Seung Woo Lee

[chemE Grad]

swlee18 @mit.edu

ANALYSIS OF TRANSPORT PHENOMENA

TOPICS IN CHEMICAL ENGINEERING

A SERIES OF TEXTBOOKS AND MONOGRAPHS

SERIES EDITOR
    Keith E. Gubbins *Cornell University*

ASSOCIATE EDITORS
    Mark A. Barteau, *University of Delaware*
    Edward L. Cussler, *University of Minnesota*
    Klavs F. Jensen, *Massachusetts Institute of Technology*
    Douglas A. Lauffenburger, *Massachusetts Institute of Technology*
    Manfred Morari, *ETH*
    W. Harmon Ray, *University of Wisconsin*
    William B. Russel, *Princeton University*

SERIES TITLES
    Receptors: Models for Binding, Trafficking, and Signalling
        D. Lauffenburger and J. Linderman
    Process Dynamics, Modeling, and Control
        B. Ogunnaike and W. H. Ray
    Microstructures in Elastic Media
        N. Phan-Thien and S. Kim
    Optical Rheometry of Complex Fluids
        G. Fuller
    Nonlinear and Mixed Integer Optimization: Fundamentals and Applications
        C. A. Floudas
    Mathematical Methods in Chemical Engineering
        A. Varma and M. Morbidelli
    The Engineering of Chemical Reactions
        L. D. Schmidt
    Analysis of Transport Phenomena
        W. M. Deen

ANALYSIS OF TRANSPORT PHENOMENA

WILLIAM M. DEEN
Massachusetts Institute of Technology

New York   Oxford
OXFORD UNIVERSITY PRESS
1998

OXFORD UNIVERSITY PRESS

Oxford   New York
Athens   Auckland   Bangkok   Bogotá   Bombay   Buenos Aires
Calcutta   Cape Town   Dar es Salaam   Delhi   Florence   Hong Kong
Istanbul   Karachi   Kuala Lumpur   Madras   Madrid   Melbourne
Mexico City   Nairobi   Paris   Singapore   Taipei   Tokyo   Toronto   Warsaw

and associated companies in
Berlin   Ibadan

Published by Oxford University Press, Inc.,
198 Madison Avenue, New York, New York 10016

Oxford is a registered trademark of Oxford University Press.

Library of Congress Cataloging-in-Publication Data
Deen, William M. (William Murray), 1947–
    Analysis of transport phenomena / William M. Deen.
        p.   cm. — (Topics in chemical engineering)
    Includes bibliographical references and index.
    ISBN 0-19-508494-2
    1. Transport theory.   2. Chemical engineering.   I. Title.
    II. Series: Topics in chemical engineering (Oxford University Press)
    TP156.T7D44   1998
    660'.2842—dc21   97-31237
                                                                CIP

9 8 7

Printed in the United States of America
on acid-free paper

To
Meredith and Michael

CONTENTS

Chapter 4

SOLUTION METHODS FOR CONDUCTION AND DIFFUSION PROBLEMS   *132*

Chapter 5

FUNDAMENTALS OF FLUID MECHANICS   *208*

Chapter 6

UNIDIRECTIONAL AND NEARLY UNIDIRECTIONAL FLOW   *251*

Chapter 7

CREEPING FLOW   *291*

Chapter 8

LAMINAR FLOW AT HIGH REYNOLDS NUMBER   *332*

Chapter 9

FORCED-CONVECTION HEAT AND MASS TRANSFER IN CONFINED LAMINAR FLOWS   *370*

Chapter 10

FORCED-CONVECTION HEAT AND MASS TRANSFER IN UNCONFINED LAMINAR FLOWS

Chapter 11

MULTICOMPONENT ENERGY AND MASS TRANSFER

Chapter 12

TRANSPORT IN BUOYANCY-DRIVEN FLOW

Chapter 13

TRANSPORT IN TURBULENT FLOW

Appendix

VECTORS AND TENSORS   *551*

INDEX   *583*

PREFACE

SCOPE AND PURPOSE

This book is intended as a text for graduate-level courses in transport phenomena for chemical engineers. It is a byproduct of nearly 20 years of teaching a one-semester course that is one of three "core" offerings required of all first-year graduate students in chemical engineering at MIT. For both master's and doctor's degree candidates, any additional courses in transport are taken as electives. Thus, the selection of material for this course has represented an attempt to define the elements of the subject that anyone with a graduate degree in chemical engineering should know. Although the book tries to bridge the gap between introductory texts and the research literature, it is intended for all students at the graduate level, not just those who will pursue transport-related research. To provide a more rounded view of the subject and allow more teaching flexibility, the book goes beyond the bare essentials and also beyond what can be presented in a one-semester course.

Transport is one of the more mathematical subjects in engineering, and to prepare students to solve a wider variety of problems and to understand the literature, the mathematical level is more advanced than in most undergraduate texts. It is assumed that the reader has a good grounding in multivariable calculus and ordinary differential equations, but not necessarily any background in solving partial differential equations. To make the book as self-contained as possible, a significant fraction (about 20%) is devoted to a presentation of the additional methods which are needed. This integration of mathematical methods with transport reflects the situation at MIT, where most chemical engineering students have not had a prior graduate-level course in applied mathematics.

ORGANIZATION AND USE

This book is based on the premise that there are certain concepts and techniques which apply almost equally to momentum, heat, and mass transfer and that at the graduate level these common themes deserve emphasis. Chapter 1 highlights the similarities among the "molecular" or "diffusive" transport mechanisms: heat conduction, diffusion of chemical species, and viscous transfer of momentum. The conservation equations for scalar quantities are derived first in general form in Chapter 2, and then they are used to obtain the governing equations for total mass, energy, and chemical species. The scaling and order-of-magnitude concepts which are crucial in modeling have received little attention in previous texts. These concepts, along with certain key methods for obtaining solutions (including similarity, perturbation, and finite Fourier transform techniques), are introduced early, in Chapters 3 and 4, using conduction and diffusion problems as examples. Thus, the first few chapters establish the tools needed for later analyses, while also covering heat and mass transfer in stationary media. Chapters 5–8 are concerned entirely with fluid mechanics. That part of the book begins with the fundamental equations for momentum transfer, and then it discusses unidirectional flow, nearly unidirectional (lubrication) flow, creeping flow, and laminar boundary layer flow, in that order. Forced-convection heat and mass transfer in laminar flow, which builds on all of the preceding material, is the subject of Chapters 9 and 10. The remaining topics can be covered in any order. The material on multicomponent energy and mass transfer is presented next, as Chapter 11, because I feel that simultaneous heat and mass transfer, at least, must be given a high priority for chemical engineering students. Chapters 12 (free convection) and 13 (turbulence) extend the discussion of convective heat and mass transfer to other types of flow.

In a one-semester course with 37 lecture hours available, I assign as reading the Appendix (vector–tensor analysis), Chapter 1, most of Chapter 2 (through Section 2.8), Chapter 3, most of Chapter 4 (through Section 4.8), Chapters 5 and 6, the first parts of Chapters 7 and 8 (through Sections 7.4 and 8.4), Chapters 9 and 10, and the first part of Chapter 11 (through Section 11.3). This leaves a few class hours at the end for other subjects, which I vary. There is usually time for one of the following: multicomponent diffusion and transport in electrolyte solutions (the rest of Chapter 11), transport in buoyancy-driven flow (Chapter 12), or transport in turbulent flow (Chapter 13). As suggested by this synopsis, I have tried to place the less essential material at the ends of chapters. I spend only a few minutes of class time on the Appendix and Chapter 1; after a brief review of vector operations, I usually begin with the conservation equations in Chapter 2. With the classroom in mind, no symbol is used which is difficult to write on a blackboard. Vectors and second-order tensors, which appear in the book as boldface letters, can be written using single and double underlines, respectively.

The book can be used in other ways. Supplemented with additional material and/ or covered at a less intense pace than that described above, it could serve as the basis for a two-semester transport sequence. If the students already have the needed backgound in similarity and perturbation methods, as well as in the use of eigenfunction expansions to solve partial differential equations, much of the material in the last half of Chapter 3 and in Chapter 4 need not be covered. Concerning the mathematics, one idiosyncratic aspect of this book is the use of the finite Fourier transform technique instead of separation of variables. To my knowledge, there are no previous textbooks which focus on the finite Fourier transform method, so that it will probably be unfamiliar not just to the students but to the instructor in a course for which this book is intended. I too learned separation of variables first, and for years I taught that method for deriving eigenfunction expansions. What I eventually found is that the finite Fourier transform approach enables students to more quickly reach the point where they can tackle interesting problems. My experience is that those who have had prior exposure to separation of variables tend to appreciate the flexibility of the finite Fourier transform method, whereas those without that background suffer no disadvantage.

ACKNOWLEDGMENTS

I owe a debt to many colleagues at MIT, from whom I have learned a great deal about transport phenomena and about teaching. My greatest debt is to Robert A. Brown, with whom I shared the teaching of our graduate transport course for a number of years. He has influenced my thinking about the subject in ways that are too numerous to list. We planned this book as a joint project and pursued it together through its early stages, but unfortunately the press of other responsibilities required that he withdraw. I have learned also from joint teaching with Robert C. Armstrong and Howard Brenner, each of whom has provided unique perspectives. Kenneth A. Smith shared his time freely in numerous discussions and never failed to give good advice. Robert C. Reid kindly reviewed an early draft of Chapter 1, and T. Alan Hatton, Jack B. Howard, Jefferson W. Tester, and Preetinder S. Virk also shared information in their areas of expertise.

I appreciate the assistance of Glorianne Collver-Jacobson and Peter Romanow in typing the manuscript and Long P. Le in preparing several of the figures.

Some very specific contributions of my colleagues need to be mentioned. A number of the examples and problems in this book are derived from "solution books" maintained by the MIT chemical engineering faculty as records of past course offerings or from doctoral qualifying exams. In some cases the history of a problem was such that its source(s) could no longer be identified. Where possible, I have tried to credit the originator of an idea. Having revised most problems to some degree, I alone am responsible for any errors or confusion. The phrase "this problem was suggested by '' is used to indicate credit without blame.

I also want to acknowledge the contributions of colleagues elsewhere. Antony N. Beris of the University of Delaware and Roger T. Bonnecaze of the University of Texas at Austin each reviewed large parts of the manuscript and made many helpful suggestions. Their efforts are greatly appreciated. Going back in time, Andreas Acrivos exposed me to the elegance of asymptotic analysis when I was a graduate student at Stanford University, and his course in viscous flow theory left an indelible impression. Influences on this book from the "transport schools" in chemical engineering at the University of Wisconsin and the University of Minnesota will be readily apparent to anyone who has followed the development of the field.[1] I hope that I have succeeded in blending those and other traditions in a useful manner.

Finally, I want to acknowledge all of the "10.50" students over the years—you know who you are! They suffered through the early drafts of this book and ultimately helped to make it better. It was their questions which led me always to a deeper understanding of the subject and thereby furthered my own education.

W.M.D.

---

[1] The 1960 text, *Transport Phenomena,* by R. B. Bird, W. E. Stewart, and E. N. Lightfoot of the University of Wiscconsin, firmly established the logic for the unified study of momentum, heat, and mass transfer in continua. More than any other book, it led to the widespread introduction of transport courses into chemical engineering curricula. Of the many contributions to transport analysis from the University of Minnesota, the most visible in the present book is the finite Fourier transform method; see footnote 1 of Chapter 4.

LIST OF SYMBOLS

Following is a list of symbols used frequently in this book. Many which appear only in one chapter are not included. In general, scalars are italic Roman (e.g., f), vectors are bold Roman (e.g., $\mathbf{v}$), and tensors are bold Greek (e.g., $\boldsymbol{\tau}$). The magnitude of a vector or tensor is usually represented by the corresponding italic letter (e.g., v for $\mathbf{v}$ and τ for $\boldsymbol{\tau}$). Vector and tensor components are identified by subscripts (e.g., v_x for the x component of $\mathbf{v}$). For more information on vector and tensor notation and on coordinate systems, see the Appendix.

Roman Letters

Bi	Biot number.
Br	Brinkman number.
C	Total molar concentration.
C_i	Molar concentration of species i.
$\hat{C}_p$	Heat capacity at constant pressure, per unit mass.
d	Diameter.
Da	Damköhler number for reaction relative to diffusion.
D_i	Pseudobinary diffusivity for species i in a dilute mixture.
D_{ij}	Binary diffusivity for species i and j.
$\mathbf{e}_i$	Unit (base) vector associated with coordinate i.
$\mathbf{g}$	Gravitational acceleration.
Gr	Grashof number.
h	Heat transfer coefficient.
$\hat{H}$	Enthalpy per unit mass.
$\overline{H}_i$	Partial molar enthalpy for species i.
H_V	Rate of energy input per unit volume from an external power source.
$\mathbf{j}_i$	Mass flux of species i relative to the mass-average velocity.
$\mathbf{J}_i$	Molar flux of species i relative to the mass-average velocity.
k	Thermal conductivity.
k_{ci}	Mass transfer coefficient for species i, based on molar concentrations.
L	Characteristic length.
M_i	Molecular weight of species i.
$\mathbf{n}$	Unit vector normal to a surface (outward from a control volume).
$\mathbf{n_I}$	Unit vector normal to an interface (phase boundary), directed from 1 to 2 or A to B.
$\mathbf{n}_i$	Mass flux of species i relative to fixed coordinates.
N_{Av}	Avogadro's number.
$\mathbf{N}_i$	Molar flux of species i relative to fixed coordinates.

Nu	Nusselt number.
P	Thermodynamic pressure.
$\mathscr{P}$	Dynamic pressure.
Pe	Péclét number.
Pr	Prandtl number.
q	Energy flux relative to the mass-average velocity.
Q	Volumetric flow rate.
r	Position vector.
R	Radius (usually) or universal gas constant (Chapters 1 and 11).
Ra	Rayleigh number.
Re	Reynolds number.
R_{Si}	Rate of formation of species i per unit area (heterogeneous reactions).
R_{Vi}	Rate of formation of species i per unit volume (homogeneous reactions).
s	Stress vector.
S	Surface (especially that enclosing a control volume).
Sc	Schmidt number.
Sh_i	Sherwood number for species i.
t	Time.
T	Temperature.
u	In boundary layer theory, the outer velocity evaluated at the surface (Chapters 8 and 10); in turbulence, the fluctuating part of the velocity (Chapter 13).
U	Characteristic velocity (often the mean velocity).
$\hat{U}$	Internal energy per unit mass.
v	Mass-average velocity.
$\mathbf{v}_i$	Velocity of species i.
$\mathbf{v_I}$	Velocity of a point on an interface (phase boundary).
$\mathbf{v_S}$	Velocity of a point on the surface of a control volume.
V	Volume (especially that of a control volume).
$\overline{V}_i$	Partial molar volume for species i.
w	Vorticity.
x_i	Mole fraction of species i in a mixture.

Greek Letters

α	Thermal diffusivity, $k / \rho \hat{C}_p$.
$\boldsymbol{\Gamma}$	Rate-of-strain tensor.
δ	Penetration depth or boundary layer thickness.
$\boldsymbol{\delta}$	Identity tensor.
$\boldsymbol{\varepsilon}$	Alternating unit tensor.
$\hat{\lambda}$	Latent heat per unit mass.
μ	Viscosity.
ν	Kinematic viscosity, μ/ρ.
ρ	Total mass density.
ρ_i	Mass concentration of species i.
$\boldsymbol{\sigma}$	Total stress tensor.
$\boldsymbol{\tau}$	Viscous stress tensor.
Φ	Viscous dissipation function.
ψ	Stream function.
ω_i	Mass fraction of species i in a mixture.

Special Symbols

D/Dt Material derivative.
∇ Gradient operator.
∇^2 Laplacian operator.
$\sim$ Order-of-magnitude equality.

Chapter 1

DIFFUSIVE FLUXES AND MATERIAL PROPERTIES

1.1 INTRODUCTION

If the temperature or the concentrations of individual chemical species in a material are perturbed so that they become nonuniform, the resulting gradients tend to disappear over time. Equilibrium states of matter are therefore characterized by the absence of spatial gradients in temperature and species concentration. Within a fluid, the same is true for gradients in velocity. The spontaneous dissipation of such gradients is a fundamental and far-reaching observation, in that it implies that energy, matter, and momentum tend to move from regions of higher to lower concentration. The purpose of *constitutive equations* is to relate these fluxes to local material properties. Accordingly, the establishment of suitable constitutive equations is one of the central aspects of the analysis of transport phenomena. The fluxes of energy, matter, and momentum each have two principal components: the *convective* transport, which automatically accompanies any bulk motion, and the *molecular* or *diffusive* transport, which is related to small-scale molecular displacements. Whereas convection is relatively easy to describe in a general manner, the molecular transport processes are specific to a given material or class of materials and therefore require much more attention. Although the forms of various constitutive equations, and in some cases the numerical values of the coefficients, can be rationalized using molecular models, these relationships remain largely empirical.

The types of constitutive equations which arise in energy and mass transfer are summarized in Table 1-1. There is, of course, an energy flux which results from a temperature gradient, as well as a flux of a given species in a mixture which results from a gradient in the concentration of that species. The constitutive equations which are ordinarily used to describe these fluxes are the "laws" named after Fourier and Fick, respectively. These basic constitutive equations, along with the analogous relationship for vis-

TABLE 1-1
Energy and Species Fluxes Resulting from Gradients in Various Quantities

Gradient	Energy flux	Flux of species i
Temperature	Conduction (Fourier)	Thermal diffusion (Soret)
Concentration of species i	Diffusion–thermo effect (Dufour)	Diffusion (Fick)
Concentration of species j ($\neq i$)		Multicomponent diffusion (Stefan–Maxwell)
Electrical potential		Ion migration (Nernst–Planck)
Pressure		Pressure diffusion

cous transfer of momentum, are the focus of this chapter. Typical values of the transport coefficients are given, and simplified molecular models are presented to provide a semi-quantitative explanation for the magnitudes of the coefficients and their dependence on temperature, pressure, and composition. Although Fourier's law and Fick's law are used to model energy and mass transfer throughout most of this book, it must be emphasized that there are many other possible gradient-flux combinations. As shown in Table 1-1, these include energy transport due to concentration gradients, and diffusion of a given species due to gradients in temperature, pressure, electrical potential, or the concentrations of other species. The general description of energy and mass transfer in multicomponent, nonisothermal systems, encompassing all of the entries in Table 1-1, is presented in Chapter 11. Additional information on the stress tensor and constitutive equations for momentum transfer is in Chapter 5.

The constitutive equations are presented using the general vector–tensor notation described in the Appendix. The reader who is unfamiliar with this notation should first see Sections A.2 through A.4.

1.2 BASIC CONSTITUTIVE EQUATIONS

Heat Conduction

The constitutive equation which is used to describe heat conduction can be traced to the work of J. Fourier in the early nineteenth century.[1] According to *Fourier's law*, the heat flux $\mathbf{q}$ (energy flow per unit cross-sectional area) is given by

$$\mathbf{q} = -k\boldsymbol{\nabla}T, \tag{1.2-1}$$

---

[1] Joseph Fourier (1768–1830) was a French civil administrator and diplomat who pursued mathematical physics in his spare time. His masterpiece was a paper submitted in 1807 to the Institut de France. In it he not only gave the first correct statement of the partial differential equation for transient conduction in a solid, but also determined the temperature in several problems using the technique of eigenfunction expansions (Fourier series), which he had developed for that purpose (see Chapter 4). Moreover, he showed how to exploit coordinate systems suited to particular geometries. This paper established the main concepts of linear boundary value problems, which dominated the study of differential equations for the next century. Among his other contributions to mathematics and physics were the Fourier integral, modern ideas concerning the dimensions and units of physical quantities, and the integral symbol (Ravetz and Grattan-Guinness, in Gillispie, Vol. 5, 1972, pp. 93–99).

where k is the *thermal conductivity*, ∇ is the gradient operator, and T is the temperature. In a moving fluid, $\mathbf{q}$ only represents energy transfer relative to the local mass-average velocity; the energy flux relative to fixed coordinates has a contribution also from convection (bulk flow). In a mixture, there are contributions to $\mathbf{q}$ beyond that shown in Eq. (1.2-1) (see Chapter 11).

The use of a single scalar coefficient, k, in Eq. (1.2-1) implies that there is no preferred direction for heat conduction. Accordingly, the flux parallels the temperature gradient. Materials for which this is true, including many solids and almost all fluids, are termed *isotropic*. Materials which are *anisotropic* have an internal structure which makes the thermal conductivity depend on direction. An example is wood, where conduction parallel to the grain is often several times faster than that perpendicular to it. For anisotropic materials, the scalar conductivity must be replaced by a conductivity tensor ($\boldsymbol{\kappa}$), giving

$$\mathbf{q} = -\boldsymbol{\kappa} \cdot \nabla T. \tag{1.2-2}$$

A schematic representation of the heat flux vector is shown in Fig. 1-1. Consider an imaginary surface which is perpendicular to a vector $\mathbf{n}$ of unit magnitude. That is, $\mathbf{n}$ is the *unit normal vector* which describes the orientation of the surface. For a surface of arbitrary orientation, as in Fig. 1-1(a), the heat flux normal to the surface (i.e., in the direction of $\mathbf{n}$) is

$$q_n = \mathbf{n} \cdot \mathbf{q} = -k(\mathbf{n} \cdot \nabla T). \tag{1.2-3}$$

If, as shown in Fig. 1-1(b), the surface happens to be perpendicular to the y axis of a rectangular coordinate system, so that $\mathbf{n} = \mathbf{e}_y$ (unit vector in y direction), then the flux normal to the surface is

$$q_y = \mathbf{e}_y \cdot \mathbf{q} = -k\frac{\partial T}{\partial y}, \tag{1.2-4}$$

which is a one-dimensional form of Fourier's law. Notice that the subscript y identifies q_y as the y component of the vector $\mathbf{q}$; in this book, subscripts are *not* used to indicate differentiation.

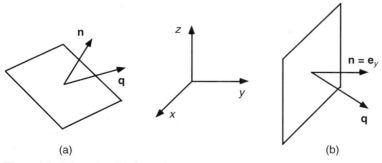

(a)   (b)

Figure 1-1. Schematic of the heat flux $\mathbf{q}$ at a surface with a unit normal $\mathbf{n}$ for (a) a surface with arbitrary orientation and (b) a surface perpendicular to the y axis.

Diffusion of Chemical Species

The molar flux of species i relative to fixed coordinates is denoted as $\mathbf{N}_i$; the corresponding flux in mass units is $\mathbf{n}_i = M_i \mathbf{N}_i$, where M_i is the molecular weight of i. The part of the species flux which is attributed to bulk flow and the part which is due to diffusion depend on the definition of the mixture velocity. Any number of average velocities can be computed for a mixture, depending on how one chooses to weight the velocities of the individual species. Again relative to fixed coordinates, the velocity $\mathbf{v}_i$ of species i is defined as

$$\mathbf{v}_i = \frac{\mathbf{N}_i}{C_i} = \frac{\mathbf{n}_i}{\rho_i}, \tag{1.2-5}$$

where C_i and ρ_i are the concentrations of i expressed in molar and mass units, respectively. Notice that $\mathbf{v}_i$ is the same for any consistent choice of flux and concentration units, so that species velocities are unambiguous.

Two useful reference frames are the *mass-average velocity*, $\mathbf{v}$, and the *molar-average velocity*, $\mathbf{v}^{(M)}$, which are defined as

$$\mathbf{v} \equiv \sum_{i=1}^{n} \omega_i \mathbf{v}_i, \qquad \mathbf{v}^{(M)} \equiv \sum_{i=1}^{n} x_i \mathbf{v}_i, \tag{1.2-6}$$

where ω_i and x_i are the mass fraction and mole fraction of species i, respectively, and n is the number of chemical species in the mixture. The mass and mole fractions are related to the total mass density (ρ) and total molar concentration (C) by

$$\omega_i = \frac{\rho_i}{\rho} = \frac{\rho_i}{\sum\limits_{i=1}^{n} \rho_i}, \qquad x_i = \frac{C_i}{C} = \frac{C_i}{\sum\limits_{i=1}^{n} C_i}. \tag{1.2-7}$$

The various diffusional fluxes which are defined using $\mathbf{v}$ or $\mathbf{v}^{(M)}$ are as follows: $\mathbf{J}_i$, the molar flux relative to the mass-average velocity; $\mathbf{J}_i^{(M)}$, the molar flux relative to the molar-average velocity; $\mathbf{j}_i$, the mass flux relative to the mass-average velocity; and $\mathbf{j}_i^{(M)}$, the mass flux relative to the molar-average velocity. The relationships among these fluxes are summarized in Table 1-2. Another reference frame which is sometimes used is the volume-average velocity (see Problem 1-2).

A constitutive equation to describe diffusion in a binary mixture was proposed by A. Fick in the mid-nineteenth century.[2] Four forms of *Fick's law* are shown in Table 1-3 (see also Problem 1-2). All of these forms of Fick's law are equivalent, and each contains the same *binary diffusivity* for species A and B, denoted as D_{AB}. It can be shown that $D_{AB} = D_{BA}$ (Problem 1-1), so that diffusion of any pair of species is characterized by only one diffusion coefficient. In a multicomponent mixture, there are as many indepen-

---

[2] Adolph Fick (1829–1901), a German physiologist, was an early proponent of the rigorous application of physics to medicine and physiology (Rothschuh, in Gillispie, Vol. 4, 1971, pp. 614–617). In addition to his law for diffusion (1855), he made pioneering theoretical and experimental contributions in biomechanics, membrane biophysics, bioenergetics, and electrophysiology. The "Fick principle" for computing cardiac output is still taught in modern courses in physiology. Based on a steady-state mass balance for a solute (oxygen) in a continuous-flow system (the blood circulation), it was a remarkably advanced concept when conceived. For a discussion of the early work on diffusion of Fick and of Thomas Graham, see Cussler (1984, pp. 16–20).

TABLE 1-2
Flux of Species i in Various Reference Frames and Units for a Mixture of n Components

Definitions of fluxes:

Reference velocity	Molar units	Mass units
$\mathbf{0}$	$\mathbf{N}_i$	$\mathbf{n}_i$
$\mathbf{v}$	$\mathbf{J}_i$	$\mathbf{j}_i$
$\mathbf{v}^{(M)}$	$\mathbf{J}_i^{(M)}$	$\mathbf{j}_i^{(M)}$

Flux relationships:

$$\mathbf{N}_i = C_i \mathbf{v} + \mathbf{J}_i = C_i \mathbf{v}^{(M)} + \mathbf{J}_i^{(M)}, \qquad \sum_{i=1}^{n} \mathbf{N}_i = C\mathbf{v}^{(M)}, \qquad \sum_{i=1}^{n} \mathbf{J}_i^{(M)} = 0$$

$$\mathbf{n}_i = \rho_i \mathbf{v} + \mathbf{j}_i = \rho_i \mathbf{v}^{(M)} + \mathbf{j}_i^{(M)}, \qquad \sum_{i=1}^{n} \mathbf{n}_i = \rho\mathbf{v}, \qquad \sum_{i=1}^{n} \mathbf{j}_i = 0$$

dent diffusion coefficients as there are pairs of species. Moreover, the fluxes and concentration gradients of *all* species are interdependent (see Chapter 11). Accordingly, Fick's law is not applicable to multicomponent systems, except in special circumstances (e.g., dilute solutions).

Stress and Momentum Flux

The linear momentum of a fluid element of mass m and velocity $\mathbf{v}$ is $m\mathbf{v}$. In a pure fluid, $\mathbf{v}$ is defined merely by selecting a fixed reference frame; in a mixture, $\mathbf{v}$ must be interpreted as the mass-average velocity, as given by Eq. (1.2-6). The local rates of momentum transfer in a fluid are determined in part by the stresses. Indeed, just as a force represents an overall rate of transfer of momentum, a stress (force per unit area) represents a *flux* of momentum.

Consider some point within a fluid, through which passes an imaginary surface having an orientation described by the unit normal $\mathbf{n}$. The stress at that point, with reference to the orientation of that test surface, is given by the *stress vector*, $\mathbf{s(n)}$. The stress vector is defined as the force per unit area on the test surface, *exerted by the fluid toward which $\mathbf{n}$ points*. As shown in Chapter 5, an equal and opposite stress is exerted by the fluid on the opposite side of the test surface. Accordingly, defining $\mathbf{s(n)}$ as the stress exerted by the fluid on a particular side of the surface establishes an algebraic

TABLE 1-3
Fick's Law for Binary Mixtures of A and B

Reference velocity	Mass units		Molar units	
$\mathbf{v}$	$\mathbf{j}_A = -\rho D_{AB}\nabla\omega_A$	(A)	$\mathbf{J}_A = -\dfrac{\rho D_{AB}}{M_A}\nabla\omega_A$	(B)
$\mathbf{v}^{(M)}$	$\mathbf{j}_A^{(M)} = -CM_A D_{AB}\nabla x_A$	(C)	$\mathbf{J}_A^{(M)} = -CD_{AB}\nabla x_A$	(D)

sign convention for each of the components of this vector. As shown in Fig. 1-2(a), for a surface of arbitrary orientation, $s(n)$ can be resolved into components which are normal (s_n) and tangent (s_1, s_2) to the surface. The normal component represents a normal stress or "pressure" on the surface, and the tangential components are shear stresses. Tensile and compressive forces correspond to $s_n > 0$ and $s_n < 0$, respectively.

As shown in Chapter 5, the stress vector is related to the *stress tensor*, σ, by

$$s(n) = n \cdot \sigma. \tag{1.2-8}$$

The stress tensor, which may be represented in component form as

$$\sigma = \begin{bmatrix} \sigma_{xx} & \sigma_{xy} & \sigma_{xz} \\ \sigma_{yx} & \sigma_{yy} & \sigma_{yz} \\ \sigma_{zx} & \sigma_{zy} & \sigma_{zz} \end{bmatrix}, \tag{1.2-9}$$

gives the information needed to compute the stress vector for a surface with *any* orientation. This is illustrated most simply by considering a surface which is normal to the y axis of a rectangular coordinate system, as shown in Fig. 1-2(b). In that case

$$s(n) = s(e_y) = \sigma_{yx}e_x + \sigma_{yy}e_y + \sigma_{yz}e_z, \tag{1.2-10}$$

so that, for example, σ_{yx} is seen to be the *force per unit area on a plane perpendicular to the y axis, acting in the x direction, and exerted by the fluid at greater y*. Note that the definition of σ_{ij} contains three parts, which specify a reference plane, a direction, and a sign convention, respectively. The orientation of the reference plane is specified by the first subscript, and the direction of the force is indicated by the second subscript.

A stress may be interpreted as a flux of momentum, as already noted. Because we have defined σ_{yx} in terms of the force exerted on a plane of constant y by fluid at greater y, positive values of σ_{yx} correspond to transfer of x momentum in the $-y$ direction. The opposite sign convention ("exerted by the fluid at lesser y") is sometimes chosen, in which case positive values of σ_{yx} correspond to momentum transfer in the $+y$ direction. The sign convention adopted here is the one used most commonly in the fluid mechanics literature; the opposite sign convention (as used by Bird et al., 1960) emphasizes the analogy between momentum transport in simple flows and the transport of heat or chemical species. It should be mentioned that some authors assign the opposite meanings to the subscripts in σ_{ij}, so that $s = \sigma \cdot n$ instead of $n \cdot \sigma$.

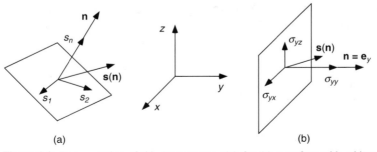

(a)   (b)

Figure 1-2. Representation of the stress vector $s(n)$ for (a) a surface with arbitrary orientaton, and (b) a surface normal to the y naxis. In the latter case the scalar components of $s(n)$ equal components of the stress tensor σ.

As described in more detail in Chapter 5, the part of the stress which is caused solely by fluid motion is termed the *viscous stress*. This contribution to the stress tensor is denoted as τ. For a *Newtonian fluid* with constant density, it is related to velocity gradients by

$$\tau = \mu(\nabla\mathbf{v} + (\nabla\mathbf{v})^{\mathrm{t}}), \qquad (1.2\text{-}11)$$

where μ is the *viscosity*, $\nabla\mathbf{v}$ is the velocity gradient tensor, and $(\nabla\mathbf{v})^{t}$ is the transpose of $\nabla\mathbf{v}$. There is a qualitative analogy among Eq. (1.2-11), Fourier's law, and Fick's law, in that each of the fluxes is proportional to the gradient of some quantity. The analogy becomes more precise if one considers the viscous shear stress in a *unidirectional flow*, where $\mathbf{v} = v_x(y)\mathbf{e}_x$. That is, the only nonvanishing component of the velocity is in the x direction, and it depends only on y. For that special case, Eq. (1.2-11) reduces to

$$\tau_{yx} = \mu\frac{dv_x}{dy}. \qquad (1.2\text{-}12)$$

The correspondence between Eq. (1.2-12) and Eq. (1.2-4), the one-dimensional form of Fourier's law, is obvious. If we had chosen the opposite sign convention for the stress, then a minus sign would have appeared in Eq. (1.2-12) and the analogy would have been exact.

Gases and low-molecular-weight liquids are ordinarily Newtonian; everyday examples are air and water. Various high-molecular-weight liquids (e.g., polymer melts), certain suspensions (e.g., blood), and complex fluid mixtures (e.g., microemulsions and foams) are non-Newtonian. Such fluids are characterized by structural features which influence, and also are influenced by, their flow. Although non-Newtonian fluids do not satisfy Eq. (1.2-11), a viscosity can still be defined for a unidirectional flow. The experimental and modeling studies which seek to establish constitutive equations for complex fluids comprise the field of *rheology;* we return to this subject briefly in Chapter 5.

1.3 DIFFUSIVITIES FOR ENERGY, SPECIES, AND MOMENTUM

The basic constitutive relations given in Section 1.2 are similar in that each relates a flux to the gradient of a particular quantity, evaluated locally in a fluid. The analogies among the constitutive equations are strengthened by expressing each as a flux that is proportional to the gradient of a concentration. The proportionality constant in each case is a type of diffusivity. In this context, "concentration" is the amount per unit volume of *any* quantity, not only chemical species.

In systems where thermal effects are more important than mechanical forms of energy, a useful measure of energy is the enthalpy. Considering variations in temperature only (i.e., neglecting the effects of pressure), the changes in the enthalpy per unit mass $(\hat{H})$ are given by

$$d\hat{H} = \left(\frac{\partial\hat{H}}{\partial T}\right)_P dT = \hat{C}_p dT, \qquad (1.3\text{-}1)$$

where $\hat{C}_p$ is the heat capacity per unit mass, at constant pressure. For constant $\hat{C}_p$ and ρ, Eq. (1.2-4) becomes

$$q_y = -\alpha \frac{\partial}{\partial y}(\rho \hat{C}_p T),$$

(1.3-2)

$$\alpha \equiv \frac{k}{\rho \hat{C}_p}.$$

(1.3-3)

The term in parentheses in Eq. (1.3-2) is a concentration of energy, and α is the *thermal diffusivity*.

If the density (ρ) of a mixture is constant, the y component of Eq. (B) of Table 1-3 becomes

$$J_{Ay} = -D_{AB} \frac{\partial C_A}{\partial y}.$$

(1.3-4)

A similar result, but with $J_{Ay}^{(M)}$ instead of J_{Ay}, is obtained if the total molar concentration (C) is constant. Either is a good approximation for a dilute liquid solution, but constant C is often more accurate for gases (e.g., for an ideal gas at constant temperature and pressure).

The concentration of linear momentum directed along the x axis is ρv_x. For a constant-density, Newtonian fluid with $v_x = v_x(y)$ and $v_y = 0$, Eq. (1.2-12) rearranges to

$$\tau_{yx} = \nu \frac{d(\rho v_x)}{dy},$$

(1.3-5)

$$\nu \equiv \frac{\mu}{\rho}.$$

(1.3-6)

The diffusivity for momentum is ν, the *kinematic viscosity*.

Equations (1.3-2), (1.3-4), and (1.3-5) each describe a process in which something is transferred from regions of high to regions of low concentration. [As already mentioned, the difference in algebraic sign in Eq. (1.3-5), as compared with Eqs. (1.3-2) and (1.3-4), is incidental, being related to the sign convention chosen for the stress.] Each of these constitutive equations is therefore consistent with the basic observation cited in Section 1.1, that systems tend to relax toward equilibrium states in which gradients in temperature, species concentration, and velocity are absent.

The diffusivities for energy, species, and momentum all have the same units (m²/s), permitting comparisons of the intrinsic rates of these transport processes. The rate of viscous momentum transfer relative to heat conduction is given by the *Prandtl number*,

$$\text{Pr} \equiv \frac{\nu}{\alpha} = \frac{\mu \hat{C}_p}{k}.$$

(1.3-7)

The rate of viscous momentum transfer relative to species diffusion is indicated by the *Schmidt number*,

$$\text{Sc} \equiv \frac{\nu}{D_{AB}} = \frac{\mu}{\rho D_{AB}}.$$

(1.3-8)

The magnitudes of the Prandtl and Schmidt numbers are extremely important in convective heat transfer and mass transfer, respectively, as discussed in Chapters 9, 10, 12, and 13. A ratio of diffusivities which is sometimes used in the analysis of simultaneous heat and mass transfer is the *Lewis number*, $\text{Le} = \alpha/D_{AB} = \text{Sc}/\text{Pr}$.

1.4 MAGNITUDES OF TRANSPORT COEFFICIENTS

In judging which physical effects are important in a given type of situation, it is helpful to have in mind the values of viscosity, thermal conductivity, and species diffusivity which are typical of various kinds of materials. The purpose of this section is to summarize some representative data. Many methods are available for estimating the values of transport coefficients in materials where specific data are limited or absent, but a discussion of those methods is beyond the scope of this book. A good source of information on property estimation is Reid et al. (1987).

Figure 1-3 shows the approximate ranges of viscosity seen for gases and ordinary (nonpolymeric) liquids, at room temperature and pressure. The S.I. unit for stress (pressure) is the pascal (Pa), so that $\mu\ [=]\ Pa \cdot s = 1\ kg\ m^{-1}\ s^{-1} = 10$ poise. The viscosities of gases fall in a narrow range around $10^{-5}\ Pa \cdot s$. The viscosities of liquid water and of many organic solvents are about $10^{-3}\ Pa \cdot s$, or about 10^2 times those typical of gases. The range of liquid viscosities is even wider than the four orders of magnitude suggested by Fig. 1-3. Depending on the forces and time scales of interest, materials ordinarily considered to be solids (e.g., window glass) may exhibit liquid-like deformation. Thus, liquid viscosities do not have an upper bound.

Ranges of thermal conductivities are depicted in Fig. 1-4. The S.I. unit for power (energy/time) is the watt (W), so that $k[=]W\ m^{-1}\ K^{-1} = 1\ kg\ m\ s^{-3}\ K^{-1} = 4.184 \times 10^2$ cal $cm^{-1}\ s^{-1}\ K^{-1}$. For gases, k ranges from about 10^{-2} to $10^{-1}\ W\ m^{-1}\ K^{-1}$, and for nonmetallic liquids is about a factor of 10 larger. Most common organic liquids have thermal conductivities within a very narrow range, 0.10–$0.17\ W\ m^{-1}\ K^{-1}$; water and other highly polar molecules have values of k several times larger. The largest values of k are for pure metals, where conductivities for heat parallel those for electric current. This correspondence between electrical and thermal conductivity arises from the fact that, in pure metals, free electrons are the major carriers of heat; in other materials, the concentration of free electrons is low, and energy is transmitted primarily by atomic or molecular motions. Nonmetallic solids range from good thermal insulators to good heat conductors.

Values of μ, k, and Pr for several gases and liquids are shown in Tables 1-4 and 1-5. For gases, μ and k both increase with temperature, whereas the opposite temperature dependence is seen for liquids (except k for water). For gases or liquids at moderate pressures, the effect of pressure on μ and k is generally negligible. The Prandtl numbers for gases are typically near unity, indicating that the intrinsic rates of heat conduction

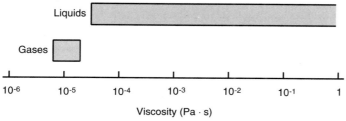

Figure 1-3. Approximate ranges for the viscosity of gases and nonpolymeric liquids, at ambient temperature and pressure.

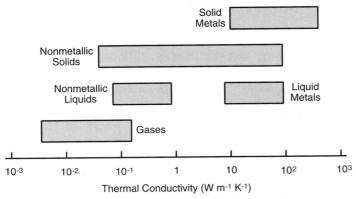

Figure 1-4. Approximate ranges for the thermal conductivities of various classes of materials.

and viscous momentum transfer are roughly the same. Values of Pr for liquids typically fall between 1 and 10. The exceptions include liquid metals (e.g., Na and Si), where Pr is small because of the large k, and high-molecular-weight organic liquids (e.g., lubricating oils and polymer melts), where Pr is large because of the large μ.

Ranges of binary diffusivities characteristic of gases, liquids, and solids are illustrated in Fig. 1-5. Diffusivities of low-molecular-weight solutes in common liquids are typically about 10^{-9} m^2/s, or roughly four orders of magnitude smaller than typical values in gases (about 10^{-5} m^2/s). The lower extreme shown for liquids, 10^{-11} m^2/s, is representative of certain large proteins in water. For high-molecular-weight solutes, D_{AB}

TABLE 1-4
Viscosities, Thermal Conductivities, and Prandtl Numbers of Gases[a]

Gas	T (°C)	$\mu(10^{-5}\text{Pa·s})$	$k(10^{-2}\text{W m}^{-1}\text{ K}^{-1})$	Pr
Air	22	1.82	2.59	0.72
	102	2.18	3.19	
NH_3	37	1.06	2.78	0.85
	147	1.46	4.20	
CO_2	37	1.54	1.79	0.78
	127	1.94	2.51	
C_2H_6	47	1.00	2.45	0.77
	117	1.20	3.47	
N_2	20	1.76	2.52	0.73
	100	2.10	3.07	
O_2	20	2.03	2.59	0.74
H_2O	100	1.27	2.39	0.94
	200	—	3.32	

[a]Most of the data are from tables or correlations in Reid et al. (1987), for atmospheric pressure. The values for air are from Gebhart et al. (1988). The Prandtl numbers for the other gases, as well as the viscosities of oxygen and water vapor, are taken from Bird et al. (1960). All Prandtl numbers shown are for 0°C.

TABLE 1-5
Viscosities, Thermal conductivities, and Prandtl Numbers of Liquids[a]

Liquid	T (°C)	$\mu(10^{-3}\text{Pa}\cdot\text{s})$	$k(\text{W m}^{-1}\text{K}^{-1})$	Pr
C_6H_6	20	0.65	0.147	7.7
	100	0.25	0.127	—
$CHCl_3$	20	0.54	0.120	4.4
	100	0.28	0.102	—
C_2H_5OH	20	1.14	0.169	16
	100	0.32	0.150	—
C_3H_8	−100	0.42	0.167	5.2
	20	0.097	0.098	—
H_2O	20	1.00	0.609	6.9
	100	0.282	0.690	1.7
Engine	0	3.9×10^3	0.15	4.7×10^4
Oil	100	17	0.14	2.8×10^2
LDPE[b]	180	1×10^6	0.2	1×10^7
Na	104	0.69	86.0	1.1×10^{-2}
	400	0.27	71.2	4.4×10^{-3}
Si	1411	0.7	64.0	1.1×10^{-2}

[a]Most of the data are from Reid et al. (1987), for atmospheric pressure. The viscosity of water and the heat capacities needed to calculate Pr for water, benzene (C_6H_6), and chloroform ($CHCl_3$) are from Weast (1968). The data for liquid sodium and engine oil were taken from Rohsenow and Hartnett (1973). The values for molten silicon are from Touloukian and Ho (1970) and Glazov et al. (1969).
[b]Low-density polyethylene. The property values for molten LDPE are order-of-magnitude estimates from Tadmor and Gogos (1979) and depend on the flow conditions.

in dilute liquid solutions is inversely proportional to the molecular radius of the solute and is also inversely proportional to the viscosity of the solvent. Thus, for very large polymeric solutes, for colloidal particles, and/or for very viscous solvents, D_{AB} may be much lower than the minimum value shown. Values of D_{AB} for solids are the most difficult to anticipate, but usually do not exceed 10^{-12} m²/s. This excludes porous materials (e.g., wood, catalyst supports, certain membranes) in which diffusion may proceed through a gas or liquid phase within the pores, yielding correspondingly larger values of D_{AB}. Diffusivities in solids may be many orders of magnitude lower than those shown in Fig. 1-5.

Values of D_{AB} for several gas pairs are shown in Table 1-6. As with μ and k, D_{AB} for gases increases with temperature. As shown in Table 1-7, D_{AB} for dilute liquid solutions also increases with temperature. At moderate pressures (P), D_{AB} in gases varies inversely with P. The pressure dependence of D_{AB} in liquids is negligible in most applications.

For gases, D_{AB} is normally assumed to be independent of mole fraction. As indicated by the data in Table 1-7 for n-heptane/benzene and water/ethyl acetate, this is not true for liquids. Depending on which component is in abundance (the "solvent"), very different values can be observed for D_{AB}. Further complicating the situation for miscible liquids, the dependence of D_{AB} on mole fraction is not necessarily monotonic. Also, for

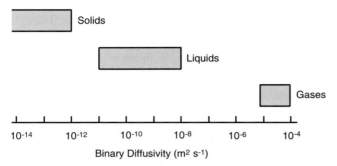

Figure 1-5. Approximate ranges for binary diffusivities in gases, liquids, and solids.

systems in which only one component may act as the solvent, D_{AB} may either decrease or increase with solute concentration.

The dependence of D_{AB} on composition leads to certain distinctions among measured values of diffusivity. The binary diffusivity D_{AB} defined in Table 1-3 is termed the *mutual diffusion coefficient;* it is the coefficient measured in the presence of macroscopic concentration gradients. Using radioisotopes, certain light scattering techniques, or other methods, it is also possible to measure diffusion within a mixture of macroscopically uniform composition; a diffusivity determined in this way is termed a *tracer diffusion coefficient.* In a mixture containing only A and B, $D_{AB} \rightarrow D_{AB}^0$ as $x_A \rightarrow 0$, where D_{AB}^0 is the tracer diffusivity for molecules of A in nearly pure B. Likewise, $D_{AB} \rightarrow D_{BA}^0$ as $x_B \rightarrow 0$, where D_{BA}^0 is the tracer diffusivity for B in nearly pure A. As exemplified by the infinite-dilution values of D_{AB} for water/ethyl acetate and *n*-heptane/benzene in Table 1-7, $D_{AB}^0 \neq D_{BA}^0$ in general. The inequality of these tracer diffusivities reflects the dependence of D_{AB} on composition, rather than any failure of the relation $D_{AB} = D_{BA}$. That

TABLE 1-6
Binary Diffusivities for Gases at
Atmospheric Pressure[a]

Gas pair	T (K)	D_{AB} (10^{-5} m^2/s)
CO_2/N_2	298	1.69
CO_2/He	298	6.20
	498	14.3
H_2/NH_3	263	5.8
	358	11.1
	473	18.9
N_2/NH_3	298	2.33
	358	3.32
N_2/H_2O	308	2.59
	352	3.64
O_2/H_2O	352	3.57

[a] The data are from Reid et al. (1987) for $P = 1$ bar $= 10^5$ Pa.

TABLE 1-7
Binary Diffusivities for Liquids at Infinite Dilution[a]

Solute	Solvent	T (K)	$D_{AB}(10^{-9}$ m²/s)
n-Heptane	Benzene	298	2.10
		353	4.25
Benzene	n-Heptane	298	3.40
		372	8.40
Water	Ethyl acetate	298	3.20
Ethyl acetate	Water	293	1.00
Methane	Water	275	0.85
		333	3.55
Serum albumin	Water	293	0.061[b]
Boron	Silicon	1411	24[c]

[a] The data are from Reid et al. (1987), except where noted.
[b] From Johnson et al. (1995). Serum albumin, which has a molecular weight of 7×10^4, is the predominant protein in blood plasma.
[c] From Kodera (1963).

is, $D_{AB} = D_{BA}$ at any given composition, but $D_{AB}{}^0$ and $D_{BA}{}^0$ correspond to two very different compositions ($x_A \rightarrow 0$ and $x_B \rightarrow 0$, respectively). As discussed in Chapter 11, an activity-coefficient correction can be introduced into Fick's law to greatly reduce the dependence of D_{AB} on composition.

In a binary system containing species A and a chemically identical isotope A^*, the diffusivity obtained is D_{AA}, termed the *self-diffusion coefficient*. This is actually a special kind of tracer diffusivity, referring to conditions where species A is surrounded by itself. Differences in the intermolecular forces cause the mobility of A surrounded by A to differ in general from that of A surrounded by B, or B surrounded by A, so that $D_{AA} \neq D_{AB}{}^0 \neq D_{BA}{}^0$. What are loosely termed "tracer diffusivities" are also sometimes reported for isotope A^* in a mixture of A and B. It is not always recognized that because this is a ternary system (A^*, A, B), these quantities differ from the binary diffusivities discussed above. A careful interpretation of such results requires a multicomponent diffusion formulation (see Chapter 11).

Representative values of the Schmidt number can be estimated from the data in the tables. For gases at room temperature and pressure, typical viscosities and densities are $\mu \cong 1 \times 10^{-5}$ Pa·s and $\rho \cong 1$ kg m⁻³, respectively, so that $\nu \cong 1 \times 10^{-5}$ m²/s. For common liquids, $\mu \cong 1 \times 10^{-3}$ Pa·s and $\rho \cong 1 \times 10^3$ kg m⁻³, yielding $\nu \cong 1 \times 10^{-6}$ m²/s. With $D_{AB} \cong 1 \times 10^{-5}$ m²/s for gases and 1×10^{-9} m²/s for liquids, Sc is roughly unity for gases and about 10^3 for liquids. For solutes in water at room temperature, Sc ranges from about 200 for H_2 (Gebhart et al., 1988) to 10^5 for large proteins. Even larger values of Sc result for polymeric solutes in very viscous liquids, such as polymer melts. In summary, species diffusion and viscous transfer of momentum have comparable intrinsic rates in gases, but diffusion of chemical species is by far the slower process in liquids.

1.5 MOLECULAR INTERPRETATION OF TRANSPORT COEFFICIENTS

The experimental trends summarized in Section 1.4 can be understood to a large extent through the use of simple molecular models. The models discussed here are chosen for their illustrative value, rather than for their quantitative accuracy in predicting viscosities, thermal conductivities, or diffusivities. The objective is to understand the orders of magnitude of the transport coefficients, and to some extent their dependence on temperature, pressure, and composition. For a more rigorous treatment of the molecular interpretation of transport coefficients in gases or liquids, a standard reference is Hirschfelder et al. (1954).

Lattice Model

The "molecular" or "diffusive" fluxes of energy, species, and momentum are based ultimately on the random motions of molecules. An elementary model to relate random motions of molecules to diffusive fluxes is developed by assuming molecular movements to be confined to a cubic lattice, as shown in Fig. 1-6. According to this model, each discrete location (lattice site) is occupied at any instant by a collection of molecules, each of which is able to jump to any of the six adjacent sites. The lattice spacing, ℓ, corresponds to the distance moved by a molecule in one "jump." All molecules are assumed to move at speed u along one of the three coordinate axes, with no preference of direction (i.e., the material is assumed to be isotropic). Thus, it is equally likely for a molecule to be moving at speed u in the $+x$ direction, $-x$ direction, $+y$ direction, and so on. It is assumed further that the molecules leaving a given position carry amounts of energy or momentum representative of that position. Implicit in this assumption is that exchanges of energy and momentum among molecules at a given position allow them to equilibrate rapidly with one another. Convective contributions to the fluxes are not considered. Other restrictions are the same as those used to obtain Eqs. (1.3-2), (1.3-4), and (1.3-5).

To exploit the analogies among transport of energy, species, and momentum, we use $\mathbf{f}$ to denote any of the flux vectors (molecular contributions only) and use b to represent any of the corresponding concentrations, as summarized in Table 1-8. Concen-

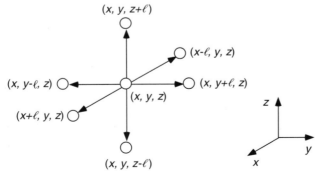

Figure 1-6. Cubic-lattice model for random molecular motion. Molecules are assumed to make jumps of length ℓ parallel to one of the coordinate axes, moving at speed u.

TABLE 1-8
Fluxes, Concentrations, and Diffusivities for Use in Random-Jump Model

Process	Flux (f_y)	Concentration (b)	Diffusivity ($\mathcal{D}$)
Heat transfer	q_y	$\rho \hat{C}_p T$	α
Species transfer	J_{Ay}	C_A	D_{AB}
Momentum transfer	$-\tau_{yx}$	ρv_x	ν

trations are obtained by dividing the total amount of a quantity at a particular lattice site by the fluid volume per site (ℓ^3).

The flux in the y direction (f_y) is evaluated by considering the net movement of molecules between the lattice sites (x, y, z) and $(x, y+\ell, z)$. Including the molecules moving along the y axis which originated at either of those sites, we obtain

$$f_y = \frac{u}{6}[b(x, y, z) - b(x, y+\ell, z)]. \tag{1.5-1}$$

The factor $\frac{1}{6}$ comes from the equal probability of jumps in any of six directions. We assume now that the jump distance ℓ is much smaller than any macroscopic or system dimension, L. This enables us to treat b as a continuous rather than a discrete function of position and allows us to represent it using a Taylor series expansion about the reference point,

$$b(x, y+\ell, z) = b(x, y, z) + \ell \left.\frac{\partial b}{\partial y}\right|_{(x,\, y,\, z)} + \frac{\ell^2}{2}\left.\frac{\partial^2 b}{\partial y^2}\right|_{(x,\, y,\, z)} + \cdots . \tag{1.5-2}$$

We will retain only the first two terms, which is valid if $\partial^2 b/\partial y^2 << \ell^{-1}\,\partial b/\partial y$. If the length scale for significant variations in b is L, the assumption that $\ell/L << 1$ is sufficient to justify this truncation of the series. Substituting Eq. (1.5-2) into Eq. (1.5-1), we find that

$$f_y = -\mathcal{D}\frac{\partial b}{\partial y}, \tag{1.5-3}$$

$$\mathcal{D} = \frac{u\ell}{6}. \tag{1.5-4}$$

Equation (1.5-3) is of the same form as Eqs. (1.3-2), (1.3-4), and (1.3-5). It is seen that the diffusivity parameter, $\mathcal{D}$, is proportional to the product of molecular speed and jump distance.

Application of Lattice Model to Gases

Any attempt to relate this elementary model to the properties of real materials depends on the interpretation of u and ℓ. The situation is simplest for gases where, according to kinetic theory, transport results from exchanges that occur during molecular collisions. The concept of a collision is meaningful when the time spent during encounters with other molecules is much smaller than that spent between such encounters. According to elementary kinetic theory, low-density gases are characterized by a *mean molecular velocity* (c) and a *mean free path* (or mean distance between collisions, λ) which are given by

$$c = \left(\frac{8RT}{\pi M}\right)^{1/2}, \tag{1.5-5}$$

$$\lambda = \frac{1}{\sqrt{2}\pi d^2 N_{Av} C} = \frac{RT}{\sqrt{2}\pi d^2 N_{Av} P}, \tag{1.5-6}$$

where R is the gas constant, N_{Av} is Avogadro's number, M is molecular weight (molar), and d is molecular diameter. For a mixture of A and B it is necessary to use average values of M and d in Eqs. (1.5-5) and (1.5-6). The appropriate averages are

$$M = 2\left(\frac{1}{M_A} + \frac{1}{M_B}\right)^{-1}, \tag{1.5-7}$$

$$d = \frac{d_A + d_B}{2}. \tag{1.5-8}$$

Using $u = c$ and $\ell = \lambda$ in Eq. (1.5-4), we obtain an estimate of $\mathcal{D}$ for gases. Translating this into thermal conductivity, binary diffusivity, and viscosity, respectively, the results are

$$k = \rho \hat{C}_p \mathcal{D} = \frac{1}{3\pi^{3/2}} \frac{(MRT)^{1/2}\hat{C}_p}{N_{Av} d^2}, \tag{1.5-9}$$

$$D_{AB} = \mathcal{D} = \frac{1}{3\pi^{3/2}} \frac{(RT)^{3/2}}{N_{Av} d^2 M^{1/2} P}, \tag{1.5-10}$$

$$\mu = \rho \mathcal{D} = \frac{1}{3\pi^{3/2}} \frac{(MRT)^{1/2}}{N_{Av} d^2}. \tag{1.5-11}$$

This model implies that $\nu = \alpha = D_{AB}$, so that the Prandtl and Schmidt numbers are simply

$$\text{Pr} = \text{Sc} = 1. \tag{1.5-12}$$

To calculate a representative value of D_{AB} from Eq. (1.5-10), suppose that $T = 293$ K, $P = 1$ atm $= 1.0133 \times 10^5$ Pa, $M = 30$ g mol^{-1}, and $d = 3$ Å $= 3 \times 10^{-10}$ m. With $R = 8.314$ Pa m^3 mol^{-1} K^{-1}, this yields $u = c = 4.5 \times 10^2$ m/s, $\ell = \lambda = 1.0 \times 10^{-7}$ m, and $D_{AB} = 0.8 \times 10^{-5}$ m^2/s. This value of D_{AB} is of the same order of magnitude as most of the gas diffusivities in Table 1-6. The prediction from Eq. (1.5-12) that Pr and Sc are roughly unity is also in agreement with the findings for gases discussed in Section 1.4. Moreover, Eqs. (1.5-9) and (1.5-11) correctly predict that k and μ increase with T, and (at low densities) are independent of P. Finally, Eq. (1.5-10) is correct in that D_{AB} increases with T and varies inversely with P. We conclude that this elementary model does remarkably well in predicting the main features of the transport properties of gases.

More careful comparisons with measurements in gases reveal that the numerical coefficients in Eqs. (1.5-9)–(1.5-11) are not very accurate and that the temperature dependence of each transport coefficient is underestimated. Although precise power-law relationships are not obeyed, it is roughly true that $\mu \propto k \propto T$ (rather than $T^{1/2}$) and that $D_{AB} \propto T^{7/4}$ (rather than $T^{3/2}$). These discrepancies are largely corrected by the rigorous

Figure 1-7. Qualitative dependence of intermolecular potential energy on separation distance, according to Eq. (1.5-13).

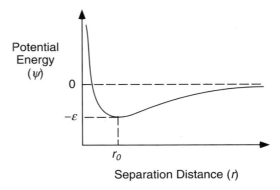

kinetic theory of Chapman and Enskog (Chapman and Cowling, 1951), which considers in some detail the effect of intermolecular potential energies on the interactions between colliding molecules. The Chapman–Enskog theory assumes that all collisions are binary and elastic and that molecular motion during collisions can be described by classical mechanics. It also treats the molecules as spherically symmetric. It successfully describes the transport properties of gases at low pressures and high temperatures, except for the thermal conductivities of polyatomic gases. In that case a correction must be included to account for transfer of internal (i.e., rotational and vibrational) energy.

An example of an intermolecular potential for uncharged, spherically symmetric molecules is the Lennard-Jones 12-6 potential,

$$\psi(r) = 4\varepsilon \left[\left(\frac{d^*}{r} \right)^{12} - \left(\frac{d^*}{r} \right)^{6} \right]. \tag{1.5-13}$$

In this expression, ψ is the intermolecular potential energy, which is assumed to depend only on the distance between molecules (r) and the parameters ε and d^*. This function is depicted qualitatively in Fig. 1-7, which shows that intermolecular forces are weakly attractive ($d\psi/dr > 0$) at large separations and strongly repulsive ($d\psi/dr < 0$) at small separations. The depth of the energy well is ε, and the transition from attraction to repulsion occurs at $r = r_0 = 2^{1/6}d^* \cong 1.1d^*$. Molecular interactions are important over distances which are a small multiple of d^*; according to Eq. (1.5-13), ψ is only 6% of ε when $r = 2d^*$. The effective diameter d^* in Eq. (1.5-13) is not identical to d in Eq. (1.5-6), but the two quantities have the same order of magnitude.

A consideration of characteristic lengths and times provides insight into why kinetic theory works as well as it does in describing the transport properties of gases. At room conditions (1 atm and 293 K) the number density of molecules in an ideal gas is $n = 2.5 \times 10^{25}$ m^{-3}. This corresponds to a mean distance between molecules of $\ell_0 = n^{-1/3} = 3.4 \times 10^{-9}$ m. For simple, polyatomic gases a typical value of d^* is 4×10^{-10} m (Hirschfelder et al., 1954). Accordingly, ℓ_0 is roughly tenfold larger than d^*, and intermolecular forces are weak except when rare dynamical events bring pairs of molecules very close together. Because the length scale of the forces is d^*, a measure of the duration of a "collision" is the time during which two molecules are separated by a distance comparable to d^*. Thus, the time a molecule spends in a collision is of the order of $d^*/v \cong 10^{-12}$ s, which is much shorter than the time it spends between collisions, $\ell/u \cong 10^{-10}$ s. Neglecting the time spent during a collision was one of several assumptions used to derive Eq. (1.5-4). In essence, the limiting factor for transport in gases is the frequency of collisions. It is this fact which leads to the very similar diffusivities for energy, species, and momentum.

Diffusivities for Liquids

A consideration of length scales for liquids yields a very different picture than for gases. For a liquid with $M = 30$ g mol^{-1} and $\rho = 1.0 \times 10^3$ kg m^{-3}, the number density and average spacing of molecules are $n = 2.0 \times 10^{28}$ m^{-3} and $\ell_0 = 3.7 \times 10^{-10}$ m. Thus, ℓ_0 for liquids is of the same order of magnitude as d^*. This indicates that significant intermolecular forces are present at all times, so that the concept of distinct molecular collisions is much less meaningful for liquids than it is for gases. The transport coefficient which is related most directly to molecular displacements is D_{AB}, and a displacement distance or jump length ℓ can be estimated using Eq. (1.5-4) as $\ell = 6D_{AB}/u$. Lacking a kinetic theory for liquids comparable to that available for gases, we equate u with the speed of sound. For $D_{AB} = 1 \times 10^{-9}$ m^2/s and $u = 1.5 \times 10^3$ m/s (the speed of sound in water), we obtain $\ell = 4.0 \times 10^{-12}$ m, or roughly $10^{-2} d^*$. Thus, whereas the typical jump length in gases ($\ell = 1 \times 10^{-7}$ m) is three orders of magnitude larger than the effective molecular diameter, the jump length in liquids is much smaller than d^*. The relatively close packing of molecules in liquids requires the coordinated motion of many molecules to permit single displacements on the order of d^* or larger, so that typical displacements are much smaller.

The observation that Pr and Sc in liquids often differ widely from unity (Section 1.4) suggests that momentum, energy, and mass transfer do not share a common mechanism, as they do in gases. The absence of a common mechanism is suggested further by the markedly different dependencies of k, D_{AB}, and μ on temperature. Although D_{AB} increases with temperature in liquids (as it does in gases), k and μ generally decrease (Section 1.4). Moreover, k typically follows a relation of the form $k = k_0(1 - AT)$, whereas μ generally varies as $\mu = \mu_0 \exp(B/T)$, with k_0, A, μ_0, and B being positive constants. Because ν and α greatly exceed D_{AB}, momentum and energy transfer in liquids must not depend entirely (or even primarily) on net molecular displacements. The basis for the enhancement of momentum and energy transfer can be understood qualitatively in terms of the intermolecular forces. As already noted, liquids are dense enough for molecules to be affected at all times by interactions with their neighbors. Rotational and vibrational motions can be transmitted through the intermolecular forces, augmenting any transport that occurs through the net displacement (translation) of molecules.

Stokes–Einstein Model

We now discuss the hydrodynamic model for diffusion of chemical species in liquids proposed by Einstein in 1906 (see Einstein, 1956). In the derivation which follows, a form of Fick's law will be obtained from thermodynamic and mechanical arguments. Consider a very dilute liquid solution, and suppose that the solute molecules are much larger in diameter than the solvent molecules. On the scale of a solute molecule, the solvent will then resemble more a continuum than a collection of discrete molecules, and the resistance to the motion of the solute will be similar to the hydrodynamic drag experienced by a solid body moving through the same solvent. A diffusive flux of a solute is the result of gradients in its concentration, or, more generally, gradients in its chemical potential. The gradient in chemical potential may be interpreted as a force promoting diffusion; this force is opposed by the hydrodynamic drag.

A force balance is written for a representative solute molecule by considering only the *average* motion of the solute, assumed to be steady. Neglecting inertia, the y component of the force balance is

$$\kappa T \frac{\partial \ln C_A}{\partial y} + f v_{Ay} = 0, \tag{1.5-14}$$

where κ is Boltzmann's constant, f is the drag coefficient, and v_{Ay} is the solute velocity, defined as in Eq. (1.2-5). The first term is the chemical potential gradient per molecule of solute A, which is assumed to act as a body force. This expression for the chemical potential gradient applies to an ideal solution at constant T and P. The second term represents the hydrodynamic force on the solute molecule. The solvent velocity far from the solute is taken to be zero.

In a very dilute solution, the mass-average velocity is essentially the same as the solvent velocity. With no bulk motion, $J_{Ay} = N_{Ay} = C_A v_{Ay}$, and Eq. (1.5-14) is rearranged to

$$J_{Ay} = -D_{AB} \frac{\partial C_A}{\partial y}, \tag{1.5-15}$$

$$D_{AB} = \frac{\kappa T}{f}. \tag{1.5-16}$$

Thus, the diffusivity varies inversely with the drag coefficient. If the solute is assumed to be a solid sphere of radius r_A, then f is evaluated using Stokes' equation for the drag on a sphere at low Reynolds number (see Chapter 7). The result is the well-known *Stokes–Einstein equation,*

$$D_{AB} = \frac{\kappa T}{6\pi\mu r_A}. \tag{1.5-17}$$

The diffusivity is seen now to be inversely proportional to both the solute radius and the solvent viscosity. Equation (1.5-17) is often used to calculate an effective solute radius from a measured value of D_{AB}.

The Stokes–Einstein equation is more general than the restrictive conditions of this derivation imply. In particular, other approaches show that this result is valid for transient as well as steady-state conditions. This relation has found widespread use in interpreting diffusion data for dilute liquid solutions. Unexpectedly, Eq. (1.5-17) is often very accurate for solute radii as small as 2–3 times that of the solvent. Empirical correlations for liquid diffusivities often use it as a starting point, because it predicts the correct limiting behavior for large solutes. It explains very simply the observation that D_{AB} declines as solute size increases. The Stokes–Einstein equation also describes well the increases in liquid diffusivity which occur with increasing temperature. The temperature dependence is contained in the ratio T/μ, with the effects of temperature on μ often being the more important factor.

Hydrodynamic models for diffusion have been developed also for solutions which are not infinitely dilute. For mutual (or gradient) diffusion of large solutes which act as hard spheres, Batchelor (1976) found that

$$D_{AB} = D_0(1 + 1.45\phi), \tag{1.5-18}$$

where D_0 is the diffusivity from the Stokes–Einstein equation and ϕ is the volume fraction of solute. In this expression, which is for moderately dilute solutions, terms containing ϕ^2 (or higher powers) have been neglected. As shown, the net effect of the thermodynamic (excluded volume) and hydrodynamic interactions between hard spheres is an increase in the mutual diffusion coefficient. Molecules which are more strongly repulsive (i.e., due to long-range electrostatic interactions) may have diffusivities which are much more sensitive to concentration than is indicated in Eq. (1.5-18) (Russel et al., 1989, Chapter 13).

Lattice Model for Diffusivities in Solids[3]

Solid-state diffusion also can be described using lattice models. In crystalline solids, especially, the regularity of the structures allows a lattice model to provide a fairly realistic representation of the geometry of random walks. A feature not encountered in our application of the lattice model to gas diffusivities is the close proximity of the atoms in a solid lattice, which introduces energy barriers that must be overcome for net displacement of a diffusing solute. Diffusion in crystalline solids can occur by *substitutional diffusion,* in which the solute moves only along paths which connect lattice sites, or by *interstitial diffusion,* in which the paths are not restricted in this manner. The latter mechanism is only feasible for atoms which are much smaller than those which compose the solid. In the discussion of elementary jump-rate theory that follows, we consider only substitutional diffusion of a dilute solute; more complete descriptions of this classic analysis are available elsewhere (Girifalco, 1964; Borg and Dienes, 1988).

Consider an atom which moves along a simple cubic lattice, as shown in Fig. 1-6. The frequency of jumps between adjacent lattice points is written in terms of the velocity of the diffusing species u and the lattice spacing ℓ as $\omega \equiv u/\ell$. Then, the diffusion coefficient in the absence of any barrier to a jump is given by rewriting Eq. (1.5-4) as

$$D = \frac{\omega \ell^2}{6}. \tag{1.5-19}$$

This expression assumes that the free energies of all lattice sites are equivalent and that movement of the diffusing molecule does not cause an increase in the free energy of the system. Now assume that movement from one lattice site to another causes the diffusing molecule to pass through a position where there is an increased free energy, as shown in Fig. 1-8. The maximum free energy along this path, denoted as ΔG^*, corresponds to a *saddle point* in a complete map of energy versus position. At equilibrium, elementary statistical mechanics indicates that the concentration of atoms in the saddle-point states (C^*) is related to that in the lattice states (C_0) by

---

[3] This analysis was suggested by R. A. Brown.

Figure 1-8. Schematic representation of a saddle point on an energy diagram for solid-state diffusion.

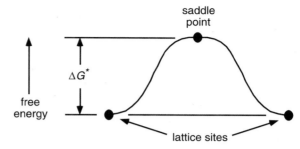

$$\frac{C^*}{C_o} = \exp\left(-\frac{\Delta G^*}{\kappa T}\right). \qquad (1.5\text{-}20)$$

The concentration ratio in Eq. (1.5-20) is just the relative probability of finding a solute atom in the saddle-point configuration. The fraction of attempted jumps which are successful is assumed to be proportional to this relative probability, so that the frequency of successful jumps from one lattice site to another is $\omega C^*/C_0$. It follows that the diffusivity is given by

$$D = \frac{\omega \ell^2}{6} \exp\left(-\frac{\Delta G^*}{\kappa T}\right). \qquad (1.5\text{-}21)$$

This is an Arrhenius-type expression for the dependence of the diffusion coefficient on temperature, similar to that used to describe the temperature dependence of reaction rate constants. Diffusion coefficients in many solids follow this type of expression.

Equation (1.5-21) is usually rewritten by separating the free energy difference into entropic and enthalpic parts as $\Delta G^* = T\,\Delta S^* - \Delta H^*$ to give

$$D = \frac{\omega \ell^2}{6} \exp\left(-\frac{\Delta S^*}{\kappa}\right) \exp\left(\frac{\Delta H^*}{\kappa T}\right) \equiv D_o \exp\left(\frac{\Delta H^*}{\kappa T}\right). \qquad (1.5\text{-}22)$$

Equation (1.5-22) is correct for a solute diffusing among totally unoccupied lattice sites; however, in a given solid all sites are not necessarily vacant. The concentration of vacancies relative to the total concentration of lattice sites can be expressed in a manner analogous to Eq. (1.5-20), by letting ΔG^+ be the free energy associated with the formation of a vacancy in a perfect crystal. Following the same reasoning as before, Eqs. (1.5-21) and (1.5-22) become

$$D = \frac{\omega \ell^2}{6} \exp\left(-\frac{\Delta G^*}{\kappa T}\right) \exp\left(-\frac{\Delta G^+}{\kappa T}\right), \qquad (1.5\text{-}23)$$

$$D = D_0 \exp\left(\frac{\Delta H^* + \Delta H^+}{\kappa T}\right). \qquad (1.5\text{-}24)$$

Equation (1.5-24) indicates that measuring the diffusion coefficient as a function of temperature yields only the combined enthalpic contributions from the saddle-point and vacancy-formation free energies.

Examples of solid-state diffusion data as a function of temperature are shown in

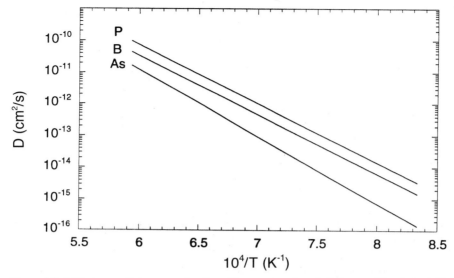

Figure 1-9. Diffusion coefficients as a function of temperature for selected group III and group V elements in silicon, based on data summarized by Casey and Pearson (1975). Results for As, B, and P are shown for temperatures ranging from 1200 to 1685 K (the melting point of Si).

Fig. 1-9, which gives diffusivities for selected group III and group V elements in silicon. The original data were correlated using Eq. (1.5-24), yielding values of D_0 ranging from 5 to 60 cm²/s and $(\Delta H^* + \Delta H^+)$ from 3.7 to 4.2 eV, where eV = electron volt = 1.602×10^{-19} J. Even near the melting point of silicon (1685 K), the diffusivities are extremely small. Taking boron at this temperature as an example, $D = 4.4 \times 10^{-11}$ cm²/s = 4.4×10^{-15} m²/s.

1.6 CONTINUUM APPROXIMATION

The constitutive equations for energy, species, and momentum transport are based on the assumption that variables such as temperature, species concentration, and velocity can be regarded as continuous functions of position. The existence of discrete molecules is ignored. This assumption is not automatically valid, and it is important to be aware of its limitations. In evaluating the continuum hypothesis it is helpful to keep in mind the length and time scales discussed in Section 1.5. These are summarized in Table 1-9.

The larger the minimum linear dimension (L) of a system, the more likely it is that a continuum model will be valid. The requirements for L may be considered from two perspectives. One consideration is the need to minimize random fluctuations in state variables over time or over position. If a system contains too few molecules, properties such as density, concentration, and velocity are not well-defined at a mathematical "point." The second consideration relates to the importance of interactions with other molecules within a given fluid, compared with interactions with molecules in other phases which act as boundaries. In molecular interpretations of μ, k, and D_{AB} such as those discussed in Section 1.5, it is usual to regard the fluid as infinite and to ignore

TABLE 1-9
Molecular Length and Time Scales Representative
of Gases and Liquids

Quantity	Gases[a]	Liquids[b]
Molecular diameter, d (m)	3×10^{-10}	3×10^{-10}
Number density, n (m^{-3})	3×10^{25}	2×10^{28}
Intermolecular spacing, ℓ_0 (m)	3×10^{-9}	4×10^{-10}
Displacement distance, ℓ (m)	1×10^{-7}	$\sim 10^{-12}$
Molecular velocity, u (m/s)	5×10^2	$\sim 10^3$
Displacement time, ℓ/u (s)	$\sim 10^{-10}$	$\sim 10^{-15}$
Duration of collisions, $\sim d/u$ (s)	$\sim 10^{-12}$	$\sim 10^{-13}$

[a]Based on an ideal gas at $P = 1$ atm and $T = 293$ K, with $M = 30$ g/mol.
[b]Based on $\rho = 1 \times 10^3$ kg/m^3 and $M = 30$ g/mol.

any effects of boundaries on these coefficients. We should have some idea of the conditions under which these "bulk" transport coefficients remain valid. It is difficult to answer either kind of question with precision, but we can present some useful guidelines.

To illustrate the concept of fluctuations, suppose that one has a "snapshot" showing the positions of all molecules in a fluid sample at a given instant in time. If one computed a mass density by counting molecules in various sample volumes (δV) centered about some fixed location, the results might resemble those in Fig. 1-10. If δV is relatively large, the calculated value of ρ is influenced by macroscopic variations in density, such as those occurring in a vertical column of air. As δV is reduced, we expect ρ to approach a local value representative of the reference "point" we have chosen in the fluid. However, as δV becomes still smaller, reductions in δV eventually lead to wide swings in ρ, as the boundary of δV passes the locations of the few molecules remaining within the sampling volume. Large and apparently random variations in ρ would be expected also for a small and constant value of δV if the reference point were moved slightly or if observations were made at the same location over time. Each of these variations in ρ at small δV represents a kind of fluctuation. Because fluctuations arise from the discrete nature of molecules, they occur not only in ρ but in other local quantities.

To assess the expected magnitude of fluctuations in relation to δV, we need the

Figure 1-10. Density as a function of sample volume, at a fixed location and a given instant in time.

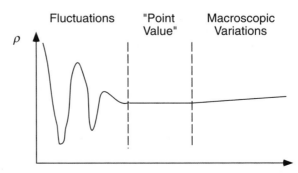

concept of an ensemble, as used in statistical mechanics. An ensemble consists of a number of hypothetical replicas of a system, all identical in their macroscopic thermodynamic properties but differing in molecular-level details. The individual systems in an ensemble are chosen to represent all possible molecular states of the real system, so that the number of systems in an ensemble is ordinarily very large. A fundamental postulate is that long-time averages of mechanical variables (e.g., pressure, energy, volume, number of molecules) in a thermodynamic system are equivalent to the corresponding ensemble averages, provided that the number of systems in the ensemble is suitably large. According to the *ergodic hypothesis,* all systems in an ensemble should be weighted equally when computing ensemble averages, because the real system spends equal amounts of time in each of the available states. We can consider a macroscopic system at equilibrium to be divided into a large number of subsystems. Whether we move from one subsystem to another, or follow the evolution over time of a single subsystem, the statistical behavior of all properties should be equivalent. The implied equivalence of spatial and temporal fluctuations simplifies our problem.

Different types of ensembles are distinguished by which variables are held fixed. In a closed system where the number of molecules (N), the volume (V), and T are held constant, the probability distribution of energies is Gaussian. A measure of the magnitude of fluctuations is the standard deviation divided by the mean, which for energy is found to be of the order of $N^{-1/2}$. In an open, isothermal system, where V, T, and chemical potential are fixed, the magnitudes of fluctuations in N and P are of the order of $\overline{N}^{-1/2}$, where $\overline{N}$ is the mean value of N.

These results suggest a criterion for the existence of "point" quantities having negligible fluctuations. Namely, we must be able to choose a length scale δL which is small enough to define local variables and to perform limiting mathematical operations such as differentiation, but at the same time large enough to make those local variables well-behaved. This can be accomplished by requiring that $\delta L \ll L$ and that the "point volume" δV contain a sufficient number of molecules. The latter requirement can be written as $\delta L > (N_{min}/n)^{1/3}$, where n is the average (macroscopic) number density of molecules and N_{min} is the minimum acceptable number of molecules in the volume $\delta V = (\delta L)^3$. Choosing $L > 10 \ \delta L$ and $N_{min} = 10^4$, and using the typical number densities in Table 1-9, gives $L > 7 \times 10^{-7}$ m for gases and $L > 8 \times 10^{-8}$ m for liquids. These results suggest that to make fluctuations in state variables acceptably small, all system dimensions should exceed roughly 1 μm for gases (at room conditions) and 0.1 μm for liquids. The exact numerical values given are of course arbitrary. For ideal gases, $n \propto P$ at any given temperature, so that the minimum value of L will vary as $P^{-1/3}$. Thus, for gases under vacuum, the minimum value of L may greatly exceed 1 μm.

We turn now to the second major consideration regarding L. For the values of μ, k, and D_{AB} determined in the bulk fluid to be meaningful, interactions with other molecules in the fluid must be much more important than those with the boundaries. This will be true if L greatly exceeds the length scale of the intermolecular interactions. In that case, the molecules in the fluid will interact with one another much more frequently than with those at the boundaries. The characteristic length for interactions in gases is the mean free path, given by Eq. (1.5-6), whereas that for liquids (excluding ion–ion interactions) is of the order of the molecular diameter. Requiring L to exceed these interaction lengths by a factor of 10, and using the values given in Table 1-9, gives

$L > 1 \times 10^{-6}$ m for gases and $L > 3 \times 10^{-9}$ m for liquids. These results suggest that bulk transport properties should be applicable ordinarily to system dimensions of about 1 μm in gases and about 3 nm in liquids.

The various estimates of minimum system dimensions are compared in Table 1-10. The fluctuation and mean-free-path criteria for gases at room conditions yield essentially identical lower bounds for L, about 1 μm in each case. However, the fluctuation-based bound for L varies only as $P^{-1/3}$, whereas Eq. (1.5-6) shows that the mean free path varies as P^{-1}. This indicates that at low pressures, the mean-free-path criterion will be the more stringent one. For liquids, the more restrictive criterion is that used to guarantee small fluctuations in point variables.

A well-known situation where the criteria of Table 1-10 are not met involves gas flow through small pores. When the mean free path exceeds (or is comparable to) the pore diameter, transport is governed by collisions with the pore wall, rather than by collisions with other gas molecules (Mason and Marrero, 1970). This type of transport is called *Knudsen flow*. Low pressures alone may cause the mean free path to become comparable to the system dimensions, as in high-vacuum equipment and in the upper atmosphere.

In liquid-filled pores of molecular dimensions, D_{AB} is typically reduced well below its bulk-solution value. When the pore radius is much larger than that of the solvent but comparable to that of the solute, this finding is attributable in part to the enhanced hydrodynamic drag experienced by a particle moving in a confined fluid. In terms of Eq. (1.5-16), f is increased by the proximity of the pore walls (Deen, 1987). If the solute, solvent, and pore are all of comparable size, a hydrodynamic explanation of diffusional hindrances is less appropriate.

Measurements of ionic conductivities in membranes which have straight, liquid-filled pores of constant cross section suggest that the value of μ in bulk water is applicable to pores with radii as small as 3 nm (Anderson and Quinn, 1972). This finding is consistent with our rough estimate of what is needed to obtain bulk property values for low-molecular-weight liquids ($L > 3$ nm). Similarly, measurements of the viscosity of thin films of n-tetradecane (a Newtonian liquid) sheared between molecularly smooth surfaces have shown that μ is within 10% of its bulk value for film thicknesses as small as 10 molecular diameters (Israelachvili and Kott, 1989). Both sets of results suggest that our criterion based on fluctuations ($L > 0.1$ μm) may be too conservative. A factor which tends to mitigate any effects of fluctuations in transport studies with porous media, especially, is that the macroscopic quantities measured are averages over large numbers of pores. With regard to heat transfer, Flik et al. (1992) discuss the limitations of

TABLE 1-10
Estimates of Minimum System Dimensions for Continuum Transport Models Using Bulk Properties

Type of fluid	Small fluctuations	Bulk properties
Gas[a]	1 μm	1 μm
Liquid[b]	0.1 μm	3 nm

[a]Based on an ideal gas at $P = 1$ atm and $T = 293$ K, with $M = 30$ g/mol.
[b]Based on $\rho = 1 \times 10^3$ kg/m^3 and $M = 30$ g/mol.

applying bulk values of thermal conductivities to microstructures composed of various materials.

There are limits on the time scales, as well as the length scales, to which the basic constitutive equations can be applied. Because these constitutive equations do not contain time as a variable, the implicit assumption is that fluxes result instantaneously from the imposition of the corresponding gradients. For all transport in gases and for species diffusion in liquids, one characteristic time scale is that corresponding to a single molecular displacement, ℓ/u. As shown in Table 1-10, this is ordinarily about 10^{-10} s in gases and 10^{-15} s in liquids. A second time scale is given by d/u, which may be interpreted roughly as the duration of a collision; this time is 10^{-12} s in gases and 10^{-13} s in liquids. The longer of the respective time estimates, namely 10^{-10} s in gases and 10^{-13} s in liquids, are representative of how fast a flux can result from an instantaneously applied gradient. These times are so small that it is not surprising that there is no need in most applications to be concerned with relaxation (time-dependent) effects in the constitutive equations. An important exception to this conclusion arises in the flow of many polymeric liquids, either solutions or molten polymers. This is because large, flexible polymers may require relatively long times to relax to equilibrium from a deformed and/ or oriented state. Since the rheological properties of the fluid depend on the configuration of the polymer molecules, the constitutive equations relating stress to velocity gradients must take the polymer relaxation time into account. These times for molten polymers and polymers in solution may be as large as several seconds.

References

Anderson, J. L., and J. A. Quinn. Ionic mobility in microcapillaries. A test for anomalous water structures. *J. Chem. Soc. Faraday Trans. I* 68: 744–748, 1972.

Batchelor, G. K. Brownian diffusion of particles with hydrodynamic interaction. *J. Fluid Mech.* 74: 1–29, 1976.

Bird, R. B., W. E. Stewart, and E. N. Lightfoot. *Transport Phenomena.* Wiley, New York, 1960.

Borg, R. J., and G. J. Dienes. *An Introduction to Solid State Diffusion.* Academic Press, New York, 1988.

Casey, H. C., Jr. and G. L. Pearson. Diffusion in semiconductors. In *Point Defects in Solids,* Vol. 2, J. H. Crawford, Jr. and L. M. Slifkin, Eds. Plenum Press, New York, 1975, pp. 163–255.

Chapman, S. and T. G. Cowling. *Mathematical Theory of Non-Uniform Gases,* second edition. Cambridge University Press, Cambridge, 1951.

Cussler, E. L. *Diffusion.* Cambridge University Press, Cambridge, 1984.

Deen, W. M. Hindered transport of large molecules in liquid-filled pores. *AIChE J.* 33: 1409–1425, 1987.

Einstein, A. *Investigations on the Theory of the Brownian Movement.* Dover, New York, 1956. [This is an edited English translation of Einstein's early work on diffusion and molecular dimensions; the original papers appeared between 1905 and 1911.]

Flik, M. I., B. I. Choi, and K. E. Goodson. Heat transfer regimes in microstructures. *J. Heat Transfer* 114: 666–674, 1992.

Gebhart, B., Y. Jaluria, R. L. Mahajan, and B. Sammakia. *Buoyancy-Induced Flows and Transport.* Hemisphere, New York, 1988.

Gillispie, C. G. (Ed.). *Dictionary of Scientific Biography.* Scribner's, New York, 1970–1980.

Girifalco, L. A. *Atomic Migration in Crystals.* Blaisdell, New York, 1964.

Glazov, V. M., S. N. Chizhevskaya, and N. N. Glagoleva. *Liquid Semiconductors.* Plenum Press, New York, 1969.

Hirschfelder, J. O., C. F. Curtiss, and R. B. Bird. *Molecular Theory of Gases and Liquids.* Wiley, New York, 1954.

Israelachvili, J. N. and S. J. Kott. Shear properties and structure of simple liquids in molecularly thin films: the transition from bulk (continuum) to molecular behavior with decreasing film thickness. *J. Colloid Interface Sci.* 129: 461–467, 1989.

Johnson, E. M., D. A. Berk, R. K. Jain, and W. M. Deen. Diffusion and partitioning of proteins in charged agarose gels. *Biophys. J.* 68: 1561–1568, 1995.

Kodera, H. Diffusion coefficients of impurities in silicon melt. *Jap. J. Appl. Phys.* 2: 212–219, 1963.

Mason, E. A. and T. R. Marrero. The diffusion of atoms and molecules. In *Advances in Atomic and Molecular Physics,* Vol. 6, D. R. Bates and I. Esterman, Eds. Academic Press, New York, 1970.

Reid, R. C., J. M. Prausnitz, and B. E. Poling. *The Properties of Gases and Liquids,* fourth edition. McGraw-Hill, New York, 1987.

Rohsenow, W. M. and J. P. Hartnett. *Handbook of Heat Transfer.* McGraw-Hill, New York, 1973.

Russel, W. B., D. A. Saville, and W. R. Schowalter. *Colloidal Dispersions.* Cambridge University Press, Cambridge, 1989.

Tadmor, Z. and C. Gogos. *Principles of Polymer Processing.* Wiley-Interscience, New York, 1979.

Touloukian, Y. S. and C. Y. Ho (Eds.). *Thermophysical Properties of Matter,* Vol. 6. IFI/Plenum, New York, 1970.

Weast, R. C., Ed. *Handbook of Chemistry and Physics,* 49th edition. Chemical Rubber Co., Cleveland, 1968.

Problems

1-1. Uniqueness of Binary Diffusivity

Show that for any binary mixture, $D_{AB} = D_{BA}$. Begin by evaluating the diffusive fluxes of species A and B using Eq. (A) or Eq. (D) of Table 1-3.

1-2. Diffusion Relative to Volume-Average Velocity

A reference frame sometimes chosen for diffusion is the *volume-average velocity,* $\mathbf{v}^{(V)}$. For a binary mixture this velocity is given by

$$\mathbf{v}^{(V)} = \mathbf{N}_A \bar{V}_A + \mathbf{N}_B \bar{V}_B,$$

where $\bar{V}_A$ and $\bar{V}_B$ are the partial molar volumes of species A and B, respectively. The total molar concentration is related to the partial molar volumes and mole fractions by

$$C = (x_A \bar{V}_A + x_B \bar{V}_B)^{-1}.$$

The molar flux of A relative to the volume-average velocity, $\mathbf{J}_A^{(V)}$, is defined as

$$\mathbf{J}_A^{(V)} = \mathbf{N}_A - C_A \mathbf{v}^{(V)}.$$

(a) Show that

$$\mathbf{J}_A^{(V)} \bar{V}_A + \mathbf{J}_B^{(V)} \bar{V}_B = 0.$$

(b) Show that

$$\mathbf{J}_A^{(V)} = C \bar{V}_B \mathbf{J}_A^{(M)}.$$

(c) Using the result in (b), show that

$$J_A^{(V)} = -D_{AB}\nabla C_A.$$

This indicates that for diffusion relative to the volume-average velocity, the "natural" driving force is the gradient in molar concentration; compare with Eqs. (A) and (D) of Table 1-3.

1-3. Equivalence of Alternate Forms of Fick's Law

The objective of this problem is to demonstrate the equivalence of Eqs. (A) and (D) of Table 1-3; obtaining Eqs. (B) and (C) from these requires only that one multiply or divide by M_A to obtain the required molar or mass units.

(a) Noting that $j_A + j_B = J_A M_A + J_B M_B = 0$ (Table 1-2), show that

$$J_A = \frac{M_B C}{\rho} J_A^{(M)}.$$

(b) Use the result in part (a) to derive Eq. (A) of Table 1-2 from Eq. (D). This shows that the various forms of Fick's law are equivalent.

1-4. Flux Normal to a Surface[*]

Consider a point $P = (x, y, z) = (2a/3, b/3, 2c/3)$, which is on the surface of an ellipsoid defined by

$$\left(\frac{x}{a}\right)^2 + \left(\frac{y}{b}\right)^2 + \left(\frac{z}{c}\right)^2 = 1.$$

Suppose that there is a flux $\mathbf{f}$ which has the components

$$\mathbf{f} = (f_x, f_y, f_z) = \left(\frac{2a}{3}, \frac{b}{3}, \frac{2c}{3}\right).$$

Notice that $\mathbf{f}$ is parallel to a line connecting the origin with point P. Compute $f_n = \mathbf{n}\cdot\mathbf{f}$ at point P, where $\mathbf{n}$ is the outward-pointing unit vector which is normal to the surface of the ellipsoid.

[*]This problem was suggested by R. A. Brown.

1-5. Computation of Stress on a Surface[*]

Suppose that the stress in a fluid at point P is measured by placing a very small transducer at that point and determining the force per unit area for different orientations of the transducer surface. The following results are obtained, independent of time:

Orientation of Surface	Measured Stress Vector
$\mathbf{e}_x$	$2\mathbf{e}_x + \mathbf{e}_y$
$\mathbf{e}_y$	$\mathbf{e}_x + 3\mathbf{e}_y + \mathbf{e}_z$
$\mathbf{e}_z$	$\mathbf{e}_y + 5\mathbf{e}_z$

(a) Determine the rectangular components of the stress tensor, σ_{ij}.

(b) Compute the stress vector s at point P for a test surface with the orientation

$$\mathbf{n} = \frac{2\mathbf{e}_x + \mathbf{e}_y + 2\mathbf{e}_z}{3}.$$

(c) For the test surface of part (b), determine the <u>normal component</u> of the stress vector.

(d) Again for the test surface of part (b), compute two tangential components of the stress vector. (*Hint:* First determine two unit vectors that are tangent to the surface. They need not be orthogonal to one another.)

1-6. Stress-Free Surface*

Suppose that the stress tensor at point P in a fluid is given by

$$\boldsymbol{\sigma} = \begin{bmatrix} \sigma_{11} & 4 & 2 \\ 4 & 0 & 3 \\ 2 & 3 & 0 \end{bmatrix},$$

where only σ_{11} is unknown. Determine σ_{11} so that there is a plane with unit normal **n** on which the stress vector $\mathbf{s} = \mathbf{n} \cdot \boldsymbol{\sigma}$ vanishes. Determine **n**.

*This problem was suggested by R. A. Brown.

Chapter 2

CONSERVATION EQUATIONS AND THE FUNDAMENTALS OF HEAT AND MASS TRANSFER

2.1 INTRODUCTION

There are two major parts to any transport model. One is the *constitutive equation* used to relate the diffusive flux of a quantity to the local material properties, as discussed in Chapter 1. The other is the *conservation equation,* which relates the rate of accumulation of the quantity to the rates at which it enters, or is formed within, a specified region. Whereas constitutive equations are largely empirical and material-specific, conservation equations are based more on first principles and have certain universal forms. In this chapter we begin by deriving a number of conservation equations which apply to any scalar quantity, including the amounts of mass, energy, and individual chemical species. The general results are then used to obtain the equations governing heat transfer in a pure fluid and mass transfer in a dilute, isothermal mixture. Simple applications of those equations are illustrated using several examples. Finally, both the species conservation equation and the interpretation of the diffusivity are reconsidered from a molecular viewpoint.

Heat transfer in pure fluids and mass transfer in isothermal mixtures are treated in parallel throughout most of this book because the differential equations and boundary conditions describing them are largely analogous. Nonetheless, there are important situations where transport processes are coupled in such a way that analogies are of little use. Examples include heat transfer in reactive mixtures and multicomponent mass transfer in gases, as discussed in Chapter 11.

In this chapter there is extensive use of vector–tensor notation, and the reader may find it helpful to review the entire Appendix before proceeding.

2.2 GENERAL FORMS OF CONSERVATION EQUATIONS

The equations describing the conservation of scalar quantities are derived in this section using the concept of a *control volume*. A control volume is any closed region in space, selected for its usefulness in formulating a desired balance equation. It is a region in which the rate of accumulation of some quantity is equated with the net rate at which that quantity enters by crossing the boundaries and/or is formed by internal sources. The size and shape of a control volume may vary with time, and the control volume boundaries need not correspond to physical interfaces.

Conservation Equations for Finite Volumes

As shown in Fig. 2-1, a control volume is assumed to have a volume $V(t)$ and surface area $S(t)$, where t is time. Differential elements of volume and surface area are denoted by dV and dS, respectively. A unit vector $\mathbf{n}$ defined at every point on the surface is normal to the surface and directed outward. The fluid velocity and the velocity of the control surface are denoted as $\mathbf{v}(\mathbf{r}, t)$ and $\mathbf{v}_S(\mathbf{r}, t)$, respectively, where $\mathbf{r}$ is the position vector. For a given point on the control surface, the fluid velocity *relative to the surface* is $\mathbf{v} - \mathbf{v}_S$, and the component of the relative velocity that is perpendicular to the control surface and directed outward is $(\mathbf{v} - \mathbf{v}_S) \cdot \mathbf{n}$. Only if $\mathbf{v} = \mathbf{v}_S$ is there no fluid flow across the surface.

    We begin by considering a *fixed control volume*, in which $\mathbf{v}_S = \mathbf{0}$ and both V and S are constant. Let $b(\mathbf{r}, t)$ denote the concentration of some quantity (i.e., its amount per unit volume), and let $\mathbf{F}(\mathbf{r}, t)$ be the total flux of that quantity (i.e., the flux relative to some fixed reference point or origin). To allow for chemical reactions and certain energy inputs from external sources, let $B_V(\mathbf{r}, t)$ denote the rate of formation of the quantity per unit volume. The amount of the quantity contained in a differential volume element centered about the point $\mathbf{r}$ is $b(\mathbf{r}, t)\, dV$. The rate at which the quantity *enters* the control volume across a differential element of surface is $-(\mathbf{F} \cdot \mathbf{n})\, dS$; the minus sign is needed because $\mathbf{n}$ points outward. The rate at which the quantity is formed within a volume element is $B_V\, dV$. Equating the overall rate of accumulation to the rates of entry and formation gives

$$\frac{d}{dt} \int_V b\, dV = -\int_S \mathbf{F} \cdot \mathbf{n}\, dS + \int_V B_V\, dV. \qquad (2.2\text{-}1)$$

Figure 2-1. Control volume. The outward normal $(\mathbf{n})$ and the fluid $(\mathbf{v})$ and surface $(\mathbf{v}_S)$ velocities all vary with position on the control surface.

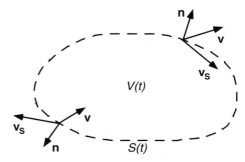

This is the most basic form of integral (or macrosopic) conservation equation. The notation $\int_V dV$ and $\int_S dS$ indicates integration over the control volume and control surface, respectively. Once these integrations over position are performed, the resulting quantities are functions only of time. Accordingly, the time derivative outside the volume integral is written as the total derivative (d/dt), not the partial derivative $(\partial/\partial t)$.

A *moving control volume* is one which is translating, rotating, and/or deforming, such that V and/or S depend on time and $\mathbf{v_S} \neq \mathbf{0}$. For such cases an additional term must be added to Eq. (2.2-1) to account for the fact that any surface motion creates a flux relative to the surface. The term $\mathbf{v_S} \cdot \mathbf{n}\, dS$ represents the rate at which volume is swept out by an element of the control surface; the product of this and the local concentration gives the rate of entry of a quantity into the control volume due to movement of the surface. Thus, the conservation equation for a moving control volume is

$$\frac{d}{dt} \int_{V(t)} b\, dV = -\int_{S(t)} \mathbf{F} \cdot \mathbf{n}\, dS + \int_{V(t)} B_V\, dV + \int_{S(t)} b\, \mathbf{v_S} \cdot \mathbf{n}\, dS, \qquad (2.2\text{-}2)$$

where the effects of surface motion are included now in the last term. The Leibniz formula for differentiating a volume integral, Eq. (A.5-9), can be written as

$$\frac{d}{dt} \int_{V(t)} b\, dV = \int_{V(t)} \frac{\partial b}{\partial t}\, dV + \int_{S(t)} b\, \mathbf{v_S} \cdot \mathbf{n}\, dS. \qquad (2.2\text{-}3)$$

This allows Eq. (2.2-2) to be simplified to

$$\int_{V(t)} \frac{\partial b}{\partial t}\, dV = -\int_{S(t)} \mathbf{F} \cdot \mathbf{n}\, dS + \int_{V(t)} B_V\, dV. \qquad (2.2\text{-}4)$$

This is a very general conservation statement, requiring only that $b(\mathbf{r}, t)$ be continuous within the volume $V(t)$. Notice that the time derivative is now inside the volume integral and is therefore a partial derivative.

If the *interior* of a control volume contains a moving interface at which the concentration is discontinuous, Eqs. (2.2-3) and (2.2-4) must each contain an additional term. Consider the control volume shown in Fig. 2-2, which encloses part of an interface between phases A and B. (A "phase" is regarded here as any region in which b is a continuous function of position.) The interface divides the control volume into two regions, with volumes V_A and V_B. The corresponding external surfaces, S_A and S_B, have

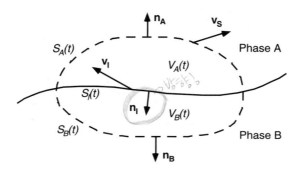

Figure 2-2. Control volume enclosing part of an interface between phase A and phase B.

unit outward normals $\mathbf{n_A}$ and $\mathbf{n_B}$, respectively. The interfacial surface (inside the control volume) is denoted by S_I and has a unit normal $\mathbf{n_I}$ which points from phase A toward phase B. Accordingly, V_A is bounded by S_A plus S_I, and V_B is bounded by S_B plus S_I. The interfacial velocity is denoted by $\mathbf{v_I}$ $(\mathbf{r}, t)$ and the velocity of any *external* surface is represented as usual by $\mathbf{v_S}(\mathbf{r}, t)$.

To obtain the required generalization of the Leibniz formula, we first apply Eq. (2.2-3) separately to V_A and V_B, which gives

$$\frac{d}{dt} \int_{V_A(t)} b \, dV = \int_{V_A(t)} \frac{\partial b}{\partial t} \, dV + \int_{S_A(t)} b \mathbf{v_S} \cdot \mathbf{n_A} \, dS + \int_{S_I(t)} b_A \mathbf{v_I} \cdot \mathbf{n_I} \, dS, \qquad (2.2\text{-}5a)$$

$$\frac{d}{dt} \int_{V_B(t)} b \, dV = \int_{V_B(t)} \frac{\partial b}{\partial t} \, dV + \int_{S_B(t)} b \mathbf{v_S} \cdot \mathbf{n_B} \, dS - \int_{S_I(t)} b_B \mathbf{v_I} \cdot \mathbf{n_I} \, dS. \qquad (2.2\text{-}5b)$$

In the integrals over the surface S_I, subscripts have been added to indicate the side of the interface on which b is evaluated. Adding these equations, we obtain

$$\frac{d}{dt} \int_{V(t)} b \, dV = \int_{V(t)} \frac{\partial b}{\partial t} \, dV + \int_{S(t)} b \mathbf{v_S} \cdot \mathbf{n} \, dS + \int_{S_I(t)} (b_A - b_B) \mathbf{v_I} \cdot \mathbf{n_I} \, dS, \qquad (2.2\text{-}6)$$

where V is the sum of V_A and V_B, S is the sum of S_A and S_B, and $\mathbf{n}$ equals either $\mathbf{n_A}$ or $\mathbf{n_B}$. In other words, V, S, and $\mathbf{n}$ have their usual meanings in connection with the *complete* control volume. Equation (2.2-6), which contains a new term to account for the discontinuity of b at the interface, is the desired generalization of the Leibniz formula, Eq. (2.2-3).

Equation (2.2-2) does not require b to be continuous within V, and therefore applies directly to the complete control volume in Fig. 2-2. Using Eq. (2.2-6) to evaluate the left-hand side of Eq. (2.2-2) results in

$$\int_{V(t)} \frac{\partial b}{\partial t} \, dV + \int_{S_I(t)} (b_A - b_B) \mathbf{v_I} \cdot \mathbf{n_I} \, dS = - \int_{S(t)} \mathbf{F} \cdot \mathbf{n} \, dS + \int_{V(t)} B_V \, dV \qquad (2.2\text{-}7)$$

which is the desired generalization of Eq. (2.2-4) for a control volume with an internal interface. Notice that the new term will vanish unless *two* conditions are present: The interface must be moving, and b must be discontinuous there.

As one more extension of the integral conservation equations, we allow for the possibility of a source term at an internal interface. Denoting the rate of formation per unit area of interface as B_S $(\mathbf{r}, t)$, the addition of this source term to Eq. (2.2-7) results in

$$\int_{V(t)} \frac{\partial b}{\partial t} \, dV + \int_{S_I(t)} (b_A - b_B) \mathbf{v_I} \cdot \mathbf{n_I} \, dS = - \int_{S(t)} \mathbf{F} \cdot \mathbf{n} \, dS + \int_{V(t)} B_V \, dV + \int_{S_I(t)} B_S \, dS.$$

$$(2.2\text{-}8)$$

Conservation Equations for Points

Conservation equations valid at a given *point* in a continuum are obtained from control volume balances by taking the limit $V \rightarrow 0$. Whether we begin with a fixed or moving control volume, we must ultimately obtain the same result when the volume is reduced to a point. Accordingly, we will employ the simplest control volume relation, Eq. (2.2-1). As the first step, the surface integral in Eq. (2.2-1) is converted to a volume integral by application of the divergence theorem, Eq. (A.5-2), which is written as

$$\int_S \mathbf{F} \cdot \mathbf{n} \, dS = \int_V \mathbf{\nabla} \cdot \mathbf{F} \, dV. \tag{2.2-9}$$

A conservation equation equivalent to Eq. (2.2-1) is then

$$\int_V \left[\frac{\partial b}{\partial t} + \mathbf{\nabla} \cdot \mathbf{F} - B_V \right] dV = 0. \tag{2.2-10}$$

Now, any integral is equal to the magnitude of the region of integration times the mean value of the integrand within that region. Thus,

$$\left\langle \frac{\partial b}{\partial t} + \mathbf{\nabla} \cdot \mathbf{F} - B_V \right\rangle V = 0, \tag{2.2-11}$$

where the angle brackets denote an average over the control volume. Because the magnitude of V is arbitrary, the bracketed quantity must vanish for any choice of control volume. A limiting process whereby $V \to 0$, with V centered on the point $\mathbf{r}$, reduces the bracketed quantity to its value at that point. Hence, Eq. (2.2-11) is satisfied in this limit only if

$$\frac{\partial b}{\partial t} = -\mathbf{\nabla} \cdot \mathbf{F} + B_V. \tag{2.2-12}$$

This relation holds at every point within a given phase. Equation (2.2-12) is the differential form of the general conservation equation and provides the basis for "microscopic" or local analyses of transport phenomena. In this book it is employed numerous times in various forms.

Conservation Equations for Interfaces

The integral forms of the conservation equations are also used to establish conservation relations at interfaces. Consider the control volume shown in Fig. 2-3, which is like that in Fig. 2-2 except that the surfaces S_A and S_B are each a constant distance ℓ from S_I. The edges of the control volume are bounded by the surface S_E, which has the outward-pointing normal vector $\mathbf{n}_E$. Applying Eq. (2.2-8) to this control volume, we obtain

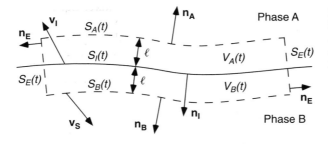

Figure 2-3. Control volume enclosing part of an interface between phase A and phase B. The surfaces S_A and S_B are each a constant distance ℓ from the interface.

$$\int_{V(t)} \frac{\partial b}{\partial t} dV + \int_{S_I(t)} (b_A - b_B)\mathbf{v_I} \cdot \mathbf{n_I} \, dS = -\int_{S_A(t)} [\mathbf{F} \cdot \mathbf{n_A}]_A \, dS - \int_{S_B(t)} [\mathbf{F} \cdot \mathbf{n_B}]_B \, dS$$

$$-\int_{S_E(t)} [\mathbf{F} \cdot \mathbf{n_E}] \, dS + \int_{V(t)} B_V \, dV + \int_{S_I(t)} B_S \, dS. \tag{2.2-13}$$

We consider now the limit $\ell \rightarrow 0$, in which the surfaces S_A and S_B approach S_I. In this limit $V \rightarrow 0$ and $S_E \rightarrow 0$, so that the corresponding integrals vanish. Moreover, $-\mathbf{n_A} \rightarrow \mathbf{n_B} \rightarrow \mathbf{n_I}$. Accordingly, Eq. (2.2-13) simplifies to

$$\int_{S_I(t)} [(\mathbf{F} - b\mathbf{v_I})_B - (\mathbf{F} - b\mathbf{v_I})_A] \cdot \mathbf{n_I} \, dS = \int_{S_I(t)} B_S \, dS. \tag{2.2-14}$$

A relation valid for any point on the interface is obtained by considering S_I to be an arbitrarily small surface containing that point. Arguments analogous to those used to derive Eq. (2.2-12) lead to a pointwise interfacial balance valid at any instant in time,

$$[(\mathbf{F} - b\mathbf{v_I})_B - (\mathbf{F} - b\mathbf{v_I})_A] \cdot \mathbf{n_I} = B_S. \tag{2.2-15}$$

Recalling that $\mathbf{F}$ is the flux relative to fixed coordinates, $\mathbf{F} - b\mathbf{v_I}$ is seen to be the flux *relative to the interface*. Thus, Eq. (2.2-15) shows that an interfacial source creates a difference in the normal components of the fluxes relative to the interface. In general, all of the quantities involved are functions of position on the interface, so that the balance changes from point to point. This interfacial balance is a key result, because it is the basis for many of the boundary conditions in heat and mass transfer problems.

All of the derivations here have neglected the possibility of interfacial accumulation and of transport within the plane of the interface. These effects may arise in problems involving adsorption at solid surfaces, or surfactants at fluid–fluid interfaces, and may be included using additional accumulation and flux terms; see Edwards et al. (1991) for the more general equations.

Convective and Diffusive Fluxes

It is common to express the total flux $\mathbf{F}$ as the sum of convective and diffusive contributions. Defining the convective part as $b\mathbf{v}$, we obtain

$$\mathbf{F} \equiv b\mathbf{v} + \mathbf{f}, \tag{2.2-16}$$

where $\mathbf{f}$ is the diffusive part of the flux. As discussed in Chapter 1, the diffusive flux for energy is the conduction flux, or $\mathbf{q}$. For a mixture, where $\mathbf{v}$ is interpreted as the mass-average velocity, the diffusive flux for species A is $\mathbf{J}_A$ (molar units) or $\mathbf{j}_A$ (mass units). Because average velocities other than $\mathbf{v}$ might be selected for a mixture, Eq. (2.2-16) does not represent the only possible decomposition of $\mathbf{F}$ into convective and diffusive parts. For example, if we were to choose $\mathbf{v}^{(M)}$ (the molar-average velocity) instead of $\mathbf{v}$, the diffusive flux $\mathbf{f}$ would be replaced by a different quantity, $\mathbf{f}^{(M)}$. For modeling most transport processes, including energy and species transport in dilute solutions, the preferred mixture velocity is $\mathbf{v}$. Accordingly, Eq. (2.2-16) is used in the relations which follow.

Combining Eqs. (2.2-12) and (2.2-16), the general conservation equation at an interior point is rewritten as

$$\boxed{\frac{\partial b}{\partial t} + \nabla \cdot (b\mathbf{v}) = -\nabla \cdot \mathbf{f} + B_V.}$$ (2.2-17)

Combining Eqs. (2.2-15) and (2.2-16), the balance at an interface is

$$[(\mathbf{f} + b(\mathbf{v} - \mathbf{v_I}))_B - (\mathbf{f} + b(\mathbf{v} - \mathbf{v_I}))_A] \cdot \mathbf{n_I} = B_S.$$ (2.2-18)

Equations (2.2-17) and (2.2-18) are simply alternative ways to write the general balances, and contain no new physical information or assumptions.

Flux Continuity and Symmetry Conditions within a Given Phase

In formulating transport models it is often helpful to have a mathematical boundary pass through the interior of a given phase. For this reason, it is important to establish certain properties of the fluxes at interior points. These are derived by again considering a control volume of vanishingly small size. Dividing each term of Eq. (2.2-4) by S and rearranging, we obtain

$$\frac{1}{S} \int_S \mathbf{F} \cdot \mathbf{n} \, dS = \frac{V}{S} \left[\frac{1}{V} \int_V \left(B_V - \frac{\partial b}{\partial t} \right) dV \right].$$ (2.2-19)

For a control volume of a given shape, $V = c_V L^3$ and $S = c_S L^2$, where L is the characteristic linear dimension and c_V and c_S are constants which depend on the shape. Accordingly, for any control volume, $V/S \to 0$ as $L \to 0$. The bracketed term in Eq. (2.2-19) remains finite in this limit, indicating that

$$\lim_{S \to 0} \frac{1}{S} \int_S \mathbf{F} \cdot \mathbf{n} \, dS = 0.$$ (2.2-20)

The analogous result for the stress in a fluid is given in Section 5.3.

One important implication of Eq. (2.2-20) is that $\mathbf{F}$ is a continuous function of position at all interior points. This is seen by applying this relation to the control volume in Fig. 2-3, but with the interface replaced by an imaginary, fixed surface. For $\ell \to 0$, the edge contributions will vanish, leading to the conclusion that the normal component of $\mathbf{F}$ evaluated at one side of the imaginary surface equals the normal component evaluated at the other side. In other words, the normal component of $\mathbf{F}$ must be continuous. [The same conclusion is reached by setting $B_S = 0$ and $\mathbf{v_I} = \mathbf{0}$ in Eq. (2.2-15).] Because the imaginary surface can have any orientation in space, *all* components of $\mathbf{F}$ must be continuous at interior points, and the flux vector must be a continuous function of position. Moreover, because b and $\mathbf{v}$ are continuous functions of position within a given phase, it follows that the diffusive flux $\mathbf{f}$ is also continuous at interior points. As shown in Chapter 5, analogous arguments lead to the conclusion that the stress in a fluid is a continuous function of position.

The usual reason for placing a mathematical boundary in the interior of a phase is to exploit some form of symmetry which is present. If there is a *plane* of symmetry with unit normal $\mathbf{n}$, then the flux component $F_n = \mathbf{n} \cdot \mathbf{F}$ must change sign at the symmetry plane. The only way F_n can change sign yet still be continuous is if it vanishes, so that

$$F_n = 0 \qquad \text{(symmetry plane)}.$$ (2.2-21)

Rotational symmetry is best described using polar coordinates. Of particular interest are situations where the field variables are independent of the angle θ in cylindrical coordi-

TABLE 2-1
General Conservation Equations for Interior Points and Interfaces

Interior points		Points at interfaces	
$\dfrac{\partial b}{\partial t} = -\nabla \cdot \mathbf{F} + B_V$	(A)	$\left[(\mathbf{F} - b\mathbf{v_I})_B - (\mathbf{F} - b\mathbf{v_I})_A\right] \cdot \mathbf{n_I} = B_S$	(B)
$\dfrac{\partial b}{\partial t} + \nabla \cdot (b\mathbf{v}) = -\nabla \cdot \mathbf{f} + B_V$	(C)	$\left[(\mathbf{f} + b(\mathbf{v} - \mathbf{v_I}))_B - (\mathbf{f} + b(\mathbf{v} - \mathbf{v_I}))_A\right] \cdot \mathbf{n_I} = B_S$	(D)

nates, or independent of both angles θ and ϕ in spherical coordinates (see Fig. A-3). Such problems are termed *axisymmetric* or *spherically symmetric,* respectively. In these cases it is found that the radial component of the flux must vanish at the origin, or that

$$F_r = 0 \quad \text{at } r = 0 \qquad \text{(axisymmetric or spherically symmetric).} \qquad (2.2\text{-}22)$$

This result is derived for the spherical case by considering a small, spherical control volume centered at $r = 0$. The outward-normal component of the flux at the control surface is $F_r = \mathbf{e_r} \cdot \mathbf{F}$. If $F_r = F_r(r, t)$ only, then Eq. (2.2-20) leads directly to Eq. (2.2-22). For the cylindrical case with $F_r = F_r(r, z, t)$ we choose a cylindrical control volume centered on the line $r = 0$ and recognize that the contributions of the ends become negligible as $r \to 0$. Thus, the integrand in the dominant surface integral is F_r, and Eq. (2.2-22) is obtained once again. Equations (2.2-21) and (2.2-22) apply also to the diffusive fluxes, f_n or f_r.

Summary

The most important results of this section are the balances derived for interior points or points on interfaces, using either the total flux or the diffusive flux. These are summarized in Table 2-1.

2.3 CONSERVATION OF MASS

Continuity

The results of Section 2.2 are applied first to total mass. The concentration variable in this case is the total mass density, or $b = \rho$. By definition, there is no net mass flow relative to the mass-average velocity, and thus there is no diffusive flux for total mass (i.e., $\mathbf{f} = 0$). In addition, there are no sources or sinks ($B_V = 0$). Thus, Eq. (C) of Table 2-1 becomes

$$\frac{\partial \rho}{\partial t} + \nabla \cdot (\rho \mathbf{v}) = 0. \qquad (2.3\text{-}1)$$

This statement of local conservation of mass is called the *continuity equation.* It is used in this book to derive a number of other general relationships.

If the fluid density is constant, the continuity equation simplifies to

$$\nabla \cdot \mathbf{v} = 0 \qquad (\text{constant } \rho). \tag{2.3-2}$$

The density of any pure fluid is related to the pressure and temperature by an equation of state of the form $\rho = \rho(P, T)$. Liquids are virtually incompressible, so that the effects of pressure on ρ can be ignored. Although gases are not incompressible in the thermodynamic sense, the effect of pressure on density in gas flows is often negligible. As discussed in Schlicting (1968, pp. 10–11), gases tend to behave as if they are incompressible if the velocity is much smaller than the speed of sound. In that the *Mach number* (Ma) is the ratio of the flow velocity to the speed of sound, this criterion for applying Eq. (2.3-2) may be stated as Ma$\ll$1. (For dry air at 0°C, the speed of sound is 331 m/s.) For both liquids and gases, spatial variations in density due to temperature may be quite important, in that they cause buoyancy-driven flows (free convection). However, as discussed in Chapter 12, even in such flows Eq. (2.3-2) is usually an excellent approximation. For these reasons, this form of the continuity equation is employed when solving all problems involving fluid mechanics or convective heat and mass transfer in this book.

Material Derivative

Using conservation of mass, we can write the general conservation equations for scalar quantities in other forms. These are expressed most easily in terms of the differential operator,

$$\frac{D}{Dt} \equiv \frac{\partial}{\partial t} + \mathbf{v} \cdot \nabla, \tag{2.3-3}$$

which is called the *material derivative* (or *substantial derivative*).

The material derivative has a specific physical interpretation: It is the rate of change as seen by an observer moving with the fluid. This interpretation can be understood by considering the rate of change of a scalar $b(\mathbf{r}, t)$ perceived by an observer moving at any velocity $\mathbf{u}$, written as $(db/dt)_\mathbf{u}$. At any fixed position $\mathbf{r}$, the time dependence of b will cause a variation equal to $(\partial b/\partial t)\Delta t$ over a small time interval Δt. An additional change is caused by movement of the observer, provided that there are spatial variations in b. In particular, a small displacement described by a vector $\Delta \mathbf{r}$, tangent to $\mathbf{u}$, causes a change in b equal to $(\Delta \mathbf{r}) \cdot (\nabla b)$ (see Section A.6). Given that $\mathbf{u} = \Delta \mathbf{r}/\Delta t$ for small time intervals, the resulting change in b over the time Δt is $(\mathbf{u} \cdot \nabla b)\Delta t$. Adding the two contributions to the change in b and letting $\Delta t \to 0$, the rate of change seen by the moving observer is

$$\left(\frac{db}{dt}\right)_\mathbf{u} = \frac{\partial b}{\partial t} + \mathbf{u} \cdot \nabla b. \tag{2.3-4}$$

For an observer moving with the fluid, $\mathbf{u} = \mathbf{v}$ and

$$\left(\frac{db}{dt}\right)_\mathbf{v} = \frac{\partial b}{\partial t} + \mathbf{v} \cdot \nabla b \equiv \frac{Db}{Dt}, \tag{2.3-5}$$

as indicated above.

Alternative Conservation Equations

One useful equation involves the amount of any quantity per unit mass, denoted as $\hat{B} \equiv b/\rho$. Changing variables from b to $\hat{B}$, the left-hand side of Eq. (C) of Table 2-1 becomes

$$\frac{\partial b}{\partial t} + \nabla \cdot (b\mathbf{v}) = \hat{B}\left[\frac{\partial \rho}{\partial t} + \nabla \cdot (\rho \mathbf{v})\right] + \rho\left[\frac{\partial \hat{B}}{\partial t} + \mathbf{v} \cdot \nabla \hat{B}\right]. \tag{2.3-6}$$

According to Eq. (2.3-1), the first bracketed term in Eq. (2.3-6) is exactly zero. Thus, Eq. (C) of Table 2-1 is equivalent to

$$\rho \frac{D\hat{B}}{Dt} = -\nabla \cdot \mathbf{f} + B_V. \tag{2.3-7}$$

This is employed, for example, in Chapter 11 to obtain a conservation equation for entropy.

   An especially useful form of conservation equation applies to fluids at constant density. In this case, it follows from Eq. (2.3-2) that $\nabla \cdot (b\mathbf{v}) = \mathbf{v} \cdot \nabla b + b(\nabla \cdot \mathbf{v}) = \mathbf{v} \cdot \nabla b$. Using this to introduce the material derivative, Eq. (C) of Table 2-1 becomes

$$\frac{Db}{Dt} = -\nabla \cdot \mathbf{f} + B_V \qquad \text{(constant } \rho\text{)}. \tag{2.3-8}$$

This result is employed in Sections 2.4 and 2.6 to obtain the most common forms of the energy and species conservation equations. If the diffusive flux is of the form $\mathbf{f} = -\mathscr{D}\nabla b$ (Chapter 1) and if the diffusivity $\mathscr{D}$ is constant, then Eq. (2.3-8) becomes

$$\frac{Db}{Dt} = \mathscr{D}\nabla^2 b + B_V \qquad \text{(constant } \rho \text{ and } \mathscr{D}\text{)}. \tag{2.3-9}$$

This is the conservation equation for a "constant-property fluid."

2.4 CONSERVATION OF ENERGY: THERMAL EFFECTS

In many problems which require an energy equation, thermal effects are more important than mechanical ones. This is usually true when there are large temperature variations, or when heats of reaction or latent heats are involved. An approximate equation for conservation of energy that is suitable for such applications is obtained from Eq. (2.3-8) by setting $b = \rho \hat{C}_p T$, $\mathbf{f} = \mathbf{q}$, and $B_V = H_V$. Thus, assuming that ρ and $\hat{C}_p$ are constant, we have

$$\rho \hat{C}_p \frac{DT}{Dt} = -\nabla \cdot \mathbf{q} + H_V \qquad \text{(constant } \rho \text{ and } \hat{C}_p\text{)}. \tag{2.4-1}$$

The source term H_V represents the rate of energy input from external power sources, per unit volume. The most common example is resistive heating due to the passage of an electric current.

   The heat flux in a solid or pure fluid is evaluated using Fourier's law,

$$\mathbf{q} = -k\nabla T. \tag{2.4-2}$$

Assuming that k is constant, Eqs. (2.4-1) and (2.4-2) are combined to give

$$\rho \hat{C}_p \frac{DT}{Dt} = k \nabla^2 T + H_V \qquad \text{(constant } \rho, \hat{C}_p, \text{ and } k\text{).} \qquad \text{(2.4-3)}$$

This energy equation is the one used most frequently in this book. Notice that it is of the same form as Eq. (2.3-9). Equation (2.4-3) is given in rectangular, cylindrical, and spherical coordinates in Table 2-2.

Two kinds of mechanical effects are ignored in Eqs. (2.4-1) and (2.4-3), both of which would appear as additional source terms. One is viscous dissipation, which is the conversion of kinetic energy to heat due to the internal friction in a fluid. This is usually negligible, with the exception of certain high-speed flows (with large velocity gradients) or flows of extremely viscous fluids. The other term not included, which involves pressure variations, is related to the work done in compressing a fluid. It is identically zero for fluids which are thermodynamically incompressible; specifically, it vanishes if $\partial \rho / \partial T$ at constant pressure is zero. Accordingly, this term is ordinarily negligible for liquids. For gases, it will be negligible if $\Delta P \ll \rho \hat{C}_p \Delta T$, where ΔP and ΔT are the magnitudes of the pressure and temperature variations. For air at ambient conditions, a temperature change of 8 K is energetically equivalent to a pressure change of 1×10^4 Pa (i.e., 0.1 atm). Thus, this effect is negligible also for many applications involving gases.

In Chapter 9 there is a rigorous derivation of conservation of energy for a pure fluid, including the mechanical terms just discussed. A general treatment of energy conservation in a mixture is given in Chapter 11.

TABLE 2-2
Approximate Forms of the Energy Conservation Equation in Rectangular, Cylindrical, and Spherical Coordinates (Thermal Effects Only)[a]

Rectangular: $T = T(x, y, z, t)$

$$\frac{\partial T}{\partial t} + v_x \frac{\partial T}{\partial x} + v_y \frac{\partial T}{\partial_y} + v_z \frac{\partial T}{\partial z} = \alpha \left[\frac{\partial^2 T}{\partial x^2} + \frac{\partial^2 T}{\partial y^2} + \frac{\partial^2 T}{\partial z^2} \right] + \frac{H_V}{\rho \hat{C}_P}$$

Cylindrical: $T = T(r, \theta, z, t)$

$$\frac{\partial T}{\partial t} + v_r \frac{\partial T}{\partial r} + \frac{v_\theta}{r} \frac{\partial T}{\partial \theta} + v_z \frac{\partial T}{\partial z} = \alpha \left[\frac{1}{r} \frac{\partial}{\partial r} \left(r \frac{\partial T}{\partial r} \right) + \frac{1}{r^2} \frac{\partial^2 T}{\partial \theta^2} + \frac{\partial^2 T}{\partial z^2} \right] + \frac{H_V}{\rho \hat{C}_P}$$

Spherical: $T = T(r, \theta, \phi, t)$

$$\frac{\partial T}{\partial t} + v_r \frac{\partial T}{\partial r} + \frac{v_\theta}{r} \frac{\partial T}{\partial \theta} + \frac{v_\phi}{r \sin \theta} \frac{\partial T}{\partial \phi} = \alpha \left[\frac{1}{r^2} \frac{\partial}{\partial r} \left(r^2 \frac{\partial T}{\partial r} \right) + \frac{1}{r^2 \sin \theta} \frac{\partial}{\partial \theta} \left(\sin \theta \frac{\partial T}{\partial \theta} \right) + \frac{1}{r^2 \sin^2 \theta} \frac{\partial^2 T}{\partial \phi^2} \right] + \frac{H_V}{\rho \hat{C}_P}$$

[a] It is assumed that ρ, $\hat{C}_p$, and k are constant and that certain mechanical effects are negligible. The thermal diffusivity is $\alpha = k/(\rho \hat{C}_p)$. More general energy equations are given in Section 9.6 for pure fluids and in Section 11.2 for mixtures.

2.5 HEAT TRANSFER AT INTERFACES

In this section we derive the conditions satisfied by the temperature and the heat flux at boundaries between pure materials (i.e., when mass transfer is absent). Either material may be a fluid or a solid, but the two phases are understood to be mutually insoluble or immiscible. These interfacial conditions involving temperatures and temperature gradients form the basis for the boundary conditions used most commonly with Eq. (2.4-3). The interfacial conditions for simultaneous heat and mass transfer are discussed in Chapter 11.

Interfacial Energy Balance

Consider a point $\mathbf{r}_s$ on the boundary between phases 1 and 2. As shown in Fig. 2-4, the orientation of the interface is described by the unit normal $\mathbf{n}$ directed from phase 1 to phase 2. The phases might be a solid and a liquid (as indicated in the figure), two solids, or two fluids. If there is no bulk flow across the interface, then the normal component of the velocity in either phase equals the normal component of the interfacial velocity. That is, in both materials $v_n = v_{In}$, where $v_n = \mathbf{n} \cdot \mathbf{v}$ and $v_{In} = \mathbf{n} \cdot \mathbf{v_I}$. Consequently, with $\mathbf{f} = \mathbf{q}$ and $B_S = H_S$, Eq. (D) of Table 2-1 becomes

$$q_n(\mathbf{r}_s, t)|_2 - q_n(\mathbf{r}_s, t)|_1 = H_S(\mathbf{r}_s,\ t), \qquad (2.5\text{-}1)$$

where the subscripts 1 and 2 indicate the side of the interface on which the normal heat flux q_n is evaluated. The source term H_S represents the rate of energy input from an external power source, per unit surface area. Such energy inputs are rare in chemical processes, although examples are given in Problems 2–13 and 2–15. For the usual situation with $H_S = 0$, Eq. (2.5-1) becomes

$$q_n(\mathbf{r}_s,\ t)|_1 = q_n(\mathbf{r}_s, t)|_2. \qquad (2.5\text{-}2)$$

The arguments "$(\mathbf{r}_s, t)$" are included in Eqs. (2.5-1) and (2.5-2) to emphasize that these relations are valid instantaneously at each point on the interface. The same is true for all of the interfacial conditions to be presented, although the arguments are omitted hereafter to simplify the notation. Thus, Eq. (2.5-2) is written more simply as $q_n|_1 = q_n|_2$.

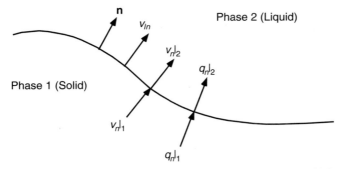

Figure 2-4. Normal components of velocities and heat fluxes at a solid–liquid interface.

Thermal Equilibrium at an Interface

Two materials are usually assumed to have equal temperatures at any point of contact, so that

$$T_1 = T_2 \tag{2.5-3}$$

at each point on the interface. The use of Eq. (2.5-3) in heat transfer analysis implies that thermal equilibrium is maintained across the phase boundary, even though the heat flux is not zero. This is equivalent to assuming that the thermal resistance of the interface itself is so small that it can be neglected in comparison with the resistances of the bulk phases. When at least one of the phases is a fluid, Eq. (2.5-3) is an excellent approximation. However, in certain situations, such as the point of contact between two metals with rough or oxidized surfaces, Eq. (2.5-3) may have to be replaced by a boundary condition involving a contact resistance (or conductance):

$$q_n|_1 = h_s(T_1 - T_2) = q_n|_2. \tag{2.5-4}$$

For a given heat flux, if the surface conductance h_s is sufficiently large, then $T_1 \rightarrow T_2$, as in Eq. (2.5-3).

Convection Boundary Condition

Equation (2.5-2), with q_n in each phase evaluated using Fourier's law, remains valid whether or not there is any fluid flow parallel to the interface. It assumes only that heat transfer *normal* to the interface is entirely by conduction. Nonetheless, in a flowing fluid it is often convenient to express the conduction heat flux in terms of a *heat transfer coefficient, h*. Taking phase 2 to be a fluid, the interfacial heat flux in the fluid can be written as

$$q_n|_2 \equiv h(T_2 - T_b), \tag{2.5-5}$$

where T_2 is still the temperature in fluid 2 *at the interface* and T_b is the *bulk temperature* in fluid 2. For heat transfer to or from a solid immersed in a large volume of fluid, the fluid temperaure far from the solid is often assumed to be constant, and that value is used as the bulk temperature. The definition of the bulk temperature for fluids in pipes or other confined geometries is discussed in Chapter 9. It should be emphasized that Eq. (2.5-5) contains no new physical information, but rather is simply a definition of the heat transfer coefficient h in fluid 2.

Suppose now that phase 1 is a solid and phase 2 is a liquid, as shown in Fig. 2-4. Using Fourier's law to evaluate $q_n|_1$ and using Eq. (2.5-5) to evaluate $q_n|_2$, Eq. (2.5-2) becomes what is called a *convection boundary condition*,

$$-k_1(\mathbf{n} \cdot \nabla T)_1 = h(T_2 - T_b). \tag{2.5-6}$$

If h and T_b are known, this type of boundary condition is convenient for an analysis which focuses on the temperature distribution in the solid. The reason that Eqs. (2.5-5) and (2.5-6) are useful is that a great deal of experimental and theoretical information on heat transfer in flowing fluids, in various geometries and flow regimes, is summarized in terms of values of h. Many analyses of convective heat transfer have as their objective the evaluation of h for a particular system; we return to this subject in Chapters 9, 10, 12, and 13.

Radiation Heat Transfer

Heat transfer by radiation is important in combustion and other high-temperature processes, particularly as a mechanism for the transfer of heat between solid surfaces separated by gases. The theory for radiation heat transfer is beyond the scope of this book; for thorough treatments of this subject see Sparrow and Cess (1978) and Siegel and Howell (1981). Here we introduce only the most basic form of the Stefan–Boltzmann law for application in simple transport models. Assuming that solid surfaces 1 and 3 are separated by gas 2, which is transparent to radiation, the radiant heat flux at surface 1 due to surface 3 is expressed as

$$q_n^{(\text{rad})}\big|_1 = \sigma_{\text{SB}} F_{13}(T_1^4 - T_3^4) \tag{2.5-7}$$

where σ_{SB} is the Stefan–Boltzmann constant ($\sigma_{\text{SB}} = 5.67 \times 10^{-12}$ W cm^{-2} K^{-4}) and F_{13} is the "view factor" for this pair of surfaces ($0 \le F_{13} \le 1$). The view factor is the fraction of radiation leaving surface 1 that is directlly intercepted by surface 3; it is a function of both the shapes and the emissivities of the surfaces. Except for very simple geometries the evaluation of F_{ij} may become quite involved. A particularly simple case is when both surfaces behave as blackbodies and thus have unit emissivities. If, in addition, surface 1 is convex and completely enclosed by surface 3, then $F_{13} = 1$.

The assumption that a gas is transparent to radiation is often satisfactory, in which case convective and radiant heat transfer from a surface are independent and additive. The relative importance of convection and radiation is assessed by defining an effective heat transfer coefficient for radiation, h_{rad}, and comparing it with h. If the radiative temperature is equal to the bulk temperature of the gas (i.e., $T_3 = T_b$ in the gas), then the radiant heat flux is expressed as

$$q_n^{(\text{rad})}\big|_1 = h_{\text{rad}}(T_1 - T_b), \tag{2.5-8}$$

$$h_{\text{rad}} = \sigma_{\text{SB}} F_{13}(T_1^2 + T_b^2)(T_1 + T_b). \tag{2.5-9}$$

Noting that $T_1 = T_2$, Eq. (2.5-8) is of the same form as Eq. (2.5-5). The importance of radiation in a given situation can be assessed from the approximate magnitude of h_{rad} computed from estimates of the absolute temperatures in the system.

Phase Change

Processes such as melting and evaporation are characterized by bulk flow across a phase boundary, so that the interfacial energy balance given by Eq. (2.5-2) is not applicable. In what follows we assume that phase 1 is the more condensed form of the material. In other words, phase 1 might be a solid and phase 2 a liquid (as in Fig. 2-4), or phase 1 might be a liquid and phase 2 a gas. Conservation of mass at the interface is expressed by setting $b = \rho$, $\mathbf{f} = \mathbf{0}$, and $B_S = 0$ in Eq. (D) of Table 2-1. This gives

$$\rho_1(v_n\big|_1 - v_{In}) = \rho_2(v_n\big|_2 - v_{In}). \tag{2.5-10}$$

Thus, with different densities in the two phases, the velocities normal to the interface are *not* equal. Latent heats are ordinarily measured at constant pressure, so that the appropriate "energy concentration" in each phase is the enthalpy per unit volume, or $\rho\hat{H}$. Setting $b = \rho\hat{H}$, $\mathbf{f} = \mathbf{q}$, and $B_S = 0$ in the general interfacial balance, we obtain

$$\rho_1 \hat{H}_1(v_n\big|_1 - v_{In}) + q_n\big|_1 = \rho_2 \hat{H}_2(v_n\big|_2 - v_{In}) + q_n\big|_2. \tag{2.5-11}$$

Equation (2.5-11) differs from Eq. (2.5-2) because of the addition of convective contri-butions to the energy fluxes across the interface. Kinetic energy terms, which may be important occasionally in evaporation or sublimation, have been neglected in Eq. (2.5-11).

Combining Eqs. (2.5-10) and (2.5-11) yields

$$q_n|_1 - q_n|_2 = \hat{\lambda}\rho_1(v_n|_1 - v_{In}) = \hat{\lambda}\rho_2(v_n|_2 - v_{In}). \tag{2.5-12}$$

$$\hat{\lambda} = \hat{H}_2 - \hat{H}_1, \tag{2.5-13}$$

where $\hat{\lambda}$ is the latent heat per unit mass. Because $\hat{\lambda} > 0$, Eq. (2.5-12) indicates that, depending on the nature of the two phases, there will be melting or evaporation when $q_n|_1 > q_n|_2$, and solidification or condensation when $q_n|_1 < q_n|_2$. In the first case, the mass flow across the interface is seen to be directed toward phase 2, such that $v_n|_1 > v_{In}$ and $v_n|_2 > v_{In}$. In the second case the mass flow is directed toward phase 1, so that $v_n|_1 < v_{In}$ and $v_n|_2 < v_{In}$.

Symmetry Conditions

It is often helpful to treat a plane of symmetry as a boundary, thereby reducing the size of the domain in which one seeks to determine the temperature. From the discussion of symmetry in Section 2.2 it follows that

$$q_n = 0 \qquad \text{(symmetry plane).} \tag{2.5-14}$$

In other words, a plane of symmetry acts as an insulated surface. When the temperature and velocity are independent of the angle θ in cylindrical coordinates or of the angles θ and ϕ in spherical coordinates, we obtain

$$q_r = 0 \quad \text{at } r = 0 \qquad \text{(axisymmetric or spherically symmetric).} \tag{2.5-15}$$

2.6 CONSERVATION OF CHEMICAL SPECIES

It was pointed out in Chapter 1 that the concentration of species i can be expressed in molar units (C_i) or mass units (ρ_i). Problems involving chemical reactions are handled most easily with molar units, so that C_i is generally preferred. The molar flux of i relative to fixed coordinates is $\mathbf{N}_i$. Thus, setting $b = C_i$, $\mathbf{F} = \mathbf{N}_i$, and $B_V = R_{Vi}$ in Eq. (A) of Table 2-1, the basic form of the conservation equation for species i is

$$\frac{\partial C_i}{\partial t} = -\nabla \cdot \mathbf{N}_i + R_{Vi}. \tag{2.6-1}$$

The source term R_{Vi} represents the net rate of formation of species i by chemical reac-tions, per unit volume. Reactions which appear in this manner (i.e., ones which occur throughout the volume of a gas mixture or liquid solution) are called *homogeneous*. Reactions which occur at catalytic surfaces are called *heterogeneous*, and their rates (per unit area) are denoted as R_{Si}. The units of R_{Vi} and R_{Si} are mol m^{-3} s^{-1} and mol m^{-2} s^{-1}, respectively. With either type of reaction the sign convention is that the rate is positive if there is net formation of species i, and negative if there is net consumption.

The rates of heterogeneous reactions appear only in interfacial conditions, as described in Section 2.7.

In addition to the choice of units, in defining the convective and diffusive fluxes of species i one must select an average velocity for the mixture. Various average velocities were defined in Chapter 1. Because the *total* mass flux is given by $\rho \mathbf{v}$, it is the mass-average velocity ($\mathbf{v}$) which must appear in the continuity equation for a mixture. Similarly, the mass-average velocity is needed when considering conservation of momentum, because $\rho \mathbf{v}$ also represents a concentration of linear momentum (see Chapter 5). Accordingly, when there is fluid flow the preferred way to write the total flux is

$$\mathbf{N}_i = C_i \mathbf{v} + \mathbf{J}_i, \tag{2.6-2}$$

where $\mathbf{J}_i$ is the molar flux of i relative to the mass-average velocity. Although Eq. (2.6-2) is preferred for most applications, in some situations (e.g., with stagnant gases) it is advantageous to use the molar-average or volume-average velocity.

Several forms of Fick's law for binary mixtures were presented in Chapter 1. It follows from Eq. (B) of Table 1-3 that for a binary mixture of A and B at *constant density*, the molar flux of A relative to the mass-average velocity is

$$\mathbf{J}_A = -D_{AB} \nabla C_A. \tag{2.6-3}$$

This is an excellent approximation for most liquid solutions, but less accurate for gas mixtures. Although strictly valid only for a binary system, Eq. (2.6-3) can be applied also to certain multicomponent mixtures. In a multicomponent liquid solution or gas mixture, there are diffusional interactions between each pair of components. However, if all components but one are present in small amounts, interactions among the minor components tend to be negligible, and only the binary diffusivity of each minor component i with the abundant component is important. This diffusivity is abbreviated as D_i, the second subscript (corresponding to the abundant gas species or liquid solvent) being dropped for simplicity. For this *pseudobinary* situation at constant density, the flux equation for each minor component is

$$\mathbf{J}_i = -D_i \nabla C_i. \tag{2.6-4}$$

Equation (2.6-4) is the most frequently used flux equation for liquid solutions, even when there are more than two components. For multicomponent aqueous solutions, in particular, its applicability can be appreciated by noting that the concentration of pure water is 56 molar (mol/liter). Thus, even if the total solute concentration is several molar, the mole fraction of water is still almost unity. The pseudobinary approximation tends to be less useful for multicomponent gas mixtures, where it is less likely that there will be a dominant component. The general equations for multicomponent diffusion in gases and liquids are presented in Chapter 11. The basis for the pseudobinary approximation for dilute mixtures is examined in more detail in Example 11.8-2.

Based on the foregoing, when all components of a liquid solution or gas mixture except one are present in small concentrations, and the density and diffusivities are constant, the conservation equations for the minor components may be written as

$$\frac{DC_i}{Dt} = D_i \nabla^2 C_i + R_{Vi} \qquad \text{(constant } \rho \text{ and } D_i \text{)} \tag{2.6-5}$$

TABLE 2-3
Species Conservation Equations for a Binary or Pseudobinary Mixture in
Rectangular, Cylindrical, and Spherical Coordinates[a]

Rectangular: $C_i = C_i(x, y, z, t)$

$$\frac{\partial C_i}{\partial t} + v_x \frac{\partial C_i}{\partial x} + v_y \frac{\partial C_i}{\partial y} + v_z \frac{\partial C_i}{\partial z} = D_i \left[\frac{\partial^2 C_i}{\partial x^2} + \frac{\partial^2 C_i}{\partial y^2} + \frac{\partial^2 C_i}{\partial z^2} \right] + R_{Vi}$$

Cylindrical: $C_i = C_i(r, \theta, z, t)$

$$\frac{\partial C_i}{\partial t} + v_r \frac{\partial C_i}{\partial r} + \frac{v_\theta}{r} \frac{\partial C_i}{\partial \theta} + v_z \frac{\partial C_i}{\partial z} = D_i \left[\frac{1}{r} \frac{\partial}{\partial r} \left(r \frac{\partial C_i}{\partial r} \right) + \frac{1}{r^2} \frac{\partial^2 C_i}{\partial \theta^2} + \frac{\partial^2 C_i}{\partial z^2} \right] + R_{Vi}$$

Spherical: $C_i = C_i(r, \theta, \phi, t)$

$$\frac{\partial C_i}{\partial t} + v_r \frac{\partial C_i}{\partial r} + \frac{v_\theta}{r} \frac{\partial C_i}{\partial \theta} + \frac{v_\phi}{r \sin \theta} \frac{\partial C_i}{\partial \phi} = D_i \left[\frac{1}{r^2} \frac{\partial}{\partial r} \left(r^2 \frac{\partial C_i}{\partial r} \right) + \frac{1}{r^2 \sin \theta} \frac{\partial}{\partial \theta} \left(\sin \theta \frac{\partial C_i}{\partial \theta} \right) + \frac{1}{r^2 \sin^2 \theta} \frac{\partial^2 C_i}{\partial \phi^2} \right] + R_{Vi}$$

---

[a] It is assumed that ρ and D_i are constant, where D_i is the binary or pseudobinary diffusivity.

This equation, which is again like Eq. (2.3-9), is given for rectangular, cylindrical, and spherical coordinates in Table 2-3. In a binary system, Eq. (2.6-5) may be applied to both components, whether or not one of them is present in abundance. However, the requirement for constant density is most easily realized in liquid solutions. For gas mixtures at uniform temperature and pressure, constant total molar concentration (C) is a more accurate assumption than is constant mass density (ρ).

2.7 MASS TRANSFER AT INTERFACES

The discussion here of interfacial conditions for mass transfer parallels that presented for heat transfer in Section 2.5.

Interfacial Species Balance

Once again consider the boundary between phase 1 and phase 2, with the unit vector **n** normal to the interface and directed from 1 to 2. Assuming that there is no bulk flow across the interface (i.e., $v_{In} = v_n$ in either phase), the normal component of the flux of species i relative to the interface is given by J_{in}. Accordingly, Eq. (D) of Table 2-1 becomes

$$J_{in}|_2 - J_{in}|_1 = R_{Si}, \tag{2.7-1}$$

where R_{Si} is the molar rate of formation of species i per unit surface area. Unlike the corresponding energy source term in Eq. (2.5-1), nonzero values of R_{Si} are encountered frequently in chemical engineering and are of great practical importance. Heterogeneous reactions often occur at solid surfaces which are catalytically active but impermeable to reactants and products. If phase 1 is an impermeable solid, then $J_{in}|_1 = 0$ and

$$J_{in}|_2 = R_{Si} \qquad (2.7\text{-}2)$$

for all reactants and products.

The fluxes of the reactants and products of a heterogeneous reaction are related by the reaction stoichiometry. Let ξ_i be the stoichiometric coefficient for substance i in the reaction, with $\xi_i < 0$ for reactants and $\xi_i > 0$ for products. For example, with a reaction written as

$$C_6H_{12} \rightarrow C_6H_6 + 3H_2$$

$\xi_1 = -1$ for cyclohexane (C_6H_{12}), $\xi_2 = 1$ for benzene (C_6H_6), and $\xi_3 = 3$ for hydrogen. The advantage of defining stoichiometric coefficients in this manner is that a species-independent reaction rate can be employed, such that $R_S \equiv R_{Si}/\xi_i$. This allows Eq. (2.7-2) to be rewritten as

$$J_{in}|_2 = \xi_i R_S. \qquad (2.7\text{-}3)$$

Thus, once the reaction rate is known, the fluxes of all reactants and products can be calculated. A kinetic expression relating R_S to the concentrations at the surface is required to complete the formulation of a problem of this type.

Species Equilibrium at Interface

As with heat transfer, it is usually assumed that the resistance of the interface itself to mass transfer is negligible. However, in the analysis of diffusion across a phase boundary it is important to bear in mind that local equilibrium does *not* imply equal interfacial concentrations. Instead, the concentrations of species i at the interface are related by an *equilibrium partition coefficient, K_i*. For example,

$$C_i|_1 = K_i C_i|_2. \qquad (2.7\text{-}4)$$

For gas–liquid systems, K_i is derived from the Henry's law constant. For liquid–liquid systems, it is determined by the relative solubilities of i in the two liquids; large deviations of K_i from unity (e.g., for solutes in water versus nonpolar organic solvents) form the basis for separations by solvent extraction. The partition coefficient is a thermodynamic quantity, and for a given solute it is expected to depend on the temperature, pressure, and composition of the two phases. In this book we will ordinarily assume that K_i is independent of the solute concentration, which is true for ideal solutions. Values of K_i for various gas–liquid, liquid–liquid, gas–solid, and liquid–solid systems span many orders of magnitude, emphasizing the importance of including this factor in Eq. (2.7-4).

Convection Boundary Condition

Assume now that phase 2 is a fluid and that there is no bulk flow across the interface. The *mass transfer coefficient (k_{ci})* for species i is defined as

$$J_{in}|_2 \equiv k_{ci}(C_i|_2 - C_{ib}), \qquad (2.7\text{-}5)$$

where C_{ib} is the *bulk concentration* of solute i in phase 2. Note that the driving force is based on concentrations only within phase 2, the fluid to which k_{ci} applies. Because the mass transfer coefficient for a given species depends in part on its diffusivity, the value of k_{ci} will tend to be different for each component of a mixture. Mass transfer coefficients are defined sometimes using other driving forces (e.g., a difference in mole fraction rather than molar concentration).

The analogy between heat and mass transfer coefficients is made evident by comparing Eq. (2.7-5) with Eq. (2.5-5). When mass transfer is accompanied by bulk flow normal to the interface, this analogy is best preserved if the mass transfer coefficient continues to refer only to the *diffusive* part of the solute flux (i.e., J_{in} rather than N_{in}). This is because heat transfer coefficients have been defined to represent only the *conduction* heat flux (q_n), and not energy transfer by bulk flow. Restricting heat and mass transfer coefficients to the representation of conduction and diffusion, respectively, is logical for another reason. Namely, these transport mechanisms require gradients in the corresponding field variable (temperature or concentration), whereas convection does not. Equations (2.5-5) and (2.7-5) imply that the fluxes being represented vanish when temperature or concentration gradients are absent.

Symmetry Conditions

The symmetry conditions for mass transfer which are analogous to Eqs. (2.5-14) and (2.5-15) are

$$J_{in} = 0 \quad \text{(symmetry plane)}, \qquad (2.7\text{-}6)$$

$$J_{ir} = 0 \quad \text{at } r = 0 \quad \text{(axisymmetric or spherically symmetric)}. \qquad (2.7\text{-}7)$$

2.8 ONE-DIMENSIONAL EXAMPLES

The examples in this section illustrate the application of conservation equations and interfacial conditions for heat or mass transfer in relatively simple situations. The problems all involve steady states, and they are "one-dimensional" in the sense that the temperature or concentration depends on a single coordinate. The field variables are therefore governed by ordinary differential equations. Three important dimensionless groups are introduced: the Biot number (Bi), the Damköhler number (Da), and the Péclet number (Pe).

Example 2.8-1 Heat Transfer in a Wire Consider the steady-state temperature in a cylindrical wire of radius R that is heated by passage of an electric current and cooled by convective heat transfer to the surrounding air. The local heating rate, H_V, is assumed to be independent of position. This is equivalent to assuming a uniform current density and electrical resistance. For steady conduction in a solid with such a heat source, Eq. (2.4-3) reduces to

$$\nabla^2 T = -\frac{H_V}{k}. \qquad (2.8\text{-}1)$$

Cylindrical coordinates (r, θ, z) are the natural choice for this problem. The convection boundary condition at the surface of the wire is written as

$$\frac{\partial T}{\partial r} = -\frac{h}{k}(T - T_\infty) \qquad \text{at } r = R, \tag{2.8-2}$$

where T_∞ is the ambient temperature. For simplicity, we assume that h is independent of position. With H_V, k, and h all assumed to be constant, there is nothing to cause the temperature to depend on the angle θ. Thus, we conclude that the temperature field is axisymmetric. It follows that Eq. (2.5-15) is applicable and that the second boundary condition in r is

$$\frac{\partial T}{\partial r} = 0 \qquad \text{at } r = 0. \tag{2.8-3}$$

If the wire is very long, and nothing is done to cause the temperature at the ends to differ, then there is also no reason for the temperature to depend on z. It is apparent now that all of the physical conditions can be satisfied by a temperature field which depends only on r.

Assuming now that $T = T(r)$ only, Table 2-2 is used to rewrite Eq. (2.8-1) as

$$\frac{1}{r}\frac{d}{dr}\left(r\frac{dT}{dr}\right) = -\frac{H_V}{k}. \tag{2.8-4}$$

This second-order equation requires two boundary conditions in r, which are given already by Eqs. (2.8-2) and (2.8-3). The temperature is determined by first integrating Eq. (2.8-4) to give

$$r\frac{dT}{dr} = -\frac{H_V r^2}{2k} + C_1. \tag{2.8-5}$$

The symmetry condition [Eq. (2.8-3)] indicates that the constant C_1 must be zero. A second integration yields

$$T = -\frac{H_V r^2}{4k} + C_2, \tag{2.8-6}$$

where C_2 is another constant. Substituting this result into the convective boundary condition at the surface [Eq. (2.8-2)] gives

$$-\frac{H_V R}{2k} = -\frac{h}{k}\left(-\frac{H_V R^2}{4k} + C_2 - T_\infty\right). \tag{2.8-7}$$

Solving for C_2, the temperature is found to be

$$T - T_\infty = \frac{H_V R^2}{4k}\left[1 - \left(\frac{r}{R}\right)^2\right] + \frac{H_V R}{2h}. \tag{2.8-8}$$

Thus, the temperature at the surface of the wire exceeds the ambient value by the amount $H_V R/2h$, and the temperature at the center of the wire is elevated further by an amount $H_V R^2/4k$.

The behavior of the temperature is revealed more clearly by using dimensionless quantities defined as

$$\Theta \equiv \frac{T - T_\infty}{T_c - T_\infty}, \qquad \eta \equiv \frac{r}{R}, \qquad \text{Bi} \equiv \frac{hR}{k}, \tag{2.8-9}$$

where $T_c = T(0)$ is the temperature at the center of the wire and Bi is the *Biot number*. By definition, Θ ranges from unity at the center of the wire to zero in the bulk air. Equation (2.8-8) is rewritten now as

$$\Theta = \frac{2 + \text{Bi}(1 - \eta^2)}{2 + \text{Bi}}. \tag{2.8-10}$$

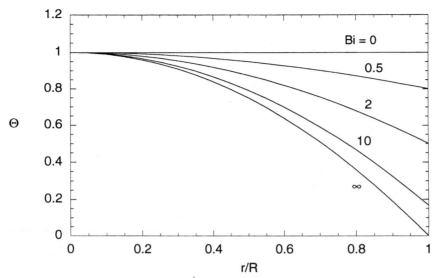

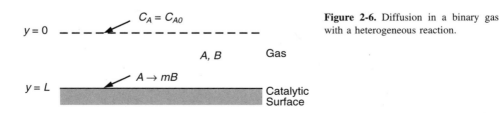

Figure 2-5. Temperature profile in an electrically heated wire, as a function of the Biot number. The plot is based on Eq. (2.8-10).

Figure 2-5 shows the dimensionless temperature profile for several values of Bi. For $Bi \ll 1$, radial heat conduction in the wire is so fast that the wire is nearly isothermal, and the main temperature drop is in the air. For $Bi \gg 1$, convective heat transfer in the air is so rapid that the external temperature drop is negligible, and the temperature at the wire surface is very close to the ambient value. Thus, the Biot number represents the ratio of the heat transfer resistance within the wire to that within the surrounding air. The significance of Biot numbers for heat or mass transfer is discussed further in Chapter 3.

Example 2.8-2 Diffusion in a Binary Gas with a Heterogeneous Reaction   This example illustrates the use of Fick's law for a binary gas, and it also shows how the reaction rate can influence the boundary condition used at a catalytic surface. The system to be considered is shown in Fig. 2-6. A stagnant gas film of thickness L is in contact with a surface which catalyzes the irreversible reaction, $A \rightarrow mB$. The reaction rate follows nth-order kinetics ($n > 0$), as given by

$$R_{SA} = -k_{sn}C_A^n, \tag{2.8-11}$$

where k_{sn} is a constant. It is assumed that C_A depends on y only and that its value at $y = 0$ is fixed at C_{A0}. It is assumed also that the gas is isothermal and isobaric, so that the total molar concentra-

Figure 2-6. Diffusion in a binary gas with a heterogeneous reaction.

tion (C) is constant. Unless the molecular weights of A and B are identical (i.e., unless $m=1$), the total mass density (ρ) will not be constant under these conditions.

Before using species conservation or Fick's law, we first see what can be learned from the continuity equation. For this steady, one-dimensional system with variable ρ, Eq. (2.3-1) becomes

$$\frac{d(\rho v_y)}{dy}=0. \tag{2.8-12}$$

Thus, ρv_y is independent of y. Because the catalytic surface is assumed to be impermeable (i.e., $v_y=0$ at $y=L$), we conclude that the mass-average velocity is zero throughout the gas film. The main consequence of this is that $J_{iy}=N_{iy}$ for both species.

Most of the results in Section 2.6 cannot be used here because ρ is not constant. However, we can apply Eq. (2.6-1) to both species. It follows from the stated assumptions that

$$\frac{dN_{Ay}}{dy}=0=\frac{dN_{By}}{dy}. \tag{2.8-13}$$

There is no reaction term in Eq. (2.8-13) because there is no *homogeneous* reaction. This equation indicates that both fluxes are independent of position, so that evaluating the flux ratio at any location determines the ratio for all y. Using Eq. (2.7-3) together with $J_{iy}=N_{iy}$, we obtain

$$N_{By}=-mN_{Ay}. \tag{2.8-14}$$

No further consideration of species B is necessary, because N_{By} can be obtained from N_{Ay} and because C_B was assumed to have no effect on the reaction kinetics.

In selecting a form of Fick's law it is advantageous to employ an expression which involves C rather than ρ, because it is C which is assumed to be constant. Thus, we adopt Eq. (D) of Table 1-3, which requires that we use the *molar-average* velocity as the reference frame. From Tables 1-2 and 1-3, the total flux of A is given by

$$N_{Ay}=x_A(N_{Ay}+N_{By})-CD_{AB}\frac{dx_A}{dy}. \tag{2.8-15}$$

Using Eq. (2.8-14) to eliminate N_{By} from this expression, we obtain

$$N_{Ay}=x_A N_{Ay}(1-m)-CD_{AB}\frac{dx_A}{dy}. \tag{2.8-16}$$

The convective flux of A in this reference frame is defined as $C_A v_y^{(M)}$, so that Eq. (2.8-16) implies that $v_y^{(M)}=N_{Ay}(1-m)/C$. Thus, the molar-average velocity does not vanish unless $m=1$, even though the mass-average velocity is zero for all stoichiometries. This indicates that there is a convective flux here when using the molar-average velocity, but not when using the mass-average velocity!

Rearranging Eq. (2.8-16) to solve for N_{Ay} gives

$$N_{Ay}=-\frac{CD_{AB}}{[1-x_A(1-m)]}\frac{dx_A}{dy}. \tag{2.8-17}$$

Taking advantage now of the assumed constancy of C, Eq. (2.8-17) becomes

$$N_{Ay}=-\frac{D_{AB}}{[1-(C_A/C)(1-m)]}\frac{dC_A}{dy}. \tag{2.8-18}$$

The flux of A at the catalytic surface is directly related to the reaction rate at the surface. From Eq. (2.7-2),

$$N_{Ay}(L) = -R_{SA} = k_{sn}[C_A(L)]^n. \tag{2.8-19}$$

Because N_{Ay} has been shown to be independent of position, Eqs. (2.8-18) and (2.8-19) can be equated to give

$$\frac{dC_A}{dy} = -\left(\frac{k_{sn}}{D_{AB}}\right)[C_A(L)]^n\left[1 - \frac{C_A}{C}(1-m)\right], \qquad C_A(0) = C_{A0}. \tag{2.8-20}$$

Before integrating Eq. (2.8-20) to determine the concentration profile, we introduce the dimensionless quantities

$$\theta \equiv \frac{C_A}{C_{A0}}, \qquad \eta \equiv \frac{y}{L}, \qquad \text{Da} \equiv \frac{k_{sn}C_{A0}^{n-1}L}{D_{AB}}, \qquad \phi \equiv \frac{C_A}{C_{A0}}\bigg|_{y=L}. \tag{2.8-21}$$

The parameter Da is the *Damköhler number*, which is seen to be the ratio of a reaction velocity $(k_{sn}C_{A0}^{n-1})$ to a diffusion velocity (D_{AB}/L). Thus, it is a measure of the intrinsic rate of reaction relative to that of diffusion.[1] The reactant concentration at the catalytic surface, which is an unknown constant, is denoted as ϕ. The governing equation in dimensionless form is then

$$\frac{d\theta}{d\eta} = -\text{Da}\,\phi^n[1 - x_{A0}(1-m)\theta], \qquad \theta(0) = 1, \tag{2.8-22}$$

where x_{A0} is the (known) mole fraction of A at $\eta = 0$.

Equation (2.8-22) is separable and can be integrated from $\eta = 0$ to $\eta = 1$ to obtain implicit expressions for $\phi = \theta(1)$, the surface concentration of the reactant. The results, which depend on the reaction stoichiometry, are

$$\text{Da}\phi^n = \begin{cases} \dfrac{1}{x_{A0}(1-m)}\ln\left[\dfrac{1 - x_{A0}(1-m)\phi}{1 - x_{A0}(1-m)}\right], & m \neq 1, \\ 1 - \phi, & m = 1. \end{cases} \tag{2.8-23}$$

Inspection of Eq. (2.8-23) reveals that $\phi \to 1$ as $\text{Da} \to 0$. In this case the reaction is slow relative to diffusion, so that the reaction is the controlling step and the reactant concentration is nearly uniform throughout the film. At the other extreme, as $\text{Da} \to \infty$, the process is controlled entirely by mass transfer and $\phi \to 0$. That is, the concentration at the surface approaches zero.

Concentration profiles for an equimolar, second-order reaction ($m = 1$, $n = 2$) at several values of Da are shown in Fig. 2-7. The transition from kinetic to diffusion control as Da is increased is evident. Also noteworthy is the qualitative similarity between this plot and Fig. 2-5. In both situations the parameter can be interpreted as the ratio of two resistances in series, those for internal and external heat transfer (Bi) or those for diffusion and reaction (Da).

The foregoing results indicate that if we were interested only in diffusion-controlled conditions ($\text{Da} \to \infty$), then we could replace Eq. (2.8-19) by $C_A = 0$ at $y = L$, or (in dimensionless form)

$$\theta(1) = 0. \tag{2.8-24}$$

This type of *fast-reaction boundary condition* holds for any irreversible, diffusion-controlled, heterogeneous reaction. When this simple condition applies, the reaction rate law does not enter into the problem. The accuracy of Eq. (2.8-24) for the present problem is judged most easily for an equimolar reaction ($m = 1$), in which case Eq. (2.8-23) indicates that

---

[1] There are no fewer than five dimensionless groups named after Damköhler, two of which involve reaction rates. The one which compares rates of reaction and diffusion is sometimes called "Damköhler group II." Because it is the only type of Damköhler number used in this book, no other identifier is added to the symbol Da. An extensive tabulation of named dimensionless groups pertinent to chemical engineering is given in Catchpole and Fulford (1966).

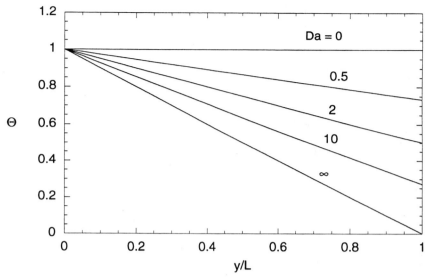

Figure 2-7. Concentration profiles for diffusion through a gas film with a heterogeneous reaction, showing the effect of the Damköhler number. The curves were obtained by solving Eq. (2.8-22) with $m=1$ and $n=2$.

$$\theta(1) \rightarrow \mathrm{Da}^{-1/n} \qquad (\mathrm{Da} \rightarrow \infty, \; m=1). \tag{2.8-25}$$

For first-order kinetics ($n=1$) and $\mathrm{Da} > 10^2$, this implies that $\theta(1) < 0.01$. Because the reactant flux (and therefore the reaction rate) varies as $1 - \theta(1)$, the simplified boundary condition will lead to an error of $<1\%$ under these conditions. For second-order kinetics ($n=2$), $\mathrm{Da} > 10^4$ is needed to maintain this small level of error.

Example 2.8-3 Diffusion in a Dilute Liquid Solution with a Heterogeneous Reaction We reconsider the situation of Example 2.8-2, but with a dilute liquid solution in place of the gas. One consequence of having a liquid is that we can assume constant ρ. Another key feature of the dilute liquid solution is that it is pseudobinary from a diffusional standpoint. Thus, the product B will have negligible influence on the diffusion of the reactant A for any stoichiometry. Because species B is assumed not to affect the reaction kinetics either, it need not be considered at all.

The previous conclusion that $v_y = 0$ remains valid for the liquid. Using this information and Eq. (2.6-4), the flux of A is given by

$$N_{Ay} = J_{Ay} = -D_A \frac{dC_A}{dy}. \tag{2.8-26}$$

This may be contrasted with Eqs. (2.8-16) or (2.8-18) for the gas-phase problem; there is no convection now. From Table 2-3, the conservation equation for species A is

$$\frac{d^2 C_A}{dy^2} = 0. \tag{2.8-27}$$

Accordingly, C_A is linear in y for all values of m. Converting to the dimensionless variables defined by Eq. (2.8-21) and proceeding much as before, it is found that the concentration of A at the catalytic surface is governed by

$$Da\phi'' = 1 - \phi. \tag{2.8-28}$$

For the gas, this result held only for $m=1$, corresponding to equimolar counterdiffusion [see Eq. (2.8-23)]. For the liquid, it is valid for all stoichiometries.

Example 2.8-4 Diffusion in a Liquid with a Reversible Homogeneous Reaction   This example illustrates the influence of homogeneous reactions on concentration profiles and also demonstrates a technique for dealing with reversible reactions. Consider the reversible conversion of A to B in a stagnant liquid film at steady state, as shown in Fig. 2-8. The concentrations at $y=0$ are known (C_{A0}, C_{B0}), and the inert solid surface at $y=L$ acts only as an impermeable barrier. The reaction kinetics are first order and reversible. From the kinetics and stoichiometry,

$$R_{VA} = k_{-1}C_B - k_1C_A = k_1(KC_B - C_A) = -R_{VB}, \tag{2.8-29}$$

$$K = k_{-1}/k_1. \tag{2.8-30}$$

At equilibrium, $R_{VA} = R_{VB} = 0$ and $C_A = KC_B$, with K the equilibrium constant defined by Eq. (2.8-30). Whether or not reaction equilibrium is actually achieved in any part of the liquid film under steady-state conditions is determined by the relative rates of reaction and diffusion, as will be seen.

For solutes A and B in a dilute liquid, the steady-state conservation equations (from Table 2-3) are

$$0 = D_A \frac{d^2C_A}{dy^2} + R_{VA}, \tag{2.8-31}$$

$$0 = D_B \frac{d^2C_B}{dy^2} - R_{VA}. \tag{2.8-32}$$

With the specified concentrations at $y=0$ and an impermeable and inert surface at $y=L$, the boundary conditions are

$$C_A = C_{A0}, \ C_B = C_{B0} \qquad \text{at } y=0, \tag{2.8-33}$$

$$\frac{dC_A}{dy} = \frac{dC_B}{dy} = 0 \qquad \text{at } y=L. \tag{2.8-34}$$

This completes the formulation of the problem.

The solution to this set of equations is complicated by the fact that Eqs. (2.8-31) and (2.8-32) are coupled through the reaction terms. However, a differential equation which is independent of the reaction rates is obtained by adding Eqs. (2.8-31) and (2.8-32). The result is

$$0 = D_A \frac{d^2C_A}{dy^2} + D_B \frac{d^2C_B}{dy^2}. \tag{2.8-35}$$

Integrating Eq. (2.8-35) once yields

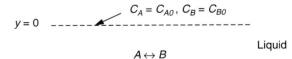

$y = 0$

$C_A = C_{A0}, \ C_B = C_{B0}$

$A \leftrightarrow B$

Liquid

$y = L$

Inert
Surface

Figure 2-8. Diffusion in a stagnant liquid film with a reversible homogeneous reaction.

$$D_A \frac{dC_A}{dy} + D_B \frac{dC_B}{dy} = a_1, \tag{2.8-36}$$

where a_1 is a constant. From the boundary conditions at $y = L$, we have $a_1 = 0$. Integrating again,

$$D_A C_A + D_B C_B = a_2. \tag{2.8-37}$$

From the boundary conditions at $y = 0$, we have $a_2 = D_A C_{A0} + D_B C_{B0}$. Thus,

$$C_B = \frac{D_A}{D_B}(C_{A0} - C_A) + C_{B0}. \tag{2.8-38}$$

Inserting Eq. (2.8-38) in Eq. (2.8-31) will yield a single differential equation involving only C_A. The solution is completed using dimensionless variables defined as

$$\eta \equiv \frac{y}{L}, \qquad \theta_A \equiv \frac{C_A}{C_{A0}}, \qquad \theta_B \equiv \frac{K C_B}{C_{A0}}. \tag{2.8-39}$$

With this choice of dimensionless concentrations, $\theta_A = \theta_B$ at equilibrium. Combining Eqs. (2.8-31) and (2.8-38) as mentioned above, and converting to dimensionless variables, gives

$$\frac{d^2 \theta_A}{d\eta^2} = (\alpha + \beta)\theta_A - (\alpha\gamma + \beta), \tag{2.8-40}$$

$$\theta_A(0) = 1, \qquad \frac{d\theta_A}{d\eta}(1) = 0. \tag{2.8-41}$$

The dimensionless parameters which appear are

$$\alpha \equiv \frac{k_1 L^2}{D_A}, \qquad \beta \equiv \frac{k_{-1} L^2}{D_B}, \qquad \gamma \equiv \frac{K C_{B0}}{C_{A0}}. \tag{2.8-42}$$

This problem is characterized by *two* Damköhler numbers, α and β; α involves the forward rate constant and D_A, whereas β involves the reverse rate constant and D_B. The third parameter determines the direction of the reaction in the liquid film: The net reaction is $A \rightarrow B$ for $\gamma < 1$ and $B \rightarrow A$ for $\gamma > 1$.

The solution to Eqs. (2.8-40) and (2.8-41) for the concentration of species A is

$$\theta_A = \left[\frac{\alpha\gamma + \beta}{\alpha + \beta}\right] + \left[\frac{\alpha(1 - \gamma)}{\alpha + \beta}\right][\cosh(\sqrt{\alpha + \beta}\,\eta) - \tanh(\sqrt{\alpha + \beta})\sinh(\sqrt{\alpha + \beta}\,\eta)]. \tag{2.8-43}$$

The corresponding result for species B is

$$\theta_B = \left[\frac{\alpha\gamma + \beta}{\alpha + \beta}\right] - \left[\frac{\beta(1 - \gamma)}{\alpha + \beta}\right][\cosh(\sqrt{\alpha + \beta}\,\eta) - \tanh(\sqrt{\alpha + \beta})\sinh(\sqrt{\alpha + \beta}\,\eta)]. \tag{2.8-44}$$

These results are valid for any values of the parameters.

It is informative to examine the concentration profiles for a very fast reaction. Suppose that $\alpha \rightarrow \infty$ and $\beta \rightarrow \infty$ with β/α fixed, which corresponds to increasing the reaction rate constants to very high values without altering the equilibrium constant. For large arguments, $\tanh x \rightarrow 1$ and $\cosh x - \sinh x \rightarrow e^{-x}$, so that Eqs. (2.8-43) and (2.8-44) reduce to

$$\theta_A = \left[\frac{\gamma + (\beta/\alpha)}{1 + (\beta/\alpha)}\right] + \left[\frac{1 - \gamma}{1 + (\beta/\alpha)}\right]e^{-\sqrt{\alpha + \beta}\,\eta}, \tag{2.8-45}$$

$$\theta_B = \left[\frac{\gamma + (\beta/\alpha)}{1 + (\beta/\alpha)}\right] - \left[\frac{(\beta/\alpha)(1 - \gamma)}{1 + (\beta/\alpha)}\right]e^{-\sqrt{\alpha + \beta}\,\eta}. \tag{2.8-46}$$

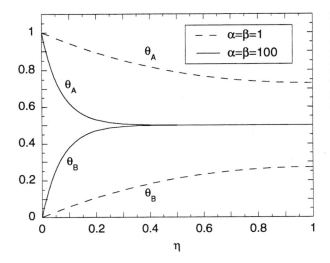

Figure 2-9. Concentration profiles for diffusion in a stagnant film with a reversible homogeneous reaction, based on Eqs. (2.8-43) and (2.8-44). Results are shown for moderate ($\alpha = \beta = 1$) and fast ($\alpha = \beta = 100$) reaction kinetics. In both cases it is assumed that $\gamma = 0$.

Over *most* of the film the exponential terms are negligible, so that

$$\theta_A \cong \left[\frac{\gamma + (\beta/\alpha)}{1 + (\beta/\alpha)} \right] \cong \theta_B. \tag{2.8-47}$$

Thus, when the reaction is fast the two species are in equilibrium throughout most of the film. Notice that the equilibrium concentration is determined in part by the relative diffusivities, which affect the value of β/α. Near the exposed surface ($\eta = 0$), where A and B are added to or removed from the liquid, there may be very large concentration gradients. In this nonequilibrium region, θ_A and θ_B each change from the surface value to the equilibrium value over a dimensionless distance $\Delta \eta$ of order of magnitude $(\alpha + \beta)^{-1/2}$. The dimensional thickness δ of this nonequilibrium layer is therefore

$$\frac{\delta}{L} \sim (\alpha + \beta)^{-1/2}. \tag{2.8-48}$$

Thus, even as the reaction rate constants reach arbitrarily high levels, the reactants never achieve equilibrium throughout the film; the thickness of the nonequilibrium layer merely diminishes, eventually varying as the inverse square root of the sum of the Damköhler numbers.

   Plots of Eqs. (2.8-43) and (2.8-44) for moderate and fast reaction kinetics and $\gamma = 0$ are shown in Fig. 2-9. For $\alpha = \beta = 1$ the reaction $A \rightarrow B$ proceeds at significant rates throughout the film, whereas for $\alpha = \beta = 100$ there are distinct reaction and equilibrium zones, as discussed.

Example 2.8-5 Directional Solidification of a Dilute Binary Alloy[2]   The partitioning of a dopant or impurity during the solidification of a melt in a uniaxial temperature field is fundamental to most methods for growth of single-crystal semiconductors and metals. This situation is simply described by the steady-state, one-dimensional model shown in Fig. 2-10. The melt–crystal interface is located at $y = 0$. Solidification is assumed to occur at a constant rate, such that the velocities of the melt and solid relative to the interface are both equal to U; this implies that the melt and solid densities are equal. The material is a dilute alloy with dopant concentration $C_i(y)$. The bulk of the melt is assumed to be well-mixed with a dopant concentration C_∞, whereas the dopant

---

[2] This example was suggested by R. A. Brown.

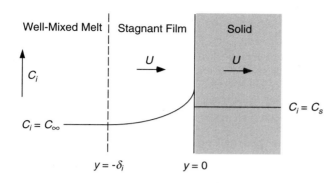

concentration in the solid is C_s. It is assumed that there is a "stagnant film" of melt of thickness δ_i in which there is no flow parallel to the interface. That is, according to the model, transport of the dopant in this layer is only by diffusion and convection *normal* to the interface. Given values for C_∞, δ_i, and the dopant properties, it is desired to predict C_s.

For this one-dimensional, steady-state problem with $C_i = C_i(y)$ and $v_y = U$, the conservation equation for the dopant (Table 2-3) becomes

$$D_i \frac{d^2 C_i}{dy^2} - U \frac{dC_i}{dy} = 0. \tag{2.8-49}$$

At the edge of the stagnant layer the concentration must match the bulk value, so that

$$C_i(-\delta_i) = C_\infty. \tag{2.8-50}$$

At the melt–solid interface, the flux of dopant in the melt must match the flux in the solid. This condition is written as

$$N_{iy}(0) = UC_i(0) - D_i \frac{dC_i}{dy}(0) = UC_s. \tag{2.8-51}$$

We have neglected diffusion of the dopant in the solid because of the extremely low diffusivity in the solid phase. For the low solidification rates which are characteristic of crystal growth, the melt and solid can be assumed to be in chemical equilibrium at the interface. This equilibrium is typically modeled for dilute alloys by the linear relationship

$$C_s = K_i C_i(0), \tag{2.8-52}$$

where K_i is the equilibrium segregation coefficient for solute i. For most impurities, such as metals in silicon, $K_i < 0.1$. Using Eq. (2.8-52) the interfacial balance is rewritten as

$$\frac{dC_i}{dy}(0) = \frac{U(1 - K_i)}{D_i} C_i(0). \tag{2.8-53}$$

The solution of Eq. (2.8-49) that satisfies Eqs. (2.8-50) and (2.8-53) is

$$\frac{C_i(y)}{C_\infty} = \frac{K_i + (1 - K_i)\exp(Uy/D_i)}{K_i + (1 - K_i)\exp(-U\delta_i/D_i)}. \tag{2.8-54}$$

This can be rewritten as

$$\frac{C_i(y)}{C_\infty} = \frac{K_i + (1 - K_i)\exp(Pe_i y/\delta_i)}{K_i + (1 - K_i)\exp(-Pe_i)}, \tag{2.8-55}$$

$$\mathrm{Pe}_i \equiv \frac{U\delta_i}{D_i}, \quad \text{film } \text{sml} \tag{2.8-56}$$

where Pe_i is the *Péclet number* for solute i. In general, the Péclet number for mass transfer is a measure of the importance of convection relative to diffusion. As exemplified by Eq. (2.8-56), it is a ratio of a flow velocity (U) to a characteristic velocity for diffusion (D_i/δ_i). For $\mathrm{Pe}_i \to 0$ in the solidification model, diffusion in the film of melt is sufficiently fast that no significant concentration gradient is able to develop. Thus, Eq. (2.8-55) indicates that $C_i(y) = C_\infty$, a constant. At the other extreme, where $\mathrm{Pe}_i \gg 1$, Eq. (2.8-55) becomes

$$\frac{C_i(y)}{C_\infty} = 1 + \frac{(1 - K_i)}{K_i} \exp\left(\frac{Uy}{D_i}\right) \quad \text{(no bulk mixing)}. \tag{2.8-57}$$

This result is the same as if we had assumed that there was no well-mixed region in the melt and had simply set $\delta_i = -\infty$. Accordingly, we refer to it as the case for "no bulk mixing."

Burton et al. (1953) first used this analysis to deduce an expression for the effective segregation coefficient for the dopant between the bulk melt and crystal, K_{eff}, defined as

$$K_{\mathrm{eff}} \equiv \frac{C_s}{C_\infty} = \frac{K_i C_i(0)}{C_\infty}. \tag{2.8-58}$$

Thus, K_{eff} is like K_i, except that K_{eff} is based on the bulk rather than the interfacial concentration of the dopant. Whereas K_i is determined only by thermodynamics, K_{eff} depends on the solidification rate and other process variables. Using Eq. (2.8-55) it is found that

$$K_{\mathrm{eff}} = \frac{K_i}{K_i + (1 - K_i)\exp(-\mathrm{Pe}_i)}. \tag{2.8-59}$$

If Eq. (2.8-57) is used for the dopant concentration profile, then it is found that $K_{\mathrm{eff}} = 1$ and there is no segregation of dopant between the bulk melt and the crystal. Hence, removal of impurities during solidification relies on vigorous mixing of the melt. Equation (2.8-59) has found widespread use in research on crystal growth from melts, as reviewed in Brown (1988).

Stagnant Film Models

The foregoing analysis is an example of the use of a *stagnant film model* to represent convective heat or mass transfer in what is in reality a very complex situation. In directional solidification, buoyancy-driven flow will usually result in two- or three-dimensional velocity fields in the melt, precluding an analytical solution to the "real" mass transfer problem. Likewise, flow patterns in systems with forced convection (e.g., stirred reactors) are often quite complex. For this reason, stagnant film models as used in this example and in Examples 2.8-2 and 2.8-3 find frequent use as first approximations in engineering calculations. In any stagnant film model, convection and diffusion parallel to the interface are ignored, and all effects of mixing in the fluid are embodied in the effective film thickness, δ_i. The film thickness is related to the mass transfer coefficient and diffusivity of i by

$$\delta_i = \frac{D_i}{k_{ci}}. \tag{2.8-60}$$

Thus, if k_{ci} and D_i are known, δ_i may be computed using Eq. (2.8-60). In many applications, such as in most experimental studies of directional solidification, δ_i is used as an adjustable parameter. For example, if one chooses the value of δ_i to match the observed value of K_{eff} for a given dopant at one solidification velocity (U), then one can use Eq. (2.8-59) to estimate K_{eff} for that dopant at some other velocity.

One of the limitations of the stagnant film approach is that, in general, the effective film thickness needed to give the correct value of the mass transfer coefficient is different for each component of a mixture. If it is assumed that the film thickness has the same value for all solutes (i.e., $\delta_i = \delta$ for all i), then Eq. (2.8-60) implies that $k_{ci} \propto D_i$. In reality, it is found that for given flow conditions we have $k_{ci} \propto D_i^n$, where n is a constant which depends on the type of flow and which is typically < 1. The prediction of n for various situations is discussed in Chapters 9, 10, 12, and 13.

2.9 SPECIES CONSERVATION FROM A MOLECULAR VIEWPOINT

The conservation equations for chemical species developed in this chapter are based on the concept of the local *concentration* of species i in a mixture. As discussed in Section 1.6, this requires that there be a large number of molecules in any point-size volume under consideration. Having a large number of molecules at every such point averages out the randomness in the instantaneous positions of individual molecules, so that the species concentrations are deterministic quantities. However, in efforts to relate molecular-level phenomena to observed rates of transport, it is natural to focus on the motion of a single molecule. The position of a given molecule can be described only in terms of probabilities, so that its calculation is a stochastic rather than a deterministic problem. Nonetheless, under the proper conditions, the equation describing the probability of finding a diffusing molecule at a given position and time has the same form as a conservation equation. The objective of this section is to derive the equation for the probability distribution, and thereby to illustrate the relationship between the molecular and the more macroscopic viewpoints.

The following derivation concerns the motion of a single molecule (or Brownian particle) relative to the surrounding fluid; in this reference frame, convection is excluded. The molecule is assumed to move independently of other solute molecules, which restricts the derivation to dilute solutions. Chemical reactions are not considered. The probability density for finding the molecule at position $\mathbf{r}$ at time t is denoted by $p(\mathbf{r}, t)$. Suppose that after being "released" from some location at $t = 0$, the molecule takes a number of discrete steps or "jumps" of uniform length ℓ in random directions. The time required for a single step is τ. For the situation illustrated in Fig. 2-11, the position at time t is $\mathbf{r} - \mathbf{R}$. The next step terminates at some position $\mathbf{r}$ at time $t + \tau$. Thus, $\mathbf{R}$ is a vector with length ℓ and random direction.

Working backward from time $t + \tau$ to time t, we recognize that $p(\mathbf{r}, t + \tau)$ is the product of $p(\mathbf{r} - \mathbf{R}, t)$ and the probability of a jump described by a specific vector $\mathbf{R}$. The probability density governing $\mathbf{R}$ is denoted by $\Lambda(\mathbf{R})$. Considering all possible jumps,

$$p(\mathbf{r}, t + \tau) = \int p(\mathbf{r} - \mathbf{R}, t) \Lambda(\mathbf{R}) \, d\mathbf{R}, \qquad (2.9\text{-}1)$$

$$\Lambda(\mathbf{R}) = \frac{\delta(|\mathbf{R}| - \ell)}{4\pi\ell^2}, \qquad (2.9\text{-}2)$$

where $\delta(x)$ is the *Dirac delta function* and the integration is over all space. Equation (2.9-1) is an example of a *Markoff integral equation*, describing a stepwise stochastic

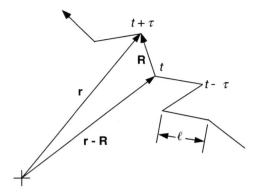

Figure 2-11. Positions of a molecule making randomly directed jumps of length ℓ, each in time τ.

process. Equation (2.9-2) ensures that the only allowed jumps are of length ℓ, because $\delta(x-x_0)=0$ for $x \neq x_0$. The Dirac delta function has the additional property that $\int \delta(x-x_0) f(x) \, dx = f(x_0)$. Using spherical coordinates to integrate $\Lambda(\mathbf{R})$ yields

$$\int \Lambda \, d\mathbf{R} = \int_0^{2\pi} \int_0^{\pi} \int_0^{\infty} \frac{\delta(r-\ell)}{4\pi\ell^2} r^2 \sin \theta \, dr \, d\theta \, d\phi = \frac{1}{\ell^2} \int_0^{\infty} \delta(r-\ell) r^2 \, dr = 1. \qquad (2.9\text{-}3)$$

That is, $\Lambda(\mathbf{R})$ is normalized so that the probability is unity that a given jump will end somewhere on a spherical surface of radius ℓ.

Two other integrals which will be needed are stated now. Using Eqs. (A.6-12) and (A.6-13), it is found that

$$\int \Lambda \mathbf{R} \, d\mathbf{R} = \mathbf{0} \qquad (2.9\text{-}4)$$

$$\int \Lambda \mathbf{R}\mathbf{R} \, d\mathbf{R} = \frac{\ell^2}{3} \boldsymbol{\delta}. \qquad (2.9\text{-}5)$$

Equation (2.9-4) gives the average value of the single-step displacement vector $\mathbf{R}$; because $\mathbf{R}$ has no directional bias, that average vanishes. Equation (2.9-5) gives the average value of the dyad $\mathbf{R}\mathbf{R}$, and the result is expressed in terms of the identity tensor $\boldsymbol{\delta}$.

After a large number of jumps, $p(\mathbf{r}, t)$ may be approximated as a continuous function of position and time. With ℓ and τ assumed to be small relative to the length and time scales of interest, the probabilities in Eq. (2.9-1) can be written as Taylor expansions about position $\mathbf{r}$ and time t. The first two terms for the expansion in time are

$$p(\mathbf{r}, t+\tau) = p(\mathbf{r}, t) + \frac{\partial p}{\partial t} \tau + \cdots \qquad (2.9\text{-}6)$$

and, using Eq. (A.6-6), the first three terms for the expansion in position are

$$p(\mathbf{r} - \mathbf{R}, t) = p(\mathbf{r}, t) - \mathbf{R} \cdot \nabla p + \frac{1}{2} \mathbf{R}\mathbf{R} : \nabla\nabla p + \cdots, \qquad (2.9\text{-}7)$$

where all derivatives are evaluated at $\mathbf{r}$ and t. Substituting these leading terms in Eq. (2.9-1) gives

$$p(\mathbf{r},\ t) + \frac{\partial p}{\partial t}\tau = p(\mathbf{r},\ t)\int \Lambda\ d\mathbf{R} - \nabla p \cdot \int \Lambda\mathbf{R}\ d\mathbf{R} + \frac{1}{2}\nabla\nabla p : \int \Lambda\mathbf{R}\mathbf{R}\ d\mathbf{R}. \quad (2.9\text{-}8)$$

Using Eqs. (2.9-3)–(2.9-5) to evaluate the integrals, we find that

$$p(\mathbf{r},\ t) + \frac{\partial p}{\partial t}\tau = p(\mathbf{r},\ t) - 0 + \frac{\ell^2}{6}\nabla\nabla p : \boldsymbol{\delta}. \quad (2.9\text{-}9)$$

Finally, applying the identity $\nabla\nabla p : \boldsymbol{\delta} = \nabla^2 p$ and rearranging yields

$$\frac{\partial p}{\partial t} = \left(\frac{\ell^2}{6\tau}\right)\nabla^2 p. \quad (2.9\text{-}10)$$

This is the desired equation governing the probability density $p(\mathbf{r},\ t)$ for finding a given solute molecule at position $\mathbf{r}$ at time t.

A deterministic result comparable to Eq. (2.9-10) is obtained by setting $\mathbf{v} = 0$ and $R_{Vi} = 0$ in Eq. (2.6-5) to give

$$\frac{\partial C_i}{\partial t} = D_i\nabla^2 C_i. \quad (2.9\text{-}11)$$

The similarities between Eqs. (2.9-10) and (2.9-11) are not coincidental. Equation (2.9-10) governs the probability density for each of N independent solute molecules in a dilute solution. The overall distribution of solute molecules in position and time can be described using a sum of such probabilities or, if N is large, a concentration. Using the index j to denote a particular solute molecule, the probability density $p_j(\mathbf{r},\ t)$ is normalized such that

$$\int p_j(\mathbf{r},\ t)\ d\mathbf{r} = 1. \quad (2.9\text{-}12)$$

That is, the probability is unity that the molecule is located somewhere in space. If there are N molecules of solute i in a volume V, then the molar concentration is related to the probability densities by

$$C_i(\mathbf{r},\ t) = \frac{1}{N_{\text{Av}}V}\sum_{j=1}^{N} p_j(\mathbf{r},\ t). \quad (2.9\text{-}13)$$

The significance of Eq. (2.9-13) is that it implies that Eq. (2.9-11) could have been derived by summing N equations like Eq. (2.9-10), one for each solute molecule. The coefficients appearing in Eqs. (2.9-10) and (2.9-11) must then be identical. We conclude that

$$D_i = \frac{\ell^2}{6\tau} = \frac{u\ell}{6}, \quad (2.9\text{-}14)$$

where $u = \ell/\tau$ is a "jump velocity" like that used in the lattice model for diffusion in Section 1.5. The diffusivity in Eq. (2.9-14) is the same as the lattice result; compare with Eq. (1.5-4).

We conclude this discussion by showing how the diffusivity D_i can be calculated from information on the movement of a single molecule or particle. Consider the molecule to be at the origin at $t = 0$. At later times its position is described by

$\mathbf{r}(t) = x(t)\ \mathbf{e}_x + y(t)\ \mathbf{e}_y + z(t)\ \mathbf{e}_z$, corresponding to a radial coordinate $r(t) = |\mathbf{r} \cdot \mathbf{r}|^{1/2} = (x^2 + y^2 + z^2)^{1/2}$. In an isotropic medium, $p = p(r, t)$ only; that is, in the absence of any directional bias, the probability density is spherically symmetric. Using spherical coordinates (Table 2-3) and $D_i = \ell^2/(6\tau)$, Eq. (2.9-10) becomes

$$\frac{\partial p}{\partial t} = \frac{D_i}{r^2} \frac{\partial}{\partial r}\left(r^2 \frac{\partial p}{\partial r}\right). \tag{2.9-15}$$

Taking as unity the probability that the molecule is located somewhere in space, the normalization condition for p [similar to Eq. (2.9-12)] is

$$\int p\ dV = \int\limits_{-\infty}^{\infty}\int\int p\ dx\ dy\ dz = 4\pi\int\limits_0^\infty p\ r^2\ dr = 1. \tag{2.9-16}$$

The solution to Eq. (2.9-15) that satisfies Eq. (2.9-16) is (see Section 4.9)

$$p(r, t) = \frac{1}{8(\pi D_i t)^{3/2}}\ e^{-r^2/4D_i t}. \tag{2.9-17}$$

Thus, $p \to 0$ as $r \to \infty$, as expected.

To relate D_i to measured molecule or particle positions (e.g., as obtained from the experimental visualization of particles or molecular dynamics simulations), we now use Eq. (2.9-17) to evaluate certain average properties of the position. These are averages which would be observed by recording the results of a large number of random walks. The *mean displacement* of the molecule,

$$\langle \mathbf{r} \rangle = \int \mathbf{r} p\ dV, \tag{2.9-18}$$

vanishes as a result of the spherical symmetry, and therefore contains no information on D_i.

The *mean-square displacement* is given by

$$\langle r^2 \rangle = \int r^2\ p\ dV. \tag{2.9-19}$$

Using Eq. (2.9-17) in Eq. (2.9-19), we find that

$$\langle r^2 \rangle = \frac{4\pi}{8(\pi D_i t)^{3/2}}\int\limits_0^\infty r^4 e^{-r^2/4D_i t}\ dt = 6D_i t. \tag{2.9-20}$$

This result can be used to evaluate D_i by measuring displacements during many identical time intervals, each of duration t. The average of the squares of these displacements will yield $\langle r^2 \rangle$, allowing the diffusivity to be calculated as $D_i = \langle r^2 \rangle/6t$.

The calculation of the mean-square displacement as a function of time is the basis for the evaluation of diffusion coefficients using molecular dynamics simulations for liquids. In such simulations the instantaneous positions of N molecules or particles are computed by solving Newton's law of motion with an interatomic or interparticle potential field which describes their interactions. If the intitial position of the jth particle is denoted as $\mathbf{r}_j(0)$, then Eq. (2.9-20) can be written for the collection of particles as

$$6D_i t = \langle | \mathbf{r}_j(t) - \mathbf{r}_j(0) |^2 \rangle, \tag{2.9-21}$$

where the angle brackets indicate averages taken over all N particles. See Allen and Tildesley (1987) for a thorough discussion of molecular dynamics methods.

References

Allen, M. P. and D. J. Tildesley. *Computer Simulation of Liquids.* Clarendon Press, Oxford, 1987.

Brown, R. A. Theory of transport processes in single crystal growth from the melt. *AIChE J.* 34: 881–911, 1988.

Burton, J. A., R. C. Prim, and W. P. Slichter. The distribution of solute in crystals grown from the melt. Part I. Theoretical. *J. Chem. Phys.* 21: 1987–1991, 1953.

Catchpole, J. P. and G. Fulford. Dimensionless groups. *Ind. Eng. Chem.* 58(3): 46–60, 1966.

Edwards, D. A., H. Brenner, and D. T. Wasan. *Interfacial Transport Processes and Rheology.* Butterworth-Heinemann, Boston, 1991.

Nye, J. F. *Physical Properties of Crystals.* Oxford University Press, Oxford, 1957.

Schlicting, H. *Boundary-Layer Theory,* sixth edition. McGraw-Hill, New York, 1968.

Siegel, R. and J. R. Howell. *Thermal Radiation Heat Transfer,* second edition. McGraw-Hill, New York, 1981.

Sparrow, E. M. and R. D. Cess. *Radiation Heat Transfer.* McGraw-Hill, New York, 1978.

Problems

2-1. Charge Conservation and Electric Current

For an electrolyte solution of uniform composition ($\nabla C_i = 0$), the current density $\mathbf{i}$ and electrical potential ϕ are related as

$$\mathbf{i} = -\kappa_e \nabla \phi,$$

where κ_e is the electrical conductance of the solution (a constant). The usual units of $\mathbf{i}$ are C s^{-1} m^{-2} or A m^{-2}. The charge density in an electrolyte solution (ρ_e) is given by

$$\rho_e = F \sum_i z_i C_i,$$

where F is Faraday's constant (9.652×10^4 C mol^{-1}), z_i is the valence of ion i, and the summation is over all ions in the solution. Except near charged surfaces, electroneutrality ($\rho_e \cong 0$) is a good approximation. Assume that this holds, and that there are no charge-producing reactions taking place in the solution.

(a) What can be learned by applying conservation of charge to an arbitrary control volume in the solution?

(b) Derive the partial differential equation which governs $\phi(\mathbf{r}, t)$ in the solution.

2-2. Flow in Porous Media: Darcy's Law

A relationship often used to model fluid flow in porous media is *Darcy's law,*

$$\mathbf{v} = -\frac{\kappa}{\mu}(\nabla P - \rho \mathbf{g}) \equiv -\frac{\kappa}{\mu} \nabla \mathcal{P},$$

where κ is the Darcy permeability, $\mathbf{g}$ is the gravitational acceleration vector, and $\mathcal{P}$ is the dynamic pressure (Chapter 5). The Darcy permeability has units of m^2 and is a constant for a given material. The velocity and pressure in Darcy's law are averaged over a length scale large compared to the size of individual pores, but small compared to the dimensions of the object. That is, microstructural details are not considered. Pore structure enters only through its effect on the value of κ, which is usually determined by experiment.

(a) For a material of average porosity ε (volume fraction of pores), use Darcy's law to obtain expressions for conservation of mass in the fluid, first in integral (control volume) form and then in differential form.

(b) Show that for an incompressible fluid, the differential equation from part (a) reduces to

$$\nabla^2 \mathcal{P} = 0.$$

(c) Consider an idealized porous material consisting of straight, cylindrical pores of diameter d. Assume that all pores are parallel to the x, y, or z axes and that they intersect at points described by a simple cubic lattice of dimension $\ell \gg d$. The overall dimensions of a sample of this material are of order of magnitude L, where $L \gg \ell$. The volume flow rate Q in a single pore segment is described by *Poiseuille's law* [Eq. (6.2-23)],

$$Q = \frac{\pi |\Delta \mathcal{P}_e| d^4}{128 \mu \ell},$$

where $|\Delta \mathcal{P}_e|$ is the pressure drop for a segment of length ℓ. Evaluate κ for this material.

(d) Suppose now that the pores run only along the x and y axes. What change is required in the form of Darcy's law? Once again evaluate the permeability.

2-3. Diffusion in a Gas with a Fast Heterogeneous Reaction

For steady diffusion through a gas film with a heterogeneous reaction, as described in Example 2.8-2, determine the concentration profile and flux for the reactant in the limit Da→∞. Consider both $m=1$ and $m \neq 1$. Compare plots of $\theta(\eta)$ for $m = \frac{1}{2}$, 1, and 2.

2-4. Entropy Conservation and the Second Law

Consider energy transfer in a stationary fluid or solid. The internal energy per unit mass is denoted by $\hat{U}$.

(a) State the differential conservation equation for $\hat{U}$.

(b) Derive the conservation equation for entropy per unit mass, $\hat{S}$, assuming that ρ is constant and $H_V = 0$. (Begin with the equation for $\hat{U}$, and then use the thermodynamic relationship between $\hat{U}$ and $\hat{S}$.)

(c) Suppose that it is desired to write the result of part (b) in the form of Eq. (2.3-7), with the entropy flux (relative to $\mathbf{v}$) denoted by $\mathbf{j_s}$ and the volumetric rate of entropy production denoted by σ. If the entropy flux is defined as $\mathbf{j_s} \equiv \mathbf{q}/T$, determine σ.

(d) According to the second law of thermodynamics, the total entropy of a system undergoing a spontaneous change must increase with time. Thus, for a fixed control volume,

$$\frac{d}{dt} \int_V \rho \hat{S} \, dV > 0.$$

Using the results of part (c), show that this requires that the thermal conductivity (k) be a positive number. To do this, consider a system which is nonisothermal and insulated from its surroundings.

2-5. Diffusivities for Two- or One-Dimensional Random Walks

(a) Using the approach of Section 2.9, show that for a *two-dimensional* random walk with steps of length ℓ in time τ, the molecular diffusivity is given by

$$D = \frac{\ell^2}{4\tau}.$$

(b) Show that for a *one-dimensional* random walk the corresponding result is

$$D = \frac{\ell^2}{2\tau}.$$

2-6. Diffusion into an Open Cone

Assume that a solute is released at a constant rate $\dot{m}$ (moles/time) into the open conical region shown in Fig. P2-6. The solute source is located at the origin (cone vertex) and the cone angle is Θ_0. There is no flux across the surface defined by $\theta = \theta_0$. Determine the steady-state concentration in the conical region.

2-7. Pressure Measurement with a Gas Purge*

It is desired to monitor the pressure in a vessel containing pure fluorine (F_2), using the gauge arrangement shown in Fig. P2-7. Because F_2 is highly corrosive, a nitrogen purge is to be used to minimize the exposure of the gauge to F_2. The N_2 flux through the tube connecting the gauge and vessel may be adjusted as needed by varying the N_2 pressure in the purge line. Assume that the purge rate is slow enough, and the vessel large enough, that the amount of N_2 in the vessel (at $y = L$) remains negligible.

Find an expression for the minimum nitrogen flux which will make the steady-state fluorine flux zero and which will also keep the mole fraction of fluorine at the gauge below some specified value (i.e., $x_F(0) \leq x_F^*$). Assume that transport in the tube is one-dimensional.

*This problem was suggested by D. I. C. Wang.

2-8. Sieving Coefficient in Ultrafiltration

Ultrafiltration is a process in which pressure-driven flow of filtrate across a semipermeable membrane causes macromolecules or colloidal particles to be concentrated in the retentate. The ability to retain a given solute depends on factors such as molecular size and pore size, but is also influenced a great deal by the operating conditions. An undesired consequence of the filtration is that the retained macromolecules tend to accumulate near the upstream surface of the membrane, a phenomenon called *concentration polarization*. The concentration of a given solute at the membrane surface may greatly exceed that in the bulk retentate, causing membrane fouling by precipitation, gel formation, or other mechanisms. Even moderate levels of solute accumulation at the upstream surface of the membrane are undesirable, in that they will tend to increase the solute concentration in the filtrate. Dynamic conditions within the membrane also affect the separation.

Figure P2-6. Diffusion from a point source of solute into an open conical region.

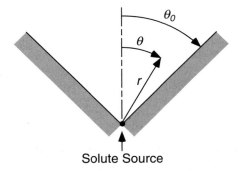

Solute Source

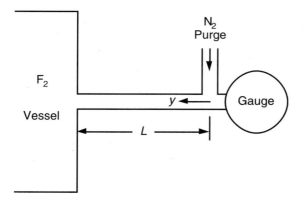

Figure P2-7. Pressure monitoring arrangement for vessel containing fluorine gas.

A simple but useful model for steady-state ultrafiltration, based on the stagnant-film concept, is shown in Fig. P2-8. The solute concentration in the retentate film $C_R(x)$ is assumed to reach the bulk value C_0 at a distance δ from the membrane; the more effective the agitation of the retentate, the smaller the value of δ. The diffusivity in the retentate is D_R and the filtrate velocity is v_F. In the membrane the solute flux (N) is given by

$$N = -D_M \frac{dC_M}{dx} + v_F(1-\sigma)C_M,$$

where $C_M(x)$ is a "liquid-equivalent" concentration and D_M is the corresponding solute diffusivity in the membrane phase. (By "liquid-equivalent" it is meant that $C_M(x)$ is defined so that the concentration variable is continuous, even at the phase boundaries.) The intrinsic selectivity of the membrane toward the solute is expressed by the diffusivity ratio D_M/D_R and the *reflection coefficient*, σ. For an ideal semipermeable membrane, $\sigma = 1$, and for a completely nonselective membrane, $\sigma = 0$. The filtrate concentration is not known in advance; it is determined by the solute flux and filtrate velocity, such that $C_F = N/v_F$. In other words, transport of solute in the filtrate is assumed to be purely convective.

The *sieving coefficient*, $\Theta \equiv C_F/C_0$, measures the overall effectiveness of the separation for a particular solute. Show that

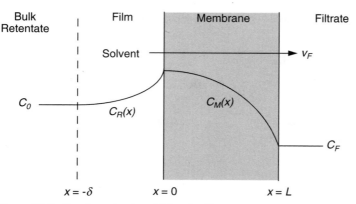

Figure P2-8. One-dimensional model for ultrafiltration.

$$\Theta = \frac{(1-\sigma)e^{\mathrm{Pe}_R}}{\sigma(1-e^{-\mathrm{Pe}_M})+(1-\sigma)e^{\mathrm{Pe}_R}},$$

$$\mathrm{Pe}_R \equiv \frac{v_F \delta}{D_R}, \qquad \mathrm{Pe}_M \equiv \frac{v_F(1-\sigma)L}{D_M}.$$

Discuss the behavior of Θ for extreme values of σ and the retentate and membrane Péclet numbers, Pe_R and Pe_M. Why is it desirable to operate with a small Pe_R and a large Pe_M?

2-9. Reversible Heterogeneous Reaction in a Liquid

Consider the situation depicted in Fig. 2-8, except assume that the reversible reaction between A and B is heterogeneous (at $y=L$) instead of homogeneous. Determine the steady-state concentration profiles $C_A(y)$ and $C_B(y)$ and the reaction rate. Use rate and equilibrium expressions analogous to those in Eqs. (2.8-29) and (2.8-30).

2-10. Oxygen Diffusion in Tissues

Oxygen is consumed in body tissues, or by cells maintained in vitro, at a rate which is often nearly independent of the O_2 concentration. As a model for a tissue region or aggregate of cells, consider steady-state O_2 diffusion in a sphere of radius r_0, with zeroth-order consumption of O_2. Assume that the O_2 concentration at the outer surface $(r=r_0)$ is maintained constant at C_0. Determine the O_2 concentration profile, $C(r)$.

If r_0 and the rate of O_2 consumption are sufficiently large, no O_2 will reach a central core, defined by $r<r_c$. For the central core the assumption of zeroth-order kinetics is no longer valid, because no O_2 is available to react. Determine when an oxygen-free central core will exist and find an expression for r_c. This situation occurs in certain solid tumors where, as the tumor grows, the cells in the center are killed by lack of O_2.

2-11. Facilitated Transport

A reversible chemical reaction can be used to augment steady-state transport of a solute across a liquid film or membrane, with no net consumption of that solute. Such a reaction permits the solute to exist in either of two forms, both of which can diffuse, and thereby increases the effective concentration of the transported species. This augmentation is termed "facilitated transport."

As shown in Fig. P2-11, assume that a stagnant liquid film (e.g., a liquid in a highly porous membrane) separates gas compartments containing different partial pressures (P_A) of species A. The solubility of A in the liquid is α, such that $C_A=\alpha P_A$ at equilibrium; C_A refers to the liquid. In the liquid, A is converted reversibly to B, which is nonvolatile and therefore does not escape. The rate and equilibrium expressions are as in Eqs. (2.8-29) and (2.8-30). For algebraic simplicity, assume that $D_A=D_B=D$. A steady state exists.

(a) For the special case of no reaction, relate the flux of A to the partial pressures and other system parameters.

(b) Determine the flux of A under reactive conditions. (It is helpful to first add the conservation equations for A and B, as in Example 2.8-4.)

(c) Determine the enhancement factor E, defined as the ratio of the flux of A with reaction to that without reaction. One would expect that for a fixed value of the equilibrium constant (K), the flux enhancement would be greatest for very fast reaction kinetics. Examine the behavior of E in this limit.

2-12. Bimolecular Reaction in a Liquid Film

Assume that the bimolecular, irreversible reaction $A+B\to C$ takes place at steady state in a liquid film of thickness L. As shown in Fig. P2-12, reactant A is introduced at $x=0$ and reactant

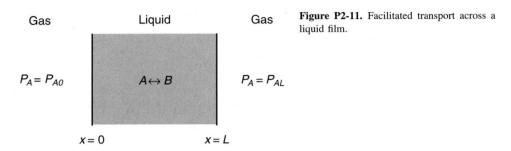

Figure P2-11. Facilitated transport across a liquid film.

B at $x=L$. The liquid solution is dilute everywhere with respect to solutes A, B, and C. The volumetric rate of formation of C is given by

$$R_{VC}=kC_A C_B,$$

where k is the second-order, homogeneous rate constant.

If the reaction kinetics are sufficiently fast, A and B cannot coexist in the liquid. In the limit $k\to\infty$ the reaction occurs at a plane $x=x_R$, which forms a boundary between the part of the film that contains A (but not B) and the part that contains B (but not A). Determine x_R, $C_A(x)$, $C_B(x)$, $C_C(x)$, and the reaction rate. Are the concentration gradients continuous or discontinuous at $x=x_R$?

2-13. Peltier Effect*

As shown in Fig. P2-13, molten and solid silicon are maintained in contact between two graphite plates separated by a distance L. Both phases occur because the melting temperature of silicon (T_m) is between the temperatures of the graphite plates (i.e., $T_1<T_m<T_2$). Assume that temperature variations in the x and y directions can be neglected.

(a) Calculate the steady-state temperature profiles in both silicon phases, and the height H of the melt–solid interface. The thermal conductivities of the melt and solid are k_m and k_s, respectively, and $k_m \neq k_s$. The heat of fusion of silicon is $\hat{\lambda}$.

(b) When a melt is an electrical conductor and its solid is a semiconductor (as is the case for silicon), passing a current from the solid to the melt releases heat at the interface, a phenomenon called the *Peltier effect* (Nye, 1957). The rate of energy release at the interface is given by

$$H_S=\beta i_z,$$

where β is the Peltier coefficient (units of volts) and i_z is the current density in the z direction (units of A/m²). Determine the effect of H_S on the interface position.

*This problem was suggested by R. A. Brown.

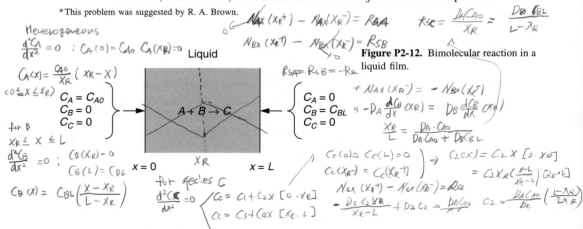

Figure P2-12. Bimolecular reaction in a liquid film.

Figure P2-13. A system with a silicon melt–solid interface.

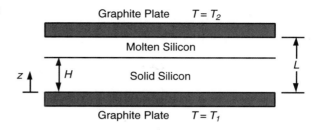

2-14. Mass Fluxes for a Growing Bubble

A vapor bubble of radius $R(t)$ is growing in a supersaturated liquid. Assume that transport is spherically symmetric and that the gas density (ρ_G) and liquid density (ρ_L) are constant. The center of the bubble is stationary (i.e., coordinates are chosen so that the origin is at the center of the bubble).

   (a) Show that the vapor in the bubble is at rest. To do this, choose a control volume of constant radius which, at a given instant, is slightly smaller than $R(t)$.
   (b) Evaluate the mass flux into the bubble (i.e., across the vapor–liquid interface).
   (c) Evaluate the mass flux in the liquid at a fixed position just outside the bubble.

2-15. Heat Conduction in Solids with Friction

Two solid plates of differing thermal properties are in contact, as shown in Fig. P2-15. The bottom surface of plate 1 is at constant temperature T_0, and at the top surface of plate 2 there is convective heat transfer to the ambient air (temperature T_∞, heat transfer coefficient h). The plate dimensions in the x and z directions are sufficiently large that $T = T(y)$ only. The rate of frictional heating at the contact surface between two solids may be expressed as

$$H_S = c\gamma U,$$

where U is the relative velocity of the solids, γ is the force per unit area holding the solids in contact (i.e., normal to the interface), and c is the coefficient of dry friction for the materials involved.

   (a) Determine the steady-state temperature profile in the plates when both are at rest.
   (b) Assume now that plate 2 moves horizontally at speed U, while plate 1 remains stationary. The plates are held in contact only by gravity. Determine the rate of interfacial energy generation and its effect on the steady-state temperature profile in the plates.

2-16. Freezing of a Liquid Outside a Tube

A pure liquid is being frozen outside a refrigerated tube, as shown in Fig. P2-16. The bulk liquid is at its freezing temperature (T_F), and the inner wall of the tube is cooled by convective heat transfer to a gas at bulk temperature T_G. The internal heat transfer coefficient is h and the heat of fusion is $\hat{\lambda}$. For simplicity, assume that the freezing occurs slowly enough that the time derivative in the energy equation is negligible; this is called a *pseudosteady approximation* (Chapter 3). Assume further that T_G is independent of axial position. Finally, assume that the tube wall is thin $(R_o \cong R_i)$ and has negligible thermal resistance. The layer of solid outside the tube is not necessarily thin.

   (a) Determine the pseudosteady temperature profile in the solid at a given instant.
   (b) Determine the rate of freezing at a given instant, expressed as dR_s/dt.
   (c) Find $R_s(t)$, assuming that $R_s(0) = R_o$.

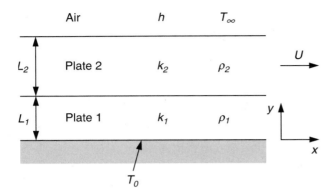

2-17. Melting of a Candle

After a candle is lit the wax next to the flame soon reaches its melting temperature (T_m), and the candle begins to melt. As shown in Fig. P2-17, assume that after a certain time t a candle of radius R has a length $L(t)$. There is a net heat flux q_0 to the top surface of the candle, which represents radiant energy transfer from the flame minus convective losses. Assume that q_0 is constant. There is also convective heat transfer from the sides of the candle to the surrounding air (ambient temperature T_∞, heat transfer coefficient h). The base of the candle is at the ambient temperature. For simplicity, assume that the melting occurs slowly enough that the time derivative in the energy conservation equation is negligible; this is a *pseudosteady approximation* (Chapter 3).

(a) Find the pseudosteady temperature profile $T(z, t)$ in the candle, assuming that the temperature is approximately independent of radial position.
(b) Evaluate dL/dt at a given instant. Assume that the layer of melted wax on top of the candle is thin enough that the heat flux toward the candle on the *liquid* side of the melt–solid interface is approximately equal to q_0.
(c) Assuming that the initial length is $L(0) = L_0$, what is the time (t_m) required for the candle to melt completely?

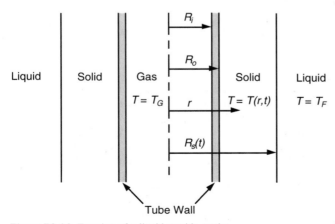

Figure P2-16. Freezing of a liquid outside a tube.

Figure P2-17. Melting of a candle.

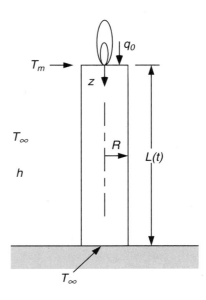

2-18. Vessel Wall with a Heater

To prevent heat loss from a fluid-filled vessel, it is decided to attach an electrical heater to the outside of the wall, as shown in Fig. P2-18. The passage of a current through the heater material yields a constant rate of energy generation, H_V. The inside bulk temperature and heat transfer coefficient are T_i and h_i, respectively. The outside values are T_o and h_o, where $T_o < T_i$. Assume that the steady-state temperature is a function of x only.

Calculate the heating rate that is just sufficient to eliminate any heat loss from the inside of the container.

2-19. Membrane Reactor

The enzyme-catalyzed reaction $A \rightarrow B$ is to be carried out under steady-state conditions with the enzyme immobilized in a membrane. The membrane has two purposes: to prevent loss of the expensive enzyme, and to help separate any residual reactant from the product. Consider the arrangement shown in Fig. P2-19. One surface of the membrane (at $x = L$) has a thin coating that permits rapid diffusion of A from the feed stream to the membrane, such that $C_A(L) \cong C_0$. A key

Figure P2-18. Vessel wall with a heater.

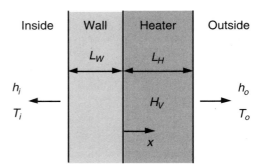

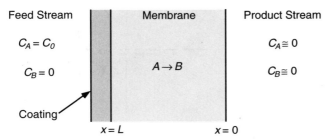

Figure P2-19. Membrane reactor. The concentrations shown for the feed stream and product stream are assumed to apply just inside the coating and membrane, respectively.

feature of the system is that the coating restricts (but does not eliminate) the passage of B. With negligible B present in the feed stream, the flux of B across the coating is described by

$$N_{Bx}(L) = k_c C_B(L),$$

where k_c is the *permeability coefficient* of the coating for B (similar to a mass transfer coefficient). For simplicity, assume that the concentrations of A and B in the product stream are maintained at low levels and that the diffusivities in the membrane are equal, $D_A = D_B = D$. The first-order, homogeneous reaction rate constant in the membrane is k_r.

(a) State the equations governing the concentrations of reactant and product in the membrane. Use the following dimensionless quantities:

$$\eta \equiv \frac{x}{L}, \qquad \Theta_A \equiv \frac{C_A}{C_0}, \qquad \Theta_B \equiv \frac{C_B}{C_0},$$

$$\lambda \equiv \left(\frac{k_r L^2}{D}\right)^{1/2} = Da^{1/2}, \qquad \gamma \equiv \frac{k_c L}{D}.$$

(b) Determine $\Theta_A(\eta)$ and $\Theta_B(\eta)$.

(c) Two measures of reactor performance are the purity of the product stream,

$$\phi_1 \equiv \frac{N_{Bx}(0)}{N_{Ax}(0) + N_{Bx}(0)},$$

and the fractional recovery of B in the product stream,

$$\phi_2 \equiv \frac{N_{Bx}(0)}{N_{Bx}(0) - N_{Bx}(L)}.$$

Derive general expressions for ϕ_1 and ϕ_2, and determine the limiting values of both quantities for very fast or very slow reactions (large or small λ, with fixed γ). Discuss conditions which will maximize ϕ_1 and/or ϕ_2.

Chapter 3

SCALING AND APPROXIMATION TECHNIQUES

3.1 INTRODUCTION

This chapter has a dual purpose: to discuss the simplification of transport models and to describe methods for obtaining approximate solutions. Finding the simplest differential equations and boundary conditions which adequately represent a real system is something of an art. A qualitative understanding of what is being modeled is essential, and in complex situations some trial and error may be required. Nonetheless, there are systematic ways to approach the formulation process. Certain solution methods are based on many of the same ideas, and they are included here for that reason. The examples in this chapter involve conduction and diffusion only, but the methods are more general. Indeed, they are used to greater advantage in many of the convective transport problems later in this book.

The examples in Section 2.8 employed various assumptions to yield steady-state, one-dimensional problems. When are such assumptions valid? The key to model simplification is the concept of scales, discussed in Section 3.2. The *scale* of a variable is an estimate of its maximum order of magnitude, and *scaling* is the process of identifying the correct orders of magnitude of the various unknowns which one confronts in a new problem. Several types of simplifications which can be made in steady or time-dependent problems are illustrated in Sections 3.3 and 3.4. The discussion of solution techniques begins with the similarity method in Section 3.5, continues with perturbation methods in Sections 3.6 and 3.7, and concludes with integral approximation methods in Section 3.8. The similarity method applies only when a dynamically determined length scale is much smaller than the system dimensions; perturbation methods are useful when a problem with scaled variables contains a small parameter; and integral methods, like the scaling process itself, depend strongly on a qualitative understanding of the system being modeled.

3.2 SCALING

Orders of Magnitude

In discussing which quantities are large enough to be important in a given problem and which are negligible, we need a symbol to indicate order of magnitude. Throughout this book, "~" is used to denote an *order-of-magnitude equality*.[1] If x and y are variables, then "$x \sim y$" indicates that the *maximum* values of x and y differ by less than a factor of ten; this is said as "x is of order y" or just "x is order y." In order-of-magnitude equalities algebraic signs are ignored. When one quantity is a constant, a rough guideline is that a threefold discrepancy is tolerated in either direction. Using this guideline, "$x \sim 1$" means that the maximum value of x is between 0.3 and 3. The three main classifications for a dimensionless quantity are that it is small ($x \ll 1$), large ($x \gg 1$), or order one ($x \sim 1$). What is effectively small or large depends on the problem, but $x < 0.1$ and $x > 10$ are reasonable starting points when no other information is available.

Scaled Variables

Putting the mathematical description of a physical process in dimensionless form ("non-dimensionalization") is a standard procedure for minimizing the number of variables and parameters. The heat and mass transfer examples in Chapter 2 all involved some use of dimensionless quantities. In all but the simplest problems, two or more alternative sets of such quantities can be constructed, each with the same number of variables and parameters. *Scaling* is a special type of nondimensionalization in which the independent and dependent variables are chosen so that they are order one. When this is done, the magnitudes of various terms in an equation are revealed by the dimensionless parameters which appear as coefficients, possibly suggesting terms which may be neglected.

The *scale* of a variable is an estimate of its maximum order of magnitude, as already mentioned. Length and time scales are also called *characteristic lengths* and *characteristic times*. *Scaled variables* are created by dividing dimensional variables by their respective scales. Dimensionless variables also have scales, in this case pure numbers. Because we are concerned mainly with differential equations, we must find the scales for derivatives. Thus, we need to consider the scales for *changes* in the variables, not just for the variables themselves. In situations where a variable changes by a small fraction of its absolute value, the distinction between the two types of scales may be important. If Θ and η are scaled dependent and independent variables, respectively, then by definition,

$$\Theta \sim 1, \qquad \frac{d\Theta}{d\eta} \sim 1, \qquad \frac{d^2\Theta}{d\eta^2} \sim 1. \tag{3.2-1}$$

In scaling the variables in differential equations, making the derivatives order one is the main priority. That is, the changes in variables are usually more important than their absolute magnitudes.

The scales in some transport problems are easily identified, but in others they are far from obvious. Indeed, just finding the proper scales can be the main goal in analyz-

---

[1]As used in this book, "$x \sim 1$" is not the same as "$x = O(1)$" (see Section 3.6).

ing a complex problem. Doing this may require considerable insight into which rate processes are important, and this is where trial and error is sometimes involved. The scale of any variable may be influenced by the values of dimensionless parameters, and the scale of a dependent variable may differ in various domains of the independent variables(s). For example, the concentration scale for a solute in one region of a liquid may differ from that in another. Or, the concentration scale at early times may differ from that at later times.

To scale transport equations we need reliable order-of-magnitude estimates of various first and second derivatives. For these estimates we assume that, for a function $f(x)$ which varies *gradually* by an amount Δf over an interval Δx, the orders of magnitude of the first and second derivatives are given by

$$\frac{df}{dx} \sim \frac{\Delta f}{\Delta x}, \tag{3.2-2}$$

$$\frac{d^2f}{dx^2} \sim \frac{(\Delta f/\Delta x)-0}{\Delta x} = \frac{\Delta f}{(\Delta x)^2}. \tag{3.2-3}$$

The finite-difference expression in Eq. (3.2-2), which is applied to both temporal and spatial derivatives, will be accurate if f varies smoothly and monotonically over the interval Δx. In estimating spatial derivatives we are aided by the fact that the molecular transport processes (conduction, diffusion, viscous stresses) have the effect of smoothing the corresponding field variables (temperature, concentration, velocity). Thus, the smoothness of a function is ordinarily taken for granted and the main problem is to identify Δf and Δx. As indicated, Eq. (3.2-3) assumes that the first derivative changes from roughly zero to $\Delta f/\Delta x$ over the interval Δx. For a function with very little curvature, the actual second derivative may be much smaller than this estimate. However, because we are usually trying to identify terms which can be neglected, this tendency to overestimate the second derivative is rarely a concern. If f and x have been scaled, then $\Delta f \sim 1$, $\Delta x \sim 1$, and the first and second derivatives are order one, as in Eq. (3.2-1).

Scales for Known Functions

The concept of scales will be illustrated using three functions selected from the heat and mass transfer examples in Chapter 2. First, consider the electrically heated wire of Example 2.8-1. For $\mathrm{Bi} \gg 1$, the expression for the temperature profile reduces to

$$\frac{T-T_\infty}{H_V R^2/k} = \frac{1}{4}\left[1-\left(\frac{r}{R}\right)^2\right], \tag{3.2-4}$$

where H_V is the heating rate and R is the radius of the wire. It is seen that $T-T_\infty$ (the temperature increment above the ambient) varies smoothly by an amount $H_V R^2/4k$ over the distance R. The factor "$\frac{1}{4}$" in the temperature change is of marginal significance for order-of-magnitude purposes and will be ignored. Thus, the temperature and length scales are $\Delta T = H_V R^2/k$ and $\Delta r = R$. Using these scales in Eqs. (3.2-2) and (3.2-3), the derivatives are estimated as

$$\frac{dT}{dr} \sim \frac{H_V R}{k}, \qquad \frac{d^2T}{dr^2} \sim \frac{H_V}{k}. \tag{3.2-5}$$

It is easily confirmed that both estimates are within a factor of 2 of the true maximum values. Thus, Eqs. (3.2-2) and (3.2-3) provide the desired level of accuracy. If $T - T_\infty$ and r are divided by their respective scales to give Θ and η, then Eq. (3.2-4) becomes

$$\Theta = \frac{1}{4}(1 - \eta^2). \tag{3.2-6}$$

The first and second derivatives of this scaled function are

$$\frac{d\Theta}{d\eta} = -\frac{\eta}{2}, \quad \frac{d^2\Theta}{d\eta^2} = -\frac{1}{2}. \tag{3.2-7}$$

These exact values are consistent with Eq. (3.2-1). In other words, both derivatives are order one.

As a second case consider diffusion across a liquid film with an irreversible, heterogeneous reaction, as in Example 2.8-3. For $Da \gg 1$, the concentration profile is simply

$$\frac{C_A}{C_{A0}} = 1 - \frac{y}{L}. \tag{3.2-8}$$

Here the concentration scale is C_{A0} and the length scale is L, the film thickness. The scaled function and its derivatives are

$$\Theta = 1 - \eta, \tag{3.2-9}$$

$$\frac{d\Theta}{d\eta} = -1, \quad \frac{d^2\Theta}{d\eta^2} = 0. \tag{3.2-10}$$

The discrepancy between the actual second derivative and the expectation from Eq. (3.2-1) is due not to an error in scaling, but to the peculiarity of the linear function.

The third case, derived from Example 2.8-4, involves a homogeneous reaction in a liquid film. For simplicity, we assume here that the reaction is irreversible and very fast. The concentration of the reactant for this situation is obtained by letting $\alpha \to \infty$, $\beta \to 0$, and $\gamma \to 0$ in Eq. (2.8-43). The result expressed in dimensional variables is

$$\frac{C_A}{C_{A0}} = \exp\left(-\frac{y}{\sqrt{D_A/k_1}}\right), \tag{3.2-11}$$

where D_A is the diffusivity and k_1 is the first-order rate constant. The concentration scale is again C_{A0}, but the length scale is now $\sqrt{D_A/k_1}$. Because most of the concentration change occurs between $y = 0$ and $y = \sqrt{D_A/k_1}$, it is irrelevant that the liquid film actually extends much farther, to $y = L$. (In this problem the key parameter is $\alpha = k_1 L^2/D_A \gg 1$, so that $L \gg \sqrt{D_A/k_1}$.) The length scale here is determined by the balance between two rate processes (reaction and diffusion), rather than by the geometry. In other words, it is *dynamic* rather than *geometric*. The scaled function and its derivatives for this case are

$$\Theta = e^{-\eta}, \tag{3.2-12}$$

$$\frac{d\Theta}{d\eta} = -e^{-\eta}, \quad \frac{d^2\Theta}{d\eta^2} = e^{-\eta}. \tag{3.2-13}$$

Again, the exact derivatives are consistent with Eq. (3.2-1).

Scales for Unknown Functions

If scales are to be useful in formulating and simplifying models, then some means is needed to identify them *before* obtaining a solution. In convective transport problems, especially, using some prior knowledge of scales to simplify the governing equations is often the only way to obtain an analytical solution (i.e., a closed-form mathematical expression involving known functions). Sometimes, it is the only practical path to a numerical solution. Thus, we need ways to find the scales for *unknown* functions.

A surprisingly effective method for scaling unknown functions is to define the dimensionless variables in such a way that no physically important term in the governing equations is multiplied by a large or small parameter. That is, the presence of such parameters in strategic places usually means that the variables are improperly scaled, and eliminating those parameters by embedding them in the variables corrects the scaling. To illustrate this approach, we again use two of the examples from Chapter 2. In these particular examples the scaling adjustments accomplish very little, because the original set of equations was already simple enough to solve exactly. However, to illustrate the thought process, we will pretend that no solutions are available. The solutions are used only at the end, to confirm the scaling results.

Example 3.2-1 Temperature Scale for an Electrically Heated Wire Again consider the physical situation in Example 2.8-1. For $Bi \gg 1$, where the surface temperature equals that of the surrounding air, the dimensional problem is stated as

$$\frac{k}{r} \frac{d}{dr}\left(r \frac{dT}{dr}\right) + H_V = 0,$$ (3.2-14)

$$\frac{dT}{dr}(0) = 0, \qquad T(R) = T_\infty.$$ (3.2-15)

Given that the heat source is distributed evenly throughout the wire, and that heat can be lost only from the surface, the temperature must vary gradually over the entire cross section. Thus, the length scale must be R, and $\eta = r/R$ must be an adequately scaled coordinate. The temperature scale is not obvious at the outset. For the dimensionless temperature we let $\Theta = (T - T_\infty)/\Delta T$, where ΔT is to be determined. The heat transfer problem expressed in terms of $\Theta(\eta)$ is

$$\frac{1}{\eta} \frac{d}{d\eta}\left(\eta \frac{d\Theta}{d\eta}\right) + \left(\frac{H_V R^2}{k \Delta T}\right) = 0$$ (3.2-16)

$$\frac{d\Theta}{d\eta}(0) = 0, \qquad \Theta(1) = 0.$$ (3.2-17)

The only parameter which remains is the dimensionless source term. By setting $\Delta T = H_V R^2/k$, we convert Eq. (3.2-16) to

$$\frac{1}{\eta} \frac{d}{d\eta}\left(\eta \frac{d\Theta}{d\eta}\right) + 1 = 0.$$ (3.2-18)

There are no large or small parameters in Eqs. (3.2-17) or (3.2-18); indeed, we have eliminated the parameters altogether! This indicates that our choice of temperature scale is appropriate. As confirmation, recall that $H_V R^2/k$ is the temperature scale found earlier from the solution.

Example 3.2-2 Length Scale for a Fast, Reversible Reaction in a Liquid Film From Example 2.8-4, the dimensionless equations governing the concentration of species A are

$$\frac{d^2\theta_A}{d\eta^2} = (\alpha+\beta)\theta_A - (\alpha\gamma+\beta), \tag{3.2-19}$$

$$\theta_A(0) = 1, \tag{3.2-20}$$

$$\frac{d\theta_A}{d\eta}(1) = 0. \tag{3.2-21}$$

The parameters α and β are Damköhler numbers for the forward and reverse reactions. We will assume that $\gamma < 1$, which is to say that the equilibrium is such that species A is the reactant and species B is the product.

The dimensionless reactant concentration, $\theta_A = C_A/C_{A0}$, is presumed to be adequately scaled. This is because the actual concentration is fixed at C_{A0} on one side and must decline to some fraction of that within the film. Our concern here is with the length scale. The dimensionless (but not necessarily scaled) coordinate is $\eta = y/L$. For slow or moderate reaction kinetics (i.e., small or moderate α and β), we expect the reaction to proceed over the entire film. The length scale is then the film thickness, L, and the coordinate is properly scaled as is. More interesting is the case of a very fast reaction, such that $\alpha \to \infty$ and $\beta \to \infty$ with β/α fixed. For this situation, Eq. (3.2-19) indicates that $\theta_A \to (\alpha\gamma+\beta)/(\alpha+\beta)$ for all parts of the film where $d^2\theta_A/d\eta^2$ remains finite. From the boundary conditions, this can be true only at some distance from $\eta=0$, because in general $(\alpha\gamma+\beta)/(\alpha+\beta) \neq 1$. In other words, this value for the reactant concentration violates Eq. (3.2-20) but not Eq. (3.2-21). If rewritten in terms of dimensional quantities, it is seen that this concentration corresponds to chemical equilibrium between A and B (see Example 2.8-4). Thus, there is evidently a nonequilibrium region only in the vicinity of $\eta=0$, and in this region $d^2\theta_A/d\eta^2$ does *not* remain finite for $\alpha \to \infty$ and $\beta \to \infty$. The failure of this derivative to be ~ 1 is evidence that the coordinate scaling is incorrect *for the nonequilibrium region*. Put another way, the thickness of the nonequilibrium region is much smaller than the assumed scale, L. Because $d^2\theta_A/d\eta^2$ becomes infinite, the reaction zone evidently becomes thinner as $\alpha \to \infty$ and $\beta \to \infty$.

The objective now is to estimate the thickness of the reaction zone when the reaction is fast. To do this, a new coordinate is defined as

$$Y \equiv (\alpha+\beta)^m \eta. \tag{3.2-22}$$

In the reaction zone $\eta \ll 1$, so that it is multiplied by a "stretching factor" to obtain a new coordinate which is properly scaled for that region. (The idea of coordinate stretching, which is especially important in boundary layer theory, is used repeatedly in this book.) By definition, $Y \sim 1$ in the nonequilibrium region. The stretching factor is based on the Damköhler numbers, because they are the parameters which created the scaling difficulty. In Eq. (3.2-22) we have chosen $(\alpha+\beta)$ as the large parameter to use in the stretching, but the analysis would be much the same if we used either α or β alone. What must be determined is the exponent, m. Following the rule stated earlier concerning large or small parameters, we will choose m so that no key terms in the governing equations are multiplied by α or β.

Using Eq. (3.2-22) in Eq. (3.2-19), the differential equation becomes

$$\frac{d^2\theta_A}{dY^2} = (\alpha+\beta)^{1-2m}\theta_A - (\alpha\gamma+\beta)(\alpha+\beta)^{-2m}. \tag{3.2-23}$$

Choosing $m = \frac{1}{2}$ yields

$$\frac{d^2\theta_A}{dY^2} = \theta_A - \frac{(\alpha\gamma+\beta)}{(\alpha+\beta)} = \theta_A - \frac{[\gamma+(\beta/\alpha)]}{[1+(\beta/\alpha)]}. \tag{3.2-24}$$

In this problem γ and β/α are assumed to be fixed, so that Eq. (3.2-24) no longer contains parameters which are indefinitely large or small. This indicates that the coordinate is now properly scaled for the nonequilibrium region. The thickness of the nonequilibrium part of the film is inversely proportional to the stretching factor. In dimensional terms, the length scale for the reaction zone, or characteristic length, is

$$\delta = \frac{L}{(\alpha + \beta)^{1/2}}.\tag{3.2-25}$$

It is important also to see what happens to the boundary condtions. In terms of Y, Eqs. (3.2-20) and (3.2-21) become

$$\theta_A(0) = 1,\tag{3.2-26}$$

$$\frac{d\theta_A}{dY}(\infty) = 0.\tag{3.2-27}$$

The liquid film actually extends from $Y=0$ to $Y=(\alpha+\beta)^{1/2}L$, but the upper limit of Y has been replaced by ∞ in Eq. (3.2-27). For a fast reaction, the impermeable boundary in this problem is many characteristic lengths away from the nonequilibrium region (i.e., $L/\delta \gg 1$). Accordingly, the geometric length scale is effectively infinite, and does not influence the solution. Using the concentration deduced above for the equilibrium region, we could replace Eq. (3.2-27) with

$$\theta_A(\infty) = \frac{(\alpha\gamma+\beta)}{(\alpha+\beta)} = \frac{[\gamma+(\beta/\alpha)]}{[1+(\beta/\alpha)]}.\tag{3.2-28}$$

The solution to this set of equations for fast reactions was presented in Example 2.8-4. For the special case of an irreversible reaction, where $\beta/\alpha \to 0$, the length scale given in Eq. (3.2-25) is identical to that found from Eq. (3.2-11).

To summarize these last two examples, in both cases it was possible to determine an unknown scale without solving the differential equation. To do this, we had to realize that the correct length scale in Example 3.2-1 was R and that the correct concentration scale in Example 3.2-2 was C_{A0}. Without this qualitative understanding of the physical situations we could not have made any progress. Thus, in scaling and order-of-magnitude analysis it is essential to develop that kind of understanding early in the process. A good practice is to begin by sketching the expected temperature, concentration, or velocity profile, using as guidance the boundary conditions, types of source terms, and other information in the governing equations. Obviously, the more experience one has with related problems, the easier it is to see how to proceed. Although not always discussed explicitly, information on the scales in transport problems can be gleaned from almost all of the examples in this book.

Scaling concepts are important in areas of science and engineering beyond the study of transport phenomena. A more general discussion of scaling and simplification in applied mathematics is provided in Segel (1972) and Lin and Segel (1974).

3.3 REDUCTIONS IN DIMENSIONALITY

The effort needed to solve a differential equation, either by analytical or numerical methods, increases sharply with the number of independent variables. Thus, in formulating a model we want to exclude any variables which are not needed to describe the real system at the desired level of precision. Of course, the physical world consists of objects or fluid regions which are three-dimensional. What we need to know, for example, are the conditions which allow heat conduction in an object to be modeled as if it were two-dimensional or one-dimensional. The number of spatial coordinates, or the "dimensionality" of a problem, can be reduced using three basic strategies, which are discussed here under the headings of *symmetry, aspect ratio,* and *series resistances.* Circumstances under which time can be eliminated as an independent variable, even in unsteady processes, are discussed in Section 3.4.

Symmetry

Simplifications based on various types of symmetry are the easiest to recognize and require the least explanation. A major part of exploiting symmetry is in choosing the best coordinate system for a given problem. It is an elementary fact, for example, that cylindrical coordinates allow one to describe the curved surface of a cylinder using one variable (e.g., $r=R$), instead of the two variables needed in a rectangular system (e.g., $x^2 + y^2 = R^2$). Likewise, a temperature field with circular symmetry can be described using just the coordinates (r, z), instead of (x, y, z). This is why special coordinate systems have been developed (see Section A.7).

A related strategy is the exclusion of any spatial variable which is not required by the conservation equation or interfacial conditions. A simple illustration of this was given in Example 2.8-1, in which the forms of the volumetric source term and boundary conditions were used to conclude that there was no reason for the temperature to depend on any variable other than r. Thus, we routinely seek the *simplest* form of solution which is consistent with the data given. There are obvious benefits in this approach, but also some hazards. For example, the simplest functional form of the velocity consistent with steady flow in a cylindrical tube is $v_z(r)$. This is all that the continuity equation, the Navier–Stokes equation, and the boundary conditions seem to require (see Chapter 6). However, it is well known that laminar flow in a tube becomes unstable beyond a certain Reynolds number and that turbulence results. As discussed in Chapter 13, the local velocity in turbulent flow is three-dimensional and time-dependent, even in symmetric conduits. Such issues of stability and multiple solutions arise when the governing equations are strongly nonlinear, as is true for the Navier–Stokes equation at high Reynolds number. They are rarely a concern in conduction and diffusion problems, where the equations are usually linear or nearly so. The key features of linear differential equations are reviewed in Section 4.1.

Aspect Ratio

The ratio of two linear dimensions of an object (e.g., length/width) is called an *aspect ratio.* A square is a rectangle with an aspect ratio of one. There are a number of potential simplifications when an object or fluid region has a large (or small) aspect ratio. One of the more important of these is when one dimension is much smaller than the other two. For example, consider conduction in the square plate shown in Fig. 3-1, where $L/W >>$ 1. Suppose that there is a temperature difference between the edges at constant z and that the top, bottom, and remaining edges exchange heat with the surrounding air. Conduction in the z direction will be important because of the applied temperature gradient, and conduction in the $\pm x$ direction will be significant because of the large surface area

Figure 3-1. Heat conduction in a square plate.

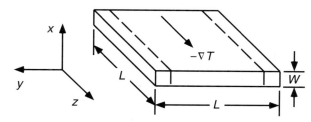

and short conduction path. Must the y direction be considered? The answer is no, provided that L/W is large enough. A rule of thumb in steady conduction and diffusion is that *edge effects*, such as the effects here of the surfaces at constant y, are insignificant beyond about one thickness from the edges. This is shown in Examples 3.7-4 and 4.5-2. Thus, for most of the plate, as indicated roughly by the region between dashed lines in Fig. 3-1, we can assume that $T = T(x, z)$ only.

We have mentioned two different reasons why we might model a temperature field as being two-dimensional. One is that a large aspect ratio makes the edge effects negligible, as just discussed. Referring again to Fig. 3-1, the other possibility is that the surfaces at constant y are well insulated, such that $\partial T/\partial y = 0$. If, in addition, the boundary conditions at the x and z surfaces are independent of y (i.e., the applied temperatures and/or heat transfer coefficients are constant), it follows that there is an exact solution with $T = T(x, z)$. This was classified as a symmetry argument.

There is a similar criterion for neglecting edge effects in viscous flows (see Example 6.5-2). Various other simplifications based on the thinness of a fluid region arise in lubrication theory, creeping flow, and boundary-layer flow, as discussed in Chapters 6–8.

Series Resistances

It is sometimes possible to reduce the dimensionality of a problem by determining which of two rate processes in series is the controlling step. In conduction or diffusion, this consideration most often involves the Biot number. Following are two examples of how the Biot number can be used in simplifying models.

Example 3.3-1 Heat Transfer from an Object Immersed in a Fluid Consider a solid object which is completely surrounded by a fluid. Heat transfer from the surface of the object to the fluid is described using the heat transfer coefficient, h. Equating conduction on the solid side of the interface to convective heat transfer on the fluid side, we obtain

$$-k\frac{\partial T}{\partial n}\bigg|_s = h(T_s - T_\infty), \qquad (3.3\text{-}1)$$

where T_s is the temperature at the surface, T_∞ is the temperature in the bulk fluid, and $\partial/\partial n \equiv \mathbf{n} \cdot \nabla$, where $\mathbf{n}$ is the outward unit normal. Now let T_c be the temperature at the center of the object. The characteristic length L is the "half-width," or shortest distance from the center to the surface (e.g., half the thickness of a plate, the radius of a sphere or long cylinder, etc.). The order of magnitude of the derivative in Eq. (3.3-1) is then

$$\frac{\partial T}{\partial n}\bigg|_s \sim \frac{T_c - T_s}{L}. \qquad (3.3\text{-}2)$$

Combining Eqs. (3.3-1) and (3.3-2) gives

$$\text{Bi} \equiv \frac{hL}{k} \sim \frac{T_c - T_s}{T_s - T_\infty} = \frac{\text{solid resistance}}{\text{fluid resistance}}. \qquad (3.3\text{-}3)$$

As shown, the ratio of the temperature differences indicates the relative thermal resistances of the two phases. Note that the thermal conductivity in Bi is that of the solid, not the fluid.
        Helpful simplifications result when Bi is either large or small. For large Bi, the *fluid* is nearly isothermal and

$$T_s = T_\infty \qquad (\text{Bi} \gg 1) \qquad (3.3\text{-}4)$$

This simplification was used already in one of the scaling examples in Section 3.2. Viewing Eq. (3.3-1) as the more general boundary condition at a fluid–solid interface, we see that when a surface temperature is specified the implicit assumption is that $\text{Bi} \gg 1$. A practical advantage of Eq. (3.3-4) is that the exact value of h need not be known to determine T in the solid (provided it can be shown that Bi is large).
        For small Bi, the fluid resistance dominates and the *solid* is nearly isothermal. For the heat transfer problem in the solid, this allows the ultimate reduction in dimensionality: from three-dimensional to "zero-dimensional"! The term *lumped model* is often used to indicate the absence of spatial variations; the contrasting term is *distributed model*. In a lumped model, the differential form of the energy equation is replaced by an integral balance; the control volume corresponds to the object. Applying Eq. (2.2-1) to steady heat transfer from an object with a heat source, we obtain

$$\int_S \mathbf{q} \cdot \mathbf{n} \, dS = \int_V H_V \, dV. \qquad (3.3\text{-}5)$$

Letting $\bar{h}$ be the heat transfer coefficient averaged over the surface, the integral balance reduces to

$$\bar{h}S(T_s - T_\infty) = H_V V. \qquad (3.3\text{-}6)$$

Noting that $T \cong T_s$ throughout the solid for $\text{Bi} \ll 1$, the solid temperature is given by

$$T = T_\infty + \frac{H_V V}{\bar{h}S} \qquad (\text{Bi} \ll 1). \qquad (3.3\text{-}7)$$

This simple result is valid for objects of any shape. For the heated wire in Example 2.8-1, an equivalent expression is obtained by letting $\text{Bi} \to 0$ in the solution for the temperature.
        Biot number for mass transfer. In defining Bi for mass transfer, it is necessary to ensure that the concentration differences in the two phases are compared on a common basis. To preserve the analogy with electrical resistances in series, the "potential" (the measure of concentration) must be continuous at the point where the "resistors" join (the interface). This requirement is met by using the partition coefficient to modify the concentration difference in one of the phases. For species i at a liquid–solid interface, the interfacial balance analogous to Eq. (3.3-1) is

$$-D_i^{(S)} \left. \frac{\partial C_i^{(S)}}{\partial n} \right|_s = k_{ci}[C_{is}^{(L)} - C_{i\infty}^{(L)}] = \frac{k_{ci}}{K_i}[C_{is}^{(S)} - K_i C_{i\infty}^{(L)}], \qquad (3.3\text{-}8)$$

where the superscripts refer to the phase and K_i is the solid-to-fluid concentration ratio at equilib-rium. Estimating the order of magnitude of the derivative as before and expressing the liquid concentrations as on the far right of Eq. (3.3-8), we obtain

Figure 3-2. Solute concentrations near a solid–liquid interface, assuming $K_i < 1$. The solid lines depict true concentrations. The dashed lines show "liquid-equivalent" concentrations in the solid (left) or "solid-equivalent" concentrations in the liquid (right).

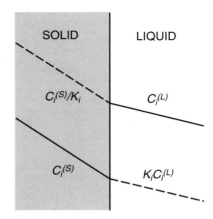

$$\mathrm{Bi} \equiv \frac{k_{ci} L}{K_i D_i^{(S)}} \sim \frac{C_{ic}^{(S)} - C_{is}^{(S)}}{C_{is}^{(S)} - K_i C_{i\infty}^{(L)}} = \frac{\text{solid resistance}}{\text{fluid resistance}}. \tag{3.3-9}$$

Only solid-phase or "solid-equivalent" concentrations are involved here, as in the lower curves in Fig. 3-2. Dividing the numerator and denominator by K_i gives liquid-phase or "liquid-equivalent" concentrations, as in the upper curves. With either choice, Bi is the same. The modeling simplifications in mass transfer for extreme values of Bi are analogous to those for heat transfer.

Example 3.3-2 Fin Approximation Consider steady heat transfer from an extended surface or "fin" to the surrounding air. As shown in Fig. 3-3, the base of the fin is at temperature T_0. The ambient temperature is T_∞, and the heat transfer coefficient is a constant, h. The length L and half-thickness W of the fin are such that $\gamma \equiv L/W \gg 1$. Of particular importance, it is assumed that $\mathrm{Bi} \equiv hW/k \ll 1$. The third dimension (in the y direction) is assumed to be large enough to make the problem two-dimensional. Thus, we start by assuming that $T = T(x, z)$.

The difference between a fin and a fully submerged object is that the temperature at one end of the fin is fixed. It will be shown that, unlike heat transfer from a submerged object, the small Biot number based on the half-thickness is insufficient to make the fin isothermal. Nonetheless, the small value of Bi allows us to eliminate one of the independent variables. The importance of the resulting "fin approximation" is that it is a prototype for reducing a two-dimensional model to a one-dimensional one. This type of approximation is applicable to a variety of situations involving conduction or diffusion in thin solids or thin layers of fluid.

Figure 3-3. Two-dimensional heat transfer from a rectangular fin.

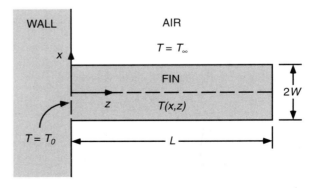

From Table 2-2, the energy equation for the fin is

$$\frac{\partial^2 T}{\partial x^2} + \frac{\partial^2 T}{\partial z^2} = 0. \tag{3.3-10}$$

The boundary conditions are

$$T(x, 0) = T_0, \tag{3.3-11}$$

$$\frac{\partial T}{\partial z}(x, L) = -\frac{h}{k}[T(x, L) - T_\infty], \tag{3.3-12}$$

$$\frac{\partial T}{\partial x}(0, z) = 0, \tag{3.3-13}$$

$$\frac{\partial T}{\partial x}(W, z) = -\frac{h}{k}[T(W, z) - T_\infty]. \tag{3.3-14}$$

Assuming that the base temperature and heat transfer coefficient are uniform makes the temperature field symmetric about $x = 0$, as expressed by Eq. (3.3-13).

The boundary condition at the top surface, Eq. (3.3-14), is used now to show that temperature variations in the x direction are negligible. Estimating the order of magnitude of the derivative and rearranging the result as in Eq. (3.3-3), we obtain

$$\frac{T(0, z) - T(W, z)}{T(W, z) - T_\infty} \sim \frac{hW}{k} = \text{Bi}. \tag{3.3-15}$$

Thus, with $\text{Bi} \ll 1$, the temperature variations over x are much smaller than the temperature scale, $T_0 - T_\infty$. This indicates that $T = T(z)$ only, to good approximation.

Given that the temperature field is approximately one-dimensional, the local value can be replaced by the cross-sectional average (at constant z). This average is

$$\overline{T}(z) \equiv \frac{1}{W} \int_0^W T(x, z) \, dx. \tag{3.3-16}$$

To obtain an energy equation involving $\overline{T}$, we average each term in Eq. (3.3-10) over the cross-section. The nonzero terms are

$$\frac{1}{W} \int_0^W \frac{\partial^2 T}{\partial x^2} \, dx = \frac{1}{W} \frac{\partial T}{\partial x} \Big|_{x=0}^{x=W} = -\frac{h}{Wk}(\overline{T} - T_\infty), \tag{3.3-17}$$

$$\frac{1}{W} \int_0^W \frac{\partial^2 T}{\partial z^2} \, dx = \frac{d^2}{dz^2}\left(\frac{1}{W} \int_0^W T \, dx\right) = \frac{d^2 \overline{T}}{dz^2}. \tag{3.3-18}$$

In Eq. (3.3-17) we have used the boundary conditions in x and have approximated the surface temperature as $\overline{T}$. The energy equation is now

$$\frac{d^2 \overline{T}}{dz^2} - \frac{h}{Wk}(\overline{T} - T_\infty) = 0. \tag{3.3-19}$$

Notice that the term representing conduction in the x direction ($\partial^2 T/\partial x^2$) has not gone away; it has only changed form. That is, temperature variations in that direction have been neglected not because there is little or no heat transfer at the top and bottom surfaces of the fin, but rather because the internal heat transfer resistance is small. Averaging in a similar manner the boundary conditions involving z, we obtain

$$\overline{T}(0) = T_0, \qquad \frac{d\overline{T}}{dz}(L) = -\frac{h}{k}[\overline{T}(L) - T_\infty]. \tag{3.3-20}$$

The cross-sectional average temperature is governed entirely by Eqs. (3.3-19) and (3.3-20).

A dimensionless temperature and z-coordinate are defined now as

$$\Theta(\zeta) \equiv \frac{\overline{T} - T_\infty}{T_0 - T_\infty}, \qquad \zeta \equiv \frac{z}{W}. \tag{3.3-21}$$

This temperature variable is scaled but the coordinate is not. We have chosen W for the nondimensionalization because the cross-sectional dimension of a thin object is often more representative than is the length. It will be shown, however, that neither W nor L is the correct length scale for temperature variations in the z direction. Using these variables, the energy equation and boundary conditions are

$$\frac{d^2\Theta}{d\zeta^2} - \mathrm{Bi}\ \Theta = 0, \tag{3.3-22}$$

$$\Theta(0) = 1, \qquad \frac{d\Theta}{d\zeta}(\gamma) = -\mathrm{Bi}\ \Theta(\gamma). \tag{3.3-23}$$

These equations, or the dimensional ones given above, can be solved for the average temperature. However, scaling the z coordinate first will sharpen our insight and, incidentally, lead to the simplest form of solution. Following the approach of Example 3.2-2, we define the *scaled* coordinate as

$$Z \equiv \mathrm{Bi}^m\ \zeta, \tag{3.3-24}$$

where the constant m is to be determined. In terms of Z, Eq. (3.3-22) becomes

$$\mathrm{Bi}^{2m}\frac{d^2\Theta}{dZ^2} - \mathrm{Bi}\ \Theta = 0. \tag{3.3-25}$$

To eliminate the small parameter from the differential equation, we choose $m = \frac{1}{2}$. The scaled coordinate is then

$$Z = \left(\frac{h}{Wk}\right)^{1/2} z. \tag{3.3-26}$$

It is seen now that the scale for z is $(Wk/h)^{1/2}$. This *dynamic* length scale reflects the competition between heat conduction from the base to the tip and heat loss from the top and bottom.

For $\Theta(Z)$, the boundary condition at the tip of the fin becomes

$$\frac{d\Theta}{dZ}(\Lambda) = -\mathrm{Bi}^{1/2}\Theta(\Lambda), \tag{3.3-27}$$

$$\Lambda \equiv \left(\frac{h}{Wk}\right)^{1/2} L. \tag{3.3-28}$$

Thus, if Bi is sufficiently small, the tip will act as if it were insulated. We conclude that the governing equations for $\Theta(Z)$ in the limit $\mathrm{Bi} \to 0$ are

$$\frac{d^2\Theta}{dZ^2} - \Theta = 0, \tag{3.3-29}$$

$$\Theta(0) = 1, \qquad \frac{d\Theta}{dZ}(\Lambda) = 0. \tag{3.3-30}$$

The solution is

$$\Theta(Z) = \cosh Z - \tanh \Lambda \sinh Z. \tag{3.3-31}$$

A fin that is dynamically or functionally "short" is one where $\Lambda \ll 1$, or $L \ll (Wk/h)^{1/2}$. That is, the geometric length is much less than the characteristic length. Conversely, a dynamically "long" fin is one where $\Lambda \gg 1$. In neither case is the aspect ratio, $\gamma = L/W$, the controlling parameter. The limiting solutions for short and long fins are

$$\Theta(Z) \to 1 \qquad (\Lambda \to 0), \tag{3.3-32}$$

$$\Theta(Z) \to e^{-Z} \qquad (\Lambda \to \infty). \tag{3.3-33}$$

The short fin is essentially isothermal, whereas the temperature in the long fin reaches the ambient value well before the tip. These results resemble those for slow and fast homogeneous reactions, respectively, in a liquid film (see Section 3.2).

For $\mathrm{Bi} \ll 1$ and $\gamma \gg 1$, but with more general boundary conditions, the approach used here will yield a temperature profile which is a good approximation in most of the fin, but *not* near the base ($z = 0$). If the base temperature is not constant (i.e., if $T_0 = T_0(x)$), the imposed values of dT_0/dx may make it impossible to neglect temperature variations in the x direction in the vicinity of the base. In effect, the specified function $T_0(x)$ introduces an additional length scale which may nullify the use of the low-Biot-number approximation near the base. Because the base temperature is constant in the present problem, Eq. (3.3-31) is a good approximation throughout the fin. The more general situation is discussed in Example 3.7-4, using the method of matched asymptotic expansions. The basic issue addressed in Example 3.7-4 is how to combine a two-dimensional solution in one region with a one-dimensional solution in an adjacent region. As with the elementary model of a fin presented here, that analysis serves as a prototype for other conduction and diffusion problems.

3.4 SIMPLIFICATIONS BASED ON TIME SCALES

If the temperature or concentration is suddenly perturbed at some location, a finite time is required for the temperature or concentration changes to be noticed a given distance away from the original disturbance. In a stagnant medium the time involved is the characteristic time for conduction or diffusion. This characteristic time is a key factor in formulating conduction or diffusion models, in that it determines how fast a system can respond to changes imposed at a boundary. A fast response may justify the use of a steady-state or pseudo-steady-state model. A slow response may allow one to model a region as infinite or semi-infinite, because the effects of one or more distant boundaries are never "felt" on the time scales of interest.

The two examples in this section involve transient diffusion across a membrane of thickness L, as shown in Fig. 3-4. The basic model formulation common to both is described first. The solute concentration and diffusivity within the membrane are $C(x, t)$ and D, respectively. (To simplify the notation, the subscripts usually used to identify individual species are omitted here.) The external solutions are assumed to be perfectly mixed, with solute concentrations $C_1(t)$ and $C_2(t)$, respectively. Each solution has a volume V, and the exposed area of the membrane is A. No solute is present initially, and at

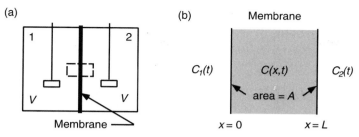

Figure 3-4. Diffusion through a membrane separating well-mixed solutions. (a) Overall view; (b) enlargement of the area indicated by the dashed rectangle in (a).

$t=0$ the concentration in one external solution is suddenly changed to C_0. The objective is to determine how the various concentrations evolve over time.

We focus first on the membrane, which is modeled as a homogeneous material. The species conservation equation from Table 2-3 reduces to

$$\frac{\partial C}{\partial t}=D\frac{\partial^2 C}{\partial x^2}.$$ (3.4-1)

The initial and boundary conditions for $C(x, t)$ are

$$C(x,\,0)=0,$$ (3.4-2)

$$C(0,\,t)=KC_1(t),$$ (3.4-3)

$$C(L,\,t)=KC_2(t),$$ (3.4-4)

where K is the partition coefficient.

Lumped models for the external compartments are derived from integral (control-volume) statements of solute conservation based on Eq. (2.2-1). Assuming that the stirred compartments are closed except for mass transfer to or from the membrane, we obtain

$$V\frac{dC_1}{dt}=-AN_x|_{x=0}, \qquad C_1(0)=C_0,$$ (3.4-5)

$$V\frac{dC_2}{dt}=+AN_x|_{x=L}, \qquad C_2(0)=0,$$ (3.4-6)

where N_x represents the solute flux in the x direction. Evaluating the solute fluxes just inside the membrane, we have

$$N_x|_{x=0}=-D\frac{\partial C}{\partial x}(0,\,t), \qquad N_x|_{x=L}=-D\frac{\partial C}{\partial x}(L,\,t).$$ (3.4-7)

This completes the basic problem statement. Equations (3.4-1)–(3.4-7) are coupled through the concentration and flux conditions at $x=0$ and $x=L$, making this a difficult problem to solve in a completely general manner.

The behavior of the concentration profile in the membrane at short times is quite different than that at long times, as shown qualitatively in Fig. 3-5. For small t, the concentration changes resulting from the step change at $x=0$ spread over only a fraction

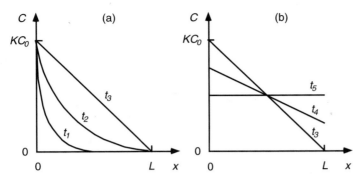

Figure 3-5. Qualitative behavior of the solute concentration in a membrane after the solute is suddenly added to one of the external solutions. Concentration profiles are shown for several distinct times, such that $t_1 < t_2 < t_3 < t_4 < t_5$.

of the membrane thickness, as indicated by the curve for $t = t_1$ in Fig. 3-5(a). The characteristic length for small t is therefore the *penetration depth*, $\delta(t)$. This time-dependent length scale is the distance over which significant concentration changes have occurred at any instant. If the external concentrations were to remain at their initial values, then the steady state corresponding to the curve labeled t_3 eventually would be achieved. This would occur after the changes had spread over the entire thickness L and after additional time had elapsed to allow the concentration profile to adjust to the presence of the right-hand boundary. If the external compartments are large enough that their concentrations change *slowly*, a curve very close to that for $t = t_3$ will in fact be achieved. The second series of changes, depicted in Fig. 3-5(b), will ordinarily occur over a much longer time scale than the first. During this second period the external concentrations gradually equilibrate, until the final state corresponding to $t = t_5$ is reached. The purpose of the following examples is to establish the time scales for the two kinds of behavior and to show how the diffusion problem can be simplified when certain criteria are met.

Example 3.4-1 Penetration Analysis for Short Times   We begin with an approximation to the membrane diffusion problem which is valid for small t (i.e., for $\delta << L$). As suggested above, suppose that V is large enough that $C_1(t)$ remains very close to its initial value, C_0, for these short times. The original boundary conditions, Eqs. (3.4-3) and (3.4-4), are then replaced by

$$C(0, t) = KC_0, \tag{3.4-8}$$

$$C(\infty, t) = 0. \tag{3.4-9}$$

Because C_0 is a known constant, the membrane diffusion problem is now uncoupled from the external mass balances, Eqs. (3.4-5) and (3.4-6). Applying the second boundary condition at $x = \infty$ instead of $x = L$ assumes that $L >> \delta$, as in Example 3.2-2. The new feature here is that δ increases with t, so that this assumption eventually fails.

The dependence of the penetration depth on time is deduced now from order-of-magnitude estimates of the two terms in Eq. (3.4-1), using Eqs. (3.2-2) and (3.2-3) to estimate the derivatives. At any representative position x, the concentration will change from zero to a significant fraction of KC_0 over the time t that has elapsed since the step change at the boundary. At any time t, the concentration will change from KC_0 to zero over the distance δ. The order-of-magnitude estimates of the derivatives are then

$$\frac{\partial C}{\partial t} \sim \frac{KC_0}{t}, \tag{3.4-10}$$

$$\frac{\partial^2 C}{\partial x^2} \sim \frac{KC_0}{\delta^2}. \tag{3.4-11}$$

Substituting these estimates into Eq. (3.4-1) and solving for $\delta(t)$ gives

$$\delta \sim (Dt)^{1/2}. \tag{3.4-12}$$

This shows that the penetration depth increases as the square root of time, a well-known feature of diffusion (or conduction) problems. It also provides an order-of-magnitude estimate of the time required for diffusion to occur over a specified distance. Setting $\delta(t) = L$ and solving for t yields the characteristic time t_D for diffusion over the distance L,

$$t_D = \frac{L^2}{D}. \tag{3.4-13}$$

Analogous arguments apply to transient heat conduction, for which the characteristic time is

$$t_H = \frac{L^2}{\alpha}, \tag{3.4-14}$$

where α is the thermal diffusivity. These time scales for diffusion and conduction are insensitive to the problem geometry. Moreover, as evidenced by the cancelation of the factor KC_0, they are not affected by the concentration or temperature scales. Thus, these results can be used to estimate diffusion or conduction times in a wide variety of situations.

Returning to the membrane diffusion example, we wish to estimate the time interval for which Eq. (3.4-9), the boundary condition applied at $x = \infty$, will be valid. To ensure that $\delta(t) \ll L$, it follows from Eq. (3.4-13) that we must have

$$t \ll \frac{L^2}{D}. \tag{3.4-15}$$

Given that V is large, this condition will be violated long before $C_1(t)$ changes appreciably from C_0. Thus, it is the failure of Eq. (3.4-9), rather than Eq. (3.4-8), which most severely limits the applicability of the penetration analysis.

The penetration problem defined by Eqs. (3.4-1), (3.4-2), (3.4-8), and (3.4-9) is much simpler than that stated originally. The solution for $C(x, t)$ is completed in section 3.5 using the method of similarity.

Example 3.4-2 Pseudosteady Analysis for Long Times   We now seek an approximation for the same membrane diffusion problem which is valid for long times. If the external concentrations change slowly enough, then we expect the instantaneous profile in the membrane to resemble that for a steady state (i.e., as if C_1 and C_2 were constant). Neglecting the time derivative in Eq. (3.4-1), the differential equation for $C(x, t)$ is

$$\frac{\partial^2 C}{\partial x^2} = 0, \tag{3.4-16}$$

similar to that for a steady state. The solution to Eq. (3.4-16) for given external concentrations $C_1(t)$ and $C_2(t)$ is

$$C(x, t) = KC_1(t) + K[C_2(t) - C_1(t)]\frac{x}{L}. \tag{3.4-17}$$

Linear profiles of this type are shown in Fig. 3-5(b). Neglecting the time derivative in the differential equation, and letting time enter only as a parameter through the boundary conditions, as done here, constitutes a *pseudosteady approximation.*

Assuming that the pseudosteady approximation is valid, the determination of $C_1(t)$ and $C_2(t)$ is straightforward. The solute fluxes computed from Eqs. (3.4-7) and (3.4-17) are

$$N_x\big|_{x=0} = \frac{DK[C_1(t) - C_2(t)]}{L} = N_x\big|_{x=L}. \tag{3.4-18}$$

Using Eq. (3.4-18) in Eqs. (3.4-5) and (3.4-6) gives

$$\frac{dC_1}{dt} = -\frac{ADK}{VL}(C_1 - C_2), \qquad C_1(0) = C_0, \tag{3.4-19}$$

$$\frac{dC_2}{dt} = +\frac{ADK}{VL}(C_1 - C_2), \qquad C_2(0) = 0. \tag{3.4-20}$$

Adding these differential equations leads to the conclusion that $C_1 + C_2 = \text{constant} = C_0$, reflecting the fact that the total amount of solute is fixed. Subtracting Eq. (3.4-20) from Eq. (3.4-19) yields a differential equation for the concentration difference, $C_1 - C_2$, which is easily solved. The final expressions for C_1 and C_2 are

$$C_1 = \frac{C_0}{2}(1 + e^{-t/\tau}), \tag{3.4-21}$$

$$C_2 = \frac{C_0}{2}(1 - e^{-t/\tau}), \tag{3.4-22}$$

$$\tau \equiv \frac{VL}{2ADK}. \tag{3.4-23}$$

The rate of response of the external concentrations is seen to be governed by the time constant τ defined by Eq. (3.4-23).

The pseudosteady approximation clearly is a major simplification. What remains to be established are the quantitative criteria which must be satisfied for this approximation to be valid. The concentration disturbance must first penetrate from $x=0$ to $x=L$, and then $C(x, t)$ must evolve to an approximation of the linear profile given by Eq. (3.4-17). This will require several characteristic diffusion times. The first requirement for pseudosteady behavior is then

$$t \gg \frac{L^2}{D}. \tag{3.4-24}$$

A second requirement is related to the rate at which the membrane concentration profile can adjust, relative to the rate at which the external concentrations change. Provided that enough time has elapsed to satisfy Eq. (3.4-24), the changes in the external concentrations will be controlled by the time constant τ. The characteristic response time for the membrane is the diffusion time scale, L^2/D. If $L^2/D \ll \tau$, the membrane concentration profile will respond almost immediately to changes in the external concentrations, as assumed in the pseudosteady analysis. Thus, the second requirement for pseudosteady behavior is

$$\frac{L^2}{D\tau} = \frac{2ALK}{V} \ll 1. \tag{3.4-25}$$

That is, the product of the membrane volume (AL) and partition coefficient (K) must be much less than the volume of the external solutions (V). The product ALK represents the capacity of the membrane for solute, expressed as an effective volume. For example, if entry of the solute into the membrane is thermodynamically favorable, then $K>1$ and the effective volume of the membrane exceeds its actual volume. Thus, Eq. (3.4-25) is equivalent to stating that the effective membrane volume must be much smaller than the external volumes. Given that membranes are ordinarily very thin, both of the requirements stated here are usually satisfied in practical situations.

To see more clearly how it is that these conditions lead to a linear (pseudosteady) concentration profile in the membrane, we estimate the individual terms in Eq. (3.4-1). Once Eq. (3.4-24) has been satisfied, the time derivative is

$$\frac{\partial C}{\partial t} \sim \frac{KC_0}{\tau}, \tag{3.4-26}$$

which is consistent with Eqs. (3.4-17), (3.4-21), and (3.4-22). Note that the time scale which appears on the right-hand side of Eq. (3.4-26) is τ, not t; compare with Eq. (3.4-10). In terms of the scaled variables $X \equiv x/L$ and $\Theta \equiv C/(KC_0)$, the magnitude of the second spatial derivative is

$$\frac{\partial^2 \Theta}{\partial X^2} = \frac{L^2}{DKC_0} \frac{\partial C}{\partial t} \sim \frac{L^2}{D\tau}. \tag{3.4-27}$$

The concentration profile will be approximately linear provided that $\partial^2\Theta/\partial X^2 \ll 1$. Thus, we arrive again at Eq. (3.4-25).

In summary, a pseudosteady approximation is justified only if the response time of the system being modeled is much shorter than the time required for significant changes in the boundary data. When this is true, the system will respond almost immediately to changes in the boundary conditions, and time will enter the problem only as a parameter. In certain situations involving step changes at the boundaries, such as the example just discussed, a minimum time period must elapse before the solution can achieve a pseudosteady form.

3.5 SIMILARITY METHOD

The similarity method is a technique which reduces a partial differential equation in two independent variables to an ordinary differential equation involving a single composite variable. Sometimes called "combination of variables," similarity analysis is applicable to certain problems in which the characteristic lengths are determined by rate processes, rather than by the geometric dimensions. Such problems generally involve regions which are regarded as being semi-infinite. Examples include (a) the penetration regime for transient diffusion or conduction introduced in Section 3.4 and (b) various steady, two-dimensional flow problems involving velocity, temperature, and/or concentration boundary layers. For this approach it is necessary that the field variable (velocity, temperature, or concentration) have profiles which are identical in shape for all positions or times, differing only by the length scale over which the variations occur. In other words, the various profiles must be sufficiently *self-similar* that they can be superimposed by introducing a time-dependent or position-dependent scale factor. For transient conduction or diffusion, the time-dependent scale factor is equivalent to the penetration depth discussed in Section 3.4; for steady boundary layer problems, the scale factor is the boundary layer thickness, which is typically a function of position along a surface. Unlike the methods for solving partial differential equations which are discussed in Chapter 4, the similarity method may be used with nonlinear as well as linear problems. The technique is illustrated by an example in this section and by several other applications throughout the book.

A similarity solution is developed here for the short-time, transient diffusion problem described in Example 3.4-1, involving a step change in concentration at $x=0$ and $t=0$. Using the dimensionless concentration $\Theta \equiv C/(KC_0)$, the governing equations are

$$\frac{\partial \Theta}{\partial t} = D\frac{\partial^2 \Theta}{\partial x^2},$$
(3.5-1)

$$\Theta(x, 0) = 0,$$
(3.5-2)

$$\Theta(0, t) = 1,$$
(3.5-3)

$$\Theta(\infty, t) = 0.$$
(3.5-4)

One requirement of the similarity method is seen to be satisfied, in that this problem contains no fixed characteristic length. That is, there is not a boundary condition imposed at some finite value of x, such as $x=L$.

Assume now that Θ can be expressed as a function of a single independent variable η, where

$$\eta \equiv \frac{x}{g(t)}$$
(3.5-5)

and $g(t)$ is a scale factor which is to be determined. The only difference between $g(t)$ and the penetration depth $\delta(t) \sim (Dt)^{1/2}$ in section 3.4 is that, whereas $\delta(t)$ was used only to describe orders of magnitude, $g(t)$ will have specific numerical values. Although our knowledge of $\delta(t)$ indicates that $g(t) \propto (Dt)^{1/2}$, we will show how to derive $g(t)$ without such prior information.

The derivatives in Eq. (3.5-1) are expressed now in terms of η. In this introductory example, subscripts are used as reminders of which variable is being held constant in each partial derivative. The derivatives are given by

$$\left(\frac{\partial \Theta}{\partial t}\right)_x = \left(\frac{\partial \eta}{\partial t}\right)_x \frac{d\Theta}{d\eta} = \frac{-\eta g'}{g}\frac{d\Theta}{d\eta},$$
(3.5-6)

$$\left(\frac{\partial \Theta}{\partial x}\right)_t = \left(\frac{\partial \eta}{\partial x}\right)_t \frac{d\Theta}{d\eta} = \frac{1}{g}\frac{d\Theta}{d\eta},$$
(3.5-7)

$$\left(\frac{\partial^2 \Theta}{\partial x^2}\right)_t = \frac{1}{g}\left(\frac{\partial \eta}{\partial x}\right)_t \frac{d}{d\eta}\left(\frac{d\Theta}{d\eta}\right) = \frac{1}{g^2}\frac{d^2\Theta}{d\eta^2},$$
(3.5-8)

where $g' \equiv dg/dt$.

In terms of the similarity variable, Eq. (3.5-1) becomes

$$\frac{d^2\Theta}{d\eta^2} + \eta\left(\frac{gg'}{D}\right)\frac{d\Theta}{d\eta} = 0.$$
(3.5-9)

If the assumption that $\Theta = \Theta(\eta)$ is correct, then neither t nor x can appear separately in the differential equation and other conditions for Θ. Because $g = g(t)$, this will be true for Eq. (3.5-9) only if the product gg' is a constant. The exact value of the constant is immaterial, except that a positive constant is needed to have $g > 0$ and $g' > 0$. The simplest equation for $\Theta(\eta)$ is obtained by setting

$$gg' = 2D. \tag{3.5-10}$$

With this choice, Eq. (3.5-9) reduces to

$$\frac{d^2\Theta}{d\eta^2} + 2\eta\frac{d\Theta}{d\eta} = 0. \tag{3.5-11}$$

Integrating Eq. (3.5-11) once yields

$$\frac{d\Theta}{d\eta} = ae^{-\eta^2}, \tag{3.5-12}$$

where a is a constant. Notice that the particular constant chosen in Eq. (3.5-10) has simplified the argument of the exponential in Eq. (3.5-12).

   Turning now to the initial and boundary conditions, it is evident that Eqs. (3.5-2) and (3.5-4) will be equivalent to one another if

$$g(0) = 0. \tag{3.5-13}$$

That is, $t = 0$ and $x = \infty$ will both correspond to $\eta = \infty$ if Eq. (3.5-13) is satisfied. Requiring this, the boundary conditions for Eq. (3.5-11) become

$$\Theta(0) = 1, \tag{3.5-14}$$

$$\Theta(\infty) = 0. \tag{3.5-15}$$

Thus, only two boundary conditions must be satisfied, which is the number that can be accommodated by a second-order, ordinary differential equation such as Eq. (3.5-11). The similarity transformation has evidently been successful.

   Integrating Eq. (3.5-12) and applying the boundary conditions yields

$$\Theta(\eta) = 1 - \frac{2}{\sqrt{\pi}}\int_0^\eta e^{-s^2}\,ds \equiv 1 - \mathrm{erf}(\eta) \equiv \mathrm{erfc}(\eta), \tag{3.5-16}$$

where erf and erfc are the *error function* and *complementary error function,* respectively. These functions are plotted in Fig. 3-6. As η varies from 0 to ∞, $\mathrm{erf}(\eta)$ increases from 0 to 1, whereas $\mathrm{erfc}(\eta)$ decreases from 1 to 0. The properties of the error function and related functions are discussed in many mathematical handbooks, including Abramowitz and Stegun (1970). Values of the functions are tabulated in such books, and they are available also from a variety of software packages (e.g., spreadsheet programs for personal computers).

   To complete the solution, we need to determine $g(t)$. Equation (3.5-10) is nonlinear in g, but noticing that $(g^2)' = 2gg'$, it is a linear differential equation for g^2, namely,

$$(g^2)' = 4D. \tag{3.5-17}$$

Integrating Eq. (3.5-17), with Eq. (3.5-13) as the initial condition, we obtain

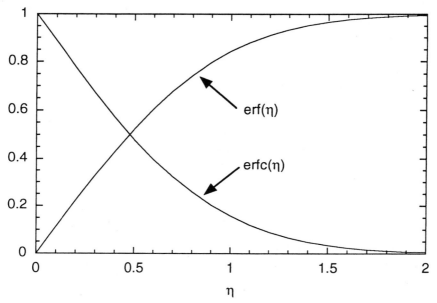

Figure 3-6. Error function and complementary error function.

$$g = 2(Dt)^{1/2}, \qquad \eta = \frac{x}{2(Dt)^{1/2}}. \qquad (3.5\text{-}18)$$

This completes the solution for the concentration.

It is found that $\mathrm{erfc}(1.4) = 0.05$, indicating that the concentration change is 95% complete at $\eta = 1.4$. Thus, a reasonable (albeit arbitrary) value for the penetration depth is $1.4g$, or $2.8\ (Dt)^{1/2}$. This is consistent with the order-of-magnitude estimate, $\delta(t) \sim (Dt)^{1/2}$. The solute flux at $x = 0$ is given by

$$N_x|_{x=0} = -D\frac{\partial C}{\partial x}\bigg|_{x=0} = -\frac{DKC_0}{g}\frac{d\Theta}{d\eta}\bigg|_{\eta=0} = \left(\frac{D}{\pi t}\right)^{1/2} KC_0. \qquad (3.5\text{-}19)$$

The concentration gradient is inversely proportional to the penetration depth. Thus, the flux varies as $t^{-1/2}$ because the penetration depth varies as $t^{1/2}$.

More extensive discussions of similarity analysis are provided in the monographs by Barenblatt (1979), Dresner (1983), and Hansen (1964).

3.6 REGULAR PERTURBATION ANALYSIS

The differential equations in transport problems are often very difficult to solve, due to nonlinearities or other complications. Nonetheless, when the governing equations contain a small parameter, it is often possible to use perturbation techniques to obtain a good approximation to the solution. An equation with a *large* parameter can always be rewritten by inverting that quantity, thereby creating a *small* parameter. Denoting the small parameter by ε, the basic strategy is to obtain a solution valid for $\varepsilon = 0$ and then to add correction terms to account for the fact that ε is not identically zero, but merely small. This procedure replaces the original (possibly intractable) problem by a sequence of simpler ones. Perturbation problems may be *regular* or *singular,* depending on what happens for $\varepsilon = 0$. Regular problems are discussed here, and singular problems are introduced in Section 3.7.

In constructing perturbation solutions it is necessary to keep track of which terms are negligible at a given level of approximation. There is a standard notation for this. The statement "$f(\varepsilon) = O(\varepsilon^n)$ as $\varepsilon \to 0$," read as "$f(\varepsilon)$ is of order ε^n as $\varepsilon \to 0$," means that $|f(\varepsilon)|/\varepsilon^n \leq m$ as $\varepsilon \to 0$, where m is any positive constant. In other words, $f(\varepsilon)/\varepsilon^n$ remains bounded as $\varepsilon \to 0$. Stated yet another way, $f(\varepsilon)$ vanishes at least as fast as ε^n as $\varepsilon \to 0$. When using the "O" notation we will usually omit the stipulation "$\varepsilon \to 0$," the limit involved being clear from the context.[2]

There is a distinction between *mathematical order* (as expressed by "O") and *order of magnitude* ("$\sim$" in this book). This is because the mathematical order is determined only by whether a function is bounded in a certain limit, without regard for the magnitude of any proportionality constant. For example, if $f = k\varepsilon$, where k is a constant, $f = O(\varepsilon)$ whether $k = 1$ or $k = 100$. Nonetheless, when the variables in a problem are properly scaled, the order of magnitude of each term is similar to its mathematical order. That is, scaling tends to keep proportionality constants from being large or small.

Example 3.6-1 Solution of an Algebraic Equation   Following Bender and Orszag (1978), the perturbation approach is illustrated first by using it to find the roots of an algebraic equation,

$$x^3 - (4 + \varepsilon)x + 2\varepsilon = 0, \tag{3.6-1}$$

where $\varepsilon \ll 1$. Each root may be regarded as a function of the parameter ε. Using a Maclaurin series to express the dependence of the root x on ε yields

$$x = \sum_{n=0}^{\infty} a_n \varepsilon^n, \tag{3.6-2}$$

where the coefficients a_n are constants. Because ε is small, the series in Eq. (3.6-2) should converge rapidly, and only a few terms should be needed to obtain an excellent approximation to any given root of Eq. (3.6-1).

Substituting Eq. (3.6-2) into Eq. (3.6-1) and collecting like powers of ε gives

$$(a_0^3 - 4a_0) + (3a_0^2 a_1 - a_0 - 4a_1 + 2)\varepsilon + (3a_0 a_1^2 + 3a_0^2 a_2 - a_1 - 4a_2)\varepsilon^2 + O(\varepsilon^3) = 0. \tag{3.6-3}$$

Note that "$O(\varepsilon^3)$" encompasses *all* of the higher-order terms, those containing ε^n with $n \geq 3$.

The strategy now is to use Eq. (3.6-3) to generate a sequence of equations, each simpler to solve than Eq. (3.6-1). To obtain first approximations to the roots of Eq. (3.6-1), we consider only the $O(1)$ terms in Eq. (3.6-3); this gives

$$a_0^3 - 4a_0 = a_0(a_0^2 - 4) = 0. \tag{3.6-4}$$

---

[2]There is another standard symbol for expressing mathematical order. If $|f(\varepsilon)|/\varepsilon^n \to 0$ as $\varepsilon \to 0$, then "$f(\varepsilon) = o(\varepsilon^n)$." If enough is known about the function f to use either symbol, "O" is preferred over "o." This is because "O" provides a closer bound on the limiting behavior of f. For example, if $f = k\varepsilon^2$, with k a nonzero constant, the statements $f = O(\varepsilon^2)$, $f = o(\varepsilon)$, and $f = o(1)$ are all correct, but $f = O(\varepsilon^2)$ is the most informative. Whether one writes $f = O[g(\varepsilon)]$ or $f = o[g(\varepsilon)]$, $g(\varepsilon)$ is termed the *gauge function*. Gauge functions are typically polynomials (as above), but sometimes functions other than polynomials are needed; see Van Dyke (1964) for examples.

This first approximation is equivalent to setting $\varepsilon = 0$ in the original equation, Eq. (3.6-1). The three roots of Eq. (3.6-4) are $a_0^{(1)} = -2$, $a_0^{(2)} = 0$, and $a_0^{(3)} = 2$, so that the first approximations to the desired roots are

$$x_1 = -2 + O(\varepsilon), \tag{3.6-5}$$

$$x_2 = O(\varepsilon), \tag{3.6-6}$$

$$x_3 = 2 + O(\varepsilon), \tag{3.6-7}$$

where x_i denotes the ith root of Eq. (3.6-1).

Improved approximations to the roots are obtained now by considering the $O(\varepsilon)$ terms in Eq. (3.6-3). Essentially, we adopt the view that the $O(1)$ terms have already been satisfied, and that the $O(\varepsilon^2)$ terms are still negligible. The $O(\varepsilon)$ problem is

$$3a_0^2 a_1 - a_0 - 4a_1 + 2 = 0. \tag{3.6-8}$$

Because the three $a_0^{(i)}$ are already known, Eq. (3.6-8) is a *linear* equation which may be solved for the corresponding $a_1^{(i)}$. It is found that $a_1^{(1)} = -\frac{1}{2}$, $a_1^{(2)} = \frac{1}{2}$, and $a_1^{(3)} = 0$. Accordingly, the roots of the original equation are now

$$x_1 = -2 - \frac{1}{2}\varepsilon + O(\varepsilon^2), \tag{3.6-9}$$

$$x_2 = \frac{1}{2}\varepsilon + O(\varepsilon^2), \tag{3.6-10}$$

$$x_3 = 2 + O(\varepsilon^2). \tag{3.6-11}$$

Continuing this procedure to $O(\varepsilon^2)$ gives the problem,

$$3a_0 a_1^2 + 3a_0^2 a_2 - a_1 - 4a_0 = 0. \tag{3.6-12}$$

Equation (3.6-12) is linear in a_2, just as Eq. (3.6-8) was linear in a_1. Solving for $a_2^{(i)}$ yields

$$x_1 = -2 - \frac{1}{2}\varepsilon + \frac{1}{8}\varepsilon^2 + O(\varepsilon^3), \tag{3.6-13}$$

$$x_2 = \frac{1}{2}\varepsilon - \frac{1}{8}\varepsilon^2 + O(\varepsilon^3), \tag{3.6-14}$$

$$x_3 = 2 + O(\varepsilon^3). \tag{3.6-15}$$

A key feature of this example was that the governing equations for $O(\varepsilon^n)$ with $n \geqslant 1$ were linear, even though the original equation was not. This linearity starting with $O(\varepsilon)$, which is a feature of all regular perturbation problems, greatly facilitates continuing the procedure to the desired level of accuracy.

Example 3.6-2 Electrically Heated Wire with Variable Thermal and Electrical Conductivities The problem in Example 2.8-1 is modified now by assuming that the thermal conductivity and source terms in an electrically heated wire are both functions of temperature. The energy equation is

$$\frac{1}{r}\frac{d}{dr}\left[rk(T)\frac{dT}{dr}\right] + H_V(T) = 0. \tag{3.6-16}$$

Assume that over some range of temperatures the thermal conductivity can be expressed as a linear function of the temperature,

$$k = k_\infty[1 + a(T - T_\infty)],$$ (3.6-17)

where k_∞ is the thermal conductivity at the reference temperature T_∞ (the ambient temperature) and a is a constant. The local rate of heat generation is given by $H_V = i^2/\kappa$, where i is the current density and κ is the electrical conductivity of the wire. From Ohm's law, $i = \kappa V/L$, where V is the applied voltage (assumed to be independent of r) and L is the length of the wire. It follows that $H_V = \kappa V^2/L^2$. Because conduction of heat and electricity in metals is by the same mechanism, we assume that $\kappa \propto k$. Thus, the source term in Eq. (3.6-16) is written as

$$H_V = H_\infty[1 + a(T - T_\infty)],$$ (3.6-18)

where H_∞ is a constant. To simplify the analysis, it is assumed that $Bi = hR/k_\infty \gg 1$, so that the surface temperature is T_∞. A convenient set of dimensionless quantities is

$$\eta \equiv \frac{r}{R}, \qquad \Theta \equiv \frac{T - T_\infty}{H_\infty R^2/k_\infty}, \qquad \varepsilon \equiv \frac{aH_\infty R^2}{k_\infty},$$ (3.6-19)

where the temperature scale is $\Delta T = H_\infty R^2/k_\infty$. With these definitions, Eq. (3.6-16) is transformed to

$$\frac{1}{\eta}\frac{d}{d\eta}\left[\eta(1 + \varepsilon\Theta)\frac{d\Theta}{d\eta}\right] + 1 + \varepsilon\Theta = 0$$ (3.6-20)

with the boundary conditions

$$\frac{d\Theta}{d\eta}(0) = 0,$$ (3.6-21)

$$\Theta(1) = 0.$$ (3.6-22)

Note that Eq. (3.6-20) is nonlinear, because the coefficient of $d\Theta/d\eta$ depends on Θ. There appears to be no general solution which is valid for all ε.

Suppose now that the temperature dependence of k and H_V is relatively weak, such that $\varepsilon \ll 1$. Treating the temperature as a function of ε and expanding the solution as a power series, we obtain

$$\Theta(\eta) = \sum_{n=0}^{\infty} \Theta_n(\eta)\varepsilon^n.$$ (3.6-23)

Because Θ depends on the radial position as well as on the parameter ε, the coefficients in the expansion are not constants, as they were in the previous example. The problem now is to determine the coefficient functions, $\Theta_n(\eta)$. We will be satisfied with the first correction to the temperature profile caused by the variable thermal properties. In other words, we will compute only $\Theta_0(\eta)$ and $\Theta_1(\eta)$.

Substituting Eq. (3.6-23) into Eq. (3.6-20) yields

$$\frac{1}{\eta}\frac{d}{d\eta}\left[\eta\left(1 + \varepsilon\Theta_0 + O(\varepsilon^2)\right)\left(\frac{d\Theta_0}{d\eta} + \varepsilon\frac{d\Theta_1}{d\eta} + O(\varepsilon^2)\right)\right] + 1 + \varepsilon\Theta_0 + O(\varepsilon^2) = 0.$$ (3.6-24)

The boundary conditions are expanded in a similar manner to give

$$\frac{d\Theta_0}{d\eta}(0) + \varepsilon\frac{d\Theta_1}{d\eta}(0) + O(\varepsilon^2) = 0,$$ (3.6-25)

$$\Theta_0(1) + \varepsilon\Theta_1(1) + O(\varepsilon^2) = 0.$$ (3.6-26)

The $O(1)$ problem governing Θ_0, which is the same as that obtained by setting $\varepsilon = 0$ in the original equations, is

$$\frac{1}{\eta}\frac{d}{d\eta}\left[\eta\frac{d\Theta_0}{d\eta}\right] + 1 = 0,\tag{3.6-27}$$

$$\frac{d\Theta_0}{d\eta}(0) = 0, \qquad \Theta_0(1) = 0.\tag{3.6-28}$$

These are the equations for constant values of k and H_V, solved earlier. From Eq. (3.2-6), the solution is

$$\Theta_0 = \frac{1}{4}(1 - \eta^2).\tag{3.6-29}$$

The $O(\varepsilon)$ problem for $\Theta_1(\eta)$ is

$$\frac{1}{\eta}\frac{d}{d\eta}\left[\eta\left(\Theta_0\frac{d\Theta_0}{d\eta} + \frac{d\Theta_1}{d\eta}\right)\right] + \Theta_0 = 0,\tag{3.6-30}$$

$$\frac{d\Theta_1}{d\eta}(0) = 0, \qquad \Theta_1(1) = 0.\tag{3.6-31}$$

Equation (3.6-30) is a linear differential equation for $\Theta_1(\eta)$, with nonhomogeneous terms arising from the $O(1)$ solution, $\Theta_0(\eta)$. The solution is found to be

$$\Theta_1(\eta) = \frac{1}{64}(1 - \eta^4).\tag{3.6-32}$$

Thus, the first two terms of the expansion for Θ are

$$\Theta(\eta) = \frac{1}{4}(1 - \eta^2) + \frac{\varepsilon}{64}(1 - \eta^4) + O(\varepsilon^2).\tag{3.6-33}$$

The procedure followed in obtaining Eq. (3.6-33) can be extended indefinitely (as time and patience allow), to derive successively better approximations to Θ.

It was noted in the previous example that the perturbation procedure always yields a sequence of linear equations beginning at $O(\varepsilon)$. Although not true in general, in the present example the $O(1)$ equation (that governing Θ_0) happened to be linear also. The fact that Θ_0 was a simple polynomial greatly facilitated the determination of Θ_1. As a rule, the success of the perturbation approach in solving a differential equation hinges on whether the first term in the perturbation expansion (e.g., Θ_0) can be expressed in terms of elementary functions.

3.7 SINGULAR PERTURBATION ANALYSIS

An important feature of regular perturbation solutions to differential equations is that the expansions are uniformly valid for all values of the independent variable. Thus, the solution in Example 3.6-2 is a good approximation to the temperature profile throughout the wire, in the limit $\varepsilon \to 0$. Singular perturbation methods are needed to treat problems where uniformly valid solutions cannot be found. The key characteristic of such problems is that each of two or more regions requires a different approximation. Singular perturbation analysis is relatively new, having been employed first in the 1950s for the solution of boundary layer problems in fluid mechanics. Out of this research developed what is now called the method of *matched asymptotic expansions* [see Chapter 4 of Van Dyke (1964)]. Matched asymptotic expansions have been described in many texts on applied mathematics. For information beyond the introduction given here, see Bender and Orszag (1978), Kevorkian and Cole (1981), Lin and Segel (1974), Nayfeh (1973), or Van Dyke (1964).

As shown in the examples which follow, a common feature of singular perturbation problems is that the small parameter, ε, multiplies the highest-order term in the equation (e.g., the highest derivative). Consequently, setting $\varepsilon = 0$ reduces the order of the equation. This reduction of order is sufficient to invalidate a regular perturbation expansion. However, not all singular perturbation problems have this feature (see Section 7.7, for example).

Example 3.7-1 Solution of an Algebraic Equation   As in the discussion of regular perturbation methods, it is instructive to begin with an algebraic equation. Consider the roots of

$$\varepsilon x^2 + x - 1 = 0, \tag{3.7-1}$$

where $\varepsilon \ll 1$. We will proceed as if the roots of this quadratic equation could not be found exactly, and use the exact answer only to check the results of the perturbation analysis. In regular perturbation analysis, as discussed in Section 3.6, the first approximation to the solution is found by setting $\varepsilon = 0$ in the governing equation. Doing that in Eq. (3.7-1) gives

$$x - 1 = 0 \tag{3.7-2}$$

or $x = 1$. However, there is an immediate difficulty: what happened to the second root? It was inadvertently lost because setting $\varepsilon = 0$ reduced the equation from second order to first order. As already mentioned, a reduction in the order of the governing equation for $\varepsilon = 0$ is a hallmark of singular perturbation problems. Such problems are singular in the sense that the solution obtained by setting $\varepsilon = 0$ is radically different than the asymptotic solution for $\varepsilon \to 0$. Thus, the solution for $\varepsilon = 0$ is not an acceptable starting point for constructing better approximations. In the present example the singularity manifests itself in the loss of one root; with differential equations it usually leads to the inability to satisfy one or more boundary conditions for $\varepsilon = 0$.

The flaw in the reasoning leading to Eq. (3.7-2) was the implicit assumption that $x = O(1)$ for both roots. This led to the expectation that $\varepsilon x^2 = O(\varepsilon)$. Resolving the difficulty requires that we pay careful attention to the scaling. As will be shown, the two roots of Eq. (3.7-1) have very different scales.

To correct the scaling problem, a new variable is defined by setting

$$X \equiv \varepsilon^a x, \tag{3.7-3}$$

similar to the transformations used to scale the coordinates in Examples 3.2-2 and 3.3-2. The constraints which determine a are that the governing equation be second order (allowing us to find both roots) and that $X = O(1)$ at most. Using Eq. (3.7-3) in Eq. (3.7-1) yields

$$X^2 + \varepsilon^{a-1} X - \varepsilon^{2a-1} = 0. \tag{3.7-4}$$

What is needed now is to identify the most important terms for $\varepsilon \to 0$, through what is called a *dominant balance*. Aside from X^2, which is mandatory, either or both of the other terms on the left-hand side might be important. Let us assume that the dominant terms are those involving X^2 and X. This implies that $\varepsilon^{a-1} = 1$ or $a = 1$, so that Eq. (3.7-4) becomes

$$X^2 + X - \varepsilon = 0. \tag{3.7-5}$$

The term that does not involve X is $O(\varepsilon)$, consistent with our assumption that the others are dominant. Equation (3.7-5) provides the basis for the perturbation solution, as shown below.

What if we had guessed the wrong set of dominant terms? If X^2 and the term involving ε were assumed to be dominant in Eq. (3.7-4), we would have concluded that $\varepsilon^{2a-1} = 1$ or $a = 1/2$. This would have given

$$X^2 + \varepsilon^{-1/2} X - 1 = 0. \tag{3.7-6}$$

The remaining term is now $O(\varepsilon^{-1/2})$, which contradicts the assumed dominant balance. This illustrates how trial and error is sometimes involved in determining scales.

Equation (3.7-5) is solved by representing each root as a power series expansion in ε,

$$X = \sum_{n=0}^{\infty} a_n \varepsilon^n, \tag{3.7-7}$$

and computing successive terms as in Example 3.6-1. The $O(1)$ problem is

$$a_0^2 + a_0 = 0, \tag{3.7-8}$$

so that the roots at leading order are $a_0^{(1)} = -1$ and $a_0^{(2)} = 0$. The $O(\varepsilon)$ problem is

$$2a_1 a_0 + a_1 - 1 = 0, \tag{3.7-9}$$

which gives $a_1^{(1)} = -1$ and $a_1^{(2)} = 1$. The roots X_i, accurate up to $O(\varepsilon^2)$, are then

$$X_1 = -1 - \varepsilon + O(\varepsilon^2), \tag{3.7-10}$$

$$X_2 = \varepsilon + O(\varepsilon^2). \tag{3.7-11}$$

In terms of the original variable x, the roots are

$$x_1 = -\varepsilon^{-1} - 1 + O(\varepsilon), \tag{3.7-12}$$

$$x_2 = 1 + O(\varepsilon). \tag{3.7-13}$$

The naive approach of setting $\varepsilon = 0$ in Eq. (3.7-1) gave us an approximation to x_2 but missed x_1 entirely. As already mentioned, the difficulty came from assuming that $x = O(1)$ for both roots. As seen now in Eqs. (3.7-12) and (3.7-13), $x_2 = O(1)$ but $x_1 = O(\varepsilon^{-1})$.

The exact roots of Eq. (3.7-1) are, of course,

$$x = \frac{-1 \pm \sqrt{1 + 4\varepsilon}}{2\varepsilon}. \tag{3.7-14}$$

Expanding the numerator for $\varepsilon \ll 1$ gives

$$\sqrt{1 + 4\varepsilon} = 1 + 2\varepsilon + O(\varepsilon^2). \tag{3.7-15}$$

It can be seen from Eqs. (3.7-14) and (3.7-15) that the "minus" and "plus" roots correspond to x_1 and x_2, respectively. Thus, the singular perturbation analysis has provided the first terms in an expansion of the exact solution.

Example 3.7-2 Diffusion in a Cylinder with a Fast, Homogeneous Reaction   In this example the singular perturbation approach is applied to a boundary-value problem. Consider a cylindrical catalyst pellet in which a fast, first-order reaction occurs at sites which are uniformly distributed throughout the pellet. The pellet is modeled as a permeable medium with an effective diffusivity D for the reactant and an effective homogeneous rate constant k_v. Both the diffusivity and the rate constant depend on the catalyst material; see Aris (1975) for a discussion of these parameters. The reactant concentration at steady state, $C(r)$, is shown qualitatively in Fig. 3-7. The fast, irreversible reaction leads to a sharp drop in concentration near the outer surface, as for the analogous problem involving a planar liquid film discussed in Section 3.2. The region in which the concentration declines sharply is called the *concentration boundary layer.*

The dimensionless reactant concentration and radial coordinate are given by

$$\Theta \equiv \frac{C}{C_0}, \qquad \eta \equiv \frac{r}{R}, \qquad (3.7\text{-}16)$$

where C_0 is the concentration at the outer surface and R is the pellet radius. Using cylindrical coordinates, the species conservation equation is

$$\varepsilon\left(\frac{d^2\Theta}{d\eta^2} + \frac{1}{\eta}\frac{d\Theta}{d\eta}\right) - \Theta = 0, \qquad (3.7\text{-}17)$$

$$\varepsilon \equiv \mathrm{Da}^{-1} = \frac{D}{k_v R^2}. \qquad (3.7\text{-}18)$$

Because the reaction is assumed to be fast, the small parameter is $\varepsilon = \mathrm{Da}^{-1}$. The boundary conditions are

$$\frac{d\Theta}{d\eta}(0) = 0, \qquad (3.7\text{-}19)$$

$$\Theta(1) = 1. \qquad (3.7\text{-}20)$$

The fact that ε multiplies the second derivative in Eq. (3.7-17) is an immediate indication that the perturbation problem is singular. If we set $\varepsilon = 0$, the differential equation reduces to $\Theta = 0$. Although accurate for the "core" region in the center of the pellet, this solution is obviously incorrect near the outer surface. This is made plain by the fact that $\Theta = 0$ is consistent with Eq. (3.7-19) but not Eq. (3.7-20). We conclude that whereas η is scaled adequately for the core region, a different variable is needed for the boundary layer. The different coordinate scalings will lead to different approximations for the reactant concentration in the two regions.

To rescale the radial coordinate for the boundary layer, let

$$\xi \equiv (1 - \eta)\varepsilon^b. \qquad (3.7\text{-}21)$$

The quantity $(1 - \eta)$, which is small in the boundary layer, is "stretched" using the factor ε^b to obtain a coordinate which is $O(1)$. That is, $\xi = O(1)$ in the boundary layer, by definition. Because ε is small, making the stretching factor large will require that $b < 0$. Changing to the new independent variable, Eq. (3.7-17) becomes

$$\varepsilon^{2b+1}\frac{d^2\Theta}{d\xi^2} - \frac{\varepsilon^{b+1}}{1 - \xi\varepsilon^{-b}}\frac{d\Theta}{d\xi} - \Theta = 0. \qquad (3.7\text{-}22)$$

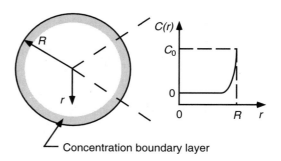

Figure 3-7. Schematic of a cylindrical catalyst pellet, showing the concentration boundary layer which results for $\mathrm{Da} \gg 1$.

Concentration boundary layer

Given that $\xi\varepsilon^{-b}$ will be small in the boundary layer (i.e., $b<0$), the coefficient of the first derivative can be simplified using an expansion of the form $1/(1-x)=1+x+O(x^2)$ for $x\to0$. The result is

$$\varepsilon^{2b+1}\frac{d^2\Theta}{d\xi^2}-\varepsilon^{b+1}[1+\xi\varepsilon^{-b}+O(\varepsilon^{-2b})]\frac{d\Theta}{d\xi}-\Theta=0. \qquad (3.7\text{-}23)$$

The required value of b is determined by the dominant balance. The equation must remain second order, and the reaction term is a key part of the physical situation. Accordingly, the first and last terms are balanced to give $2b+1=0$, or $b=-1/2$. Thus, the boundary layer coordinate is

$$\xi\equiv(1-\eta)\varepsilon^{-1/2} \qquad (3.7\text{-}24)$$

and the corresponding species conservation equation is

$$\frac{d^2\hat\Theta}{d\xi^2}-\varepsilon^{1/2}[1+\xi\varepsilon^{1/2}+O(\varepsilon)]\frac{d\hat\Theta}{d\xi}-\hat\Theta=0. \qquad (3.7\text{-}25)$$

To distinguish between the two regions, the concentration in the boundary layer is denoted now as $\hat\Theta=\hat\Theta(\xi)$.

The fact that ε appears only as $\varepsilon^{1/2}$ in Eq. (3.7-25) suggests an expansion for the boundary layer region of the form

$$\hat\Theta=\hat\Theta_0(\xi)+\varepsilon^{1/2}\hat\Theta_1(\xi)+O(\varepsilon). \qquad (3.7\text{-}26)$$

Substituting Eq. (3.7-26) into Eq. (3.7-25) gives

$$\frac{d^2\hat\Theta_0}{d\xi^2}+\varepsilon^{1/2}\frac{d^2\hat\Theta_1}{d\xi^2}-\varepsilon^{1/2}[1+\xi\varepsilon^{1/2}+O(\varepsilon)]\left[\frac{d\hat\Theta_0}{d\xi}+\varepsilon^{1/2}\frac{d\hat\Theta_1}{d\xi}+O(\varepsilon)\right]$$
$$-\hat\Theta_0-\varepsilon^{1/2}\hat\Theta_1+O(\varepsilon)=0. \qquad (3.7\text{-}27)$$

The boundary condition at the surface, Eq. (3.7-20), is expanded in a similar manner. From Eq. (3.7-27) and that boundary condition, the $O(1)$ problem for the boundary layer is

$$\frac{d^2\hat\Theta_0}{d\xi^2}-\hat\Theta_0=0, \qquad (3.7\text{-}28)$$

$$\hat\Theta_0(0)=1. \qquad (3.7\text{-}29)$$

Notice that Eq. (3.7-28) is the same as the equation which would have been written in rectangular coordinates. The curvature of the cylinder surface has no effect at $O(1)$ because the boundary layer thickness is much smaller than the cylinder radius. That is, the boundary layer is thin enough to make the surface look flat. The solution to Eq. (3.7-28) that satisfies Eq. (3.7-29) is

$$\hat\Theta_0=a_0e^{-\xi}+(1-a_0)e^{\xi}, \qquad (3.7\text{-}30)$$

where a_0 is a constant which remains to be determined. Of importance, it is *not* legitimate to use Eq. (3.7-19) here, because the boundary layer does not extend to the center of the cylinder. The boundary condition at the center applies to the *core* region, which must be handled separately.

To evaluate a_0, it is necessary to match Eq. (3.7-30) to the core solution, denoted by $\tilde\Theta=\tilde\Theta(\eta)$. Unlike problems involving distinct regions separated by a definite surface, such as a phase boundary, the matching is not accomplished at a fixed location. Instead, the matching must be done in an asymptotic sense, expressed as

$$\lim_{\xi\to\infty}\hat\Theta(\xi)=\lim_{\eta\to1}\tilde\Theta(\eta), \qquad \varepsilon\to0. \qquad (3.7\text{-}31)$$

Equation (3.7-31) is based on the observation that for $\varepsilon \ll 1$, large values of the boundary layer coordinate ξ correspond to values of the original variable η near unity. In general, the matching must be accomplished using the successive levels of approximation to $\hat{\Theta}(\xi)$ and $\tilde{\Theta}(\eta)$, as defined by expansions for each of these concentration variables in ε. In the present problem the matching requirement is trivial, because it is readily shown that $\tilde{\Theta}(\eta) = 0$ at all levels of approximation. Thus, Eq. (3.7-31) gives $\hat{\Theta}_0(\infty) = 0$, which requires that $a_0 = 1$ and

$$\hat{\Theta}_0 = e^{-\xi}. \tag{3.7-32}$$

This completes the $O(1)$ solution for the boundary layer.

The $O(\varepsilon^{1/2})$ problem for the boundary layer, which will give the first correction due to the curvature of the surface, is given by

$$\frac{d^2\hat{\Theta}_1}{d\xi^2} - \frac{d\hat{\Theta}_0}{d\xi} - \hat{\Theta}_1 = 0 \tag{3.7-33}$$

$$\hat{\Theta}_1(0) = 0. \tag{3.7-34}$$

Substituting Eq. (3.7-32) into Eq. (3.7-33) and solving for $\hat{\Theta}_1$ gives

$$\hat{\Theta}_1 = \frac{\xi e^{-\xi}}{2}, \tag{3.7-35}$$

which satisfies the (trivial) matching requirement that $\hat{\Theta}_1(\infty) = 0$.

In general, an overall or composite solution can be constructed by adding the solutions for the two regions and subtracting the part of the answer that is common to both solutions:

$$\Theta = \hat{\Theta}(\xi) + \tilde{\Theta}(\eta) - \lim_{\xi \to \infty} \hat{\Theta}(\xi) = \hat{\Theta}(\xi) + \tilde{\Theta}(\eta) - \lim_{\eta \to 1} \tilde{\Theta}(\eta). \tag{3.7-36}$$

The last term in either form of Eq. (3.7-36) corrects for the "double counting" which would otherwise occur in the matching region defined either by $\xi \to \infty$ or $\eta \to 1$. The composite solution for the present example is

$$\Theta = \hat{\Theta} = e^{-\xi}\left[1 + \frac{\xi}{2}\varepsilon^{1/2} + O(\varepsilon)\right]. \tag{3.7-37}$$

This completes the solution for the concentration field up to $O(\varepsilon)$.

It is informative to compute the flux of reactant into the cylinder, which is given by

$$-N_r|_{r=R} = D\frac{dC}{dr}\bigg|_{r=R} = \frac{DC_0}{R}\left(-\varepsilon^{-1/2}\frac{d\Theta}{d\xi}\bigg|_{\xi=0}\right) = \frac{DC_0}{R}\varepsilon^{-1/2}\left[1 - \frac{\varepsilon^{1/2}}{2} + O(\varepsilon)\right]. \tag{3.7-38}$$

The first term in the final expression comes from $\hat{\Theta}_0(\xi)$ and is the same as the flux into a planar slab of material. The second term comes from $\hat{\Theta}_1(\xi)$ and is a first approximation for the effects of surface curvature. As can be seen, the convex shape of the cylindrical surface reduces the flux.

This example illustrates many features of problems which can be solved only by singular perturbation, but it also happens to have an exact solution valid for the entire cylinder. The exact solution is

$$\Theta(\eta) = \frac{I_0(\eta\varepsilon^{-1/2})}{I_0(\varepsilon^{-1/2})}, \tag{3.7-39}$$

where $I_0(x)$ is the modified Bessel function of the first kind, of order zero (see Section 4.7). Using Eq. (3.7-39) to calculate the flux into the cylinder gives

$$-N_r|_{r=R} = \frac{DC_0}{R} \varepsilon^{-1/2} \frac{I_1(\varepsilon^{-1/2})}{I_0(\varepsilon^{-1/2})}, \tag{3.7-40}$$

where $I_1(x)$ is the modified Bessel function of the first kind, of order one. The Bessel functions in Eq. (3.7-40) have the following asymptotic expansions for large arguments (Abramowitz and Stegun, 1970):

$$I_0(x) = \frac{e^x}{(2\pi x)^{1/2}} \left[1 + \frac{1}{8x} + O(x^{-2}) \right], \tag{3.7-41}$$

$$I_1(x) = \frac{e^x}{(2\pi x)^{1/2}} \left[1 - \frac{3}{8x} + O(x^{-2}) \right]. \tag{3.7-42}$$

Using these results to evaluate the first two terms of the expansion for $I_1(\varepsilon^{-1/2})/I_0(\varepsilon^{-1/2})$, the flux given by Eq. (3.7-40) is found to be identical to the result in Eq. (3.7-38). Thus, as seen in the previous example, the singular perturbation analysis yields an expansion of the exact solution.

Example 3.7-3 Quasi-Steady-State Approximation in Chemical Kinetics A key feature of Example 3.7-2 was the emergence of two very different length scales for concentration variations, one for the core region and another, much smaller scale, for the boundary layer. The singular perturbation approach also applies to problems involving two or more divergent *time* scales, as will be illustrated now using a problem in chemical kinetics. Consider the sequential homogeneous reactions

$$A \underset{k_{-1}}{\overset{k_1}{\rightleftarrows}} B \overset{k_2}{\rightarrow} C.$$

For a well-stirred batch reactor of fixed volume, the species concentrations are governed by the following first-order differential equations and initial conditions:

$$\frac{dC_A}{dt} = -k_1 C_A + k_{-1} C_B, \qquad C_A(0) = C_0, \tag{3.7-43}$$

$$\frac{dC_B}{dt} = k_1 C_A - (k_{-1} + k_2) C_B, \qquad C_B(0) = 0, \tag{3.7-44}$$

$$\frac{dC_C}{dt} = k_2 C_B, \qquad C_C(0) = 0. \tag{3.7-45}$$

It is assumed here that only species A is present initially, at concentration C_0.

If the second reaction is fast enough to maintain the intermediate (species B) at relatively low concentrations, it is common practice in kinetics to assume that $dC_B/dt \cong 0$. This is called the *quasi-steady-state approximation,* abbreviated here as QSSA. Using this approach, Eq. (3.7-44) gives

$$C_B \cong C_A \left(\frac{k_1}{k_{-1} + k_2} \right). \tag{3.7-46}$$

Using Eq. (3.7-46) to eliminate $C_B(t)$ in favor of $C_A(t)$, Eq. (3.7-43) is readily integrated to obtain $C_A(t)$. Evaluating $C_B(t)$ using that result and Eq. (3.7-46), it is straightforward to compute the rate of formation of the product (species C) from Eq. (3.7-45).

To examine the limitations of this very useful approach, we introduce the dimensionless variables

$$\theta(\tau) \equiv \frac{C_A}{C_0}, \tag{3.7-47}$$

$$\phi(\tau) \equiv \frac{C_B}{C_0}\left(\frac{k_{-1}+k_2}{k_1}\right) = \frac{C_B}{\varepsilon C_0}, \tag{3.7-48}$$

$$\tau \equiv k_1 t \tag{3.7-49}$$

and the dimensionless parameters

$$\varepsilon \equiv \frac{k_1}{k_{-1}+k_2}, \tag{3.7-50}$$

$$\lambda \equiv \frac{k_2}{k_{-1}}. \tag{3.7-51}$$

The definition of θ is motivated by the fact that C_0 is the scale for C_A. When the QSSA is even roughly correct, Eq. (3.7-46) provides a suitable scale for C_B, thereby motivating the definition of ϕ. Making the time dimensionless using k, as in the definition of τ, is equivalent to adopting a time scale representative of changes in C_A, but not necessarily changes in C_B and C_C. The QSSA is expected to work best when $C_B \ll C_A$, and so the analysis focuses on small values of ε. The remaining parameter, λ, is not necessarily large or small.

With these definitions, Eqs. (3.7-43) and (3.7-44) become

$$\frac{d\theta}{d\tau} = -\theta + \left(\frac{1}{1+\lambda}\right)\phi, \qquad \theta(0) = 1, \tag{3.7-52}$$

$$\varepsilon \frac{d\phi}{d\tau} = \theta - \phi, \qquad \phi(0) = 0. \tag{3.7-53}$$

Equation (3.7-45) is not considered further because C_A and C_B (or θ and ϕ) can be obtained independently of C_C.

In terms of the dimensionless variables, adopting the QSSA would lead us to neglect the time derivative in Eq. (3.7-53), with the result that

$$\phi = \theta. \tag{3.7-54}$$

One of the differential equations, Eq. (3.7-53), would then be replaced by this algebraic equation. Consequently, the ability to satisfy the initial condition for the intermediate, $\phi(0) = 0$, would be lost. It is evident then that the QSSA is not *uniformly valid* for all τ. Further reflection indicates that the QSSA fails at *small* τ, when ϕ increases rapidly from 0 to approximately θ. Because the right-hand side of Eq. (3.7-53) is always $O(1)$, even for small τ, it must be that $\varepsilon \, d\phi/d\tau = O(1)$ for sufficiently small τ. This fact motivates the choice of a new time variable ω for small times,

$$\omega \equiv \tau/\varepsilon. \tag{3.7-55}$$

With time rescaled in this manner, the governing equations for *small* times are

$$\frac{d\theta}{d\omega} = -\varepsilon\theta + \left(\frac{\varepsilon}{1+\lambda}\right)\phi, \qquad \theta(0) = 1, \tag{3.7-56}$$

$$\frac{d\phi}{d\omega} = \theta - \phi, \qquad \phi(0) = 0. \tag{3.7-57}$$

The rescaling of time used here is analogous to the rescaling of distance used in steady-state boundary layer problems. In singular perturbation parlance, Eqs. (3.7-56) and (3.7-57) are the governing equations for an *inner region,* while Eqs. (3.7-52) and (3.7-53) apply to an *outer region.*

We proceed now to solve for θ and ϕ for short times (the inner region, dependent variables $\hat{\theta}$ and $\hat{\phi}$) and for long times (the outer region, dependent variables $\tilde{\theta}$ and $\tilde{\phi}$). The two solutions are matched asymptotically. Because the solutions in neither region are identically zero, this provides a better illustration of the mechanics of asymptotic matching than does Example 3.7-2.

We begin with the outer problem. Expanding $\tilde{\theta}(\tau)$ and $\tilde{\phi}(\tau)$ in integral powers of ε as

$$\tilde{\theta}(\tau) = \tilde{\theta}_0(\tau) + \tilde{\theta}_1(\tau)\varepsilon + \tilde{\theta}_2(\tau)\varepsilon^2 + O(\varepsilon^3), \tag{3.7-58}$$

$$\tilde{\phi}(\tau) = \tilde{\phi}_0(\tau) + \tilde{\phi}_1(\tau)\varepsilon + \tilde{\phi}_2(\tau)\varepsilon^2 + O(\varepsilon^3), \tag{3.7-59}$$

the governing equations for the outer region become

$$\frac{d}{d\tau}(\tilde{\theta}_0 + \varepsilon\tilde{\theta}_1) = -(\tilde{\theta}_0 + \varepsilon\tilde{\theta}_1) + \left(\frac{1}{1+\lambda}\right)(\tilde{\phi}_0 + \varepsilon\tilde{\phi}_1) + O(\varepsilon^2), \tag{3.7-60}$$

$$\varepsilon\frac{d}{d\tau}(\tilde{\phi}_0 + \varepsilon\tilde{\phi}_1) = (\tilde{\theta}_0 + \varepsilon\tilde{\theta}_1) - (\tilde{\phi}_0 + \varepsilon\tilde{\phi}_1) + O(\varepsilon^2). \tag{3.7-61}$$

It follows that the $O(1)$ outer problem is

$$\frac{d\tilde{\theta}_0}{d\tau} = -\tilde{\theta}_0 + \left(\frac{1}{1+\lambda}\right)\tilde{\phi}_0, \tag{3.7-62}$$

$$\tilde{\phi}_0 = \tilde{\theta}_0. \tag{3.7-63}$$

Because Eq. (3.7-63) is equivalent to the QSSA, it is evident that the QSSA is simply a first approximation to the outer or long-time behavior. The solution to Eqs. (3.7-62) and (3.7-63) is

$$\tilde{\theta}_0 = a_0 e^{-\lambda\tau/(1+\lambda)} = \tilde{\phi}_0, \tag{3.7-64}$$

where a_0 is a constant to be determined later by matching. It would be incorrect to try to evaluate a_0 by using the initial conditions, because Eq. (3.7-64) is not valid for small times.

It follows from Eqs. (3.7-60) and (3.7-61) that all higher-order ($n \geq 1$) corrections in the outer solution are governed by

$$\frac{d\tilde{\theta}_n}{d\tau} = -\tilde{\theta}_n + \left(\frac{1}{1+\lambda}\right)\tilde{\phi}_n, \tag{3.7-65}$$

$$\tilde{\phi}_n = \tilde{\theta}_n - \frac{d\tilde{\phi}_{n-1}}{d\tau}. \tag{3.7-66}$$

Thus, each pair of terms in the outer expansions happens to be governed by one differential equation and one algebraic equation. The differential equations are Eq. (3.7-62) for $n=0$ and Eq. (3.7-65) for $n \geq 1$; the algebraic equations are Eq. (3.7-63) for $n=0$ and Eq. (3.7-66) for $n \geq 1$. Solving the $n=1$ outer equations for $\tilde{\theta}_1$ and $\tilde{\phi}_1$ and using Eq. (3.7-64) yields

$$\tilde{\theta} = a_0 e^{-\lambda\tau/(1+\lambda)} + \varepsilon\left[a_1 e^{-\lambda\tau/(1+\lambda)} + \frac{a_0\lambda}{(1+\lambda)^2}\tau e^{-\lambda\tau/(1+\lambda)}\right] + O(\varepsilon^2), \tag{3.7-67}$$

$$\tilde{\phi} = a_0 e^{-\lambda\tau/(1+\lambda)} + \varepsilon\left[a_1 e^{-\lambda\tau/(1+\lambda)} + \frac{a_0\lambda}{(1+\lambda)}\left(1 + \frac{\tau}{(1+\lambda)}\right)e^{-\lambda\tau/(1+\lambda)}\right] + O(\varepsilon^2), \tag{3.7-68}$$

where a_1 is another constant.

Consider now the inner problem. Expanding $\hat{\theta}(\omega)$ and $\hat{\phi}(\omega)$ in integer powers of ε as

$$\hat{\theta}(\omega) = \hat{\theta}_0(\omega) + \hat{\theta}_1(\omega)\varepsilon + \hat{\theta}_2(\omega)\varepsilon^2 + O(\varepsilon^3), \tag{3.7-69}$$

$$\hat{\phi}(\omega) = \hat{\phi}_0(\omega) + \hat{\phi}_1(\omega)\varepsilon + \hat{\phi}_2(\omega)\varepsilon^2 + O(\varepsilon^3) \tag{3.7-70}$$

gives the inner equations,

$$\frac{d\hat{\theta}_0}{d\omega} = 0, \qquad\qquad \hat{\theta}_0(0) = 1, \qquad\qquad (3.7\text{-}71)$$

$$\frac{d\hat{\theta}_n}{d\omega} = -\hat{\theta}_{n-1} + \left(\frac{1}{1+\lambda}\right)\hat{\phi}_{n-1}, \qquad \hat{\theta}_n(0) = 0, \qquad (3.7\text{-}72)$$

$$\frac{d\hat{\phi}_0}{d\omega} = \hat{\theta}_0 - \hat{\phi}_0, \qquad\qquad \hat{\phi}_0(0) = 0, \qquad\qquad (3.7\text{-}73)$$

$$\frac{d\hat{\phi}_n}{d\omega} = \hat{\theta}_n - \hat{\phi}_n, \qquad\qquad \hat{\phi}_n(0) = 0, \qquad\qquad (3.7\text{-}74)$$

where $n \geqslant 1$. In this case there are two differential equations at each successive level of approximation. Because these equations are valid for small times, it is now correct to invoke the initial conditions for θ and ϕ, as shown. Solving again only the $O(1)$ and $O(\varepsilon)$ problems gives

$$\hat{\theta} = 1 - \frac{\varepsilon}{(1+\lambda)}(1 + \lambda\omega - e^{-\omega}) + O(\varepsilon^2), \qquad (3.7\text{-}75)$$

$$\hat{\phi} = 1 - e^{-\omega} - \frac{\varepsilon}{(1+\lambda)}[(1-\lambda)(1 - e^{-\omega}) + \omega(\lambda - e^{-\omega})] + O(\varepsilon^2). \qquad (3.7\text{-}76)$$

To complete the solution through $O(\varepsilon)$ it is necessary to evaluate the constants a_0 and a_1 which appear in the outer solution, Eqs. (3.7-67) and (3.7-68). To do this, the inner and outer expansions must be matched asymptotically. Because the same two constants are involved, it is not necessary to work with both θ and ϕ; it is sufficient to match the expansions for either. Choosing ϕ, the matching requirement is

$$\lim_{\tau \to 0} \tilde{\phi}(\tau) = \lim_{\omega \to \infty} \hat{\phi}(\omega). \qquad (3.7\text{-}77)$$

The next step is to examine the limiting behavior of $\tilde{\phi}$ and $\hat{\phi}$ and to convert the resulting expressions to a common independent variable; either τ or ω could be used. Considering first the limiting behavior of $\tilde{\phi}$ and expanding the exponentials in Eq. (3.7-68) for small τ gives

$$\lim_{\tau \to 0} \tilde{\phi} = a_0\left[1 - \frac{\lambda\tau}{1+\lambda} + O(\tau^2)\right] + \varepsilon a_1\left[1 - \frac{\lambda\tau}{1+\lambda} + O(\tau^2)\right]$$

$$+ \varepsilon a_0\left[\frac{\lambda}{1+\lambda}\right]\left[1 + \frac{\tau}{1+\lambda}\right]\left[1 - \frac{\lambda\tau}{1+\lambda} + O(\tau^2)\right] + O(\varepsilon^2). \qquad (3.7\text{-}78)$$

Recalling that $\tau = \omega\varepsilon$, Eq. (3.7-78) reduces to

$$\lim_{\tau \to 0} \tilde{\phi} = a_0 + \varepsilon\left[a_1 - \frac{a_0\lambda}{1+\lambda}(\omega - 1)\right] + O(\varepsilon^2). \qquad (3.7\text{-}79)$$

Notice the regrouping of terms caused by the variable change; one of the a_0 terms is now multiplied by ε.

Neglecting the exponential in Eq. (3.7-76), the limiting behavior of $\hat{\phi}$ is

$$\lim_{\omega \to \infty} \hat{\phi} = 1 - \frac{\varepsilon}{1+\lambda}[1 + \lambda(\omega - 1)] + O(\varepsilon^2). \qquad (3.7\text{-}80)$$

Because of the form of the $O(\varepsilon)$ term in the outer expansion, Eq. (3.7-79), the 1 has not been neglected relative to $\lambda\omega$, although the term $e^{-\omega}$ was neglected. For Eq. (3.7-79) to match Eq. (3.7-80) up to $O(\varepsilon^2)$, it is necessary that

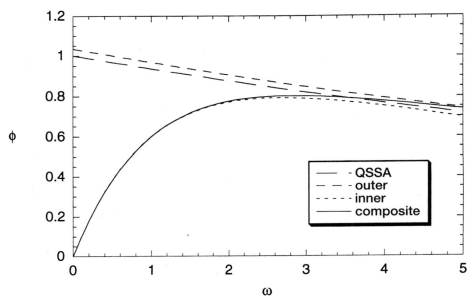

Figure 3-8. Dimensionless concentrations of species B obtained using the following: the QSSA, equivalent to Eq. (3.7-64); the outer expansion, Eq. (3.7-68); the inner expansion, Eq. (3.7-76); and the composite solution from the singular perturbation analysis, Eq. (3.7-82). The results are for $\varepsilon = 0.1$ and $\lambda = 2$.

$$a_0 = 1, \qquad a_1 = -\frac{1}{1+\lambda}. \tag{3.7-81}$$

By analogy with Eq. (3.7-36), the composite solution for ϕ is

$$\phi = \tilde{\phi} + \hat{\phi} - 1 + \frac{\varepsilon}{1+\lambda}[1 + \lambda(\omega - 1)] + O(\varepsilon^2), \tag{3.7-82}$$

where $\tilde{\phi}$ and $\hat{\phi}$ are evaluated from Eqs. (3.7-68) and (3.7-76), respectively. This completes the solution through $O(\varepsilon)$.

The various results for the dimensionless concentration of species B, expressed as $\phi(\omega)$, are compared in Fig. 3-8 for $\varepsilon = 0.1$ and $\lambda = 2$. Most accurate is the composite solution from the singular perturbation analysis, which interpolates smoothly between the inner and outer solutions. The solution obtained using the QSSA is inaccurate for short times, as mentioned earlier, but is very close to the composite result for $\omega > 4$.

In summary, this example demonstrates that the QSSA provides an approximation to the true "long-time" or "outer" solution that is accurate at $O(1)$. The $O(\varepsilon)$ correction obtained here provides an explicit indication of the error involved for small (but nonzero) ε. The singular perturbation methodology has been applied successfully to several other reaction schemes; for examples see Bowen et al. (1963) and Segel and Slemrod (1989).

Example 3.7-4 Limitations of the Fin Approximation   We return to the problem of heat transfer in a thin object, the fin depicted in Fig. 3-3. The temperature at the base of the fin ($z = 0$) is assumed now to be a function of the transverse coordinate, so that $T_0 = T_0(x)$. In Example 3.3-2 it was mentioned that when the conditions for the fin approximation are satisfied, namely small Biot number and large aspect ratio, the approximate temperature profile will be accurate for most of the fin, but not necessarily near the base. The difficulty is that a given function $T_0(x)$ may be inconsistent with the fin approximation's neglect of temperature variations in the x direction. In other words, the fin approximation will not in general satisfy the boundary condition at $z=0$. We have seen that the failure of an approximate solution to satisfy a boundary condition is typical of problems where singular perturbation analyses are useful, and that is true here. By treating the part of the fin near the base as an inner (boundary layer) region, and the rest of the fin as an outer region, the method of matched asymptotic expansions gives rigorous bounds on the accuracy of the fin approximation and provides a procedure for calculating corrections to the approximate solution obtained previously.

The dimensionless temperature and coordinates which will be used are

$$\Theta \equiv \frac{T-T_\infty}{\bar{T}_0-T_\infty}, \qquad \eta \equiv \frac{x}{W}, \qquad \zeta \equiv \frac{z}{W}, \tag{3.7-83}$$

$$\bar{T}_0 \equiv \frac{1}{W}\int_0^W T_0(x)dx. \tag{3.7-84}$$

This notation is similar to that in Example 3.3-2, except that Θ is now the *local* temperature, rather than the cross-sectional average. For simplicity, it is assumed that $T_0(x)$ is symmetric about $x=0$.

The small parameter here is the Biot number, so that $\varepsilon = \mathrm{Bi} = hW/k$. The analysis is simplified somewhat by assuming that the fin is dynamically long, corresponding to $\Lambda \gg 1$ in Example 3.3-2. Thus, the aspect ratio γ will not appear. The problem formulation for $\Theta(\eta, \zeta)$ is

$$\frac{\partial^2\Theta}{\partial\eta^2}+\frac{\partial^2\Theta}{\partial\zeta^2}=0, \tag{3.7-85}$$

$$\Theta(\eta, 0)=f(\eta)\equiv\frac{T_0-T_\infty}{\bar{T}_0-T_\infty}, \tag{3.7-86}$$

$$\Theta(\eta, \infty)=0, \tag{3.7-87}$$

$$\left.\frac{\partial\Theta}{\partial\eta}\right|_{\eta=0}=0, \tag{3.7-88}$$

$$\left.\frac{\partial\Theta}{\partial\eta}\right|_{\eta=1}=-\varepsilon\Theta(1, \zeta). \tag{3.7-89}$$

Outer solution. The z coordinate appropriate for the outer region was identified in Example 3.3-2 as

$$Z \equiv \varepsilon^{1/2}\zeta. \tag{3.7-90}$$

Denoting the outer solution as $\tilde{\Theta}(\eta, Z)$ and changing the z coordinate in Eq. (3.7-85) from ζ to Z, the governing equations for the outer problem are

$$\frac{\partial^2\tilde{\Theta}}{\partial\eta^2}+\varepsilon\frac{\partial^2\tilde{\Theta}}{\partial Z^2}=0, \tag{3.7-91}$$

$$\tilde{\Theta}(\eta, \infty)=0, \tag{3.7-92}$$

$$\left.\frac{\partial\tilde{\Theta}}{\partial\eta}\right|_{\eta=0}=0, \tag{3.7-93}$$

$$\left.\frac{\partial\tilde{\Theta}}{\partial\eta}\right|_{\eta=1}=-\varepsilon\tilde{\Theta}(1, Z). \tag{3.7-94}$$

The boundary condition at the base has been omitted, because it does not apply to the outer problem.

The absence of any fractional powers of ε in the outer problem suggests that we assume an outer expansion of the form

$$\tilde{\Theta}(\eta, Z) = \tilde{\Theta}_0(\eta, Z) + \varepsilon\tilde{\Theta}_1(\eta, Z) + \varepsilon^2\tilde{\Theta}_2(\eta, Z) + O(\varepsilon^3). \tag{3.7-95}$$

The governing equations for $\tilde{\Theta}_i$ are then

$$\frac{\partial^2\tilde{\Theta}_0}{\partial\eta^2} = 0; \qquad \frac{\partial^2\tilde{\Theta}_i}{\partial\eta^2} = -\frac{\partial^2\tilde{\Theta}_{i-1}}{\partial Z^2}, \quad i \geq 1, \tag{3.7-96}$$

$$\tilde{\Theta}_i(\eta, \infty) = 0, \quad i \geq 0, \tag{3.7-97}$$

$$\left.\frac{\partial\tilde{\Theta}_i}{\partial\eta}\right|_{\eta=0} = 0, \quad i \geq 0, \tag{3.7-98}$$

$$\left.\frac{\partial\tilde{\Theta}_0}{\partial\eta}\right|_{\eta=1} = 0; \qquad \left.\frac{\partial\tilde{\Theta}_i}{\partial\eta}\right|_{\eta=1} = -\tilde{\Theta}_{i-1}(1, Z), \quad i \geq 1. \tag{3.7-99}$$

Equations (3.7-96), (3.7-98), and (3.7-99) indicate that the solution of the $O(1)$ problem is independent of η, so that

$$\tilde{\Theta}_0 = A_0(Z). \tag{3.7-100}$$

All that is known about A_0 at this point is that, from Eq. (3.7-97), $A_0(\infty) = 0$. Solving the differential equation for the $O(\varepsilon)$ problem, we find that

$$\tilde{\Theta}_1 = -\frac{d^2A_0}{dZ^2}\frac{\eta^2}{2} + A_1(Z). \tag{3.7-101}$$

Equation (3.7-99) for $i = 1$ reveals now that

$$\frac{d^2A_0}{dZ^2} = A_0, \tag{3.7-102}$$

from which we obtain

$$A_0(Z) = C_0e^{-Z}, \tag{3.7-103}$$

where C_0 is a constant. It follows from Eqs. (3.7-97), (3.7-101), and (3.7-103) that $A_1(\infty) = 0$. Proceeding to $O(\varepsilon^2)$, Eq. (3.7-96) for $i = 2$ gives

$$\tilde{\Theta}_2 = C_0e^{-Z}\frac{\eta^4}{24} - \frac{d^2A_1}{dZ^2}\frac{\eta^2}{2} + A_2(Z). \tag{3.7-104}$$

Equation (3.7-99) for $i = 2$ provides the differential equation for A_1, which is

$$\frac{d^2A_1}{dZ^2} - A_1 = -\frac{C_0}{3}e^{-Z}. \tag{3.7-105}$$

Equation (3.7-105) has the particular solution $A_1 = (C_0/6) Ze^{-Z}$, so that

$$A_1(Z) = C_1e^{-Z} + C_0\frac{Ze^{-Z}}{6}, \tag{3.7-106}$$

where C_1 is another constant.

The outer solution is now determined up to $O(\varepsilon^2)$ as

$$\tilde{\Theta} = e^{-Z}\left[C_0 + \varepsilon\left(-C_0\frac{\eta^2}{2} + C_1 + C_0\frac{Z}{6}\right)\right] + O(\varepsilon^2).$$ (3.7-107)

The constants in Eq. (3.7-107) must be determined by matching $\tilde{\Theta}$ with the inner solution, which is denoted by $\hat{\Theta}(\eta, \zeta)$. From Eq. (3.7-90), in the limit $\varepsilon \to 0$, large values of ζ correspond to small values of Z. Accordingly, the asymptotic matching is expressed as

$$\lim_{z \to 0} \tilde{\Theta}(\eta, Z) = \lim_{\zeta \to \infty} \hat{\Theta}(\eta, \zeta), \qquad \varepsilon \to 0.$$ (3.7-108)

Evaluating the required limit of the outer solution gives

$$\lim_{Z \to 0} \tilde{\Theta} = \left[1 - Z + \frac{Z^2}{2} + O(Z^3)\right]\left[C_0 + \varepsilon\left(-C_0\frac{\eta^2}{2} + C_1 + C_0\frac{Z}{6}\right)\right] + O(\varepsilon^2)$$

$$= C_0 - \varepsilon^{1/2}C_0\zeta + \varepsilon\left(C_0\frac{\zeta^2}{2} - C_0\frac{\eta^2}{2} + C_1\right) + O(\varepsilon^{3/2}).$$ (3.7-109)

The change from Z to ζ in the second form of Eq. (3.7-109) will be needed later to complete the matching. The importance of this result right now is that it indicates that the inner expansion, for $\hat{\Theta}(\eta, \zeta)$, must contain half-integer powers of ε. Otherwise, it will be impossible to match the terms involving $\varepsilon^{1/2}$, $\varepsilon^{3/2}$, and so forth, which are implied by Eq. (3.7-109).

Inner solution and matching. The scaled variables for the inner problem are the same as those in the original problem statement, Eqs. (3.7-85)–(3.7-89). Based on Eq. (3.7-109), we assume an inner expansion of the form

$$\hat{\Theta}(\eta, \zeta) = \hat{\Theta}_0(\eta, \zeta) + \varepsilon^{1/2}\hat{\Theta}_1(\eta, \zeta) + \varepsilon\hat{\Theta}_2(\eta, \zeta) + O(\varepsilon^{3/2}).$$ (3.7-110)

Substituting Eq. (3.7-110) into the original problem statement yields the governing equations for $\hat{\Theta}_i$, which are

$$\frac{\partial^2\hat{\Theta}_i}{\partial\eta^2} + \frac{\partial^2\hat{\Theta}_i}{\partial\zeta^2} = 0, \quad i \geq 0,$$ (3.7-111)

$$\hat{\Theta}_0(\eta, 0) = f(\eta); \qquad \hat{\Theta}_i(\eta, 0) = 0, \quad i \geq 1,$$ (3.7-112)

$$\left.\frac{\partial\hat{\Theta}_i}{\partial\eta}\right|_{\eta=0} = 0, \quad i \geq 0,$$ (3.7-113)

$$\left.\frac{\partial\hat{\Theta}_0}{\partial\eta}\right|_{\eta=1} = 0; \qquad \left.\frac{\partial\hat{\Theta}_1}{\partial\eta}\right|_{\eta=1} = 0; \qquad \left.\frac{\partial\hat{\Theta}_i}{\partial\eta}\right|_{\eta=1} = -\hat{\Theta}_{i-2}(1, \zeta), \quad i \geq 2.$$ (3.7-114)

The solution to the $O(1)$ inner problem is obtained using the methods of Chapter 4. The result is

$$\hat{\Theta}_0(\eta, \zeta) = 1 + b_0\zeta + \sqrt{2}\sum_{n=1}^{\infty}[(f_n - b_n)e^{-n\pi\zeta} + b_n e^{n\pi\zeta}]\cos n\pi\eta,$$ (3.7-115)

$$f_n = \sqrt{2}\int_0^1 f(\eta)\cos n\pi\eta \, d\eta,$$ (3.7-116)

where the b_n are constants which remain to be determined. The $O(\varepsilon^{1/2})$ inner problem has the simple solution

$$\hat{\Theta}_1 = a\zeta,$$ (3.7-117)

where a is another constant to be found from matching. To obtain the correct behavior for $\zeta \to \infty$, it is evident that the $e^{n\pi\zeta}$ terms in Eq.(3.7-115) cannot be present; compare with Eq. (3.7-109). Accordingly, $b_n = 0$ for all $n > 0$, and

$$\lim_{\zeta \to \infty} \hat{\Theta}(\eta, \zeta) = 1 + b_0 \zeta + \varepsilon^{1/2} a \zeta + O(\varepsilon). \tag{3.7-118}$$

A comparison of Eqs. (3.7-118) and (3.7-109) indicates now that $C_0 = 1$, $b_0 = 0$, and $a = -1$. An additional term in the inner expansion would be needed to evaluate the remaining constant in the outer solution, C_1.

With the constants evaluated as just described, the inner and outer solutions are

$$\tilde{\Theta}(\eta, Z) = e^{-Z}\left[1 + \varepsilon\left(-\frac{\eta^2}{2} + C_1 + \frac{Z}{6}\right)\right] + O(\varepsilon^2), \tag{3.7-119}$$

$$\hat{\Theta}(\eta, \zeta) = 1 + \sqrt{2} \sum_{n=1}^{\infty} f_n e^{-n\pi\zeta} \cos n\pi\eta - \varepsilon^{1/2}\zeta + O(\varepsilon). \tag{3.7-120}$$

As expected, the leading term in the outer solution (e^{-Z}) is identical to the result from the fin approximation [see Eq. (3.3-33)]. Moreover, Eq. (3.7-119) shows that the error in the fin approximation is $O(\varepsilon)$ in the outer region, as suggested by Eq. (3.3-15). The magnitude of the error in the inner region depends on the temperature profile specified at the base. If T_0 is constant, corresponding to $f(\eta) = 1$, it is readily confirmed from Eq.(3.7-116) that $f_n = 0$ for all $n > 0$. In that case, the infinite series in Eq. (3.7-120) vanishes and the fin approximation is as accurate in the inner region as it is in the outer region; that is, the inner temperature equals e^{-Z} up to $O(\varepsilon)$. However, if T_0 is not constant, the infinite series does not vanish and the errors in the fin approximation in the inner region are $O(1)$.

The scaling used for the inner region assumes that it has a dimensionless thickness of ~ 1 in terms of ζ, or a dimensional thickness of $\sim W$. An estimate of the actual thickness is obtained by examining the behavior of the infinite series in Eq. (3.7-120), because that is the largest term which appears only in the inner solution. The most slowly decaying term in the sum, $e^{-\pi\zeta}$, is already quite small for $\zeta = 1$. This is consistent with the scaling assumption. It also illustrates the rule of thumb for modeling given in Section 3.3, which is that edge effects in conduction and diffusion are important over only about one thickness of the region being modeled. That is, deviations from the fin approximation near the base are a type of edge effect.

In summary, singular perturbation problems differ from regular ones in that there is no first approximation which is uniformly valid throughout the domain of interest. Singular problems are created typically when ε multiplies the highest-order derivative in a differential equation. Therefore, setting $\varepsilon = 0$ in these problems reduces the order of the equation and makes it impossible to satisfy all of the boundary conditions. The usual consequence is that two or more regions must be considered separately, with a different scaling of the governing equations and different dominant terms in each region. The solutions for the various regions are joined by asymptotic matching, as illustrated by the examples.

3.8 INTEGRAL APPROXIMATION METHOD

As already noted, transport problems frequently possess no exact, closed-form solution, even after the governing equations have been simplified as much as possible. In such cases the perturbation methods discussed in Sections 3.6 and 3.7 may be useful, provided that the problem contains a small parameter and provided that a suitable analytical solution exists as a leading approximation. If these conditions are not met, it is still possible to find approximate solutions using the *integral approximation method.* This approach is based on a very different philosophy, in that one does not seek to satisfy the governing differential equation (e.g., the energy conservation equation) locally at every point in the region of interest. Rather, the differential equation is integrated over a control volume, and it is only the resulting integral equation which is satisfied, together with certain boundary conditions.

The integration is accomplished by *assuming* a simple functional form for the solution (e.g., a form for the temperature profile). The assumed form contains unknown constants or functions which are determined by the integral equation, the boundary conditions, and sometimes other conditions which are imposed. The simplification results from the fact that an ordinary differential equation is reduced by this procedure to a set of algebraic equations, and a partial differential equation is reduced either to algebraic equations or to ordinary differential equations, depending on the form of the assumed solution. The final solution satisfies the conservation equation in a global or averaged sense, but its local behavior is only approximate. For example, energy is conserved for the control volume as a whole, but not at every point in the region being modeled.

The application of the integral approximation method to boundary layer flows was introduced by T. von Kármán in the 1920s. As low-cost computing became widely available, beginning in the 1960s, its use in transport research began to decline. It is included here not for its importance in current research, but for its educational value. Reasonable solutions for difficult problems can be obtained with an expenditure of effort appropriate for a homework assignment. Equally important, the method reinforces scaling concepts by requiring that one think carefully about the form of the temperature, concentration, or velocity field before attempting a solution.

The integral approach will be illustrated by two examples involving the diffusion of a solute into a solid sheet or liquid film, where it reacts. The dimensionless problem statement common to both examples is presented first. It is assumed that the reaction is homogeneous and irreversible, with nth-order kinetics. Solute enters the film only at $x=0$, the surface at $x=L$ being impermeable. The dimensionless variables and Damköhler number are defined as

$$\Theta(\eta, \tau) \equiv \frac{C}{C_0}, \qquad \eta \equiv \frac{x}{L}, \qquad \tau \equiv \frac{tD}{L^2}, \tag{3.8-1}$$

$$\mathrm{Da} \equiv \frac{k_{vn} C_0^{n-1} L^2}{D}, \tag{3.8-2}$$

where C_0 is the concentraton at the exposed surface and k_{vn} is the rate constant. The full problem statement is then

$$\frac{\partial \Theta}{\partial \tau} = \frac{\partial^2 \Theta}{\partial \eta^2} - \mathrm{Da}\, \Theta^n, \tag{3.8-3}$$

$$\Theta(0, \tau) = 1, \tag{3.8-4}$$

$$\frac{\partial \Theta}{\partial \eta}(1, \tau) = 0, \tag{3.8-5}$$

$$\Theta(\eta, 0) = 0. \tag{3.8-6}$$

This problem can be solved exactly for $n=0$ or $n=1$ by the techniques presented in Chapter 4, but not for other values of n. This is true not only for the transient problem shown, but also for the corresponding steady-state problem.

Example 3.8-1 Approximate Steady-State Solution   We begin with the steady-state version of the reaction-diffusion problem described above. Integrating the time-independent form of Eq. (3.8-3) over the thickness of the film yields the desired integral equation, which is

$$\left.\frac{d\Theta}{d\eta}\right|_{\eta=0} + Da \int_0^1 \Theta^n \, d\eta = 0. \tag{3.8-7}$$

Equation (3.8-5) has been used, but no approximations have yet been made. The approximation comes in picking a concentration profile which will allow us to estimate the derivative and integral in Eq. (3.8-7). A reasonable choice is

$$\Theta(\eta) = e^{-\eta/\delta}, \tag{3.8-8}$$

where δ is a constant. With $\delta > 0$, Eq. (3.8-8) gives the necessary decrease in Θ with increasing distance into the film, and it is consistent with the boundary condition in Eq. (3.8-4). Thus, Eq. (3.8-8) is *qualitatively* correct, although in general it will not satisfy Eq. (3.8-5). Of course, Eq. (3.8-8) is not unique, in that many other functions (e.g., various polynomials) exhibit similar qualitative behavior. Substituting Eq. (3.8-8) into Eq. (3.8-7), and performing the differentiation and integration, yields $\delta = Da^{-1}$ for $n=0$. For $n \neq 0$, δ must satisfy

$$\delta^2 = \frac{n}{Da}(1 - e^{-n/\delta})^{-1} \qquad (n \neq 0). \tag{3.8-9}$$

Thus, the original differential equation has been reduced to an algebraic equation for δ. For specified values of Da and n, an iterative solution of Eq. (3.8-9) will complete the approximate solution for $\Theta(\eta)$. Notice from Eq. (3.8-8) that δ^{-1} is equal to the dimensionless solute flux entering the film at $\eta=0$. That is, for $\eta=0$, $-d\Theta/d\eta = \delta^{-1}$. The values of δ^{-1} are compared later with exact results for the flux, for special cases.

   Explicit expressions for δ are obtained from Eq. (3.8-9) in the limits of very slow or fast reactions, Da $\rightarrow 0$ and Da $\rightarrow \infty$, respectively. These results are

$$\delta \rightarrow \frac{1}{Da} \qquad (Da \rightarrow 0, \, n \neq 0), \tag{3.8-10}$$

$$\delta \rightarrow \left(\frac{n}{Da}\right)^{1/2} \qquad (Da \rightarrow \infty, \, n \neq 0). \tag{3.8-11}$$

There are no slow or fast reaction limits for $n=0$, because (as already mentioned) $\delta = Da^{-1}$ in that case for all Da.

   The accuracy of the integral solution is judged by comparisons with the exact analytical results obtainable for $n=0$ and $n=1$. For $n=0$, the exact concentration profile and flux are

$$\Theta = 1 - Da \; \eta\left(1 - \frac{\eta}{2}\right), \tag{3.8-12}$$

$$-\left.\frac{d\Theta}{d\eta}\right|_{\eta=0} = Da. \tag{3.8-13}$$

TABLE 3-1
**Comparison of Exact and Integral-Approximate Fluxes for a
Steady, First-Order Reaction in a Film**[a]

Da	Exact flux	Approximate flux	% Error
0.1	0.09679	0.09538	-1.5
1	0.7616	0.7146	-6.2
10	3.151	3.089	-2.0
100	10.00	10.00	0

[a] The dimensionless fluxes shown are the values of $(-\partial\Theta/\partial\eta)$ at $\eta=0$.

Thus, the approximate flux for $n=0$ (i.e., $\delta^{-1}=\mathrm{Da}$) turns out to be exact, although the concentration profile, $\Theta = e^{-\mathrm{Da}\eta}$, is not! For $n=1$ the exact results are

$$\Theta = \cosh\sqrt{\mathrm{Da}}\,\eta - \tanh\sqrt{\mathrm{Da}}\,\sinh\sqrt{\mathrm{Da}}\,\eta, \qquad (3.8\text{-}14)$$

$$-\frac{d\Theta}{d\eta}\bigg|_{\eta=0} = \sqrt{\mathrm{Da}}\,\tanh\sqrt{\mathrm{Da}}. \qquad (3.8\text{-}15)$$

A comparison of the exact and approximate fluxes for $n=1$ is given in Table 3-1. The maximum error of about 6% is at $\mathrm{Da}=1$, whereas the asymptotic results for $\mathrm{Da}\to0$ and $\mathrm{Da}\to\infty$, Eqs. (3.8-10) and (3.8-11), turn out to be exact.

A result for $\mathrm{Da}\to0$ and any n can be obtained using a regular perturbation scheme. Although a singular perturbation approach applies, in principle, to the opposite extreme of $\mathrm{Da}\to\infty$, the governing equation for the $O(1)$ boundary layer problem has the same form as the original, intractable differential equation. Thus, no progress can be made in that case, other than inferring (from scaling arguments) that the flux must be proportional to $\mathrm{Da}^{1/2}$.

Example 3.8-2 Approximate Transient Solution We return to the transient reaction-diffusion problem defined by Eqs. (3.8-3)–(3.8-6). The integral equation to be solved is now

$$\int_0^1 \frac{\partial\Theta}{\partial\tau}\,d\eta + \frac{\partial\Theta}{\partial\eta}\bigg|_{\eta=0} + \mathrm{Da}\int_0^1 \Theta^n\,d\eta = 0. \qquad (3.8\text{-}16)$$

The key step, as before, is the selection of a suitable functional form for the concentration. For consistency with Eq. (3.8-8), the approximate transient profile is chosen as

$$\Theta(\eta,\,\tau) = e^{-\eta/\delta(\tau)}. \qquad (3.8\text{-}17)$$

The only difference from Eq. (3.8-8) is that the dimensionless penetration depth, δ, is now a function of time. The penetration depth at steady state, $\delta\,(\infty)$, is what was determined in Example 3.8-1. By requiring that $\delta(0)=0$, the assumed profile is made to satisfy Eq. (3.8-6).

Substituting Eq. (3.8-17) into Eq. (3.8-16) leads to

$$\frac{d\delta}{d\tau}\left[1 - e^{-1/\delta}\left(1 + \frac{1}{\delta}\right)\right] - \frac{1}{\delta} + \frac{\mathrm{Da}}{n}\delta(1 - e^{-n/\delta}) = 0, \qquad \delta(0)=0. \qquad (3.8\text{-}18)$$

Thus, the partial differential equation for $\Theta(\eta,\,\tau)$ has been reduced to an ordinary differential equation for $\delta(\tau)$. Although it cannot be solved analytically, Eq. (3.8-18) can be integrated numerically using standard software.

Further analytical progress can be made for fast reactions, or $\mathrm{Da}\gg1$. In this case, $\delta\ll1$;

the upper bound for δ is given by Eq. (3.8-11). It is permissible then to neglect all terms in Eq. (3.8-18) which are multiplied by $e^{-1/\delta}$, resulting in

$$\frac{d\delta}{d\tau} - \frac{1}{\delta} + \frac{Da}{n}\delta = 0, \qquad \delta(0) = 0. \tag{3.8-19}$$

This nonlinear differential equation is solved by first multiplying by δ and then recognizing that $\delta \, (d\delta/d\tau) = (1/2) \, d(\delta^2)/d\tau$. The result is a first-order, linear equation for δ^2. The solution is

$$\delta(\tau) = \left[\frac{n}{Da}(1 - e^{-2Da\,\tau/n})\right]^{1/2} \qquad (Da \gg 1). \tag{3.8-20}$$

At steady state, we recover the result given in Eq. (3.8-11).

The time required to reach a steady state (t_s) is estimated from Eq. (3.8-20) by seeing when the exponential term becomes small. Considering $e^{-3} = 0.05$ to be small enough, we set $2 \, Da \, \tau_s/n = 3$, or

$$t_s = \frac{3}{2} \frac{n}{k_{vn}C_0^{n-1}}. \tag{3.8-21}$$

Alternatively, an order-of-magnitude estimate for t_s could be based on the characteristic time for diffusion over the maximum (steady state) penetration distance. In this problem the dimensionless penetration distance is $\delta(\tau)$ and the dimensional distance is $\delta(\tau)L$. Accordingly, from Eq. (3.4-13), another estimate for t_s is

$$t_s \sim \frac{(\delta(\infty)L)^2}{D} = \frac{n}{k_{vn}C_0^{n-1}}. \tag{3.8-22}$$

The two estimates of t_s are obviously similar, emphasizing once again that the duration of transients can be estimated without knowing the transient solution.

The integral method is impressive in that it yields approximate solutions with relatively little mathematical effort, even for problems which are intractable using other analytical methods. However, this approach has two very important limitations. One is that it is very difficult to predict the accuracy of the approximation, which may be very sensitive to the form of the trial function used—that is, the assumed form of the concentration field (or other field variable). The second, closely related limitation is that there is no systematic way to improve the accuracy of the approximate solution. Perturbation methods, by contrast, provide (a) an orderly way of keeping track of the size of neglected terms and (b) a systematic procedure for increasing accuracy (adding another term to the series expansion). However, those methods are limited to problems which have a large or small parameter!

Many examples of the integral method applied to heat conduction problems are given in Arpaci (1966). Conceptually related to the integral technique is a class of methods based on weighted residual equations; the underlying principles and numerous examples are provided by Finlayson (1980). Among these methods are collocation and the Galerkin method of weighted residuals; a discussion of these techniques is beyond the scope of this book. In general, for a similar form of trial function (e.g., a polynomial of a given order), weighted residual methods are more accurate than the integral method. Inherent also in weighted residual methods is a procedure for improving accuracy by adding more functions to the approximate solutions. Weighted residuals form the basis for certain widely used numerical methods for solving differential equations, including orthogonal collocation and finite elements.

References

Abramowitz, M. and I. A. Stegun. *Handbook of Mathematical Functions.* U.S. Department of Commerce, National Bureau of Standards, Washington, DC, 1970.

Aris, R. *The Mathematical Theory of Diffusion and Reaction in Permeable Catalysts,* Vol. 1. Clarendon Press, Oxford, 1975.

Arnold, J. H. Studies in diffusion: III. Unsteady-state vaporization and absorption. *Trans. A.I.Ch.E.* 40: 361–378, 1944.

Arpaci, V. S. *Conduction Heat Transfer.* Addison-Wesley, Reading, MA, 1966.

Barenblatt, G. I. *Similarity, Self-Similarity, and Intermediate Asymptotics.* Consultants Bureau, New York, 1979.

Bender, C. M. and S. A. Orszag. *Advanced Mathematical Methods for Scientists and Engineers.* McGraw-Hill, New York, 1978.

Bowen, J. R., A. Acrivos, and A. K. Oppenheim. Singular perturbation refinement to quasi-steady state approximation in chemical kinetics. *Chem. Eng. Sci.* 18: 177–188, 1963.

Casey, H. C., Jr. and G. L. Pearson. Diffusion in semiconductors. In *Point Defects in Solids,* Vol. 2, J. H. Crawford, Jr. and L. M. Slifkin, Eds. Plenum Press, New York, 1975, pp. 163–255.

Dresner, L. *Similarity Solutions of Nonlinear Partial Differential Equations.* Pitman, Boston, 1983.

Finlayson, B. A. *Nonlinear Analysis in Chemical Engineering.* McGraw-Hill, New York, 1980.

Hansen, A. G. *Similarity Analyses of Boundary Value Problems in Engineering.* Prentice-Hall, Englewood Cliffs, NJ, 1964.

Kamke, E. *Differentialgleichungen,* Vol. 1. Akademische Verlagsgesellschaft Becker & Erler Kom.-Ges., Leipzig, Germany, 1943, p. 475.

Kevorkian, J. and J. D. Cole. *Perturbation Methods in Applied Mathematics.* Springer-Verlag, New York, 1981.

Lifshitz, I. M. and V. V. Slyozov. The kinetics of precipitation from supersaturated solid solutions. *J. Phys. Chem. Solids* 19: 35–50, 1961.

Lin, C. C. and L. A. Segel. *Mathematics Applied to Deterministic Problems in the Natural Sciences.* Macmillan, New York, 1974.

Nayfeh, A. H. *Perturbation Methods.* Wiley-Interscience, New York, 1973.

Segel, L. A. Simplification and scaling. *SIAM Rev.* 14: 547–571, 1972.

Segel, L. A. and M. Slemrod. The quasi-steady-state assumption: A case study in perturbation. *SIAM Rev.* 31: 446–477, 1989.

Van Dyke, M. *Perturbation Methods in Fluid Mechanics.* Academic Press, New York, 1964.

Problems

3-1. Absorption from a Bubble with Reaction

Consider a spherical bubble of radius $R(t)$ containing pure gas A. The gas diffuses into the surrounding liquid, where it is consumed by a first-order, irreversible, homogeneous reaction (rate constant k). The liquid concentration of A at the gas–liquid interface is C_0. Assume that $C_A = 0$ far from the bubble and that the liquid is stagnant.

(a) If $R(t)$ varies slowly enough and if enough time elapses following introduction of the bubble, a pseudosteady analysis is justified for the liquid. Derive expressions for the liquid concentration profile, $C_A(r, t)$, and the flux of A at the gas–liquid interface, N_{Ar} (R, t), under these conditions. Determine the ratio of the flux with reaction to that without reaction, as a function of the Damköhler number. (*Hint:* The transformation $C_A(r, t) = \psi(r, t)/r$ will be helpful.)

(b) In part (a), convective transport caused by bubble shrinkage was neglected. Identify the conditions under which this is a good approximation. Specifically, what must be the relationship between C_0 and the concentration of A in the gas bubble (C_b) ? (*Hint:* Use an interfacial balance to relate C_b, C_0, dR/dt, and the flux of A in the liquid.)

(c) Use order-of-magnitude reasoning to identify the range of times for which a pseudo-steady analysis is justified. Assume that the conditions of part (b) are satisfied (i.e., convection is negligible).

3-2. Heat Transfer to Solder

Solder is made from various alloys designed to melt upon contact with a very hot surface (i.e., a metal piece which is to be joined to another). As shown in Fig. P3-2, assume that a solder wire of radius R is melting at a constant rate, yielding a constant wire velocity v_w. The melting temperature of the solder is T_m, the ambient temperature is T_∞, and the latent heat of fusion is $\hat{\lambda}$. There is a constant heat transfer coefficient h from the wire to the ambient air. The wire is long enough that it eventually reaches the ambient temperature far from the surface.

(a) Under what conditions is a one-dimensional analysis valid for the temperature profile in the wire (i.e., $T \cong T(z)$ only)?

(b) Determine $T(z)$ for the conditions of part (a).

(c) Determine the required rate of heat conduction from the surface to the wire, as a function of v_w and other parameters.

3-3. Concentration-Dependent Diffusivity

Suppose that it is desired to determine the extent to which the diffusivity (D_A) of As in solid Si depends on the concentration (C_A) of As. An experiment is performed in which one surface of a sample of pure Si is contacted with As over a period $0 \le t \le t_0$. At the end of the experiment the sample is sectioned and analyzed to determine $C_A(x, t_0)$, where x is distance from the exposed surface. The concentration profile is found to be like that shown in Fig. P3-3.

(a) Show that the conservation equation governing $C_A(x, t)$ with $D_A = D_A(C_A)$ can be transformed to an ordinary differential equation by introducing a similarity variable.

(b) Describe how the result of (a) can be used to determine $D_A(C_A)$ for $0 \le C_A \le C_0$ from a single concentration profile like that in Fig. P3-3. (*Hint:* Assume that the slope and the area under the curve can be calculated accurately. The diffusivity can be evaluated in

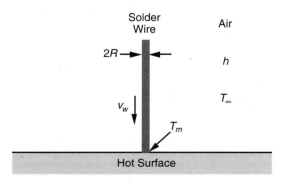

Figure P3-2. Heat transfer to solder.

Figure P3-3. Qualitative behavior of the concentration of As in solid Si, at some time t_0 after the surface at $x=0$ is contacted with As.

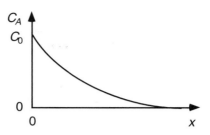

terms of those quantities.)
This problem is based on an analysis given in Casey and Pearson (1975).

3-4. Similarity Solution with a Constant-Flux Boundary Condition

Consider a semi-infinite slab initially at temperature T_∞. A constant heat flux q_0 is imposed at $x=0$ beginning at $t=0$. Rewrite the energy equation in terms of the local heat flux $q_x(x, t)$ rather than $T(x, t)$, and show that q_x can be evaluated by the method of similarity. Use this approach to obtain an expression for $T(x, t)$.

3-5. Transient Diffusion in a Permeable Tube with Open Ends

The open-ended cylindrical tube of radius R and length $2L$ shown in Fig. P3-5 is immersed initially in a fluid with solute concentration C_0. At $t=0$ the concentration in the surrounding fluid is changed rapidly to a new constant value, $C_1<C_0$. The tube wall is permeable to the solute, the rate of transmural diffusion being governed by the solute permeability, k_s (same units as a mass transfer coefficient). After the reduction in the outside concentration, solute leaves the tube both by diffusion through the open ends and by diffusion across the tube wall. Assume that $L/R\gg1$.

(a) Identify the conditions under which radial variations in the internal concentration will be negligible. Under these conditions $C(r, z, t)\cong\overline{C}(z, t)$, where $\overline{C}(z, t)$ is the cross-sectional average concentration. Are different sets of conditions required for $t\ll R^2/D$ and $t\gg R^2/D$?

(b) Show that when the conditions of part (a) are satisfied, we obtain

$$\frac{\partial\overline{C}}{\partial t}=D\frac{\partial^2\overline{C}}{\partial z^2}-\frac{2k_s}{R}(\overline{C}-C_1),$$

$$\overline{C}(z, 0)=C_0,\ \overline{C}(\pm L, t)=C_1.$$

(c) Sketch qualitatively the concentration profiles for various values of t. Under what conditions will the solute concentration in much of the tube be independent of z? Show that under these conditions the dimensionless concentration in the central part of the tube is given by

Figure P3-5. Transient diffusion in a permeable tube with open ends.

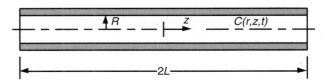

(i) very small K

(2) moderate K

(3) larger

$$\psi \equiv \frac{\overline{C}-C_1}{C_0-C_1} \cong e^{-2k_s t/R}.$$

(d) When the conditions of part (c) are satisfied, a similarity solution can be obtained for the entire tube. To do this, define a new axial coordinate x such that $x=0$ at one end of the tube, and use a new dimensionless concentration

$$\Theta(x, t) = e^{2k_s t/R}\psi(x, t).$$

Complete the solution. Why does similarity work for the variable Θ but not for ψ? Why is it necessary to use x rather than z?

3-6. Scaling and Perturbation Analyses for a Bimolecular Reaction

Consider an irreversible, bimolecular reaction in a liquid film at steady state, as shown in Fig. P2-12. The objective here is to extend the analysis of Problem 2-12, which was limited to "infinitely fast" reaction kinetics.

$D_A \dfrac{d^2 C_A}{dx^2} = -R_{VA}$

$D_B \dfrac{d^2 C_B}{dt^2} = -R_{VB}$

$= -R_{VA}$

$\dfrac{d^2(C_A - C_B)}{dx^2} = 0$ ⟹ $C_A - C_B = ax + b$.

$C_A(0) - C_B(0) = b = C_{A0}$

$C_A(L) - C_B(L) = aL + b = -C_{A0}$

$C_B(x) = C_A(x) - C_{A0}\left(1 - \dfrac{2x}{L}\right)$

$D_A \dfrac{d^2 C_A}{dx^2} = k \cdot C_A \cdot C_B$

$= k \cdot C_A \left[C_A - C_{A0}\left(1 - \dfrac{2x}{L}\right)\right]$

(a) Sketch qualitatively $C_A(x)$, $C_B(x)$, and $C_C(x)$ for a very small value of the reaction rate constant k, a moderate value of k, and a large (but finite) value of k.

(b) For the special case of equal reactant diffusivities ($D_A = D_B$) and equal boundary concentrations of A and B ($C_{A0} = C_{BL}$), show that the concentration of A is governed by

$$\frac{d^2\theta}{d\eta^2} = \text{Da } \theta(\theta + 2\eta - 1),$$

$$\theta(0) = 1, \qquad \theta(1) = 0,$$

where $\theta = C_A/C_{A0}$, $\eta = x/L$, and $\text{Da} = kC_{A0}L^2/D_A$. [*Hint:* First combine the conservation equations for A and B and solve for $C_B(x)$ in terms of $C_A(x)$.]

$\dfrac{d^2\theta_0}{d\eta^2} + \epsilon \dfrac{d^2\theta_1}{d\eta^2} + O(\epsilon^2)$

$= \epsilon(\theta_0 + \epsilon\theta_1 + O(\epsilon^2))[\theta_0 + \epsilon\theta_1 + O(\epsilon^2)$

$+ 2\eta - 1]$

$\theta_0(0) + \epsilon\theta_1(0) + O(\epsilon^2) = 1$

$\theta_0(1) + \epsilon\theta_1(1) + O(\epsilon^2) = 0$

(c) For slow reaction kinetics ($\text{Da} \ll 1$), a regular perturbation scheme can be used. Determine $\theta(\eta)$ for slow kinetics, including terms of $O(\text{Da})$.

(d) For fast (but not infinitely fast) reaction kinetics the reaction will be confined largely to a zone of thickness δ at the center of the liquid film. In this region, $C_A \sim C_B \sim C^*$. Use order-of-magnitude considerations to determine δ and C^*. Specifically, determine the dependence of δ/L and C^*/C_{A0} on Da, for $\text{Da} \gg 1$. These results show when the approximation used for infinitely fast kinetics ($\delta = 0$) is reasonable.

3-7. Diffusion with a Slow Second-Order Reaction

Consider a second-order, irreversible reaction occurring at steady state in a liquid film of thickness L. The concentration of the reactant (species A) at $x=0$ is C_0, and there is no transport across the surface at $x=L$. The volumetric reaction rate is given by

$$R_{VA} = -kC_A^2.$$

Assuming that the reaction is slow relative to diffusion, use a regular perturbation scheme to determine $C_A(x)$ and the flux of A at $x=0$. The small parameter is the Damköhler number, $\text{Da} = kC_0L^2/D_A$. Obtain three terms in the expansion for $\Theta = C_A/C_0$. That is, include terms which are $O(\text{Da}^2)$.

3-8. Perturbation Analysis of Diffusion and Reaction with Saturable Kinetics

Assume that the reaction $A \rightarrow B$ follows the rate expression

$$R_{VA} = -\frac{k_0 C_A}{K_m + C_A},$$

$O(1)$

$\dfrac{d^2\theta_0}{d\eta^2} = 0$, $\theta_0(0) = 1$, $\theta_0(1) = 0$

$\theta_0 = 1 - \eta$

$O(\epsilon)\dfrac{d^2\theta_1}{d\eta^2} = \theta_0(\theta_0 + 2\eta - 1) = \eta - \eta^2$ ⟹

$\theta_0(0) = 0$, $\theta_1(0) \approx 0$

$\theta = 1 - \eta - \dfrac{\epsilon}{12}[\eta - 2\eta^3 + \eta^4] + O(\epsilon^2)$

where k_0 and K_m are constants. Enzymatically catalyzed reactions often have rate laws of this form. The reaction is carried out in a long, cylindrical region of radius R under steady-state conditions, with $C_A = C_0$ at the surface $(r = R)$.

Assuming that the reaction is slow relative to diffusion, use a regular perturbation scheme to determine $C_A(r)$ and the flux of A at the surface. The small parameter is the Damköhler number, $Da = k_0 R^2 / D_A C_0$. The second parameter which arises is $\gamma = K_m / C_0$; assume that γ is neither large nor small compared to unity. Obtain three terms in the expansion for $\Theta = C_A / C_0$. That is, include terms which are $O(Da^2)$.

3-9. Ostwald Ripening[*]

In a supersaturated solid solution, precipitation may lead to the formation of grains or particles of a new phase. It is found that after an initial stage of this process that involves nucleation, there is a second stage in which the particles grow to much larger sizes. In the second stage the degree of supersaturation is slight, and the rate of particle growth is controlled by diffusion. This coarsening of particles, known as *Ostwald ripening,* has important effects on the physical properties of various alloys. An extensive analysis of the kinetics of the second stage of this process was given by Lifshitz and Slyozov (1961). The objective here is to derive one of their simpler results.

Consider a solid solution consisting of components A and B, with B the abundant species. The particles being formed by precipitation are composed of pure B. Assume that at any given time the precipitate consists of widely dispersed spherical particles of radius $R(t)$, which do not interact with one another. Accordingly, it is sufficient to consider one representative particle for which $C_A = C_A(r, t)$ in the surrounding solution. Far from the particle the concentration of A is constant at $C_{A\infty}$. A key aspect of this problem is that the concentration of A at the particle surface is influenced by the interfacial energy, and therefore it depends on particle size. The relationship is

$$C_A [R(t),\ t] = C_{A\infty} + \frac{\Gamma}{R(t)},$$

where the constant Γ is proportional to the interfacial energy. It is the interfacial energy term which provides the driving force for diffusion.

(a) Derive an expression for the growth rate of the particle, dR/dt, valid when growth is slow enough that diffusion in the solution is pseudosteady. (Assume that the densities of the particle and solution are the same, and show that as a consequence there is no convection.)

(b) Show that eventually $R(t) \propto t^{1/3}$, which is the result of Lifshitz and Slyozov.

(c) Use order-of-magnitude reasoning to determine when the pseudosteady assumption is valid.

[*]This problem was suggested by R. A. Brown.

3-10. Reaction and Diffusion with a Nonuniform Catalyst Distribution[*]

A certain catalyst is prepared by immersing a thin sheet of inert porous material in a solution, from which the catalytic agent is deposited in the porous medium. After deposition the catalyst concentration $C_C(x)$ within the porous slab is found to be nonuniform. Assume that the catalyst concentration is given by

$$C_C(x) = C_{C0}[1 - (x/L)^2],$$

where L is the slab thickness and C_{C0} is a constant. The slab is to be used in a laboratory experiment involving the irreversible reaction $A \rightarrow B$, which follows the rate expression

$$R_{VA} = -kC_C C_A.$$

Entry of A will occur only at one surface, where the concentration of A will be maintained at C_{A0}; the opposite surface of the slab is to be placed next to an impermeable barrier. The conditions are such that Da will be very large.

(a) It is intended that the side with depleted catalyst $(x=L)$ be placed next to the impermeable barrier. For this arrangement, use a singular perturbation analysis to determine the
steady-state flux of A entering the slab. Evaluate the first two terms in the expansion for
the flux.

(b) Suppose that the slab is accidentally reversed, so that the side with high catalyst concentration is placed next to the barrier. Show that the first term in a perturbation expansion
for C_A is governed by the *Airy equation*, which is of the form

$$\frac{d^2W}{dz^2} = zW.$$

The solutions to this equation, called Airy functions, are given in Abramowitz and Stegun (1970,
p. 446).

*This problem was suggested by R. A. Brown.

3-11. Heat Transfer in a Packed Bed Reactor

A catalytic reaction is to be carried out in a long, packed-bed reactor of radius R. A component supplied by the flowing gas reacts exothermically at the surface of the catalyst pellets, releas-
ing heat at a volumetric rate H_V. The catalyst surface area per unit volume is a, and heat transfer
from the catalyst pellets to the gas is characterized by a heat transfer coefficient, h. For the
purposes of this problem, the bulk gas temperature T_G will be assumed to be constant. The reactor
wall is maintained at a constant temperature, T_W. The packing has an effective thermal conductivity k_e.

(a) Show that if axial conduction is neglected, a steady-state energy balance for the packing
leads to

$$\frac{k_e}{r}\frac{d}{dr}\left(r\frac{dT}{dr}\right) - ha(T-T_G) + H_V = 0,$$

$$T(R) = T_W,$$

where $T(r)$ is the catalyst temperature. (Temperature variations within a catalyst pellet
have been neglected; the coordinate r refers to radial position in the reactor.)

(b) Convert the energy equation of part (a) to dimensionless form, and identify the conditions under which there will be a thermal boundary layer. Where are the large temperature gradients located? What are the physical reasons for the boundary layer?

(c) For the boundary layer conditions of part (b), use the singular perturbation approach to
obtain the first two terms of a series expansion for the temperature field.

3-12. Heat Transfer in Unidirectional Solidification*

One method for establishing the steep temperature gradients needed for directional solidification is to interface two heat pipes (constant temperature pieces) with temperatures T_1 and T_2
which bracket the melting temperature T_m of the material. This arrangement is shown for a cylindrical geometry in Fig. P3-12. Melt and solid are contained within a long ampoule of radius R.
Assume that the thickness of the ampoule wall is negligible and that the ampoule wall, melt, and
solid all have the same thermal conductivity k. The heat transfer coefficient between the ampoule
and furnace wall is h.

Figure P3-12. Heat transfer in unidirectional solidification.

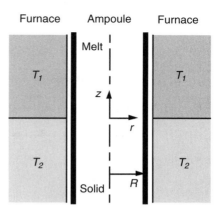

(a) For order-of-magnitude estimates, representative values are $R=1$ cm, $k=0.5$ W K^{-1} cm^{-1}, $h=0.05$ W K^{-1} cm^{-2}, $T_1=600$ K, $T_m=500$ K, and $T_2=400$ K. Show that radiative heat transfer from the furnace to the ampoule is negligible compared to convective heat transfer. Also show that radial temperature gradients within the melt and solid are negligible.

(b) For the one-dimensional conditions suggested by part (a), determine the axial temperature profile $\bar{T}(z)$ in the melt and solid, assuming that the ampoule and its contents are stationary. [Derive a general expression, not using the specific numerical values of part (a).] Find the location of the interface, which is not necessarily at $z=0$.

(c) Repeat part (b), assuming that the ampoule and its contents are moving at a constant velocity $v_z=-U$. Assume that the melt and solid have the same density ρ. The latent heat of fusion is $\hat{\lambda}$.

*This problem was suggested by R. A. Brown.

3-13. Time Response of a Thermocouple*

A thermocouple is to be used to measure temperature fluctuations in an air stream. The variations in air temperature (T_A) are expected to be of the form

$$T_A(t)=T_0+T_1 \sin 2\pi\omega t,$$

where T_0 and T_1 are constants and ω is the frequency of the temperature oscillations. The thermocouple bead can be modeled as a sphere of radius R. Denoting the thermal conductivity of the air as k_A, the heat transfer coefficient between the sphere and the air is given by $h=k_A/R$.

(a) The thermocouple diameter is 0.010 in. Other representative data are as follows:

Parameter	Air	Thermocouple
$\rho(\text{lb}_m\ \text{ft}^{-3})$	0.07	500
$\hat{C}_p(\text{Btu lb}_m^{-1}\ {}^\circ\text{F}^{-1})$	0.24	0.1
$k(\text{Btu hr}^{-1}\ \text{ft}^{-1}\ {}^\circ\text{F}^{-1})$	0.015	50

Show that the thermocouple will be nearly isothermal at any given instant, justifying a lumped model. Explain why this will be true even for large ω, when the characteristic length may be much smaller than R.

(b) Derive a general expression for the thermocouple temperature response, $T_T(t)$. Notice the effect of ω on the magnitude and delay of the response. Aside from proper calibra-

$T = T_\infty$

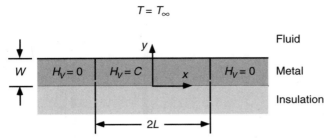

Figure P3-14. Heat transfer from a heated section of a wall.

tion, what is needed to make the thermocouple reading accurately reflect the air temperature?

*This problem was suggested by R. A. Brown.

3-14. Heat Transfer from a Heated Section of a Wall

For a laboratory heat transfer experiment it is desired to heat one section of a flat wall in
such a way that the heated section is nearly isothermal at steady state, with sharp transitions in
the temperature at the edges of the heated section. As shown in Fig. P3-14, it is proposed that the
wall be a metal sheet of thickness W backed by a rigid material which provides good thermal and
electrical insulation; the other dimensions of the metal sheet will greatly exceed W. Resistive
heating at a constant rate $H_V = C$ will be confined to a strip of width $2L$ by attaching a pair of
electrical leads at that spacing (running parallel to the z axis). Assume that convective heat transfer
at the top surface of the metal sheet is characterized by a constant heat transfer coefficient h; the
bulk temperature of the fluid above the wall is T_∞.

Starting with the energy conservation equation for the metal sheet, use scaling arguments
to identify ranges for the dimensionless parameters which will accomplish the aforementioned
design objectives. Determine the temperature in the *central* part of the heated section (i.e., not too
close to $x = \pm L$). Also, obtain an order-of-magnitude estimate of the distance required for the
surface temperature to decline from the heated value to T_∞.

3-15. Velocity of a Solidification Front

Assume that a pure liquid, initially at temperature T_∞, occupies the space $x > 0$. At $t = 0$ the
temperature at $x = 0$ is suddenly lowered to T_0, which is below the freezing temperature T_F of the
liquid. Solidification will begin, and the solid–liquid interface will advance into the region for-
merly occupied by liquid. As shown in Fig. P3-15, the interface location at time t is given by
$x = \delta(t)$. Assume that the liquid and solid have the same density, but different thermal conductivi-

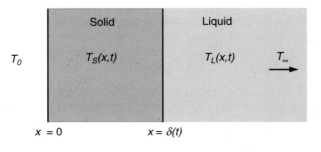

Figure P3-15. Moving boundary in
solidification.

ties (k_L and k_S, respectively). The latent heat of fusion is $\hat{\lambda}$. In this situation, $\delta(t)$ may be determined exactly, without invoking pseudosteady or other approximations.

(a) State the differential equations and boundary conditions governing the solid and liquid temperatures, $T_S(x, t)$ and $T_L(x, t)$, respectively.
(b) Taking into account the fixed temperatures at $x=0$ and $x\to\infty$ but ignoring for the moment the interfacial conditions at $x = \delta(t)$, show that both the solid and liquid temperatures may be expressed in terms of error functions. The variables involved are $\eta_i = x/(2(\alpha_i t)^{1/2})$, where $i = S$ or L.
(c) Use the results of part (b) and the requirement that $T = T_F$ at the solid–liquid interface to show that $\delta \propto t^{1/2}$. To define the proportionality constant, use

$$\delta(t) = 2\gamma(\alpha_S t)^{1/2},$$

where γ is a numerical constant which remains to be determined.
(d) Use the interfacial energy balance to show that γ is given implicitly by

$$\frac{e^{-\gamma^2}}{\mathrm{erf}\ \gamma} - \frac{k_L \alpha_S^{1/2}(T_\infty - T_F)e^{-\gamma^2 \alpha_S/\alpha_L}}{k_S \alpha_L^{1/2}(T_F - T_0)\mathrm{erfc}\,[\gamma(\alpha_S/\alpha_L)^{1/2}]} = \frac{\gamma\pi^{1/2}\lambda}{\hat{C}_{PS}(T_F - T_0)}.$$

Complete the solution for $T_S(x, t)$ and $T_L(x, t)$ in terms of γ.
(e) Consider the special case where the initial liquid temperature is at the freezing point, $T_\infty = T_F$. Show that a *pseudosteady analysis* involving $T_S(x, t)$ leads to

$$\delta(t) = \left(\frac{2k_S(T_F - T_0)t}{\lambda\rho}\right)^{1/2}.$$

Comparing this with the exact results of parts (c) and (d), show that the pseudosteady and exact results for $\delta(t)$ are in good agreement, provided that $\hat{C}_{PS}(T_F - T_0)/\hat{\lambda} \ll 1$. (*Hint:* erf $x \to 2x/\sqrt{\pi}$ as $x\to 0$.) Use an order-of-magnitude analysis of the time-dependent energy equation and boundary conditions to explain why a pseudosteady approximation is accurate when this requirement is met.

3-16. Integral Analysis of Diffusion and Reaction with Saturable Kinetics

An enzyme-catalyzed reaction with kinetics as in Problem 3-8 is being carried out at steady state in a liquid film of thickness L. At one surface ($x=0$) the concentration of the reactant is maintained at C_0, whereas the other bounding surface ($x=L$) is inert and impermeable. Use the integral approximation method to estimate the flux of reactant at $x=0$. (*Hint:* A reasonable form for the concentration profile is provided by an expression analogous to Eq. (3.8-8).]

3-17. Time to Reach a Steady State in Transient Diffusion

In the discussion of the membrane diffusion problem in Section 3.4 it was suggested from order-of-magnitude considerations that a pseudosteady concentration profile would be achieved after a time $t \gg L^2/D$. The objective here is to obtain a more specific estimate of the time to reach a steady state, using the integral approximation method. Assume now that the boundary concentrations, $C = KC_0$ at $x=0$ and $C=0$ at $x=L$, are held constant for $t \gg 0$, so that a true steady state will be achieved for $t\to\infty$. Let t_s be the time required to approach the steady concentration profile to within 5%.

(a) During the time period $0 < t < t_L$ when the concentration change at $x=0$ has not yet penetrated over the entire thickness L of the membrane, a reasonable approximation to the concentration profile is

$$\frac{C}{KC_0}=(1-\eta)^2, \qquad \eta=\frac{x}{\delta(t)}.$$

The penetration depth, $\delta(t)$, remains to be determined; as usual, $\delta(0)=0$. What is realistic about this assumed concentration profile and what is not? If the time for the concentration change to penetrete the full thickness of the membrane is defined by $\delta(t_L)=L$, determine t_L.

(b) For $t>t_L$, a reasonable approximation to the concentration profile is

$$\frac{C}{KC_0}=a(t)(1-\zeta)^2+\left(1-a(t)\right)(1-\zeta), \qquad \zeta=\frac{x}{L}.$$

Remaining to be determined is the function $a(t)$. For consistency with part (a) and with the steady-state solution, what must be the values of $a(t_L)$ and $a(\infty)$? Use this assumed profile to calculate t_s.

3-18. Transport in a Membrane with Oscillatory Flow

A membrane of thickness L in contact with aqueous solutions is subjected to a pulsatile pressure on one side, leading to time-periodic flow across the membrane. The transmembrane velocity is given by

$$v_x=U(t)=U_0+U_1 \sin \pi\omega t,$$

where U_0, U_1, and ω are constants. Consider convective and diffusive transport of some dilute solute i, where the flux is given by

$$N_{ix}(x, t)=-D_i\frac{\partial C_i}{\partial x}+U(t)C_i.$$

The solute concentration in the membrane is $C_i(x, t)$, and the concentrations at the boundaries are constant, such that $C_i(0, t)=C_0$ and $C_i(L, t)=0$. The pressure pulses are relatively slow, so that

$$\varepsilon \equiv \frac{\omega L^2}{D_i}\ll 1.$$

It is desired to investigate solute transport after many pressure pulses have occurred (i.e., ignoring any initial transients).

(a) Of the various possible time scales, identify the best choice for defining a scaled, dimensionless time, τ. Letting $\Theta\equiv C_i/C_0$ and $X\equiv x/L$, state the equations governing $\Theta(X, \tau)$.

(b) Show that this may be approached as a regular perturbation problem involving the small parameter ε. Write the differential equations and boundary conditions governing the first two terms of the expansion.

(c) Determine the first term in the expansion, $\Theta_0(X, \tau)$.

3-19. Transient Diffusion in a Porous Medium with Reversible Adsorption

Assume that species A is able to diffuse within some porous material, where it is reversibly adsorbed on the solid phase. Let S denote the adsorbed form of A. The rate of adsorption is given by

$$R_{VS}=k_A C_A-k_S C_S,$$

where the concentrations C_A and C_S are based on total volume (fluid plus solid). With the porous material modeled as a homogeneous medium, the effective diffusivity of A is D_A. The adsorbed species is assumed to be immobile, so that $D_S = 0$. The porous medium, which occupies the space $x > 0$, initially contains no A or S. Starting at $t = 0$, $C_A = C_0$ at $x = 0$.

(a) Use an order-of-magnitude analysis to show that for short contact times the concentration profile for A is approximately the same as if there were no adsorption. Suitable dimensionless variables are $\theta = C_A/C_0$, $\varphi = C_S/KC_0$, $\tau = tk_A$, and $X = x(k_A/D_A)^{1/2}$, where $K = k_A/k_S$. What is a sufficiently small value of τ? Obtain expressions for θ and φ valid for short times.

(b) Show that for large τ, when changes in θ are relatively slow, a pseudoequilibrium is reached where $\theta \cong \varphi$. (This is not a true equilibrium because the adsorption rate is not zero.) What is a sufficiently large value of τ? Show that the functional form of θ in this case is the same as for no adsorption, except that the apparent diffusivity is reduced from D_A to $D_A/(1 + K)$. This difference in apparent diffusivity can be of great importance in systems designed to promote adsorption, where $K \gg 1$.

3-20. Similarity Solutions for Time-Dependent Boundary Concentrations

Consider a transient diffusion problem like that in Section 3.5, but with a partition coefficient of unity ($K = 1$) and the boundary concentration a given function of time, $C_0(t)$. One might attempt a similarity solution by using $C_0(t)$ as a time-dependent concentration scale. That is, one could assume a solution of the form

$$\frac{C(x, t)}{C_0(t)} = \theta(\eta),$$

where $\eta = x/g(t)$ as before.

(a) Show that this approach will work only if $C_0(t) = at^\gamma$, where a and γ are constants.
(b) With $C_0(t)$ as stated in part (a) and $\eta = x/[2(Dt)^{1/2}]$, the governing equation becomes

$$\frac{d^2\theta}{d\eta^2} + 2\eta\frac{d\theta}{d\eta} - 4\gamma\theta = 0; \qquad \theta(0) = 1, \quad \theta(\infty) = 0.$$

Show that the transformation

$$\theta = \eta^{-1/2} e^{-\eta^2/2} w(\eta^2)$$

yields the differential equation

$$\frac{d^2w}{dz^2} + \left[-\frac{1}{4} + \frac{\kappa}{z} + \frac{\left(\frac{1}{4} - \mu^2\right)}{z^2} \right] w = 0,$$

where $z = \eta^2$, $\kappa = -[\gamma + (1/4)]$, and $\mu = 1/4$.

Note: The transformation used here is given in Kamke (1943). The differential equation for $w(z)$ is called *Whittaker's equation,* and its solutions (Whittaker functions) are expressible in terms of the *confluent hypergeometric functions* M and U (Abramowitz and Stegun, 1970, p. 505). Using this information, the solute concentration and the flux at the boundary are found to be (for $\gamma \geq 0$)

$$\theta = \frac{\Gamma(1 + \gamma)}{\sqrt{\pi}} e^{-\eta^2} U\left(\frac{1}{2} + \gamma, \frac{1}{2}, \eta^2\right),$$

$$N_x|_{x=0} = \frac{\Gamma(1 + \gamma)}{\Gamma\left(\frac{1}{2} + \gamma\right)} aD^{1/2}t^{\gamma - 1/2} = \frac{\Gamma(1 + \gamma)}{\Gamma\left(\frac{1}{2} + \gamma\right)} \left(\frac{D}{t}\right)^{1/2} C_0(t),$$

where $\Gamma(x)$ is the gamma function (Abramowitz and Stegun, 1970, p. 255). Notice that for $\gamma = \frac{1}{2}$ the flux is independent of time. For $\gamma = 0$ the concentration at the boundary is constant, and these results reduce to those derived in Section 3.5 for constant C_0 and $K = 1$; this is seen by noting that $U(\frac{1}{2}, \frac{1}{2}, \eta^2) = \sqrt{\pi} \exp(\eta^2) \operatorname{erfc} \eta$, $\Gamma(1) = 1$, and $\Gamma(\frac{1}{2}) = \sqrt{\pi}$.

3-21. Diffusion and Reaction in the Bleaching of Wood Pulp[*]

Chlorine dioxide (ClO_2) is commonly used to bleach the aqueous suspension of wood pulp which is produced by the digester in the kraft process. The ClO_2 reacts rapidly and irreversibly with lignin, which constitutes about 5% of the pulp; the remainder of the pulp (predominantly cellulose) is inert with respect to ClO_2. In addition to its reaction with lignin, ClO_2 also undergoes a slow, spontaneous decomposition. Thus, the reactions during the bleaching process may be written as

$$A + B \xrightarrow{k_1} C,$$

$$A \xrightarrow{k_2} P,$$

where A is ClO_2, B is unbleached lignin, C is bleached lignin, P represents products of the ClO_2 decomposition, and k_1 and k_2 are the homogeneous reaction rate constants.

The kinetics of the bleaching process may be studied under simplified conditions by exposing a water-filled mat of pulp fibers to a dilute aqueous solution of ClO_2. As shown in Fig. P3-21, the solution–mat interface is at $x = 0$, and the thickness of the mat is taken to be infinite. As ClO_2 diffuses in and reacts, the boundary between brown (unbleached) and white (bleached) pulp moves away from the interface, the location of that boundary being denoted by $x = \delta(t)$. The large magnitude of k_1 is what maintains a visually sharp boundary. Assume that the ClO_2 concentration at $x = 0$ remains constant at C_{A0}, that the unbleached lignin concentration at $x > \delta(t)$ is constant at C_{B0}, and that $C_{A0} \ll C_{B0}$. Note that, because it is a component of the pulp fibers, lignin is immobile.

(a) Derive the conservation statement which applies at $x = \delta(t)$, assuming that $k_1 \to \infty$ and $k_2 = 0$.

(b) For the conditions of part (a), show that diffusion in the region $0 < x < \delta(t)$ can be modeled as pseudo-steady. [*Hint:* Assume that the overall rate, and therefore the process time scale, is controlled by the interfacial balance from part (a).]

(c) Use the results of parts (a) and (b) to solve for $\delta(t)$.

(d) If k_2 is small but not identically zero, how must the above results be modified? Derive the differential equation which governs $\delta(t)$ for this case.

Figure P3-21. Bleaching of pulp fibers.

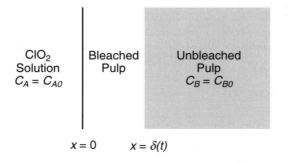

ClO_2
Solution
$C_A = C_{A0}$

Bleached
Pulp

Unbleached
Pulp
$C_B = C_{B0}$

$x = 0$ $x = \delta(t)$

(e) Of course, k_1 is not truly infinite. Derive an order-of-magnitude estimate for the thickness (ℓ) of the reaction zone.

(f) For large but finite k_1, give an order-of-magnitude estimate for the amount of time which must elapse before the concept of a pseudosteady reaction front at $x = \delta(t)$ becomes valid.

*This problem was suggested by K. A. Smith.

3-22. Concentrations in a Stagnant-Film Model of a Bioreactor

The liquid phase in a stirred-batch reactor containing a cell suspension is to be modeled as a stagnant film of thickness δ in contact with a well-mixed bulk solution, as shown in Fig. P3-22. Beginning at $t = 0$, species A is synthesized by the cells and enters the extracellular medium at $x = 0$ with a constant flux N_0; no A is present initially. Throughout the liquid, A undergoes an irreversible reaction with first-order rate constant k. Additional quantities are as follows: S, the total surface area at $x = 0$ or $x = \delta$; V, the volume of the bulk solution; and $\varepsilon V \equiv S\delta$, the volume of the stagnant film. The film is assumed to be thin, such that $\varepsilon \ll 1$.

(a) Formulate the complete transient problem, including the equations which govern the solute concentration in the bulk solution as well as in the stagnant film.

(b) Consider the steady state with $\mathrm{Da} \equiv kL^2/D_A \ll 1$. Use order-of-magnitude reasoning to estimate the concentration scales. The scales for C_A, ΔC_A, and $\hat{C}_A$ are denoted by C_A^0, ΔC_A^0, and C_A^+, respectively. Note that the scales for C_A and for the spatial variation in C_A (written as ΔC_A) are not necessarily equal. Also note that whereas Da and ε are both small, their ratio may be large or small. Distinguish between situations where $\mathrm{Da} \gg \varepsilon$ and $\mathrm{Da} \ll \varepsilon$.

(c) Calculate the steady-state concentrations in the film and bulk solution for $\mathrm{Da} \ll 1$. Are the results consistent with the scales estimated in part (b)?

(d) Repeat parts (b) and (c) for $\mathrm{Da} \gg 1$. That is, estimate the magnitudes of the concentration scales and then calculate the steady-state concentrations.

(e) Obtain order-of-magnitude estimates of the time required to reach a steady state for $\mathrm{Da} \ll 1$ and $\mathrm{Da} \gg 1$.

3-23. Absorption with a Nearly Irreversible Reaction

Assume that gaseous species A is absorbed by a stagnant liquid, where it reacts to form species B according to

$$A \underset{\leftarrow}{\overset{\rightarrow}{}} 2B.$$

The reaction rate is given by

$$R_{VA} = -k_V(C_A - KC_B^2)$$

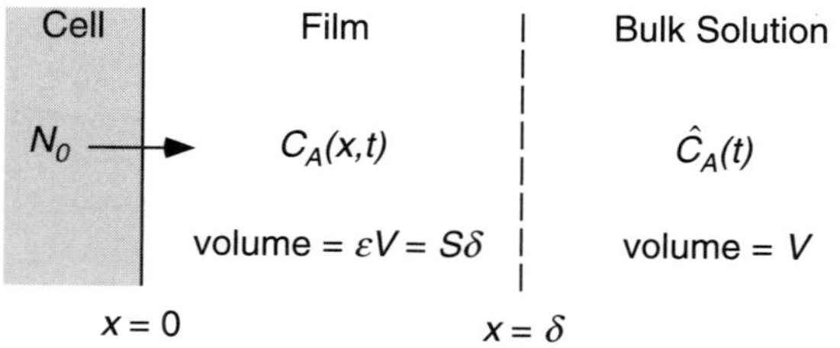

Figure P3-22. Stagnant-film model of a bioreactor. Species A enters the extracellular solution with a constant flux at $x = 0$ and is consumed by a first-order, irreversible reaction in the film and bulk solution.

Cell	Film		Bulk Solution
$N_0 \rightarrow$	$C_A(x,t)$		$\hat{C}_A(t)$
	volume $= \varepsilon V = S\delta$		volume $= V$
$x = 0$		$x = \delta$	

where the equilibrium constant K has units of inverse concentration. The reaction equilibrium strongly favors B, but not quite to the extent that the reaction can be considered irreversible. The system is at steady state and the liquid layer, which occupies the space $x>0$, is effectively infinite in thickness. The liquid-phase concentrations at the gas-liquid interface ($x=0$) are $C_A(0)=C_O$ and $C_B(0)=0$. For simplicity, assume equal diffusivities, $D_A=D_B=D$.

(a) State the differential equations and boundary conditions governing the liquid-phase concentrations, with dimensionless concentrations defined as $\Theta(\eta)\equiv C_A/C_0$ and $\Psi(\eta)\equiv C_B/C_0$. What is the proper length scale for converting x to the dimensionless coordinate η? What is needed for the liquid thickness to be considered infinite?

(b) Use perturbation methods to calculate the flux of A into the liquid film, $N_{Ax}(0)$, for $\varepsilon\equiv KC_0\ll1$. Determine the first two terms in the expansion for the flux. Discuss the effect of the reaction reversibility on the flux. (*Hint:* Expand both of the concentration variables as regular perturbation series.)

3-24. Unsteady-State Evaporation

A pure liquid composed of species A is brought into contact with a stagnant gas which initially contains only species B. Evaporation of the liquid begins at $t=0$, leading to a mixture of A and B in the gas; B is insoluble in the liquid and therefore remains in the gas phase. The gas–liquid interface is maintained at $z=0$, and the gas occupies the space $z>0$. The mole fraction of A in the gas at $z=0$ is x_{A0}. The system is at constant temperature and pressure and the gas mixture is ideal, so that the total molar concentration in the gas is constant.

(a) Show that the mole fraction of A in the gas, $x_A(z, t)$, is governed by

$$\frac{\partial x_A}{\partial t}=D_{AB}\frac{\partial^2 x_A}{\partial z^2}+\left[\frac{D_{AB}}{1-x_{A0}}\frac{\partial x_A}{\partial z}(0,\,t)\right]\frac{\partial x_A}{\partial z},$$

$$x_A(z,\,0)=0,\qquad x_A(0,\,t)=x_{A0},\qquad x_A(\infty,\,t)=0.$$

(b) Using a similarity variable (η) analogous to that in Section 3.5, show that

$$\frac{d^2\Theta}{d\eta^2}+2(\eta-\phi)\frac{d\Theta}{d\eta}=0,$$

$$\Theta(0)=1,\qquad \Theta(\infty)=0,$$

$$\phi(x_{A0})\equiv-\frac{1}{2}\left(\frac{x_{A0}}{1-x_{A0}}\right)\frac{d\Theta}{d\eta}(0),$$

where $\Theta(\eta)\equiv x_A/x_{A0}$. The unknown constant ϕ is determined later [see part (d)].

(c) Determine $\Theta(\eta)$.

(d) Derive an expression which could be used to compute ϕ from x_{A0}. It will be of the implicit form, $x_{A0}=f(\phi)$.

The similarity solution for this problem was given originally by Arnold (1944).

3-25. Melting of a Small Crystal[*]

A small, spherical crystal of initial radius R_0 and initial temperature T_0 is placed in a large volume of the same liquid, which is at temperature T_∞. The melting temperature T_m is such that $T_0<T_m<T_\infty$, so that the crystal immediately begins to melt. Assume that the bulk liquid is stagnant and that the solid and liquid densities are equal. The latent heat for melting (per unit mass) is denoted as $\hat{\lambda}$.

(a) Show that $\mathbf{v}=\mathbf{0}$ throughout the liquid, despite the changing crystal radius, $R(t)$.

(b) Assuming that the temperature field is pseudosteady, derive expressions for $R(t)$ and for the time t_m required for the crystal to melt completely.

(c) State the order-of-magnitude criteria which must be satisfied for the pseudosteady analysis to be valid.

*This problem was suggested by K. A. Smith.

Chapter 4

SOLUTION METHODS FOR CONDUCTION AND DIFFUSION PROBLEMS

4.1 INTRODUCTION

The preceding chapters were concerned mainly with the *formulation* of transport models. Some solution methods were introduced in Chapter 3, but they are approximate and/or fairly specialized. Many problems in transport involve linear partial differential equations, and this chapter is concerned with methods which apply to this broad class of problems. The focus on conduction and diffusion in solids and stagnant fluids is continued from Chapter 3. In addition to their many applications, the partial differential equations in such problems are the simplest ones we will encounter, making them ideal for illustrating unfamiliar methods. With $\mathbf{v} = \mathbf{0}$ and constant thermophysical properties, Eq. (2.4-3) for conservation of energy reduces to

$$\rho \hat{C}_p \frac{\partial T}{\partial t} = k\nabla^2 T + H_V. \tag{4.1-1}$$

For species i in a dilute solution, again with $\mathbf{v} = \mathbf{0}$, Eq. (2.6-5) gives

$$\frac{\partial C_i}{\partial t} = D_i \nabla^2 C_i + R_{Vi}. \tag{4.1-2}$$

One or the other of these equations is the starting point for almost all problems in this chapter.

Most of this chapter is devoted to the *finite Fourier transform* (FFT) method. The FFT method is one of many analytical or numerical techniques in which exact or approximate solutions to partial differential equations are found by expanding the solution in terms of a set of known functions, called *basis functions*, and then determining the unknown coefficients in the expansion. Applying the FFT method to partial differential equations like Eqs. (4.1-1) and (4.1-2) reduces the number of spatial variables until only a two-point boundary-value problem or initial-value problem remains, which is solved by standard methods. The FFT method is basically equivalent to the technique of *separation of variables* discussed in many traditional texts on mathematical methods in physics and engineering; for example, see Arpaci (1966), Butkov (1968), Carslaw and Jaeger (1959), Churchill (1963), and Morse and Feshbach (1953). The FFT method is more flexible, however, and permits a more direct attack on many problems. The reader who is familiar with separation of variables is in a position to notice these improvements, but prior exposure to that method is unnecessary for what is presented here.[1]

After the FFT method is discussed in some detail, there is an introduction to another approach for solving linear problems, which employs "point-source" solutions of the differential equations (Green's functions). Whereas the FFT method requires that at least one spatial dimension be finite, Green's functions are used most simply in problems with unbounded domains. Thus, the two methods are complementary.

The methods presented here were chosen because of their broad applicability and because exposure to them also provides a helpful foundation for understanding modern numerical techniques for solving partial differential equations (e.g., weighted residual methods). Space limitations preclude a discussion of the many other analytical methods applicable to linear problems. For discussions of other classical techniques see, for example, Morse and Feshbach (1953) and Carslaw and Jaeger (1959). Useful compilations of the solutions to many conduction and diffusion problems are provided by Carslaw and Jaeger (1959) and Crank (1975).

4.2 FUNDAMENTALS OF THE FINITE FOURIER TRANSFORM (FFT) METHOD

Linearity

The FFT method is applicable to linear boundary-value problems in domains where at least one of the spatial dimensions is finite. To clarify what is meant by that statement, a precise definition of linearity is needed. Let $\Theta = \Theta(\mathbf{r}, t)$ be the field variable (e.g., temperature or concentration), and let $\mathcal{L}$ be a differential operator which contains one or more spatial derivatives and perhaps also a time derivative. Then, a wide variety of partial differential equations can be represented as

$$\mathcal{L}\Theta = S(\mathbf{r}), \qquad (4.2\text{-}1)$$

where $S(\mathbf{r})$ is a specified function of position. The differential operator is said to be *linear* if

---

[1] The use of finite Fourier transforms in the solution of conduction and diffusion problems has received little attention in standard texts. It is mentioned in Carslaw and Jaeger (1959) and Butkov (1968), but only briefly. This transform methodology was introduced to chemical engineering by N. R. Amundson, while teaching at the University of Minnesota. The author was exposed to this approach by R. A. Brown, who learned it as a graduate student at Minnesota in the 1970s. Incidentally, the reader should not confuse this method with *fast Fourier transform* methods, which, unfortunately, are also abbreviated as FFT. Widely used in certain areas of engineering and physical science, those are numerical methods for computing Fourier transforms of discretely sampled data. The "ordinary" Fourier transform applies to an infinite domain.

$$\mathcal{L}(a_1\Theta_1 + a_2\Theta_2) = a_1\mathcal{L}\Theta_1 + a_2\mathcal{L}\Theta_2, \qquad (4.2\text{-}2)$$

where a_1 and a_2 are any constants and $\Theta_1(\mathbf{r}, t)$ and $\Theta_2(\mathbf{r}, t)$ are any functions [not necessarily solutions of Eq. (4.2-1)].

As an example, consider Eq. (4.1-2) for the case of an irreversible, first-order reaction,

$$\frac{\partial C_i}{\partial t} = D_i \nabla^2 C_i - k_1 C_i, \qquad (4.2\text{-}3)$$

where k_1 is the homogeneous reaction rate constant. Putting Eq. (4.2-3) in the form of Eq. (4.2-1) gives

$$\left(D_i\nabla^2 - k_1 - \frac{\partial}{\partial t}\right)C_i = 0, \qquad (4.2\text{-}4)$$

so that $\mathcal{L} = D_i\nabla^2 - k_1 - \partial/\partial t$ and $S = 0$. It is easily confirmed that this differential operator is linear. For a second-order reaction, with $R_{Vi} = -k_2 C_i^2$, the corresponding operator is $\mathcal{L} = D_i\nabla^2 - k_2 C_i - \partial/\partial t$. In this case, the dependence of $\mathcal{L}$ on the field variable (C_i) makes it impossible to satisfy Eq. (4.2-2), so that the operator is seen to be nonlinear. Any differential operator containing products of derivatives will also be nonlinear. For the conduction and diffusion problems governed by Eqs. (4.1-1) or (4.1-2), linearity demands that the source term (H_V or R_{Vi}) either must be independent of the field variable or must be a linear function of T or C_i. If any part of the source term is independent of T or C_i, then the function S in Eq. (4.2-1) will be nonzero.

A *boundary-value problem* consists of a differential equation together with certain boundary conditions, which may also be written in terms of differential operators. (For time-dependent problems, an initial condition is also needed.) A *linear* boundary-value problem is one in which all of the differential operators are linear. There are only three types of linear boundary conditions for the second-order differential equations with which we are concerned. They are

$\Theta = f(\mathbf{r_s})$	(Dirichlet)	(4.2-5a)
$\mathbf{n}\cdot\nabla\Theta = g(\mathbf{r_s})$	(Neumann)	(4.2-5b)
$\mathbf{n}\cdot\nabla\Theta + h_1(\mathbf{r_s})\Theta = h_0\ (\mathbf{r_s})$	(Robin)	(4.2-5c)

Equation (4.2-5a), where Θ itself is a specified function of position $\mathbf{r_s}$ on a bounding surface, is called an "essential" or "Dirichlet" condition; Eq. (4.2-5b), where the normal component of the gradient is specified, is termed a "natural" or "Neumann" condition; and Eq. (4.2-5c) is a "mixed" or "Robin" condition.

Uniqueness

A very important characteristic of linear boundary-value problems is that, with minor exceptions, the solutions are unique. Thus, once a solution is found by any method, it must be the only solution. This does not mean that a solution for, say, $T(\mathbf{r}, t)$ cannot be written in more than one form. Indeed, this chapter contains many examples of problems where the solutions can be written as two or more infinite series involving different sets of elementary functions. The uniqueness property does require, however, that the alternative expressions yield the same values of $T(\mathbf{r}, t)$ once all summations and other operations have been performed. The exceptions to uniqueness are steady-state problems involving Neumann (gradient) conditions on *all* boundaries. In such problems the solution is unique only to within an additive constant.

Example 4.2-1 Uniqueness Proof for Laplace's Equation   For steady conduction or diffusion without volumetric source terms, Eqs. (4.1-1) and (4.1-2) both reduce to *Laplace's equation*. As an example of a uniqueness proof for the solution of a linear partial differential equation, we assume that the variable $\psi(\mathbf{r})$ is subject to Laplace's equation in a volume V with a nonhomogeneous Dirichlet condition on the bounding surface S:

$$\nabla^2\psi=0 \text{ in } V, \qquad \psi=f(\mathbf{r_s}) \text{ on } S. \tag{4.2-6}$$

Setting $\phi=\psi$ in Green's first identity [Eq. (A.5–4)] and using $\nabla^2\psi=0$, we obtain

$$\int_V |\nabla\psi|^2 \, dV = \int_S \psi\frac{\partial\psi}{\partial n} \, dS. \tag{4.2-7}$$

Suppose now that there are two different solutions to Eq. (4.2-6), denoted by ψ_1 and ψ_2, and let $\Theta\equiv\psi_1-\psi_2$. It follows that Θ is also a solution to Laplace's equation and

$$\int_V |\nabla\Theta|^2 \, dV = \int_S \Theta\frac{\partial\Theta}{\partial n} \, dS. \tag{4.2-8}$$

Because, by definition, $\Theta=0$ on S, the right-hand side vanishes and

$$\int_V |\nabla\Theta|^2 dV = 0. \tag{4.2-9}$$

The integrand here cannot be negative, so that Eq. (4.2-9) is satisfied only if $|\nabla\Theta|=0$ everywhere. This requires that Θ be constant throughout V. However, Θ can be constant in V and zero on S only if $\Theta=0$ everywhere. We conclude that ψ_1 and ψ_2 cannot differ; that is, the solution to Eq. (4.2-6) must be unique. A similar approach yields proofs for the Neumann and Robin boundary conditions.

Domains

Aside from the requirement of linearity, the other restriction of the FFT method is that the boundaries must correspond to constant values of the coordinates, or *coordinate surfaces*. Consider, for example, the three regions depicted in Fig. 4-1. Each might represent the cross section of a long, solid rod for which it is desired to determine the

$$R^2 = x^2 + y^2$$

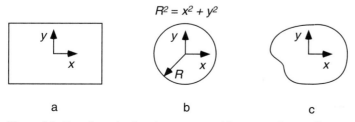

a   b   c

Figure 4-1. Two-dimensional regions represented by rectangular coordinates.

steady-state, two-dimensional temperature field, expressed as $T(x, y)$. In the rectangular domain of case (a) the boundaries each correspond to constant x or constant y, as required. In case (b) the location of the circular boundary is still readily described using rectangular coordinates, but neither x nor y is constant on the boundary. The best approach here, of course, is to use cylindrical coordinates, as done in several of the examples of Chapters 2 and 3; then the boundary corresponds to a constant value of r, namely $r = R$. In case (c) there is no obvious coordinate system which would describe the boundary. Thus, the requirement that the boundaries be coordinate surfaces restricts us to certain regular shapes. This restriction is not as severe as it first seems, because orthogonal, curvilinear coordinate systems have been developed for numerous geometries (see Section A.7). The problems in this book involve only rectangular, cylindrical, or spherical coordinates.

Homogeneity

An additional consideration is whether a differential equation or boundary condition is *homogeneous* or *nonhomogeneous*. Equation (4.2-1) is said to be homogeneous in the dependent variable Θ if $S = 0$; Eq. (4.2-3) is an example of a linear, homogeneous equation. The homogeneous forms of the boundary conditions in Eq. (4.2-5) result when the functions on the right-hand sides are zero (i.e., $f = 0$, $g = 0$, or $h_0 = 0$). A partial differential equation which is homogeneous always has as one possible solution the trivial solution, $\Theta = 0$. Whether or not this is a solution for the boundary-value problem will depend, of course, on the boundary conditions. If the partial differential equation and auxiliary conditions (the boundary conditions and, for transient problems, the initial condition) are *all* linear and homogeneous, then $\Theta = 0$ is indeed the solution for the problem. In other words, if there are no nonhomogeneities, nothing happens. Thus, the nonhomogeneous terms are sometimes called "forcing functions." The reader who is familiar with the separation of variables technique for solving partial differential equations will recall that it is directly applicable only to problems where the differential equation is homogeneous, and there is only one nonhomogeneous boundary (or initial) condition. If the differential equation is nonhomogeneous and/or there is more than one nonhomogeneous boundary condition, two or more subsidiary problems must be solved, and the results must be added to construct the desired overall solution. The number of subsidiary problems which must be solved equals the number of nonhomogeneities. As will be seen, the FFT approach is more flexible.

FFT Method

The objective of the introductory FFT example which follows is to illustrate the mechanics of the method; many other important considerations are deferred until later in the chapter. Such topics include the identification of suitable sets of basis functions, the convergence properties of the series solutions, and certain other issues relating to how and why the method works.

Example 4.2-2 Steady Conduction in a Square Rod   Consider the problem depicted in Fig. 4-2, in which the dimensionless coordinates are x and y and the dimensionless temperature is $\Theta(x, y)$. (For simplicity, most examples in this chapter begin with a problem already stated in dimensionless form; see Chapters 2 and 3 for examples of problem formulation and nondimen-

Figure 4-2. Steady heat conduction in a square rod with specified surface temperatures.

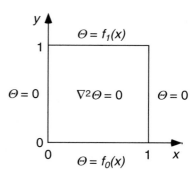

sionalization.) The temperature field is nonuniform as a consequence of the specified temperature variations along the boundaries at $y=0$ and $y=1$. There is no volumetric energy-source term. The problem statement for $\Theta(x, y)$ is

$$\frac{\partial^2 \Theta}{\partial x^2} + \frac{\partial^2 \Theta}{\partial y^2} = 0, \tag{4.2-10}$$

$$\Theta(0, y)=0, \qquad \Theta(1, y)=0, \tag{4.2-11}$$

$$\Theta(x, 0)=f_0(x), \qquad \Theta(x, 1)=f_1(x). \tag{4.2-12}$$

We start by assuming that the solution can be written in a series expansion as

$$\Theta(x, y)= \sum_{n=1}^{\infty} C_n(y)\Phi_n(x), \tag{4.2-13}$$

$$\Phi_n(x)=\sqrt{2}\, \sin n\pi x, \qquad n=1, 2, \dots. \tag{4.2-14}$$

Notice that each term in Eq. (4.2-13) is a product of two functions, each involving only one spatial variable. This feature is common to all FFT and separation of variables solutions. The functions $\Phi_n(x)$ specified by Eq. (4.2-14) are the *basis functions*. What constitutes an appropriate set of basis functions is described in Section 4.3, and the conditions under which an expansion such as Eq. (4.2-13) will be valid are discussed in Section 4.4. For now, the main thing to notice is that each of the basis functions vanishes at $x=0$ and $x=1$, so that Eq. (4.2-13) satisfies the homogeneous boundary conditions in Eq. (4.2-11). The $\sqrt{2}$ factor in Eq. (4.2-14), which looks extraneous, has the effect of *normalizing* the basis functions; the advantage in doing this will be evident when we are finished. To complete the solution it is necessary to determine the coefficients which multiply the basis functions in Eq. (4.2-13), namely $C_n(y)$.

For this problem the finite Fourier transform (FFT) of the temperature is defined as

$$\Theta_n(y)= \int_0^1 \Phi_n(x)\Theta(x, y)\, dx, \tag{4.2-15}$$

where $\Theta_n(y)$ is referred to simply as the "transformed temperature." Equation (4.2-15) defines an integral transform in which the original function is multiplied by a basis function, and integrated over the finite interval $0\le x\le 1$. The limits of integration correspond to the range of x in the problem; if the x boundaries were at locations other than $x=0$ and $x=1$, the definition of the transform would be modified accordingly. It will be shown that $\Theta_n(y)$ is identical to the expansion coefficient $C_n(y)$, so that determining $\Theta_n(y)$ for all n is sufficient to solve the problem.

A differential equation governing $\Theta_n(y)$ is obtained by transforming the original set of equations as in Eq. (4.2-15). Applying the FFT to both sides of Eq. (4.2-10) gives

$$\int_0^1 \Phi_n(x) \left[\frac{\partial^2 \Theta}{\partial x^2} + \frac{\partial^2 \Theta}{\partial y^2} \right] dx = 0. \tag{4.2-16}$$

Using Eq. (4.2-15), the second term is rewritten in terms of the transformed temperature as

$$\int_0^1 \Phi_n \frac{\partial^2 \Theta}{\partial y^2} dx = \frac{d^2}{dy^2} \int_0^1 \Phi_n \Theta \, dx = \frac{d^2 \Theta_n}{dy^2}. \tag{4.2-17}$$

The first term in Eq. (4.2-16) is rearranged using two integrations by parts to give

$$\int_0^1 \Phi_n \frac{\partial^2 \Theta}{\partial x^2} dx = \left(\Phi_n \frac{\partial \Theta}{\partial x} \right) \Big|_{x=0}^{x=1} - \int_0^1 \frac{\partial \Phi_n}{\partial x} \frac{\partial \Theta}{\partial x} dx$$

$$= \left(\Phi_n \frac{\partial \Theta}{\partial x} \right) \Big|_{x=0}^{x=1} - \left(\Theta \frac{\partial \Phi_n}{\partial x} \right) \Big|_{x=0}^{x=1} + \int_0^1 \Theta \frac{d^2 \Phi_n}{dx^2} dx. \tag{4.2-18}$$

The first of the "boundary terms" in Eq. (4.2-18), that involving $\Phi_n(\partial \Theta / \partial x)$, vanishes because $\Phi_n(x)$ was defined such that $\Phi_n(0) = \Phi_n(1) = 0$. The second boundary term, involving $\Theta(d\Phi_n/dx)$, vanishes because of the homogeneous boundary conditions satisfied by Θ, Eq. (4.2-11). Concerning the final term in Eq. (4.2-18), it follows from the definition of the basis functions that $d^2\Phi_n / dx^2 = -(n\pi)^2 \sqrt{2} \sin n\pi x = -(n\pi)^2 \Phi_n$. Accordingly, Eq. (4.2-18) becomes

$$\int_0^1 \Phi_n \frac{\partial^2 \Theta}{\partial x^2} dx = \int_0^1 \Theta \frac{d^2 \Phi_n}{dx^2} dx = -(n\pi)^2 \Theta_n. \tag{4.2-19}$$

These results are combined to give the differential equation for $\Theta_n(y)$,

$$\frac{d^2 \Theta_n}{dy^2} - (n\pi)^2 \Theta_n = 0. \tag{4.2-20}$$

Equation (4.2-20) requires two boundary conditions, which are obtained by transforming the original boundary conditions involving y, Eq. (4.2-12). Thus,

$$\Theta_n(0) = \int_0^1 \Phi_n(x) f_0(x) \, dx \equiv A_n, \tag{4.2-21}$$

$$\Theta_n(1) = \int_0^1 \Phi_n(x) f_1(x) \, dx \equiv B_n, \tag{4.2-22}$$

where A_n and B_n are constants. This completes the information required to calculate $\Theta_n(y)$. Linearly independent solutions of Eq. (4.2-20) include [$\exp(n\pi y)$, $\exp(-n\pi y)$], [$\sinh n\pi y$, $\cosh n\pi y$], [$\cosh n\pi y$, $\cosh n\pi(c-y)$], and [$\sinh n\pi y$, $\sinh n\pi(c-y)$], where c is any constant. Choosing the last of these pairs (with $c=1$) and satisfying Eqs. (4.2-21) and (4.2-22), the transformed temperature is found to be

$$\Theta_n(y) = \frac{A_n \sinh n\pi(1-y) + B_n \sinh n\pi y}{\sinh n\pi}. \tag{4.2-23}$$

What remains is to establish the relationship between $\Theta_n(y)$ and $C_n(y)$. This is done by first transforming Eq. (4.2-13) to give

$$\int_0^1 \Phi_m(x)\Theta(x, y) \, dx = \Theta_m(y) = \sum_{n=1}^{\infty} C_n(y) \int_0^1 \Phi_m(x) \Phi_n(x) \, dx. \tag{4.2-24}$$

The integral involving the product of the basis functions is evaluated as

$$\int_0^1 \Phi_m(x) \Phi_n(x) \, dx = 2 \int_0^1 \sin m\pi x \sin n\pi x \, dx = \delta_{mn}, \tag{4.2-25}$$

where δ_{mn} is the Kronecker delta. Given that $\delta_{mn} = 0$ for $m \neq n$ and $\delta_{mn} = 1$ for $m = n$, Eqs. (4.2-24) and (4.2-25) indicate that $\Theta_m(y) = C_m(y)$. Notice that the $\sqrt{2}$ factor included in the definition of $\Phi_n(x)$ was needed to make $\Theta_n(y)$ and $C_n(y)$ identical. If the basis functions had not been normalized in this manner, the integral in Eq. (4.2-25) would have equaled $\delta_{mn}/2$ instead of δ_{mn}, giving the less convenient result that $\Theta_n(y) = C_n(y)/2$. The overall solution is

$$\Theta(x, y) = \sqrt{2} \sum_{n=1}^{\infty} \left[\frac{A_n \sinh n\pi(1-y) + B_n \sinh n\pi y}{\sinh n\pi} \right] \sin n\pi x. \tag{4.2-26}$$

It is worth noting that obtaining a solution for nonhomogeneous boundary conditions at both $y = 0$ and $y = 1$ was not significantly more difficult than if there had been only one nonhomogeneous boundary condition.

In summary, the FFT method first involved transforming the original problem in x and y to one involving y only. The transformed problem was solved using elementary methods to obtain $\Theta_n(y)$. To complete the overall series solution, it was necessary only to recognize that $\Theta_n(y)$ and $C_n(y)$ were identical. This situation contrasts with that encountered when using other integral transforms (e.g., the Laplace transform) to solve partial differential equations, where deriving and solving the transformed problem may be straightforward, but inverting the transform to recover the final solution may be extremely difficult. With the FFT method the inversion step, in this case going from $\Theta_n(y)$ to $\Theta(x,y)$, is automatic!

4.3 BASIS FUNCTIONS AS SOLUTIONS TO EIGENVALUE PROBLEMS

Sets of functions which can be used as basis functions in the FFT method are the solutions to certain types of eigenvalue problems. These solutions are called *eigenfunctions*. This section begins by reviewing the formulation and solution of eigenvalue problems, and then it discusses the properties of eigenfunctions which make them suitable basis functions.

Eigenvalue Problems

Eigenvalue problems arise when certain homogeneous differential equations are subject only to homogeneous boundary conditions. Of concern here are functions $\Phi(x)$ which satisfy differential equations of the form

$$\mathcal{L}_x \Phi = -\lambda^2 \Phi, \tag{4.3-1}$$

where $\mathcal{L}_x$ is a second-order differential operator involving x, and λ is a constant which can have only certain values, termed *characteristic values* or *eigenvalues*. For the problems of interest here, the values of λ (and all other constants) are real. The operator $\mathcal{L}_x$ that corresponds to a given conduction or diffusion problem usually may be identified simply by inspection of the conservation equation. With the partial differential equation written in the form of Eq. (4.2-1), one finds $\mathcal{L}_x$ by selecting only those terms in $\mathcal{L}$ which involve x. To maintain the correct algebraic signs in Eq. (4.3-1), the coefficient of d^2/dx^2 in $\mathcal{L}_x$ must be positive.

For example, consider steady-state heat conduction in two dimensions, with temperature $T(x, y)$. Including the possibility of a volumetric heat source, the energy conservation equation is

$$\nabla^2 T = \frac{\partial^2 T}{\partial x^2} + \frac{\partial^2 T}{\partial y^2} = -\frac{H_v}{k}. \tag{4.3-2}$$

Equation (4.3-2) is rewritten in the form of Eq. (4.2-1) as

$$\mathscr{L}T = \left(\frac{\partial^2}{\partial x^2} + \frac{\partial^2}{\partial y^2}\right) T = -\frac{H_v}{k}. \tag{4.3-3}$$

The operator $\mathscr{L}_x$ obtained from Eq. (4.3-3) is just $\mathscr{L}_x = d^2/dx^2$. Referring to Eq. (4.3-1), the differential equation for the eigenvalue problem is then

$$\frac{d^2\Phi}{dx^2} = -\lambda^2\Phi. \tag{4.3-4}$$

The same eigenvalue equation is obtained for a transient conduction problem involving $T(x, t)$. Indeed, *any* conduction or diffusion problem in rectangular coordinates, involving a stationary medium with constant thermophysical properties, yields Eq. (4.3-4).

The boundary conditions for eigenvalue problems must be of certain types. In general, they may be any combination of the *homogeneous* forms of the three types of conditions given by Eq. (4.2-5); certain other allowable types of boundary conditions are discussed in Section 4.6. Thus, the boundary conditions for Eq. (4.3-4) are usually some combination of

$$\Phi = 0, \tag{4.3-5a}$$

$$\Phi' = 0, \tag{4.3-5b}$$

$$\Phi' + A\Phi = 0, \tag{4.3-5c}$$

where primes are used to denote differentiation with respect to x and A is a constant.

In the following three examples, solutions are derived for eigenvalue problems of the type just discussed. In each case x is a dimensionless variable defined so that the boundaries are at $x = 0$ and $x = 1$.

Example 4.3-1 Eigenvalue Problem with Two Dirichlet Conditions   If both boundary conditions are of the form of Eq. (4.3-5a), then the problem statement is

$$\frac{d^2\Phi}{dx^2} = -\lambda^2\Phi, \qquad \Phi(0) = \Phi(1) = 0. \tag{4.3-6}$$

The differential equation has the general solution

$$\Phi = a \sin \lambda x + b \cos \lambda x, \tag{4.3-7}$$

where a and b are constants. Applying the boundary conditions, these constants must satisfy a pair of algebraic equations, which can be written as a single matrix equation,

$$\begin{bmatrix} 0 & 1 \\ \sin \lambda & \cos \lambda \end{bmatrix} \begin{bmatrix} a \\ b \end{bmatrix} = \begin{bmatrix} 0 \\ 0 \end{bmatrix}. \tag{4.3-8}$$

A set of homogeneous equations such as that given by Eq. (4.3-8) always has the trivial solution $a=b=0$. For there to be nontrivial solutions, the determinant of the coefficient matrix must be zero. This requirement leads to the *characteristic equation*, which restricts the possible values of λ. From Eq. (4.3-8), the characteristic equation is

$$\sin \lambda = 0. \tag{4.3-9}$$

Accordingly, λ can only have certain discrete values λ_n, corresponding to the zeros of the function $\sin \lambda$. These are given by

$$\lambda_n = n\pi, \tag{4.3-10}$$

where n is any integer. Returning to Eq. (4.3-8) and noting that $\cos \lambda_n = \cos(n\pi) = (-1)^n$, it is seen from either of the algebraic equations that $b=0$ for all n, while the value of a is undetermined. The eigenfunctions are then

$$\Phi_n(x) = a_n \sin n\pi x, \tag{4.3-11}$$

where a_n is any constant. Notice that $\Phi_0(x)$ can be dropped from the set of eigenfunctions, because it is just the trivial solution $\Phi_0(x) = 0$. It is shown later that the eigenfunctions corresponding to $n<0$ are also not needed, in that they are redundant.

Example 4.3-2 Eigenvalue Problem with One Neumann and One Dirichlet Condition The problem to be considered is now

$$\frac{d^2\Phi}{dx^2} = -\lambda^2\Phi, \qquad \Phi'(0) = \Phi(1) = 0. \tag{4.3-12}$$

The general solution is again Eq. (4.3-7), where the constants now must satisfy

$$\begin{bmatrix} \lambda & 0 \\ \sin \lambda & \cos \lambda \end{bmatrix}\begin{bmatrix} a \\ b \end{bmatrix} = \begin{bmatrix} 0 \\ 0 \end{bmatrix}. \tag{4.3-13}$$

Equation (4.3-13) yields the characteristic equation,

$$\lambda \cos \lambda = 0, \tag{4.3-14}$$

which leads to the eigenvalues,

$$\lambda_n = \left(n + \frac{1}{2}\right)\pi, \tag{4.3-15}$$

where n is any integer. Another possible eigenvalue, not included in Eq. (4.3-15), is $\lambda = 0$. Ignoring for the moment the possibility that $\lambda = 0$ and noting that $\sin \lambda_n = (-1)^n$, either algebraic relation in Eq. (4.3-13) implies that $a = 0$. The value of b is undetermined. The resulting eigenfunctions are

$$\Phi_n = b_n \cos\left(n + \frac{1}{2}\right)\pi x, \tag{4.3-16}$$

where b_n is any constant.

The special case of $\lambda = 0$ always requires separate consideration. Setting $\lambda = 0$ in Eq. (4.3-7) yields $\Phi = b$. In other words, the solution corresponding to $\lambda = 0$ is simply a constant. Checking the boundary conditions in Eq. (4.3-12), it is seen that a nonzero, constant value for Φ will satisfy $\Phi'(0) = 0$ but not $\Phi(1) = 0$. This excludes the possibility that $\lambda = 0$, so that Eq. (4.3-16) does in fact represent the complete set of eigenfunctions.

Example 4.3-3 Eigenvalue Problem with One Robin and One Dirichlet Condition   The problem statement is

$$\frac{d^2\Phi}{dx^2} = -\lambda^2\Phi, \qquad \Phi'(0) + A\Phi(0) = 0, \qquad \Phi(1) = 0, \qquad (4.3\text{-}17)$$

where A is a constant. For applications involving conduction or diffusion, usually $A<0$ for a "left-hand" boundary. Proceeding as before, the constants in the general solution must satisfy

$$\begin{bmatrix} \lambda & A \\ \sin \lambda & \cos \lambda \end{bmatrix}\begin{bmatrix} a \\ b \end{bmatrix} = \begin{bmatrix} 0 \\ 0 \end{bmatrix}. \qquad (4.3\text{-}18)$$

The characteristic equation is $\lambda \cos \lambda - A \sin \lambda = 0$, or

$$\lambda = A \tan \lambda. \qquad (4.3\text{-}19)$$

Unlike the preceding examples, the eigenvalues must be determined numerically. For $A = -1$, the first three positive roots of Eq. (4.3-19) are $\lambda_1 = 2.029$, $\lambda_2 = 4.913$, and $\lambda_3 = 7.979$. Equation (4.3-18) provides one relationship between the constants in the general solution, namely $\lambda a + Ab = 0$. Retaining a_n as the undetermined constant, the solution is

$$\Phi_n = a_n\left(\sin \lambda_n x - \frac{\lambda_n}{A} \cos \lambda_n x\right). \qquad (4.3\text{-}20)$$

Because $\Phi = 0$ for $\lambda = 0$, the zero eigenvalue can be ignored.

Notice that in the last example the eigenfunctions do not correspond to either elementary function in the general solution, sine or cosine. Rather, the eigenfunctions are a specific linear combination of $\sin \lambda_n x$ and $\cos \lambda_n x$, as dictated by the boundary conditions. For eigenvalue problems involving Eq. (4.3-4), this is always the case. Sometimes, as in the first two examples, we may get $a_n = 0$ or $b_n = 0$, so that the linear combination happens to be "all cosine" or "all sine", respectively.

Inner Product and Orthogonality

Having illustrated the formulation and solution of eigenvalue problems, the remaining task in this section is to identify the relationship between sets of eigenfunctions and the sequences of functions which can be used as basis functions in the FFT method. For a series expansion like Eq. (4.2-13) to be valid, the basis functions must be *orthogonal*. For the definition of orthogonality, as well as for subsequent discussion of the FFT method, it is helpful to introduce the *inner product* notation. The inner product of any functions $f(x)$ and $g(x)$ with respect to an interval $a \leq x \leq b$ and a weighting function $w(x)$ is defined by

$$\langle f, g \rangle \equiv \int_a^b f(x)g(x)w(x)\,dx. \qquad (4.3\text{-}21)$$

The weighting function is always such that $w(x) > 0$ for $a < x < b$. For problems leading to Eq. (4.3-4), the weighting function is $w = 1$; the identification of weighting functions for more general types of problems is discussed in Section 4.6. For any proper weighting function, a direct consequence of the definition of the inner product is that

$$\langle f, f \rangle > 0. \qquad (4.3\text{-}22)$$

That is, the inner product of any function with itself is positive definite. As discussed in Section 4.4, this feature of the inner product provides a useful measure of the magnitude of a function within a stated interval. It is readily shown that the inner product also has the properties

$$\langle f, g \rangle = \langle g, f \rangle, \tag{4.3-23}$$

$$\langle f, g+h \rangle = \langle f, g \rangle + \langle f, h \rangle, \tag{4.3-24}$$

$$\langle cf, g \rangle = c \langle f, g \rangle, \tag{4.3-25}$$

where $h(x)$ is some other function and c is a constant.

Two functions $\Phi_n(x)$ and $\Phi_m(x)$ are said to be *orthogonal* if

$$\langle \Phi_n, \Phi_m \rangle = C_{nm} \delta_{nm}, \tag{4.3-26}$$

where C_{nm} is a constant. Thus, for any set of orthogonal functions, the inner product vanishes unless $n = m$. Based on the definition of the inner product, any statement concerning orthogonality necessarily refers to a specific interval and a specific weighting function. If $C_{nm} = 1$, then the functions are said to be *orthonormal*. Referring back to Eq. (4.2-25), it is seen that the functions $\Phi_n(x) = \sqrt{2} \sin n\pi x$ are orthonormal, whereas the functions $\Phi_n(x) = \sin n\pi x$ are merely orthogonal. In applying the FFT method the basis functions used will always be normalized, so that

$$\langle \Phi_n, \Phi_m \rangle = \delta_{nm}. \tag{4.3-27}$$

The formal resemblance between the inner product of two orthonormal functions and the scalar or dot product of a pair of orthogonal unit vectors is noteworthy. Indeed, as discussed in Section 4.4, there is a strong analogy between representing a given vector as a sum of components involving orthogonal unit vectors and representing a given function as a sum of terms involving orthonormal functions.

Functions cannot be orthogonal unless they are *linearly independent*. Linear independence of a set of N functions is related to what is required to satisfy the equation

$$\sum_{n=1}^{N} c_n \Phi_n(x) = 0, \tag{4.3-28}$$

where the c_n are constants. The set of functions $\Phi_n(x)$ ($n = 1, 2, \ldots, N$) is linearly independent if Eq. (4.3-28) can be satisfied only by choosing $c_n = 0$ for *all* n. Suppose, for example, that for two distinct values i and j, $\Phi_i(x) = \pm \Phi_j(x)$. Then, the terms corresponding to $n = i$ and $n = j$ in the sum could be made to cancel by setting $c_i = \mp c_j$. Because Eq. (4.3-28) could be satisfied without making either of those constants zero, the set of functions would not be linearly independent. Moreover, if $\Phi_i(x) = \pm \Phi_j(x)$, Eqs. (4.3-22) and (4.3-25) indicate that the inner product of Φ_i and Φ_j cannot vanish. Thus, this set of functions does not satisfy Eqs. (4.3-26) or (4.3-27).

A key feature of certain types of eigenvalue problems is that they yield eigenfunctions which form orthogonal sequences. Eigenvalue problems which are *self-adjoint* have this feature. If an eigenvalue problem is self-adjoint, then it is also true that the eigenvalues are real. The definition of self-adjointness, along with a broad class of problems which has this important property ("Sturm–Liouville problems"), is discussed in Section 4.6. For now, it suffices to state that all eigenvalue problems of the type discussed in this section, involving Eq. (4.3-4) and boundary conditions chosen from Eq. (4.3-5), are self-adjoint.

To obtain an orthogonal set of basis functions from a set of eigenfunctions like that generated in any of the three examples in this section, it is necessary only to ensure linear independence. Considering the eigenfunctions given by Eq. (4.3-11), note that $\sin(x) = -\sin(-x)$. Because the eigenfunctions with $n < 0$ differ only in sign from those with $n > 0$, including negative as well as positive values of n would result in a set of functions which is not linearly independent. Accordingly, we choose to exclude the negative integers. As mentioned below Eq. (4.3-11), $n = 0$ is already excluded because it corresponds to the trivial solution. With regard to the eigenfunctions in Eq. (4.3-16), note that $\cos(x) = \cos(-x)$. Consequently, the eigenfunctions for $n < 0$ duplicate those for $n > 0$, so that once again we exclude those for $n < 0$. In this case we must retain $n = 0$, because it gives a nontrivial solution, $\cos(\pi x/2)$. Similar reasoning applies to the eigenfunctions given by Eq. (4.3-20). For each positive root of Eq. (4.3-19) there is a "mirror image" negative root. This, coupled with the fact that Eq. (4.3-20) involves only sines and cosines, indicates that we should discard all eigenfunctions corresponding to negative values of λ.

For convenient reference, the orthonormal sequences obtained from four common eigenvalue problems are given in Table 4-1. These four problems include all combinations of Dirichlet and Neumann boundary conditions applied at $x = 0$ and $x = \ell$. Notice that in case IV one of the eigenfunctions is a constant, $\Phi_0(x) = 1/\sqrt{\ell}$, corresponding to $\lambda = 0$. In each of these cases there are simple, explicit expressions for the eigenvalues. When one or more Robin (mixed) conditions is present, as in Example 4.3-3, the eigenvalues must be found numerically.

With this additional information about basis functions, the reader may find it helpful to review Example 4.2-2. The basis functions defined by Eq. (4.2-14) are seen now to be normalized and linearly independent versions of the eigenfunctions derived in Example 4.3-1. Those eigenfunctions were useful because they satisfy homogeneous

Table 4-1
Orthonormal Sequences of Functions from Certain Eigenvalue Problems in Rectangular Coordinates[a]

Case	Boundary conditions	Basis functions
I	$\Phi(0) = 0,\ \Phi(\ell) = 0$	$\Phi_n(x) = \sqrt{\dfrac{2}{\ell}}\,\sin\dfrac{n\pi x}{\ell},\ n = 1, 2, \ldots$
II	$\Phi'(0) = 0,\ \Phi(\ell) = 0$	$\Phi_n(x) = \sqrt{\dfrac{2}{\ell}}\,\cos\left(n + \dfrac{1}{2}\right)\dfrac{\pi x}{\ell},\ n = 0, 1, 2, \ldots$
III	$\Phi(0) = 0,\ \Phi'(\ell) = 0$	$\Phi_n(x) = \sqrt{\dfrac{2}{\ell}}\,\sin\left(n + \dfrac{1}{2}\right)\dfrac{\pi x}{\ell},\ n = 0, 1, 2, \ldots$
IV	$\Phi'(0) = 0,\ \Phi'(\ell) = 0$	$\Phi_n(x) = \sqrt{\dfrac{2}{\ell}}\,\cos\dfrac{n\pi x}{\ell},\ n = 1, 2, \ldots$
		$\Phi_0(x) = \dfrac{1}{\sqrt{\ell}}$

[a] All of the functions shown satisfy Eq. (4.3-4) in the interval $[0, \ell]$.

boundary conditions of the same type (Dirichlet) as the homogeneous boundary conditions in the heat transfer example. Notice also that in Eq. (4.2-19) it was important for the basis functions to satisfy Eq. (4.3-4), the differential equation of the eigenvalue problem. Finally, note the crucial role that orthogonality played in relating $C_n(y)$ to $\Theta_n(y)$, using Eqs. (4.2-24) and (4.2-25).

4.4 REPRESENTATION OF AN ARBITRARY FUNCTION USING ORTHONORMAL FUNCTIONS

A cornerstone of the FFT method (and the separation of variables method) is the assumption that an arbitrary function, such as the temperature field in a heat transfer problem, can be represented as a series expansion involving some set of orthogonal functions. Thus, it was assumed in Eq. (4.2-13) that a certain two-dimensional temperature field, $\Theta(x, y)$, could be expressed as the sum of an infinite number of terms, each consisting of a basis function $\Phi_n(x)$ multiplied by a coefficient $C_n(y)$. Although Example 4.2-2 illustrated the mechanics of evaluating the coefficients for one such series, it left some fundamental questions unanswered. For what kinds of functions is such a series valid? As more and more terms are evaluated, in what manner will the series converge to the function it is supposed to represent? This section addresses those questions, and it also presents a general procedure for evaluating the coefficients in such a series. The results presented here apply not just to series constructed using the trigonometric basis functions described in Section 4.3, but also to expansions employing any other basis functions derived from the general Sturm-Liouville problem described in Section 4.6. Those include Bessel functions (Section 4.7) and Legendre polynomials (Section 4.8).

Generalized Fourier Series

To address the main issues, it is sufficient to consider arbitrary functions of a single variable, $f(x)$. It is desired to represent such a function using a series of the form

$$f(x) = \sum_{n=1}^{\infty} c_n \Phi_n(x), \tag{4.4-1}$$

where the functions $\Phi_n(x)$ form a complete orthonormal sequence, as discussed in Section 4.3. (The starting value of n has no special significance; sometimes it will be convenient to label the first term with $n=0$.) Any such expansion involving orthonormal functions is referred to here as a *Fourier series,* and the constant coefficients c_n are called *spectral coefficients.* When reference is made to expansions involving specific basis functions, adjectives such as "Fourier-sine," "Fourier-Bessel," or "Fourier-Legendre" are sometimes used.

To use a Fourier series to represent a function $f(x)$ over some closed interval $[a, b]$ in x (i.e., for $a \le x \le b$), it is sufficient that $f(x)$ be *piecewise continuous* in that interval [see, for example, Churchill (1963)]. A piecewise continuous function is one which is continuous throughout the interval, or which exhibits a finite number of jump discontinuites. At a jump discontinuity the function has different limits evaluated from the left and right, but both limits are finite. It is helpful to bear in mind that the physical processes of conduction or diffusion (and the second derivatives which are the mathematical embodiments of those processes) tend to smooth the temperature or concentration profile, usually ensuring that the field variable and its first derivatives are continuous within a given phase. Discontinuities in the field variable may exist at internal interfaces or along boundaries, but such discontinuities ordinarily consist of finite jumps. Accordingly, almost any conduction or diffusion problem will involve only piecewise continuous functions. A more general classification of functions which can be represented using Eq. (4.4-1) has been developed using linear operator theory [see Naylor and Sell (1982) and Ramkrishna and Amundson (1985)].

Calculation of Spectral Coefficients

Assuming that a given function $f(x)$ can be represented by a Fourier series, the spectral coefficients in Eq. (4.4-1) are given by

$$c_n = \langle \Phi_n(x), f(x) \rangle. \tag{4.4-2}$$

The inner product notation is used here, as defined by Eq. (4.3-21). The fundamental justification for Eq. (4.4-2) is the Fourier Series Theorem (Naylor and Sell, 1982); its basis in functional analysis is beyond the scope of this book. A heuristic derivation of Eq. (4.4-2) is obtained by forming inner products of $\Phi_m(x)$ with both sides of Eq. (4.4-1), and then using the orthonormality of the basis functions, to give

$$\langle \Phi_m(x), f(x) \rangle = \sum_{n=1}^{\infty} c_n \langle \Phi_m(x), \Phi_n(x) \rangle = \sum_{n=1}^{\infty} c_n \delta_{mn} = c_m. \tag{4.4-3}$$

Equation (4.4-2), without the inner product notation, was used indirectly in Example 4.2-2. That is, the constants A_n and B_n defined by Eqs. (4.2-21) and (4.2-22) are the spectral coefficients for the functions $f_0(x)$ and $f_1(x)$, respectively, the temperatures specified along two of the boundaries.

It is noteworthy that Eq. (4.4-1) is analogous to the more familiar representation of an arbitrary vector as a sum of scalar components, each multiplied by a corresponding base vector. The spectral coefficients correspond to the components and the basis functions are analogous to the base vectors. Just as the base vectors for a given coordinate system are (usually) defined so that they are mutually orthogonal and normalized to unit magnitude, the basis functions in Eq. (4.4-1) are orthonormal. The inner product operation in Eq. (4.4-3) is like forming the scalar or dot product of a base vector with some other vector. The scalar product selects the component of the second vector corresponding to the particular base vector, just as Eq. (4.4-3) isolates the corresponding spectral coefficient. The main difference is that the number of base vectors needed in a physical problem is ordinarily just three, whereas the number of basis functions in Eq. (4.4-1) is infinite. The fact that there is an infinite number of basis functions in any set tempts one to think that it would not hurt to omit one. However, just as it is impossible to represent an arbitrary three-dimensional vector using only e_x and e_y (while omitting e_z), it is impossible to accurately represent an arbitrary function using an incomplete set of basis functions. A very readable introduction to the concept of orthogonal bases in finite-dimensional (vector) space versus infinite-dimensional (function) space is given by Churchill (1963).

The Fourier series representations of functions of two or more variables follow immediately from Eqs. (4.4-1) and (4.4-2), by simply treating the additional variables as parameters. For example, an expansion for a function $\Theta(x, y)$ can be written as

$$\Theta(x, y) = \sum_{n=1}^{\infty} \Theta_n(y)\Phi_n(x), \tag{4.4-4}$$

$$\Theta_n(y) = \langle \Phi_n(x), \Theta(x, y) \rangle. \tag{4.4-5}$$

Equation (4.4-4) is the same as the expansion for the temperature in Example 4.2-2, whereas Eq. (4.4-5) restates the definition of the FFT of $\Theta(x, y)$, given previously by Eq. (4.2-15). It is seen that the transformed temperature, $\Theta_n(y)$, is a spectral coefficient. Thus, the strategy in the FFT method was to compute the spectral coefficients for the two-dimensional temperature field, which in turn depended on the spectral coefficients for certain boundary data (i.e., A_n and B_n).

Convergence

To discuss the convergence of Fourier series we return to functions of one variable. Let $f_N(x)$ represent the partial sum involving the first N terms in the series for $f(x)$, or

$$f_N(x) \equiv \sum_{n=1}^{N} c_n\Phi_n(x). \tag{4.4-6}$$

The extent to which these partial sums converge as N is increased may be described both in a local and in an average sense. An important result pertaining to local or pointwise convergence applies when both $f(x)$ and $f'(x)$ are piecewise continuous on the interval $[a, b]$ (Hildebrand, 1976; Greenberg, 1988). If that is true, then

$$\lim_{N \to \infty} f_N(x) = \frac{f(x^+) + f(x^-)}{2} \tag{4.4-7}$$

at each point in the open interval (a, b) (i.e., for $a < x < b$). In Eq. (4.4-7), $f(x^+)$ and $f(x^-)$ represent the right- and left-hand limits of the function, respectively, evaluated at x. Thus, at all points in (a, b) where the function is continuous, the series converges to $f(x)$; at a jump discontinuity, convergence is to the arithmetic mean of the limits. A subtle but important distinction is that the pointwise convergence described by Eq. (4.4-7) is guaranteed at all *interior points,* including those arbitrarily close to the endpoints, but not at the endpoints themselves.

　　The overall or average convergence properties of a Fourier series are best described using the concept of a *norm.* The norm of a function $f(x)$, written as $\|f(x)\|$, is defined as

$$\|f(x)\| \equiv \langle f(x), f(x) \rangle^{1/2} = \left[\int_a^b f^2(x)w(x) \, dx \right]^{1/2}, \tag{4.4-8}$$

which involves the inner product of the function with itself. This inner product provides a measure of the magnitude of the function, with respect to a specific interval $[a, b]$ and weighting function $w(x)$. It is analogous to forming the scalar (dot) product of a vector with itself, and then taking the square root to evaluate the magnitude of the vector. For Fourier series it is found that

$$\lim_{N \to \infty} \| f(x) - f_N(x) \| = 0. \tag{4.4-9}$$

In words, the *norm of the error* involved in approximating $f(x)$ with $f_N(x)$ vanishes as $N \to \infty$. This behavior is referred to as *convergence in the mean* (Churchill, 1963).

An interesting result which is related to Eq. (4.4-9) is that, for a fixed number of terms N, choosing the coefficients in the Fourier series in the manner prescribed by Eq. (4.4-2) has the effect of minimizing the norm of the error (Churchill, 1963). In other words, a Fourier series containing a finite number of terms provides the best "least squares" approximation to the function $f(x)$. Another result, which concerns the spectral coefficients themselves, is that

$$\lim_{n \to \infty} c_n = 0. \tag{4.4-10}$$

Thus, if an infinite series involving a set of basis functions has coefficients which do not exhibit this behavior, then it is not a legitimate Fourier series.

Following are some examples of Fourier series representations for simple functions, which illustrate typical convergence behavior and other features of such series. The selection of examples was influenced by the fact that, when constructing a Fourier series for boundary or initial data as part of an FFT solution, the choice of basis functions is ordinarily dictated by other parts of the problem. Thus, the basis functions which must be used to expand some function $f(x)$ are not necessarily those which yield the simplest Fourier series for that function. All of these examples employ trigonometric basis functions chosen from Table 4-1, with the interval [0, 1].

Example 4.4-1 Fourier–Sine Series for f(x) = 1   With the FFT method it is often necessary to represent a constant by a Fourier series. Suppose it is desired to represent $f(x) = 1$ using the sines given as case I in Table 4-1. From Eq. (4.4-2), the coefficients are given by

$$c_n = \sqrt{2} \int_0^1 \sin n\pi x \, dx = \frac{\sqrt{2}}{n\pi} [1 - (-1)^n]. \tag{4.4-11}$$

It follows that the desired sine series is

$$1 = 2 \sum_{n=1}^{\infty} [1 - (-1)^n] \frac{\sin n\pi x}{n\pi}. \tag{4.4-12}$$

Although it may seem quite improbable that the right-hand side equals unity, this result is correct.[2] Plots of this series are shown in Fig. 4-3, based on partial sums involving either 4 or 20 terms. As expected, $f_{20}(x)$ is a much better approximation to $f(x) = 1$ than is $f_4(x)$. However, no matter

---

[2] The contributions in the remarkable paper submitted by Fourier in 1807 to the Institut de France were described in the first footnote of Chapter 1. What was not mentioned is that, after review by Laplace, Lagrange, Monge, and Lacroix, the paper was rejected. Although the other reviewers favored acceptance, Lagrange was adamantly opposed to the idea that trigonometric series could be used to represent arbitrary functions. Thus, any skepticism which the reader might have when first encountering an identity like Eq. (4.4-12) is excusable. Although they were known to other prominent mathematicians, Fourier's results were not actually published until the appearance of his book in 1822. The 1807 manuscript was rediscovered many years after his death and, with biographical information, published after a delay of 165 years (Grattan-Guinness, 1972)!

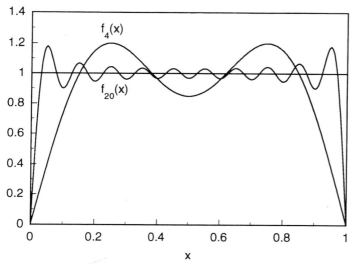

Figure 4-3. Fourier-sine series representations of $f(x) = 1$, based on 4 or 20 terms in Eq. (4.4-12). The exact function is also shown.

how many terms are used, this particular series fails at the endpoints, because the basis functions vanish there and $f(x)$ does not. This illustrates how increasing the number of terms gives convergence in the mean and local convergence at interior points, including points progressively closer to the ends of the interval, but does not necessarily help at the endpoints themselves. Incidentally, notice that only the odd-numbered "harmonics" (odd values of n) contribute to the sum in Eq. (4.4-12). This occurs whenever the function $f(x)$ is symmetric about the midpoint of the interval, as is the case for any constant.

If the cosines in case IV of Table 4-1 had been chosen to represent $f(x) = 1$, the problem would have been exceptionally easy, because it would have been sufficient to set $c_0 = 1$ and $c_n = 0$ for $n > 0$. This underscores the point that if $f(x)$ happens to have the same functional form as one of the basis functions, a one-term representation is exact and an infinite series is not required. That situation is analogous to a vector which parallels one of the coordinate axes and which therefore has a single nonvanishing component.

Example 4.4-2 Fourier-Sine Series for $f(x) = 1 - x$ Again choosing the sines in case I of Table 4-1, the coefficients for $f(x) = 1 - x$ are

$$c_n = \sqrt{2} \int_0^1 \sin(n\pi x)(1 - x)\, dx = \frac{\sqrt{2}}{n\pi} \tag{4.4-13}$$

and the complete series is given as

$$1 - x = 2 \sum_{n=1}^{\infty} \frac{\sin n\pi x}{n\pi}. \tag{4.4-14}$$

In this case there is no symmetry about $x = \frac{1}{2}$, and all harmonics contribute. Plots of this series are shown in Fig. 4-4. The behavior near $x = 0$ is similar to that in the previous example, but in this case the series is able to give the exact value of zero at $x = 1$.

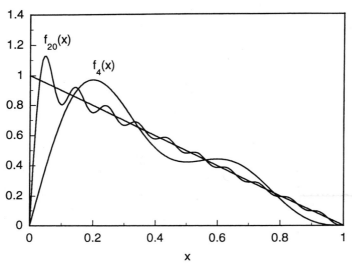

Figure 4-4. Fourier-sine series representations of $f(x)=1-x$, based on 4 or 20 terms in Eq. (4.4-14). The exact function is also shown.

Example 4.4-3 Fourier-Cosine Series for a Step Function Consider the step function defined by

$$f(x) = \begin{cases} 0, & 0 \le x < \tfrac{1}{2}, \\ 1, & \tfrac{1}{2} < x \le 1, \end{cases} \qquad (4.4\text{-}15)$$

which will be represented now using the cosines in case II of Table 4-1. A discontinuous function like this is handled by performing the integration for the inner product in a piecewise manner. In this case the piece for $[0, \tfrac{1}{2}]$ vanishes, leaving

$$c_n = \sqrt{2}\int_{1/2}^{1} \cos\!\left(n+\tfrac{1}{2}\right)\pi x\, dx = \frac{\sqrt{2}}{(n+\tfrac{1}{2})\pi}\left[(-1)^n \pm \frac{\sqrt{2}}{2}\right], \qquad (4.4\text{-}16)$$

where the minus sign applies for $n=0$, 1, 4, 5, ... and the plus sign is for $n=2$, 3, 6, 7, ... A computationally convenient representation of the series is obtained by using four sums involving a new index:

$$f(x) = 2\!\left(1 - \frac{\sqrt{2}}{2}\right)\sum_{m=0}^{\infty}\frac{\cos(4m+\tfrac{1}{2})\pi x}{(4m+\tfrac{1}{2})\pi} + 2\!\left(-1 - \frac{\sqrt{2}}{2}\right)\sum_{m=0}^{\infty}\frac{\cos[4m+\tfrac{3}{2}]\pi x}{[4m+\tfrac{3}{2}]\pi},$$

$$+ 2\!\left(1 + \frac{\sqrt{2}}{2}\right)\sum_{m=0}^{\infty}\frac{\cos(4m+\tfrac{5}{2})\pi x}{(4m+\tfrac{5}{2})\pi} + 2\!\left(-1 + \frac{\sqrt{2}}{2}\right)\sum_{m=0}^{\infty}\frac{\cos[4m+\tfrac{7}{2}]\pi x}{[4m+\tfrac{7}{2}]\pi}. \qquad (4.4\text{-}17)$$

Results from this series are plotted in Fig. 4-5. Notice that Eq. (4.4-17) yields 8 terms when the upper limit for m is 1, and 40 terms when the upper limit is 9. Once again, the approximation fails at one of the boundaries $(x=1)$, no matter how many terms are used. In accordance with Eq. (4.4-7), the series evidently converges to $f=\tfrac{1}{2}$ at $x=\tfrac{1}{2}$ (i.e., the arithmetic mean of the right- and left-hand limits at the discontinuity). The actual computed values are $f_8(\tfrac{1}{2})=0.464$ and $f_{40}(\tfrac{1}{2})=0.493$.

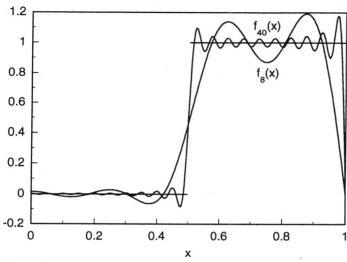

Figure 4-5. Fourier-cosine series representations of a step function, based on 8 or 40 terms in Eq. (4.4-17). The exact function is also shown.

In Fig. 4-5 there is a distinct overshoot evident near the discontinuity at $x=\frac{1}{2}$, even for a series with as many as 40 terms. An unfortunate property of Fourier series is that the amplitude of the overshoot near any discontinuity eventually fails to decay further as more and more terms are added. This is called the *Gibbs phenomenon;* see, for example, Morse and Feshbach (1953). With sines or cosines the overshoot reaches an asymptotic limit of about 9% of the size of the step. Similar overshoots are also seen in Figs. 4-3, 4-4, and 4-5 near the boundaries where point-wise convergence fails. Convergence in the mean is not violated, because the overshoots are confined to progressively thinner regions as the number of terms is increased.

Differentiation and Integration of Fourier Series

Differentiation of a Fourier series may be needed, for example, to compute the heat flux from a representation of the temperature field. The obvious approach is to differentiate each term, but caution is needed because the resulting series does not always converge to the derivative. If, however,

(1) $f(x)$ is continuous in $[a, b]$,
(2) $f'(x)$ is piecewise continuous in $[a, b]$, and
(3) $f(x)$ satisfies the same (homogeneous) boundary conditions as $\Phi_n(x)$,

then termwise differentiation of the Fourier series will yield a series which converges to $f'(x)$ at all points in (a, b) where $f'(x)$ is continuous (Hildebrand, 1976). These three conditions are *sufficient* for termwise differentiation, but they are not all *necessary*. In the case of trigonometric Fourier series, termwise differentiation is sometimes valid even if condition (3) is not satisfied (Churchill, 1963; Hildebrand, 1976).

Termwise integration of a Fourier series may be desired, for example, to determine the average temperature of an object. It is sufficient for this that the function be piecewise continuous (Churchill, 1963), so that the validity of termwise integration is rarely (if ever) an issue.

4.5 FFT METHOD FOR PROBLEMS IN RECTANGULAR COORDINATES

The examples in this section illustrate the application of the finite Fourier transform method to a variety of problems involving rectangular coordinates. Included are transient as well as steady-state problems.

Example 4.5-1 Steady Conduction with Multiple Nonhomogeneous Terms Consider the steady, two-dimensional heat conduction problem shown in Fig. 4-6, which is a more complicated version of Example 4.2-2. The dimensionless temperature has specified, nonzero values on all boundaries, and there is a volumetric heat source term, $H(x, y)$. Thus, the partial differential equation and all four boundary conditions are nonhomogeneous. Using the FFT method, we seek a solution of the form

$$\Theta(x, y) = \sum_{n=1}^{\infty} \Theta_n(y)\Phi_n(x),\tag{4.5-1}$$

$$\Theta_n(y) = \langle \Phi_n(x), \Theta(x, y) \rangle.\tag{4.5-2}$$

In the absence of any homogeneous boundary conditions, there is no preferred "direction" for the expansion of $\Theta(x, y)$. For the general case depicted in Fig. 4-6, basis functions involving y would serve as well as ones involving x, so that the coordinate preference expressed by Eq. (4.5-1) is arbitrary. Unlike the situation in Example 4.2-2, no choice of basis functions will satisfy any of the boundary conditions exactly, term by term, because none of the conditions are homogeneous. Nonetheless, we choose as basis functions ones which satisfy homogeneous boundary conditions of the same *type* (Dirichlet) as the nonhomogeneous conditions in the actual problem. From case I in Table 4-1, we have

$$\Phi_n(x) = \sqrt{2} \sin n\pi x, \qquad n = 1,2, \dots .\tag{4.5-3}$$

The energy conservation equation for this problem is written as

$$\frac{\partial^2 \Theta}{\partial x^2} + \frac{\partial^2 \Theta}{\partial y^2} + H(x, y) = 0.\tag{4.5-4}$$

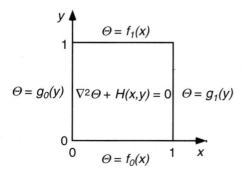

Figure 4-6. Steady heat conduction in a square rod with specified surface temperatures and a volumetric heat source.

Applying the FFT to the first term we obtain, with integration by parts,

$$\int_0^1 \Phi_n \frac{\partial^2 \Theta}{\partial x^2} dx = \left(\Phi_n \frac{\partial \Theta}{\partial x}\right) \Big|_{x=0}^{x=1} - \left(\Theta \frac{d\Phi_n}{dx}\right) \Big|_{x=0}^{x=1} + \int_0^1 \Theta \frac{d^2 \Phi_n}{dx^2} dx. \tag{4.5-5}$$

The first boundary term in Eq. (4.5-5) vanishes, because $\Phi_n(0) = \Phi_n(1) = 0$. Evaluating the derivative of Φ_n and using the boundary temperatures specified at $x=0$ and $x=1$, the second boundary term becomes

$$\left(\Theta \frac{d\Phi_n}{dx}\right) \Big|_{x=0}^{x=1} = \sqrt{2} n\pi [g_1(y)(-1)^n - g_0(y)] \equiv C_n(y). \tag{4.5-6}$$

Thus, the absence of homogeneous boundary conditions in the x direction requires that one of the boundary terms in Eq. (4.5-5) be retained. The remaining integral in Eq. (4.5-5) is evaluated as

$$\int_0^1 \Theta \frac{d^2 \Phi_n}{dx^2} dx = -(n\pi)^2 \int_0^1 \Theta \Phi_n \, dx = -(n\pi)^2 \Theta_n. \tag{4.5-7}$$

The other terms from the original differential equation, Eq. (4.5-4), are transformed in a straightforward manner as

$$\int_0^1 \Phi_n \frac{\partial^2 \Theta}{\partial y^2} dx = \frac{d^2}{dy^2} \int_0^1 \Phi_n \Theta \, dx = \frac{d^2 \Theta_n}{dy^2}, \tag{4.5-8}$$

$$\int_0^1 \Phi_n H \, dx = \sqrt{2} \int_0^1 \sin n\pi x \, H(x, y) \, dx \equiv H_n(y). \tag{4.5-9}$$

Collecting all of the terms, the transformed differential equation is

$$\frac{d^2 \Theta_n}{dy^2} - (n\pi)^2 \Theta_n = C_n(y) - H_n(y). \tag{4.5-10}$$

It is seen that having one or more nonhomogeneous boundary conditions in the basis function coordinate (in this case, x) and/or a nonhomogeneous term in the original partial differential equation causes the differential equation for the transformed temperature to be nonhomogeneous [compare Eq. (4.5-10) with Eq. (4.2-20)]. The boundary conditions satisfied by Eq. (4.5-10) are

$$\Theta_n(0) = A_n, \qquad \Theta_n(1) = B_n, \tag{4.5-11}$$

where the constants A_n and B_n are defined by Eqs. (4.2-21) and (4.2-22), respectively. As discussed in Example 4.2-2, nonzero values of these constants are the consequence of nonhomogeneous boundary conditions in the y coordinate.

As a special case, assume that $\Theta = 2$ at $x=0$, $\Theta = 0$ at $x=1$, $\Theta = 1$ at $y=0$ and $y=1$, and $H=0$. Equations (4.5-10) and (4.5-11) become

$$\frac{d^2 \Theta_n}{dy^2} - (n\pi)^2 \Theta_n = -2\sqrt{2} n\pi, \tag{4.5-12}$$

$$\Theta_n(0) = \Theta_n(1) = \frac{\sqrt{2}}{n\pi} [1 - (-1)^n]. \tag{4.5-13}$$

Equation (4.5-12) remains nonhomogeneous because of the nonhomogeneous boundary condition at $x=0$. The boundary values for Θ_n in Eq. (4.5-13) are just the spectral coefficients for the function $f(x)=1$ (see Example 4.4-1). Equation (4.5-12) has the particular solution $\Theta_n = (2\sqrt{2})/(n\pi)$, so that, choosing $\sinh n\pi y$ and $\sinh n\pi(1-y)$ as the independent solutions of the homogeneous equation, the general solution is

$$\Theta_n = a_n \sinh n\pi y + b_n \sinh n\pi(1-y) + \frac{2\sqrt{2}}{n\pi}. \qquad (4.5\text{-}14)$$

Evaluating the constants a_n and b_n using Eq. (4.5-13), the overall solution is

$$\Theta(x, y) = 2\sum_{n=1}^{\infty}\left\{2 - [1 + (-1)^n]\left[\frac{\sinh n\pi y + \sinh n\pi(1-y)}{\sinh n\pi}\right]\right\}\frac{\sin n\pi x}{n\pi}. \qquad (4.5\text{-}15)$$

The form of the temperature field is clarified by noticing that, from a comparison with Eq. (4.4-14), the first part of the sum in Eq. (4.5-15) (the part independent of y) is the Fourier series representation of $f(x) = 2(1-x)$. Also, in the y-dependent part of Eq. (4.5-15), all odd-numbered terms drop out. Accordingly, Eq. (4.5-15) is rewritten as

$$\Theta(x, y) = 2(1-x) - 4\sum_{\substack{n=1 \\ n \text{ even}}}^{\infty}\left[\frac{\sinh n\pi y + \sinh n\pi(1-y)}{\sinh n\pi}\right]\frac{\sin n\pi x}{n\pi}. \qquad (4.5\text{-}16)$$

This expression has two distinct parts. There is a linear function of x that matches the specified temperatures at $x=0$ and $x=1$, and which is the solution to the one-dimensional problem that would result if the boundaries at $y=0$ and $y=1$ were insulated. There is also a Fourier series involving functions of both x and y, which is needed to match the actual boundary conditions at $y=0$ and $y=1$. Because the part of $\Theta(x, y)$ represented by the Fourier sine series is antisymmetric about the midplane, $x=\frac{1}{2}$, only the even-numbered harmonics contribute.

The advantage of writing the solution as Eq. (4.5-16) is that the Fourier series now represents a function which vanishes at the endpoints, $x=0$ and $x=1$, as do the basis functions. Consequently, as discussed in Section 4.4, termwise differentiation of the series is valid if it is desired to calculate $\partial\Theta/\partial x$. Moreover, the solution yields the exact temperatures at $x=0$ and $x=1$. The series in Eq. (4.5-15) has neither of these attributes; it will converge to the correct values at $x=1$ and at points arbitrarily close to $x=0$, but not at $x=0$ itself.

This problem could have been solved also using an expansion in y. The same basis functions (but written in y) would be used and the amount of effort required would be similar. The alternate forms of the solution just discussed, and the possibility of other series representations involving $\Phi_n(y)$, emphasize that a given temperature field can often be represented by several different Fourier series. Domains which have two (or three) finite dimensions usually offer a choice of expansion coordinate. Experience is the best guide as to which expansion will yield the simplest result.

Example 4.5-2 Edge Effects in Steady Conduction   This example illustrates certain new mathematical features, while also leading to an important physical conclusion. The mathematical objectives are to illustrate how to handle convective boundary conditions and a semi-infinite domain, whereas the physical point concerns the limitations of one-dimensional models for steady heat conduction. The problem to be considered is shown in Fig. 4-7. There is convective heat transfer at two boundaries (involving Bi as a parameter) and a specified temperature at a third boundary. The fourth boundary of the two-dimensional domain is assumed to be indefinitely far from the region of interest, and no data are given there.

The only finite dimension is in the x direction, so that the solution must be expanded using basis functions involving x. In general, to identify the boundary conditions which must be satisfied by $\Phi_n(x)$, one substitutes into (the homogeneous form of) the x boundary conditions an assumed solution of the form $\Theta(x, y) = \Phi_n(x)Y(y)$. The function $Y(y)$ factors out, revealing the required boundary conditions. For the present example, the associated eigenvalue problem is found to be

$$\frac{d^2\Phi}{dx^2} = -\lambda^2\Phi, \qquad \Phi(0)=0, \qquad \Phi'(1) + \text{Bi}\ \Phi(1)=0. \qquad (4.5\text{-}17)$$

Figure 4-7. Steady heat conduction in a semi-infinite domain. The right-hand boundary is assumed to be at $y = \infty$.

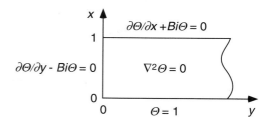

This problem is similar to Eq. (4.3-17) with $A = $ Bi, the Biot number, except that the boundary conditions are reversed. Following the same procedure as in Example 4.3-3, it is found that

$$\Phi_n = a_n \sin\lambda_n x, \qquad n = 1, 2, \ldots, \tag{4.5-18}$$

$$\lambda_n = -\text{Bi} \tan \lambda_n, \tag{4.5-19}$$

$$a_n = \left(\frac{2\lambda_n}{\lambda_n - \cos \lambda_n \sin \lambda_n}\right)^{1/2}. \tag{4.5-20}$$

The eigenvalues are the positive roots of Eq. (4.5-19), and Eq. (4.5-20) is the normalization condition. For Bi $= 1$, the first three eigenvalues are $\lambda_1 = 2.029$, $\lambda_2 = 4.913$, and $\lambda_3 = 7.979$.

In transforming the differential equation for $\Theta(x, y)$ the key step is the integration by parts, given again by Eq. (4.5-5). Regrouping the boundary terms, those for $x = 1$ are evaluated as

$$\left(\Phi_n \frac{\partial \Theta}{\partial x} - \Theta \frac{d\Phi_n}{dx}\right)\bigg|_{x=1} = [\Phi_n(-\text{Bi } \Theta) - \Theta(-\text{Bi } \Phi_n)]\bigg|_{x=1} = 0, \tag{4.5-21}$$

whereas those for $x = 0$ are given by

$$\left(\Phi_n \frac{\partial \Theta}{\partial x} - \Theta \frac{d\Phi_n}{dx}\right)\bigg|_{x=0} = -\frac{d\Phi_n}{dx}\bigg|_{x=0} = -a_n\lambda_n. \tag{4.5-22}$$

As seen here and in Example 4.2-2, when the boundary condition in the physical problem is homogeneous, the corresponding boundary term in the integration by parts will vanish. It is important to recognize that if we had chosen a different set of basis functions, it would not have been possible to evaluate these boundary terms. If, for example, we tried to use the sine functions for which $\Phi_n(1) = 0$, we would fail to reach any conclusion from Eq. (4.5-21) because the problem does not specify a value for $\Theta(1, y)$. This emphasizes the importance of choosing basis functions which satisfy types of boundary conditions (Dirichlet, Neumann, or Robin) which correspond to those in the problem at hand.

Proceeding as in the previous FFT examples, the differential equation for the transformed temperature is found to be

$$\frac{d^2\Theta_n}{dy^2} - \lambda_n^2\Theta_n = -a_n\lambda_n. \tag{4.5-23}$$

Transforming the boundary condition at $y = 0$, we find that

$$\frac{d\Theta_n}{dy}(0) - \text{Bi } \Theta_n(0) = 0. \tag{4.5-24}$$

Equation (4.5-23) has the particular solution $\Theta_n = a_n/\lambda_n$. The semi-infinite domain makes it advantageous to express the homogeneous solution in terms of exponentials. Thus, the general solution is written as

$$\Theta_n = b_n e^{-\lambda_n y} + c_n e^{\lambda_n y} + \frac{a_n}{\lambda_n},\qquad(4.5\text{-}25)$$

where b_n and c_n are the unknown constants. For the temperature to be *finite* at $y=\infty$, it is necessary that $c_n=0$. Thus, information on the mysterious fourth boundary is not needed. Evaluating b_n using Eq. (4.5-24), the final solution for the transformed temperature is

$$\Theta_n = \frac{a_n}{\lambda_n}\left[1 - \left(\frac{\text{Bi}}{\lambda_n + \text{Bi}}\right)e^{-\lambda_n y}\right].\qquad(4.5\text{-}26)$$

Using Eqs. (4.5-26) and (4.5-18) in Eq. (4.5-1), the overall solution is found to be

$$\Theta(x, y) = 2\sum_{n=1}^{\infty}\left[1 - \left(\frac{\text{Bi}}{\lambda_n + \text{Bi}}\right)e^{-\lambda_n y}\right]\left[\frac{\sin \lambda_n x}{\lambda_n - \cos \lambda_n \sin \lambda_n}\right].\qquad(4.5\text{-}27)$$

The decaying exponentials indicate that for large y, the temperature is a function of x only. In other words, far from the left-hand edge of the object in Fig. 4-7 the solution is one-dimensional. The form of the solution far from the edge is identified by substituting $\Theta = f(x)$ in the original problem statement. It is straightforward to show that the solution of the resulting one-dimensional problem is $f(x) = 1 - [\text{Bi}/(1 + \text{Bi})]x$. Using this result to simplify Eq. (4.5-27), the final form of the temperature field is

$$\Theta(x, y) = 1 - \left(\frac{\text{Bi}}{1 + \text{Bi}}\right)x - 2\sum_{n=1}^{\infty}\left(\frac{\text{Bi}}{\lambda_n + \text{Bi}}\right)e^{-\lambda_n y}\left(\frac{\sin \lambda_n x}{\lambda_n - \cos \lambda_n \sin \lambda_n}\right).\qquad(4.5\text{-}28)$$

We wish now to estimate how near the left-hand edge the one-dimensional solution is valid. It is sufficient to consider the most slowly decaying exponential, which is that containing λ_1. Accepting $\exp(-\lambda_1 y) = 0.05$ as being close enough to zero and using $\lambda_1 = 2.03$ (as stated above for $\text{Bi} = 1$), we find that $y = 1.5$. In other words, the thermal effects of the edge are "felt" over about 1.5 thicknesses. (The thickness is the dimension in the x direction, which is used to make both coordinates dimensionless.) As a general rule, edge effects in steady conduction or diffusion in a thin object become negligible after one to two thicknesses. This is a valuable guideline in modeling, as mentioned in Section 3.3.

Example 4.5-3 Transient Diffusion Approaching a Steady State   We return to the membrane diffusion problem introduced in Section 3.4; we now consider times short enough that the concentrations in the external solutions remain approximately constant, but long enough that the similarity solution derived in Section 3.5 is not necessarily valid. With reference to Fig. 3-5(a), we wish to compute the solution for $t \le t_3$, whereas the similarity result is valid at most for $t \le t_2$. The dimensionless problem for $\Theta(x, t)$ is depicted in Fig. 4-8.

The FFT procedure for transient problems is very similar to that for steady-state situations. We seek a solution of the form

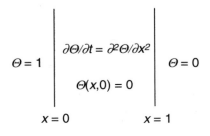

Figure 4-8. Transient diffusion in a membrane, with constant concentrations at the boundaries and an initial concentration of zero.

$$\Theta(x, t) = \sum_{n=1}^{\infty} \Theta_n(t) \Phi_n(x),$$ (4.5-29)

where the coefficients are now functions of time. The main distinction is that in a one-dimensional transient problem there is never a choice in the basis function variable. The basis functions cannot involve time because the differential equation has only a first derivative in t; there is no way to formulate an eigenvalue problem in t. The basis functions needed here are those satisfying Dirichlet boundary conditions (case I of Table 4-1), namely

$$\Phi_n(x) = \sqrt{2} \sin n\pi x, \qquad n = 1, 2, \dots .$$ (4.5-30)

Transforming the time and space derivatives in the partial differential equation, we obtain

$$\int_0^1 \Phi_n \frac{\partial \Theta}{\partial t} dx = \frac{d\Theta_n}{dt},$$ (4.5-31)

$$\int_0^1 \Phi_n \frac{\partial^2 \Theta}{\partial x^2} dx = \left(\Phi_n \frac{\partial \Theta}{\partial x} - \Theta \frac{d\Phi_n}{dx} \right) \Big|_{x=0}^{x=1} + \int_0^1 \Theta \frac{d^2\Phi_n}{dx^2} dx$$

$$= \left(\Theta \frac{d\Phi_n}{dx} \right) \Big|_{x=0} - \lambda_n^2 \Theta_n = \sqrt{2} n\pi - (n\pi)^2 \Theta_n.$$ (4.5-32)

The initial condition, $\Theta = 0$ at $t = 0$, transforms simply to $\Theta_n(0) = 0$. Accordingly, the complete transformed problem is

$$\frac{d\Theta_n}{dt} + (n\pi)^2 \Theta_n = \sqrt{2} n\pi, \qquad \Theta_n(0) = 0.$$ (4.5-33)

The solution to Eq. (4.5-33) is

$$\Theta_n = \frac{\sqrt{2}}{n\pi} \left(1 - e^{-(n\pi)^2 t} \right)$$ (4.5-34)

and the overall solution is then

$$\Theta(x, t) = 2 \sum_{n=1}^{\infty} \left(1 - e^{-(n\pi)^2 t} \right) \frac{\sin n\pi x}{n\pi}.$$ (4.5-35)

The solution in Eq. (4.5-35) consists of two main parts. There is a transient part which decays exponentially with time, and a part which is independent of time and evidently represents a steady state achieved at large t. Using Eq. (4.4-13), the steady-state solution is identified as the Fourier expansion for the function $f(x) = 1 - x$. Thus, the solution is rewritten more simply as

$$\Theta(x, t) = 1 - x - 2 \sum_{n=1}^{\infty} e^{-(n\pi)^2 t} \frac{\sin n\pi x}{n\pi}.$$ (4.5-36)

Not only does Eq. (4.5-36) more clearly reveal the form of the steady-state solution, it allows for termwise differentiation in x and provides an exact represention of the values of Θ at the boundaries, as discussed in connection with Eq. (4.5-17).

The series in Eq. (4.5-36) converges very rapidly for large t, so that the time needed to reach a steady state can be estimated simply by looking at the first term. The first exponential reaches 5% of its initial value at $\pi^2 t = 3$, or $t = 3/\pi^2 = 0.3$. The corresponding dimensional time is $0.3L^2/D$, where L is the membrane thickness and D is the solute diffusivity. Thus, the time to reach a steady state is of the same order of magnitude as the characteristic time for diffusion over the membrane thickness, $t_D = L^2/D$ (see Section 3.4). In contrast to the situation for large times, for small t the series in Eq. (4.5-36) converges very slowly, eventually requiring an impractical number of terms as $t \to 0$. Alternative approaches are needed for calculations involving very small t, such as the similarity method. Thus, the similarity and FFT methods may be applied to the same problem in a complementary manner.

Example 4.5-4 Transient Conduction with No Steady State   We now consider a problem similar to the last example, but with Neumann (constant flux) conditions at both boundaries. Suppose that for $t>0$ there is a constant radiant heat flux imposed on one surface of a solid slab, while the other surface is thermally insulated. The problem is illustrated in Fig. 4-9.

   With two Neumann conditions, the basis functions are those from case IV of Table 4-1:

$$\Phi_0(x)=1; \qquad \Phi_n(x)=\sqrt{2}\,\cos n\pi x, \quad n=1,\,2,\,\ldots. \tag{4.5-37}$$

This is the first FFT example in which it has been necessary to include the constant basis function, Φ_0, which corresponds to the eigenvalue $\lambda_0=0$. The constant basis function requires special treatment and has an important effect on the structure of the solution, as will be seen. The expansion for the dimensionless temperature Θ is of the usual form,

$$\Theta(x,\,t)=\sum_{n=0}^{\infty}\Theta_n(t)\Phi_n(x), \tag{4.5-38}$$

the only difference being that the summation index now must start at zero. Transforming the spatial derivative in the usual manner, the integration by parts leads to

$$\int_0^1 \Phi_n \frac{\partial^2 \Theta}{\partial x^2}\,dx = \left(\Phi_n \frac{\partial \Theta}{\partial x} - \Theta \frac{d\Phi_n}{dx}\right)\Big|_{x=0}^{x=1} + \int_0^1 \Theta \frac{d^2\Phi_n}{dx^2}\,dx$$

$$= -\left(\Phi_n \frac{\partial \Theta}{\partial x}\right)\Big|_{x=0} - \lambda_n^2 \Theta_n = \Phi_n(0) - (n\pi)^2 \Theta_n. \tag{4.5-39}$$

Equation (4.5-39), the transformed time derivative [i.e., Eq. (4.5-31)], and the transformed initial condition yield the relations governing Θ_n. The results are written separately for $n=0$ and $n>0$ as

$$\frac{d\Theta_0}{dt}=1, \qquad \Theta_0(0)=0, \tag{4.5-40}$$

$$\frac{d\Theta_n}{dt}+(n\pi)^2\Theta_n=\sqrt{2}, \qquad \Theta_n(0)=0, \qquad n=1,2,\,\ldots. \tag{4.5-41}$$

Solving Eqs. (4.5-40) and (4.5-41) for Θ_0 and Θ_n, respectively, gives

$$\Theta_0=t, \tag{4.5-42}$$

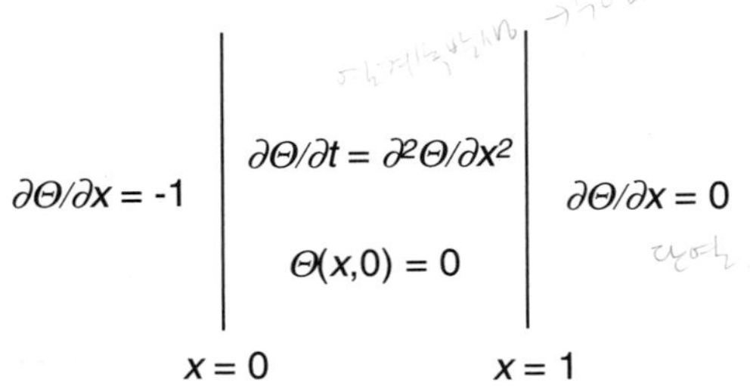

Figure 4-9. Transient heat conduction in a solid slab, with a constant heat flux at one boundary and the other boundary insulated.

$\partial\Theta/\partial t = \partial^2\Theta/\partial x^2$

$\partial\Theta/\partial x = -1$

$\Theta(x,0) = 0$

$\partial\Theta/\partial x = 0$

$x = 0$ $x = 1$

$$\Theta_n = \frac{\sqrt{2}}{(n\pi)^2}\left(1 - e^{-(n\pi)^2 t}\right), \qquad n = 1, 2, \ldots . \tag{4.5-43}$$

The overall solution is

$$\Theta(x, t) = t + 2\sum_{n=1}^{\infty}\left(1 - e^{-(n\pi)^2 t}\right)\frac{\cos n\pi x}{(n\pi)^2}. \tag{4.5-44}$$

Focusing now on the behavior of the solution at long times, we see that

$$\lim_{t\to\infty}\Theta(x, t) = t + 2\sum_{n=1}^{\infty}\frac{\cos n\pi x}{(n\pi)^2} \equiv t + f(x). \tag{4.5-45}$$

In contrast to Example 4.5-3, where the solution eventually became a function of x only, here there is no steady state. The physical reason for this is the imbalance in the heat fluxes at the two surfaces. Because the boundary conditions imply a continual net input of energy, the temperature increases indefinitely; the constant rate of temperature increase achieved at large times [the part of Eq. (4.5-45) proportional to t] is simply a reflection of the constant rate of heating at the $x=0$ surface. (In reality, such temperature increases could occur only for a limited time, because a constant net heat flux could not be sustained once the surface became sufficiently hot.)

The form of the function $f(x)$ in Eq. (4.5-45) is inferred now from a combination of local (differential) and overall (integral) energy balances. We begin by stating some general relations, and then we apply them to this particular problem. Integrating the local conservation equation over x gives the integral balance for a control volume extending from $x=0$ to $x=1$:

$$\int_0^1 \frac{\partial\Theta}{\partial t}\, dx = \int_0^1 \frac{\partial^2\Theta}{\partial x^2}\, dx = \left.\frac{\partial\Theta}{\partial x}\right|_{x=0}^{x=1}. \tag{4.5-46}$$

Equation (4.5-46) is integrated over time to give

$$\overline{\Theta}(t) - \overline{\Theta}(0) = \int_0^t \left.\frac{\partial\Theta}{\partial x}\right|_{x=0}^{x=1} ds, \tag{4.5-47}$$

$$\overline{\Theta}(t) \equiv \int_0^1 \Theta(x, t)\, dx, \tag{4.5-48}$$

where $\overline{\Theta}(t)$ is the average temperature at a given instant and s is a dummy variable. The foregoing relations are valid for all t. For a system with *constant fluxes* at both boundaries, we assume that the long-time solution is of the form

$$\lim_{t\to\infty}\Theta(x, t) = Bt + f(x), \tag{4.5-49}$$

where B is a constant; compare this with Eq. (4.5-45). It follows that

$$\lim_{t\to\infty}\overline{\Theta}(t) = Bt + \int_0^1 f(x)\, dx. \tag{4.5-50}$$

Using Eq. (4.5-49) to evaluate the terms in the differential energy equation at large times, we obtain

$$\lim_{t\to\infty}\frac{\partial\Theta}{\partial t} = \frac{\partial}{\partial t}[Bt + f(x)] = B, \tag{4.5-51}$$

$$\lim_{t\to\infty}\frac{\partial^2\Theta}{\partial x^2} = \frac{\partial^2}{\partial x^2}[Bt + f(x)] = \frac{d^2 f}{dx^2}. \tag{4.5-52}$$

This confirms that Eq. (4.5-49) is the correct limiting form for the solution, provided that $f(x)$ satisfies the differential equation,

$$\frac{d^2f}{dx^2} = B. \tag{4.5-53}$$

Using the initial and boundary conditions of the present problem in Eq. (4.5-47), we find that for all t,

$$\overline{\Theta}(t) = t. \tag{4.5-54}$$

Comparing this result with Eq. (4.5-50), it is seen that $B = 1$ [in agreement with Eq. (4.5-45)] and that

$$\int_0^1 f(x)\,dx = 0. \tag{4.5-55}$$

Equation (4.5-53) and the boundary conditions indicate that $f(x)$ also must satisfy

$$\frac{d^2f}{dx^2} = 1, \qquad f'(0) = -1, \qquad f'(1) = 0. \tag{4.5-56}$$

At first glance it appears that with Eq. (4.5-55) in addition to the two boundary conditions in Eq. (4.5-56), the problem for $f(x)$ is overspecified. That is not true, because the information gained from the boundary condition at $x = 0$ turns out to be equivalent to that at $x = 1$; the integral constraint in Eq. (4.5-55) is needed to obtain a unique solution. The solution for $f(x)$ is

$$f(x) = \frac{x^2}{2} - x + \frac{1}{3}. \tag{4.5-57}$$

Thus, the cosine series in Eq. (4.5-45) is found to be an expansion of the polynomial given by Eq. (4.5-57); an independent calculation of the Fourier expansion of Eq. (4.5-57) confirms that this is the case. The complete solution [Eq. (4.5-44)] is then rewritten more clearly as

$$\Theta(x,\ t) = t + \frac{x^2}{2} - x + \frac{1}{3} - 2\sum_{n=1}^{\infty} e^{-(n\pi)^2 t}\frac{\cos n\pi x}{(n\pi)^2}. \tag{4.5-58}$$

Once again, it is only this last form of the solution which represents the boundary data exactly, and in which termwise differentiation of the Fourier series may be used to calculate $\partial\Theta/\partial x$.

The approach just described is useful also for transient problems which have steady states, the main difference being that for such cases $B = 0$. Thus, for Example 4.5-3,

$$\frac{d^2f}{dx^2} = 0, \qquad f(0) = 1, \qquad f(1) = 0. \tag{4.5-59}$$

The solution obtained from Eq. (4.5-59),

$$f(x) = 1 - x, \tag{4.5-60}$$

is that already incorporated in Eq. (4.5-36).

Example 4.5-5 Transient Diffusion with a Time-Dependent Boundary Condition   We now consider a problem similar to that in Example 4.5-3, except that one of the boundary conditions is a function of time for $t > 0$. As shown in Fig. 4-10, it is assumed that after the initial step change at $t = 0$, the concentration at $x = 0$ decays exponentially with time, with a (dimensionless) time constant κ^{-1}.

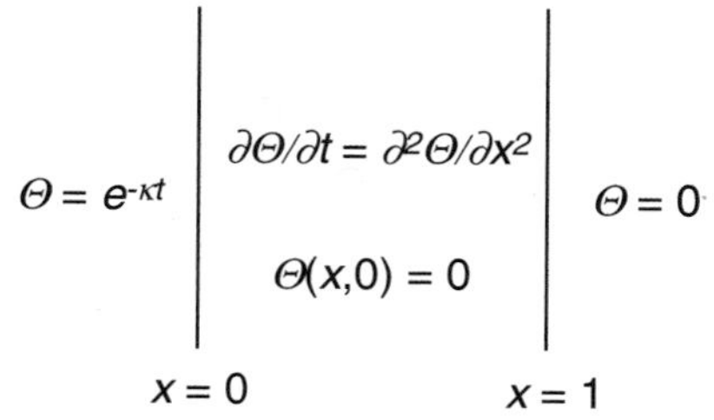

Figure 4-10. Transient diffusion in a membrane, with a time-dependent concentration at one boundary.

$$\Theta = e^{-\kappa t} \quad \left| \begin{array}{c} \partial\Theta/\partial t = \partial^2\Theta/\partial x^2 \\ \\ \Theta(x,0) = 0 \end{array} \right| \quad \Theta = 0$$

$$x = 0 \qquad\qquad x = 1$$

The basis functions are the same as those used in Example 4.5-3, and the approach is very similar. Equations (4.5-32) and (4.5-33) become

$$\int_0^1 \Phi_n \frac{\partial^2\Theta}{\partial x^2}\,dx = \left(\Theta\,\frac{d\Phi_n}{dx}\right)\Bigg|_{x=0} - \lambda_n^2\Theta_n = \sqrt{2}\,n\pi e^{-\kappa t} - (n\pi)^2\Theta_n, \tag{4.5-61}$$

$$\frac{d\Theta_n}{dt} + (n\pi)^2\Theta_n = \sqrt{2}\,n\pi e^{-\kappa t}, \qquad \Theta_n(0) = 0. \tag{4.5-62}$$

Equation (4.5-62) differs from Eq. (4.5-33) only in that the nonhomogeneous term in the differential equation is now a function of time. This first-order, linear differential equation has the general solution

$$\Theta_n = c_n e^{-(n\pi)^2 t} + \frac{\sqrt{2}}{n\pi}\left[1 - \frac{\kappa}{(n\pi)^2}\right]^{-1} e^{-\kappa t}. \qquad (?) \tag{4.5-63}$$

Using the initial condition to evaluate the constant c_n, the transformed concentration is found to be

$$\Theta_n = \frac{\sqrt{2}}{n\pi}\left[1 - \frac{\kappa}{(n\pi)^2}\right]^{-1} e^{-\kappa t}\left[1 - e^{-[(n\pi)^2 - \kappa]t}\right]. \tag{4.5-64}$$

Combining the transformed concentration from Eq. (4.5-64) with the basis functions from Eq. (4.5-30), the overall solution is

$$\Theta(x, t) = 2e^{-\kappa t}\sum_{n=1}^{\infty}\left[1 - \frac{\kappa}{(n\pi)^2}\right]^{-1}\left[1 - e^{-[(n\pi)^2 - \kappa]t}\right]\frac{\sin n\pi x}{n\pi}, \tag{4.5-65}$$

which reduces to Eq. (4.5-35) for $\kappa = 0$ (constant boundary concentration), as it should.

Using the separation-of-variables method to solve problems with time-dependent boundary conditions, such as this one, it is necessary to employ a special technique called *Duhamel superposition* [see Arpaci (1966)]. As seen here, the FFT method requires no modification for such problems.

The solution given by Eq. (4.5-65) provides insight into the pseudosteady approximation used for the membrane diffusion example in Section 3.4. If κ is small, then the changes imposed on the boundary at $x=0$ are relatively slow. In terms of Example 3.4-2, small κ is similar to having $L^2/(D\tau) \ll 1$. In other words, the characteristic time for diffusion is much smaller than the external time constant which influences the boundary condition(s), so that the diffusional response of the system is comparatively rapid. From Eq. (4.5-65), a more precise criterion for slow changes at the boundary is $\kappa/\pi^2 \ll 1$. When that condition is satisfied, the solution reduces to

$$\Theta(x,\ t) = 2e^{-\kappa t} \sum_{n=1}^{\infty} [1 - e^{-(n\pi)^2 t}] \frac{\sin n\pi x}{n\pi} = e^{-\kappa t} \Theta^*(x,\ t), \tag{4.5-66}$$

where $\Theta^*(x,t)$ denotes the solution obtained by specifying the constant concentration $\Theta = 1$ at $x = 0$; compare Eq. (4.5-66) with Eq. (4.5-35). For $t \gg 1/\pi^2$ (or $t \gg 0.1$), all of the transient terms in the summation are negligible and Eq. (4.5-66) reduces to

$$\Theta(x,\ t) = e^{-\kappa t} \left(2 \sum_{n=1}^{\infty} \frac{\sin n\pi x}{n\pi} \right) = e^{-\kappa t}(1 - x). \tag{4.5-67}$$

The last form of Eq. (4.5-67) is exactly what we would have obtained, much more simply, from a pseudosteady analysis of the present problem (i.e., setting $\partial\Theta/\partial t = 0$ in the partial differential equation). As discussed in Section 3.4, it is seen that the pseudosteady approximation has two requirements: The boundary conditions must change slowly compared to the diffusional response of the system, and a certain amount of time must elapse before the concentration profile can achieve its pseudosteady form.

Example 4.5-6 Steady Conduction in Three Dimensions Problems involving three (or four) independent variables are solved by the application of two (or three) finite Fourier transforms. The steady-state heat conduction problem in three dimensions shown in Fig. 4-11 will illustrate the approach. It is assumed that $\Theta = 1$ on the top surface of a rectangular solid of dimensions $a \times b \times c$, and $\Theta = 0$ on all other surfaces; there is no volumetric heat source term.
The full problem statement is

$$\frac{\partial^2\Theta}{\partial x^2} + \frac{\partial^2\Theta}{\partial y^2} + \frac{\partial^2\Theta}{\partial z^2} = 0, \tag{4.5-68}$$

$$\Theta(0,\ y,\ z) = 0, \qquad \Theta(a,\ y,\ z) = 0, \tag{4.5-69}$$

$$\Theta(x,\ 0,\ z) = 0, \qquad \Theta(x,\ b,\ z) = 0, \tag{4.5-70}$$

$$\Theta(x,\ y,\ 0) = 0, \qquad \Theta(x,\ y,\ c) = 1. \tag{4.5-71}$$

The expansion for $\Theta(x,\ y,\ z)$ will be written using two sets of basis functions, one involving x and the other y. Because of the Dirichlet boundary conditions, we choose (from case I of Table 4-1) the basis functions

$$\Phi_n(x) = \sqrt{\frac{2}{a}} \sin \frac{n\pi x}{a}, \qquad n = 1,2, \dots, \tag{4.5-72}$$

$$\Psi_m(y) = \sqrt{\frac{2}{b}} \sin \frac{m\pi y}{b}, \qquad m = 1,2, \dots. \tag{4.5-73}$$

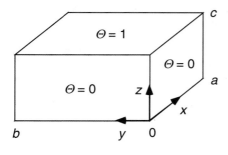

Figure 4-11. Steady heat conduction in a rectangular solid. There is a constant temperature, $\Theta = 1$, at the top surface $(z = c)$ and $\Theta = 0$ on all other surfaces.

The temperature is transformed first using the basis functions in x to give

$$\Theta_n(y, z) = \int_0^a \Phi_n(x) \Theta(x, y, z) \, dx. \tag{4.5-74}$$

At this stage the expansion is written as

$$\Theta(x, y, z) = \sum_{n=1}^{\infty} \Theta_n(y, z) \Phi_n(x), \tag{4.5-75}$$

which is a straightforward extension of Eq. (4.4-4). Transforming a second time using the basis functions in y results in

$$\Theta_{nm}(z) = \int_0^b \Psi_m(y) \Theta_n(y, z) \, dy = \int_0^b \int_0^a \Phi_n(x) \Psi_m(y) \Theta(x, y, z) \, dx \, dy, \tag{4.5-76}$$

$$\Theta(x, y, z) = \sum_{n=1}^{\infty} \sum_{m=1}^{\infty} \Theta_{nm}(z) \Phi_n(x) \Psi_m(y). \tag{4.5-77}$$

Clearly, the order in which the transforms were applied was immaterial. The problem is reduced now to that of computing the coefficient functions, $\Theta_{nm}(z)$.

In transforming the boundary-value problem we apply both FFT's (x and y) at once, thereby proceeding directly to the governing equations for $\Theta_{nm}(z)$. Applying both transforms to the x derivative in Eq. (4.5-68) gives

$$\int_0^b \int_0^a \Phi_n \Psi_m \frac{\partial^2 \Theta}{\partial x^2} \, dx \, dy = \int_0^b \left[\left(\Phi_n \frac{\partial \Theta}{\partial x} - \Theta \frac{d\Phi_n}{dx} \right) \Big|_{x=0}^{x=a} + \int_0^a \Theta \frac{d^2 \Phi_n}{dx^2} \, dx \right] \Psi_m \, dy$$

$$= -(n\pi/a)^2 \Theta_{nm}. \tag{4.5-78}$$

Doing the same for the y and z derivatives, we obtain

$$\int_0^b \int_0^a \Phi_n \Psi_m \frac{\partial^2 \Theta}{\partial y^2} \, dx \, dy = -(m\pi/b)^2 \Theta_{nm}, \tag{4.5-79}$$

$$\int_0^b \int_0^a \Phi_n \Psi_m \frac{\partial^2 \Theta}{\partial z^2} \, dx \, dy = \frac{d^2 \Theta_{nm}}{dz^2}. \tag{4.5-80}$$

Combining Eqs. (4.5-78)–(4.5-80), we obtain the differential equation for $\Theta_{nm}(z)$,

$$\frac{d^2 \Theta_{nm}}{dz^2} - [(n/a)^2 + (m/b)^2] \pi^2 \Theta_{nm} = 0. \tag{4.5-81}$$

Applying the two integral transforms to the nonhomogeneous boundary condition at $z = c$ gives

$$\int_0^b \int_0^a \Phi_n \Psi_m \Theta(x, y, c) \, dx \, dy = \int_0^b \int_0^a \Phi_n \Psi_m \, dx \, dy$$

$$= \left(-\frac{\sqrt{2a}}{n\pi} \cos \frac{n\pi x}{a} \right) \Big|_{x=0}^{x=a} \left(-\frac{\sqrt{2b}}{m\pi} \cos \frac{m\pi x}{b} \right) \Big|_{y=0}^{y=b}$$

$$= \frac{\sqrt{2a}}{n\pi} [1 - (-1)^n] \frac{\sqrt{2b}}{m\pi} [1 - (-1)^m]. \tag{4.5-82}$$

The boundary conditions for Eq. (4.5-81) are then

$$\Theta_{nm}(0) = 0, \qquad \Theta_{nm}(c) = \frac{2\sqrt{ab}}{nm\pi^2} [1 - (-1)^n][1 - (-1)^m]. \tag{4.5-83}$$

The solution for the (doubly) transformed temperature is

$$\Theta_{nm} = \frac{2\sqrt{ab}}{nm\pi^2} [1-(-1)^n][1-(-1)^m] \frac{\sinh\{[(n/a)^2+(m/b)^2]^{1/2}\pi z\}}{\sinh\{[(n/a)^2+(m/b)^2]^{1/2}\pi c\}}. \tag{4.5-84}$$

Using Eq. (4.5-77), and recognizing from Eq. (4.5-84) that only the odd values of n and m will contribute, the overall solution is found to be

$$\Theta(x, y, z) = \frac{16}{\pi^2} \sum_{\substack{n=1 \\ n \text{ odd}}}^{\infty} \sum_{\substack{m=1 \\ m \text{ odd}}}^{\infty} \frac{1}{nm} \frac{\sinh\{[(n/a)^2+(m/b)^2]^{1/2}\pi z\}}{\sinh\{[(n/a)^2+(m/b)^2]^{1/2}\pi c\}} \sin\left(\frac{n\pi x}{a}\right) \sin\left(\frac{m\pi y}{b}\right). \tag{4.5-85}$$

In general, if there are k independent variables in a partial differential equation, then $k-1$ finite Fourier transforms are needed to obtain the solution.

4.6 SELF-ADJOINT EIGENVALUE PROBLEMS AND STURM–LIOUVILLE THEORY

It was mentioned in Section 4.3 that if an eigenvalue problem is self-adjoint, then the eigenfunctions are orthogonal (with the proper weighting function) and the eigenvalues are real. It was also stated that eigenvalue problems based on Eq. (4.3-4) are self-adjoint. To understand such statements and to extend the FFT method to a wider variety of problems, including conduction or diffusion in cylindrical or spherical coordinates, it is necessary to define what is a self-adjoint eigenvalue problem. An eigenvalue problem involving the independent variable x ordinarily consists of a second-order differential equation in the form of Eq. (4.3-1), along with homogeneous boundary conditions of the types given in Eq. (4.3-5). That problem is said to be *self-adjoint* if the operator $\mathcal{L}_x$ is such that

$$\langle \mathcal{L}_x u, v \rangle = \langle u, \mathcal{L}_x v \rangle, \tag{4.6-1}$$

where $u(x)$ and $v(x)$ are functions which satisfy the *boundary conditions* of the eigenvalue problem, but not necessarily the differential equation. Implicit in Eq. (4.6-1) is the specification of a certain interval $[a, b]$ and weighting function $w(x)$.

To illustrate the application of Eq. (4.6-1), consider the familiar operator $\mathcal{L}_x = d^2/dx^2$, for which $w(x) = 1$. Starting with the left-hand side of Eq. (4.6-1) and integrating by parts, we find that

$$\langle \mathcal{L}_x u, v \rangle = \int_a^b v \frac{d^2u}{dx^2} dx = \left(v \frac{du}{dx} - u \frac{dv}{dx} \right) \Big|_{x=a}^{x=b} + \int_a^b u \frac{d^2v}{dx^2} dx = \langle u, \mathcal{L}_x v \rangle, \tag{4.6-2}$$

which confirms that problems involving this operator are self-adjoint. The boundary terms vanish as a consequence of the homogeneous boundary conditions satisfied by u and v. This is obvious when the boundary conditions are of the Dirichlet or Neumann types. If, say, there is a Robin boundary condition at $x=b$, then $u'(b)+Au(b)=v'(b)+Av(b)=0$ and

$$\left(v \frac{du}{dx} - u \frac{dv}{dx} \right) \Big|_{x=b} = v(b)[-Au(b)] - u(b)[-Av(b)] = 0. \tag{4.6-3}$$

Thus, with $\mathscr{L}_x = d^2/dx^2$, any combination of the usual three types of boundary conditions gives a self-adjoint eigenvalue problem.

It is shown now that a broad class of differential equations yields self-adjoint eigenvalue problems. Consider the general, second-order, linear, homogeneous equation,

$$\frac{d^2u}{dx^2} + f_1(x)\frac{du}{dx} + f_0(x)u = 0. \tag{4.6-4}$$

Setting $p(x) = \exp(\int f_1(x)\,dx)$ and $q(x) = p(x)f_0(x)$, Eq. (4.6-4) is rewritten as

$$\frac{d}{dx}\left(p\frac{du}{dx}\right) + qu = 0. \tag{4.6-5}$$

Motivated by the form of Eq. (4.6-5), we choose the operator $\mathscr{L}_x$ associated with some weighting function $w(x)$ as

$$\mathscr{L}_x \equiv \frac{1}{w}\left[\frac{d}{dx}\left(p\frac{d}{dx}\right) + q\right], \tag{4.6-6}$$

where it is assumed that p, dp/dx, q, and w are all continuous in the interval of interest, $a \le x \le b$, and that $p > 0$ and $w > 0$ for $a < x < b$. Assuming again that $u(x)$ and $v(x)$ satisfy any combination of the three types of homogeneous boundary conditions, the left-hand side of Eq. (4.6-1) becomes

$$\langle \mathscr{L}_x u, v\rangle = \int_a^b \frac{1}{w}\left[\frac{d}{dx}\left(p\frac{du}{dx}\right) + qu\right]vw\,dx$$

$$= \int_a^b v\frac{d}{dx}\left(p\frac{du}{dx}\right)dx + \int_a^b quv\,dx$$

$$= \left(pv\frac{du}{dx}\right)\Bigg|_{x=a}^{x=b} - \int_a^b p\frac{dv}{dx}\frac{du}{dx}\,dx + \int_a^b quv\,dx$$

$$= p\left(v\frac{du}{dx} - u\frac{dv}{dx}\right)\Bigg|_{x=a}^{x=b} + \int_a^b u\frac{d}{dx}\left(p\frac{dv}{dx}\right)dx + \int_a^b quv\,dx$$

$$= \int_a^b u\frac{d}{dx}\left(p\frac{dv}{dx}\right)dx + \int_a^b quv\,dx. \tag{4.6-7}$$

The boundary terms vanish as before, provided that $p(a)$ and $p(b)$ are finite. Evaluating the right-hand side of Eq. (4.6-1) gives

$$\langle u, \mathscr{L}_x v\rangle = \int_a^b u\frac{1}{w}\left[\frac{d}{dx}\left(p\frac{dv}{dx}\right) + qv\right]w\,dx$$

$$= \int_a^b u\frac{d}{dx}\left(p\frac{dv}{dx}\right)dx + \int_a^b quv\,dx = \langle \mathscr{L}_x u, v\rangle. \tag{4.6-8}$$

Accordingly, eigenvalue problems based on the differential equation,

$$\frac{1}{w}\left[\frac{d}{dx}\left(p\frac{d\Phi}{dx}\right) + q\Phi\right] = -\lambda^2\Phi \tag{4.6-9}$$

with boundary conditions as in Eq. (4.3-5), are self-adjoint with respect to the weighting function $w(x)$. Eigenvalue problems of this form are called *Sturm–Liouville* problems[3] [see, for example, Churchill (1963) and Greenberg (1988)].

The boundary conditions given in Eq. (4.3-5) are not the only ones which yield self-adjoint eigenvalue problems. Also leading to such problems is the *periodicity* requirement,

$$\Phi(a) = \Phi(b), \qquad \Phi'(a) = \Phi'(b). \qquad (4.6\text{-}10)$$

Inspection of Eq. (4.6-7) indicates that if u and v satisfy Eq. (4.6-10), then the boundary terms will vanish in the integration by parts, as required for a self-adjoint problem. Equation (4.6-10) arises, for example, with problems in cylindrical coordinates involving the angle θ, in which the domain is a complete circle. One other way a self-adjoint eigenvalue problem is obtained is if p vanishes at one or both ends of the interval. If, for example, $p(a) = 0$ and $\Phi(a)$ and $\Phi'(a)$ are both finite, then the corresponding boundary term in Eq. (4.6-7) will again vanish. This type of condition arises in certain problems in cylindrical or spherical coordinates (see Sections 4.7 and 4.8).

All Sturm–Liouville problems yield eigenvalues which are real and eigenfunctions which are orthogonal. A proof that λ is real is given, for example, in Churchill (1963). The orthogonality of the eigenfunctions is a direct consequence of the fact that the Sturm–Liouville operator is self-adjoint, as is shown now. Considering two different eigenfunctions Φ_n and Φ_m, self-adjointness implies that

$$\langle \mathcal{L}_x\Phi_n, \Phi_m \rangle = \langle \Phi_n, \mathcal{L}_x\Phi_m \rangle. \qquad (4.6\text{-}11)$$

From the definition of the eigenvalue problem and the properties of the inner product (Section 4.3) we obtain

$$\langle \mathcal{L}_x\Phi_n, \Phi_m \rangle = \langle -\lambda_n^2\Phi_n, \Phi_m \rangle = -\lambda_n^2\langle \Phi_n, \Phi_m \rangle, \qquad (4.6\text{-}12)$$

$$\langle \Phi_n, \mathcal{L}_x\Phi_m \rangle = \langle \Phi_n, -\lambda_m^2\Phi_m \rangle = -\lambda_m^2\langle \Phi_n, \Phi_m \rangle. \qquad (4.6\text{-}13)$$

Subtracting Eq. (4.6-13) from Eq. (4.6-12) leads to

$$0 = (\lambda_m^2 - \lambda_n^2)\langle \Phi_n, \Phi_m \rangle. \qquad (4.6\text{-}14)$$

Because $\lambda_m \neq \lambda_n$, this implies that

$$\langle \Phi_n, \Phi_m \rangle = \int_a^b \Phi_n\Phi_m w \, dx = 0 \qquad (n \neq m). \qquad (4.6\text{-}15)$$

In other words, the solutions to Sturm–Liouville problems are orthogonal.

In Sturm–Liouville problems with boundary conditions from Eq. (4.3-5), or with $p = 0$ at one of the endpoints, the eigenfunctions are unique. That is, there is only one linearly independent eigenfunction corresponding to a given eigenvalue. A proof of this, along with a discussion of the exception which occurs with periodic boundary conditions, is given by Churchill (1963).

When applying the FFT method, the most awkward term to transform in the partial differential equation is that involving $\mathcal{L}_x$, the second-order operator in the basis function

---

[3] This general theory is due to C. Sturm (1803–1855) and J. Liouville (1809–1882), French mathematicians in the generation folowing Fourier.

coordinate. Taking advantage of the integration by parts performed for the Sturm–Liouville operator in Eq. (4.6-7), the transform of that term is written generally as

$$\langle \mathcal{L}_x \Theta, \Phi_n \rangle = p \left(\Phi_n \frac{\partial \Theta}{\partial x} - \Theta \frac{d\Phi_n}{dx} \right) \Bigg|_{x=a}^{x=b} - \lambda_n^2 \Theta_n. \tag{4.6-16}$$

The eigenvalue problems introduced in Section 4.3 and applied in Section 4.5 correspond to $w=1$, $p=1$, and $q=0$. The operators and eigenfunctions in some of the more common problems in cylindrical and spherical coordinates are discussed in Sections 4.7 and 4.8. Eigenvalue problems not encompassed by classical Sturm–Liouville theory, such as those where w and p are discontinuous, are discussed in Ramkrishna and Amundson (1974a,b, 1985) and Hatton et al. (1979). A few such situations are examined in the problems at the end of this chapter.

4.7 FFT METHOD FOR PROBLEMS IN CYLINDRICAL COORDINATES

For the most general type of problem in cylindrical coordinates, $\Theta = \Theta(r, z, \theta, t)$. When there is no volumetric source term, the dimensionless energy or species conservation equation is

$$\frac{\partial \Theta}{\partial t} = \frac{1}{r} \frac{\partial}{\partial r} \left(r \frac{\partial \Theta}{\partial r} \right) + \frac{\partial^2 \Theta}{\partial z^2} + \frac{1}{r^2} \frac{\partial^2 \Theta}{\partial \theta^2}. \tag{4.7-1}$$

We will be concerned only with steady two-dimensional and unsteady one-dimensional problems involving $\Theta(r, z)$, $\Theta(r, \theta)$, or $\Theta(r, t)$. For $\Theta(r, z)$, the eigenvalue problems in z will have the same differential operator as in rectangular coordinates, and thus they will be identical to those considered in Section 4.3. For $\Theta(r, \theta)$, the eigenvalue problems in θ again have the same differential operator; the only distinctive feature of these nonaxisymmetric problems has to do with periodic boundary conditions, as discussed at the end of this section. However, with $\Theta(r, z)$ or $\Theta(r, t)$, the eigenvalue problems in r involve a differential equation not yet considered, Bessel's equation. Because they are the most unique aspect of cylindrical problems, the solutions of this equation are discussed first.

Bessel Functions

Eigenvalue problems in r result in

$$\frac{1}{r} \frac{d}{dr} \left(r \frac{d\Phi}{dr} \right) = -\lambda^2 \Phi, \tag{4.7-2}$$

which is a Sturm–Liouville equation with $w(r)=p(r)=r$ and $q(r)=0$ [compare with Eq. (4.6-9)]. For problems involving annular regions, such that $a \leq r \leq b$ with $a \neq 0$, a self-adjoint eigenvalue problem results when any of the three types of boundary conditions from Eq. (4.3-5) are applied at $r=a$ and $r=b$. For domains containing the origin, all that is required at $r=0$ is that Φ and Φ' be finite, as discussed in Section 4.6.

Equation (4.7-2) is a particular form of *Bessel's equation*, which is written more generally as

$$x \frac{d}{dx}\left(x \frac{df}{dx}\right) + (m^2 x^2 - v^2) f = 0, \tag{4.7-3}$$

where m is a parameter and v is any real constant. In Eq. (4.7-2), $m = \lambda$ and $v = 0$. The properties of the solutions to Bessel's equation have been studied extensively (Watson, 1944). The two linearly independent solutions are usually written as $J_v(mx)$ and $Y_v(mx)$, and are known as *Bessel functions* of order v of the first and second kind, respectively. Values of Bessel functions of integer order are available from numerous sources, including software written for personal computers, making calculations involving these functions quite routine. The solutions to Eq. (4.7-2) are the Bessel functions of order zero, $J_0(\lambda r)$ and $Y_0(\lambda r)$. The first derivatives and the integrals of $J_0(\lambda r)$ and $Y_0(\lambda r)$ can both be expressed in terms of the corresponding Bessel functions of order one, $J_1(\lambda r)$ and $Y_1(\lambda r)$, so that we need be concerned here only with the properties of J_0, J_1, Y_0, and Y_1. Graphs of these functions are shown in Fig. 4-12. As the plots suggest, all of the functions have an infinite number of roots. Two values worth noting are $J_0(0) = 1$ and $J_1(0) = 0$. An important distinction between Bessel functions of the first and second kinds is that, whereas $J_0(0)$ and $J_1(0)$ are finite, $Y_0(0)$ and $Y_1(0)$ are not.

Of the many identities involving Bessel functions, ones which are particularly helpful to us in evaluating derivatives and integrals are

$$\frac{dJ_0(mx)}{dx} = -mJ_1(mx), \qquad \frac{d}{dx}[xJ_1(mx)] = mxJ_0(mx), \tag{4.7-4a,b}$$

$$\frac{dY_0(mx)}{dx} = -mY_1(mx), \qquad \frac{d}{dx}[xY_1(mx)] = mxY_0(mx). \tag{4.7-5a,b}$$

Consider an eigenvalue problem involving Eq. (4.7-2) on the interval $0 \le r \le 1$, in which one boundary condition is $\Phi(1) = 0$. The fact that $Y_0(0)$ is unbounded requires that the solution $Y_0(\lambda r)$ be excluded, so that

$$\Phi(r) = aJ_0(\lambda r), \tag{4.7-6}$$

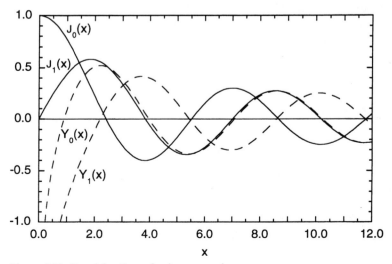

Figure 4-12. Bessel functions of orders zero and one.

where a is a constant. The boundary condition at $r=1$ yields the characteristic equation,

$$J_0(\lambda_n) = 0. \tag{4.7-7}$$

That is, the eigenvalues are the roots of J_0. The normalization condition for the eigenfunctions is

$$\langle \Phi_n, \Phi_n \rangle = 1 = a_n^2 \int_0^1 J_0^2(\lambda_n r) r \, dr. \tag{4.7-8}$$

Notice the inclusion of the weighting function $w(r) = r$ in the inner product.

To determine a_n, we need to evaluate the integral in Eq. (4.7-8). Integrals like this occur in any eigenvalue problem involving J_0, so that it is worthwhile to derive a general result for the interval $[0,\ell]$, valid for any of the three types of boundary conditions at $r = \ell$. This is done by first recalling that $J_0(\lambda_n r)$ satisfies Bessel's equation with $m = \lambda_n$ and $\nu = 0$, so that

$$r \frac{d}{dr}\left(r \frac{dJ_0}{dr}\right) = -\lambda_n^2 r^2 J_0. \tag{4.7-9}$$

The next step is to multiply both sides of Eq. (4.7-9) by $2(dJ_0/dr)$. The left-hand side gives

$$2r \frac{dJ_0}{dr} \frac{d}{dr}\left(r \frac{dJ_0}{dr}\right) = \frac{d}{dr}\left(r \frac{dJ_0}{dr}\right)^2. \tag{4.7-10}$$

Integrating Eq. (4.7-10) from $r=0$ to $r=\ell$ and using Eq. (4.7-4a) to evaluate the derivative of J_0, we obtain

$$\int_0^\ell \frac{d}{dr}\left(r \frac{dJ_0}{dr}\right)^2 dr = \left(r \frac{dJ_0}{dr}\right)^2 \bigg|_{r=0}^{r=\ell} = \ell^2 \left(\frac{dJ_0}{dr}\right)^2 \bigg|_{r=\ell} = \ell^2 \lambda_n^2 J_1^2(\lambda_n \ell). \tag{4.7-11}$$

The modified right-hand side of Eq. (4.7-9) is rearranged as

$$-2\lambda_n^2 r^2 J_0 \frac{dJ_0}{dr} = -\lambda_n^2 r^2 \frac{d}{dr}(J_0)^2. \tag{4.7-12}$$

Integrating Eq. (4.7-12) by parts gives

$$-\lambda_n^2 \int_0^\ell r^2 \frac{d}{dr}(J_0)^2 \, dr = -\lambda_n^2\left[(rJ_0)^2 \bigg|_{r=0}^{r=\ell} - 2\int_0^\ell J_0^2 r \, dr\right] = -\lambda_n^2\left[\ell^2 J_0^2(\lambda_n \ell) - 2\int_0^\ell J_0^2 r \, dr\right]. \tag{4.7-13}$$

Equating the results from Eqs. (4.7-11) and (4.7-13) leads to the desired identity, which is

$$\int_0^\ell J_0^2(\lambda_n r) r \, dr = \frac{\ell^2}{2\lambda_n^2}\left[\lambda_n^2 J_0^2(\lambda_n \ell) + \left(\frac{dJ_0}{dr}\right)^2 \bigg|_{r=\ell}\right] = \frac{\ell^2}{2}[J_0^2(\lambda_n \ell) + J_1^2(\lambda_n \ell)]. \tag{4.7-14}$$

Either form of Eq. (4.7-14) may be more convenient, depending on the boundary condition at $r = \ell$.

A second integral which may as well be evaluated now is one which arises when representing any constant as a Fourier–Bessel series involving $J_0(\lambda_n r)$. The integral is

Table 4-2
Orthonormal Sequences of Functions from Certain Eigenvalue Problems in Cylindrical Coordinates [a]

Case	Boundary condition	Characteristic equation [b]	Basis functions
I	$\Phi_n(\ell)=0$	$J_0(\lambda_n\ell)=0$	$\Phi_n(r)=\dfrac{\sqrt{2}}{\ell}\dfrac{J_0(\lambda_n r)}{J_1(\lambda_n\ell)}$
II	$\Phi'_n(\ell)=0$	$J_1(\lambda_n\ell)=0$	$\Phi_n(r)=\dfrac{\sqrt{2}}{\ell}\dfrac{J_0(\lambda_n r)}{J_0(\lambda_n\ell)}$, $n=1,2,\ldots$
			$\Phi_0(r)=\dfrac{\sqrt{2}}{\ell}$
III	$\Phi'_n(\ell)+A\Phi_n(\ell)=0$	$\lambda_n=A\dfrac{J_0(\lambda_n\ell)}{J_1(\lambda_n\ell)}$	$\Phi_n(r)=\dfrac{\sqrt{2}}{\ell}\left[1+\left(\dfrac{A}{\lambda_n}\right)^2\right]^{-1/2}\dfrac{J_0(\lambda_n r)}{J_0(\lambda_n\ell)}$

[a] All of the functions shown satisfy Eq. (4.7-2) on the interval $[0,\ \ell]$, with $\Phi'(0)=0$.
[b] The first few eigenvalues for cases I and II are as follows:
I: $\lambda_1\ell=2.405$, $\lambda_2\ell=5.520$, $\lambda_3\ell=8.654$.
II: $\lambda_0\ell=0$, $\lambda_1\ell=3.832$, $\lambda_2\ell=7.016$, $\lambda_3\ell=10.173$.

$$\int_0^\ell J_0(\lambda_n r)r\ dr=\frac{1}{\lambda_n}\int_0^\ell \frac{d}{dr}[rJ_1(\lambda_n r)]\ dr=\frac{\ell}{\lambda_n}J_1(\lambda_n\ell). \qquad (4.7\text{-}15)$$

Equation (4.7-15) makes use of the identity given as Eq. (4.7-4b).

Returning now to the eigenvalue problem for a full cylinder with a Dirichlet boundary condition at $r=1$, we find from Eqs. (4.7-6), (4.7-8), and (4.7-14) that

$$\Phi_n(r)=\sqrt{2}\,\frac{J_0(\lambda_n r)}{J_1(\lambda_n)}. \qquad (4.7\text{-}16)$$

A very similar procedure is used to derive the orthonormal basis functions corresponding to Neumann or Robin conditions. Table 4-2 summarizes the three types of basis functions arising in the analysis of conduction or diffusion inside a complete cylinder of radius ℓ. For problems involving annular regions, where $r=0$ is not in the domain and the solution $Y_0(\lambda_n r)$ cannot be excluded, the basis functions will be linear combinations of $J_0(\lambda_n r)$ and $Y_0(\lambda_n r)$.

Example 4.7-1 Transient Conduction in an Electrically Heated Wire   Consider a long, cylindrical wire, initially at ambient temperature, which is subjected to a constant rate of heating for $t>0$ due to passage of an electric current. Heat transfer from the wire to the surroundings is described using a convection boundary condition, and the Biot number is not necessarily large or small. This is a transient problem leading to the steady state analyzed in Example 2.8-1. Referring to the dimensional quantities used in that example, we choose $H_V R^2/k$ and R as the temperature and length scales, respectively. Then, the dimensionless problem formulation for $\Theta(r,\ t)$ is

$$\frac{\partial\Theta}{\partial t}=\frac{1}{r}\frac{\partial}{\partial r}\left(r\frac{\partial\Theta}{\partial r}\right)+1, \qquad (4.7\text{-}17)$$

$$\Theta(r,\ 0)=0, \qquad (4.7\text{-}18)$$

$$\frac{\partial\Theta}{\partial r}(0,\ t)=0, \qquad \frac{\partial\Theta}{\partial r}(1,\ t)+\text{Bi}\ \Theta(1,\ t)=0. \qquad (4.7\text{-}19)$$

The only eigenvalue problem derivable from Eq. (4.7-17) is one involving r. The domain includes $r=0$, so that the basis functions will contain only $J_0(\lambda_n r)$. From case III of Table 4-2, the orthonormal basis functions suitable for the Robin boundary condition in Eq. (4.7-19) are

$$\Phi_n(r) = \sqrt{2}\left[1 + \left(\frac{\text{Bi}}{\lambda_n}\right)^2\right]^{-1/2}\frac{J_0(\lambda_n r)}{J_0(\lambda_n)} \tag{4.7-20}$$

with eigenvalues given by the positive roots of

$$\lambda_n = \text{Bi}\,\frac{J_0(\lambda_n)}{J_1(\lambda_n)}. \tag{4.7-21}$$

Inspection of Eq. (4.7-21) and the graphs of the Bessel functions in Fig. 4-12 indicates that zero is not an eigenvalue for any $\text{Bi} > 0$.

The transformed temperature is defined as

$$\Theta_n(t) = \int_0^1 \Phi_n(r)\Theta(r,t)r\,dr. \tag{4.7-22}$$

Notice again the inclusion of the weighting factor r. Using Eq. (4.6-16) to transform the term in Eq. (4.7-17) containing the derivatives in r gives

$$\int_0^1 \Phi_n\left[\frac{1}{r}\frac{\partial}{\partial r}\left(r\frac{\partial\Theta}{\partial r}\right)\right]r\,dr = r\left(\Phi_n\frac{\partial\Theta}{\partial r} - \Theta\frac{d\Phi_n}{dr}\right)\Bigg|_{r=0}^{r=1} - \lambda_n^2\Theta_n = -\lambda_n^2\Theta_n. \tag{4.7-23}$$

The boundary terms vanish because of the homogeneous boundary conditions in Eq. (4.7-19). Transforming the nonhomogeneous term in Eq. (4.7-17) and making use of Eq. (4.7-15), we obtain

$$\int_0^1 \Phi_n r\,dr = \sqrt{2}\left[1 + \left(\frac{\text{Bi}}{\lambda_n}\right)^2\right]^{-1/2}\frac{J_1(\lambda_n)}{\lambda_n J_0(\lambda_n)}. \tag{4.7-24}$$

The transformation of the time derivative and the initial condition is straightforward. It is found that the transformed temperature must satisfy

$$\frac{d\Theta_n}{dt} + \lambda_n^2\,\Theta_n = \sqrt{2}\left[1 + \left(\frac{\text{Bi}}{\lambda_n}\right)^2\right]^{-1/2}\frac{J_1(\lambda_n)}{\lambda_n J_0(\lambda_n)}, \qquad \Theta_n(0) = 0. \tag{4.7-25}$$

The solution to Eq. (4.7-25) is

$$\Theta_n = \sqrt{2}\left[1 + \left(\frac{\text{Bi}}{\lambda_n}\right)^2\right]^{-1/2}\frac{J_1(\lambda_n)}{\lambda_n^3 J_0(\lambda_n)}[1 - e^{-\lambda_n^2 t}] \tag{4.7-26}$$

and the overall solution is

$$\Theta(r,t) = 2\sum_{n=1}^{\infty}\left[1 + \left(\frac{\text{Bi}}{\lambda_n}\right)^2\right]^{-1}[1 - e^{-\lambda_n^2 t}]\frac{J_1(\lambda_n)J_0(\lambda_n r)}{\lambda_n^3 J_0^2(\lambda_n)}. \tag{4.7-27}$$

The form of the overall solution is clarified by rewriting the steady-state part. From Example 2.8-1, the steady-state solution is

$$\Theta_{ss}(r) = \frac{1}{2\,\text{Bi}} + \frac{1}{4}(1 - r^2). \tag{4.7-28}$$

Substituting Θ_{ss} for the time-independent part of Eq. (4.7-27), the final solution becomes

$$\Theta(r,\ t)=\frac{1}{2\,\text{Bi}}+\frac{1}{4}(1-r^2)-2\sum_{n=1}^{\infty}\left[1+\left(\frac{\text{Bi}}{\lambda_n}\right)^2\right]^{-1}e^{-\lambda_n^2 t}\frac{J_1(\lambda_n)J_0(\lambda_n\ r)}{\lambda_n^3 J_0^2(\lambda_n)}.$$
(4.7-29)

Modified Bessel Functions

Another differential equation which commonly arises in problems involving cylindrical coordinates is the *modified Bessel's equation,* written generally as

$$x\frac{d}{dx}\left(x\frac{df}{dx}\right)-(m^2x^2+\nu^2)f=0,$$
(4.7-30)

where m is a parameter and ν is any real constant. Equations (4.7-3) and (4.7-30) differ only in the sign of the m^2x^2 term. The solutions to Eq. (4.7-30) are written as $I_\nu(mx)$ and $K_\nu(mx)$, and are called *modified Bessel functions* of order ν of the first and second kind, respectively. As with the "regular" Bessel functions, our concern is with the functions corresponding to $\nu=0$ and $\nu=1$. Graphs of these modified Bessel functions are shown in Fig. 4-13. The most obvious difference between Bessel functions and modified Bessel functions is that the latter do not display oscillatory behavior or possess multiple roots. The limiting values of the modified Bessel functions are

$$I_0(0)=1,\qquad I_1(0)=0,\qquad I_0(\infty)=\infty,\qquad I_1(\infty)=\infty,$$
(4.7-31)

$$K_0(0)=\infty,\qquad K_1(0)=\infty,\qquad K_0(\infty)=0,\qquad K_1(\infty)=0.$$
(4.7-32)

Identities which are particularly helpful in evaluating derivatives and integrals are

$$\frac{dI_0(mx)}{dx}=mI_1(mx),\qquad \frac{d}{dx}[xI_1(mx)]=mxI_0(mx),$$
(4.7-33a,b)

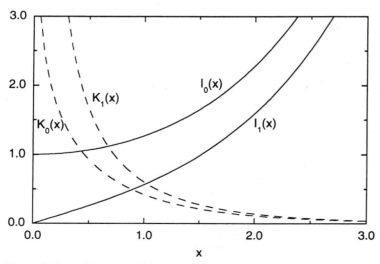

Figure 4-13. Modified Bessel functions of orders zero and one.

$$\frac{dK_0(mx)}{dx} = -mK_1(mx), \qquad \frac{d}{dx}[xK_1(mx)] = -mxK_0(mx). \qquad (4.7\text{-}34\text{a,b})$$

Example 4.7-2 Steady Diffusion with a First-Order Reaction in a Cylinder   Consider a one-dimensional, steady-state problem involving diffusion and a homogeneous reaction in a long cylinder. The reaction is first order and irreversible, and the reactant is maintained at a constant concentration at the surface. The Damköhler number (Da) is not necessarily large or small. The conservation equation and boundary condition for the reactant are written as

$$\frac{1}{r}\frac{d}{dr}\left(r\frac{d\Theta}{dr}\right) - \text{Da }\Theta = 0, \qquad \frac{d\Theta}{dr}(0)=0, \qquad \Theta(1)=1. \qquad (4.7\text{-}35)$$

A comparison of Eqs. (4.7-30) and (4.7-35) shows the latter differential equation to be a modified Bessel equation of order zero, with $m = \text{Da}^{1/2}$. Accordingly, the general solution to Eq. (4.7-35) is

$$\Theta(r) = aI_0(\text{Da}^{1/2}r) + bK_0(\text{Da}^{1/2}r), \qquad (4.7\text{-}36)$$

where a and b are constants. Given that $K_0(\text{Da}^{1/2}r)$ is not finite at $r=0$, that solution is excluded. Using the boundary condition at $r=1$ to evaluate a, the final solution is found to be

$$\Theta(r) = \frac{I_0(\text{Da}^{1/2}r)}{I_0(\text{Da}^{1/2})}. \qquad (4.7\text{-}37)$$

This solution was stated without proof in Example 3.7-2, in discussing the results of the singular perturbation analysis of this problem for $\text{Da}\rightarrow\infty$.

Example 4.7-3 Steady Conduction in a Hollow Cylinder   Modified Bessel functions arise also in FFT solutions, as this example will show. Assume that a hollow cylinder has a constant temperature at its base, and a different constant temperature at its top and its inner and outer curved surfaces (see Fig. 4-14). The steady, two-dimensional conduction problem involving $\Theta(r, z)$ is

$$\frac{1}{r}\frac{\partial}{\partial r}\left(r\frac{\partial\Theta}{\partial r}\right) + \frac{\partial^2\Theta}{\partial z^2} = 0, \qquad (4.7\text{-}38)$$

$$\Theta(r, 0) = 1, \qquad \Theta(r, \gamma) = 0, \qquad (4.7\text{-}39)$$

Figure 4-14. Steady, two-dimensional heat conduction in a hollow cylinder. The base is maintained at a different temperature than that of the top or the curved surfaces. Only half of the hollow cylinder is shown.

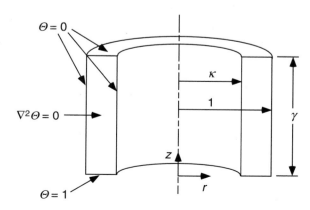

$$\Theta(\kappa, z) = 0, \qquad \Theta(1, z) = 0. \tag{4.7-40}$$

We can construct an FFT solution using basis functions either in r or in z. Basis functions in r will involve a linear combination of J_0 and Y_0; the latter solution cannot be excluded here because $r = 0$ is outside the domain. The eigenvalues will be solutions to a transcendental equation which involves the parameter κ (the ratio of the inner radius to the outer radius), and they will need to be determined numerically. Indeed, it is found that the characteristic equation is

$$J_0(\lambda_n \kappa) Y_0(\lambda_n) = J_0(\lambda_n) Y_0(\lambda_n \kappa) \tag{4.7-41}$$

and the basis functions are given by

$$\Phi_n(r) = \frac{Y_0(\lambda_n) J_0(\lambda_n r) - J_0(\lambda_n) Y_0(\lambda_n r)}{\left\{ \int_\kappa^1 [Y_0(\lambda_n) J_0(\lambda_n r) - J_0(\lambda_n) Y_0(\lambda_n r)]^2 \, r \, dr \right\}^{1/2}}. \tag{4.7-42}$$

If these basis functions are used, the coefficients in the expansion for the temperature will involve $\sinh \lambda_n z$ and $\cosh \lambda_n z$ (or the corresponding exponential functions of z). Alternatively, we can expand the temperature using basis functions in z. Given the Dirichlet boundary conditions in that coordinate, the basis functions (from Table 4-1, case I) are then

$$\Phi_n(z) = \sqrt{\frac{2}{\gamma}} \sin \frac{n\pi z}{\gamma}, \qquad n = 1, 2, \dots. \tag{4.7-43}$$

This set of basis functions has a major advantage, in that all of the eigenvalues are known explicitly. Thus, the preferred expansion is

$$\Theta(r, z) = \sum_{n=1}^{\infty} \Theta_n(r) \Phi_n(z). \tag{4.7-44}$$

Transforming the term in Eq. (4.7-38) containing the z derivative gives

$$\int_0^\gamma \Phi_n \frac{\partial^2 \Theta}{\partial z^2} \, dz = \left(\Phi_n \frac{\partial \Theta}{\partial z} - \Theta \frac{d\Phi_n}{dz} \right) \Big|_{z=0}^{z=\gamma} - \lambda_n^2 \Theta_n = \left(\frac{n\pi}{\gamma} \right) \sqrt{\frac{2}{\gamma}} - \left(\frac{n\pi}{\gamma} \right)^2 \Theta_n. \tag{4.7-45}$$

The transformation of the other term in the partial differential equation, and of the boundary conditions in r, is routine. The governing equation for the transformed temperature is

$$\frac{1}{r} \frac{d}{dr} \left(r \frac{d\Theta_n}{dr} \right) - \left(\frac{n\pi}{\gamma} \right)^2 \Theta_n = - \left(\frac{n\pi}{\gamma} \right) \sqrt{\frac{2}{\gamma}}, \qquad \Theta_n(\kappa) = \Theta_n(1) = 0. \tag{4.7-46}$$

The homogeneous form of this differential equation is seen to be the modified Bessel's equation of order zero. Including a particular solution to satisfy the nonhomogeneous term, the general solution for Eq. (4.7-46) is

$$\Theta_n = A_n I_0 \left(\frac{n\pi r}{\gamma} \right) + B_n K_0 \left(\frac{n\pi r}{\gamma} \right) + \left(\frac{\gamma}{n\pi} \right) \sqrt{\frac{2}{\gamma}}. \tag{4.7-47}$$

The domain does not include $r = 0$ or $r = \infty$, so that both of the modified Bessel functions must be retained. Applying the boundary conditions given in Eq. (4.7-46), the constants in Eq. (4.7-47) are found to be

$$A_n = - \left(\frac{\gamma}{n\pi} \right) \left(\sqrt{\frac{2}{\gamma}} \right) \left[\frac{K_0(n\pi/\gamma) - K_0(n\pi\kappa/\gamma)}{I_0(n\pi\kappa/\gamma) K_0(n\pi/\gamma) - I_0(n\pi/\gamma) K_0(n\pi\kappa/\gamma)} \right], \tag{4.7-48}$$

$$B_n = -\left(\frac{\gamma}{n\pi}\right)\left(\sqrt{\frac{2}{\gamma}}\right)\left[\frac{I_0(n\pi\kappa/\gamma) - I_0(n\pi/\gamma)}{I_0(n\pi\kappa/\gamma)K_0(n\pi/\gamma) - I_0(n\pi/\gamma)K_0(n\pi\kappa/\gamma)}\right]. \tag{4.7-49}$$

The solution is completed by using Eqs. (4.7-43) and (4.7-47)–(4.7-49) in Eq. (4.7-44).

Nonaxisymmetric Problems

It was mentioned earlier that eigenvalue problems involving the angle θ employ the same differential equation as in rectangular coordinates. Thus, $\Phi(\theta)$ satisfies

$$\frac{d^2\Phi}{d\theta^2} = -\lambda^2\Phi \tag{4.7-50}$$

together with appropriate boundary conditions. If the domain does not encompass a full circle, then the boundary conditions will be any of the usual homogeneous types. If, however, the domain is a complete circle (e.g., $0 \le \theta \le 2\pi$), there will be no true boundaries at planes of constant θ. In those situations it is sufficient to require that the field variable and its first derivative with respect to θ be periodic in θ, with a period of 2π. That is the same as stating that the field variable is a single-valued function of position; adding 2π to θ returns one to the same physical location. In such cases the conditions which accompany Eq. (4.7-50) are of the form

$$\Phi(c) = \Phi(c + 2\pi), \qquad \Phi'(c) = \Phi'(c + 2\pi), \tag{4.7-51}$$

where c is an arbitrary constant. Equations (4.7-50) and (4.7-51) yield a self-adjoint eigenvalue problem, as discussed in Section 4.6.

Solutions to Eq. (4.7-50) which satisfy Eq. (4.7-51) are $\sin \lambda_n\theta$ and $\cos \lambda_n\theta$, with $\lambda_n = n$, an integer. As usual, the negative eigenvalues are excluded to ensure orthogonality of the eigenfunctions. A peculiarity of these periodic problems is that both $\sin n\theta$ and $\cos n\theta$ are eigenfunctions individually, rather than in a certain linear combination. Thus, there are *two* linearly independent eigenfunctions corresponding to each eigenvalue (each value of n). For this reason, different numbering systems are employed for the eigenfunctions and eigenvalues. The eigenfunctions for problems which are periodic in θ are

$$\Phi_0 = \frac{1}{\sqrt{2\pi}}; \quad \Phi_n = \frac{1}{\sqrt{\pi}}\sin\frac{(n+1)\theta}{2}, \quad n = 1,3,5,\ldots; \quad \Phi_n = \frac{1}{\sqrt{\pi}}\cos\frac{n\theta}{2}, \quad n = 2,4,6,\ldots,$$

$$\tag{4.7-52}$$

which are orthonormal with respect to the weighting function $w(\theta) = 1$ and any interval $[c, c + 2\pi]$.

Another distinctive feature of problems where $\Theta = \Theta(r, \theta)$ is that, unlike most other two-dimensional problems (with both dimensions finite), there is no choice in the basis function coordinate. That is, the expansion must be constructed using $\Phi_n(\theta)$ obtained from Eq. (4.7-52). Any attempt to use $\Phi_n(r)$ fails because the resulting functions of θ do not have the required periodicity.

4.8 FFT METHOD FOR PROBLEMS IN SPHERICAL COORDINATES

The discussion here is restricted to spherical problems which are *axisymmetric* (i.e., independent of the angle ϕ). For $\Theta = \Theta(r, \theta, t)$ and no source term, the dimensionless conservation equation is

$$\frac{\partial \Theta}{\partial t} = \frac{1}{r^2}\frac{\partial}{\partial r}\left(r^2 \frac{\partial \Theta}{\partial r}\right) + \frac{1}{r^2 \sin\theta}\frac{\partial}{\partial \theta}\left(\sin\theta \frac{\partial\Theta}{\partial\theta}\right), \tag{4.8-1}$$

where $0 \le \theta \le \pi$. It is usually advantageous in spherical problems to replace the coordinate θ with

$$\eta \equiv \cos\theta, \tag{4.8-2}$$

where $-1 \le \eta \le 1$. Changing variables in Eq. (4.8-1), the differential equation for $\Theta(r, \eta, t)$ is found to be

$$\frac{\partial \Theta}{\partial t} = \frac{1}{r^2}\frac{\partial}{\partial r}\left(r^2 \frac{\partial\Theta}{\partial r}\right) + \frac{1}{r^2}\frac{\partial}{\partial\eta}\left((1-\eta^2)\frac{\partial\Theta}{\partial\eta}\right). \tag{4.8-3}$$

Thus, trigonometric functions of θ have been eliminated in favor of polynomials in η. We will be concerned only with problems involving $\Theta(r, t)$ or $\Theta(r, \eta)$. The eigenvalue problems for the former yield spherical Bessel functions in r, whereas those for the lattter give Legendre polynomials in η. Those functions are discussed now in turn.

Spherical Bessel Functions

The eigenvalue problems in r which come from Eq. (4.8-3) are based on

$$\frac{1}{r^2}\frac{d}{dr}\left(r^2 \frac{d\Phi}{dr}\right) = -\lambda^2 \Phi, \tag{4.8-4}$$

which is a Sturm–Liouville equation with $w(r) = p(r) = r^2$ and $q(r) = 0$ [compare with Eq. (4.6-9)]. Equation (4.8-4) differs from Eq. (4.7-2), which is Bessel's equation of order zero, only in the replacement of r by r^2. It is a particular form of the *spherical Bessel's equation,* which is written more generally as

$$\frac{d}{dx}\left(x^2 \frac{df}{dx}\right) + [m^2 x^2 - n(n+1)]f = 0, \tag{4.8-5}$$

where m is any real constant and n is a non-negative integer. The solutions to Eq. (4.8-5), called *spherical Bessel functions,* may be expressed in terms of Bessel functions of order $n + (1/2)$. Accordingly, their properties are covered in discussions of Bessel functions of fractional order [see Watson (1944) or Abramowitz and Stegun (1970)]. Fortunately, the solution to Eq. (4.8-4), which corresponds to $n = 0$, is also expressible as

$$\Phi(r) = a\frac{\sin\lambda r}{r} + b\frac{\cos\lambda r}{r}, \tag{4.8-6}$$

where a and b are constants. Thus, only elementary functions are needed for conduction or diffusion problems involving $\Theta(r, t)$.

In problems involving complete spheres, the requirement that the temperature or concentration be finite everywhere eliminates one solution of Eq. (4.8-4). That is, for $r \to 0$, $\sin(\lambda r)/r \to \lambda$ but $\cos(\lambda r)/r \to \infty$, so that $b = 0$ in Eq. (4.8-6) when the domain includes $r = 0$. Thus, the nth eigenfunction becomes

$$\Phi_n(r) = a_n \frac{\sin \lambda_n r}{r}. \tag{4.8-7}$$

For the interval $[0, \ell]$ the normalization requirement is

$$\langle \Phi_n, \Phi_n \rangle = 1 = a_n^2 \int_0^\ell \left(\frac{\sin \lambda_n r}{r} \right)^2 r^2 \, dr = a_n^2 \int_0^\ell \sin^2 \lambda_n r \, dr. \tag{4.8-8}$$

Notice that the weighting function for this inner product is $w(r) = r^2$. Evaluating the integral in Eq. (4.8-8), we find that for any of the three types of homogeneous boundary conditions at the sphere surface, the coefficient in Eq. (4.8-7) is given by

$$a_n = \sqrt{\frac{2}{\ell}} \left(1 - \frac{\cos \lambda_n \ell \sin \lambda_n \ell}{\lambda_n \ell} \right)^{-1/2}. \tag{4.8-9}$$

The boundary condition used will determine the specific values of λ_n and thus a_n. The eigenvalues and basis functions corresponding to the three types of boundary conditions are listed in Table 4-3. The use of these basis functions in an FFT solution is illustrated in Example 4.8-2.

Modified Spherical Bessel Functions

A differential equation closely related to Eq. (4.8-5) is the *modified spherical Bessel's equation*,

$$\frac{d}{dx} \left(x^2 \frac{df}{dx} \right) - [m^2 x^2 + n(n+1)] f = 0. \tag{4.8-10}$$

Table 4-3
Orthonormal Sequences of Functions from Certain Eigenvalue Problems in Spherical Coordinates [a]

Case	Boundary condition	Characteristic equation [b]	Basis functions
I	$\Phi_n(\ell) = 0$	$\lambda_n \ell = n\pi, \quad n = 1, 2, \dots$	$\Phi_n(r) = \sqrt{\dfrac{2}{\ell}} \dfrac{\sin(n\pi r/\ell)}{r}$
II	$\Phi_n'(\ell) = 0$	$\lambda_n \ell = \tan \lambda_n \ell$	$\Phi_n(r) = \sqrt{\dfrac{2}{\ell r}} \dfrac{\sin \lambda_n r}{\sin \lambda_n \ell}, \quad n = 1, 2, \dots$ $\Phi_0(r) = \sqrt{\dfrac{3}{\ell^3}}$
III	$\Phi_n'(\ell) + A\Phi_n(\ell) = 0$	$\lambda_n \ell = (1 - A\ell^2) \tan \lambda_n \ell$	$\Phi_n(r) = \sqrt{\dfrac{2}{\ell}} \left[\dfrac{1 - A\ell}{\sin^2 \lambda_n \ell - A\ell} \right]^{1/2} \dfrac{\sin \lambda_n r}{r}$

[a] All of the functions shown satisfy Eq. (4.8-4) on the interval $[0, \ell]$, with $\Phi'(0) = 0$.
[b] For case II, $\lambda_0 = 0$. Otherwise, $\lambda_n > 0$.

As with the corresponding equations for cylindrical problems [i.e., Eqs. (4.7-3) and (4.7-30)], Eqs. (4.8-5) and (4.8-10) differ only in the sign of the m^2x^2 term. The solutions to Eq. (4.8-10), called *modified spherical Bessel functions,* are expressible in terms of modified Bessel functions of order $n+(1/2)$. For the special case of interest here, where $n=0$, the general solution is expressed in elementary functions as

$$f(x) = A \frac{\sinh mx}{x} + B \frac{\cosh mx}{x}. \qquad (4.8\text{-}11)$$

The application of Eqs. (4.8-10) and (4.8-11) to a diffusion problem is illustrated in Example 4.8-1.

The correspondence between the solutions to the ordinary or modified spherical Bessel's equations and the solutions to the analogous equations in rectangular coordinates is noteworthy. Rectangular problems yield either trigonometric or hyperbolic functions; spherical problems give the corresponding functions divided by r. This underlies a well-known transformation used in solving spherical conduction or diffusion problems, in which there is a change in the dependent variable given by

$$\Theta(r, t) \equiv \frac{\Psi(r, t)}{r}. \qquad (4.8\text{-}12)$$

It is readily confirmed that Eq. (4.8-12) transforms a problem for $\Theta(r, t)$ involving the spherical ∇^2 operator into a problem for $\Psi(r, t)$ involving the rectangular one. An example of a situation in which this transformation is useful is given in Problem 3-1.

Legendre Polynomials

The eigenvalue problem in η which is associated with Eq. (4.8-3) has the differential equation

$$\frac{d}{d\eta}\left[(1-\eta^2)\frac{d\Phi}{d\eta}\right] = -\lambda^2\Phi, \qquad (4.8\text{-}13)$$

which is *Legendre's equation.* This is a Sturm–Liouville equation with $w(\eta)=1$, $p(\eta)=(1-\eta^2)$, and $q(\eta)=0$, and it has solutions which are discussed in detail in Hobson (1955). One key finding is that for Eq. (4.8-13) to have a solution which is bounded at $\eta=\pm1$ (corresponding to $\theta=0$ and π), the eigenvalues must be given by

$$\lambda_n^2 = n(n+1), \qquad n=0,1,2, \dots . \qquad (4.8\text{-}14)$$

Thus, for any problem on the interval $[-1, 1]$ in η, the eigenvalues are determined solely by the requirement that the solution be finite. The other key finding is that Eq. (4.8-13) has only one solution which is bounded at $\eta=\pm1$, namely

$$\Phi_n(\eta) = a_n P_n(\eta), \qquad (4.8\text{-}15)$$

where the functions $P_n(x)$ are *Legendre polynomials.* (The other linearly independent solution of Eq. (4.8-13), which is not bounded at $\eta=\pm1$, involves what are called *Legendre polynomials of the second kind;* that solution is of no physical interest and will be disregarded.) From Sturm–Liouville theory (Section 4.6), the Legendre polynomials $P_n(x)$ are orthogonal on the interval $[-1, 1]$ with respect to the weighting function

$w(x)=1$. The first two Legendre polynomials are $P_0(x)=1$ and $P_1(x)=x$, and the remaining members of this orthogonal sequence can be generated using the recursion relation,

$$P_{n+1}(x)=\frac{(2n+1)xP_n(x)-nP_{n-1}(x)}{n+1}. \qquad (4.8\text{-}16)$$

Alternatively, the Legendre polynomials can be computed using *Rodrigues' formula,*

$$P_n(x)=\frac{1}{2^n n!}\frac{d^n}{dx^n}[(x^2-1)^n]. \qquad (4.8\text{-}17)$$

It may be confirmed using either Eq. (4.8-16) or Eq. (4.8-17) that the even-numbered Legendre polynomials contain only even powers of x, whereas the odd-numbered ones contain only odd powers of x. Consequently, the even and odd-numbered Legendre polynomials are even and odd functions, respectively.

The Legendre polynomials are orthogonal but not orthonormal; they are standardized such that $P_n(1)=1$ for all n. Applying the normalization requirement to Eq. (4.8-15) leads to

$$\langle \Phi_n, \Phi_n\rangle=1=a_n^2\int_{-1}^{1}P_n^2(\eta)\,d\eta=a_n^2\left(\frac{2}{2n+1}\right). \qquad (4.8\text{-}18)$$

The integral in Eq. (4.8-18) was evaluated using Rodrigues' formula and n integrations by parts [see Arpaci (1966)]. From Eqs. (4.8-15) and (4.8-18), the orthonormal basis functions in η are

$$\Phi_n(\eta)=\left(\frac{2n+1}{2}\right)^{1/2}P_n(\eta), \qquad n=0,\ 1,\ 2,\ \cdots. \qquad (4.8\text{-}19)$$

The first four Legendre polynomials and corresponding basis functions are listed in Table 4-4. The application of Legendre polynomials to a diffusion problem is illustrated in Example 4.8-3.

Example 4.8-1 Steady Diffusion in a Sphere with a First-Order Reaction   Consider a spherical particle of unit (dimensionless) radius, within which there is a first-order, irreversible reaction. The reactant concentration at the outer surface is constant, and the system is at steady state. The resulting one-dimensional problem for the reactant concentration $\Theta(r)$ is

$$\frac{1}{r^2}\frac{d}{dr}\left(r^2\frac{d\Theta}{dr}\right)-\text{Da }\Theta=0, \qquad \frac{d\Theta}{dr}(0)=0, \qquad \Theta(1)=1, \qquad (4.8\text{-}20)$$

where the Damköhler number is based on the sphere radius. A comparison with Eq. (4.8-10) reveals that this is a modified spherical Bessel's equation. From Eq. (4.8-11), the general solution is

$$\Theta(r)=A\frac{\sinh(\sqrt{\text{Da}}\ r)}{r}+B\frac{\cosh(\sqrt{\text{Da}}\ r)}{r}. \qquad (4.8\text{-}21)$$

To keep the concentration finite at $r=0$ (and to satisfy the no-flux condition), it is necessary that $B=0$. Evaluating A using the boundary condition at the surface, the solution is

TABLE 4-4
Legendre Polynomials and Corresponding
Orthonormal Basis Functions[a]

Legendre polynomials	Basis functions
$P_0(\eta)=1$	$\Phi_0(\eta)=\dfrac{1}{\sqrt{2}}$
$P_1(\eta)=\eta$	$\Phi_1(\eta)=\sqrt{\dfrac{3}{2}}\,\eta$
$P_2(\eta)=\dfrac{1}{2}(3\eta^2-1)$	$\Phi_2(\eta)=\sqrt{\dfrac{5}{8}}(3\eta^2-1)$
$P_3(\eta)=\dfrac{1}{2}(5\eta^3-3\eta)$	$\Phi_3(\eta)=\sqrt{\dfrac{7}{8}}(5\eta^3-3\eta)$
$P_n(x)=\dfrac{1}{2^n n!}\dfrac{d^n}{dx^n}\big[(x^2-1)^n\big]$	$\Phi_n(\eta)=\left(\dfrac{2n+1}{2}\right)^{1/2}P_n(\eta)$

[a] All of the functions shown satisfy Eq. (4.8-13) on the interval $[-1, 1]$,
with λ_n given by Eq. (4.8-14).

$$\Theta(r)=\frac{\sinh(\sqrt{\mathrm{Da}}\,r)}{r\,\sinh(\sqrt{\mathrm{Da}})}. \tag{4.8-22}$$

The analogous problem for a cylinder was solved in Example 4.7-2.

Example 4.8-2 Transient Diffusion in a Sphere with a First-Order Reaction Consider a transient problem which leads to the steady state of the previous example. Specifically, assume that no reactant is present initially and that the dimensionless surface concentration is unity for $t>0$. The problem statement for $\Theta(r, t)$ is then

$$\frac{\partial\Theta}{\partial t}=\frac{1}{r^2}\frac{\partial}{\partial r}\left(r^2\frac{\partial\Theta}{\partial r}\right)-\mathrm{Da}\,\Theta, \tag{4.8-23}$$

$$\Theta(r,\,0)=0, \tag{4.8-24}$$

$$\frac{\partial\Theta}{\partial r}(0,\,t)=0, \qquad \Theta(1,\,t)=1. \tag{4.8-25}$$

The required basis functions are given by case I of Table 4-3, with $\ell=1$. The expansion and the transformed concentration are written as

$$\Theta(r,\,t)=\sum_{n=1}^{\infty}\Theta_n(t)\Phi_n(r), \tag{4.8-26}$$

$$\Theta_n(t)=\int_0^1\Phi_n(r)\Theta(r,\,t)r^2\,dr. \tag{4.8-27}$$

Using Eq. (4.6-16) to transform the r derivative term in Eq. (4.8-23), we find that

$$\int_0^1\Phi_n\left[\frac{1}{r^2}\frac{\partial}{\partial r}\left(r^2\frac{\partial\Theta}{\partial r}\right)\right]r^2\,dr=r^2\left(\Phi_n\frac{\partial\Theta}{\partial r}-\Theta\frac{d\Phi_n}{dr}\right)\bigg|_{r=0}^{r=1}-\lambda_n^2\Theta_n$$

$$= -\sqrt{2}(n\pi)(-1)^n - (n\pi)^2 \Theta_n. \tag{4.8-28}$$

The remaining terms in the partial differential equation transform in a straightforward manner, as does the initial condition. The resulting problem for Θ_n is

$$\frac{d\Theta_n}{dt} + [(n\pi)^2 + \mathrm{Da}]\Theta_n = -\sqrt{2}(n\pi)(-1)^n, \qquad \Theta_n(0) = 0. \tag{4.8-29}$$

Notice that the effect of the first-order reaction is to give a second term multiplying Θ_n on the left-hand side of the differential equation; setting $\mathrm{Da} = 0$ in Eq. (4.8-29), we obtain the relation for transient diffusion without reaction. The solution to Eq. (4.8-29) is

$$\Theta_n = -\left[\frac{\sqrt{2}(n\pi)(-1)^n}{(n\pi)^2 + \mathrm{Da}}\right]\{1 - \exp[-((n\pi)^2 + \mathrm{Da})t]\}. \tag{4.8-30}$$

The overall solution is then

$$\Theta(r, t) = -2\sum_{n=1}^{\infty}\left[\frac{(n\pi)(-1)^n}{(n\pi)^2 + \mathrm{Da}}\right]\{1 - \exp[-((n\pi)^2 + \mathrm{Da})t]\}\frac{\sin n\pi r}{r}. \tag{4.8-31}$$

As mentioned earlier, the steady-state solution for this problem is that obtained in Example 4.8-1. Rewriting Eq. (4.8-31) using Eq. (4.8-22), we obtain the final form of the solution,

$$\Theta(r, t) = \frac{\sinh(\sqrt{\mathrm{Da}}\ r)}{r\sinh(\sqrt{\mathrm{Da}})} + 2\sum_{n=1}^{\infty}\left[\frac{(n\pi)(-1)^n}{(n\pi)^2 + \mathrm{Da}}\right]\exp[-((n\pi)^2 + \mathrm{Da})t]\frac{\sin n\pi r}{r}. \tag{4.8-32}$$

Notice that for $\mathrm{Da} \ll \pi^2$, the reaction has negligible effect on the time required to reach the steady state, even though Da may still be large enough to have an important influence on the steady-state concentration profile. In this case the duration of the transient phase is governed by the time required for diffusion over the entire radius of the sphere. By contrast, for $\mathrm{Da} > \pi^2$, the time to reach the steady state is shortened appreciably by the reaction. The reason is that for large Da the relatively fast reaction prevents the reactant from penetrating beyond a small distance from the surface, even at steady state. The smaller length scale shortens the diffusion time.

Example 4.8-3 Heat Conduction in a Suspension of Spheres   This example concerns the steady temperature field in a system consisting of a sphere surrounded by some other stationary phase. The sphere and the surrounding material have different thermal conductivities, and the temperature gradient far from the sphere is assumed to be uniform. The solution of this "microscopic" problem illustrates the use of Legendre polynomials and also shows how the FFT method is applied to a situation involving two phases. In Example 4.8-4, the solution to the microscopic problem is used to derive a result useful on a more macroscopic scale, namely, the effective conductivity of a suspension or composite solid containing spherical particles.

The microscopic problem is illustrated in Fig. 4-15. A sphere of radius R, consisting of material B, is surrounded by material A. The ratio of the thermal conductivities is denoted as $\gamma \equiv k_B/k_A$. A steady heat flux results from the imposition of a uniform temperature gradient G far from the sphere (parallel to the z axis). For convenience in later applications, all quantities here are dimensional. Noting that $\partial T/\partial z = G$ is equivalent to $\partial T/\partial r = G\eta$, the steady-state temperature outside the sphere, $\Psi(r, \eta)$, is governed by

$$\frac{\partial}{\partial r}\left(r^2\frac{\partial\Psi}{\partial r}\right) + \frac{\partial}{\partial\eta}\left[(1 - \eta^2)\frac{\partial\Psi}{\partial\eta}\right] = 0, \tag{4.8-33}$$

$$\frac{\partial\Psi}{\partial r}(\infty, \eta) = G\eta. \tag{4.8-34}$$

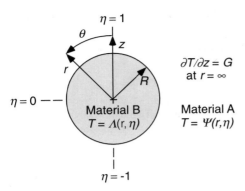

$\eta = 1$

θ

z

r

R

$\eta = 0$ ——

Material B
$T = \Lambda(r, \eta)$

$\eta = -1$

$\partial T/\partial z = G$
at $r = \infty$

Material A
$T = \Psi(r, \eta)$

Figure 4-15. Steady heat conduction in a sphere (material B) and surrounding phase (material A). There is a uniform temperature gradient G far from the sphere, and $\eta = \cos\theta$.

The temperature inside the sphere, $\Lambda(r, \eta)$, must satisfy

$$\frac{\partial}{\partial r}\left(r^2\frac{\partial\Lambda}{\partial r}\right) + \frac{\partial}{\partial\eta}\left[(1-\eta^2)\frac{\partial\Lambda}{\partial\eta}\right] = 0 \tag{4.8-35}$$

with $\Lambda(0,\eta)$ finite. (This problem is not spherically symmetric, so that we cannot require that $\partial\Lambda/\partial r = 0$ at $r=0$.) Requiring that the temperature and the normal component of the heat flux be continuous at the interface, the internal and external temperature fields are coupled by the conditions

$$\Psi(R, \eta) = \Lambda(R, \eta), \tag{4.8-36}$$

$$\frac{\partial\Psi}{\partial r}(R, \eta) = \gamma\frac{\partial\Lambda}{\partial r}(R, \eta). \tag{4.8-37}$$

The only eigenvalue problem associated with Eq. (4.8-33) is that involving η. Accordingly, the basis functions will be those obtained from Legendre polynomials, $\Phi_n(\eta)$, as given in Table 4-4. Recalling that the weighting function for inner products involving $\Phi_n(\eta)$ is unity, the transformed temperature in the surrounding material is

$$\Psi_n(r) = \int_{-1}^{1} \Phi_n(\eta)\Psi(r, \eta)\,d\eta. \tag{4.8-38}$$

Transforming the individual terms in Eq. (4.8-33) with the help of Eq. (4.6-16), we obtain

$$\int_{-1}^{1} \Phi_n\frac{\partial}{\partial r}\left(r^2\frac{\partial\Psi}{\partial r}\right)d\eta = \frac{d}{dr}\left(r^2\frac{d\Psi_n}{dr}\right), \tag{4.8-39}$$

$$\int_{-1}^{1} \Phi_n\frac{\partial}{\partial\eta}\left[(1-\eta^2)\frac{\partial\Psi}{\partial\eta}\right]d\eta = (1-\eta^2)\left[\Phi_n\frac{\partial\Psi}{\partial\eta} - \Psi\frac{d\Phi_n}{d\eta}\right]\Bigg|_{\eta=-1}^{\eta=1} - \lambda_n^2\psi_n$$

$$= -n(n+1)\Psi_n. \tag{4.8-40}$$

Notice that the elimination of the boundary terms in the integration by parts requires only that the temperature and its gradient in η be finite at $\eta = \pm1$. The last step in Eq. (4.8-40) made use of Eq. (4.8-14). From Eqs. (4.8-39) and (4.8-40), the differential equation for $\Psi_n(r)$ is

$$\frac{d}{dr}\left(r^2\frac{d\Psi_n}{dr}\right) - n(n+1)\Psi_n = 0. \tag{4.8-41}$$

Similarly, the differential equation for the transformed temperature inside the sphere is

$$\frac{d}{dr}\left(r^2\frac{d\Lambda_n}{dr}\right)-n(n+1)\Lambda_n=0. \tag{4.8-42}$$

Equations (4.8-41) and (4.8-42) are *equidimensional equations*, which have solutions of the form r^p. Substituting that trial solution into either of the differential equations, the roots of the resulting quadratic equation are found to be $p=n$ and $p=-(n+1)$. Thus, the general solution of Eq. (4.8-41) is

$$\Psi_n=A_n r^n+B_n r^{-(n+1)}, \qquad n=0,\ 1,\ 2,\ \dots\ . \tag{4.8-43}$$

The solution of Eq. (4.8-42) is similar, but it is simplified immediately by recognizing that for the temperature at $r=0$ to be finite, all negative powers of r must be eliminated. It follows that the solution inside the sphere has the form

$$\Lambda_n=C_n r^n, \qquad n=0,\ 1,\ 2,\ \dots\ . \tag{4.8-44}$$

What remains is to determine the constants A_n, B_n, and C_n using the boundary conditions at $r=R$ and $r=\infty$.

The boundary condition at $r=\infty$, Eq. (4.8-34), is transformed as

$$\frac{d\Psi_n}{dr}(\infty)=\int_{-1}^{1}\Phi_n\frac{\partial\Psi}{\partial r}(\infty,\ \eta)\ d\eta=G\int_{-1}^{1}\Phi_n\eta\ d\eta$$

$$=\sqrt{\frac{2}{3}}\,G\int_{-1}^{1}\Phi_n\Phi_1\ d\eta=\sqrt{\frac{2}{3}}\,G\,\delta_{n1}. \tag{4.8-45}$$

Thus, the fact that the temperature gradient at $r=\infty$ is proportional to Φ_1, together with the orthogonality of the basis functions, causes all terms to vanish except that for $n=1$. The gradient in the transformed temperature outside the sphere is evaluated from Eq. (4.8-43) as

$$\frac{d\Psi_n}{dr}=nA_n r^{n-1}-(n+1)B_n r^{-(n+2)}. \tag{4.8-46}$$

Comparing Eqs. (4.8-45) and (4.8-46), it is found that

$$A_1=\sqrt{\frac{2}{3}}\,G; \qquad A_n=0,\quad n\geq 2. \tag{4.8-47}$$

No conclusion is reached at this point concerning any of the B_n, because all of the corresponding terms in Eq. (4.8-46) vanish at $r=\infty$. Likewise, no information is provided concerning A_0.

The matching conditions at $r=R$, Eqs. (4.8-36) and (4.8-37), are transformed to

$$\Psi_n(R)=\Lambda_n(R), \tag{4.8-48}$$

$$\frac{d\Psi_n}{dr}(R)=\gamma\frac{d\Lambda_n}{dr}(R). \tag{4.8-49}$$

Using Eqs. (4.8-43), (4.8-44), and (4.8-47) to evaluate the terms in Eqs. (4.8-48) and (4.8-49), and solving for the constants B_n and C_n, we obtain

$$B_0=0; \qquad B_1=\sqrt{\frac{2}{3}}\,GR^3\left(\frac{1-\gamma}{2+\gamma}\right); \qquad B_n=0,\quad n\geq 2, \tag{4.8-50}$$

$$C_0 = A_0; \qquad C_1 = \sqrt{\frac{2}{3}} G\left(\frac{3}{2+\gamma}\right); \qquad C_n = 0, \quad n \geq 2. \tag{4.8-51}$$

The constant A_0 remains undetermined.

The values obtained for A_n, B_n, and C_n indicate that all terms in the Fourier series expansions corresponding to $n > 1$ will vanish. This is a consequence of the fact that the one nonhomogeneous equation in the problem, the boundary condition at $r = \infty$, is satisfied exactly by the $n = 1$ term in the expansion for Ψ. The $n = 0$ terms in Ψ and Λ both equal a constant, $A_0 \Phi_0$. It is not possible to assign a unique value to A_0 because the boundary condition at $r = \infty$ is of the Neumann type. That is, there is no information which fixes the absolute level of the temperature, so that it is known only to within an additive constant. The final solution for the temperature, constructed from the $n = 0$ and $n = 1$ terms, is

$$T(r, \eta) = \begin{cases} \Lambda(r, \eta) = A + \left(\dfrac{3}{2+\gamma}\right) Gr\eta, & 0 \leq r \leq R, \\[2mm] \Psi(r, \eta) = A + \left[1 + \left(\dfrac{1-\gamma}{2+\gamma}\right)\left(\dfrac{R}{r}\right)^3\right] Gr\eta, & R < r < \infty, \end{cases} \tag{4.8-52}$$

where $A \equiv A_0/\sqrt{2}$. The first two terms in $\Psi(r, \eta)$ correspond to the unperturbed temperature field far from the sphere; the remaining term represents the effect of the sphere on the external temperature, which is seen to decay as r^{-2}.

Example 4.8-4 Conductivity of a Suspension   The effective conductivity of a suspension or composite solid containing spherical particles is calculated now using the approach described by Jeffrey (1973). The objective is to describe heat transfer on a length scale which greatly exceeds the average spacing between spheres, but which is much smaller than the macroscopic dimensions of the material. In other words, we seek the value of thermal conductivity which must be used if the composite is modeled as a homogeneous phase. The macroscopic form of Fourier's law is written as

$$\langle \mathbf{q} \rangle = -k_{\text{eff}} \langle \nabla T \rangle, \tag{4.8-53}$$

where $\langle \mathbf{q} \rangle$ and $\langle \nabla T \rangle$ are the volume-averaged heat flux and temperature gradient, respectively, and k_{eff} is the effective thermal conductivity. The average temperature gradient is given by

$$\langle \nabla T \rangle = \lim_{V \to \infty} \frac{1}{V} \int_V \nabla T \, dV, \tag{4.8-54}$$

where the limit indicates merely that the sampling volume V contains many spheres. The average heat flux is defined in a similar manner. Distinguishing the contributions of the two phases, the flux is expressed as

$$\langle \mathbf{q} \rangle = \lim_{V \to \infty} \frac{1}{V} \left[\int_{V_A} (-k_A \nabla T) \, dV + \int_{V_B} (-k_B \nabla T) \, dV \right], \tag{4.8-55}$$

where V_A and V_B represent the total volumes occupied by phases A and B, respectively. Equation (4.8-55) is rewritten as

$$\langle \mathbf{q} \rangle = \lim_{V \to \infty} \frac{1}{V} \left[\int_V (-k_A \nabla T) \, dV + \int_{V_B} (k_A - k_B) \nabla T \, dV \right]$$

$$= -k_A \langle \nabla T \rangle - \lim_{V \to \infty} \frac{1}{V} \int_{V_B} (k_B - k_A) \nabla T \, dV, \tag{4.8-56}$$

where the integral over V_B now represents the "excess heat flux" due to the spheres. The excess flux is the flux above that which would exist if the entire material consisted of phase A. The contribution of a single sphere to the excess flux is described now using a vector $\mathbf{S}$, defined as

$$\mathbf{S} = (k_B - k_A) \int_{V_s} \nabla T \, dV, \tag{4.8-57}$$

where V_S is the volume of *one* sphere. Lettting $\bar{\mathbf{S}}$ represent the average of $\mathbf{S}$ over a statistically representative sample of spheres, the average flux is expressed as

$$\langle \mathbf{q} \rangle = -k_A \langle \nabla T \rangle - n\bar{\mathbf{S}}, \tag{4.8-58}$$

where n is the number of spheres per unit volume. This expression applies to suspensions which are not necessarily dilute.

For a dilute suspension, all spheres will act independently, and $\bar{\mathbf{S}} = \mathbf{S}_0$, the value of $\mathbf{S}$ for an isolated sphere. Thus, for a dilute suspension the average heat flux is given by

$$\langle \mathbf{q} \rangle = -k_A \langle \nabla T \rangle - n\mathbf{S}_0. \tag{4.8-59}$$

The problem is reduced now to evaluating $\mathbf{S}_0$, which requires that the temperature gradient be integrated over the volume of an isolated sphere. The temperature within a sphere under the conditons of interest is given by $\Lambda(r, \eta)$ in Eq. (4.8-52). Noting that $z = r\eta$, the internal temperature gradient is

$$\nabla T = \nabla \Lambda = \left(\frac{3}{2+\gamma}\right) G \mathbf{e}_z = \left(\frac{3}{2+\gamma}\right) \mathbf{G}. \tag{4.8-60}$$

With the unperturbed temperature gradient far from the sphere written in terms of the vector $\mathbf{G}$, as in the last form of Eq. (4.8-60), this result is independent of the coordinate system chosen. It is seen that the temperature gradient within the sphere is a constant, which makes the integration in Eq. (4.8-57) trivial. Setting $\langle \nabla T \rangle = \mathbf{G}$ for an isolated sphere, it follows that

$$n\mathbf{S}_0 = (k_B - k_A)\left(\frac{3}{2+\gamma}\right)\phi\langle \nabla T \rangle, \tag{4.8-61}$$

where $\phi = nV_S$ is the fraction of the total volume occupied by spheres. Finally, using Eq. (4.8-61) in Eq. (4.8-59) and comparing with Eq. (4.8-53), it is found that

$$\frac{k_{\text{eff}}}{k_A} = 1 + 3\left(\frac{\gamma-1}{\gamma+2}\right)\phi + O(\phi^2), \qquad \gamma \equiv \frac{k_B}{k_A}. \tag{4.8-62}$$

This result is equivalent to one derived originally by Maxwell (1873). It is seen that the first correction to the thermal conductivity is proportional to ϕ. The extension of this approach to a less dilute suspension, where sphere–sphere interactions yield a ϕ^2 contribution, is given by Jeffrey (1973). The effective diffusivity of a solute in a dilute suspension of spheres is derived in Example 4.10-2, using a different method.

4.9 POINT-SOURCE SOLUTIONS

A number of problems have been discussed in this and the preceding chapters involving volumetric heat sources or homogeneous chemical reactions, in which the source (positive or negative in algebraic sign) has been assumed to be distributed continuously throughout the region of interest. Attention is given now to situations where the source is restricted to a plane, a line, or a point within the domain, depending on whether the problem is one-, two-, or three-dimensional, respectively. For convenience, all such source terms are referred to here as *point sources*. They may be either instantaneous or continuous in time. Point-source solutions are useful for two reasons. First, the concept of a point source provides a simple but effective model for physical situations in which the actual source is highly localized. Second, the singular solutions obtained in this manner provide the basis for certain analytical and numerical methods, applicable to a wide variety of linear problems, in which the overall solution is expressed as an integral or sum involving point-source solutions. Solutions for point sources of unit magnitude are called *Green's functions*. In this section a series of diffusion examples is used to derive certain point-source solutions, each of which is valid for an unbounded domain. The utility of the corresponding Green's functions in integral representations is discussed in Section 4.10.

Example 4.9-1 Instantaneous Point Source in One Dimension　The solute concentration $C(x, t)$ resulting from an instantaneous source of solute at the plane $x=0$ at $t=0$ must satisfy

$$\frac{\partial C}{\partial t} = D\frac{\partial^2 C}{\partial x^2},$$ (4.9-1)

$$C(\pm\infty, t)=0; \qquad C(x, 0)=0 \quad \text{for } x\neq 0,$$ (4.9-2)

$$m_0 = \int_{-\infty}^{\infty} C(x, t) \, dx,$$ (4.9-3)

where m_0 is the amount of solute released (moles/area). In this and subsequent examples in this section, all variables are dimensional. Equations (4.9-2) and (4.9-3) express the fact that the source is localized, that no solute is present for $t<0$, and that the total amount of solute for $t>0$ is fixed at m_0.

　　It will be shown now that the desired concentration field can be obtained simply by differentiating the similarity solution for a sudden step change in the concentration at a boundary, as derived in Section 3.5. This approach relies on the fact that if a function $C(x, t)$ satisfies Eq. (4.9-1), as the similarity solution does, then $\partial C/\partial x$ (or any other derivative of the solution) also satisfies that differential equation. This is shown by substituting $\partial C/\partial x$ for C in Eq. (4.9-1) and interchanging the order of the partial derivatives. The use of the x derivative in particular is motivated by the respective initial and boundary conditions of the two problems. As given by Eqs. (3.5-16) and (3.5-18), the solution for a sudden increase in concentration from 0 to C_0 at $x=0$ is

$$\frac{C(\eta)}{C_0} = 1 - \frac{2}{\pi^{1/2}}\int_0^\eta e^{-s^2} \, ds = 1 - \text{erf } \eta, \qquad \eta = \frac{x}{2(Dt)^{1/2}}.$$ (4.9-4)

Although derived for the semi-infinite region $x\geq 0$, this also represents the solution to a diffusion problem for an infinite domain, in which $C=2C_0$ at $x=-\infty$ and $C=0$ at $x=\infty$. Note in this regard that erf is an odd function, so that $\text{erf}(-\infty)=-\text{erf}(\infty)=-1$, and that C and $\partial C/\partial x$ from Eq. (4.9-4) are both continuous at $x=0$. The key observation is that the values of $\partial C/\partial x$ obtained from this solution approach a "spike" at $x=0$ as $t\rightarrow 0$. This suggests that we can obtain a solution which has the desired character at $x=0$ and $t=0$ by using the x derivative of Eq. (4.9-4). Thus, the point-source solution is evidently of the form

$$C(x, t) = \frac{Ke^{-x^2/4Dt}}{(\pi Dt)^{1/2}}, \tag{4.9-5}$$

where K is a constant. It is straightforward to confirm that Eq. (4.9-5) satisfies the initial and boundary conditions given by Eq. (4.9-2). Using Eq. (4.9-3), it is found that $K = m_0/2$, so that the solution for an instantaneous source at $x=0$ and $t=0$ is

$$C(x, t) = \frac{m_0 e^{-x^2/4Dt}}{2(\pi Dt)^{1/2}}. \tag{4.9-6}$$

It is seen that the concentration profile resembles a normal or Gaussian probability distribution, with a peak centered at $x=0$ and a variance of $2Dt$. For a source at $x=x'$ and $t=t'$, the generalization of Eq. (4.9-6) is

$$C(x, t) = \frac{m_0 e^{-(x-x')^2/4D(t-t')}}{2[\pi D(t-t')]^{1/2}}. \tag{4.9-7}$$

Example 4.9-2 Instantaneous Point Source in Three Dimensions   The solution for an instantaneous point source of solute at the origin at $t=0$ must satisfy

$$\frac{\partial C}{\partial t} = D\nabla^2 C = D\left(\frac{\partial^2 C}{\partial x^2} + \frac{\partial^2 C}{\partial y^2} + \frac{\partial^2 C}{\partial z^2}\right), \tag{4.9-8}$$

$$C(\pm\infty, y, z, t) = C(x, \pm\infty, z, t) = C(x, y, \pm\infty, t) = 0; \qquad C(x, y, z, 0) = 0 \quad \text{for } r \neq 0, \tag{4.9-9}$$

$$m = \int_{-\infty}^{\infty}\int_{-\infty}^{\infty}\int_{-\infty}^{\infty} C(x, y, z, t) \, dx \, dy \, dz, \tag{4.9-10}$$

where m is the number of moles added at $t=0$ and $r = (x^2 + y^2 + z^2)^{1/2}$.

The desired three-dimensional solution is obtained from the one-dimensional result of the previous example by recognizing that the problem is separable. Let $C_1(x, t)$ be the solution given by Eq. (4.9-5), and let $C_2(y, t)$ and $C_3(z, t)$ be the corresponding solutions to one-dimensional problems in y and z. Now assume that

$$C(x, y, z, t) = C_1(x, t)C_2(y, t)C_3(z, t). \tag{4.9-11}$$

Evaluating the derivatives in Eq. (4.9-8) using Eq. (4.9-11), it is found that

$$\frac{\partial C}{\partial t} = C_2 C_3 \frac{\partial C_1}{\partial t} + C_1 C_3 \frac{\partial C_2}{\partial t} + C_1 C_2 \frac{\partial C_3}{\partial t}, \tag{4.9-12}$$

$$D\nabla^2 C = C_2 C_3 \left(D\frac{\partial^2 C_1}{\partial x^2}\right) + C_1 C_3 \left(D\frac{\partial^2 C_2}{\partial y^2}\right) + C_1 C_2 \left(D\frac{\partial^2 C_3}{\partial z^2}\right). \tag{4.9-13}$$

A term-by-term comparison of Eqs. (4.9-12) and (4.9-13) confirms that Eq. (4.9-11) is a solution to Eq. (4.9-8). It is also readily confirmed that Eq. (4.9-9) is satisfied. Moreover, the fact that C_1,

C_2, and C_3 correspond to sources at the planes $x=0$, $y=0$, and $z=0$, respectively, indicates that the product $C_1C_2C_3$ must correspond to a source at the intersection of those planes, namely the origin. It is found from Eq. (4.9-10) that

$$m = \int_{-\infty}^{\infty} C_1 \, dx \int_{-\infty}^{\infty} C_2 \, dy \int_{-\infty}^{\infty} C_3 \, dz = 8K^3. \qquad (4.9\text{-}14)$$

The result for an instantaneous point source at the origin at $t=0$ is then

$$C(x, y, z, t) = \frac{me^{-(x^2+y^2+z^2)/4Dt}}{8(\pi Dt)^{3/2}} = \frac{me^{-r^2/4Dt}}{8(\pi Dt)^{3/2}} = C(r, t). \qquad (4.9\text{-}15)$$

The form shown for spherical coordinates is the basis for Eq. (2.9-17), which was presented without proof. The generalization of Eq. (4.9-15) to a point source at position (x', y', z') at $t=t'$ is

$$C(x, y, z, t) = \frac{me^{-[(x-x')^2+(y-y')^2+(z-z')^2]/4D(t-t')}}{8[\pi D(t-t')]^{3/2}} = \frac{me^{-|\mathbf{r}-\mathbf{r'}|^2/4D(t-t')}}{8[\pi D(t-t')]^{3/2}} = C(\mathbf{r}-\mathbf{r'}, t-t'),$$

$$(4.9\text{-}16)$$

where $\mathbf{r}$ and $\mathbf{r'}$ are position vectors with the components (x, y, z) and (x', y', z'), respectively.

Example 4.9-3 Method of Images Consider an instantaneous point source of solute located a distance L from an impermeable boundary, at $t=0$. For example, the solute might be a nonvolatile contaminant buried in soil at a depth L below the surface. Referring to the (x, y, z) coordinates in Fig. 4-16, the point source is located at $(0, 0, -L)$. Simply applying the point-source solution for an unbounded domain, Eq. (4.9-16) with $x'=0$, $y'=0$, $z'=-L$, and $t'=0$, does not yield a solution which satisfies the no-flux condition at the boundary, $\partial C/\partial z=0$ at $z=0$. However, we can exploit the fact that Eq. (4.9-8) is linear and homogeneous, so that the sum of any two of its solutions is also a solution. Specifically, we add to the problem a fictitious "image" source of equal strength at the position $(0, 0, L)$, as shown. The sum of the two point-source solutions will yield a concentration field which is symmetric about the plane $z=0$, thereby ensuring that the no-flux condition is met. The requirements of the real problem that $C=0$ for $x=\pm\infty$, $y=\pm\infty$, and $z=-\infty$ will continue to be satisfied. That $C=0$ also for $z=+\infty$ is not of interest, because $z>0$ is outside the physical domain. Accordingly, the solution for m moles released at the actual source is

$$C(x, y, z, t) = \frac{m}{8(\pi Dt)^{3/2}} \left[e^{-[x^2+y^2+(z-L)^2]/4Dt} + e^{-[x^2+y^2+(z+L)^2]/4Dt} \right]. \qquad (4.9\text{-}17)$$

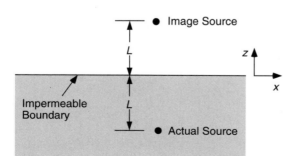

Figure 4-16. Instantaneous solute source at a distance L below an impermeable boundary. A fictitious image source of equal strength is placed an equal distance above the boundary. The impermeable boundary corresponds to the x–y plane.

Notice that the solution given by Eq. (4.9-17) was constructed by assuming that m moles were released at each source (actual and image), for a total of $2m$. With no flux across the plane $z=0$, m moles will remain in the physically relevant region, $z<0$. Thus, the total amount of solute represented by Eq. (4.9-17) is correct. The approach employed here to obtain a point-source solution for a partially bounded domain is an example of the *method of images,* a classical technique which is discussed, for example, by Morse and Feshbach (1953, Part I, pp. 812–816).

Example 4.9-4 Continuous Point Source in Three Dimensions Solutions for continuous point sources may be expressed as integrals over time of the corresponding solutions for instantaneous point sources. For example, suppose that a solute is released at the origin at a time-dependent rate, $\dot{m}(t)$ (moles/time). The continuous release of solute can be approximated as a series of discrete releases, each occuring during a small time interval $\Delta t'$. The amount of solute released between times t' and $t' + \Delta t'$ is given by $\dot{m}(t')\Delta t'$. The linearity of the governing equations indicates that all such releases will make additive contributions to the solute concentration. Letting $\Delta t' \to 0$, so that the sum of the contributions at the various release times becomes an integral, the resulting concentration is found to be

$$C(r,\ t) = \frac{1}{8(\pi D)^{3/2}} \int_0^t \frac{e^{-r^2/4D(t-t')}}{(t-t')^{3/2}}\, \dot{m}(t')\ dt'. \tag{4.9-18}$$

This expression represents a superposition of solute releases occuring at all times t' up to the current time t; in the integral, t is treated as a parameter.

For the special case of a constant release rate, the integral in Eq. (4.9-18) is evaluated with the help of the variable change, $u = [r/2][D(t-t')]^{-1/2}$. The result for constant $\dot{m}$ is

$$C(r,\ t) = \frac{\dot{m}}{4\pi Dr}\,\mathrm{erfc}\left(\frac{r}{2(Dt)^{1/2}}\right). \tag{4.9-19}$$

One way to obtain the steady-state solution is to let $t \to \infty$ and recall that $\mathrm{erfc}(0) = 1$ (see Section 3.5). The steady-state concentration resulting from a constant release rate at the origin is simply

$$C(r) = \frac{\dot{m}}{4\pi Dr}. \tag{4.9-20}$$

It is easily confirmed that this solution satisfies the steady-state problem,

$$\frac{1}{r^2}\frac{d}{dr}\left(r^2 \frac{dC}{dr}\right) = 0, \qquad C(\infty) = 0, \qquad \dot{m} = -4\pi r^2 D \frac{dC}{dr}(r), \tag{4.9-21}$$

where $\dot{m}$ is equated with the total rate of solute diffusion through any spherical shell.

4.10 INTEGRAL REPRESENTATIONS

In Section 4.9 it was shown that solutions to certain diffusion problems could be obtained by adding point-source solutions or integrating them over time. The purpose of this section is to show, in a more general way, how solutions to diffusion problems can be constructed by superposition of point-source solutions. We begin by defining the Green's functions which correspond to the basic point-source solutions derived in Section 4.9, and then show how those lead to integral representations of the solutions to problems which involve spatially distributed sources. This discussion is restricted to three-dimensional domains in which $C \to 0$ as $|\mathbf{r}| \to \infty$, where $\mathbf{r}$ is the position vector. As in Section 4.9, all variables are dimensional.

Green's Functions for Diffusion in Unbounded Domains

For steady diffusion with a constant source of solute $\dot{m}$ at position $\mathbf{r}'$, the differential conservation equation is written as

$$D\nabla^2 C = -\dot{m}\delta(\mathbf{r} - \mathbf{r}'), \tag{4.10-1}$$

where δ is the Dirac delta function. The Dirac delta for three dimensons has the properties

$$\delta(\mathbf{r}) = 0 \quad \text{for } \mathbf{r} \neq \mathbf{0}, \qquad \int_V h(\mathbf{r})\delta(\mathbf{r})dV = h(\mathbf{0}), \tag{4.10-2}$$

where $h(\mathbf{r})$ is any function and the integration is over all space. Thus, Eq. (4.10-1) reduces to Laplace's equation (no source) at all points except $\mathbf{r} = \mathbf{r}'$. The Green's function for this problem, denoted as $f(\mathbf{r} - \mathbf{r}')$, is defined as the solution for $\dot{m} = 1$. Accordingly,

$$D\nabla^2 f = -\delta(\mathbf{r} - \mathbf{r}') \tag{4.10-3}$$

and it is found from Eq. (4.9-20) that

$$f(\mathbf{r} - \mathbf{r}') = \frac{1}{4\pi D|\mathbf{r} - \mathbf{r}'|}. \tag{4.10-4}$$

Whether to include D in the Green's function or to lump the diffusivity with the source is a matter of taste. Including it in the Green's function, as done here, provides the closest parallel with the approach for transient problems discussed below.

The Green's function for transient diffusion, denoted as $F(\mathbf{r} - \mathbf{r}', t - t')$, satisfies

$$D\nabla^2 F - \frac{\partial F}{\partial t} = -\delta(\mathbf{r} - \mathbf{r}')\delta(t - t'). \tag{4.10-5}$$

As mentioned in Section 2.9, the Dirac delta with a scalar argument has properties like those stated above for $\delta(\mathbf{r})$. Thus,

$$\delta(t) = 0 \quad \text{for } t \neq 0, \qquad \int_{-\infty}^{\infty} h(t)\delta(t)\, dt = h(0). \tag{4.10-6}$$

From Eq. (4.9-16), the Green's function for transient diffusion is

$$F(\mathbf{r} - \mathbf{r}', t - t') = \begin{cases} 0 & \text{for } t < t', \\ \dfrac{e^{-|\mathbf{r} - \mathbf{r}'|^2/4D(t-t')}}{8[\pi D(t - t')]^{3/2}} & \text{for } t \geq t'. \end{cases} \tag{4.10-7}$$

Integral Representations

Solutions for spatially distributed sources may be written as integrals of Green's functions over position. The approach is analogous to that used for the time-dependent point source in Example 4.9-4, except that there the integration was over time. For example, consider a steady problem with a volumetric source of solute $\dot{m}_V(\mathbf{r}')$ (moles/volume/time). The concentration increment at position $\mathbf{r}$ due to sources in a small volume $\Delta V'$ centered at $\mathbf{r}'$ is $\dot{m}_V(\mathbf{r}')f(\mathbf{r}-\mathbf{r}')\Delta V'$. Letting $\Delta V' \to 0$ and integrating, the overall concentration is given by

$$C(\mathbf{r}) = \int_V \dot{m}_V(\mathbf{r}')f(\mathbf{r}-\mathbf{r}')\,dV', \qquad (4.10\text{-}8)$$

where the integration is over all positions $\mathbf{r}'$ in the unbounded domain V. For time-dependent sources, it is necessary also to integrate over all times t' up to the present time t. Assuming that no solute is present at $t = 0$, the transient result is

$$C(\mathbf{r}, t) = \int_0^t \int_V \dot{m}_V(\mathbf{r}', t')F(\mathbf{r}-\mathbf{r}', t-t')\,dV'\,dt'. \qquad (4.10\text{-}9)$$

It is equally straightforward to describe the effects of solute sources at the surface S of an object immersed in an unbounded fluid. For a steady problem with the source distribution $\dot{m}_S(\mathbf{r}')$ (moles/area/time), the solution is

$$C(\mathbf{r}) = \int_S \dot{m}_S(\mathbf{r}')f(\mathbf{r}-\mathbf{r}')\,dS', \qquad (4.10\text{-}10)$$

where the integration is over all positions $\mathbf{r}'$ on the surface of the object. Suppose now that there is a Dirichlet boundary condition, such that the concentration is a known function of position on the surface, $\mathbf{r}_S$. Letting $\mathbf{r} \to \mathbf{r}_S$, Eq. (4.10-10) becomes

$$C(\mathbf{r}_S) = \int_S \dot{m}_S(\mathbf{r}')f(\mathbf{r}_S-\mathbf{r}')\,dS', \qquad (4.10\text{-}11)$$

which is a *boundary integral equation* involving the function $\dot{m}_S(\mathbf{r}')$ as the only unknown. Such problems are solved numerically by discretizing the integral to yield a set of algebraic equations involving values of $\dot{m}_S(\mathbf{r}')$ at specific points. The discrete values of $\dot{m}_S(\mathbf{r}')$ obtained by solving those equations may be used in Eq. (4.10-10) to evaluate the concentration at points not on the surface. The formulation and solution of such problems is discussed by Brebbia et al. (1984).

This chapter concludes with two examples which illustrate the use of integral representations.

Example 4.10-1 Multipole Expansion The objective here is to examine the effects of steady diffusion from an arbitrary particle on the solute concentration in the surrounding fluid, far from the particle. Such effects are of interest in calculating rates of mass transfer in suspensions. As shown in Fig. 4-17, the surface of the particle is described by the position vector $\mathbf{r}'$ (with maximum magnitude $|\mathbf{r}'|_{\max}$), and there is a unit normal $\mathbf{n}$ directed toward the fluid. It is desired to determine the concentration field at positions $\mathbf{r}$ in the fluid such that $|\mathbf{r}| \gg |\mathbf{r}'|_{\max}$.

From Eq. (4.10-10), the concentration field in the fluid is given by

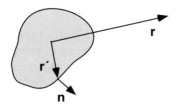

Figure 4-17. A particle of arbitrary shape immersed in a fluid. Positions within the fluid and on the particle surface are denoted by $\mathbf{r}$ and $\mathbf{r}'$, respectively, and $\mathbf{n}$ is the unit normal.

$$C(\mathbf{r}) = \int_S \mathbf{n} \cdot \mathbf{N}(\mathbf{r}') f(\mathbf{r} - \mathbf{r}') \ dS', \tag{4.10-12}$$

where the solute source term on the surface ($\dot{m}_S$) has been rewritten as the outward-normal component of the flux ($\mathbf{n} \cdot \mathbf{N}$). There is no specific consideration here of the rate processes occurring within the particle; whatever combination of diffusion and chemical reactions is present, the result is perceived in the fluid only as a flux $\mathbf{N}(\mathbf{r}')$ at the surface.

At distant positions in the fluid the particle will appear almost as a point, and the distinction between surface positions $\mathbf{r}'$ and the origin (which is within the particle) will be relatively unimportant. Accordingly, the integral in Eq. (4.10-12) is evaluated by expanding the Green's function in a Taylor series about $\mathbf{r}' = \mathbf{0}$. Using Eq. (A.6-6), we obtain

$$f(\mathbf{r} - \mathbf{r}') = f(\mathbf{r}) - \mathbf{r}' \cdot (\nabla f)|_{\mathbf{r}'=\mathbf{0}} + \frac{1}{2}\,\mathbf{r}'\mathbf{r}' : (\nabla\nabla f)|_{\mathbf{r}'=\mathbf{0}} + \cdots . \tag{4.10-13}$$

Evaluating only the first two terms, using Eq. (4.10-4), the result is

$$f(\mathbf{r} - \mathbf{r}') = \frac{1}{4\pi D r} + \frac{\mathbf{r}' \cdot \mathbf{r}}{4\pi D r^3} + \cdots . \tag{4.10-14}$$

For higher-order terms, see Morse and Feshbach (1953, Part II, pp. 1276–1283). Using Eq. (4.10-14) in Eq. (4.10-12), the result is

$$C(\mathbf{r}) = \frac{1}{4\pi D}\left[\frac{\int_S \mathbf{n} \cdot \mathbf{N}(\mathbf{r}')\ dS'}{r} + \frac{\int_S \mathbf{n} \cdot \mathbf{N}(\mathbf{r}')(\mathbf{r}' \cdot \mathbf{r})\ dS'}{r^3} + \cdots\right]$$

$$= \frac{\dot{m}}{4\pi D r} + \frac{1}{4\pi D r^3}\int_S \mathbf{n} \cdot \mathbf{N}(\mathbf{r}')(\mathbf{r}' \cdot \mathbf{r})\ dS' + \cdots . \tag{4.10-15}$$

Equation (4.10-15) is an example of a *multipole expansion;* the terminology comes from the use of such expansions in electrostatics. It is seen that the first, or "monopole" term, which decays as r^{-1}, results from net release of solute from the particle to the fluid. Indeed, as shown by the second form of the equation, the first integral is just $\dot{m}$, the total rate of release. Thus, to first approximation, the particle behaves as a point source of solute [compare with Eq. (4.9-20)]. The next term is the "dipole," which involves the first moment of the surface flux and which decays as r^{-2}. For particles which do not act as net sources (i.e., when there are no chemical reactions or solute reservoirs in the particles), the dipole is the dominant effect [see, by analogy, $\Psi(r, \eta)$ in Eq. (4.8-52)]. Multipole expansions are used to describe long-range effects in a variety of physical contexts; they are especially valuable in the analysis of momentum transfer in suspensions, as discussed in Chapter 7.

Example 4.10-2 Effective Diffusivity in a Suspension   The effective thermal conductivity of a dilute suspension of spheres was calculated in Example 4.8-4 by using the temperature gradient within a sphere. The effective diffusivity in a suspension is derived here using a different method, which is based on the form of the concentration field far from a sphere. This approach is similar to that employed originally by Maxwell (1873).

To focus on the effects of the particles, the steady-state solute concentration in the fluid is expressed as

$$C(\mathbf{r}) = C_\infty(\mathbf{r}) + C_0(\mathbf{r}), \tag{4.10-16}$$

where $C_\infty(\mathbf{r})$ is the concentration which would exist at position $\mathbf{r}$ if there were no particles present in the system, and $C_0(\mathbf{r})$ is the perturbation caused by the particles. It is assumed that the unperturbed part of the concentration corresponds to a uniform imposed gradient, such that ∇C_∞ is constant. For a dilute suspension, the contribution to C_0 from each particle is that of an isolated sphere. Assuming that there are no net sources of solute in the particles, Example 4.10-1 indicates that the single-particle contribution will be in the form of a dipole; that dipole is evaluated using the results of Example 4.8-3. Specifically, the concentration perturbation at $\mathbf{r}$ due to a *single* sphere of radius a centered at $\mathbf{r}'$ is

$$c(\mathbf{r}, \mathbf{r}') = a^3 \left(\frac{1-\gamma}{2+\gamma}\right)\left(\frac{\mathbf{r} \cdot \nabla C_\infty}{|\mathbf{r} - \mathbf{r}'|^3}\right), \qquad \gamma \equiv \frac{KD_S}{D}. \tag{4.10-17}$$

This result was obtained by equating c with the dipole part of Ψ in Eq. (4.8-52) and by making the substitutions $R = a$, $r\eta = z$, and $Gz = \mathbf{r} \cdot \nabla C_\infty$. The solute diffusivities in the sphere and the surrounding fluid are denoted by D_S and D, respectively, and K is the partition coefficient for the solute, defined as the sphere-to-fluid concentration ratio at equilibrium. The reason that K appears in the diffusivity parameter γ is that the derivation of Eq. (4.8-52) assumed that the field variable was continuous at the phase boundary. Thus, with concentrations based on values in the fluid, the apparent diffusivity within the sphere is KD_S [see Eq. (3.3-9) and related discussion].

Suppose now that a suspension with a uniform number density of n spheres per unit volume occupies a volume V, which is large enough that it contains many particles. For convenience, V is chosen to be a sphere of radius R. Summing the effects of all particles, the perturbation in the concentration far from V is given by

$$C_0(\mathbf{r}) = n \int_V c(\mathbf{r}, \mathbf{r}') \, dV'. \tag{4.10-18}$$

This expression resembles Eq. (4.10-8), although $c(\mathbf{r}, \mathbf{r}')$ is not a Green's function. The integral is simplified by recognizing that at distant positions the distinction between $\mathbf{r}'$ and the origin (which is at the center of V) is insignificant. Approximating $c(\mathbf{r}, \mathbf{r}')$ by $c(\mathbf{r}, \mathbf{0})$, it is found that

$$C_0(\mathbf{r}) = n \, V a^3 \left(\frac{1-\gamma}{2+\gamma}\right)\left(\frac{\mathbf{r} \cdot \nabla C_\infty}{r^3}\right) = \phi R^3 \left(\frac{1-\gamma}{2+\gamma}\right)\left(\frac{\mathbf{r} \cdot \nabla C_\infty}{r^3}\right), \tag{4.10-19}$$

where ϕ is the volume fraction of spheres.

The effective diffusivity of the suspension is found by comparing Eq. (4.10-19) with the result obtained by assuming that the volume V is occupied by a single phase. Based on Eq. (4.10-17), a homogeneous sphere of radius R, centered at the origin, will disturb the concentration in the surrounding fluid by the amount

$$C_0(\mathbf{r}) = R^3 \left(\frac{1-\gamma_{\text{eff}}}{2+\gamma_{\text{eff}}}\right)\left(\frac{\mathbf{r} \cdot \nabla C_\infty}{r^3}\right), \qquad \gamma_{\text{eff}} \equiv \frac{K_{\text{eff}} D_{\text{eff}}}{D}, \tag{4.10-20}$$

where K_{eff} and D_{eff} are the effective partition coefficient and effective diffusivity, respectively. Equating C_0 from Eqs. (4.10-19) and (4.10-20), it is found that

$$\frac{K_{\text{eff}} D_{\text{eff}}}{D} = 1 + 3 \left(\frac{\gamma-1}{\gamma+2}\right)\phi + O(\phi^2), \tag{4.10-21}$$

which is analogous to Eq. (4.8-62). The "$O(\phi^2)$" is a reminder that particle-particle interactions, which would yield a ϕ^2 term, were not included in the analysis.

It is evident from Eq. (4.10-21) that a definite value of D_{eff} is not obtained until K_{eff} is specified. Specifying K_{eff} is the same as defining the concentration which is to be used in conjunction with D_{eff}. There are two common approaches. One is to base the suspension concentration on that in the fluid (continuous) phase. In that case, at equilibrium, the suspension concentration will equal that in particle-free fluid (i.e., in the fluid surrounding V), and $K_{\text{eff}} \equiv 1$. The other common approach is to use the volume-average concentration in the suspension, including both phases, which gives $K_{\text{eff}} \equiv 1 + (K-1)\phi$. Unfortunately, a frequent source of confusion in the literature on diffusion in suspensions and porous media is the failure to state explicitly which choice was made.

An important special case is a suspension of impermeable spheres. Setting $\gamma = 0$ in Eq. (4.10-21) gives

$$\frac{D_{\text{eff}}}{D} = 1 - \frac{3}{2}\phi + O(\phi^2), \qquad K_{\text{eff}} \equiv 1, \tag{4.10-22}$$

$$\frac{D_{\text{eff}}}{D} = 1 - \frac{1}{2}\phi + O(\phi^2), \qquad K_{\text{eff}} \equiv 1 - \phi. \tag{4.10-23}$$

The difference between these expressions underscores the need to define K_{eff}.

The brief discussion here of Green's functions and integral representations necessarily omits many aspects of these subjects. A more extensive discussion of the use of Green's functions in conduction problems is given in Carslaw and Jaeger (1959), and a more general introduction to their application in physical problems is found in Morse and Feshbach (1953). A review of the transport properties of systems where one phase is dispersed in another, extending beyond heat conduction and diffusion, is given in Batchelor (1974).

References

Abramowitz, M. and I. A. Stegun. *Handbook of Mathematical Functions*. U.S. Department of Commerce, National Bureau of Standards, Washington, DC, 1970.

Arpaci, V. S. *Conduction Heat Transfer*. Addison-Wesley, Reading, MA, 1966.

Batchelor, G. K. Transport properties of two-phase materials with random structure. *Annu. Rev. Fluid Mech.* 6: 227–255, 1974.

Brebbia, C. A., J. C. F. Telles, and L. C. Wrobel. *Boundary Element Techniques*. Springer-Verlag, Berlin, 1984.

Butkov, E. *Mathematical Physics*. Addison-Wesley, Reading, MA, 1968.

Carslaw, H. S. and J. C. Jaeger. *Conduction of Heat in Solids*, second edition. Clarendon Press, Oxford, 1959.

Casassa, E. F. Equilibrium distribution of flexible polymer chains between a macroscopic solution phase and small voids. *J. Polym. Sci. Polym. Lett.* 5: 773–778, 1967.

Churchill, R. V. *Fourier Series and Boundary Value Problems*, second edition. McGraw-Hill, New York, 1963.

Crank, J. *The Mathematics of Diffusion*, second edition. Clarendon Press, Oxford, 1975.

Grattan-Guinness, I. *Joseph Fourier 1768–1830*. MIT Press, Cambridge, MA, 1972.

Greenberg, M. D. *Advanced Engineering Mathematics*. Prentice-Hall, Englewood Cliffs, NJ, 1988.

Hatton, T. A., A. S. Chiang, P. T. Noble, and E. N. Lightfoot. Transient diffusional interactions between solid bodies and isolated fluids. *Chem. Eng. Sci.* 34: 1339–1344, 1979.

Hildebrand, F. B. *Advanced Calculus for Applications*, second edition. Prentice-Hall, Englewood Cliffs, NJ, 1976.

Hobson, E. W. *The Theory of Spherical and Ellipsoidal Harmonics.* Chelsea, New York, 1955.

Jeffrey, D. J. Conduction through a random suspension of spheres. *Proc. R. Soc. Lond.* A 335: 355–367, 1973.

Juhasz, N. M. and W. M. Deen. Effect of local Peclet number on mass transfer to a heterogeneous surface. *Ind. Eng. Chem. Res.* 30: 556–562, 1991.

Keller, K. H. and T. R. Stein. A two-dimensional analysis of porous membrane transport. *Math. Biosci.* 1: 421–437, 1967.

Lewis, R. S., W. M. Deen, S. R. Tannenbaum, and J. S. Wishnok. Membrane mass spectrometer inlet for quantitation of nitric oxide. *Biol. Mass Spectrometry* 22: 45–52, 1993.

Maxwell, J. C. *A Treatise on Electricity and Magnetism,* Vol. 1, third edition. Clarendon Press, Oxford, 1892 [first edition published in 1873].

Morse, P. M. and H. Feshbach. *Methods of Theoretical Physics.* McGraw-Hill, New York, 1953.

Naylor, A. and G. R. Sell. *Linear Operator Theory in Engineering and Science.* Springer-Verlag, New York, 1982.

Ramkrishna, D. and N. R. Amundson. Stirred pots, tubular reactors, and self-adjoint operators. *Chem. Eng. Sci.* 29: 1353–1361, 1974a.

Ramkrishna, D. and N. R. Amundson. Transport in composite materials: reduction to a self-adjoint formalism. *Chem. Eng. Sci.* 29: 1457–1464, 1974b.

Ramkrishna, D. and N. R. Amundson. *Linear Operator Methods in Chemical Engineering.* Prentice-Hall, Englewood Cliffs, NJ, 1985.

Watson, G. N. *A Treatise on the Theory of Bessel Functions,* second edition. Cambridge University Press, London, 1944.

Problems

4-1. Steady Conduction in a Square Rod with a Heat Source

· Consider the steady, two-dimensional, heat conduction problem shown in Fig. P4-1. Two adjacent sides of a square rod are insulated, while the other two sides are maintained at the ambient temperature. There is a constant volumetric heat source, H. All quantities are dimensionless. Use the FFT method to determine $\Theta(x, y)$.

4-2. Steady Diffusion in a Square Rod with a First-Order Reaction

Assume that a first-order, irreversible reaction takes place in a long solid of square cross section, as shown in Fig. P4-2. All surfaces are exposed to a uniform concentration of reactant. Use the FFT method to determine $\Theta(x, y)$.

Figure P4-1. Steady conduction in a square rod with a constant heat source.

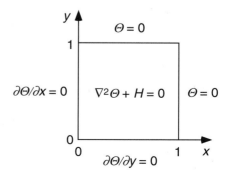

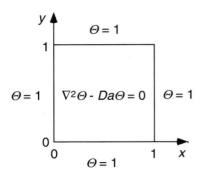

Figure P4-2. Steady diffusion with a first-order reaction in a square rod.

4-3. Steady Conduction in a Long, Rectangular Solid

Consider steady conduction in the rectangular solid shown in Fig. P4-3. The top and bottom surfaces are insulated, the left-hand surface has a specified temperature $f(x)$, and the right-hand boundary is sufficiently distant that it can be modeled as being at $y=\infty$. Use the FFT method to determine $\Theta(x, y)$. [*Hint:* Compare your result with Eq. (3.7-115).]

4-4. Steady Conduction with Neumann Boundary Conditions

Consider steady, two-dimensional conduction in the square rod shown in Fig. P4-4. Three sides are insulated and there is a specified heat flux at the fourth.

(a) Use the FFT method to determine $\Theta(x, y)$. Is there a unique solution?

(b) What happens if the flux at the top is a constant? Suppose, for example, that $\partial\Theta/\partial y = 1$ instead of $\partial\Theta/\partial y = \frac{1}{2} - x$.

4-5. Fourier Series Representations

(a) Derive the Fourier series for $f(x) = x$ for $0 \leq x \leq 1$, using the sines given as case I of Table 4-1. Make a plot which compares the exact function with the results from 4- and 20-term partial sums of the series. Compute the norm of the error for both cases.

(b) Repeat part (a) using the cosines given as case IV of Table 4-1. How do local convergence and mean convergence compare with part (a)?

(c) Does termwise differentiation of the series from (a) lead to a representation of $f'(x)$? How about the series from (b)?

4-6. Transient Diffusion from a Solid Sheet

It is desired to characterize the rate of diffusion of a contaminant (unreacted monomer) out of a sheet of polymeric material. The model to be employed assumes that the monomer is initially at uniform concentration, and that at $t=0$ the sheet is immersed in a well-stirred container of

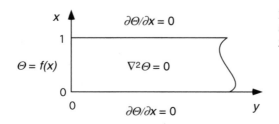

Figure P4-3. Steady conduction in a long, rectangular solid. The right-hand boundary is at $y = \infty$.

Figure P4-4. Steady conduction with Neumann boundary conditions.

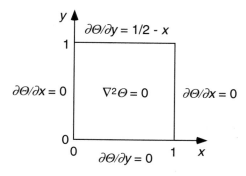

solvent where the monomer is maintained at negligible concentration. It is proposed that the dimensionless problem statement be

$$\frac{\partial \Theta}{\partial t} = \frac{\partial^2 \Theta}{\partial x^2}; \quad \Theta(x,\,0) = 1, \quad \frac{\partial \Theta}{\partial x}(0,\,t) = 0, \quad \Theta(1,\,t) = 0.$$

(a) Use the FFT method to determine $\Theta(x,\,t)$.
(b) Calculate the average monomer concentration remaining in the sheet, $\overline{\Theta}(t)$.
(c) Determine the outward flux of monomer at the surface of the sheet; express the flux in dimensional form.

4-7. Uniqueness Proofs for Laplace's Equation

In Example 4.2-1 it was shown that for Laplace's equation applied within an arbitrary volume V, with a Dirichlet boundary condition at the surface S, there is a unique solution. In this problem the Neumann and Robin boundary conditions are considered.

(a) Show that for the Neumann boundary condition, Eq. (4.2-5b), the solution is unique to within an additive constant.
(b) Show that for the Robin boundary condition, Eq. (4.2-5c), the solution is unique. [*Hint:* First show that for the situations of interest in heat or mass transfer, the coefficient function $h_1(\mathbf{r_s}) \geq 0$. What is the relationship between this function and the Biot numbers for heat or mass transfer?]

4-8. Transient Diffusion in a Film with a First-Order Reaction

Consider transient diffusion in a liquid or solid film with a first-order, homogeneous, irreversible reaction. As shown in Fig. P4-8, assume that there is no reactant present initially and that for $t > 0$ the reactant concentration is held constant at $x = 0$. The surface at $x = 1$ is impermeable. The Damköhler number, Da, is not necessarily large or small.

Figure P4-8. Transient diffusion in a film with a first-order reaction.

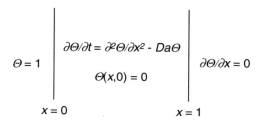

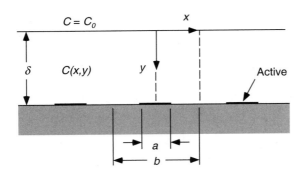

Figure P4-9. Mass transfer to a partially active catalytic surface. The reaction is confined to strips on the surface of width a.

(a) Determine $\Theta(x, t)$ using the FFT method.

(b) Use the exact solution from (a) to estimate the time required to reach a steady state, t_{ss}, for both small and large Da. Compare the estimate from the FFT solution with one obtained from order-of-magnitude and scaling considerations.

4-9. Mass Transfer to a Partially Active Catalytic Surface

Consider steady diffusion of a reactant across a liquid film to a catalytic surface which has alternating "active" and "inactive" areas, as shown in Fig. P4-9. On the active areas of width a there is a constant flux N_0 (directed toward the surface); on the inactive areas the flux is zero. At the top surface of the film the reactant concentration is a constant, C_0. Because of the symmetry it is sufficient to consider one-half of a repeating unit of width b, as indicated by the dashed lines. The fraction of the surface which is active, denoted as $\varepsilon = a/b$, is not necessarily small.

(a) Use the FFT method to determine $C(x, y)$.

(b) If the local mass transfer coefficient, $k_c(x)$, is defined as

$$k_c(x) = \frac{-D(\partial C/\partial y)|_{y=\delta}}{C_0 - C(x, \delta)},$$

then the average mass transfer coefficient, $\langle k_c \rangle$, is given by

$$\langle k_c \rangle = \frac{2}{b}\int_0^{b/2} k_c \, dx = \frac{2}{b}\int_0^{a/2} k_c \, dx.$$

Evaluate the relative mass transfer coefficient, $\langle k_c \rangle / k_c^*$, where $k_c^* = D/\delta$ is the mass transfer coefficient for a completely active surface. The relative mass transfer coefficient depends on two dimensionless parameters, ε and $K = 2\delta/b$. For what combinations of ε and K will $\langle k_c \rangle / k_c^* \to 1$?

Discussions of mass transfer at surfaces of this type are given in Juhasz and Deen (1991) and Keller and Stein (1967).

4-10. Diffusion in a Liquid Film with a Zeroth-Order Reaction

A liquid film resting on a surface is brought into contact with a gas mixture at $t = 0$. A component of the gas which is not present initially in the liquid is absorbed and undergoes a zeroth-order reaction (i.e., with a constant rate whenever the reactant is present). It is proposed that the reaction–diffusion process be modeled as shown in Fig. P4-8, except with the reaction term written as Da rather than $Da\Theta$.

(a) Define the Damköhler number used here.

(b) Evaluate the steady-state concentration profile, $\Theta_{ss}(x)$, and determine Da*, the maximum value of Da for which the proposed problem formulation is valid.

(c) Assuming that Da<Da*, use the FFT method to determine $\Theta(x, t)$.

4-11. Transient Diffusion in a Cylinder with a First-Order Reaction

Consider a transient reaction-diffusion problem in a cylinder which, at large times, leads to the steady-state problem described in Example 4.7-2.

(a) Assuming that there is no reactant within the cylinder initially, determine $\Theta(r, t)$.

(b) Evaluate the reactant flux at the surface, as a function of time.

4-12. Partitioning of Polymers Between Pores and Bulk Solution

One of the simplest models for the configuration of a linear, flexible polymer is a "random-flight chain." In this model the polymer is assumed to be a freely jointed chain consisting of N mass points connected by rectilinear segments of length ℓ. The position of each successive mass point is selected at random, with reference only to the position of the immediately preceding mass point. Accordingly, the shape of the chain is described by a random walk. The analogy between the resulting set of polymer configurations and the random walks of diffusing molecules allows the probability of certain chain configurations to be calculated using an equation like that used to describe transient diffusion.

Certain configurations of long chains are unable to fit in small pores. The extent of this steric exclusion is reflected in the partition coefficient K, which is the equilibrium ratio of the polymer concentration in the pores to the polymer concentration in bulk solution. For a random-flight chain and cylindrical pores the partition coefficient is given by

$$K = 2 \int_0^1 P(r, 1) r \, dr,$$

where $P(r, 1)$ is the probability of finding one end of the chain at dimensionless position r, once all chain configurations that intersect the pore wall have been excluded. The probability distribution $P(r, t)$ is governed by

$$\frac{\partial P}{\partial t} = \frac{\lambda^2}{r} \frac{\partial}{\partial r}\left(r \frac{\partial P}{\partial r}\right), \qquad P(r, 0) = 1, \qquad P(1, t) = 0,$$

where λ is the radius of gyration of the polymer divided by the pore radius. In this model, which assumes that N is large, the radius of gyration is given by $(N\ell^2/6)^{1/2}$. The variable t represents position along the chain, expressed as a fraction of the total contour length (i.e., $0 \leq t \leq 1$).

Show that the partition coefficient is given by

$$K(\lambda) = 4 \sum_{i=1}^{\infty} \frac{\exp(-\beta_i^2 \lambda^2)}{\beta_i^2},$$

where β_i are the roots of the Bessel function J_0. This result was derived originally by Casassa (1967).

4-13. Steady Conduction in a Fibrous Material

It is desired to characterize steady heat conduction through a composite which consists of long, parallel fibers of material B (conductivity k_B) surrounded by a continuous phase of material A (conductivity k_A). The fibers are widely spaced, so that it is sufficient to model a single fiber of radius R, as shown in Fig. P4-13. Assume that, far from a given fiber, there is a uniform temperature gradient of magnitude G directed perpendicular to the fiber axis.

$\partial T/\partial x = G$ at $r = \infty$

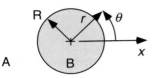

Figure P4-13. Steady conduction through a cylindrical fiber (B) and surrounding material (A), with a uniform temperature gradient far from the fiber.

Use the FFT method to determine the temperature field, $T(r, \theta)$, throughout both materials. In particular, show that within a fiber we have

$$T(r, \theta) = A + \left(\frac{2}{1+\gamma}\right) Gr \cos \theta$$

where A is a constant and $\gamma = k_B/k_A$.

4-14. Effective Thermal Conductivities of Fibrous Materials

(a) Consider a material which contains widely spaced, *parallel fibers*. Using the fiber temperature given in Problem 4-13, calculate the effective thermal conductivity for conduction *perpendicular* to the fiber axes (the x direction in Fig. P4-13). Show that

$$\frac{k_{\text{eff}}^{(x)}}{k_A} = 1 + 2\left(\frac{\gamma-1}{\gamma+1}\right)\phi + O(\phi^2),$$

where ϕ is the volume fraction of fibers.

(b) Again consider a material with *parallel fibers*. Determine the fiber temperature when the imposed gradient is *parallel* to the fiber axes (in the z direction). (*Hint:* The temperature field is one-dimensional.) For heat conduction in this direction, show that

$$\frac{k_{\text{eff}}^{(z)}}{k_A} = 1 + (\gamma-1)\phi.$$

The difference between $k_{\text{eff}}^{(x)}$ and $k_{\text{eff}}^{(z)}$ demonstrates that materials with parallel fibers are anisotropic (see Section 1.2).

(c) In general, a fiber axis will be oriented at an angle β relative to the unperturbed temperature gradient G, where $0 \le \beta \le \pi$ (see Fig. P4-14). (The second angle needed to fully define the relative orientation of the fiber axis, which describes rotations about G, need not be considered here because of symmetry.) Show that, for any fiber orientation, the solution for the temperature can be constructed by adding solutions obtained for the "perpendicular problem" [part (a)] and "parallel problem" [part (b)]. Specifically, show that within a fiber we have

$$\nabla T = \left(\frac{2}{1+\gamma}\right)(\mathbf{n} \cdot \mathbf{G})\mathbf{n} + (\mathbf{p} \cdot \mathbf{G})\mathbf{p},$$

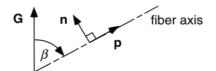

Figure P4-14. Orientation of a fiber relative to the unperturbed temperature gradient.

where **n** and **p** are unit vectors normal to and parallel to the fiber axis, respectively. Note that **n**, **p**, and **G** are all coplanar.

(d) Show that for a material with *randomly oriented fibers*, the orientation-averaged temperature gradient in the fibers is given by

$$\overline{\nabla T} = \frac{1}{2}\int_0^\pi \nabla T \sin \beta \, d\beta = \frac{1}{3}\left(\frac{5+\gamma}{1+\gamma}\right)\mathbf{G},$$

which indicates that the material is isotropic. Finally, show that the effective thermal conductivity is

$$\frac{k_{\text{eff}}}{k_A} = 1 + \frac{(\gamma-1)(\gamma+5)}{3(\gamma+1)}\phi + O(\phi^2).$$

4-15. Transient Diffusion in a Cylinder with Variable Surface Concentration

Consider the concentration $\Theta(r, t)$ of some solute in a long cylinder of unit (dimensionless) radius. Initially, the cylinder is in equilibrium with an external solution containing no solute. The solute is then introduced into the external solution, such that the surface concentration is given by

$$\Theta(1, t) = 1 - e^{-\beta t},$$

where β is a constant. That is, the surface concentration reaches a steady-state value of unity, but not instantaneously.

(a) Use the FFT method to determine $\Theta(r, t)$.

(b) Determine the fractional approach to equilibrium, $F(t)$, which is defined as the ratio of total solute in the cylinder at time t to that at $t = \infty$.

4-16. Continuous Heating of a Sphere

A solid sphere of unit radius is immersed in a fluid in which the temperature is a specified function of time only, so that the dimensionless temperature of the sphere is $\Theta(r, t)$. The initial temperatures of the sphere and the surrounding fluid are zero. For $t > 0$, the fluid is heated in such a manner that $\Theta(1, t) = Ct$, where C is a positive constant.

(a) Use the FFT method to determine $\Theta(r, t)$.

(b) Show that for large t, the solution is of the form

$$\Theta(r, t) = f(r) + Ct,$$

where C is the constant defined above. Determine $f(r)$, and use this result to rewrite the solution from part (a).

4-17. Steady-Periodic Temperature of a Sphere

Consider a situation similar to that described in Problem 4-16, except that now the fluid temperature oscillates in time with a frequency ω, such that $\Theta(1, t) = \sin(2\pi\omega t)$. For forcing functions which are periodic in time, such as this one, the solution eventually achieves a *steady-periodic* character, in which the oscillations are repeated indefinitely. Thus, after an initial transient, the sphere temperature will be periodic in time.

(a) Use the FFT method to determine $\Theta(r, t)$, including both the transient and steady-periodic parts of the solution.

(b) Under what conditions will the result from part (a) be equivalent to that obtained using a pseudosteady approximation?

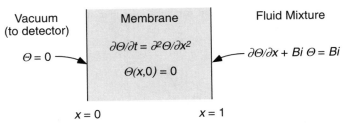

Figure P4-18. Detector with membrane inlet.

4-18. Time Response of a Detector with a Membrane Inlet
    One strategy for continuously monitoring the concentration of a gaseous species or a vola-
tile solute in a liquid is to equip a mass spectrometer (or other suitable detector) with a membrane
inlet which is permeable to the species of interest. When the inlet device contacts the mixture, a
vacuum maintained in the instrument causes the species to diffuse across the membrane and enter
the detector. An example of such a system was described by Lewis et al. (1993). Ideally, the inlet
is designed so that the rate of removal of the species is small enough to have a negligible effect
on the system being monitored.
    A schematic of a membrane inlet is shown in Fig. P4-18. Consider what happens when the
solute of interest is added suddenly to the fluid. Assume that the detector itself responds instantaneously and that its signal is proportional to the solute throughput (i.e., the flux leaving the
membrane at $x=0$). Thus, the overall response is limited by mass transfer in the fluid and diffusion through the membrane.

   (a) How must the dimensionless quantities (x, t, Θ, Bi) be defined? Pertinent dimensional
       quantities include the bulk concentration in the fluid (C_0 for $t>0$), the membrane thick-
       ness (L), the solute diffusivity in the membrane (D), and the mass transfer coefficient
       (k_c) in the fluid. The partition coefficient, or membrane-to-fluid concentration ratio at
       equilibrium, is denoted as K.
   (b) Use the FFT method to determine $\Theta(x, t)$.
   (c) Evaluate the solute flux leaving the membrane at $x=0$.
   (d) Estimate the half-time of the response (i.e., the time required to reach one-half of the
       final detector signal) for Bi $=1$.

4-19. Flow in a Porous Sphere
    As discussed in Problem 2-2, if Darcy's law is used to describe the flow of an incompress-
ible fluid in a porous material, the pressure field is governed by

$$\nabla^2 \mathcal{P} = 0.$$

Suppose that it is desired to measure the Darcy permeability (κ) in a small, porous sphere of
radius R using the arrangement shown in Fig. P4-19. The sphere will be held at the end of a
circular tube by suction. A known volumetric flow rate Q will be imposed, and the pressures $\mathcal{P}_1$
and $\mathcal{P}_0$ will be measured. At the sphere surface, it is assumed that $\mathcal{P} = \mathcal{P}_1$ for $0 \le \theta < \gamma$ and $\mathcal{P} = \mathcal{P}_0$
for $\gamma < \theta \le \pi$, where $\mathcal{P}_1$ and $\mathcal{P}_0$ are constants.

   (a) Use the FFT method to determine $\mathcal{P}(r, \theta)$.
   (b) Derive the pressure–flow relationship which would permit calculation of κ using the
       experiment described above.

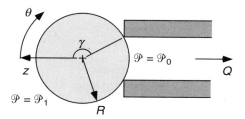

Figure P4-19. A porous sphere held by suction at the end of a circular tube.

4-20. Inner Product and Eigenfunctions for Uniform Convection*

A differential operator which arises in problems involving conduction or diffusion accompanied by a uniform bulk velocity is

$$\mathcal{L}_x = \frac{d^2}{dx^2} - \mathrm{Pe}\frac{d}{dx},$$

where Pe is the Péclet number. Equation (2.8-49) provides an example of such an operator. An examination of the self-adjoint eigenvalue problems associated with $\mathcal{L}_x$ is needed for solving certain problems involving uniform convection. An application of these results is in Problem 4-21.

(a) Consider eigenvalue problems based on $\mathcal{L}_x$, where

$$\mathcal{L}_x \Phi = \frac{d^2\Phi}{dx^2} - \mathrm{Pe}\frac{d\Phi}{dx} = -\lambda^2\Phi$$

together with any combination of the three types of homogeneous boundary conditions. Show that a weighting function $w(x) = e^{-\mathrm{Pe}\,x}$ is needed in the inner product to make such a problem self-adjoint.

(b) A useful trick for solving differential equations like that in part (a) is to modify the dependent variable so as to eliminate the first derivative. Show that this is accomplished by setting $\Phi = \exp(\mathrm{Pe}\,x/2)\Psi$ and that the general solution is

$$\Phi(x) = \exp(\mathrm{Pe}\,x/2)(a\,\sin\,\gamma x + b\,\cos\,\gamma x), \qquad \gamma^2 \equiv \lambda^2 - (\mathrm{Pe}^2/4).$$

(c) Determine the eigenvalues and orthonormal eigenfunctions for the Dirichlet boundary conditions, $\Phi(0) = \Phi(1) = 0$. What happens to the exponential factors?

*This problem was suggested by R. A. Brown.

4-21. Transient Effects in Directional Solidification*

A stagnant-film model for unidirectional solidification was presented in Example 2.8-5. The objective here is to extend that model to describe transients associated with the start-up of the process. Assume that the initial concentration of dopant is uniform at C_∞ and that solidification at velocity U begins suddenly at $t=0$ (see Fig. 2-10).

(a) Use the FFT method to determine $C(y,\,t)$ in the melt. For information on the inner product and eigenfunctions, see Problem 4-20.

(b) An important consequence of the time-dependent concentration in the melt is that the dopant concentration in the solid is not uniform; that is, $C_S = C_S\,(y,\,t)$. Determine C_S $(y,\,t)$, assuming that diffusion within the solid is negligible. If the dopant level within the solid must be within 5% of the steady-state value to give acceptable material properties, how much of the product must be discarded?

*This problem was suggested by R. A. Brown.

4-22. Point Source Near a Boundary Maintained at Constant Concentration

Assume that a reactive solute is released at a steady rate $\dot{m}$ at a point source located a distance L from a catalytic surface. The reaction rate is sufficiently fast that $C=0$ at the surface. Construct a solution for $C(\mathbf{r})$.

4-23. Inner Product and Eigenfunctions for Non-Uniform Material Properties

When the heat capacity per unit volume $(\rho \hat{C}_p)$ and/or thermal conductivity vary with position, the eigenvalue problems for conduction differ from those for constant properties. For transient, one-dimensional conduction in rectangular coordinates, with spatially varying properties, the dimensionless energy equation is written as

$$C(x)\frac{\partial \Theta}{\partial t} = \frac{\partial}{\partial x}\left(K(x)\frac{\partial \Theta}{\partial x}\right).$$

Particular care is needed when the dimensionless properties $C(x)$ and $K(x)$ are discontinuous, as when the region of interest is a layered composite consisting of two or more different materials.

(a) If $C(x)$ and $K(x)$ are *continuous* functions for $0 \leq x \leq 1$ and if the usual types of homogeneous boundary conditons apply at $x=0$ and $x=1$, show that the differential equation and inner product which yield a self-adjoint eigenvalue problem are

$$\mathcal{L}_x \Phi = \frac{1}{C(x)}\frac{d}{dx}\left(K(x)\frac{d\Phi}{dx}\right) = -\lambda^2 \Phi,$$

$$\langle u, v \rangle = \int_0^1 u(x)v(x)C(x)\ dx.$$

(b) If $C(x)$ and $K(x)$ are both *piecewise constant*, such that

$$C(x)=\begin{cases}C_1, & 0 \leq x < \gamma, \\ C_2, & \gamma < x \leq 1,\end{cases} \qquad K(x)=\begin{cases}K_1 & 0 \leq x < \gamma, \\ K_2 & \gamma < x \leq 1,\end{cases}$$

then the operator $\mathcal{L}_x$ is given by

$$\mathcal{L}_x =\begin{cases}(K_1/C_1)(d^2/dx^2), & 0 \leq x < \gamma, \\ (K_2/C_2)(d^2/dx^2), & \gamma < x \leq 1.\end{cases}$$

If the usual types of boundary conditions apply at $x=0$ and $x=1$, show that a self-adjoint eigenvalue problem is obtained with the inner product

$$\langle u, v \rangle = C_1 \int_0^\gamma u(x)v(x)\ dx + C_2 \int_\gamma^1 u(x)v(x)\ dx,$$

provided that

$$\Phi(\gamma^-)=\Phi(\gamma^+), \qquad K_1\Phi'(\gamma^-)=K_2\Phi'(\gamma^+).$$

What is the physical basis for these conditions at $x=\gamma$? The results given here apply to a composite material consisting of two layers; this approach may be extended to any number of layers, as described in Ramkrishna and Amundson (1974b).

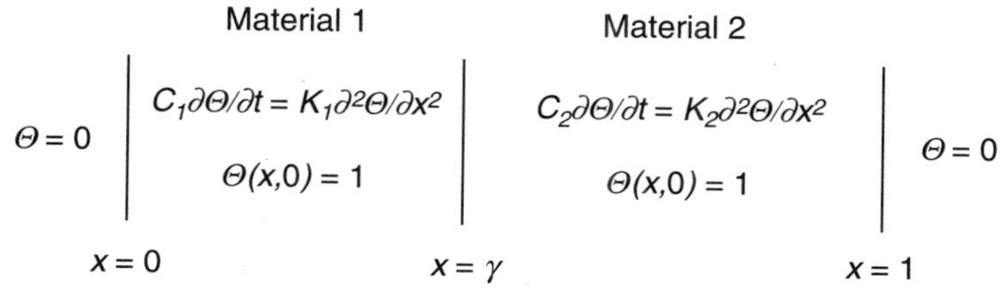

Material 1   Material 2

$$\Theta = 0 \quad \left| \begin{array}{c} C_1 \partial \Theta / \partial t = K_1 \partial^2 \Theta / \partial x^2 \\ \\ \Theta(x,0) = 1 \end{array} \right| \begin{array}{c} C_2 \partial \Theta / \partial t = K_2 \partial^2 \Theta / \partial x^2 \\ \\ \Theta(x,0) = 1 \end{array} \right| \quad \Theta = 0$$

$x = 0$ $x = \gamma$ $x = 1$

Figure P4-24. Conduction in a layered composite consisting of materials 1 and 2, which have different thermophysical properties.

4-24. Transient Conduction in a Layered Composite

Consider transient conduction in a composite solid consisting of two layers, as shown in Fig. P4-24. Using the information on the inner product and eigenfunctions in Problem 4-23, determine $\Theta(x, t)$. [*Hint:* The characteristic equation for the eigenvalues comes from the matching conditions at $x = \gamma$. Also, note that the form of the eigenfunctions changes at $x = \gamma$. It may be helpful to let $\Phi_n(x) = \Phi_n^{(1)}(x)$ for $0 \le x < \gamma$, and $\Phi_n(x) = \Phi_n^{(2)}(x)$ for $\gamma < x \le 1$.]

4-25. Green's Functions for Diffusion in Two Dimensions

(a) For steady diffusion in two dimensions, show that the Green's function may be written as

$$f(\mathbf{r} - \mathbf{r}') = -\frac{\ln|\mathbf{r} - \mathbf{r}'|}{2\pi D}.$$

Note that in this case it is not possible to satisfy the condition $C \to 0$ as $r \to \infty$, where $r = (x^2 + y^2)^{1/2}$.

(b) For transient diffusion in two dimensions, show that the Green's function is

$$F(\mathbf{r} - \mathbf{r}', t - t') = \begin{cases} 0, & \text{for } t < t', \\ \dfrac{e^{-|\mathbf{r} - \mathbf{r}'|^2 / 4D(t-t')}}{4\pi D(t - t')}, & \text{for } t \ge t'. \end{cases}$$

4-26. Transient Diffusion from a Solid to a Stirred Solution of Finite Volume

In Problem 4-6, which involves transient diffusion from a solid to a stirred liquid, it is assumed that the solute concentration in the liquid is zero at all times. The physical situation considered here is similar, except that the liquid volume is assumed to be finite. Thus, the solute concentration in the liquid increases as that in the solid decreases, until a steady state is achieved. Assuming that the mass transfer resistance in the liquid is negligible, the solute concentration in the solid, $\Theta(x, t)$, is governed by

$$\frac{\partial \Theta}{\partial t} = \frac{\partial^2 \Theta}{\partial x^2}; \quad \Theta(x, 0) = 1, \quad \frac{\partial \Theta}{\partial x}(0, t) = 0, \quad \Theta(1, t) = \Psi(t),$$

where $\Psi(t)$ denotes the concentration in the liquid.

In this problem the concentration field consists of two parts, $\Theta(x, t)$ for the solid and $\Psi(t)$ for the liquid. For such problems it is helpful to write the overall solution as a vector,

$$\boldsymbol{\Theta}(x, t) = \begin{bmatrix} \Theta(x, t) \\ \Psi(t) \end{bmatrix} = \begin{bmatrix} \Theta(x, t) \\ \Theta(1, t) \end{bmatrix},$$

where the second form is obtained by using the boundary condition at $x=1$. A vector operator $\mathcal{L}_x$ is defined so that the differential conservation equations for $\Theta(x, t)$ and $\Psi(t)$ are expressed as

$$\mathcal{L}_x \Theta = \frac{\partial \Theta}{\partial t}.$$

The nth basis function is also written as a vector, $\Phi_n(x)$, so that the differential equation for the eigenvalue problem is expressed as

$$\mathcal{L}_x \Phi_n = -\lambda_n^2 \Phi_n.$$

The expansion for $\Theta(x, t)$ is written as

$$\Theta(x, t) = \sum_{n=1}^{\infty} \langle \Theta, \Phi_n \rangle \Phi_n = \sum_{n=1}^{\infty} \Theta_n(t) \Phi_n(x).$$

It is seen that the inner product of the solution and basis function vectors is a scalar, $\Theta_n(t)$, which represents the time dependence common to the two parts of the solution. The overall approach closely parallels the usual FFT procedure, but (as given below) a special form of inner product is needed to make the approach work.

(a) Show that the liquid concentration is governed by

$$\frac{d\Psi}{dt} = -\gamma \frac{\partial \Theta}{\partial x}(1, t), \qquad \Psi(0) = 0.$$

How must $\Psi(t)$ and γ be defined?

(b) Based on the differential equations for $\Theta(x, t)$ and $\Psi(t)$, the differential operator with Θ is written as

$$\mathcal{L}_x \Theta = \begin{bmatrix} \dfrac{\partial^2 \Theta}{\partial x^2} \\ -\gamma \dfrac{\partial \Theta}{\partial x}(1, t) \end{bmatrix}$$

and the corresponding eigenvalue problem is

$$\mathcal{L}_x \Phi_n = \begin{bmatrix} \Phi_n'' \\ -\gamma \Phi_n'(1) \end{bmatrix} = -\lambda_n^2 \begin{bmatrix} \Phi_n(x) \\ \Phi_n(1) \end{bmatrix}, \qquad \Phi_n'(0) = 0.$$

Show that if the inner product of the vector functions $\mathbf{u}(x)$ and $\mathbf{v}(x)$ is defined as

$$\langle \mathbf{u}, \mathbf{v} \rangle = \int_0^1 u(x)v(x)\, dx + \frac{u(1)v(1)}{\gamma},$$

then the eigenvalue problem is self-adjoint. Note that this inner product of vectors is similar to the dot product of vectors, in that both result in a sum of scalar terms.

(c) Use the first line of the eigenvalue problem to determine $\Phi_n(x)$ to within a multiplicative constant. (The normalization condition is applied later.)

(d) Use the second line of the eigenvalue problem to find the characteristic equation governing λ_n.

(e) Use the normalization condition,

$$\langle \Phi_n, \Phi_n \rangle = \int_0^1 \Phi_n^2\, dx + \frac{\Phi_n^2(1)}{\gamma} = 1,$$

to complete the solution for Φ_n. Also, prove that the eigenfunctions are orthogonal in the inner product defined in part (b). [*Hint:* Use the approach in Section 4.6.]

(f) Show that $\Theta_n(t)$ is governed by

$$\frac{d\Theta_n}{dt} = -\lambda_n^2 \Theta_n, \qquad \Theta_n(0) = H,$$

where H is a constant. Evaluate H and solve for $\Theta_n(t)$.

(g) Complete the solution for $\Theta(x, t)$ and $\Psi(t)$.

Problems of this general type, where a solid or fluid region with a spatially distributed concentration or temperature is in contact with a stirred fluid, are discussed in Ramkrishna and Amundson (1974a) and Hatton et al. (1979).

Chapter 5

FUNDAMENTALS OF
FLUID MECHANICS

5.1 INTRODUCTION

This chapter is the first of several devoted to fluid mechanics and is intended to lay the groundwork for the later analysis of specific types of flows. The primary objective of most such analyses is to determine the fluid velocity, $\mathbf{v}(\mathbf{r}, t)$, a function of the position vector $\mathbf{r}$ and time t. Once $\mathbf{v}$ is determined, the forces exerted by the fluid on various surfaces can be calculated in a straightforward manner, and the convective transport of energy and chemical species can be described. The chapter begins with some purely descriptive aspects of fluid motion (kinematics), before turning to the forces which produce that motion (dynamics). The section on kinematics focuses mainly on the meaning of certain quantities derived from spatial derivatives of $\mathbf{v}$, including the vorticity and the rate of strain. The treatment of fluid dynamics begins with conservation of linear momentum and the general description of stress in a fluid. Following that is a brief discussion of pressure and of the resulting forces in static fluids. Presented then are several constitutive equations which relate velocity gradients to viscous stresses, for Newtonian or generalized Newtonian fluids. The dynamic pressure and the stream function, constructs which are very useful in solving flow problems, are defined. As an introduction to approximations which are applicable under particular flow conditions, the chapter closes with a discussion of scaling in connection with the Navier–Stokes equation.

5.2 FLUID KINEMATICS

Motion of Material Points

A useful definition of a fluid, which encompasses gases or liquids, is that it is *a material that deforms continuously when subjected to a shearing stress.* Whereas a solid placed under stress will achieve a new static shape (if it deforms at all), a fluid will continue to change its "shape" indefinitely. For a fluid it is not the *amount* of deformation but rather the *rate* of deformation (or rate of strain) which is crucial in the mechanics, because it is the rate of deformation which is related ultimately to the viscous stresses. For a material to deform, there must be motion of one point in the material relative to another. In a fluid, such reference points are called *material points.* A material point is an infinitesimal mass which moves always at the local fluid velocity and which retains its identity as it changes position. In other words, a material point is a small "particle" of fluid which may be tracked over time. In Chapter 2 the material derivative Db/Dt was discussed as providing the rate of change of a scalar quantity b that would be perceived by an observer moving at the local fluid velocity, $\mathbf{v}$. An equivalent statement is that the material derivative provides the rate of change evaluated at a material point (as opposed to a point at a fixed location). As seen already with the conservation equations for scalar quantities, certain relationships are stated most simply by using this material frame of reference.

The concept of a material point is important in the theory of flow visualization. A *streamline* is a curve in space which is tangent everywhere to the vector $\mathbf{v}$. Imagine having a complete "snapshot" of $\mathbf{v}$ at some instant $t = t_0$ and then selecting some fixed point. A streamline through that point is generated mathematically by identifying the curve which follows the direction of $\mathbf{v}$ at time t_0. The parameter which is varied in generating the streamline is not time, which is fixed at t_0, but rather some measure of arc length along the curve. A related (but not identical) concept is a *streakline,* a curve formed by the instantaneous positions of a succession of material points "released" continuously from a fixed location. Although streamlines and streaklines do not coincide in general, for steady flow they are the same. Thus, a way to obtain images of streamlines is to introduce lines of smoke in a steady gas flow, or streams of dye or reflective material in a liquid, and to photograph the resulting streaks under conditions where diffusion of the marker is negligible. Many excellent pictures of flows have been generated in this manner [see, for example, Van Dyke (1982)]. Streamlines are discussed further in Section 5.9. The distinctions among streamlines, streaklines, and the paths of individual particles in unsteady flow are explained in more detail in Aris (1962, pp. 76–82).

Returning to the concept of deformation, the existence of relative motion between two material points implies that there must be a velocity gradient. However, the existence of a velocity gradient does not necessarily imply deformation. The reason for this is that a purely rotational motion, such as might occur even with a rigid solid, also leads to a difference in linear velocity between any pair of material points. Even the simplest flows, such as a shear flow with a uniform velocity gradient, combine elements of rotation and deformation (see Example 5.2-2). Suppose that we choose one material point as a reference and examine the motion of some other point relative to the first. In rigid rotation, a line segment connecting the two points will rotate about the first while retaining the same length; in what may be termed "pure deformation," the segment will change in length but will not rotate. All relative motions of two material points represent combinations of these two effects. Thus, to calculate a meaningful rate of deformation in a fluid, it is necessary to account for the part of the motion that is attributable to rigid-body rotation.

Rigid-Body Rotation

Purely rotational motion, like that of a rigid solid, is considered first. In the situation illustrated in Fig. 5-1, the fluid is rotating counterclockwise about an axis which is perpendicular to the page, and points 1 and 2 are material points on two of the circular streamlines. The positions of these points relative to the coordinate origin are given by the vectors $\mathbf{r}_1$ and $\mathbf{r}_2$, whereas the vectors $\mathbf{R}_1$ and $\mathbf{R}_2$ are normal to the axis of rotation. Thus, $\mathbf{R}_1$ and $\mathbf{R}_2$ are parallel to the paper, but in general $\mathbf{r}_1$ and $\mathbf{r}_2$ are not. The angular velocity is denoted as $\boldsymbol{\omega} = \omega \mathbf{n}$, where $\mathbf{n}$ is a unit vector parallel to the axis of rotation. (Letting $\boldsymbol{\omega}$ represent a vector rather than a tensor deviates from our usual convention.) The linear and angular velocities at any point are related by

$$\mathbf{v} = \boldsymbol{\omega} \times \mathbf{R}. \tag{5.2-1}$$

According to the right-hand rule for cross (vector) products, the counterclockwise rotation shown in Fig. 5-1 requires that $\mathbf{n}$ be directed toward the reader. Letting $\Delta\mathbf{v} = \mathbf{v}_2 - \mathbf{v}_1$, $\Delta\mathbf{r} = \mathbf{r}_2 - \mathbf{r}_1$, and $\Delta\mathbf{R} = \mathbf{R}_2 - \mathbf{R}_1$, the relative velocities and relative positions of the two points are related by

$$\Delta\mathbf{v} = \boldsymbol{\omega} \times \Delta\mathbf{R} = \boldsymbol{\omega} \times \Delta\mathbf{r}. \tag{5.2-2}$$

The second equality is a consequence of the fact that any components of $\mathbf{r}_1$ and $\mathbf{r}_2$ in the direction parallel to $\boldsymbol{\omega}$ or $\mathbf{n}$ do not contribute to the cross product.

The customary measure of rotational motion at any point in a fluid is the *vorticity*,

$$\mathbf{w} \equiv \nabla \times \mathbf{v}. \tag{5.2-3}$$

For the special case of rigid-body rotation, there is a simple proportionality between $\mathbf{w}$ and $\boldsymbol{\omega}$. Taking the curl of both sides of Eq. (5.2-1) and using identity (6) of Table A-1, we find that

$$\mathbf{w} = \mathbf{R} \cdot \nabla \boldsymbol{\omega} - \boldsymbol{\omega} \cdot \nabla \mathbf{R} + \boldsymbol{\omega}(\nabla \cdot \mathbf{R}) - \mathbf{R}(\nabla \cdot \boldsymbol{\omega}). \tag{5.2-4}$$

Because $\boldsymbol{\omega}$ is a constant vector and $\mathbf{R}$ does not vary in the direction of $\mathbf{n}$, all terms vanish except that containing $\nabla \cdot \mathbf{R}$. Recognizing that $\mathbf{R}$ amounts to a two-dimensional position vector, we find that $\nabla \cdot \mathbf{R} = 2$, by analogy with Eq. (A.6-10). It follows that

$$\mathbf{w} = 2\boldsymbol{\omega}. \tag{5.2-5}$$

Rewriting Eq. (5.2-2) in terms of $\mathbf{w}$, it is found that for rigid-body rotation,

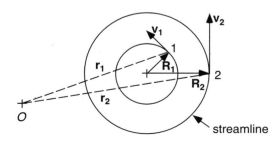

Figure 5-1. Motion of material points in a fluid undergoing rigid-body rotation. One point is labeled on each of two circular streamlines. The coordinate origin is point O.

$$\Delta \mathbf{v} = \frac{1}{2}(\mathbf{w} \times \Delta \mathbf{r}). \tag{5.2-6}$$

Although in rigid-body rotation the vorticity is a reflection merely of the (constant) angular velocity, in complex flows it provides a useful measure of the local rotational motion at any point in the fluid. Note that when we use Eq. (5.2-3) to compute $\mathbf{w}(\mathbf{r}, t)$ from $\mathbf{v}(\mathbf{r}, t)$, there is no need to specify an axis of rotation. If $\mathbf{w} = \mathbf{0}$ everywhere, the flow is said to be *irrotational*.

The *velocity gradient*, $\nabla \mathbf{v}$, is a dyad whose nine elements in rectangular coordinates consist of all possible first derivatives of the three velocity components with respect to the three coordinates. We now make use of the observation that any tensor can be written as the sum of a symmetric tensor and an antisymmetric tensor (see Section A.2). Thus, the velocity gradient can be expressed as

$$\nabla \mathbf{v} = \boldsymbol{\Gamma} + \boldsymbol{\Omega}, \tag{5.2-7}$$

$$\boldsymbol{\Gamma} \equiv \frac{1}{2}[\nabla \mathbf{v} + (\nabla \mathbf{v})^t], \tag{5.2-8}$$

$$\boldsymbol{\Omega} \equiv \frac{1}{2}[\nabla \mathbf{v} - (\nabla \mathbf{v})^t], \tag{5.2-9}$$

where $\boldsymbol{\Gamma}$ is the *rate-of-strain tensor* and $\boldsymbol{\Omega}$ is the *vorticity tensor.* The advantage of this decomposition of $\nabla \mathbf{v}$ is that, as their names suggest, the symmetric ($\boldsymbol{\Gamma}$) and antisymmetric ($\boldsymbol{\Omega}$) parts are uniquely associated with the local rates of deformation and rotation, respectively. [Some authors omit the factor $\frac{1}{2}$ from the definitions of the rate-of-strain and vorticity tensors, so that Eq. (5.2-7) would be written as $\nabla \mathbf{v} = \frac{1}{2}(\boldsymbol{\Gamma} + \boldsymbol{\Omega})$. The definitions used here are the ones more commonly employed.]

To complete the description of rigid-body rotation, we focus first on the properties of $\boldsymbol{\Omega}$. Any antisymmetric tensor has only three independent, nonzero elements. Thus, as pointed out in Aris (1962, pp. 24–25), an antisymmetric tensor can be constructed from the three components of any vector. The vorticity *tensor* $\boldsymbol{\Omega}$ is related to the vorticity *vector* $\mathbf{w}$ according to

$$\boldsymbol{\Omega} = \frac{1}{2}(\boldsymbol{\varepsilon} \cdot \mathbf{w}) = \frac{1}{2}\begin{pmatrix} 0 & w_z & -w_y \\ -w_z & 0 & w_x \\ w_y & -w_x & 0 \end{pmatrix}, \tag{5.2-10}$$

where $\boldsymbol{\varepsilon}$ is the alternating unit tensor (see Section A.3). The matrix in Eq. (5.2-10) shows the relationship between the elements of $\boldsymbol{\Omega}$ and the rectangular components of $\mathbf{w}$. Using the identity

$$\frac{1}{2}(\mathbf{w} \times \mathbf{a}) = \mathbf{a} \cdot \boldsymbol{\Omega}, \tag{5.2-11}$$

which is valid for any vector $\mathbf{a}$, we find from Eq. (5.2-6) that for any pair of material points in a fluid undergoing rigid-body rotation the relative velocity is

$$\Delta \mathbf{v} = \Delta \mathbf{r} \cdot \boldsymbol{\Omega}. \tag{5.2-12}$$

This result is used below in interpreting the difference in velocity between neighboring material points in an arbitrary flow.

Deformation

To describe the deformation of a fluid, as measured by the tendency of two material points to move closer together or farther apart, we examine the difference in velocity between nearby points in relation to their difference in position. The locations of material points 1 and 2 at time t are given by $\mathbf{r}_1(t)$ and $\mathbf{r}_2(t)$, respectively. Their velocities are

$$\mathbf{v}_1 = \frac{d\mathbf{r}_1}{dt} = \mathbf{v}(\mathbf{r}_1, t), \qquad \mathbf{v}_2 = \frac{d\mathbf{r}_2}{dt} = \mathbf{v}(\mathbf{r}_2, t). \tag{5.2-13}$$

The points are assumed to be very close to one another, so that the magnitude of $\Delta\mathbf{r} = \mathbf{r}_2 - \mathbf{r}_1$ is arbitrarily small. This allows us to relate their velocities by a Taylor series expansion about position $\mathbf{r}_1$. Using Eq. (A.6-7), the velocity at $\mathbf{r}_2$ is

$$\mathbf{v}(\mathbf{r}_2, t) = \mathbf{v}(\mathbf{r}_1 + \Delta\mathbf{r}, t) = \mathbf{v}(\mathbf{r}_1, t) + \Delta\mathbf{r}\cdot\nabla\mathbf{v} + O(|\Delta\mathbf{r}|^2). \tag{5.2-14}$$

Using Eq. (5.2-7) in (5.2-14), the relative velocity, $\Delta\mathbf{v} = \mathbf{v}_2 - \mathbf{v}_1$, is found to be

$$\Delta\mathbf{v} = (\Delta\mathbf{r}\cdot\mathbf{\Gamma}) + (\Delta\mathbf{r}\cdot\mathbf{\Omega}) + O(|\Delta\mathbf{r}|^2). \tag{5.2-15}$$

It was seen in Eq. (5.2-12) that the relative velocity due to rigid-body rotation is $\Delta\mathbf{r}\cdot\mathbf{\Omega}$, so that the relative velocity due to deformation is evidently $\Delta\mathbf{r}\cdot\mathbf{\Gamma}$.

To relate $\mathbf{\Gamma}$ more directly to changes in the distance $|\Delta\mathbf{r}|$, we note that

$$\Delta\mathbf{v}\cdot\Delta\mathbf{r} = \frac{d(\Delta\mathbf{r})}{dt}\cdot\Delta\mathbf{r} = \frac{1}{2}\frac{d}{dt}(|\Delta\mathbf{r}|^2) = |\Delta\mathbf{r}|\frac{d|\Delta\mathbf{r}|}{dt}. \tag{5.2-16}$$

Using Eq. (5.2-15) to evaluate $\Delta\mathbf{v}$ in Eq. (5.2-16), we obtain

$$|\Delta\mathbf{r}|\frac{d|\Delta\mathbf{r}|}{dt} = (\Delta\mathbf{r}\cdot\mathbf{\Gamma}\cdot\Delta\mathbf{r}) + (\Delta\mathbf{r}\cdot\mathbf{\Omega}\cdot\Delta\mathbf{r}) + O(|\Delta\mathbf{r}|^3). \tag{5.2-17}$$

Based on the association of vorticity with rotation, we expect the term involving $\mathbf{\Omega}$ to make no contribution to the rate of change in $|\Delta\mathbf{r}|$. Indeed, it is straightforward to show that for any vector $\mathbf{a}$ and any antisymmetric tensor $\mathbf{\Omega}$,

$$\mathbf{a}\cdot\mathbf{\Omega}\cdot\mathbf{a} = 0. \tag{5.2-18}$$

Using this result in Eq. (5.2-17), we obtain

$$|\Delta\mathbf{r}|\frac{d|\Delta\mathbf{r}|}{dt} = (\Delta\mathbf{r}\cdot\mathbf{\Gamma}\cdot\Delta\mathbf{r}) + O(|\Delta\mathbf{r}|^3). \tag{5.2-19}$$

This confirms the earlier statement that the rate of deformation in a fluid is related only to the part of $\nabla\mathbf{v}$ contained in $\mathbf{\Gamma}$.

More extensive discussions of fluid kinematics are given in Aris (1962), Batchelor (1970), and Serrin (1959). This section concludes with two examples.

Example 5.2-1 Rate of Strain in an Extensional Flow   If two material points are aligned with the x axis and separated by an infinitesimal (scalar) distance Δx, then

$$\Delta\mathbf{r}\cdot\mathbf{\Gamma}\cdot\Delta\mathbf{r} = (\Delta x)^2(\mathbf{e}_x\cdot\mathbf{\Gamma}\cdot\mathbf{e}_x) = (\Delta x)^2\Gamma_{xx} = (\Delta x)^2\frac{\partial v_x}{\partial x}. \tag{5.2-20}$$

Using this result in Eq. (5.2-19), we find that for $\Delta x \to 0$,

$$\frac{1}{\Delta x} \frac{d(\Delta x)}{dt} = \frac{\partial v_x}{\partial x}.$$

(5.2-21)

In general, a rate of strain is a rate of change in a length relative to the total length. Thus, Eq. (5.2-21) shows that $\Gamma_{xx} = \partial v_x / \partial x$ is precisely the local rate of strain in the fluid along the x axis. When there is a positive rate of strain in the flow direction, as in this example when $\partial v_x / \partial x > 0$, the flow is said to have an *extensional* component.

Example 5.2-2 Rotation and Deformation in a Simple Shear Flow   An example of a simple shear flow is a velocity field given by

$$v_x = cy, \qquad v_y = v_z = 0,$$

(5.2-22)

where c is a positive constant. Using Eqs. (5.2-3), (5.2-8), and (5.2-9), we have

$$\mathbf{w} = \begin{bmatrix} 0 \\ 0 \\ -c \end{bmatrix}, \quad \Gamma = \begin{bmatrix} 0 & c/2 & 0 \\ c/2 & 0 & 0 \\ 0 & 0 & 0 \end{bmatrix}, \quad \Omega = \begin{bmatrix} 0 & -c/2 & 0 \\ c/2 & 0 & 0 \\ 0 & 0 & 0 \end{bmatrix},$$

(5.2-23)

indicating that the flow involves a spatially uniform combination of deformation (straining) and rotation. The fact that $w_x = w_y = 0$ and $w_z < 0$ implies that the rotation is about the z axis and that it is clockwise when viewed from "above" the x–y plane.

Consider now two points in the x–y plane, separated by a small distance ΔL. As shown in Fig. 5-2, the points are assumed to be on a line which intersects the x axis at an angle θ. Using Γ from Eq. (5.2-23), it is found that

$$\begin{aligned}
\Delta \mathbf{r} \cdot \Gamma \cdot \Delta \mathbf{r} &= [\Delta L(\cos \theta \mathbf{e}_x + \sin \theta \mathbf{e}_y)] \cdot [(c/2)(\mathbf{e}_x \mathbf{e}_y + \mathbf{e}_y \mathbf{e}_x)] \cdot [\Delta L(\cos \theta \mathbf{e}_x + \sin \theta \mathbf{e}_y)] \\
&= [(c \Delta L/2)(\cos \theta \mathbf{e}_y + \sin \theta \mathbf{e}_x)] \cdot [\Delta L(\cos \theta \mathbf{e}_x + \sin \theta \mathbf{e}_y)] \\
&= c(\Delta L)^2 \sin \theta \cos \theta.
\end{aligned}$$

(5.2-24)

Accordingly, the rate of strain is given by Eq. (5.2-19) as

$$\frac{1}{\Delta L} \frac{d(\Delta L)}{dt} = c \sin \theta \cos \theta = \frac{c}{2} \sin 2\theta.$$

(5.2-25)

Thus, the rate of strain in this flow is independent of position, but dependent on the relative orientation of the points. The absolute value of the rate of strain is zero when the points are aligned with the x axis or y axis (i.e., when θ is a multiple of $\pi/2$).

Figure 5-2. Two material points in a simple shear flow. For the arrows depicting the velocity, point 1 is used as the origin.

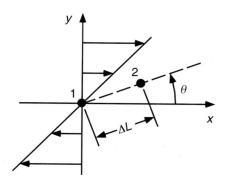

5.3 CONSERVATION OF MOMENTUM

The linear momentum of a solid body with mass m and translational (center-of-mass) velocity $\mathbf{v}$ is $m\mathbf{v}$, a vector. For a body of constant mass, Newton's second law of motion is stated as

$$m\frac{d\mathbf{v}}{dt} = \mathbf{F}, \tag{5.3-1}$$

where $\mathbf{F}$ is the net force acting on the object. Thus, the rate of change of momentum equals the net force. The objective of this section is to rewrite Eq. (5.3-1) in integral and differential forms applicable to fluids. The resulting equations describing the conservation of linear momentum are comparable in most respects to the conservation equations for mass, energy, and chemical species derived in Chapter 2. The first task is to describe the rate of change of the momentum of a body of fluid, using local velocities instead of the center-of-mass velocity. There are two basic ways to do that, both of which are informative and will be discussed. The second task is to characterize the forces exerted on a body of fluid by its surroundings, leading to an expression for $\mathbf{F}$. As in Chapter 2, the derivations begin with macroscopic (control volume) equations, which are reduced eventually to differential equations valid at any point in a fluid.

Momentum in a Material Volume

Conservation of momentum may be expressed using any choice of control volume, but the approach that adheres most closely to Eq. (5.3-1) makes use of a control volume that always contains the same mass of fluid. If the surface bounding the control volume is assumed to deform with the flow, such that the surface velocity $\mathbf{v_S}$ always equals the local fluid velocity $\mathbf{v}$, then no mass will cross any part of the surface. Accordingly, the control volume will always contain the same material. Such a control volume is called a *material volume;* its surface and volume are denoted as $S_M(t)$ and $V_M(t)$, respectively. A *material point,* as discussed in Section 5.2, is simply a material volume in the limit $V_M \to 0$.

For a material volume, Newton's second law is written as

$$\frac{d}{dt} \int_{V_M(t)} \rho\mathbf{v} \, dV = \mathbf{F}. \tag{5.3-2}$$

The product $\rho\mathbf{v}$ is the *concentration of linear momentum* (the amount of momentum per unit volume), so that the integral represents the total momentum of the fluid in the material volume and the left-hand side gives the rate of change of momentum. The center-of-mass velocity in Eq. (5.3-1) has been replaced in Eq. (5.3-2) by the local fluid velocity. What is desired now is to rewrite the left-hand side of Eq. (5.3-2) in more useful form.

Two integral transformations from the Appendix are used to obtain the desired result. The Leibniz rule for differentiating the volume integral of any vector $\mathbf{u}(\mathbf{r}, t)$, from Eq. (A.5-10), is

$$\frac{d}{dt} \int_{V(t)} \mathbf{u} \, dV = \int_{V(t)} \frac{\partial\mathbf{u}}{\partial t} \, dV + \int_{S(t)} (\mathbf{n}\cdot\mathbf{v_S})\mathbf{u} \, dS. \tag{5.3-3}$$

The divergence theorem applied to the dyadic product $\mathbf{vu}$, from Eq. (A.5-3), is

$$\int_{V(t)} \mathbf{\nabla}\cdot(\mathbf{vu})\ dV = \int_{S(t)} \mathbf{n}\cdot(\mathbf{vu})\ dS. \tag{5.3-4}$$

Combining these two results for a *material volume*, where $\mathbf{v_S} = \mathbf{v}$, gives

$$\frac{d}{dt}\int_{V_M(t)} \mathbf{u}\ dV = \int_{V_M(t)} \left[\frac{\partial \mathbf{u}}{\partial t} + \mathbf{\nabla}\cdot(\mathbf{vu})\right] dV. \tag{5.3-5}$$

This identity, which holds also if $\mathbf{u}$ is replaced by a scalar, is called *Reynolds' transport theorem.* To apply it to Eq. (5.3-2), we let $\mathbf{u} = \rho\mathbf{v}$. The integrand on the right-hand side is expanded now to obtain

$$\frac{\partial}{\partial t}(\rho\mathbf{v}) + \mathbf{\nabla}\cdot(\rho\mathbf{vv}) = \mathbf{v}\left[\frac{\partial\rho}{\partial t} + \mathbf{\nabla}\cdot(\rho\mathbf{v})\right] + \rho\left[\frac{\partial\mathbf{v}}{\partial t} + \mathbf{v}\cdot\mathbf{\nabla v}\right]. \tag{5.3-6}$$

From the continuity equation, Eq. (2.3-1), the first bracketed term is identically zero. Rewriting the remaining term using the material derivative, Eq. (5.3-5) simplifies to

$$\frac{d}{dt}\int_{V_M(t)} \rho\mathbf{v}\ dV = \int_{V_M(t)} \rho\frac{D\mathbf{v}}{Dt}\ dV. \tag{5.3-7}$$

Combining Eqs. (5.3-2) and (5.3-7), Newton's second law for a material volume becomes

$$\int_{V_M(t)} \rho\frac{D\mathbf{v}}{Dt}\ dV = \mathbf{F}. \tag{5.3-8}$$

Being able to write the rate of change of total momentum in terms of a material derivative inside a volume integral, as done here, will be very useful for obtaining a differential form of the momentum conservation equation.

Momentum in an Arbitrary Control Volume

Of importance, Eq. (5.3-8) applies not just to a material volume, but to any control volume. This will be shown by adopting a viewpoint more like that used to derive the conservation equations for scalar quantities in Chapter 2. The additional factor which must be considered with an arbitrary control volume is that any mass crossing the control surface carries with it a certain amount of momentum. In other words, there is the possibility of a net increase or decrease of momentum in the control volume due to convective transport of momentum across its boundary. For an arbitrary control volume, the momentum balance is expressed as

$$\frac{d}{dt}\int_{V(t)} \rho\mathbf{v}\ dV + \int_{S(t)} \rho\mathbf{v}[(\mathbf{v} - \mathbf{v_S})\cdot\mathbf{n}]\ dS = \mathbf{F}. \tag{5.3-9}$$

Thus, the momentum input due to the force $\mathbf{F}$ is equated with the rate of accumulation plus the convective losses. The convective transport of momentum is represented by the surface integral, which involves the product of the momentum concentration with the net outward velocity at any point on the surface. If $\mathbf{v}=\mathbf{v}_S$, so that the control volume is a material volume, then Eq. (5.3-9) reduces to Eq. (5.3-2). The key conceptual point is that the stresses which contribute to $\mathbf{F}$ correspond only to *diffusive* fluxes of momentum, ensuring that the surface integral in Eq. (5.3-9) does not involve any double-counting of momentum transfer. This will become clearer when $\mathbf{F}$ is evaluated.

Applying the Leibniz formula and divergence theorem once again, Eq. (5.3-9) is rewritten as

$$\int_{V(t)} \left[\frac{\partial}{\partial t}(\rho\mathbf{v}) + \nabla\cdot(\rho\mathbf{vv}) \right] dV = \mathbf{F}, \tag{5.3-10}$$

where we have made use of the fact that $(\rho\mathbf{v})(\mathbf{v}\cdot\mathbf{n})=\mathbf{n}\cdot(\rho\mathbf{vv})$. Expanding and simplifying the integrand as in Eq. (5.3-6), we obtain the same result as before. Thus,

$$\int_{V(t)} \rho\frac{D\mathbf{v}}{Dt}\,dV = \mathbf{F} \tag{5.3-11}$$

for *any* control volume. This completes our first main task; what remains is to evaluate $\mathbf{F}$.

Evaluation of Forces

The external forces acting on a body of fluid are of two general types. Forces which act directly on a mass (or volume) of fluid are called *body forces,* whereas those which act on surfaces are expressed in terms of *stresses* (forces per unit area). The net body force and net surface force on the control volume are denoted as $\mathbf{F}_V$ and $\mathbf{F}_S$, respectively, so that $\mathbf{F}=\mathbf{F}_V+\mathbf{F}_S$. The volumetric or body force includes contributions from gravity and, in special circumstances, from other fields (e.g., an electric field acting on a fluid which has a net charge). We confine our attention here to gravity; electrostatic forces are discussed in Chapter 11. If gravity is the only body force, then

$$\mathbf{F}_V = \int_{V(t)} \rho\mathbf{g}\,dV, \tag{5.3-12}$$

where $\mathbf{g}$ is the gravitational acceleration. The integrand, $\rho\mathbf{g}$, is the gravitational force per unit volume at any point in the fluid.

Viscous forces and pressure act on surfaces, and therefore contribute to $\mathbf{F}_S$. Surface forces are described using the stress vector $\mathbf{s}(\mathbf{n})$, which is the force per unit area on a surface with a unit normal $\mathbf{n}$. As discussed in Chapter 1, the convention used in this book is that $\mathbf{s}(\mathbf{n})$ is the stress exerted by the material that is on the side of the surface toward which $\mathbf{n}$ points. [This amounts to a sign convention for $\mathbf{s}(\mathbf{n})$, as shown below.] Thus, when $\mathbf{n}$ is the usual outward normal from the control surface, $\mathbf{s}(\mathbf{n})$ is the stress exerted on the control surface by the surroundings. Integrating the stress over the surface of any control volume gives

$$\mathbf{F}_S = \int_{S(t)} \mathbf{s}(\mathbf{n})\,dS. \tag{5.3-13}$$

Using eqs. (5.3-12) and (5.3-13), Eq. (5.3-11) becomes

$$\int_{V(t)} \rho \frac{D\mathbf{v}}{Dt}\, dV = \int_{V(t)} \rho \mathbf{g}\, dV + \int_{S(t)} \mathbf{s(n)}\, dS, \qquad (5.3\text{-}14)$$

which is a macroscopic (integral) statement of conservation of momentum for any control volume.

Some important properties of $\mathbf{s(n)}$ are revealed by applying the integral form of conservation of momentum to a control volume of small size. Equation (5.3-14) is re-arranged first to obtain

$$\frac{1}{S}\int_{S(t)} \mathbf{s(n)}\, dS = \frac{V}{S}\left[\frac{1}{V}\int_{V(t)}\left(\rho \frac{D\mathbf{v}}{Dt} - \rho \mathbf{g}\right) dV\right]. \qquad (5.3\text{-}15)$$

For any control volume with a characteristic linear dimension ℓ, $V/S = O(\ell)$ as $\ell \to 0$. The integrands in Eq. (5.3-15) remain finite in this limit, so that for a sufficiently small control volume the right-hand side vanishes and we find that

$$\lim_{S \to 0} \frac{1}{S}\int_{S(t)} \mathbf{s(n)}\, dS = 0. \qquad (5.3\text{-}16)$$

This key result indicates that the stresses on any vanishingly small volume are in balance, and therefore it describes a *stress equilibrium* which holds in the vicinity of any point. The analogous result for the flux of a scalar quantity is Eq. (2.2-20).

Stress equilibrium at a point has a number of consequences. The simplest of those is seen by considering a control volume of thickness 2ℓ centered on an imaginary planar surface within a fluid, as shown in Fig. 5-3. For a sufficiently thin control volume (i.e., $\ell \to 0$), the stresses integrated over the edges will be negligible compared to the contributions from the top and bottom surfaces (S_1 and S_2), so that Eq. (5.3-16) becomes

$$\lim_{S \to 0} \frac{1}{S}\left[\mathbf{s(n)}\big|_1 S_1 + \mathbf{s(-n)}\big|_2 S_2\right] = 0, \qquad (5.3\text{-}17)$$

where S refers to the total control surface. The areas S_1 and S_2, which are the same, are made arbitrarily small in the limit indicated, so that we conclude that at *any point* in a fluid,

$$\mathbf{s(n)} = -\mathbf{s(-n)}. \qquad (5.3\text{-}18)$$

Thus, the stress changes sign when the direction of $\mathbf{n}$ is reversed. In essence, this is just a manifestation of the action–reaction law of mechanics (Newton's third law of motion).

Figure 5-3. Control volume of thickness 2ℓ centered on an imaginary, planar test surface within a fluid. The unit normals directed toward sides 1 and 2 are $-\mathbf{n}$ and $\mathbf{n}$, respectively.

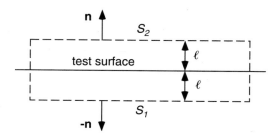

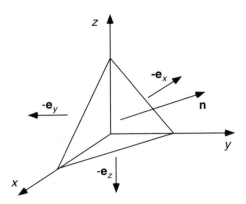

Figure 5-4. Stress tetrahedron.

Another major consequence of stress equilibrium concerns the calculation of the stress vector for a surface with an arbitrary orientation. Consider a small, tetrahedral control volume, as shown in Fig. 5-4. Three of the surfaces are normal to the rectangular coordinate axes, whereas the fourth is oriented in a direction described by the unit normal **n**. If the area of the arbitrarily oriented ("n") surface is S_0, then the areas of the x, y, and z surfaces are the projections of S_0 given by

$$S_1 = (\mathbf{n} \cdot \mathbf{e}_x)S_0, \qquad S_2 = (\mathbf{n} \cdot \mathbf{e}_y)S_0, \qquad S_3 = (\mathbf{n} \cdot \mathbf{e}_z)S_0. \qquad (5.3\text{-}19)$$

Applied to this tetrahedron, Eq. (5.3-16) becomes

$$\lim_{S \to 0} \frac{S_0}{S} [\mathbf{s}(\mathbf{n}) + (\mathbf{n} \cdot \mathbf{e}_x)\mathbf{s}(-\mathbf{e}_x) + (\mathbf{n} \cdot \mathbf{e}_y)\mathbf{s}(-\mathbf{e}_y) + (\mathbf{n} \cdot \mathbf{e}_z)\mathbf{s}(-\mathbf{e}_z)] = \mathbf{0}. \qquad (5.3\text{-}20)$$

It follows that the term in square brackets must vanish at any point in a fluid. Switching from (x, y, z) to the indices $(1, 2, 3)$ and noting from Eq. (5.3-18) that $\mathbf{s}(-\mathbf{e}_i) = -\mathbf{s}(\mathbf{e}_i)$ for $i = 1$, 2, or 3, we conclude from Eq. (5.3-20) that

$$\mathbf{s}(\mathbf{n}) = \mathbf{n} \cdot \sum_i \mathbf{e}_i \mathbf{s}(\mathbf{e}_i). \qquad (5.3\text{-}21)$$

Thus, the stress vector $\mathbf{s}(\mathbf{n})$ is expressed as the dot product of **n** with a sum of three dyads of the form $\mathbf{e}_i \mathbf{s}(\mathbf{e}_i)$. As introduced in Chapter 1 [see Fig. 1-2 and Eq. (1.2-10)], we define the jth component of the vector $\mathbf{s}(\mathbf{e}_i)$ as σ_{ij}. That is,

$$\mathbf{s}(\mathbf{e}_i) = \sum_j \sigma_{ij} \mathbf{e}_j. \qquad (5.3\text{-}22)$$

Using this definition in Eq. (5.3-21) results in

$$\mathbf{s}(\mathbf{n}) = \mathbf{n} \cdot \sum_i \sum_j \sigma_{ij} \mathbf{e}_i \mathbf{e}_j = \mathbf{n} \cdot \boldsymbol{\sigma}, \qquad (5.3\text{-}23)$$

where $\boldsymbol{\sigma}$ is the *stress tensor*. What we have just done is to derive the relationship between the stress vector and stress tensor, which was stated previously (without proof) as Eq. (1.2-8). It is seen that a knowledge of $\boldsymbol{\sigma}$ at a given point in a fluid is all that is required to calculate the stress for any specified orientation **n**. Accordingly, we will have little further need for the stress vector.

Using Eq. (5.3-23) to evaluate the stress vector in Eq. (5.3-14), the integral form of conservation of linear momentum becomes

$$\int_{V(t)} \rho \frac{D\mathbf{v}}{Dt} dV = \int_{V(t)} \rho \mathbf{g} \, dV + \int_{S(t)} \mathbf{n} \cdot \boldsymbol{\sigma} \, dS. \tag{5.3-24}$$

Applying the divergence theorem to the surface integral and collecting all terms in a single volume integral, we obtain

$$\int_{V(t)} \left[\rho \frac{D\mathbf{v}}{Dt} - \rho \mathbf{g} - \nabla \cdot \boldsymbol{\sigma} \right] dV = 0. \tag{5.3-25}$$

Because Eq. (5.3-25) applies to a control volume of any size, however small, the integrand must vanish at every point. Thus, we conclude that

$$\rho \frac{D\mathbf{v}}{Dt} = \rho \mathbf{g} + \nabla \cdot \boldsymbol{\sigma}. \tag{5.3-26}$$

This relation is valid for any fluid which behaves as a continuum, provided that the only body force is gravity. The left-hand side is the rate of change of momentum at a material point. That is, it is the mass times the acceleration, in a reference frame moving with the fluid. The right-hand side is a sum of forces. Because of the pointwise nature of Eq. (5.3-26), mass is expressed as a density and the forces are also expressed on a volumetric basis (i.e., force per unit volume).

5.4 TOTAL STRESS, PRESSURE, AND VISCOUS STRESS

The main task which remains in stating the basic equations of fluid mechanics is to relate $\boldsymbol{\sigma}$ to the dynamic conditions and to the properties of a given fluid. Certain general characteristics of the stress are discussed in this section, namely, the symmetry of $\boldsymbol{\sigma}$ and the decomposition of the total stress into pressure and viscous contributions. Pressure forces are explored further in Section 5.5, in the context of fluid statics. Constitutive equations which describe the viscous stresses in particular types of fluids are presented in Section 5.6.

Symmetry of the Stress Tensor

An important characteristic of the stress tensor is that, with exceptions for a few special fluids, it is symmetric. That is, $\sigma_{ij} = \sigma_{ji}$ for all i and j. This symmetry is a consequence of the conservation of angular momentum, as will be shown now. Conservation of angular momentum for a material volume is written as

$$\int_{V(t)} \rho \frac{D}{Dt} (\mathbf{r} \times \mathbf{v}) \, dV = \int_{V(t)} \rho (\mathbf{r} \times \mathbf{g}) \, dV + \int_{S(t)} \mathbf{r} \times (\mathbf{n} \cdot \boldsymbol{\sigma}) \, dS, \tag{5.4-1}$$

which is analogous to Eq. (5.3-24) for linear momentum. The rate of change of angular momentum is equated here to the net torque computed from the moments of the gravitational and surface forces about the origin. The main assumption in Eq. (5.4-1) is that the only torques are those associated with the macroscopic forces just mentioned; the limitations of this assumption are discussed later.

The surface integral in Eq. (5.4-1) is rewritten as a volume integral by using the divergence theorem. After considerable manipulation to express the integrand of the surface integral in the form $\mathbf{n} \cdot \boldsymbol{\alpha}$, where $\boldsymbol{\alpha}$ is a tensor, it is found that

$$\int_{S(t)} \mathbf{r} \times (\mathbf{n} \cdot \boldsymbol{\sigma}) \, dS = \int_{V(t)} [\mathbf{r} \times (\nabla \cdot \boldsymbol{\sigma}) - \boldsymbol{\varepsilon} : \boldsymbol{\sigma}] \, dV, \qquad (5.4\text{-}2)$$

where the integrand on the right-hand side equals $\nabla \cdot \boldsymbol{\alpha}$. (The tensor $\boldsymbol{\alpha}$ itself is of no particular interest.) The integral on the left-hand side of Eq. (5.4-1) is rewritten by using

$$\mathbf{c} \cdot [\nabla (\mathbf{a} \times \mathbf{b})[= (\mathbf{c} \cdot \nabla \mathbf{a}) \times \mathbf{b} - (\mathbf{c} \cdot \nabla \mathbf{b}) \times \mathbf{a}, \qquad (5.4\text{-}3)$$

which follows from identity (4) of Table A-1. Setting $\mathbf{a} = \mathbf{r}$ and $\mathbf{b} = \mathbf{c} = \mathbf{v}$, using Eq. (A.6-9), and noting that $\mathbf{v} \times \mathbf{v} = \mathbf{0}$, it is found that

$$\int_{V(t)} \rho \frac{D}{Dt} (\mathbf{r} \times \mathbf{v}) \, dV = \int_{V(t)} \rho \left(\mathbf{r} \times \frac{D\mathbf{v}}{Dt} \right) dV. \qquad (5.4\text{-}4)$$

Using Eqs. (5.4-2) and (5.4-4) in Eq. (5.4-1), and regrouping some of the integrals, results in

$$\int_{V(t)} \mathbf{r} \times \left[\rho \frac{D\mathbf{v}}{Dt} - \rho \mathbf{g} - \nabla \cdot \boldsymbol{\sigma} \right] dV = - \int_{V(t)} \boldsymbol{\varepsilon} : \boldsymbol{\sigma} \, dV. \qquad (5.4\text{-}5)$$

We note now that conservation of linear momentum, Eq. (5.3-22), implies that the bracketed term is zero. Applying Eq. (5.4-5) to an arbitrarily small control volume, it follows that

$$\boldsymbol{\varepsilon} : \boldsymbol{\sigma} = \mathbf{0} \qquad (5.4\text{-}6)$$

at any point in the fluid. The x, y, and z components of the vector $\boldsymbol{\varepsilon} : \boldsymbol{\sigma}$ are $\sigma_{zy} - \sigma_{yz}$, $\sigma_{xz} - \sigma_{zx}$, and $\sigma_{yx} - \sigma_{xy}$, respectively. Because each component must vanish, we conclude that $\sigma_{ij} = \sigma_{ji}$, as asserted above. Thus, only six of the nine elements of $\boldsymbol{\sigma}$ are independent.

The limitation on the conclusion that $\boldsymbol{\sigma}$ is symmetric is due not to any violation of the general principle of conservation of angular momentum, but rather to the fact that torques other than those included in Eq. (5.4-1) may exist. This is true for certain structured fluids (e.g., some suspensions). The conservation of angular momentum in structured fluids is analyzed in detail in Dahler and Scriven (1963), where the distinction between external and internal angular momentum for such fluids is discussed. Torques of a microscopic nature may arise, for example, from magnetic forces acting on colloidal magnetite or other magnetic particles suspended in a liquid (Rosensweig, 1985). Such materials, called *ferrofluids,* are important in certain technologies. However, for the overwhelming majority of fluids, including gases, homogeneous liquids of low molecular weight, polymeric liquids, and most suspensions, it is safe to assume that the stress tensor is symmetric.

Pressure and Viscous Stress

The distinctive mechanical properties of pressure motivate the separation of pressure from other contributions to the total stress. Pressure is the only stress in a fluid at rest. The other principal features of pressure are that it is a stress which always acts normal to a surface, is positive when it exerts a compressive force, and is isotropic. The last statement refers to the fact that, at any given point in a fluid, the same scalar value of the pressure (P) applies to hypothetical test surfaces having any orientation (see Problem 5-1). In other words, pressure acts equally in all directions. Given these characteristics of pressure, the total stress is written as

$$\boldsymbol{\sigma} = -P\boldsymbol{\delta} + \boldsymbol{\tau}, \tag{5.4-7}$$

where $\boldsymbol{\tau}$ is the *viscous stress* (or *deviatoric stress*) tensor. Multiplying P by the identity tensor ($\boldsymbol{\delta}$) ensures that pressure is a normal stress (contributes only to the diagonal elements of $\boldsymbol{\sigma}$) and is isotropic (gives diagonal elements which are equal); the minus sign is needed to make positive pressures compressive. For static fluids $\boldsymbol{\tau}$ vanishes and $\boldsymbol{\sigma} = -P\boldsymbol{\delta}$.

As used in this book, P has the same meaning as the thermodynamic pressure. That is, P is a variable which is defined locally by an equation of state of the form $P = P(\rho, T)$ (e.g., the ideal gas law). For isothermal flow of single-component fluids, the unknowns in general are $\mathbf{v}$, P, and ρ, and the governing equations are continuity (conservation of mass), conservation of momentum (including an appropriate constitutive equation), and the equation of state. Incompressible fluids, our main concern in this book, are an idealization in which ρ is a given constant. This reduces the number of unknowns by one and renders the equation of state useless for determining P, in that P is now a many-valued function of ρ. Thus, in incompressible flow P is treated simply as a mechanical variable which must help satisfy continuity and conservation of momentum. Unless P is specified at a boundary, its absolute value remains arbitrary. This is not a source of difficulty, because an incompressible flow is influenced only by the *gradient* in pressure. Static, incompressible fluids (Section 5.5) provide the simplest example of the purely mechanical treatment of P. The relationship between P and the mean normal stress in a flowing fluid is discussed in Section 5.6.

The viscous stress is defined so that it vanishes in a fluid which is at rest or which translates or rotates as a rigid body. That is, the viscous stress is associated with *deformation* of the fluid (see Section 5.2). (With viscoelastic fluids, which do not respond instantaneously to changes in imposed stresses, the phrase "fluid at rest" carries with it the understanding that sufficient time has elapsed for the fluid to reach mechanical equilibrium.) The requirement that $\boldsymbol{\tau}$ vanish for solid-body motion follows from the definition of a fluid given in Section 5.2. That is, it is a material which deforms continuously in the presence of shear stresses. No deformation occurs in solid-body motion, so that the off-diagonal (shear) components of $\boldsymbol{\sigma}$ must vanish (i.e., $\sigma_{ij} = 0$ for $i \neq j$). But, according to Eq. (5.4-7), the off-diagonal components of $\boldsymbol{\sigma}$ equal those of $\boldsymbol{\tau}$. Because the off-diagonal components must vanish and because the normal stresses in a stagnant fluid are accounted for by P, $\boldsymbol{\tau} = \mathbf{0}$ for solid-body motion. It is noteworthy that the symmetry of $\boldsymbol{\sigma}$, combined with the fact that $\sigma_{ij} = \tau_{ij}$ for $i \neq j$, shows that $\boldsymbol{\tau}$ is also symmetric.

Using Eq. (5.4-7), the divergence of the total stress is

$$\nabla \cdot \boldsymbol{\sigma} = \nabla \cdot (-P\boldsymbol{\delta}) + \nabla \cdot \boldsymbol{\tau} = -\nabla P + \nabla \cdot \boldsymbol{\tau}. \tag{5.4-8}$$

TABLE 5-1
Cauchy Momentum Equation in Rectangular Coordinates

x component:
$$\rho\left[\frac{\partial v_x}{\partial t}+v_x\frac{\partial v_x}{\partial x}+v_y\frac{\partial v_x}{\partial y}+v_z\frac{\partial v_x}{\partial z}\right]=\rho g_x-\frac{\partial P}{\partial x}+\left[\frac{\partial \tau_{xx}}{\partial x}+\frac{\partial \tau_{yx}}{\partial y}+\frac{\partial \tau_{zx}}{\partial z}\right]$$

y component:
$$\rho\left[\frac{\partial v_y}{\partial t}+v_x\frac{\partial v_y}{\partial x}+v_y\frac{\partial v_y}{\partial y}+v_z\frac{\partial v_y}{\partial z}\right]=\rho g_y-\frac{\partial P}{\partial y}+\left[\frac{\partial \tau_{xy}}{\partial x}+\frac{\partial \tau_{yy}}{\partial y}+\frac{\partial \tau_{zy}}{\partial z}\right]$$

z component:
$$\rho\left[\frac{\partial v_z}{\partial t}+v_x\frac{\partial v_z}{\partial x}+v_y\frac{\partial v_z}{\partial y}+v_z\frac{\partial v_z}{\partial z}\right]=\rho g_z-\frac{\partial P}{\partial z}+\left[\frac{\partial \tau_{xz}}{\partial x}+\frac{\partial \tau_{yz}}{\partial y}+\frac{\partial \tau_{zz}}{\partial z}\right]$$

The general statement of conservation of linear momentum, Eq. (5.3-26), is written now as

$$\rho\frac{D\mathbf{v}}{Dt}=\rho\mathbf{g}-\nabla P+\nabla\cdot\boldsymbol{\tau},\tag{5.4-9}$$

which is called the *Cauchy momentum equation*. The components of this equation in rectangular, cylindrical, and spherical coordinates are given in Tables 5-1, 5-2, and 5-3, respectively.

5.5 FLUID STATICS

The only mechanical variable in a static fluid is the pressure. Setting $\mathbf{v}=0$ and $\boldsymbol{\tau}=0$ in Eq. (5.4-9), conservation of linear momentum implies that

$$\nabla P=\rho\mathbf{g}.\tag{5.5-1}$$

This expresses the elementary fact that the pressure in a static fluid varies only with height (i.e., in the direction parallel to the gravitational vector). If the density is constant, Eq. (5.5-1) is integrated readily to determine $P(\mathbf{r})$. For example, if the $+z$ direction is taken to be "up," then $\mathbf{g}=-g\mathbf{e}_z$, $dP/dz=-\rho g$, and

TABLE 5-2
Cauchy Momentum Equation in Cylindrical Coordinates

r component:
$$\rho\left[\frac{\partial v_r}{\partial t}+v_r\frac{\partial v_r}{\partial r}+\frac{v_\theta}{r}\frac{\partial v_r}{\partial \theta}-\frac{v_\theta^2}{r}+v_z\frac{\partial v_r}{\partial z}\right]$$
$$=\rho g_r-\frac{\partial P}{\partial r}+\left[\frac{1}{r}\frac{\partial}{\partial r}(r\tau_{rr})+\frac{1}{r}\frac{\partial \tau_{\theta r}}{\partial \theta}-\frac{\tau_{\theta\theta}}{r}+\frac{\partial \tau_{zr}}{\partial z}\right]$$

θ component:
$$\rho\left[\frac{\partial v_\theta}{\partial t}+v_r\frac{\partial v_\theta}{\partial r}+\frac{v_\theta}{r}\frac{\partial v_\theta}{\partial \theta}+\frac{v_r v_\theta}{r}+v_z\frac{\partial v_\theta}{\partial z}\right]$$
$$=\rho g_\theta-\frac{1}{r}\frac{\partial P}{\partial \theta}+\left[\frac{1}{r^2}\frac{\partial}{\partial r}(r^2\tau_{r\theta})+\frac{1}{r}\frac{\partial \tau_{\theta\theta}}{\partial \theta}+\frac{\partial \tau_{z\theta}}{\partial z}\right]$$

z component:
$$\rho\left[\frac{\partial v_z}{\partial t}+v_r\frac{\partial v_z}{\partial r}+\frac{v_\theta}{r}\frac{\partial v_z}{\partial \theta}+v_z\frac{\partial v_z}{\partial z}\right]$$
$$=\rho g_z-\frac{\partial P}{\partial z}+\left[\frac{1}{r}\frac{\partial}{\partial r}(r\tau_{rz})+\frac{1}{r}\frac{\partial \tau_{\theta z}}{\partial \theta}+\frac{\partial \tau_{zz}}{\partial z}\right]$$

TABLE 5-3
Cauchy Momentum Equation in Spherical Coordinates

r component:
$$\rho\left[\frac{\partial v_r}{\partial t}+v_r\frac{\partial v_r}{\partial r}+\frac{v_\theta}{r}\frac{\partial v_r}{\partial \theta}+\frac{v_\phi}{r\sin\theta}\frac{\partial v_r}{\partial \phi}-\frac{v_\theta^2+v_\phi^2}{r}\right]$$
$$=\rho g_r-\frac{\partial P}{\partial r}+\left[\frac{1}{r^2}\frac{\partial}{\partial r}(r^2\tau_{rr})+\frac{1}{r\sin\theta}\frac{\partial}{\partial \theta}(\tau_{\theta r}\sin\theta)+\frac{1}{r\sin\theta}\frac{\partial\tau_{\phi r}}{\partial \phi}-\frac{\tau_{\theta\theta}+\tau_{\phi\phi}}{r}\right]$$

θ component:
$$\rho\left[\frac{\partial v_\theta}{\partial t}+v_r\frac{\partial v_\theta}{\partial r}+\frac{v_\theta}{r}\frac{\partial v_\theta}{\partial \theta}+\frac{v_\phi}{r\sin\theta}\frac{\partial v_\theta}{\partial \phi}+\frac{v_rv_\theta}{r}-\frac{v_\phi^2\cot\theta}{r}\right]$$
$$=\rho g_\theta-\frac{1}{r}\frac{\partial P}{\partial \theta}+\left[\frac{1}{r^2}\frac{\partial}{\partial r}(r^2\tau_{r\theta})+\frac{1}{r\sin\theta}\frac{\partial}{\partial \theta}(\tau_{\theta\theta}\sin\theta)+\frac{1}{r\sin\theta}\frac{\partial\tau_{\phi\theta}}{\partial \phi}+\frac{\tau_{r\theta}}{r}-\frac{\cot\theta}{r}\tau_{\phi\phi}\right]$$

ϕ component:
$$\rho\left[\frac{\partial v_\phi}{\partial t}+v_r\frac{\partial v_\phi}{\partial r}+\frac{v_\theta}{r}\frac{\partial v_\phi}{\partial \theta}+\frac{v_\phi}{r\sin\theta}\frac{\partial v_\phi}{\partial \phi}+\frac{v_\phi v_r}{r}+\frac{v_\theta v_\phi\cot\theta}{r}\right]$$
$$=\rho g_\phi-\frac{1}{r\sin\theta}\frac{\partial P}{\partial \phi}+\left[\frac{1}{r^2}\frac{\partial}{\partial r}(r^2\tau_{r\phi})+\frac{1}{r}\frac{\partial\tau_{\theta\phi}}{\partial \theta}+\frac{1}{r\sin\theta}\frac{\partial\tau_{\phi\phi}}{\partial \phi}+\frac{\tau_{r\phi}}{r}+\frac{2\cot\theta}{r}\tau_{\theta\phi}\right]$$

$$P(z)=P(0)-\rho gz. \tag{5.5-2}$$

In terms of the position vector **r**, the pressure in a constant-density fluid at rest is given by

$$P(\mathbf{r})=P(\mathbf{0})+\rho\mathbf{g}\cdot\mathbf{r}, \tag{5.5-3}$$

where $P(\mathbf{0})$ is the pressure at the origin.

Example 5.5-1 Buoyancy A consequence of the fact that pressure increases with depth in a static fluid is that the pressure exerts a net upward force on any submerged object. To calculate this force, consider an object of arbitrary shape submerged in a constant-density fluid, as shown in Fig. 5-5(a). The net pressure force on this object, $\mathbf{F_P}$, is given by

$$\mathbf{F_P}=-\int_S P\mathbf{n}\,dS \tag{5.5-4}$$

where the minus sign reflects the fact that positive pressures are compressive (i.e., pressure acts in the $-\mathbf{n}$ direction). The pressure force is evaluated most easily by considering the situation in Fig. 5-5(b), where the solid has been replaced by an identical volume of fluid. Because the pressure in the fluid depends on depth only, this replacement of the solid does not affect the pressure distribution in the surrounding fluid. In particular, the pressure distribution on the fluid control surface in (b) must be identical to that on the solid surface in (a). Situation (b) has the advantage that we can apply the divergence theorem to the integral in Eq. (5.5-4), because P is a continuous function of position within the volume V; this is not necessarily true for (a), because we have said nothing about the meaning of P within a solid. Applying the divergence theorem and using Eq. (5.5-1) to evaluate the pressure gradient, we find that

$$\mathbf{F_P}=-\int_V \nabla P\,dV=-\rho_f\mathbf{g}V, \tag{5.5-5}$$

where ρ_f is the (constant) density of the fluid. Thus, the upward force on the object due to the pressure equals the weight of an equivalent volume of fluid; this is *Archimedes' law*. An important implication of Eq. (5.5-5) is that $\mathbf{F_P}$ is independent of the absolute pressure. That is, adding any

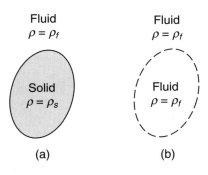

Fluid
$\rho = \rho_f$

Fluid
$\rho = \rho_f$

Solid
$\rho = \rho_s$

Fluid
$\rho = \rho_f$

(a)   (b)

Figure 5-5. Pressure force on a submerged object. The control volume in (b) is used to calculate the pressure force on the object in (a).

constant to P does not affect the net pressure force on a completely submerged object. Moreover, if P is constant, as is approximately true for a stagnant column of gas of moderate height, then $\mathbf{F_P}=\mathbf{0}$.

The *buoyancy force* is the net force on the object due to the static pressure variations and gravity. Evaluating the gravitational force on the solid body by setting $\rho=\rho_s$ in Eq. (5.3-12), the buoyancy force is found to be

$$\mathbf{F_B}=(\rho_s-\rho_f)\mathbf{g}V. \tag{5.5-6}$$

We return to the concept of buoyancy in Chapter 12, when considering flow caused by spatial variations in fluid density.

Example 5.5-2 Pressure Force on Part of an Object: Projected Areas   When only part of the surface of an object is in contact with a fluid, there is in general a net pressure force on that part of the surface. Consider the situation shown in Fig. 5-6, where it is desired to compute the x component of the pressure force on the solid surface, S_1. The orientation of $\mathbf{e}_x$ relative to $\mathbf{g}$ is considered to be arbitrary, so that the x direction is not necessarily horizontal. Denoting the desired force as F_{1x}, we have

$$F_{1x}=\mathbf{e}_x\cdot\int_{S_1} P\mathbf{n}_1 \, dS, \tag{5.5-7}$$

where $\mathbf{n_1}$ is the unit normal directed toward the surface. Applying Eqs. (5.5-4) and (5.5-5) to the control volume shown by the dashed lines in Fig. 5-5 and noticing that the pressures acting normal to the top and bottom (or front and back) surfaces in the figure make no contribution to F_{Px}, it is found that

$$F_{Px}=-(\mathbf{e}_x\cdot\mathbf{g})\rho V=\mathbf{e}_x\cdot\left[-\int_{S_1} P\mathbf{n}_1 \, dS - \int_{S_2} P\mathbf{n}_2 \, dS\right]. \tag{5.5-8}$$

Identifying the first integral with F_{1x} and setting $\mathbf{n_2}=-\mathbf{e}_x$ in the second integral, we obtain

$$F_{1x}=\langle P\rangle_2 S_2+(\mathbf{e}_x\cdot\mathbf{g})\rho V, \tag{5.5-9}$$

where $\langle P\rangle_2$ is the pressure averaged over the surface S_2. If x is directed downward (i.e., $\mathbf{e}_x\cdot\mathbf{g}=+g$), it is seen that the pressure force on S_1 equals the pressure force on S_2 plus the weight of the intervening fluid. If x is horizontal (i.e., $\mathbf{e}_x\cdot\mathbf{g}=0$) or if the weight of the fluid is negligible, then

$$F_{1x}=\langle P\rangle_2 S_2. \tag{5.5-10}$$

Figure 5-6. Use of the projected area to calculate the pressure force on part of an object.

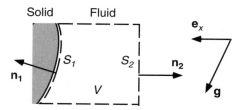

Equation (5.5-10) indicates that no matter how complex the shape of S_1, the pressure force can be obtained simply by multiplying the projected area (S_2) by the corresponding average pressure. This fact greatly simplifies force calculations in statics problems involving curved interfaces.

5.6 CONSTITUTIVE EQUATIONS FOR THE VISCOUS STRESS

Constitutive equations are necessary to relate the viscous stress to the rate of strain and to material properties such as viscosity. In this section we state the constitutive equation for a Newtonian fluid and also provide a few examples of constitutive equations for non-Newtonian fluids. The idea that the viscous stress tensor should be a function only of the rate-of-strain tensor, or $\tau = \tau(\Gamma)$, is motivated by the demonstration in Section 5.2 that Γ measures the local rate of deformation of a fluid. Viscous stresses are frictional in character, in that they reflect the molecular-level resistance to attempts to move one part of a fluid relative to another. Thus, the type of fluid motion which matters in this regard is deformation—the pulling apart or pushing together of material points—and not uniform translation or rigid-body rotation. The fact that τ must depend on Γ, and not simply ∇v, can be proven formally by considering what is necessary to ensure *material objectivity*. This is the principle that the form of a constitutive equation must not depend on the position or velocity of an observer, and therefore must be invariant to time-dependent translations or rotations of the coordinate system (Leal, 1992, pp. 39–48).

Newtonian Fluids

The concept of a Newtonian fluid was introduced in Chapter 1, and many examples of such fluids were mentioned. The defining characteristics of a Newtonian fluid are that it is homogeneous and isotropic, responds instantaneously to changes in the local rate of strain, and has a local viscous stress which is a linear function of the local rate of strain. As shown, for example, by Aris (1962, pp. 105–112), the most general expression for $\tau(\Gamma)$ for a fluid with these characteristics is

$$\tau = 2\mu\Gamma + \left(\kappa - \frac{2}{3}\mu\right)(\nabla \cdot v)\,\delta, \tag{5.6-1}$$

where μ and κ are scalars which may depend on state variables such as temperature, but which do not depend directly on Γ. The divergence of the velocity ($\nabla \cdot v$) is the sum of the diagonal elements of Γ (sometimes called the *trace* of Γ and written as $tr\Gamma$), so that both terms in Eq. (5.6-1) are indeed linear functions of Γ. The coefficients have been selected so that μ is the shear viscosity discussed in Chapter 1, whereas κ is the *dilatational viscosity* or *bulk viscosity*.

The dilatational viscosity measures the molecular-level resistance to an isotropic (i.e., spherically symmetric) expansion or compression. It is related to the rate of energy dissipation in an isothermal fluid under these conditions (Schlicting, 1968, pp. 57–60). The effects of κ on fluid dynamics are difficult to detect and are usually ignored. It has been shown that $\kappa=0$ for ideal, monatomic gases, and it is generally believed that for other fluids $\kappa<<\mu$. Moreover, for incompressible fluids, where $\nabla \cdot \mathbf{v}=0$ from continuity, the term in Eq. (5.6-1) that is multiplied by κ vanishes. For these reasons we will be concerned only with μ in the analysis of flow problems. An analysis of dilatational viscosity in a liquid containing gas bubbles is given in Batchelor (1970, pp. 253–255).

Setting $\kappa=0$ in Eq. (5.6-1), the constitutive equation for a Newtonian fluid is given by

$$\tau=2\mu\left[\Gamma-\frac{1}{3}(\nabla \cdot \mathbf{v})\,\delta\right],\tag{5.6-2}$$

where Γ is defined by Eq. (5.2-8). Equation (5.6-2) is given in component form for the three common coordinate systems in Tables 5-4, 5-5, and 5-6. For an *incompressible* Newtonian fluid, the viscous stress is simply

$$\tau=2\mu\Gamma=\mu[\nabla\mathbf{v}+(\nabla\mathbf{v})^{t}].\tag{5.6-3}$$

Note from Table 5-4 that if $v_{x}=v_{x}(y)$ and $v_{y}=v_{z}=0$, Eq. (5.6-3) reduces to Eq. (1.2-12).

As discussed in Section 5.4, the total stress is related to the pressure and viscous stress by

$$\sigma=-P\delta+\tau,\tag{5.6-4}$$

where we interpret P as the thermodynamic pressure. Of interest is the extent to which P differs from (minus) the *mean normal stress*, which may be used as a purely mechani-

---

TABLE 5-4
Viscous Stress Components for Newtonian Fluids in Rectangular Coordinates

$$\tau_{xx}=2\mu\left[\frac{\partial v_{x}}{\partial x}-\frac{1}{3}\nabla \cdot \mathbf{v}\right]$$

$$\tau_{yy}=2\mu\left[\frac{\partial v_{y}}{\partial y}-\frac{1}{3}\nabla \cdot \mathbf{v}\right]$$

$$\tau_{zz}=2\mu\left[\frac{\partial v_{z}}{\partial z}-\frac{1}{3}\nabla \cdot \mathbf{v}\right]$$

$$\tau_{xy}=\tau_{yx}=\mu\left[\frac{\partial v_{x}}{\partial y}+\frac{\partial v_{y}}{\partial x}\right]$$

$$\tau_{yz}=\tau_{zy}=\mu\left[\frac{\partial v_{y}}{\partial z}+\frac{\partial v_{z}}{\partial y}\right]$$

$$\tau_{zx}=\tau_{xz}=\mu\left[\frac{\partial v_{z}}{\partial x}+\frac{\partial v_{x}}{\partial z}\right]$$

$$\nabla \cdot \mathbf{v}=\frac{\partial v_{x}}{\partial x}+\frac{\partial v_{y}}{\partial y}+\frac{\partial v_{z}}{\partial z}$$

TABLE 5-5
Viscous Stress Components for Newtonian Fluids in Cylindrical Coordinates

$$\tau_{rr}=2\mu\left[\frac{\partial v_r}{\partial r}-\frac{1}{3}(\boldsymbol{\nabla}\cdot\mathbf{v})\right]$$

$$\tau_{\theta\theta}=2\mu\left[\frac{1}{r}\frac{\partial v_\theta}{\partial\theta}+\frac{v_r}{r}-\frac{1}{3}(\boldsymbol{\nabla}\cdot\mathbf{v})\right]$$

$$\tau_{zz}=2\mu\left[\frac{\partial v_z}{\partial z}-\frac{1}{3}(\boldsymbol{\nabla}\cdot\mathbf{v})\right]$$

$$\tau_{r\theta}=\tau_{\theta r}=\mu\left[r\frac{\partial}{\partial r}\left(\frac{v_\theta}{r}\right)+\frac{1}{r}\frac{\partial v_r}{\partial\theta}\right]$$

$$\tau_{\theta z}=\tau_{z\theta}=\mu\left[\frac{\partial v_\theta}{\partial z}+\frac{1}{r}\frac{\partial v_z}{\partial\theta}\right]$$

$$\tau_{zr}=\tau_{rz}=\mu\left[\frac{\partial v_z}{\partial r}+\frac{\partial v_r}{\partial z}\right]$$

$$\boldsymbol{\nabla}\cdot\mathbf{v}=\frac{1}{r}\frac{\partial}{\partial r}(rv_r)+\frac{1}{r}\frac{\partial v_\theta}{\partial\theta}+\frac{\partial v_z}{\partial z}$$

cal definition of pressure (e.g., Batchelor, 1970, p. 141). If defined in this way, the pressure $(\overline{P})$ is given by

$$\overline{P}\equiv-\frac{1}{3}(\sigma_{xx}+\sigma_{yy}+\sigma_{zz}).\tag{5.6-5}$$

Using Eqs. (5.6-1) and (5.6-4) to evaluate the normal stress components for a Newtonian fluid, we find that, for example,

$$\sigma_{xx}=-P+2\mu\frac{\partial v_x}{\partial x}+\left(\kappa-\frac{2}{3}\mu\right)(\boldsymbol{\nabla}\cdot\mathbf{v}).\tag{5.6-6}$$

TABLE 5-6
Viscous Stress Components for Newtonian Fluids in Spherical Coordinates

$$\tau_{rr}=2\mu\left[\frac{\partial v_r}{\partial r}-\frac{1}{3}(\boldsymbol{\nabla}\cdot\mathbf{v})\right]$$

$$\tau_{\theta\theta}=2\mu\left[\frac{1}{r}\frac{\partial v_\theta}{\partial\theta}+\frac{v_r}{r}-\frac{1}{3}(\boldsymbol{\nabla}\cdot\mathbf{v})\right]$$

$$\tau_{\phi\phi}=2\mu\left[\frac{1}{r\sin\theta}\frac{\partial v_\phi}{\partial\phi}+\frac{v_r}{r}+\frac{v_\theta\cot\theta}{r}-\frac{1}{3}(\boldsymbol{\nabla}\cdot\mathbf{v})\right]$$

$$\tau_{r\theta}=\tau_{\theta r}=\mu\left[r\frac{\partial}{\partial r}\left(\frac{v_\theta}{r}\right)+\frac{1}{r}\frac{\partial v_r}{\partial\theta}\right]$$

$$\tau_{\theta\phi}=\tau_{\phi\theta}=\mu\left[\frac{\sin\theta}{r}\frac{\partial}{\partial\theta}\left(\frac{v_\phi}{\sin\theta}\right)+\frac{1}{r\sin\theta}\frac{\partial v_\theta}{\partial\phi}\right]$$

$$\tau_{\phi r}=\tau_{r\phi}=\mu\left[\frac{1}{r\sin\theta}\frac{\partial v_r}{\partial\phi}+r\frac{\partial}{\partial r}\left(\frac{v_\phi}{r}\right)\right]$$

$$\boldsymbol{\nabla}\cdot\mathbf{v}=\frac{1}{r^2}\frac{\partial}{\partial r}(r^2v_r)+\frac{1}{r\sin\theta}\frac{\partial}{\partial\theta}(v_\theta\sin\theta)+\frac{1}{r\sin\theta}\frac{\partial v_\phi}{\partial\phi}$$

Substitution of this and the analogous expressions for σ_{yy} and σ_{zz} in Eq. (5.6-5) results in

$$\overline{P} = P - \kappa(\nabla \cdot \mathbf{v}). \tag{5.6-7}$$

Thus, P and $\overline{P}$ are identical for a stagnant fluid, an incompressible fluid, or a fluid where $\kappa = 0$. Because κ is ordinarily quite small, there is little distinction between P and $\overline{P}$ for Newtonian fluids. Incidentally, it is apparent now why the factor $\frac{2}{3}$ was included in Eq. (5.6-1): It is needed to ensure the (near) equality between P and $\overline{P}$.

Navier–Stokes Equation

An important special case is that of a Newtonian fluid with a constant density and viscosity. Using Eqs. (5.6-3) and (A.4-11), we find that

$$\nabla \cdot \boldsymbol{\tau} = 2\mu[\nabla \cdot \boldsymbol{\Gamma}] = \mu[\nabla \cdot (\nabla \mathbf{v} + (\nabla \mathbf{v})^t)] = \mu[\nabla^2 \mathbf{v} + \nabla(\nabla \cdot \mathbf{v})] = \mu \nabla^2 \mathbf{v}. \tag{5.6-8}$$

Substituting this result in Eq. (5.4-9), conservation of linear momentum is written as

$$\rho \frac{D\mathbf{v}}{Dt} = \rho \mathbf{g} - \nabla P + \mu \nabla^2 \mathbf{v}, \tag{5.6-9}$$

which is the *Navier–Stokes equation*.[1] This expression of conservation of momentum, together with the continuity equation written as

$$\nabla \cdot \mathbf{v} = 0, \tag{5.6-10}$$

provides the usual starting point for the analysis of the flow of gases at moderate velocities and of simple liquids. As mentioned in Section 5.4, the unknowns in an analysis of the flow of a pure, incompressible, isothermal fluid are just $\mathbf{v}$ and P. Counting the three scalar components of $\mathbf{v}$, there are a total of four unknown functions. Noting that the Navier–Stokes equation has three components, Eqs. (5.6-9) and (5.6-10) are seen to provide four partial differential equations in these four unknowns. The Navier–Stokes equation is given in component form for the three common coordinate systems in Tables 5-7, 5-8, and 5-9, whereas $\nabla \cdot \mathbf{v}$ for those coordinate systems is shown in Tables 5-4, 5-5, and 5-6.

Non-Newtonian Fluids

Any fluid which does not obey Eq. (5.6-1) is non-Newtonian. In general, such fluids have structural features which are influenced by the flow, and which in turn influence the relationship between the viscous stress and the rate of strain. Examples include polymer melts, certain polymer solutions, and suspensions of nonspherical and/or strongly interacting particles. Among the distinguishing characteristics of various non-Newtonian fluids are a dependence of the apparent viscosity on the rate of strain, unusually large values of the normal components of the viscous stresses, and "memory" ef-

---

[1]This equation was reported independently by the French physicist L. Navier (1785–1836) in 1822 and the Anglo-Irish physicist G. G. Stokes (1819–1903) in 1845. A discussion of Navier's contributions is in Dugas (1988, pp. 409–414); additional information on Stokes is given in Chapter 7.

TABLE 5-7
Navier–Stokes Equation in Rectangular Coordinates

x component:
$$\rho\left[\frac{\partial v_x}{\partial t}+v_x\frac{\partial v_x}{\partial x}+v_y\frac{\partial v_x}{\partial y}+v_z\frac{\partial v_x}{\partial z}\right]=\rho g_x-\frac{\partial P}{\partial x}+\mu\left[\frac{\partial^2 v_x}{\partial x^2}+\frac{\partial^2 v_x}{\partial y^2}+\frac{\partial^2 v_x}{\partial z^2}\right]$$

y component:
$$\rho\left[\frac{\partial v_y}{\partial t}+v_x\frac{\partial v_y}{\partial x}+v_y\frac{\partial v_y}{\partial y}+v_z\frac{\partial v_y}{\partial z}\right]=\rho g_y-\frac{\partial P}{\partial y}+\mu\left[\frac{\partial^2 v_y}{\partial x^2}+\frac{\partial^2 v_y}{\partial y^2}+\frac{\partial^2 v_y}{\partial z^2}\right]$$

z component:
$$\rho\left[\frac{\partial v_z}{\partial t}+v_x\frac{\partial v_z}{\partial x}+v_y\frac{\partial v_z}{\partial y}+v_z\frac{\partial v_z}{\partial z}\right]=\rho g_z-\frac{\partial P}{\partial z}+\mu\left[\frac{\partial^2 v_z}{\partial x^2}+\frac{\partial^2 v_z}{\partial y^2}+\frac{\partial^2 v_z}{\partial z^2}\right]$$

fects in the relationship between the stress and the rate of strain. A dependence of viscosity on rate of strain may reflect the ability of the flow to orient particles or macro-molecules, to break up particle aggregates, and/or to influence the conformation of individual polymer molecules. The resistance of long-chain polymers to orientation or elongation gives polymeric fluids an elastic character, which is manifested as an additional tension along streamlines (i.e., an elevated normal stress in the flow direction). The finite time required for such molecules to reach an equilibrium configuration is the source of memory effects. Fluids which exhibit elastic characteristics and have finite relaxation times are called *viscoelastic.*

The dependence of the viscosity on the rate of strain is the major concern in many applications, and it can be modeled using fairly straightforward modifications of the constitutive equation for a Newtonian fluid. The resulting equations describe what are called *generalized Newtonian fluids,* which are discussed next. The analysis of viscoelastic phenomena, including normal stress and memory effects, is beyond the scope of this book. For discussions of that and other aspects of polymer rheology see Bird et al. (1987), Pearson (1985), or Tanner (1985). Bird et al. (1987) is the source for most of the information presented below on generalized Newtonian fluids.

TABLE 5-8
Navier–Stokes Equation in Cylindrical Coordinates

r component:
$$\rho\left[\frac{\partial v_r}{\partial t}+v_r\frac{\partial v_r}{\partial r}+\frac{v_\theta}{r}\frac{\partial v_r}{\partial \theta}-\frac{v_\theta^2}{r}+v_z\frac{\partial v_r}{\partial z}\right]$$
$$=\rho g_r-\frac{\partial P}{\partial r}+\mu\left[\frac{\partial}{\partial r}\left(\frac{1}{r}\frac{\partial}{\partial r}(rv_r)\right)+\frac{1}{r^2}\frac{\partial^2 v_r}{\partial \theta^2}-\frac{2}{r^2}\frac{\partial v_\theta}{\partial \theta}+\frac{\partial^2 v_r}{\partial z^2}\right]$$

θ component:
$$\rho\left[\frac{\partial v_\theta}{\partial t}+v_r\frac{\partial v_\theta}{\partial r}+\frac{v_\theta}{r}\frac{\partial v_\theta}{\partial \theta}+\frac{v_r v_\theta}{r}+v_z\frac{\partial v_\theta}{\partial z}\right]$$
$$=\rho g_\theta-\frac{1}{r}\frac{\partial P}{\partial \theta}+\mu\left[\frac{\partial}{\partial r}\left(\frac{1}{r}\frac{\partial}{\partial r}(rv_\theta)\right)+\frac{1}{r^2}\frac{\partial^2 v_\theta}{\partial \theta^2}+\frac{2}{r^2}\frac{\partial v_r}{\partial \theta}+\frac{\partial^2 v_\theta}{\partial z^2}\right]$$

z component:
$$\rho\left[\frac{\partial v_z}{\partial t}+v_r\frac{\partial v_z}{\partial r}+\frac{v_\theta}{r}\frac{\partial v_z}{\partial \theta}+v_z\frac{\partial v_z}{\partial z}\right]$$
$$=\rho g_z-\frac{\partial P}{\partial z}+\mu\left[\frac{1}{r}\frac{\partial}{\partial r}\left(r\frac{\partial v_z}{\partial r}\right)+\frac{1}{r^2}\frac{\partial^2 v_z}{\partial \theta^2}+\frac{\partial^2 v_z}{\partial z^2}\right]$$

TABLE 5-9
Navier–Stokes Equation in Spherical Coordinates

r component:

$$\rho\left[\frac{\partial v_r}{\partial t}+v_r\frac{\partial v_r}{\partial r}+\frac{v_\theta}{r}\frac{\partial v_r}{\partial \theta}+\frac{v_\phi}{r\sin\theta}\frac{\partial v_r}{\partial \phi}-\frac{v_\theta^2+v_\phi^2}{r}\right]$$

$$=\rho g_r-\frac{\partial P}{\partial r}+\mu\left[\nabla^2 v_r-\frac{2}{r^2}v_r-\frac{2}{r^2}\frac{\partial v_\theta}{\partial \theta}-\frac{2}{r^2}v_\theta\cot\theta-\frac{2}{r^2\sin\theta}\frac{\partial v_\phi}{\partial \phi}\right]$$

θ component:

$$\rho\left[\frac{\partial v_\theta}{\partial t}+v_r\frac{\partial v_\theta}{\partial r}+\frac{v_\theta}{r}\frac{\partial v_\theta}{\partial \theta}+\frac{v_\phi}{r\sin\theta}\frac{\partial v_\theta}{\partial \phi}+\frac{v_r v_\theta}{r}-\frac{v_\phi^2\cot\theta}{r}\right]$$

$$=\rho g_\theta-\frac{1}{r}\frac{\partial P}{\partial \theta}+\mu\left[\nabla^2 v_\theta+\frac{2}{r^2}\frac{\partial v_r}{\partial \theta}-\frac{v_\theta}{r^2\sin^2\theta}-\frac{2\cos\theta}{r^2\sin^2\theta}\frac{\partial v_\phi}{\partial \phi}\right]$$

ϕ component:

$$\rho\left[\frac{\partial v_\phi}{\partial t}+v_r\frac{\partial v_\phi}{\partial r}+\frac{v_\theta}{r}\frac{\partial v_\phi}{\partial \theta}+\frac{v_\phi}{r\sin\theta}\frac{\partial v_\phi}{\partial \phi}+\frac{v_\phi v_r}{r}+\frac{v_\theta v_\phi\cot\theta}{r}\right]$$

$$=\rho g_\phi-\frac{1}{r\sin\theta}\frac{\partial P}{\partial \phi}+\mu\left[\nabla^2 v_\phi-\frac{v_\phi}{r^2\sin^2\theta}+\frac{2}{r^2\sin\theta}\frac{\partial v_r}{\partial \phi}+\frac{2\cos\theta}{r^2\sin^2\theta}\frac{\partial v_\theta}{\partial \phi}\right]$$

$$\nabla^2=\frac{1}{r^2}\frac{\partial}{\partial r}\left(r^2\frac{\partial}{\partial r}\right)+\frac{1}{r^2\sin\theta}\frac{\partial}{\partial \theta}\left(\sin\theta\frac{\partial}{\partial \theta}\right)+\frac{1}{r^2\sin^2\theta}\left(\frac{\partial^2}{\partial \phi^2}\right)$$

Generalized Newtonian Fluids

A generalized Newtonian fluid differs from a Newtonian fluid only in that the viscous stress is not a linear function of the rate of strain. Thus, restricting ourselves to incompressible liquids, the coefficient μ in Eq. (5.6-3) is not independent of Γ. A number of empirical expressions have been used to describe variations in the apparent viscosity with the rate of strain. A scalar measure of the rate of strain which is suitable for such expressions is the magnitude of the rate-of-strain tensor, defined as

$$\Gamma\equiv\left[\frac{1}{2}(\Gamma:\Gamma)\right]^{1/2}. \tag{5.6-11}$$

The sign convention for the square root is that $\Gamma>0$. Thus, for a flow where $v_x=v_x(y)$ and $v_y=v_z=0$, $\Gamma=|\Gamma_{yx}|=(1/2)|dv_x/dy|$. For shearing flows the *shear rate* equals 2Γ, or $|dv_x/dy|$ in this example. Expressions for Γ in the three common coordinate systems are given in Table 5-10. The actual entries in the table are the *viscous dissipation function*, $\Phi=(2\Gamma)^2$, which is discussed further in Chapter 9.

Most polymeric liquids (solutions or melts) exhibit constant viscosities at low shear rates, as well as viscosities that decrease as some power of Γ at higher shear rates. In some cases a constant viscosity is reached again at very high shear rates. An empirical relation which describes this behavior is the *Carreau model*,

$$\frac{\mu-\mu_\infty}{\mu_0-\mu_\infty}=[1+(2\lambda\Gamma)^2]^{(n-1)/2}, \tag{5.6-12}$$

where μ_0, μ_∞, λ, and n are constants chosen to match the behavior of a given fluid. With $n<1$, μ decreases with increasing Γ; for polymer solutions, n typically ranges between 0.2 and 0.6. The parameter λ is a characteristic time constant for the fluid, so that $2\lambda\Gamma$ is the ratio of that time constant to a characteristic time for the flow. The two

TABLE 5-10
Dissipation Function[a] and Magnitude of Rate-of-Strain Tensor

Rectangular: $\Phi = (2\Gamma)^2 = 2\left[\left(\dfrac{\partial v_x}{\partial x}\right)^2 + \left(\dfrac{\partial v_y}{\partial y}\right)^2 + \left(\dfrac{\partial v_z}{\partial z}\right)^2\right]$

$+\left[\dfrac{\partial v_y}{\partial x}+\dfrac{\partial v_x}{\partial y}\right]^2+\left[\dfrac{\partial v_z}{\partial y}+\dfrac{\partial v_y}{\partial z}\right]^2+\left[\dfrac{\partial v_x}{\partial z}+\dfrac{\partial v_z}{\partial x}\right]^2-\dfrac{2}{3}(\nabla\cdot\mathbf{v})^2$

Cylindrical: $\Phi = (2\Gamma)^2 = 2\left[\left(\dfrac{\partial v_r}{\partial r}\right)^2 + \left(\dfrac{1}{r}\dfrac{\partial v_\theta}{\partial \theta}+\dfrac{v_r}{r}\right)^2 + \left(\dfrac{\partial v_z}{\partial z}\right)^2\right]$

$+\left[r\dfrac{\partial}{\partial r}\left(\dfrac{v_\theta}{r}\right)+\dfrac{1}{r}\dfrac{\partial v_r}{\partial \theta}\right]^2+\left[\dfrac{1}{r}\dfrac{\partial v_z}{\partial \theta}+\dfrac{\partial v_\theta}{\partial z}\right]^2+\left[\dfrac{\partial v_r}{\partial z}+\dfrac{\partial v_z}{\partial r}\right]^2-\dfrac{2}{3}(\nabla\cdot\mathbf{v})^2$

Spherical: $\Phi = (2\Gamma)^2 = 2\left[\left(\dfrac{\partial v_r}{\partial r}\right)^2 + \left(\dfrac{1}{r}\dfrac{\partial v_\theta}{\partial \theta}+\dfrac{v_r}{r}\right)^2 + \left(\dfrac{1}{r\sin\theta}\dfrac{\partial v_\phi}{\partial \phi}+\dfrac{v_r}{r}+\dfrac{v_\theta\cot\theta}{r}\right)^2\right]$

$+\left[r\dfrac{\partial}{\partial r}\left(\dfrac{v_\theta}{r}\right)+\dfrac{1}{r}\dfrac{\partial v_r}{\partial \theta}\right]^2+\dfrac{\sin\theta}{r}\dfrac{\partial}{\partial \theta}\left(\dfrac{v_\phi}{\sin\theta}\right)+\dfrac{1}{r\sin\theta}\dfrac{\partial v_\theta}{\partial \phi}\right]^2$

$+\left[\dfrac{1}{r\sin\theta}\dfrac{\partial v_r}{\partial \phi}+r\dfrac{\partial}{\partial r}\left(\dfrac{v_\phi}{r}\right)\right]^2-\dfrac{2}{3}(\nabla\cdot\mathbf{v})^2$

[a] The dissipation function (Φ) appears in the general form of the energy conservation equation for a Newtonian fluid. The rate (per unit volume) at which mechanical energy is converted to heat is given by $\mu\Phi$ (see Chapter 9).

Newtonian limits are $\mu = \mu_0$ for $2\lambda\Gamma \ll 1$ and $\mu = \mu_\infty$ for $2\lambda\Gamma \gg 1$. For many concentrated polymer solutions and polymer melts, the available data can be fitted adequately with $\mu_\infty = 0$, leaving only three parameters in Eq. (5.6-12).

An expression with just two parameters, which is equivalent to Eq. (5.6-12) at intermediate shear rates, is the *power-law model*,

$$\mu = m(2\Gamma)^{n-1}, \tag{5.6-13}$$

where n has the same significance as in Eq. (5.6-12) and m is another constant. Although Eq. (5.6-13) requires Γ to be within a certain range, that range often spans several orders of magnitude. *Shear-thinning* behavior is obtained for $n<1$, Newtonian behavior is obtained for $n=1$, and *shear-thickening* behavior is obtained for $n>1$.

Some materials behave more or less as solids until they are subjected to a stress which exceeds some threshhold value. Examples include paint and various foods. The minimum stress required to induce flow, called the *yield stress*, is a property of the material. A simple representation of this kind of behavior is the *Bingham model*,

$$\mu = \begin{cases} \infty & \text{for } \tau \le \tau_0, \\ \mu_0 + \dfrac{\tau_0}{2\Gamma} & \text{for } \tau \ge \tau_0, \end{cases} \tag{5.6-14}$$

where τ_0 is the yield stress and τ is the magnitude of $\boldsymbol{\tau}$, given by

$$\tau \equiv \left[\dfrac{1}{2}(\boldsymbol{\tau}:\boldsymbol{\tau})\right]^{1/2}. \tag{5.6-15}$$

Equation (5.6-14) indicates that a Bingham fluid behaves as a solid at stresses below τ_0; because $\mu = \infty$ and the stress is finite, $\Gamma = 0$ under those conditions. That is, no deformation occurs for $\tau \leq \tau_0$. At the other extreme of large rates of strain, such that $2\mu_0 \Gamma \gg \tau_0$, the fluid becomes Newtonian with a viscosity μ_0.

Whichever of the generalized Newtonian models is used, the viscous stress is substituted into the Cauchy momentum equation to give

$$\rho \frac{D\mathbf{v}}{Dt} = \rho \mathbf{g} - \nabla P + \nabla \cdot \{\mu(\Gamma)[\nabla \mathbf{v} + (\nabla \mathbf{v})^t]\} \tag{5.6-16}$$

instead of the Navier–Stokes equation. This and the continuity equation for an incompressible fluid [Eq. (5.6-10)] provide the governing equations for a generalized Newtonian fluid. The dependence of μ on Γ makes the viscous term in Eq. (5.6-16) nonlinear. Nonetheless, a number of flow problems involving generalized Newtonian fluids in simple geometries can be solved analytically.

5.7 FLUID MECHANICS AT INTERFACES

In this section we consider the behavior of the velocity and stress at fluid-solid and fluid-fluid interfaces. The equations satisfied by velocity and stress components which are tangential to or normal to an interface provide the boundary condtions for fluid mechanics problems. Velocities are discussed first, followed by stresses at interfaces where surface tension effects are absent, and then the modifications to the stress balances required by surface tension. Finally, some of the effects of surface tension are illustrated using examples involving static fluids.

Tangential Velocity

An empirical observation, verified in countless situations, is that the velocity components tangent to a fluid–solid or fluid–fluid interface are continuous. That is, there is ordinarily no relative motion or "slip" where the materials contact one another. If $\mathbf{t}$ is any unit vector tangent to the interface, then the corresponding tangential component of velocity is $v_t = \mathbf{t} \cdot \mathbf{v}$. The matching of tangential velocities on the interface between materials 1 and 2 is written as

$$v_t|_1 = v_t|_2 \tag{5.7-1}$$

and is referred to as the *no-slip condition*. (Although not indicated explicitly, this and all other interfacial conditions discussed in this section apply locally at each point on the interface, like those discussed in Chapter 2.) At an interface between a fluid and a stationary solid, Eq. (5.7-1) reduces to the frequently used requirement that $v_t = 0$ in the fluid contacting the solid surface.

Normal Velocity

As shown by Eq. (2.5-10), the component of velocity normal to an interface ($v_n = \mathbf{n} \cdot \mathbf{v}$, where $\mathbf{n}$ is the unit normal) is governed by conservation of mass. Thus, for a stationary, impermeable, and inert solid, $v_n = 0$ in the fluid contacting the solid surface. This will be referred to as the *no-penetration* condition. In this context, "inert" means that the solid is not undergoing a phase change.

Stress Balance

At an interface where surface tension is negligible, each component of the total stress must be in equilibrium. The proof of this is basically the same as that given for Eq. (5.3-18), the imaginary test surface in Fig. 5-3 being replaced by an interface. Accordingly,

$$\mathbf{s}(\mathbf{n})_1 = \mathbf{s}(\mathbf{n})|_2, \tag{5.7-2}$$

where $\mathbf{n}$ is normal to the interface. This result applies at any point on a flat or curved interface, provided that surface tension is negligible. To satisfy Eq. (5.7-2), the component of the total stress normal to the interface ($\sigma_{nn} = \mathbf{n} \cdot \mathbf{n} \cdot \boldsymbol{\sigma}$) and any component of stress tangent to the interface ($\sigma_{nt} = \mathbf{t} \cdot \mathbf{n} \cdot \boldsymbol{\sigma}$) must each balance. In terms of the pressure and viscous stresses, these interfacial conditions become

$$\tau_{nn}|_1 - P_1 = \tau_{nn}|_2 - P_2, \tag{5.7-3}$$

$$\tau_{nt}|_1 = \tau_{nt}|_2. \tag{5.7-4}$$

Notice that for static fluids, Eq. (5.7-3) reduces simply to an equality of pressures. The equality of pressures is a good approximation also in many fluid dynamic problems, in that the normal viscous stresses are often small or even zero. In particular, for an incompressible Newtonian fluid, the normal viscous stress at any solid surface is zero (see Problem 5-4).

In analyzing liquid flows it is often possible to neglect the shear stresses at gas–liquid interfaces. This is because gas viscosities are relatively small, with a typical gas-to-liquid viscosity ratio being $\sim 10^{-2}$ (Chapter 1). Thus, unless the rates of strain in the gas greatly exceed those in the liquid, the shear stresses exerted by the gas on the liquid will be small compared to the viscous stresses within the liquid. If one is interested only in the liquid flow, treating the interface as shear-free results in major simplifications, in that it uncouples the flow problem in the liquid from that in the gas. This uncoupling is discussed in Example 6.2-4. It is important to recognize, however, that under these same conditions the shear-free interface is ordinarily *not* a good approximation when analyzing flow in the gas. That is, the stress at the interface with a flowing liquid is not usually small relative to other viscous stresses in the *gas*.

Stress Balance with Surface Tension

The surface tension (γ) at an interface between two immiscible fluids is a material property which has units of energy per unit area or force per unit length. In thermodynamics it represents the energy required to create new interfacial area, whereas in mechanics it is viewed as a force per unit length, acting within the plane of the interface. As will be shown, it can affect the mechanics of a fluid–fluid interface if one or both of the following conditions hold: if the interface is nonplanar or if the surface tension varies with position.

The effects of surface tension on the stress balance at an interface are derived by considering a control volume of thickness 2ℓ centered on an interfacial area S_I between two fluids, as shown in Fig. 5-7. Applying conservation of linear momentum to this control volume gives

$$\int_{V(t)} \rho \frac{D\mathbf{v}}{Dt} \, dV = \int_{V(t)} \rho \mathbf{g} \, dV + \int_{S(t)} \mathbf{s} \, dS + \int_{C(t)} \gamma \mathbf{m} \, dC, \qquad (5.7\text{-}5)$$

which is similar to Eq. (5.3-14), but with a term added to represent the force associated with surface tension. The new term is based on the fact that the surface tension is a force per unit length acting on the closed curve C, which represents the intersection of the control volume with the interface. Positive surface tensions act in the direction of the unit vector $\mathbf{m}$, which is outward, normal to C, and tangent to the interface. Letting $\ell \rightarrow 0$, the volume integrals in Eq. (5.7-5) will become negligible, as will the contributions of the edges to the surface (stress) integral. What remains of the surface integral are pieces which, in the limit, correspond to the top and bottom sides of the surface S_I. [The reasoning here is similar to that used to derive Eq. (2.2-15).] The contour (surface tension) integral is unaffected by ℓ, so that the momentum conservation statement becomes

$$\int_{S_I(t)} [\mathbf{s}(\mathbf{n})|_2 - \mathbf{s}(\mathbf{n})|_1] \, dS + \int_{C(t)} \gamma \mathbf{m} \, dC = 0. \qquad (5.7\text{-}6)$$

Here we have used the fact that $\mathbf{s}(-\mathbf{n}) = -\mathbf{s}(\mathbf{n})$; [see Eq. (5.3-18)].

The contour integral in Eq. (5.7-6) is converted now to a surface integral by using Eq. (A.8-30). This gives

$$\int_{C(t)} \gamma \mathbf{m} \, dC = \int_{S_I(t)} (\nabla_s \gamma + 2 \mathscr{H} \, \mathbf{n} \gamma) \, dS, \qquad (5.7\text{-}7)$$

where ∇_s is the *surface gradient* and $\mathscr{H}$ is the *mean curvature*. As its name implies, ∇_s is a differential operator which describes spatial variations of quantities within a surface. For a planar surface, it is just a two-dimensional form of the usual gradient operator, without a component normal to the surface. In general, $\mathscr{H}$ varies from point to point, according to the local shape of the surface. Its algebraic sign is such that $\mathscr{H} < 0$ when $\mathbf{n}$ points away from the local center of curvature (i.e., away from a convex surface) and $\mathscr{H} > 0$ when $\mathbf{n}$ points toward the local center of curvature. For a planar surface, $\mathscr{H} = 0$. Cylinders and spheres are special in that they have constant curvatures. For $\mathbf{n}$ directed

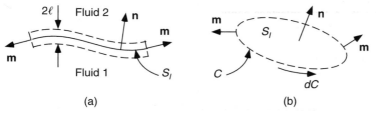

(a)   (b)

Figure 5-7. A control volume surrounding part of the interface between two immiscible fluids. Panels (a) and (b) show cross-sectional and top views, respectively. The unit vector $\mathbf{n}$ is normal to the interface. The unit vector $\mathbf{m}$ is tangent to the interface and normal to the contour C, which is the intersection of the control volume with the interface. With $\mathbf{n}$ oriented as shown, the contour integration is in the direction indicated by the arrow labeled dC. In other words, $\mathbf{n}$ points in the direction which a right-hand screw would advance if rotated in the sense of dC.

outward from a cylinder or sphere of radius R, $\mathcal{H} = -1/(2R)$ and $\mathcal{H} = -1/R$, respectively. Additional information on ∇_s and $\mathcal{H}$ is given in Section A.8.

Using Eq. (5.7-7) in Eq. (5.7-6) and letting $S_t \to 0$, we obtain the desired stress balance at any point on an interface,

$$\mathbf{s(n)}|_2 - \mathbf{s(n)}|_1 + \nabla_s \gamma + 2\mathcal{H}\,\mathbf{n}\gamma = \mathbf{0}. \tag{5.7-8}$$

Comparing this with Eq. (5.7-2), it is seen that surface tension introduces two new terms. One of those is zero if the surface tension is independent of position ($\nabla_s\gamma = \mathbf{0}$), whereas the other vanishes if the interface is flat ($\mathcal{H} = 0$).

Although sufficiently general for many fluid mechanical problems which involve surface tension, Eq. (5.7-8) is in fact a special case of a more general result given by Scriven (1960). In that analysis the interface is treated as a two-dimensional fluid, characterized not just by γ but also by *surface viscosity* coefficients analogous to μ and κ. The viscous resistance to interfacial deformation per se is an important feature of certain systems which contain surfactants. For additional information on the interfacial transport of momentum, including the measurement of interfacial viscosities, see Edwards et al. (1991).

Resolving Eq. (5.7-8) into normal and tangential components, we obtain

$$P_1 - P_2 + \tau_{nn}|_2 - \tau_{nn}|_1 + 2\mathcal{H}\,\gamma = 0, \tag{5.7-9}$$

$$\tau_{nt}|_2 - \tau_{nt}|_1 + \mathbf{t}\cdot\nabla_s\,\gamma = 0. \tag{5.7-10}$$

In these results, which are analogous to Eqs. (5.7-3) and (5.7-4), it is seen that the curvature term affects only the normal stress balance, whereas gradients in surface tension affect only the shear stress condition. The new term in the normal stress balance, Eq. (5.7-9), is sometimes called the *capillary pressure;* it may be written as $P_\gamma = -2\mathcal{H}\gamma$. As illustrated by the examples at the end of this section, nonzero values of γ do not necessarily result in any flow, even when the interface is curved and $P_\gamma \neq 0$. In contrast, nonzero *gradients* in surface tension cannot occur under static conditions. This is seen by recalling that $\tau = 0$ in static fluids and by noticing that the shear stress balance, Eq. (5.7-10), cannot be satisfied if $\nabla_s\gamma \neq \mathbf{0}$ and $\tau = \mathbf{0}$. Unlike capillary pressures, surface tension gradients act even at flat interfaces. Spatial variations in γ may arise from temperature gradients ($d\gamma/dT < 0$ for pure liquids), nonuniform concentrations of surfactants, or a nonuniform electric charge density at a liquid surface.

Places where a fluid–fluid interface contacts a solid surface, forming a three-phase contact line, require special attention. At such locations the fluid interface tends to meet the solid at a particular angle (α, the *contact angle*), which is a property of the three materials. Two types of gas–liquid–solid behavior are shown in Fig. 5-8. *Wetting* and *nonwetting* of the solid by the liquid are said to occur for $\alpha < \pi/2$ and $\alpha > \pi/2$, respectively. At the three-phase line there is an unbalanced surface tension acting toward the solid, along the tangent described by the contact angle. The result is a tendency for the liquid to move along the solid surface. This is the basis for *capillarity* phenomena, such as the rise of fluids in small vertical tubes and the spontaneous fluid uptake exhibited by many porous materials.

This section concludes with two examples which illustrate the effects of surface tension in static fluids.

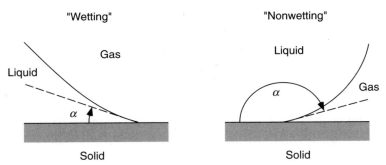

Figure 5-8. Contact angle (α) at a gas–liquid–solid boundary, for both wetting and nonwetting behavior. The dashed line indicates the tangent at the three-phase contact point.

Example 5.7-1 Pressure in a Bubble or Drop The objective is to determine the pressure difference across the surface of a gas bubble or a liquid drop immersed in some other liquid, neglecting any effects of fluid motion. The interior of the bubble or drop is denoted as fluid 1 and the surrounding phase as fluid 2, and **n** points outward. Ignoring the normal viscous stresses, Eq. (5.7-9) reduces to

$$P_1 - P_2 = -2\,\mathcal{H}\,\gamma. \tag{5.7-11}$$

Assuming that static pressure variations within either fluid are negligible on the length scale of the bubble or drop, P_1 and P_2 are constants within the respective phases. As mentioned above, for a sphere of radius R the mean curvature is a constant, $\mathcal{H} = -1/R$. Thus, we obtain

$$P_1 - P_2 = \frac{2\gamma}{R}, \tag{5.7-12}$$

which is called the *Young–Laplace equation.* Notice that Eq. (5.7-11) can be satisfied for constant P_1, P_2, and γ only if the curvature is independent of position. This is the reason that the equilibrium shape of a small bubble or drop is spherical.

Example 5.7-2 Free Surface Next to a Wetted Wall Consider a gas–liquid interface near a wetted, vertical surface, as shown in Fig. 5-9. The contact angle (α) requires that the liquid rise near the surface. The height of the gas–liquid interface is described by $z = h(x)$, where x is distance from the wall and $h(\infty) = 0$. It is desired to determine the function $h(x)$ under static conditions.

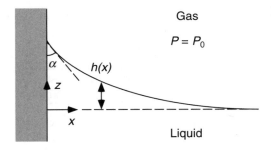

Figure 5-9. Gas–liquid interface near a vertical wall.

This is an example of a *free surface* problem, in which the location of an interface is determined by certain static or dynamic conditions, rather than being specified in advance.

The normal stress balance again reduces to Eq. (5.7-11), where now fluid 1 is taken to be the liquid and fluid 2 the gas. When **n** is directed upward (i.e., $n_z = e_z \cdot \mathbf{n} > 0$) and the surface is described as $z = h(x)$ only, the curvature is given by [see Eq. (A.8-24)]

$$2\mathcal{H} = \frac{d^2 h/dx^2}{[1 + (dh/dx)^2]^{3/2}}. \qquad (5.7\text{-}13)$$

It is assumed that the gas pressure is constant, such that $P_2 = P_0$, whereas the static pressure in the liquid varies with height. Using Eq. (5.5-2), the liquid pressure at the gas–liquid interface is

$$P_1(x) = P_0 - \rho_L g h(x), \qquad (5.7\text{-}14)$$

where the constant term (P_0) has been evaluated from the requirement that the gas and liquid pressures be equal at $x = \infty$, where the interface is flat and $h = 0$. Substituting the expressions for the curvature and pressures in Eq. (5.7-11), it is found that $h(x)$ must satisfy

$$\frac{d^2 h}{dx^2} = \left(\frac{\rho_L g}{\gamma}\right) h \left[1 + \left(\frac{dh}{dx}\right)^2\right]^{3/2} \qquad (5.7\text{-}15)$$

with the boundary conditions

$$\frac{dh}{dx}(0) = -\cot\alpha, \qquad (5.7\text{-}16)$$

$$h(\infty) = 0. \qquad (5.7\text{-}17)$$

The slope of the gas–liquid interface at the wall is dictated by the contact angle, as expressed by Eq. (5.7-16). The coefficient on the right-hand side of Eq. (5.7-15) indicates that the characteristic length for this problem is

$$\lambda \equiv \left(\frac{\gamma}{\rho_L g}\right)^{1/2}, \qquad (5.7\text{-}18)$$

which is called the *capillary length*.

The nonlinearity of Eq. (5.7-15) requires that, in general, $h(x)$ be determined by numerical integration. However, for weakly wetting surfaces, such that $\alpha = (\pi/2) - \varepsilon$ with $\varepsilon > 0$ and $\varepsilon \ll 1$, an analytical solution is obtained from a regular perturbation analysis. The cotangent in Eq. (5.7-16) is expanded in a Taylor series about the angle $\pi/2$ to give

$$\cot\alpha = \varepsilon + O(\varepsilon^3) \qquad (5.7\text{-}19)$$

and the interface height is expressed as

$$h(x) = \sum_{n=0}^{\infty} \varepsilon^n h_n(x). \qquad (5.7\text{-}20)$$

Using the methods of Section 3.6, it is found that $h_0 = 0$, $h_1 \neq 0$, $h_2 = 0$, and $h_3 \neq 0$. The solution up to $O(\varepsilon^3)$ is

$$h(x) = \varepsilon \lambda e^{-x/\lambda} + O(\varepsilon^3). \qquad (5.7\text{-}21)$$

Thus, the maximum height of the interface for small ε is $\varepsilon\lambda$.

5.8 DYNAMIC PRESSURE

In solving flow problems involving incompressible fluids with known boundaries (i.e., without any free surfaces) it is advantageous to combine the pressure and gravitational terms in the Navier–Stokes or Cauchy momentum equations by using the *dynamic pressure* (sometimes called the *modified pressure* or *equivalent pressure*). As illustrated by many examples in later chapters, this approach is extremely useful for analyzing both confined flows and flows around submerged objects. The advantage in combining the pressure and gravitational terms, when possible, is that problems can then be solved without reference to the direction of gravity. In other words, one solution suffices for all spatial orientations. For instance (as shown in Example 6.2-1), the parabolic velocity profile in a fluid-filled, parallel-plate channel is independent of whether the channel is horizontal or vertical. The modified pressure loses its utility in free-surface problems, such as those involving wave dynamics or flow in open channels. In such cases pressure and gravitational effects cannot be fully combined, because the actual pressure is needed to determine the location of the gas–liquid interface (as in Example 5.7-2).

The dynamic pressure ($\mathcal{P}$) is defined as

$$\nabla\mathcal{P} \equiv \nabla P - \rho\mathbf{g}, \tag{5.8-1}$$

so that the Navier–Stokes equation becomes

$$\rho\frac{D\mathbf{v}}{Dt} = -\nabla\mathcal{P} + \mu\nabla^2\mathbf{v}. \tag{5.8-2}$$

It is evident from Eq. (5.8-2) that $\nabla\mathcal{P}=\mathbf{0}$ in a static fluid. Thus, when the Navier–Stokes equation applies, spatial variations in $\mathcal{P}$ are always associated with fluid motion. When there are density variations or body forces other than gravity, the interpretation of $\mathcal{P}$ is less simple (see Chapters 11 and 12).

The value of $\mathcal{P}$ at any position $\mathbf{r}$ is determined only to within an additive constant, because $\mathcal{P}$ is defined only in terms of its gradient. Indeed, integration of Eq. (5.8-1) for a steady flow leads to

$$P(\mathbf{r}) = \rho\mathbf{g}\cdot\mathbf{r} + \mathcal{P}(\mathbf{r}) + c, \tag{5.8-3}$$

where c is a constant. It is seen that even if the absolute value of P is known at some location (e.g., $\mathbf{r}=\mathbf{0}$), the arbitrary nature of c allows the value of $\mathcal{P}$ there to be chosen freely. For a static fluid, where $\mathcal{P}$ is constant, Eq. (5.8-3) reduces to Eq. (5.5-3); in that case $\mathcal{P}+c=P(\mathbf{0})$, the pressure at the origin. The usual practice in problem solving is to select a reference value of $\mathcal{P}=0$ at some convenient location.

The pressure force ($\mathbf{F_P}$) on an object of volume V which is completely submerged in a static fluid was calculated in Example 5.5-1, leading to Archimedes' law. Using Eq. (5.8-3), the corresponding result for a flowing fluid is

$$\mathbf{F_P} = -\int_S P\mathbf{n}\ dS = -\rho\mathbf{g}V - \int_S \mathcal{P}\mathbf{n}\ dS, \tag{5.8-4}$$

where S denotes the entire (closed) surface of the object. The first term $(-\rho\mathbf{g}V)$ was derived previously from the static pressure variation, whereas the integral involving $\mathcal{P}$ gives the additional pressure force due to fluid motion. Thus, the contributions of pressure to the drag or lift forces on a submerged object are computed using $\mathcal{P}$, not P. As shown in Example 5.5-1, adding any constant to P or $\mathcal{P}$ does not change the values of the integrals in Eq. (5.8-4). Accordingly, the reference value chosen for $\mathcal{P}$ has no effect on drag or lift calculations.

5.9 STREAM FUNCTION

The stream function is a mathematical construct which is extremely useful for solving incompressible flow problems where only two nonvanishing velocity components and two spatial coordinates are involved. This includes flows which have a planar character and flows which are axisymmetric. In addition to being a useful vehicle for deriving the velocity and pressure fields in such problems, the stream function itself provides infor-mation which helps to visualize flow patterns. In this section the equations which govern the stream function are derived, and then the physical interpretation of the stream func-tion is discussed.

For a two-dimensional flow described by rectangular coordinates, the stream func-tion $\psi(x, y)$ is defined by

$$v_x \equiv \frac{\partial \psi}{\partial y}, \qquad v_y \equiv -\frac{\partial \psi}{\partial x}. \tag{5.9-1}$$

The rationale for this choice is seen by substituting Eq. (5.9-1) into the corresponding continuity equation for an incompressible fluid,

$$\frac{\partial v_x}{\partial x} + \frac{\partial v_y}{\partial y} = \frac{\partial}{\partial x}\left(\frac{\partial \psi}{\partial y}\right) + \frac{\partial}{\partial y}\left(-\frac{\partial \psi}{\partial x}\right) = 0. \tag{5.9-2}$$

Thus, the stream function is defined in such a way that if $\psi(x, y)$ can be determined for a given flow, the continuity equation will be satisfied automatically. Reversing the signs in Eq. (5.9-1) does not alter the ability of $\psi(x, y)$ to satisfy continuity, so it is not surprising that some authors use the opposite sign convention.

Conservation of momentum for a Newtonian fluid is ensured by rewriting the Navier–Stokes equation in terms of $\psi(x, y)$. Using the dynamic pressure, the x and y components of Eq. (5.8-2) are

$$\rho\left(\frac{\partial v_x}{\partial t} + \mathbf{v} \cdot \nabla v_x\right) = -\frac{\partial \mathcal{P}}{\partial x} + \mu \nabla^2 v_x, \tag{5.9-3}$$

$$\rho\left(\frac{\partial v_y}{\partial t} + \mathbf{v} \cdot \nabla v_y\right) = -\frac{\partial \mathcal{P}}{\partial y} + \mu \nabla^2 v_y. \tag{5.9-4}$$

The pressure is eliminated by differentiating each term in Eq. (5.9-3) by y and each term in Eq. (5.9-4) by x and then subtracting one equation from the other. The result is

$$\rho\frac{\partial}{\partial t}\left(\frac{\partial v_x}{\partial y} - \frac{\partial v_y}{\partial x}\right) + \rho\left[\frac{\partial}{\partial y}(\mathbf{v} \cdot \nabla v_x) - \frac{\partial}{\partial x}(\mathbf{v} \cdot \nabla v_y)\right] = \mu \nabla^2\left(\frac{\partial v_x}{\partial y} - \frac{\partial v_y}{\partial x}\right). \tag{5.9-5}$$

The velocity gradients in the first and last terms are related to the stream function by

$$\frac{\partial v_x}{\partial y} - \frac{\partial v_y}{\partial x} = \frac{\partial}{\partial y}\left(\frac{\partial \psi}{\partial y}\right) - \frac{\partial}{\partial x}\left(-\frac{\partial \psi}{\partial x}\right) = \nabla^2 \psi. \tag{5.9-6}$$

Expanding the quantity in square brackets in Eq. (5.9-5), it is found after some re-arrangement and cancellation of terms that

$$-\left[\frac{\partial}{\partial y}(\mathbf{v}\cdot\nabla v_x)-\frac{\partial}{\partial x}(\mathbf{v}\cdot\nabla v_y)\right]=\frac{\partial\psi}{\partial x}\frac{\partial}{\partial y}(\nabla^2\psi)-\frac{\partial\psi}{\partial y}\frac{\partial}{\partial x}(\nabla^2\psi)\equiv\frac{\partial(\psi,\ \nabla^2\psi)}{\partial(x,\ y)}, \qquad (5.9\text{-}7)$$

where the last equality defines the *Jacobian determinant*. Collecting the expressions for the individual terms, the Navier–Stokes equation is found to be equivalent to

$$\frac{\partial}{\partial t}(\nabla^2\psi)-\frac{\partial(\psi,\ \nabla^2\psi)}{\partial(x,\ y)}=\nu\nabla^4\psi. \qquad (5.9\text{-}8)$$

Thus, the three coupled differential equations involving v_x, v_y, and P (i.e., continuity and two components of the Navier–Stokes equation) have been converted to a single differential equation for ψ. Equation (5.9-8) is fourth order, but the disadvantages of dealing with a higher-order differential equation are normally outweighed by the advantages of having a single equation and single unknown. Accordingly, Eq. (5.9-8) is the preferred starting point for the analysis of many two-dimensional flow problems.

Stream functions can be defined also for planar flows in cylindrical coordinates $(r,\ \theta)$, and for axisymmetric flows in cylindrical $(r,\ z)$ or spherical $(r,\ \theta)$ coordinates. The governing equations for the various cases are given in Table 5-11. The derivation of the stream function equations for general curvilinear coordinates is described, for example, in Leal (1992, pp. 144–149). The use of ψ to solve planar or axisymmetric flow problems is illustrated in Chapters 7 and 8.

TABLE 5-11
Stream Function Equations

Coordinate system and conditions	Velocity components	Equation equivalent to Navier–Stokes equation[a]	Differential operators
Rectangular with $v_z=0$ and no z dependence	$v_x=\dfrac{\partial\psi}{\partial y}$ $v_y=-\dfrac{\partial\psi}{\partial x}$	$\dfrac{\partial}{\partial t}(\nabla^2\psi)-\dfrac{\partial(\psi,\ \nabla^2\psi)}{\partial(x,\ y)}=\nu\nabla^4\psi$	$\nabla^2\equiv\dfrac{\partial^2}{\partial x^2}+\dfrac{\partial^2}{\partial y^2}$ $\nabla^4\equiv\nabla^2(\nabla^2)$
Cylindrical with $v_z=0$ and no z dependence	$v_r=\dfrac{1}{r}\dfrac{\partial\psi}{\partial\theta}$ $v_\theta=-\dfrac{\partial\psi}{\partial r}$	$\dfrac{\partial}{\partial t}(\nabla^2\psi)-\dfrac{1}{r}\dfrac{\partial(\psi,\ \nabla^2\psi)}{\partial(r,\ \theta)}=\nu\nabla^4\psi$	$\nabla^2\equiv\dfrac{\partial^2}{\partial r^2}+\dfrac{1}{r}\dfrac{\partial}{\partial r}+\dfrac{1}{r^2}\dfrac{\partial^2}{\partial\theta^2}$
Cylindrical with $v_\theta=0$ and no θ dependence	$v_r=\dfrac{1}{r}\dfrac{\partial\psi}{\partial z}$ $v_z=-\dfrac{1}{r}\dfrac{\partial\psi}{\partial r}$	$\dfrac{\partial}{\partial t}(E^2\psi)-\dfrac{1}{r}\dfrac{\partial(\psi,\ E^2\psi)}{\partial(r,\ z)}-\dfrac{2}{r^2}\dfrac{\partial\psi}{\partial z}E^2\psi$ $=\nu E^4\psi$	$E^2\equiv\dfrac{\partial^2}{\partial r^2}-\dfrac{1}{r}\dfrac{\partial}{\partial r}+\dfrac{\partial^2}{\partial z^2}$ $E^4\equiv E^2(E^2)$
Spherical with $v_\phi=0$ and no ϕ dependence	$v_r=\dfrac{1}{r^2\sin\theta}\dfrac{\partial\psi}{\partial\theta}$ $v_\theta=-\dfrac{1}{r\sin\theta}\dfrac{\partial\psi}{\partial r}$	$\dfrac{\partial}{\partial t}(E^2\psi)-\dfrac{1}{r^2\sin\theta}\dfrac{\partial(\psi,\ E^2\ \psi)}{\partial(r,\ \theta)}$ $+\dfrac{2E^2\psi}{r^2\sin^2\theta}\left(\dfrac{\partial\psi}{\partial r}\cos\theta-\dfrac{1}{r}\dfrac{\partial\psi}{\partial\theta}\sin\theta\right)$ $=\nu E^4\psi$	$E^2\equiv\dfrac{\partial^2}{\partial r^2}+\dfrac{\sin\theta}{r^2}\dfrac{\partial}{\partial\theta}\left(\dfrac{1}{\sin\theta}\dfrac{\partial}{\partial\theta}\right)$ $E^4\equiv E^2(E^2)$

[a] Jacobians are given by

$$\frac{\partial(f,\ g)}{\partial(x,\ y)}=\begin{vmatrix}\partial f/\partial x & \partial f/\partial y\\ \partial g/\partial x & \partial g/\partial y\end{vmatrix}.$$

We turn now to the interpretation of the stream function and its use in flow visualization. The concept of a *streamline* as a line which is tangent everywhere to the instantaneous velocity vector was mentioned in Section 5.2. The most general way to determine such curves from a given velocity field is by a calculation of *trajectories*. This approach makes use of the fact that

$$\frac{d\mathbf{r}}{ds} = \mathbf{t},$$

(5.9-9)

where $\mathbf{r}$ is the position of a given point on a curve, s is arc length, and $\mathbf{t}$ is a unit vector tangent to the curve. If the curve is a streamline, then the unit tangent is (by definition) $\mathbf{t} = \mathbf{v}/|\mathbf{v}| = \mathbf{v}/v$. Replacing arc length in Eq. (5.9-9) by $p \equiv \int v^{-1} \, ds$, the trajectory equation for a streamline is

$$\frac{d\mathbf{r}}{dp} = \mathbf{v}(\mathbf{r}, t) \quad \text{or} \quad \frac{dx_i}{dp} = v_i(\mathbf{r}, t), \qquad i = 1, 2, 3.$$

(5.9-10)

The second form of Eq. (5.9-10) emphasizes that, in general, the trajectory approach yields a set of three coupled differential equations which must be solved for the coordinates $x_i(p)$. The streamline passing through a particular point in space is obtained by specifying, for example, the "initial" values $x_i(0)$ corresponding to $p=0$. In unsteady flows, streamlines change from instant to instant, so that p must not be confused with time. Indeed, in calculating a streamline at a given instant, p is varied while t (time) is held constant. This approach may be used for any flow where $\mathbf{v}(\mathbf{r}, t)$ is known; it is not necessary that the flow be two-dimensional or incompressible.

The usefulness of the stream function in flow visualization stems from the fact that ψ is constant along a given streamline. This is shown by recalling that the rate of change of any scalar function $b(\mathbf{r}, t)$ along a curve with unit tangent $\mathbf{t}$ is given by $\mathbf{t} \cdot \nabla b$ [see Eq. (A.6-3)]. Thus, the rate of change of ψ along a streamline is proportional to $\mathbf{v} \cdot \nabla \psi$, and

$$\mathbf{v} \cdot \nabla \psi = v_x \frac{\partial \psi}{\partial x} + v_y \frac{\partial \psi}{\partial y} = \frac{\partial \psi}{\partial y} \frac{\partial \psi}{\partial x} - \frac{\partial \psi}{\partial x} \frac{\partial \psi}{\partial y} = 0.$$

(5.9-11)

In other words, ψ is constant, as stated above; different streamlines correspond to different values of that constant. Accordingly, for any flow in which a stream function can be defined, calculating ψ provides an alternative to the trajectory method for locating streamlines. The stream function approach is usually simpler, especially because ψ is often computed anyway in the process of determining $\mathbf{v}$ and P.

Another interesting property of the stream function is that the flow rate across any curve connecting two streamlines equals the difference between the values of ψ. Consider, for example, the two-dimensional flow shown in Fig. 5-10. Two streamlines are shown, corresponding to the constants ψ_1 and ψ_2. There is, by definition, no flow normal to any streamline. (For this reason, an impermeable solid surface always corresponds to a streamline.) Accordingly, the flow rate across any simple curve connecting the two streamlines in Fig. 5-10 must be the same. Choosing for convenience the straight line connecting points 1 and 2, which have the coordinates (x, y_1) and (x, y_2), respectively, the volume flow rate (per unit depth) is

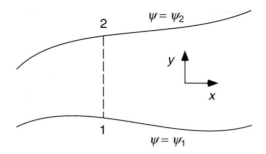

Figure 5-10. Calculation of flow rate between two streamlines.

$$q = \int_{y_1}^{y_2} v_x\, dy = \int_{y_1}^{y_2} \left(\frac{\partial \psi}{\partial y}\right) dy = \psi(x, y_2, t) - \psi(x, y_1, t) = \psi_2 - \psi_1. \tag{5.9-12}$$

Thus, the flow rate equals the difference in the values of the stream function, as stated. When streamlines are plotted using equal intervals in ψ, similar to a contour map, the local velocity is inversely proportional to the spacing between streamlines.

One other noteworthy aspect of the stream function is its relationship to the vorticity. For a planar (x, y) flow, the only nonvanishing component of the vorticity vector is w_z. Noticing that the left-hand side of Eq. (5.9-6) equals $-w_z$, we see that $\nabla^2\psi = -w_z$ for planar flow in rectangular coordinates. Similar relationships between the stream function and vorticity exist for the other cases. These relationships are

$$\nabla^2\psi = -w_z, \qquad\qquad \text{rectangular } (x, y), \qquad\qquad (5.9\text{-}13a)$$

$$\nabla^2\psi = -w_z, \qquad\qquad \text{cylindrical } (r, \theta), \qquad\qquad (5.9\text{-}13b)$$

$$E^2\psi = rw_\theta, \qquad\qquad \text{cylindrical } (r, z), \qquad\qquad (5.9\text{-}13c)$$

$$E^2\psi = -(r \sin \theta)w_\phi, \qquad \text{spherical } (r, \theta), \qquad\qquad (5.9\text{-}13d)$$

where E^2 is defined in Table 5-11. In each case there is only one nonvanishing component of vorticity, so that the component shown equals the magnitude of the vorticity, $|\mathbf{w}|$ or w. The relationships in Eq. (5.9-13) are very useful for solving problems involving irrotational flow, where $w=0$ (Chapter 8). In addition, they provide the basis for an effective solution strategy for creeping flow, where the stream-function form of the Navier–Stokes equation reduces to just $\nabla^4\psi=0$ or $E^4\psi=0$ (Chapter 7). That approach, useful for both analytical and numerical solutions, is to rewrite the fourth-order differential equation for ψ as a pair of coupled, second-order equations for ψ and w.

This section concludes with two examples which illustrate the calculation of streamlines from a known velocity field.

Example 5.9-1 Streamlines from the Trajectory Method Consider the steady, two-dimensional velocity field given by

$$v_x = x, \qquad v_y = -y. \tag{5.9-14}$$

Using Eq. (5.9-10), the differential equations governing $x(p)$ and $y(p)$ are

$$\frac{dx}{dp} = v_x = x, \qquad \frac{dy}{dp} = v_y = -y. \tag{5.9-15}$$

The general solutions are

$$\ln x = p + a, \qquad \ln y = -p + b, \tag{5.9-16}$$

where a and b are constants. Adding the solutions so as to eliminate the parameter p, the equation for a streamline is found to be

$$xy = c, \tag{5.9-17}$$

where $c = \exp(ab)$. Different streamlines correspond to different values of the constant c. For example, $c = 1$ yields the streamline which passes through the point $(1, 1)$.

Example 5.9-2 Streamlines from the Stream Function   Again consider the velocity field given by Eq. (5.9-14). Using the definition of the stream function for this geometry, Eq. (5.9-1), we obtain

$$\frac{\partial \psi}{\partial y} = v_x = x, \qquad \frac{\partial \psi}{\partial x} = -v_y = y. \tag{5.9-18}$$

Integrating these differential equations for the stream function gives

$$\psi(x, y) = xy + f(x), \qquad \psi(x, y) = xy + g(y), \tag{5.9-19}$$

where $f(x)$ and $g(y)$ are arbitrary functions. The two expressions for $\psi(x, y)$ in Eq. (5.9-19) correspond to the same function, so that we must have $f(x) = g(y) = $ constant. The value of that integration constant is arbitrary. Setting it equal to zero, we obtain

$$\psi = xy \tag{5.9-20}$$

as the equation for the streamlines. The choice made for the constant had the effect of assigning the value $\psi = 0$ to the streamlines corresponding to $x = 0$ or $y = 0$. This and the previous expression, Eq. (5.9-17), are identical if we set $c = \psi$. The streamlines for this flow are plotted in Fig. 5–11. The flow direction is downward and out to the sides.

5.10 NONDIMENSIONALIZATION AND SIMPLIFICATION OF THE NAVIER–STOKES EQUATION

The continuity and Navier–Stokes equations form a set of coupled, nonlinear, partial differential equations. The formidable problems encountered in solving the full nonlinear equations, whether in velocity or stream function form, have led to extensive investigations of simplified equations which are valid for specific geometric and/or dynamic conditions. The main types of approximations used for laminar flows are discussed in Chapters 6–8. As seen in Chapter 3, converting the governing equations to dimensionless form is usually a key step in identifying valid approximations. The purpose of this section is to introduce some of the scales and dimensionless groups which lead to special forms of the Navier–Stokes equation. The starting point is the dimensional form using the dynamic pressure,

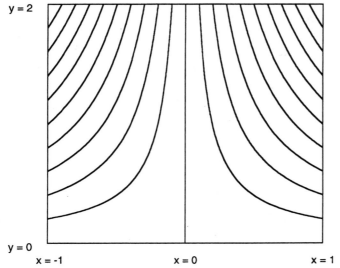

y = 2

y = 0

x = -1   x = 0   x = 1

Figure 5-11. Streamlines for the two-dimensional flow in Examples 5.9-1 and 5.9-2. The increment in ψ (or c) between any two streamlines is 0.2.

$$\rho\left(\frac{\partial \mathbf{v}}{\partial t}+\mathbf{v}\cdot\boldsymbol{\nabla}\mathbf{v}\right)=-\boldsymbol{\nabla}\mathcal{P}+\mu\nabla^2\mathbf{v}. \tag{5.10-1}$$

To facilitate the discussion of approximations, the material derivative of the velocity has been separated into its temporal and spatial parts.

Dimensionless Equations

The continuity and Navier–Stokes equations are nondimensionalized by introducing length, velocity, time, and pressure scales which are suitable for a given flow. The characteristic length (L) is a distance over which the velocity changes appreciably; in tube flow, for example, L is the tube radius or diameter. The characteristic velocity (U) is representative of the maximum velocity. The characteristic time (τ) and characteristic pressure or stress (Π) depend on the particular dynamic conditions, as discussed below. With dimensionless variables and differential operators defined as

$$\tilde{\mathbf{r}}=\frac{\mathbf{r}}{L}, \quad \tilde{\mathbf{v}}=\frac{\mathbf{v}}{U}, \quad \tilde{t}=\frac{t}{\tau}, \quad \tilde{\mathcal{P}}=\frac{\mathcal{P}}{\Pi}, \quad \tilde{\boldsymbol{\nabla}}=L\boldsymbol{\nabla}, \quad \tilde{\nabla}^2=L^2\nabla^2 \tag{5.10-2}$$

the continuity and Navier–Stokes equations become

$$\tilde{\boldsymbol{\nabla}}\cdot\tilde{\mathbf{v}}=0, \tag{5.10-3}$$

$$\mathrm{Re}\left(\frac{1}{\mathrm{Sr}}\frac{\partial\tilde{\mathbf{v}}}{\partial\tilde{t}}+\tilde{\mathbf{v}}\cdot\tilde{\boldsymbol{\nabla}}\tilde{\mathbf{v}}\right)=-\left(\frac{\Pi L}{\mu U}\right)\tilde{\boldsymbol{\nabla}}\tilde{\mathcal{P}}+\tilde{\nabla}^2\tilde{\mathbf{v}}. \tag{5.10-4}$$

The dimensionless continuity equation, Eq. (5.10-3), has the same form as its dimensional counterpart and requires no further comment. The dimensionless groups which appear now on the the left-hand side of the Navier–Stokes equation, Eq. (5.10-4), are the *Reynolds number*,[2]

$$\text{Re} \equiv \frac{UL\rho}{\mu} = \frac{UL}{\nu},$$

(5.10-5)

and the *Strouhal number*,

$$\text{Sr} \equiv \frac{\tau U}{L}.$$

(5.10-6)

The remainder of this section is devoted mainly to the significance of Re and Sr for the analysis of various types of flows.

Approximations Based on the Reynolds Number: Steady Flows

The physical significance of the Reynolds number is clearest for steady flows, for which Eq. (5.10-4) reduces to

$$\text{Re}(\tilde{\mathbf{v}} \cdot \tilde{\boldsymbol{\nabla}}\tilde{\mathbf{v}}) = -\left(\frac{\Pi L}{\mu U}\right)\tilde{\boldsymbol{\nabla}}\tilde{\mathscr{P}} + \tilde{\nabla}^2 \tilde{\mathbf{v}}.$$

(5.10-7)

The velocity terms in Eq. (5.10-7) fall into two categories. The nonlinear terms $(\tilde{\mathbf{v}} \cdot \tilde{\boldsymbol{\nabla}}\tilde{\mathbf{v}})$, which describe the fluid acceleration in a material reference frame, are called *inertial* terms. They may be interpreted also as representing *convective* transport of momentum (see section 5.3). The terms containing the second derivatives $(\tilde{\nabla}^2 \tilde{\mathbf{v}})$, which have their origin in the description of the viscous stresses, are referred to as the *viscous* terms. As discussed in Chapter 1, viscous transport of momentum is essentially a *diffusive* process. Because Re is the coefficient of the inertial terms in Eq. (5.10-7) and the viscous terms have a coefficient of unity, Re is evidently a measure of the relative importance of inertial and viscous effects in steady flows. Alternatively, it is a measure of the relative importance of the convective and diffusive transport of momentum. Inertial (or convective) effects tend to be prominent when Re >> 1 and nearly absent when Re << 1.

The pressure scales appropriate for various conditions are revealed by considering extreme values of Re. An approximation to Eq. (5.10-7) for Re →0 is

$$\mathbf{0} = -\left(\frac{\Pi L}{\mu U}\right)\tilde{\boldsymbol{\nabla}}\tilde{\mathscr{P}} + \tilde{\nabla}^2 \tilde{\mathbf{v}}.$$

(5.10-8)

If all variables are scaled properly, then the coefficient of the pressure gradient cannot be large or small (see Section 3.2). The most natural choice for that coefficient is unity, so that the pressure scale for small Re is

$$\Pi = \frac{\mu U}{L}.$$

(5.10-9)

---

[2]Osborne Reynolds (1842–1912) was a professor of engineering at Manchester, England. His experimental observations on water flow in pipes (Reynolds, 1883) established the importance of the dimensionless group which would be named in his honor and thereby provided the foundation for the concepts of dimensional analysis and dynamic similarity in fluid mechanics. As discussed in later chapters, Reynolds was also responsible for the basic theory of lubrication and for the averaging process which underlies the statistical analysis of turbulence, among other contributions.

This is called the *viscous pressure scale.* Substituting Eq. (5.10-9) in Eq. (5.10-8), we obtain

$$0 = -\tilde{\nabla}\tilde{\mathcal{P}} + \tilde{\nabla}^2\tilde{\mathbf{v}}, \qquad (5.10\text{-}10)$$

which is the dimensionless form of *Stokes' equation.* Stokes' equation is obtained also by setting $\rho = 0$ in Eq. (5.10-1); none of the scales involves the density. Equation (5.10-10) is the usual starting point for analyzing low Reynolds number flow, also called *creeping flow* or *Stokes flow.* As shown in Chapter 7, the linearity of this differential equation greatly facilitates the solution of problems involving small length and velocity scales and/or extremely viscous Newtonian fluids.

An approximation to Eq. (5.10-7) for $\mathrm{Re} \to \infty$ is

$$\mathrm{Re}(\tilde{\mathbf{v}} \cdot \tilde{\nabla}\tilde{\mathbf{v}}) = -\left(\frac{\Pi L}{\mu U}\right)\tilde{\nabla}\tilde{\mathcal{P}}, \qquad (5.10\text{-}11)$$

where the viscous terms have been neglected but the pressure term retained. The pressure term must be kept, in general, for any value of Re, because the magnitude of the pressure variations is essentially dictated by the velocity field. (Indeed, in many problems with incompressible fluids, $\mathcal{P}$ may be viewed simply as an additional degree of freedom needed to satisfy conservation of mass.) The pressure scale is identified by dividing both sides of Eq. (5.10-11) by Re and again making the coefficient of the pressure gradient unity. The result for Π is

$$\Pi = \mathrm{Re}\left(\frac{\mu U}{L}\right) = \rho U^2, \qquad (5.10\text{-}12)$$

which is called the *inertial pressure scale.* Substituting Eq. (5.10-12) into Eq. (5.10-11), we obtain

$$\tilde{\mathbf{v}} \cdot \tilde{\nabla}\tilde{\mathbf{v}} = -\tilde{\nabla}\tilde{\mathcal{P}}, \qquad (5.10\text{-}13)$$

which is the dimensionless momentum conservation equation for an *inviscid* fluid (sometimes called an *ideal* or *perfect* fluid) at steady state. This result is obtained also by setting $\mu = 0$ in the original Navier–Stokes equation; none of the scales involves the viscosity. Because Eq. (5.10-13) is a first-order differential equation and cannot satisfy all of the boundary conditions which apply to the original second-order equation, it is not a good approximation throughout any real fluid. Nonetheless, when used in conjunction with a boundary layer momentum equation, it is a key element in the modeling of high-speed flows, as discussed in Chapter 8.

When neither extreme of the Reynolds number applies, Eq. (5.10-7) can be rewritten using either pressure scale. The results are

$$\mathrm{Re}(\tilde{\mathbf{v}} \cdot \tilde{\nabla}\tilde{\mathbf{v}}) = -\tilde{\nabla}\tilde{\mathcal{P}} + \tilde{\nabla}^2\tilde{\mathbf{v}} \qquad (\Pi = \mu U/L), \qquad (5.10\text{-}14)$$

$$\mathrm{Re}(\tilde{\mathbf{v}} \cdot \tilde{\nabla}\tilde{\mathbf{v}} + \tilde{\nabla}\tilde{\mathcal{P}}) = \tilde{\nabla}^2\tilde{\mathbf{v}} \qquad (\Pi = \rho U^2). \qquad (5.10\text{-}15)$$

No terms were neglected here, so that both of these steady-state equations are exact. The viscous pressure scale is usually chosen for $\mathrm{Re} \sim 1$, leading to a preference for Eq. (5.10-14).

Approximations Based on the Strouhal Number: Unsteady Flows

In a time-dependent flow there are at least three time scales to consider: (1) the convective time scale, $\tau_c = L/U$, (2) the viscous time scale, $\tau_v = L^2/\nu$, and (3) the imposed time scale (if any), τ_i. The convective time scale is just the time required to travel a distance L at velocity U, whereas the viscous time scale is exactly analogous to the characteristic times for diffusion and conduction introduced in Section 3.4. The imposed time scale, present in some problems, is the characteristic time of a forcing function which affects the flow. Some examples are the mechanical time constants associated with turning a pump on or off, the period with which a surface is made to oscillate, and the period of a pulsatile pump such as the heart. Inspection of Eq. (5.10-6) reveals that Sr is the ratio of the actual or process time scale (τ) to the convective time scale. Likewise, Re is the ratio of the viscous time scale to the convective time scale. When the intrinsic rate of a process is large, its time constant is small. Thus, when the rate of viscous momentum transfer is very fast, τ_v will be small compared to τ_c and Re will be small. The relationships among Re, Sr, and the various time scales are summarized as

$$\text{Re} = \frac{\tau_v}{\tau_c}, \qquad \text{Sr} = \frac{\tau}{\tau_c}, \qquad \frac{\text{Re}}{\text{Sr}} = \frac{\tau_v}{\tau}. \qquad (5.10\text{-}16)$$

The Strouhal number is most meaningful when a forcing function is present and it is desired to model events on that time scale (i.e., when $\tau = \tau_i$). The most dramatic simplification in any time-dependent problem occurs when the system is pseudosteady (see Section 3.4). For Re→0, the time-dependent analog of Stokes' equation is

$$\frac{\text{Re}}{\text{Sr}} \frac{\partial \tilde{\mathbf{v}}}{\partial \tilde{t}} = -\tilde{\nabla}\tilde{\mathscr{P}} + \nabla^2 \tilde{\mathbf{v}}. \qquad (5.10\text{-}17)$$

Thus, when $\text{Re/Sr} = \tau_v/\tau_i \ll 1$, the imposed changes are slow relative to the viscous response and creeping flow is pseudosteady. Because creeping flows are characterized by very small viscous time scales, such flows are pseudosteady except when subjected to exceedingly rapid forcings. Accordingly, Stokes' equation ordinarily applies to time-dependent as well as steady flows at small Re. For Re→∞, the time-dependent analog of Eq. (5.10-13) is

$$\frac{1}{\text{Sr}} \frac{\partial \tilde{\mathbf{v}}}{\partial \tilde{t}} + \tilde{\mathbf{v}} \cdot \tilde{\nabla}\tilde{\mathbf{v}} = -\tilde{\nabla}\tilde{\mathscr{P}}. \qquad (5.10\text{-}18)$$

In this case the flow is pseudosteady if $1/\text{Sr} = \tau_c/\tau_i \ll 1$. In summary, pseudosteady behavior is exhibited by creeping flow if $\tau_i \gg \tau_v$ and by inviscid flow if $\tau_i \gg \tau_c$.
       If there is no imposed time scale, the process time scale is usually the convective time ($\tau = \tau_c$), in which case $\text{Sr} \equiv 1$. For example, consider a solid sphere, initially at rest, which is allowed to settle in a large volume of fluid. The only length scale is the radius (or diameter) of the sphere, and the velocity scale is clearly the terminal settling velocity. In analyzing the acceleration of the sphere to its terminal velocity, a reasonable time scale is the time required to fall one radius at that velocity.
       For $\text{Sr} = 1$, the time-dependent versions of Eqs. (5.10-14) and (5.10-15) are

$$\text{Re}\left(\frac{\partial \tilde{\mathbf{v}}}{\partial \tilde{t}} + \tilde{\mathbf{v}} \cdot \tilde{\nabla}\tilde{\mathbf{v}}\right) = -\tilde{\nabla}\tilde{\mathscr{P}} + \tilde{\nabla}^2\tilde{\mathbf{v}} \qquad (\Pi = \mu U/L), \qquad (5.10\text{-}19)$$

$$\text{Re}\left(\frac{\partial \tilde{\mathbf{v}}}{\partial \tilde{t}} + \tilde{\mathbf{v}} \cdot \tilde{\boldsymbol{\nabla}}\tilde{\mathbf{v}} + \tilde{\boldsymbol{\nabla}}\tilde{\mathscr{P}}\right) = \tilde{\nabla}^2\tilde{\mathbf{v}} \qquad (\Pi = \rho U^2). \qquad (5.10\text{-}20)$$

It is apparent that in such situations creeping flow will remain pseudosteady; that is, Eq. (5.10-19) will reduce once more to Stokes' equation. However, Eq. (5.10-20) indicates that no such simplification is possible for flows at moderate or large Re when Sr = 1.

References

Aris, R. *Vectors, Tensors, and the Basic Equations of Fluid Mechanics.* Prentice-Hall, Englewood Cliffs, NJ, 1962 (reprinted by Dover, New York, 1989).

Batchelor, G. K. *An Introduction to Fluid Dynamics.* Cambridge University Press, Cambridge,1970.

Bird, R. B., R. C. Armstrong, and O. Hassager. *Dynamics of Polymeric Liquids,* Vol. 1, second edition. Wiley, New York, 1987.

Dahler, J. S. and L. E. Scriven. Theory of structured continua. I. General consideration of angular momentum and polarization. *Proc. R. Soc. (Lond.)* A 275: 504–527, 1963.

Dugas, R. *A History of Mechanics.* Dover, New York, 1988. [This is a reprint of a book originally published in 1955.]

Edwards, D. A., H. Brenner, and D. T. Wasan. *Interfacial Transport Processes and Rheology.* Butterworth-Heinemann, Boston, 1991.

Leal, L. G. *Laminar Flow and Convective Transport Processes.* Butterworth-Heinemann, Boston, 1992.

Pearson, J. R. A. *Mechanics of Polymer Processing.* Elsevier, London, 1985.

Reynolds, O. An experimental investigation of the circumstances which determine whether the motion of water shall be direct or sinuous, and of the law of resistance in parallel channels. *Philos. Trans. R. Soc. Lond.* A 174: 935–982, 1883.

Rosensweig, R. E. *Ferrohydrodynamics.* Cambridge University Press, Cambridge, 1985.

Schlicting, H. *Boundary-Layer Theory,* sixth edition. McGraw-Hill, New York, 1968.

Scriven, L. E. Dynamics of a fluid interface. *Chem. Eng. Sci.* 12: 98–108, 1960.

Serrin, J. Mathematical principles of classical fluid mechanics. In *Encyclopedia of Physics (Handbuch der Physik),* Vol. 8, No. 1, S. Flügge and C. Truesdell, Eds. Springer-Verlag, Berlin, 1959, pp. 125–263.

Tanner, R. I. *Engineering Rheology.* Clarendon, Oxford, 1985.

Van Dyke, M. *An Album of Fluid Motion.* Parabolic Press, Stanford, CA, 1982.

Problems

5-1. Isotropic Property of Pressure

Given that pressure is the only stress in a static fluid and that it acts normal to all surfaces, show that it must be isotropic. That is, if at some point in a fluid P_x, P_y, and P_z are the pressures acting on imaginary test surfaces normal to $\mathbf{e}_x$, $\mathbf{e}_y$, and $\mathbf{e}_z$, respectively, and P is the pressure acting on a test surface with an arbitrary unit normal $\mathbf{n}$, show that $P_x = P_y = P_z = P$. (*Hint:* Use the stress tetrahedron shown in Fig. 5-4.)

5-2. Transport of Vorticity

The questions below pertain to an incompressible, Newtonian fluid.

(a) Show that in general (i.e., for a time-dependent, three-dimensional flow), the vorticity vector satisfies

$$\frac{D\mathbf{w}}{Dt} = \mathbf{w} \cdot \nabla \mathbf{v} + \nu \nabla^2 \mathbf{w}.$$

(*Hint:* Begin by taking the curl of each term in the Navier–Stokes equation.)

(b) Show that for any planar flow, the result in (a) reduces to

$$\frac{Dw}{Dt} = \nu \nabla^2 w.$$

Note that this vorticity equation is like the conservation equation for any other scalar quantity, but with no homogeneous source term. Thus, vorticity is created only at surfaces.

(c) Derive the result in (b) from the stream function equations.

5-3. Velocity and Pressure Calculations

Assume that for a certain steady, two-dimensional flow in a region defined by $x \geq 0$ and $y \geq 0$, one velocity component is given by

$$v_x(x, y) = C x^m,$$

where C and m are constants.

(a) Assuming that the fluid is incompressible and that $v_y(x, 0) = 0$, determine $v_y(x, y)$.
(b) For what values of m will the flow be irrotational?
(c) For what values of m can the Navier–Stokes equation be satisfied? For those cases determine the dynamic pressure, $\mathcal{P}(x, y)$, assuming that $\mathcal{P}(0, 0) = \mathcal{P}_0$.

5-4. Normal Viscous Stress at a Surface

Show that for an incompressible, Newtonian fluid, the normal component of the viscous stress vanishes at any impermeable, solid surface. It is sufficient to prove this for planar, cylindrical, and spherical surfaces. (*Hint:* Use the continuity equation.)

5-5. Soap Bubble

Soap contains surfactants which lower the surface tension of an air–water interface. The wall of a soap bubble consists of a thin film of water between two such interfaces. This problem is concerned with the equilibrium size of a spherical soap bubble, where R is the inner radius, ℓ is the thickness of the water film, and it is assumed that $\ell \ll R$. The pressure inside the bubble is P_B, the external pressure is P_0, and the surface tension is γ_s; all are assumed constant.

(a) Using the pointwise normal stress balance at an interface, Eq. (5.7-9), show that

$$P_B - P_0 = \frac{4\gamma_s}{R}.$$

This approach is like that used in Example 5.7-1.

(b) Show that the relation in (a), as well as the Young–Laplace equation, can be obtained also by performing a force balance on a hemispherical control volume corresponding to one-half of the bubble. (*Hint:* Use the results of Example 5.5-2.)

5-6. Capillary Rise

When one end of a small tube made of some wettable material (e.g., glass) is immersed in water, surface tension causes the water in the tube to rise above its level outside. Let R be the radius of the tube and let L represent the height at the bottom of the meniscus, relative to the outside water level. The surface tension is γ and the contact angle is α.

(a) Assuming that static pressure variations near the meniscus are negligible, show that

$$L = \frac{2\gamma \cos \alpha}{\rho_w g R},$$

where ρ_w is the density of water. This result will be valid provided that $L \gg R$.

(b) Suppose now that the condition $L \gg R$ does not hold, so that the expression in (a) is not a good approximation. Derive (but do not attempt to solve) the equations which govern the interface location for the general case. (The formula for the curvature of a surface with height expressed as $F(x, y)$ is given in Section A.8.)

(c) If it is desired to derive corrections to the result in (a), show that the general equations in (b) can be reformulated as a perturbation problem using the parameter

$$\varepsilon \equiv \frac{\rho_w g R^2}{\gamma}.$$

Show that the answer in (a) corresponds to the $O(1)$ solution to the perturbation problem.

5-7. Streamlines and Particle Paths

Assume that a transient, three-dimensional flow has velocity components given by

$$v_x = \frac{x}{1 + at}, \qquad v_y = \frac{y}{1 + bt}, \qquad v_z = \frac{z}{1 + ct},$$

where a, b, and c are constants. The objective here is to compare and contrast the streamlines in this flow with the paths of fluid particles.

(a) Find the equations governing the streamline which passes through the point $(1, 1, 1)$ at time t. Note that the other points included on this streamline change with time.

(b) Calculate the path of a particle which starts at $(1, 1, 1)$ at $t = 0$. Particle paths are governed by

$$\frac{d\mathbf{r}}{dt} = \mathbf{v}(\mathbf{r}, t).$$

In contrast to the calculation of streamlines, t is the variable used to trace the particle path; compare with Eq. (5.9-10). Determine the particle location at $t = 1$, denoted as $\mathbf{r}(1)$.

(c) Use the results of part (a) to show that unless $a = b = c$, neither the streamline for $t = 0$ nor that for $t = 1$ passes through the point $\mathbf{r}(1)$ found in part (b). This illustrates the fact that, in general, streamlines and particle paths do not coincide.

Chapter 6

UNIDIRECTIONAL AND NEARLY UNIDIRECTIONAL FLOW

6.1 INTRODUCTION

Two characteristics of fluid dynamics problems which are helpful in organizing solution strategies are the number of *directions* and the number of *dimensions*. As used here, the number of directions is the number of nonvanishing velocity components, whereas the number of dimensions is the number of spatial coordinates needed to describe the flow. Another pertinent characteristic, of course, is whether the flow is steady or time-dependent. In seeking solutions to flow problems involving incompressible Newtonian (or generalized Newtonian) fluids, it is the number of directions which tends to be the most crucial. There are relatively few exact, analytical solutions to the full Navier–Stokes equation, and most of those are for situations in which the flow is *unidirectional;* that is, there is only one nonzero velocity component. Some examples are: pressure-driven flow in a tube (Poiseuille flow); the flow in the gap between coaxial cylinders, one or both of which is rotating (Couette flow); and radial flow directed outward from a cylindrical or spherical surface. Whether or not a flow is unidirectional depends not just on the intrinsic properties of the motion but also on the coordinate system chosen; Couette flow, for instance, is unidirectional in cylindrical but not in rectangular coordinates. The same is true for the number of dimensions. With Poiseuille flow, for example, where the one nonzero velocity component is $v_z(r)$, it is the use of cylindrical coordinates which makes the problem two-dimensional (r, z) instead of three-dimensional (x, y, z).

The first part of this chapter illustrates exact solutions which are obtainable for unidirectional flow. The three main types of situations considered are steady flow due to a pressure gradient, steady flow due to a moving surface, and time-dependent flow due to either a pressure gradient or a moving surface. The remainder of the chapter is concerned with the limitations of those results in modeling actual flows and is also concerned with the extensions of such models to a broader class of situations. The limitations arise from entrance and edge effects and also from stability considerations. The extensions involve viscous flows which are *nearly unidirectional,* in the sense that a second velocity component, while not identically zero, is small. An example is pressure-driven flow in a gradually tapered channel. The analysis of such problems proceeds much like that for unidirectional flow, by exploiting what is termed the *lubrication approximation.* Including in the scope of our analysis nearly unidirectional as well as exactly unidirectional flow greatly increases the number of real situations which can be modeled.

The numbers of directions and dimensions and the presence or absence of time-dependence are not the only characteristics which define classes of flow problems. As mentioned in Section 5.10, another key consideration is the relative importance of inertial and viscous effects. In most of the problems in this chapter, including the nearly unidirectional flows, the viscous terms in the momentum equations are dominant and inertia is unimportant. The general features of viscous and inertially dominated flows are the focus of Chapters 7 and 8, respectively.

6.2 STEADY FLOW WITH A PRESSURE GRADIENT

The flow of an incompressible fluid in a tube or other channel of constant cross section is said to be *fully developed* if the velocity does not vary in the direction of flow (i.e., it is independent of axial position). This occurs only beyond a certain distance from the tube inlet, called the *entrance length.* The entrance length depends on the tube diameter and Reynolds number, as discussed in Section 6.5. Fully developed flows are ordinarily unidirectional; indeed, they are the most important type of unidirectional flow. The general features of such flows are discussed first, followed by specific examples.

Consider the flow of an incompressible fluid in a tube of arbitrary cross-sectional shape, as depicted in Fig. 6-1. The axial coordinate is x and the velocity field is assumed to be fully developed, so that $\mathbf{v} = \mathbf{v}(y, z, t)$ only. It is desired to identify the conditions for which the flow will be unidirectional (i.e., for which $v_y = v_z = 0$). Given that $\partial v_x / \partial x = 0$, the continuity equation for this three-dimensional situation reduces to

$$\frac{\partial v_y}{\partial y} + \frac{\partial v_z}{\partial z} = 0. \tag{6.2-1}$$

To examine the possibility of flow in the y–z plane we define a stream function $\psi(y, z, t)$ as

$$v_y \equiv \frac{\partial \psi}{\partial z}, \qquad v_z \equiv -\frac{\partial \psi}{\partial y}. \tag{6.2-2}$$

An impermeable, solid surface is always a streamline (Section 5.9), so that at the tube wall ψ is at most a function of time. Moreover, it follows from Eq. (5.9-13) that

$$\frac{\partial^2 \psi}{\partial y^2} + \frac{\partial^2 \psi}{\partial z^2} = -w_x. \tag{6.2-3}$$

Figure 6-1. Flow in a tube of constant but arbitrary cross section. (a) Section normal to the direction of flow; (b) section parallel to the direction of flow.

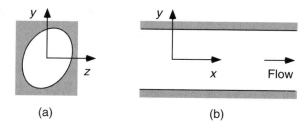

(a)

(b)

Accordingly, if the axial component of the vorticity vanishes (i.e., $w_x=0$), then ψ must satisfy a two-dimensional form of Laplace's equation and must have a value that is independent of position on the boundary. It follows from Example 4.2-1 that ψ is uniform throughout the tube cross section and, from Eq. (6.2-2), that $v_y=v_z=0$. In other words, if there is no rotational (swirling) component of the velocity, a fully developed flow in a tube of any cross-sectional shape will be unidirectional.

The Navier–Stokes equation written in terms of the dynamic pressure is

$$\rho\left(\frac{\partial \mathbf{v}}{\partial t}+\mathbf{v}\cdot\nabla\mathbf{v}\right)=-\nabla\mathcal{P}+\mu\nabla^2\mathbf{v}. \tag{6.2-4}$$

With $v_y=v_z=0$, all terms in the y and z components of this equation vanish, indicating that $\partial\mathcal{P}/\partial y=\partial\mathcal{P}/\partial z=0$ or that $\mathcal{P}=\mathcal{P}(x, t)$ only. Of importance, the inertial terms in the x component also vanish. That is, $\mathbf{v}\cdot\nabla v_x=0$ for this fully developed flow. The x component of Eq. (6.2-4) is reduced then to a *linear* partial differential equation, which for unsteady flow is

$$\rho\frac{\partial v_x}{\partial t}=-\frac{\partial\mathcal{P}}{\partial x}+\mu\left(\frac{\partial^2 v_x}{\partial y^2}+\frac{\partial^2 v_x}{\partial z^2}\right). \tag{6.2-5}$$

Because $v_x=v_x(y, z, t)$ and $\mathcal{P}=\mathcal{P}(x, t)$, $\partial\mathcal{P}/\partial x$ can depend at most on time. In other words, none of the other terms in Eq. (6.2-5) depend on x, so that $\partial\mathcal{P}/\partial x$ cannot be a function of x either. For steady, fully developed flow we obtain

$$0=-\frac{d\mathcal{P}}{dx}+\mu\left(\frac{\partial^2 v_x}{\partial y^2}+\frac{\partial^2 v_x}{\partial z^2}\right). \tag{6.2-6}$$

Here $\mathcal{P}=\mathcal{P}(x)$ only, so that $d\mathcal{P}/dx$ must be a constant.

Equations (6.2-5) and (6.2-6) are the starting points for analyses of the fully developed flow of Newtonian fluids. The simplifications in the Cauchy momentum equation for fully developed flow are similar. Applications of Eq. (6.2-6) and the analogous momentum equation for cylindrical coordinates are illustrated by the examples which follow.

Example 6.2-1 Flow in a Parallel-Plate Channel   The objective is to determine the steady, fully developed velocity field for pressure-driven flow in a channel with flat, parallel walls, as shown in Fig. 6–2. This is called *plane Poiseuille flow*. It is assumed that the z dimension is large, so that $v_x=v_x(y)$ only; see Example 6.5-2 for a discussion of edge effects in channels of rectangular cross section. Equation (6.2-6) reduces to

$$\frac{d^2 v_x}{dy^2}=\frac{1}{\mu}\frac{d\mathcal{P}}{dx} \tag{6.2-7}$$

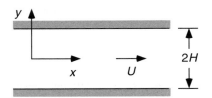

Figure 6-2. Flow in a parallel-plate channel with wall spacing $2H$ and mean velocity U.

with no-slip boundary conditions given by

$$v_x(\pm H)=0. \tag{6.2-8}$$

The integration of Eq. (6.2-7) is simplified by the fact that $d\mathcal{P}/dx$ is a constant. Integrating twice and evaluating the two integration constants using Eq. (6.2-8) gives

$$v_x(y)=-\frac{H^2}{2\mu}\frac{d\mathcal{P}}{dx}\left[1-\left(\frac{y}{H}\right)^2\right]. \tag{6.2-9}$$

The great advantage in using $\mathcal{P}$ instead of P is that Eq. (6.2-9) is valid for a channel with any spatial orientation. If the channel is horizontal (i.e., $g_x=0$), then $d\mathcal{P}/dx=dP/dx$; if the channel is vertical and there is no applied pressure (i.e., $dP/dx=0$), then $d\mathcal{P}/dx=-\rho g$. Note too that the same parabolic velocity profile results if one of the no-slip boundary conditions is replaced with a symmetry condition at $y=0$, namely $\tau_{yx}=0$ or $dv_x/dy=0$.

The mean velocity in the channel (U) is related to the pressure drop by

$$U=\frac{1}{H}\int_0^H v_x\,dy=-\frac{H^2}{3\mu}\frac{d\mathcal{P}}{dx}, \tag{6.2-10}$$

so that Eq. (6.2-9) may be rewritten as

$$v_x(y)=\frac{3}{2}U\left[1-\left(\frac{y}{H}\right)^2\right]. \tag{6.2-11}$$

It is seen that the maximum (center-plane) velocity is $\frac{3}{2}U$. The volumetric flow rate per unit width of channel is

$$q=2UH=-\frac{2H^3}{3\mu}\frac{d\mathcal{P}}{dx}. \tag{6.2-12}$$

Example 6.2-2 Flow of a Power-Law Fluid in a Circular Tube   Consider steady flow of a power-law fluid in a cylindrical tube of radius R. For this fully developed, axisymmetric flow we have $v_z=v_z(r)$, $v_r=v_\theta=0$, and $\mathcal{P}=\mathcal{P}(z)$. With such a velocity field, Γ_{rz} (or Γ_{zr}) is the only nonvanishing component of the rate-of-deformation tensor. It follows from Eq. (5.6-3) that, for a Newtonian or generalized Newtonian fluid, τ_{rz} (or τ_{zr}) is the only nonzero component of the viscous stress, and that $\tau_{rz}=\tau_{rz}(r)$ only. From the z component of the Cauchy momentum equation we obtain

$$\frac{1}{r}\frac{d}{dr}(r\tau_{rz})=\frac{d\mathcal{P}}{dz}. \tag{6.2-13}$$

With $\mathcal{P}=\mathcal{P}(z)$ and $\tau_{rz}=\tau_{rz}(r)$, Eq. (6.2-13) is satisfied only if both sides equal a constant; that is, $d\mathcal{P}/dz$ must be constant. Integrating, it is found that

$$r\tau_{rz}=\frac{d\mathcal{P}}{dz}\frac{r^2}{2}+C_1, \tag{6.2-14}$$

where C_1 is a constant. A symmetry argument like that used to derive Eq. (2.2-22) leads to the conclusion that $\tau_{rz}=0$ at $r=0$, so that $C_1=0$ and

$$\tau_{rz}=\frac{d\mathcal{P}}{dz}\frac{r}{2}. \tag{6.2-15}$$

The fact that there will be a pressure drop in the direction of flow (i.e., $d\mathcal{P}/dz<0$) indicates that $\tau_{rz}\leq0$.

For a power-law fluid it is found from Eq. (5.6-13) and Table 5-10 that

$$\tau_{rz}=-m\left|\frac{dv_z}{dr}\right|^n. \tag{6.2-16}$$

The minus sign in Eq. (6.2-16) is required because $\tau_{rz}\leq0$, and $|dv_z/dr|$ is used because $dv_z/dr\leq0$. Combining Eqs. (6.2-15) and (6.2-16) and solving for dv_z/dr gives

$$\frac{dv_z}{dr}=-\left(-\frac{1}{2m}\frac{d\mathcal{P}}{dz}\right)^{1/n}r^{1/n}. \tag{6.2-17}$$

The solution is completed by integrating Eq. (6.2-17) and using the no-slip condition, $v_z(R)=0$, to evaluate the integration constant. The result is

$$v_z(r)=\left(\frac{3n+1}{n+1}\right)U\left[1-\left(\frac{r}{R}\right)^{(n+1)/n}\right], \tag{6.2-18}$$

$$U=\frac{2}{R^2}\int_0^R v_z r\,dr=\left(\frac{n}{3n+1}\right)\left(-\frac{R^{n+1}}{2m}\frac{d\mathcal{P}}{dz}\right)^{1/n}, \tag{6.2-19}$$

where U is the mean velocity. The shape of the velocity profile in Eq. (6.2-18) is determined by the power-law exponent, n. For $n\to0$, the centerline velocity approaches the mean velocity, approximating plug flow. (As indicated in Chapter 5, polymeric fluids exhibit values of n as small as 0.2.) In general, the volumetric flow rate is given by

$$Q=\pi R^2 U=\left(\frac{n\pi}{3n+1}\right)\left(-\frac{R^{3n+1}}{2m}\frac{d\mathcal{P}}{dz}\right)^{1/n}. \tag{6.2-20}$$

For a Newtonian fluid, where $n=1$ and $m=\mu$, Eqs. (6.2-18)–(6.2-20) yield the well-known results for *Poiseuille flow*,

$$v_z(r)=2U\left[1-\left(\frac{r}{R}\right)^2\right], \tag{6.2-21}$$

$$U=-\frac{R^2}{8\mu}\frac{d\mathcal{P}}{dz}, \tag{6.2-22}$$

$$Q=-\frac{\pi R^4}{8\mu}\frac{d\mathcal{P}}{dz}. \tag{6.2-23}$$

Thus, the velocity profile for a Newtonian fluid in a circular tube is parabolic, as for a parallel-plate channel, but the maximum (centerline) velocity in this case is $2U$. Equation (6.2-23) is one form of *Poiseuille's law*.[1]

[1] This type of flow is named after Jean Poiseuille (1797–1869), a French physician whose interest in the mechanics of blood flow led him to conduct an extensive set of pressure and flow measurements in Paris in the 1830s and 1840s. He employed water and other liquids in glass capillary tubes, and was extremely meticu-

Example 6.2-3 Flow of Two Immiscible Fluids in a Parallel-Plate Channel A simple type of two-phase flow occurs when immiscible fluids occupy distinct layers in a parallel-plate channel, as depicted in Fig. 6–3. The density and viscosity of fluid 1 (ρ_1 and μ_1) may differ from those of fluid 2 (ρ_2 and μ_2). It is desired to determine the steady, fully developed velocities of the two fluids, which are denoted as $v_x^1(y)$ and $v_x^2(y)$, respectively.

The Navier–Stokes equation for each phase reduces to

$$\frac{d^2 v_x^{(i)}}{dy^2} = \frac{1}{\mu_i}\frac{d\mathcal{P}^{(i)}}{dx}. \tag{6.2-24}$$

Integrating this twice gives

$$v_x^{(i)}(y) = \frac{1}{\mu_i}\frac{d\mathcal{P}^{(i)}}{dx}\frac{y^2}{2} + a_i y + b_i, \tag{6.2-25}$$

where a_i and b_i are constants. These four constants are determined by the conditions

$$v_x^{(1)}(H_1) = 0, \tag{6.2-26}$$

$$v_x^{(2)}(-H_2) = 0, \tag{6.2-27}$$

$$v_x^{(1)}(0) = v_x^{(2)}(0), \tag{6.2-28}$$

$$\mu_1\frac{dv_x^{(1)}}{dy}(0) = \mu_2\frac{dv_x^{(2)}}{dy}(0). \tag{6.2-29}$$

Equations (6.2-26) and (6.2-27) are the usual no-slip conditions at the solid surfaces, whereas Eqs. (6.2-28) and (6.2-29) express the matching of the tangential components of velocity and stress at the fluid–fluid interface. It is convenient to introduce the constants

$$u_i \equiv -\frac{H_i^2}{2\mu_i}\frac{d\mathcal{P}^{(i)}}{dx}, \tag{6.2-30}$$

$$K \equiv \left(\frac{\mu_1}{\mu_2}\right)\left(\frac{H_2}{H_1}\right), \tag{6.2-31}$$

where u_i has units of velocity [compare with Eq. (6.2-9)] and K is dimensionless. The velocities in the two fluids are written as

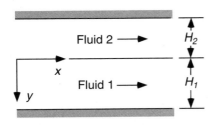

Figure 6-3. Flow of two immiscible fluids in a parallel-plate channel.

lous. At about the same time, the dependence of Q on R^4 was revealed also by a less extensive and precise set of measurements made in Berlin by a German engineer, Gotthilf Hagen (1797–1884). Thus, Eq. (6.2-23) is sometimes called the *Hagen-Poiseuille equation*. The first derivation of this result from the Navier–Stokes equation is attributed to the physicist Eduard Hagenbach (1833–1910) of Basel in 1860, although several others appear to have done this at about the same time. Incidentally, Poiseuille also invented the U-tube mercury manometer, to measure arterial pressures in animals (Sutera and Skalak, 1993).

$$v_x^{(1)}(y) = u_1 \left[1 - \left(\frac{y}{H_1} \right)^2 \right] + \left(\frac{u_2 - u_1}{1 + K} \right) \left[1 - \left(\frac{y}{H_1} \right) \right],$$

(6.2-32)

$$v_x^{(2)}(y) = u_1 \left[1 - \left(\frac{u_2}{u_1} \right) \left(\frac{y}{H_2} \right)^2 \right] + \left(\frac{u_2 - u_1}{1 + K} \right) \left[1 - K \left(\frac{y}{H_2} \right) \right],$$

(6.2-33)

where u_i and K are assumed to be known.

What remains is to relate the pressures in the two fluids. In Examples 6.2-1 and 6.2-2 it was seen that for fully developed flow of a single, incompressible fluid in a channel of known dimensions, specifying the mean velocity was the same as setting the axial gradient of the dynamic pressure [see Eqs. (6.2-10), (6.2-19), and (6.2-22)]. Extra care is needed in the present problem, because $d\mathcal{P}^{(i)}/dx$, while constant within each fluid, is generally not the same in the two phases. The constraints on the two-phase flow are revealed by considering the actual pressure, P. For the general case of a channel inclined at an arbitrary angle, $P^{(i)} = P^{(i)}(x, y, z)$ even in the absence of flow, because of static pressure variations. However, from the definition of the dynamic pressure, Eq. (5.8-1), it follows that for this flow

$$\frac{\partial P^{(i)}}{\partial x} = \frac{d\mathcal{P}^{(i)}}{dx} + \rho_i g_x.$$

(6.2-34)

Thus, the constancy of $d\mathcal{P}^{(i)}/dx$ implies that $\partial P^{(i)}/\partial x$ too is constant within each phase. The final piece of information needed comes from the normal stress balance at the fluid–fluid interface, based on Eq. (5.7-9). Given that the interface is flat and that there are no normal viscous stresses (i.e., $\tau_{yy} = 0$), the values of P there must match. We conclude that $\partial P/\partial x$ has the same constant value throughout both fluids. To emphasize that there is only one independent pressure gradient, Eq. (6.2-30) is rewritten as

$$u_i = -\frac{H_i^2}{2\mu_i} \left(\frac{\partial P}{\partial x} - \rho_i g_x \right).$$

(6.2-35)

The need to consider actual pressure, and not just dynamic pressure, is typical of problems involving fluid–fluid interfaces.

Example 6.2-4 Flow of a Liquid Film Down an Inclined Surface   With reference to Fig. 6-4, the objective is to determine the velocity of a liquid film flowing down a surface which is oriented at an angle β relative to vertical. The film thickness (H) is assumed to be constant, making it possible to have steady, fully developed flow. Thus, it is assumed that $v_x = v_x(y)$ only and $v_y = v_z = 0$, from which it follows (as before) that $\mathcal{P} = \mathcal{P}(x)$ only and that $d\mathcal{P}/dx$ is constant.

The falling liquid film can be viewed as a special case of the two-fluid problem in Example 6.2-3, in which fluid 1 is now the liquid and fluid 2 the gas. Assuming that the gas occupies a

Figure 6-4. Flow of a liquid film down an inclined surface.

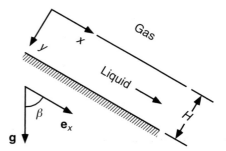

space at least as thick as the liquid film, and recalling that a typical ratio of liquid to gas viscosities is $\sim 10^2$ (Chapter 1), the parameter K defined by Eq. (6.2-31) will be extremely large. Assuming also that the gas pressure is uniform, the reasoning leading to Eq. (6.2-35) indicates that $\partial P/\partial x = 0$ in the liquid. Noting that $g_x = g \cos \beta$, it follows from Eqs. (6.2-32) and (6.2-35) that the liquid velocity is

$$v_x(y) = \frac{H^2 \rho_L g \cos \beta}{2\mu_L}\left[1 - \left(\frac{y}{H}\right)^2\right] \tag{6.2-36}$$

where the subscript L denotes liquid properties. The velocity profile can be rewritten in terms of the mean velocity (U) as

$$v_x(y) = \frac{3}{2}U\left[1 - \left(\frac{y}{H}\right)^2\right], \tag{6.2-37}$$

$$U = \frac{H^2 \rho_L g \cos \beta}{3\mu_L}. \tag{6.2-38}$$

Notice that Eq. (6.2-37) is exactly the same as the result obtained in Example 6.2-1 for flow in a parallel-plate channel of half-width H.

Assuming that $K \to \infty$, as done in deriving Eq. (6.2-36), is the same as neglecting the shear stress exerted by the gas on the liquid. As discussed in Section 5.7, this is a common approximation at gas–liquid interfaces, and it has the effect of making the liquid velocity independent of the gas properties. It is readily confirmed that Eq. (6.2-36) is obtained also by solving

$$\frac{d^2v_x}{dy^2} = \frac{1}{\mu_L}\frac{d\mathcal{P}}{dx} = -\frac{\rho_L g \cos \beta}{\mu_L} \tag{6.2-39}$$

with the boundary conditions

$$\frac{dv_x}{dy}(0) = 0, \qquad v_x(H) = 0. \tag{6.2-40}$$

This is clearly the preferred approach if one is interested only in the liquid, in that it avoids having to determine the velocity field in the gas.

6.3 STEADY FLOW WITH A MOVING SURFACE

Flow is induced not just by a gradient in dynamic pressure, but also by motion at a solid or fluid interface. Flows caused by the tangential movement of a surface provide the purest illustrations of surface-driven flow, in that there need not be any pressure gradient in the flow direction. The two examples in this section, both of which entail rotation of a solid surface, fall in that category. The main feature that distinguishes these rotational flows from the other unidirectional flows in this chapter is that the streamlines are curved, implying that all material points are undergoing spatial accelerations. A pressure gradient normal to the direction of flow (i.e., $\partial \mathcal{P}/\partial r > 0$) is needed to sustain such accelerations.

Example 6.3-1 Couette Flow Steady flow in the annular gap between long, coaxial cylinders, one or both of which is rotated at a constant angular velocity, is termed *Couette flow*. In a Couette viscometer, as shown in Fig. 6-5(a), the inner cylinder is fixed and the outer one is rotated. The

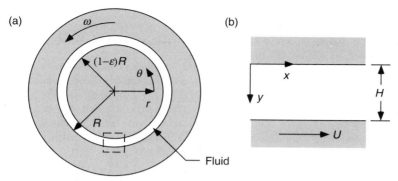

Figure 6-5. Couette viscometer. The outer cylinder is rotated at angular velocity ω and the inner cylinder is fixed. (a) Overall view; (b) enlargement of area indicated by dashed rectangle in (a) for $\varepsilon \ll 1$, with $H = \varepsilon R$ and $U = \omega R$.

fluid-filled gap is of thickness εR and the linear velocity at the wetted surface of the outer cylinder is ωR. Noting the circular symmetry and neglecting end effects, it is assumed that $\mathbf{v}$ and $\mathcal{P}$ do not depend on θ or z. Similar to fully developed flow in a channel, a velocity field of the form $v_\theta = v_\theta(r)$ and $v_r = v_z = 0$ is consistent with continuity.

From the θ component of the Navier–Stokes equation it follows that

$$0 = \frac{d}{dr}\left[\frac{1}{r}\frac{d}{dr}(rv_\theta)\right].$$

(6.3-1)

Integrating Eq. (6.3-1) twice yields terms in v_θ involving r^{-1} and r. Evaluating the integration constants using the no-slip boundary conditions,

$$v_\theta[(1-\varepsilon)R] = 0, \qquad v_\theta[R] = \omega R,$$

(6.3-2)

we obtain

$$v_\theta(r) = \omega R \frac{\left[(1-\varepsilon)^2\left(\frac{R}{r}\right) - \left(\frac{r}{R}\right)\right]}{[(1-\varepsilon)^2 - 1]}.$$

(6.3-3)

The r component of momentum serves only to determine $\mathcal{P}(r)$ from v_θ, and therefore it provides no additional information concerning the velocity. The z component only confirms that for the assumed form of $\mathbf{v}$, $\mathcal{P}$ must be independent of z.

Couette viscometers are normally designed with narrow gaps, which leads to a simpler velocity profile. For $\epsilon \ll 1$ the effects of surface curvature will be negligible, as in the first approximation to the concentration field in Example 3.7-2. Reformulating the problem using local rectangular coordinates, as shown in Fig. 6–5(b), conservation of momentum gives

$$0 = \frac{d^2 v_x}{dy^2}.$$

(6.3-4)

The velocity profile for small gaps is then simply

$$v_x = \frac{Uy}{H},$$

(6.3-5)

where $U \equiv \omega R$ and $H \equiv \varepsilon R$. Couette flow for planar surfaces, as given by Eq. (6.3-5), is referred to as *plane Couette flow*. An identical, linear velocity profile is obtained (more laboriously) by expanding the numerator and denominator of Eq. (6.3-3) in Taylor series about $\varepsilon=0$. Local rectangular coordinates find extensive use in lubrication theory (Section 6.6) and boundary layer theory (Chapter 8). As seen here, they are appropriate when the smallest radius of curvature of the bounding surface(s) greatly exceeds the other length scales governing the flow.

Measurements of the torque required to turn the outer cylinder (or to keep the inner cylinder stationary) provide an effective method to determine the viscosity of the fluid. The magnitude of the torque exerted on the fluid by the outer cylinder is $G=(2\pi R^2 L)\tau_{r\theta}(R)$, where L is the wetted length of the cylinders. The shear stress, $\tau_{r\theta}=\mu r d(v_\theta/r)/dr$ (Table 5-5), is determined in general by differentiating Eq. (6.3-3). For narrow gaps, however, $\tau_{r\theta}\cong\tau_{yx}=\mu dv_x/dy$ computed from Eq. (6.3-5). An important property of plane Couette flow is that the shear stress is independent of position, or $\tau_{yx}=\mu U/H$. Using $\tau_{r\theta}=\tau_{yx}=\mu\omega/\varepsilon$, the torque for $\varepsilon\ll1$ is

$$G=\frac{2\pi R^2 L\mu\omega}{\varepsilon}. \tag{6.3-6}$$

Example 6.3-2 Surface of a Liquid in Rigid-Body Rotation Figure 6-6 shows an example of a system with a gas–liquid interface of unknown shape, consisting of a liquid in an open container of radius R that is rotated at an angular velocity ω. If the container is rotated long enough, a steady state is reached in which the liquid is in rigid-body rotation. It is desired to determine the steady-state interface height, $h(r)$, assuming that the ambient air is at a constant pressure, P_0. Because the viscous stress vanishes for rigid-body rotation, the analysis will apply to any liquid, Newtonian or non-Newtonian. The effects of surface tension will be neglected.

The velocity of a fluid in rigid-body rotation is the same as that in a rotating solid, or

$$v_\theta(r)=\omega r \tag{6.3-7}$$

with $v_r=v_z=0$. Because of the free surface, P will be used as the pressure variable instead of $\mathcal{P}$. The circular symmetry indicates that $P=P(r, z)$ only; the θ component of the Cauchy momentum equation (Table 5-2) confirms this. Given that $g_r=0$ and $g_z=-g$, the r and z components of conservation of momentum become

$$-\frac{\rho v_\theta^2}{r}=-\frac{\partial P}{\partial r}, \tag{6.3-8}$$

$$0=-\rho g-\frac{\partial P}{\partial z}. \tag{6.3-9}$$

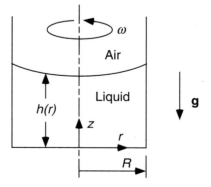

Figure 6-6. A liquid in rigid-body rotation in an open container.

Using Eq. (6.3-7) in Eq. (6.3-8) and integrating, the pressure is found to be of the form

$$P(r, z) = \frac{\rho \omega^2 r^2}{2} + g(z),$$ (6.3-10)

where $g(z)$ is an unknown function. Integrating Eq. (6.3-9), another expression for the pressure is

$$P(r, z) = f(r) - \rho g z,$$ (6.3-11)

where $f(r)$ is an unknown function. The two unknown functions are identified by comparing Eqs. (6.3-10) and (6.3-11), leading to the conclusion that

$$P(r, z) = \frac{\rho \omega^2 r^2}{2} - \rho g z + c,$$ (6.3-12)

where c is an unknown constant.

To determine the height of the interface, we note that with viscous stresses absent and surface tension assumed to be negligible, the normal stress balance requires that the liquid pressure at the interface equal P_0. Setting $z = h(r)$ and $P = P_0$ in Eq. (6.3-12) gives

$$h(r) = \frac{\omega^2 r^2}{2g} + \frac{(c - P_0)}{\rho g}.$$ (6.3-13)

Thus, the interface is parabolic in shape, with the lowest point at the center. The additional condition needed to determine c is a specification of the total volume of fluid in the container, or

$$V = 2\pi \int_0^R h(r) r \, dr.$$ (6.3-14)

Using Eq. (6.3-13) in (6.3-14) to calculate c, the final result is

$$h(r) = h_0 + \frac{\omega^2 R^2}{2g} \left[\left(\frac{r}{R} \right)^2 - \frac{1}{2} \right],$$ (6.3-15)

where $h_0 \equiv V/(\pi R^2)$ is the liquid height under static conditions.

6.4 TIME-DEPENDENT FLOW

Exact solutions for a number of time-dependent, unidirectional flow problems can be found by using the similarity method (Chapter 3) or the finite Fourier transform method (Chapter 4), as illustrated by the two examples in this section.

Example 6.4-1 Flow Near a Flat Plate Suddenly Set in Motion   A flat plate at $y = 0$ is in contact with a Newtonian fluid, initially at rest, which occupies the space $y > 0$. At $t = 0$ the plate is suddenly set in motion in the x direction at a velocity U, and that plate velocity is maintained indefinitely. The objective is to determine the fluid velocity as a function of time and position. This problem may be viewed, for example, as representing the early time (or penetration) phase in the start-up of a Couette viscometer.

Assuming that $v_x = v_x(y, t)$ only and that $\partial \mathcal{P}/\partial x = 0$, the problem formulation is

$$\frac{\partial v_x}{\partial t} = \nu \frac{\partial^2 v_x}{\partial y^2}, \tag{6.4-1}$$

$$v_x(y, 0) = 0, \qquad v_x(0, t) = U, \qquad v_x(\infty, t) = 0, \tag{6.4-2}$$

where $\nu \equiv \mu/\rho$ is the kinematic viscosity. Setting $\Theta = v_x/U$ and replacing D with ν, this is the same as the transient diffusion problem solved using the similarity method in Section 3.5. Thus, with the similarity variable defined as

$$\eta \equiv \frac{y}{2(\nu t)^{1/2}} \tag{6.4-3}$$

the solution is

$$\frac{v_x}{U} = \operatorname{erfc}(\eta). \tag{6.4-4}$$

The analogy between this problem and the transient diffusion problem of Section 3.5 underscores the fact that the kinematic viscosity is the diffusivity for momentum transfer. Thus, as mentioned in the discussion of the Strouhal number in Section 5.10, the characteristic time for viscous momentum transfer over a distance L is

$$t_V = \frac{L^2}{\nu}. \tag{6.4-5}$$

It follows that the time to achieve steady flow in the Couette viscometer of Example 6.3-1 is $\sim H^2/\nu$, or $\sim(\varepsilon R)^2/\nu$.

Example 6.4-2 Pulsatile Flow in a Circular Tube   It is desired to determine the velocity of a Newtonian fluid in a circular tube of radius R, when the fluid is subjected to a time-periodic pressure gradient. It is assumed that the fluid is initially at rest and that for $t>0$ the axial pressure gradient is given by

$$\frac{\partial \mathcal{P}}{\partial z} = -\rho a(1 + \gamma \sin \omega t), \tag{6.4-6}$$

where a, γ, and ω are positive constants. The time-averaged pressure gradient is $-\rho a$, where a has units of m s^{-2}; γ is dimensionless and ω has units of s^{-1}.

For a unidirectional flow with $v_z = v_z(r, t)$ and $v_r = v_\theta = 0$, the dimensional problem to be solved is

$$\frac{\partial v_z}{\partial t} = a(1 + \gamma \sin \omega t) + \frac{\nu}{r} \frac{\partial}{\partial r}\left(r \frac{\partial v_z}{\partial r}\right), \tag{6.4-7}$$

$$v_z(r, 0) = 0, \qquad v_z(R, t) = 0. \tag{6.4-8}$$

As will be seen, a sufficient condition at the tube centerline is that $v_z(0, t)$ be finite.

It is advantageous here to use dimensionless variables. The length, time, and velocity scales are chosen as the tube radius (R), the imposed time scale (ω^{-1}), and the mean velocity corresponding to the time-averaged pressure gradient (U_0), respectively. From Eqs. (6.2-22) and (6.4-6) we find that $U_0 = R^2 a/(8\nu)$. With the dimensionless variables defined as

$$\eta \equiv \frac{r}{R}, \qquad \tau \equiv \omega t, \qquad \Theta \equiv \frac{8\nu}{R^2 a} v_z \tag{6.4-9}$$

the problem is rewritten as

$$\beta\frac{\partial\Theta}{\partial\tau}=\frac{1}{\eta}\frac{\partial}{\partial\eta}\left(\eta\frac{\partial\Theta}{\partial\eta}\right)+8(1+\gamma\sin\,\tau), \qquad (6.4\text{-}10)$$

$$\Theta(\eta,\,0)=0, \qquad \Theta(1,\,\tau)=0, \qquad (6.4\text{-}11)$$

$$\beta\equiv\frac{\omega R^2}{\nu}. \qquad (6.4\text{-}12)$$

The parameter β is the ratio of the viscous time scale (R^2/ν) to the imposed time scale (ω^{-1}). Pseudosteady behavior is expected for $\beta\ll1$ (see Section 5.10).

In applying the finite Fourier transform method the solution is expanded as

$$\Theta(\eta,\,\tau)=\sum_{n=1}^{\infty}\Theta_n(\tau)\Phi_n(\eta), \qquad (6.4\text{-}13)$$

$$\Phi_n(\eta)=\sqrt{2}\,\frac{J_0(\lambda_n\eta)}{J_1(\lambda_n)}, \qquad J_0(\lambda_n)=0, \qquad (6.4\text{-}14)$$

$$\Theta_n(\tau)=\langle\Theta,\,\Phi_n\rangle=\int_0^1\Theta\,\Phi_n\eta\,d\eta. \qquad (6.4\text{-}15)$$

The Bessel function J_0 satisfies the requirement that the velocity be finite at $\eta=0$ (see Section 4.7). The differential equation and initial condition are transformed to

$$\beta\frac{d\Theta_n}{d\tau}+\lambda_n^2\,\Theta_n=\frac{8\sqrt{2}}{\lambda_n}(1+\gamma\sin\,\tau), \qquad \Theta_n(0)=0, \qquad (6.4\text{-}16)$$

which has the solution

$$\Theta_n(\tau)=\frac{8\sqrt{2}}{\lambda_n^3}\left[\left(1-e^{-\lambda_n^2\tau/\beta}\right)+\left(\frac{\gamma\lambda_n^2}{\lambda_n^4+\beta^2}\right)\left(\beta e^{-\lambda_n^2\tau/\beta}+\lambda_n^2\,\sin\,\tau-\beta\,\cos\,\tau\right)\right]. \qquad (6.4\text{-}17)$$

Using Eqs. (6.4-14) and (6.4-17) in Eq. (6.4-13), the overall solution is found to be

$$\Theta(\eta,\tau)=16\sum_{n=1}^{\infty}\left[\left(1-e^{-\lambda_n^2\tau/\beta}\right)+\left(\frac{\gamma\lambda_n^2}{\lambda_n^4+\beta^2}\right)\left(\beta e^{-\lambda_n^2\tau/\beta}+\lambda_n^2\,\sin\,\tau-\beta\,\cos\,\tau\right)\right]\frac{J_0(\lambda_n\eta)}{\lambda_n^3\,J_1(\lambda_n)}. \qquad (6.4\text{-}18)$$

It is seen that the velocity contains transient contributions, which decay with time, and time-periodic contributions, which continue indefinitely.

Examining in more detail the individual parts of Eq. (6.4-18), we observe that the part that is independent of γ (the amplitude of the pressure pulses) must be the response to a step change in the pressure gradient, from zero to some final value. Thus, for a step change followed by a constant pressure gradient (i.e., for $\gamma=0$), the dimensionless velocity is

$$\Theta(\eta,\tau)=16\sum_{n=1}^{\infty}\left(1-e^{-\lambda_n^2\tau/\beta}\right)\frac{J_0(\lambda_n\eta)}{\lambda_n^3\,J_1(\lambda_n)}. \qquad (6.4\text{-}19)$$

Based on the slowest-decaying term, an approach to within 5% of the steady-state solution for a step change requires that $\tau/\beta=3/\lambda_1^2=3/(2.405)^2=0.52$, or $t=0.52\,R^2/\nu=0.52\,t_V$. Thus, the transients disappear in less than one characteristic time for viscous momentum transfer. The steady-state part of Eq. (6.4-19) must correspond to the familiar, parabolic velocity profile in a tube.

Comparing Eqs. (6.4-19) and (6.2-21), we infer that

$$2(1 - \eta^2) = 16 \sum_{n=1}^{\infty} \frac{J_0(\lambda_n \eta)}{\lambda_n^3 J_1(\lambda_n)} \tag{6.4-20}$$

so that Eq. (6.4-19) is rewritten as

$$\Theta(\eta, \tau) = 2(1 - \eta^2) - 16 \sum_{n=1}^{\infty} e^{-\lambda_n^2 \tau/\beta} \frac{J_0(\lambda_n \eta)}{\lambda_n^3 J_1(\lambda_n)}. \tag{6.4-21}$$

This is a clearer form of the step-change solution.

Returning to Eq. (6.4-18), we examine now what happens at times larger than the viscous time scale, when all transients have disappeared. Neglecting all of the exponential terms and also using Eq. (6.4-20) to simplify Eq. (6.4-18), we find that

$$\Theta(\eta, \tau) = 2(1 - \eta^2) + 16\gamma \sum_{n=1}^{\infty} \left[\left(\frac{\lambda_n^2 \sin \tau - \beta \cos \tau}{\lambda_n^4 + \beta^2} \right) \right] \frac{J_0(\lambda_n \eta)}{\lambda_n J_1(\lambda_n)}. \tag{6.4-22}$$

This result indicates that, in general, the steady-periodic velocity field is not in phase with the pressure oscillations. However, if $\beta \ll \lambda_1^2 = 5.78$, so that the cosine terms are negligible, Eq. (6.4-22) reduces to

$$\Theta(\eta, \tau) = 2(1 - \eta^2)(1 + \gamma \sin \tau). \tag{6.4-23}$$

This result shows that if the viscous response time is short compared to the period of the pressure oscillations (i.e., if β is less than about 0.3), then the velocity and pressure pulses will be in phase. Equation (6.4-23) (but not the other results) could have been obtained much more simply by using a pseudosteady analysis.

6.5 LIMITATIONS OF EXACT SOLUTIONS

There are four types of limitations which one faces in using exact solutions such as those in the preceding sections to model real flows. The limitations arise from (1) geometric nonidealities, (2) flow instabilities, (3) entrance effects, and (4) edge effects. The geometric restrictions are perhaps the most obvious. If, for example, the tube radius is not exactly constant in Example 6.2-2, or the cylinders are not exactly concentric in Example 6.3-1, then the assumptions of the Poiseuille and Couette flow analyses, respectively, are violated. Fortunately, small geometric nonidealities tend to have little influence on laminar flow, and their effects can be analyzed using the methods of Section 6.6.

The most familiar example of a flow instability is the transition from laminar to turbulent flow in a pipe that occurs usually at $\text{Re} = 2UR/\nu \cong 2.1 \times 10^3$. From a mathematical viewpoint, it is the nonlinearity of the Navier–Stokes equation, with the resulting possibility of multiple solutions for a given set of dynamic conditions, which allows such instabilities to exist. The steady, fully developed solution for tube flow given by Eq. (6.2-21) is exact for all values of Re, but (unfortunately) the corresponding flow is not stable at high Re and thus is not seen experimentally if Re is too large. Instabilities can result not just in laminar-turbulent transitions but also in transitions from one mode of laminar flow to another. One of the most thoroughly analyzed situations is Couette flow with both cylinders rotating, where it is found that the instability leads first to toroidal vortices which are regularly spaced along the axis of the cylinders. The methodology of a linear stability analysis is illustrated in Example 12.3-2, which concerns the flow induced in a stagnant layer of fluid by heating it from below. Research on these and other problems in hydrodynamic stability is reviewed in Chandrasekhar (1961) and in Drazin and Reid (1981).

Entrance effects, which are a concern in channel flows, have been studied extensively. As mentioned in Section 6.2, fully developed flow is achieved only after a certain minimum distance downstream from the inlet, the entrance length. Atkinson et al. (1969) correlated theoretical and experimental entrance lengths for tubes and parallel-plate channels as

$$\frac{L_V}{R} = 1.18 + 0.112 \text{ Re}, \qquad \text{Re} = \frac{2UR}{\nu} \qquad \text{(cylindrical tube)}, \qquad (6.5\text{-}1)$$

$$\frac{L_V}{H} = 1.25 + 0.088 \text{ Re}, \qquad \text{Re} = \frac{2UH}{\nu} \qquad \text{(parallel-plate channel)}, \qquad (6.5\text{-}2)$$

where L_V is the distance needed to approach to within 1% of the fully developed velocity at the centerline. The inlet velocity was assumed to be uniform in each case, equal to the mean velocity U. If L_0 is defined as the half-width (R or H), then the results for both geometries are summarized conveniently as

$$\frac{L_V}{L_0} \cong 1 + 0.1 \text{ Re}, \qquad L_0 = R \text{ or } H. \qquad (6.5\text{-}3)$$

Another useful compilation of theoretical and experimental results, including the excess pressure drops associated with entrance regions in tubes, is given in Christiansen and Lemmon (1965). A derivation of the entrance length in a parallel-plate channel at high Re is presented below as Example 6.5-1.

The possibility of edge effects must be considered whenever a three-dimensional problem is reduced to two dimensions by neglecting one or more bounding surfaces. In Example 6.2-1 the sides of the channel were neglected so that v_x did not depend on z; in Example 6.3-1 the top and bottom of the annular gap were ignored so that v_θ did not depend on z. It was mentioned in Section 3.3 that for viscous flows, as well as for conduction and diffusion in stationary materials, a useful rule of thumb is that edge effects in domains of rectangular cross-section extend only about one thickness in from the sides. Thus, for a parallel-plate channel of height $2H$ and width $2W$, the flow is approximately two-dimensional if $H/W \ll 1$. This result is derived in Example 6.5-2.

Example 6.5-1 Velocity Entrance Length in a Parallel-Plate Channel   The objective is to estimate the entrance length in a parallel-plate channel of height $2H$, for large values of Re. It is convenient to set $y = 0$ at the lower wall, as shown in Fig. 6-7(a), and to take advantage of the symmetry by restricting the analysis to the bottom half of the channel. As indicated by the profiles in Fig. 6-7(b)–(d), the velocity is assumed to be uniform at the inlet ($x = 0$), whereas for $x > 0$ the fluid is affected by the no-slip condition at the wall. It is assumed that for small x there are nonzero values of $\partial v_x / \partial y$ only in a region of thickness $\delta(x)$ near the wall, which will be called the "wall layer." The velocity in the central part of the channel, termed the "core," is assumed to remain independent of y in the developing region, such that $v_x = u(x)$. Viscous momentum transfer causes the wall layer to grow, so that the core must accelerate to conserve mass; that is, $du/dx > 0$

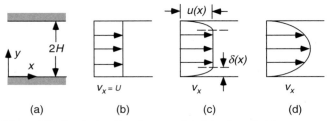

Figure 6-7. Developing flow in a parallel-plate channel. (a) Coordinates; (b) v_x for $x=0$; (c) v_x for $0<x<L_V$; (d) v_x for $x \geqslant L_V$.

because $d\delta/dx>0$. When the entire cross section has adjusted, at $x=L_V$, the flow is fully developed. The qualitative features of the velocity profiles shown in Fig. 6-7 are borne out by measurements in developing flows (Atkinson et al., 1967).

As indicated by Eq. (6.5-2) and as the results of this analysis will confirm, a large value of Re ensures that $H/L_V \ll 1$. It follows that $\delta/x \ll 1$ throughout most of the entrance region, which has important implications for momentum transfer in the wall layer. One implication is that $\partial^2 v_x/\partial x^2 \ll \partial^2 v_x/\partial y^2$. This is seen by noting that the scales for x, y, and v_x in the wall layer are x, δ, and U, respectively, so that $\partial^2 v_x/\partial x^2 \sim U/x^2$ and $\partial^2 v_x/\partial y^2 \sim U/\delta^2$. Thus, the $\partial^2 v_x/\partial x^2$ term can be neglected in the x component of the Navier–Stokes equation. Another implication of the smallness of δ/x is that $V/U \ll 1$, where V is the scale for v_y. This is seen from the continuity equation, where

$$\frac{\partial v_x}{\partial x}=-\frac{\partial v_y}{\partial y} \quad \text{or} \quad \frac{U}{x} \sim \frac{V}{\delta} \quad \text{or} \quad \frac{V}{U} \sim \frac{\delta}{x} \ll 1. \tag{6.5-4}$$

The smallness of v_y implies that all terms in the y component of the Navier–Stokes equation will be much smaller than the dominant terms in the x component. It follows that $\partial\mathcal{P}/\partial y \ll \partial\mathcal{P}/\partial x$, or that $\mathcal{P}$ is approximately a function of x only. Using these approximations, the x component of the Navier–Stokes equation for the wall layer is written as

$$v_x\frac{\partial v_x}{\partial x}+v_y\frac{\partial v_x}{\partial y}=-\frac{1}{\rho}\frac{d\mathcal{P}}{dx}+\nu\frac{\partial^2 v_x}{\partial y^2}. \tag{6.5-5}$$

This is identical to the *boundary layer approximation*, which is discussed in more detail in Chapter 8.

In the core region it is assumed that the large value of Re makes the viscous terms in the Navier–Stokes equation negligible (see Section 5.10 and also Chapter 8). With $v_x=u(x)$, the x component of the Navier–Stokes equation for the core reduces to

$$u\frac{du}{dx}=-\frac{1}{\rho}\frac{d\mathcal{P}}{dx}, \tag{6.5-6}$$

where again $\mathcal{P}=\mathcal{P}(x)$ only. Because $\mathcal{P}$ does not vary appreciably with y, Eq. (6.5-6) can be used to evaluate the pressure gradient in the wall layer in terms of $u(x)$. With regard to v_y, the continuity equation is integrated to give

$$v_y(x, y)=-\int_0^y \frac{\partial v_x}{\partial x}\, dY, \tag{6.5-7}$$

where Y is a dummy variable. Equation (6.5-7) makes use of the fact that $v_y = 0$ at $y = 0$. Using Eqs. (6.5-6) and (6.5-7) to replace the pressure term and v_y, respectively, in Eq. (6.5-5), the momentum equation for the wall layer becomes

$$\nu \frac{\partial^2 v_x}{\partial y^2} = v_x \frac{\partial v_x}{\partial x} - \left(\int_0^y \frac{\partial v_x}{\partial x} dY \right) \frac{\partial v_x}{\partial y} - u \frac{du}{dx}. \tag{6.5-8}$$

An approximate solution for the velocity in the developing region is sought now using the integral method (see Section 3.8). An integral form of Eq. (6.5-8) is obtained by integrating each term over the thickness of the wall layer. The left-hand side becomes

$$\nu \int_0^\delta \frac{\partial^2 v_x}{\partial y^2} dy = \nu \left(\frac{\partial v_x}{\partial y} \right) \Big|_{y=0}^{y=\delta} = - \nu \frac{\partial v_x}{\partial y} \Big|_{y=0}. \tag{6.5-9}$$

The term involving $\partial v_x / \partial y$ in Eq. (6.5-8) is evaluated using integration by parts, which gives

$$\int_0^\delta \left(\int_0^y \frac{\partial v_x}{\partial x} dY \right) \frac{\partial v_x}{\partial y} dy = \left(v_x \int_0^y \frac{\partial v_x}{\partial x} dY \right) \Big|_{y=0}^{y=\delta} - \int_0^\delta v_x \frac{\partial v_x}{\partial x} dy = \int_0^\delta (u - v_x) \frac{\partial v_x}{\partial x} dy. \tag{6.5-10}$$

Combining Eqs. (6.5-9) and (6.5-10) with the integrals of the other terms in Eq. (6.5-8) leads to

$$\nu \frac{\partial v_x}{\partial y} \Big|_{y=0} = \int_0^\delta \frac{\partial}{\partial x} [v_x (u - v_x)] \, dy + \frac{du}{dx} \int_0^\delta (u - v_x) \, dy. \tag{6.5-11}$$

Noticing that the term in square brackets vanishes at both limits of the integral, Eq. (6.5-11) is rewritten with the help of Eq. (A.5-8) as

$$\nu \frac{\partial v_x}{\partial y} \Big|_{y=0} = \frac{d}{dx} \int_0^\delta v_x (u - v_x) \, dy + \frac{du}{dx} \int_0^\delta (u - v_x) \, dy. \tag{6.5-12}$$

This relation, which has been used widely in boundary layer theory, is called the *momentum-integral equation* or the *Kármán integral equation* [see Schlicting (1968, p. 145)].

In terms of the coordinates in Fig. 6-7, the fully developed velocity profile from Example 6.2-1 is

$$v_x(y) = \frac{3}{2} U \left[2 \left(\frac{y}{H} \right) - \left(\frac{y}{H} \right)^2 \right]. \tag{6.5-13}$$

For consistency with this, it is assumed that the velocity in the developing wall layer is of the form

$$v_x(x,y) = u(x) \left[2 \left(\frac{y}{\delta(x)} \right) - \left(\frac{y}{\delta(x)} \right)^2 \right]. \tag{6.5-14}$$

From Fig. 6–7 and a comparison of Eqs. (6.5-13) and (6.5-14) we infer that

$$u(0) = U, \quad \delta(0) = 0; \quad u(L_V) = \frac{3}{2} U, \quad \delta(L_V) = H. \tag{6.5-15}$$

Using Eq. (6.5-14) to evaluate the terms in Eq. (6.5-12), it is found that

$$v = \frac{1}{15} u\delta \frac{d\delta}{dx} + \frac{3}{10} \delta^2 \frac{du}{dx} \tag{6.5-16}$$

or, in dimensionless form,

$$\frac{1}{Re} = \frac{1}{30} \tilde{u}\tilde{\delta} \frac{d\tilde{\delta}}{d\tilde{x}} + \frac{3}{20} \tilde{\delta}^2 \frac{d\tilde{u}}{d\tilde{x}} \tag{6.5-17}$$

$$\tilde{x} \equiv \frac{x}{H}, \qquad \tilde{\delta} \equiv \frac{\delta}{H}, \qquad \tilde{u} \equiv \frac{u}{U}, \qquad Re \equiv \frac{2UH}{\nu}. \tag{6.5-18}$$

Equation (6.5-17) provides one of the two relationships needed to determine the unknown functions $u(x)$ and $\delta(x)$.

Another relationship between $u(x)$ and $\delta(x)$ is obtained from conservation of mass. For an incompressible fluid the volumetric flow rate in a channel of constant cross section is independent of axial position. Using the variable $\eta \equiv y/\delta(x)$ to integrate over the channel cross section, this implies that

$$UH = \int_0^H v_x \, dy = u\delta \left[\int_0^1 (2\eta - \eta^2) \, d\eta + \int_1^{H/\delta} d\eta \right] = u\left(H - \frac{\delta}{3} \right). \tag{6.5-19}$$

This relationship between u and δ is used to eliminate u from the momentum result. It is found from Eq. (6.5-19) that

$$\tilde{u} = \left(1 - \frac{\tilde{\delta}}{3} \right)^{-1}, \qquad \frac{d\tilde{u}}{d\tilde{x}} = \frac{1}{3}\left(1 - \frac{\tilde{\delta}}{3} \right)^{-2} \frac{d\tilde{\delta}}{d\tilde{x}}, \tag{6.5-20}$$

so that Eq. (6.5-17) becomes

$$\frac{1}{Re} = \left[\frac{\tilde{\delta}}{30}\left(1 - \frac{\tilde{\delta}}{3} \right)^{-1} + \frac{\tilde{\delta}^2}{20}\left(1 - \frac{\tilde{\delta}}{3} \right)^{-2} \right] \frac{d\tilde{\delta}}{d\tilde{x}}. \tag{6.5-21}$$

This first-order differential equation for $\tilde{\delta}$, which is subject to the initial condition $\tilde{\delta}(0) = 0$, is nonlinear but separable. Integrating over the entire entrance region, we find that

$$\frac{1}{Re} \int_0^{L_V/H} d\tilde{x} = \frac{L_V}{HRe} = \frac{1}{30} \int_0^1 \left[\tilde{\delta}\left(1 - \frac{\tilde{\delta}}{3} \right)^{-1} + \frac{3}{2}\tilde{\delta}^2\left(1 - \frac{\tilde{\delta}}{3} \right)^{-2} \right] d\tilde{\delta} = \frac{41}{40} - \frac{12}{5} \ln \frac{3}{2}, \tag{6.5-22}$$

where the integral involving $\tilde{\delta}$ has been evaluated using the substitution $t = 1 - (\tilde{\delta}/3)$. The final expression for the velocity entrance length is

$$\frac{L_V}{H} = 0.052 \, Re. \tag{6.5-23}$$

This is very similar to the high-Re asymptote of Eq. (6.5-2), although the numerical coefficient is smaller (0.052 versus 0.088). Given that the coefficient 0.088 corresponds to 99% of the transition from a uniform to a parabolic velocity profile and that the deviation from fully developed flow decays exponentially as x becomes large (Atkinson et al., 1969), the coefficient 0.052 corresponds to about 93% of the transition. Thus, the integral analysis provides a reasonable estimate of the entrance length for large Re.

Example 6.5-2 Edge Effects for Flow in a Rectangular Channel   Consider fully developed flow in a channel with the rectangular cross section shown in Fig. 6-8(a). The objective is to determine how far one must be from the side walls to obtain the parabolic velocity profile of Example 6.2-1, where the channel edges were ignored. The analysis focuses on a region near one edge, as shown in Fig. 6-8(b).

The dimensional form of the momentum equation which applies here is Eq. (6.2-6). Dimensionless variables are defined as

$$Y \equiv \frac{y}{H}, \qquad Z \equiv \frac{z}{H}, \qquad \Theta \equiv \frac{v_x}{U} = \frac{3\mu v_x}{H^2(-d\mathcal{P}/dx)}, \tag{6.5-24}$$

where U is the mean velocity in a channel without edge effects [see Eq. (6.2-10)]. The dimensionless problem statement becomes

$$\frac{\partial^2 \Theta}{\partial Y^2} + \frac{\partial^2 \Theta}{\partial Z^2} = -3, \tag{6.5-25}$$

$$\Theta(Y, 0) = 0, \qquad \Theta(Y, \infty) \text{ finite}, \tag{6.5-26}$$

$$\frac{\partial \Theta}{\partial Y}(0, Z) = 0, \qquad \Theta(1, Z) = 0. \tag{6.5-27}$$

It is assumed in the second boundary condition in Eq. (6.5-26) that the channel is wide enough that the flow is influenced only by the nearest channel edge. That is, the opposing edge is effectively at $Z = \infty$.

The finite Fourier transform method is applied with an expansion of the form

$$\Theta(Y, Z) = \sum_{n=0}^{\infty} \Theta_n(Z)\Phi_n(Y), \tag{6.5-28}$$

$$\Phi_n(Y) = \sqrt{2}\cos\lambda_n Y, \qquad \lambda_n = \left(n + \frac{1}{2}\right)\pi, \tag{6.5-29}$$

$$\Theta_n(Z) = \langle \Theta, \Phi_n \rangle = \int_0^1 \Theta\Phi_n \, dY, \tag{6.5-30}$$

which yields the transformed problem

$$\frac{d^2\Theta_n}{dZ^2} - \lambda_n^2\Theta_n = -\frac{3\sqrt{2}}{\lambda_n}(-1)^n; \qquad \Theta_n(0) = 0, \quad \Theta_n(\infty) \text{ finite}. \tag{6.5-31}$$

The solution for the transformed velocity is

$$\Theta_n(Z) = \frac{3\sqrt{2}}{\lambda_n^3}(-1)^n(1 - e^{-\lambda_n Z}), \tag{6.5-32}$$

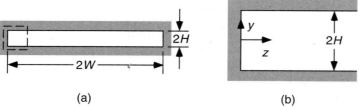

(a)   (b)

Figure 6-8. Channel of rectangular cross section: (a) overall view of cross section; (b) enlargement of area indicated by the dashed rectangle in (a). The flow is in the x direction (not shown).

which gives as the overall solution

$$\Theta(Y, Z) = 6 \sum_{n=0}^{\infty} \frac{(-1)^n}{\lambda_n^3}(1 - e^{-\lambda_n Z})\cos \lambda_n Y. \tag{6.5-33}$$

It is seen that the solution consists of two parts, one of which is independent of Z. That part evidently corresponds to the parabolic velocity profile obtained by neglecting edge effects [Eq. (6.2-11)]. Thus, the solution is rewritten more explicitly as

$$\Theta(Y, Z) = \frac{3}{2}(1 - Y^2) - 6 \sum_{n=0}^{\infty} \frac{(-1)^n}{\lambda_n^3}e^{-\lambda_n Z}\cos \lambda_n Y. \tag{6.5-34}$$

The distance over which the velocity field depends on Z as well as Y is determined by examining the most slowly decaying term in the Fourier series in Eq. (6.5-34), that with $n=0$ and $\lambda_0 = \pi/2$. In particular, for $Z \geqslant 2$ that term is found to influence the maximum velocity (that at $Y=0$) by $\leqslant 4\%$. It is concluded that edge effects are negligible for dimensional distances of $z \geqslant 2H$. In other words, as stated earlier, the velocity field is influenced by the channel edges only over distances equivalent to about one channel thickness.

6.6 LUBRICATION APPROXIMATION

In Sections 6.2 through 6.4 a number of examples were presented in which there was only one nonzero velocity component, and in which an exact solution could be obtained. Most of those problems were greatly simplified by the fact that the nonlinear inertial terms $(\mathbf{v} \cdot \nabla \mathbf{v})$ in the Navier–Stokes or Cauchy momentum equations were identically zero. We consider now a closely related type of flow, in which a second velocity component and the inertial terms are small, but not exactly zero. Liquid flows in long, narrow channels and in thin films often have these characteristics of being nearly unidirectional and dominated by viscous stresses. The entrance-region flow in Example 6.5-1 was nearly unidirectional, in that $v_y \ll v_x$, but the assumption of large Re in that analysis made inertial effects prominent. Thus, entrance-region (or boundary-layer) flows have one of the main characteristics of the problems considered here, but lack the other.

The prototype for nearly unidirectional, viscous flows is a steady, two-dimensional (x, y) flow in a thin channel or a narrow gap between solid objects. The channel height or gap width varies with position, and there may be relative motion of the solid surfaces. If the main flow is in the x direction and if the conditions discussed below are met, the Navier–Stokes equation reduces to

$$\frac{\partial^2 v_x}{\partial y^2} = \frac{1}{\mu}\frac{d\mathcal{P}}{dx}, \qquad \mathcal{P} = \mathcal{P}(x) \text{ only}, \tag{6.6-1}$$

which is called the *lubrication approximation*. This is basically the same as the x-momentum equation used for fully developed flow (e.g., plane Poiseuille flow). The only differences are that now $d\mathcal{P}/dx$ depends on x instead of being constant, and $v_x = v_x(x, y)$ instead of $v_x(y)$. The integration of Eq. (6.6-1) over y remains straightforward, despite these differences, because the right-hand side is a function of x only. Thus, if $d\mathcal{P}/dx$ is given, and v_x (or $\partial v_x/\partial y$) is specified at two values of y, then $v_x(x, y)$ can be readily determined. As illustrated by the examples later in this section, the additional information needed to calculate $v_y(x, y)$ and $\mathcal{P}(x)$ is provided by the continuity equation,

$$\frac{\partial v_x}{\partial x} + \frac{\partial v_y}{\partial y} = 0, \tag{6.6-2}$$

together with boundary conditions involving $\mathcal{P}$ (or the mean velocity). No use of the y momentum equation is made in this type of problem, aside from showing that $\mathcal{P} = \mathcal{P}(x)$ only.

A comparison of Eq. (6.6-1) with the complete x and y components of the Navier–Stokes equation for steady, two-dimensional flow,

$$\rho \left(v_x \frac{\partial v_x}{\partial x} + v_y \frac{\partial v_x}{\partial y} \right) = -\frac{\partial \mathcal{P}}{\partial x} + \mu \left(\frac{\partial^2 v_x}{\partial x^2} + \frac{\partial^2 v_x}{\partial y^2} \right), \tag{6.6-3}$$

$$\rho \left(v_x \frac{\partial v_y}{\partial x} + v_y \frac{\partial v_y}{\partial y} \right) = -\frac{\partial \mathcal{P}}{\partial y} + \mu \left(\frac{\partial^2 v_y}{\partial x^2} + \frac{\partial^2 v_y}{\partial y^2} \right), \tag{6.6-4}$$

reveals four requirements which must be met for the lubricaton approximation to be valid. These are: (1) $\partial \mathcal{P}/\partial y \ll \partial \mathcal{P}/\partial x$, so that $\mathcal{P}$ is approximately a function of x only; (2) $\partial^2 v_x/\partial x^2 \ll \partial^2 v_x/\partial y^2$; (3) $\rho v_x \partial v_x/\partial x \ll \mu \partial^2 v_x/\partial y^2$; and (4) $\rho v_y \partial v_x/\partial y \ll \mu \partial^2 v_x/\partial y^2$. It will be seen that (1) and (2) are both consequences of having a thin film or narrow channel, and thus they amount to a single geometric requirement. Moreover, it will be shown that the two inertial terms in the x-momentum equation are comparable, so that (3) and (4) reduce to a single requirement involving the Reynolds number. Thus, it will be seen that the lubrication approximation depends on two basic conditions, one geometric and one dynamic.

The geometric requirement is revealed by the continuity equation. Let L_x and L_y represent the length scales for velocity variations in the x and y directions, respectively, and let U and V be the respective scales for v_x and v_y. It follows from continuity that $U/L_x \sim V/L_y$, or

$$\frac{V}{U} \sim \frac{L_y}{L_x}. \tag{6.6-5}$$

As discussed in Example 6.5-1, to neglect pressure variations in the y direction it is sufficient that all terms in the y momentum equation be small, or that $V/U \ll 1$. Thus, to satisfy requirement (1) it is usually sufficient that

$$\frac{L_y}{L_x} \ll 1, \tag{6.6-6}$$

which holds for thin films or channels. Because $(\partial^2 v_x/\partial x^2)/(\partial^2 v_x/\partial y^2) \sim (U/L_x^2)/(U/L_y^2) \sim (L_y/L_x)^2$, Eq. (6.6-6) also satisfies requirement (2). Another important inference from the continuity equation is that the two inertial terms in Eq. (6.6-3) are of similar magnitude. That is, $v_y \partial v_x/\partial y \sim VU/L_y \sim U^2/L_x \sim v_x \partial v_x/\partial x$. Consequently, requirements (3) and (4) are both satisfied if $\rho U^2/L_x \ll \mu U/L_y^2$ or

$$\left(\frac{UL_y}{\nu} \right) \left(\frac{L_y}{L_x} \right) \equiv \mathrm{Re} \left(\frac{L_y}{L_x} \right) \ll 1, \tag{6.6-7}$$

which is the dynamic requirement for the lubrication approximation. Equation (6.6-7) shows that the inertial terms are made negligible by any suitable combination of Reynolds number and aspect ratio; having Re itself be small (as in creeping flow, Chapter 7) is helpful but not absolutely necessary. Notice that the Reynolds number is based on the cross-sectional (smaller) dimension. Equations (6.6-6) and (6.6-7) are the two conditions needed to make the lubrication approximation valid.

Following are four examples of the analysis of nearly unidirectional, viscous flows.

Example 6.6-1 Flow in a Tapered Channel Consider steady flow in a two-dimensional channel with a wall spacing which decreases gradually in the direction of flow, as shown in Fig. 6-9. (For clarity in labeling, the channel height has been greatly exaggerated.) The walls are not necessarily planar. Given the volumetric flow rate and the local half-height $h(x)$, the objective is to determine the velocity and pressure. Denoting the local mean velocity as $u(x)$, it follows from conservation of mass that the product $u(x)h(x)$ for an incompressible fluid must be a constant, such that

$$q = 2u(x)h(x), \tag{6.6-8}$$

where q is the volumetric flow rate per unit width. Thus, specifying q and $h(x)$ determines $u(x)$.

We begin by defining the conditions under which the lubrication approximation can be used. To apply Eqs. (6.6-6) and (6.6-7) it is necessary to identify the appropriate length and velocity scales. As the scales for v_x and y we arbitrarily choose the downstream values of u and h, which are denoted by u_L and h_L, respectively. The length scale for velocity changes in the x direction (L_x) is not the channel length (L); it is determined by the magnitude of $\partial v_x / \partial x$. Using Eq. (6.6-8), we find that

$$\frac{\partial v_x}{\partial x} \sim \frac{du}{dx} = -\frac{u}{h}\frac{dh}{dx} \sim \frac{u_L}{h_L}\frac{(h_0 - h_L)}{L} \equiv \frac{u_L}{L_x} \tag{6.6-9}$$

or $L_x = h_L L/(h_0 - h_L)$. In other words, L_x is defined as the distance over which v_x changes by an amount comparable to u_L, and that distance will greatly exceed L if the fractional change in the channel height is small. With $L_y = h_L$ and this result for L_x, Eqs. (6.6-6) and (6.6-7) become

$$\frac{(h_0 - h_L)}{L} \ll 1, \qquad \left(\frac{q}{2\nu}\right)\frac{(h_0 - h_L)}{L} \ll 1, \tag{6.6-10}$$

where $\mathrm{Re} \equiv 2uh\rho/\mu = q/\nu$ is independent of position along the channel. It is readily confirmed that the same criteria result if the upstream values (u_0 and h_0) are used as scales.

For h_0 or h_L to be suitable scales for y it is necessary that entrance effects be negligible (see Section 6.5). Indeed, it will be shown that the lubrication approximation yields a velocity profile which is parabolic in y, like that for fully developed flow in a parallel-plate channel. Accordingly, an additional requirement is that $L \gg L_V$, where L_V may be estimated from Eq. (6.5-2) by using $H = h_0$. This completes the set of conditions.

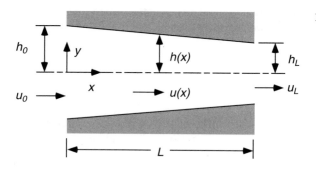

Figure 6-9. Flow in a tapered channel.

To determine $v_x(x, y)$ we integrate Eq. (6.6-1) twice to give

$$v_x(x, y) = \frac{d\mathcal{P}}{dx} \frac{y^2}{2\mu} + c_1(x)y + c_2(x),$$
(6.6-11)

where $c_1(x)$ and $c_2(x)$ are unknown functions. Evaluating those functions using the boundary conditions

$$v_x[x, \pm h(x)] = 0,$$
(6.6-12)

we find that, analogous to Eqs. (6.2-10) and (6.2-11),

$$\frac{d\mathcal{P}}{dx} = -\frac{3}{2} \frac{\mu q}{h^3(x)}, \qquad q = 2uh$$
(6.6-13)

$$v_x(x, y) = \frac{3}{4} \frac{q}{h(x)} \left[1 - \left(\frac{y}{h(x)}\right)^2 \right] = \frac{3}{2} u(x) \left[1 - \left(\frac{y}{h(x)}\right)^2 \right].$$
(6.6-14)

Thus, a channel with gradually tapering walls yields a velocity profile like that in a parallel-plate channel, the only difference being that the channel height and mean velocity vary with axial position.

To evaluate $v_y(x, y)$ we integrate the continuity equation to obtain

$$v_y(x, y) = v_y(x, 0) - \int_0^y \frac{\partial v_x}{\partial x} dY,$$
(6.6-15)

where Y is a dummy variable. Using Eq. (6.6-14) to evaluate the integrand, and recognizing that symmetry requires v_y to vanish at $y=0$, the result is

$$v_y(x, y) = \frac{3}{4} \frac{q}{h(x)} \frac{dh}{dx} \left[\left(\frac{y}{h(x)}\right) - \left(\frac{y}{h(x)}\right)^3 \right].$$
(6.6-16)

As one might expect, v_y is seen to be directly proportional to the slope of the channel walls, dh/dx.

Finally, the pressure is determined by integrating Eq. (6.6-13). The total pressure drop (neglecting the entrance region) is

$$\mathcal{P}_0 - \mathcal{P}_L = \frac{3}{2} \mu q \int_0^L h^{-3}(x) \, dx,$$
(6.6-17)

where $\mathcal{P}_0 = \mathcal{P}(0)$ and $\mathcal{P}_L = \mathcal{P}(L)$. To proceed further it is necessary to specify the function $h(x)$. Assuming planar walls, the half-height is

$$h(x) = h_0 + (h_L - h_0) \frac{x}{L}$$
(6.6-18)

and the total pressure drop is

$$\mathcal{P}_0 - \mathcal{P}_L = \frac{3}{2} \mu q L \left(\frac{h_0 + h_L}{2h_0^2 \, h_L^2}\right).$$
(6.6-19)

For $h_0 = h_L = H$, we recover the result for a parallel-plate channel of constant half-height H, which is

$$\mathcal{P}_0 - \mathcal{P}_L = \frac{3}{2}\frac{\mu q L}{H^3}. \qquad (6.6\text{-}20)$$

Example 6.6-2 Flow in a Permeable Tube   Membrane filtration modules are often configured as bundles of hollow fibers, and they are frequently operated with a low external pressure to cause filtrate to pass outward through the walls of the fibers. To model the flow in one such fiber, consider steady flow in a cylindrical tube of radius R and length L. The mean axial velocity at the tube inlet is given as u_0. For simplicity, assume that the fluid velocity normal to the tube wall (the membrane) is a known constant, v_w. The objective is to determine the velocity and pressure in the tube.

The main difference between this and the preceding example is that the axial variations in the velocity are caused by the loss of fluid through the permeable wall, rather than a change in cross section. Analogous to Eq. (6.6-5), the continuity equation in cylindrical coordinates indicates that $v_w/u_0 \sim L_r/L_z$, where v_w is the scale for v_r and u_0 is the scale for v_z. Neglecting entrance effects, the scale for r is $L_r = R$. Accordingly, Eqs. (6.6-6) and (6.6-7) become

$$\frac{v_w}{u_0} \ll 1, \qquad \left(\frac{u_0 R}{\nu}\right)\left(\frac{v_w}{u_0}\right) = \left(\frac{v_w R}{\nu}\right) \equiv \mathrm{Re}_w \ll 1. \qquad (6.6\text{-}21)$$

The aspect ratio in the geometric condition has been replaced by the ratio of velocity scales, and the dynamic condition has been reduced to the requirement that the Reynolds number based on the *wall velocity* (Re_w) be small. No explicit consideration of L_z was needed because the ratio of velocity scales was known directly.

The lubrication approximation for axisymmetric flow in cylindrical coordinates, analogous to Eq. (6.6-1), is

$$\frac{1}{r}\frac{\partial}{\partial r}\left(r\frac{\partial v_z}{\partial r}\right) = \frac{1}{\mu}\frac{d\mathcal{P}}{dz}, \qquad \mathcal{P} = \mathcal{P}(z) \text{ only}, \qquad (6.6\text{-}22)$$

where $d\mathcal{P}/dz$ is a function of z. Integrating twice and applying the no-slip condition at the tube wall (which usually remains valid despite the presence of flow normal to the wall), we obtain

$$\frac{d\mathcal{P}}{dz} = -\frac{8\mu u(z)}{R^2} \qquad u(z) = \frac{2}{R^2}\int_0^R v_z\, r\, dr \qquad (6.6\text{-}23)$$

$$v_z(r, z) = 2u(z)\left[1 - \left(\frac{r}{R}\right)^2\right], \qquad (6.6\text{-}24)$$

where $u(z)$ is the local mean velocity. The last result is analogous to Eq. (6.2-21).

The next step is to determine $u(z)$ and $v_r(r, z)$ from conservation of mass. Combining Eq. (6.6-24) with the continuity equation gives

from A-3 (2)

$$\frac{1}{r}\frac{\partial}{\partial r}(rv_r) = -\frac{\partial v_z}{\partial z} = -2\frac{du}{dz}\left[1 - \left(\frac{r}{R}\right)^2\right]. \qquad (6.6\text{-}25)$$

Integrating Eq. (6.6-25) and requiring that v_r vanish at $r=0$, we find that

$$v_r(r, z) = -R\frac{du}{dz}\left[\left(\frac{r}{R}\right) - \frac{1}{2}\left(\frac{r}{R}\right)^3\right]. \qquad (6.6\text{-}26)$$

The mean axial velocity and wall velocity are related now by evaluating Eq. (6.6-26) at $r=R$, where $v_r = v_w$. The result is a differential equation governing $u(z)$,

$$\frac{du}{dz} = -\frac{2v_w}{R}, \qquad u(0) = u_0, \tag{6.6-27}$$

which is readily integrated (for constant v_w) to give

$$u(z) = u_0 - \frac{2v_w z}{R}. \tag{6.6-28}$$

Substituting Eq. (6.6-27) in Eq. (6.6-26) yields the final expression for v_r,

$$v_r(r, z) = 2v_w \left[\left(\frac{r}{R}\right) - \frac{1}{2}\left(\frac{r}{R}\right)^3 \right]. \tag{6.6-29}$$

Assuming that the excess pressure drop in the entrance region is negligible, the overall pressure drop is computed by integrating Eq. (6.6-23) to give

$$\mathcal{P}_0 - \mathcal{P}_L = \frac{8\mu u_0 L}{R^2}\left(1 - \frac{v_w L}{u_0 R}\right). \tag{6.6-30}$$

The first term corresponds to the pressure drop in an impermeable tube of length L. As shown by the second term, the effect of the fluid loss through the wall is to reduce the axial pressure drop.

As a final remark, note from Eq. (6.6-28) that the axial velocity will fall to zero at $z = u_0 R / (2v_w)$. However, for the scaling assumptions embodied in Eq. (6.6-21) to be valid, u must remain of the same order of magnitude as u_0. Thus, with v_w constant, the tube cannot be too long. This additional restriction on the analysis is expressed as

$$L \ll \frac{u_0 R}{2v_w}. \tag{6.6-31}$$

The restriction due to the neglect of the entrance region is $L \gg L_V$, where L_V may be estimated using Eq. (6.5-1).

Example 6.6-3 Lubrication Theory The classical application of the lubrication approximation, originating with Reynolds (1886), is to situations in which two solids in relative motion are separated by a thin film of fluid (e.g., machine parts with lubricating oil). The objective in this example is to determine the velocity, pressure, and forces for a class of two-dimensional, steady problems in which one surface slides along the other, without rotation. As shown in Fig. 6-10, two solid surfaces are assumed to be separated by a distance $h(x)$. The coordinates are chosen so that the upper surface is stationary, whereas the lower surface has a tangential velocity, $v_x = U$. As a result of the motion of the lower surface and/or an applied pressure difference, there is a volumetric flow rate q (per unit width). It is assumed that the pressure is specified at two locations,

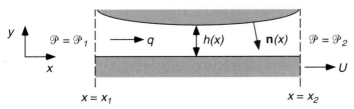

Figure 6-10. General lubrication problem in two dimensions.

such that $\Delta\mathcal{P} = \mathcal{P}(x_2) - \mathcal{P}(x_1) = \mathcal{P}_2 - \mathcal{P}_1$ is known. The interpretation of the pressure boundary conditions is discussed later.

As the first step, the no-slip boundary conditions

$$v_x(x, 0) = U, \qquad v_x(x, h) = 0 \tag{6.6-32}$$

are applied to the integrated form of the lubrication equation, Eq. (6.6-11). (The argument of the function h is omitted now, with the understanding that $h = h(x)$ throughout the analysis.) The result is

$$v_x(x, y) = U\left[1 - \left(\frac{y}{h}\right)\right] - \frac{h^2}{2\mu}\frac{d\mathcal{P}}{dx}\left[\left(\frac{y}{h}\right) - \left(\frac{y}{h}\right)^2\right], \tag{6.6-33}$$

which resembles a linear combination of plane Couette flow (the term containing U) and plane Poiseuille flow (the term containing $d\mathcal{P}/dx$). The next step is to relate $d\mathcal{P}/dx$, which is an unknown function of x, to the constant flow rate q. It is found using Eq. (6.6-33) that

$$q \equiv \int_0^h v_x\,dy = \frac{Uh}{2} - \frac{h^3}{12\mu}\frac{d\mathcal{P}}{dx}. \tag{6.6-34}$$

Solving for $d\mathcal{P}/dx$, we obtain

$$\frac{d\mathcal{P}}{dx} = \frac{6\mu U}{h^2} - \frac{12\mu q}{h^3}. \tag{6.6-35}$$

Using Eq. (6.6-35) to eliminate $d\mathcal{P}/dx$, Eq. (6.6-33) becomes

$$v_x(x, y) = U\left[1 - 4\left(\frac{y}{h}\right) + 3\left(\frac{y}{h}\right)^2\right] + \frac{6q}{h}\left[\left(\frac{y}{h}\right) - \left(\frac{y}{h}\right)^2\right], \tag{6.6-36}$$

which is the final expression for $v_x(x, y)$.

The velocity derivative of most interest is

$$\frac{\partial v_x}{\partial y} = \frac{U}{h}\left[-4 + 6\left(\frac{y}{h}\right)\right] + \frac{6q}{h^2}\left[1 - 2\left(\frac{y}{h}\right)\right], \tag{6.6-37}$$

which is needed to calculate the shear stress. The derivative needed to determine v_y is

$$\frac{\partial v_x}{\partial x} = \frac{2}{h}\frac{dh}{dx}\left[U - \frac{3q}{h}\right]\left[2\left(\frac{y}{h}\right) - 3\left(\frac{y}{h}\right)^2\right] = -\frac{\partial v_y}{\partial y}, \tag{6.6-38}$$

from which it follows that

$$v_y(x, y) = 2\frac{dh}{dx}\left(\frac{3q}{h} - U\right)\left[\left(\frac{y}{h}\right)^2 - \left(\frac{y}{h}\right)^3\right]. \tag{6.6-39}$$

As in Example 6.6-1, it is seen that $v_y \propto dh/dx$.

Integrating Eq. (6.6-35), the pressure is calculated as

$$\mathcal{P}(x) - \mathcal{P}(x_1) = 6\mu U\int_{x_1}^x h^{-2}\,dX - 12\mu q\int_{x_1}^x h^{-3}\,dX. \tag{6.6-40}$$

Taking the term containing U as being representative of the right-hand side, it is apparent that $\mathcal{P} \sim \mu UL/H^2$, where $L \equiv x_2 - x_1$ and H is the characteristic gap height. For comparison, Eqs. (6.6-37) and (6.6-39) indicate that $\tau_{yx} = \tau_{xy} \cong \mu\partial v_x/\partial y \sim \mu U/H$. Thus, the ratio of the pressure to the shear stress is large, $\sim L/H$. As discussed in more detail in the next example, it is this large ratio which leads to favorable lubrication performance.

To solve a problem of this type it is necessary to specify $h(x)$ and two of the following three quantities: U, q, and $\Delta\mathcal{P}$. Evaluating Eq. (6.6-40) at $x=x_2$ gives a relationship among U, q, and $\Delta\mathcal{P}$,

$$\Delta\mathcal{P} = 6\mu U \int_{x_1}^{x_2} h^{-2}\, dx - 12\mu q \int_{x_1}^{x_2} h^{-3}\, dx, \qquad (6.6\text{-}41)$$

which allows the third quantity to be calculated. The velocity, pressure, and viscous stresses are then computed as needed from the other relationships given above.

In principle, the boundary conditions for the pressure require an asymptotic matching of the pressure variable used here with the pressures computed in the bulk fluid on either side of the narrow gap. Thus, strictly speaking, the lubrication problem is coupled to two external flow problems, both of which are characterized by length scales much larger than the gap thickness, which is the scale here. The widely different length scales lead to different dominant terms in the Navier–Stokes equation for the three regions, so that the overall problem has a singular perturbation character; if the gap length is also the external length scale, then the small parameter is H/L. Fortunately, in lubrication problems the external pressure variations are very small compared to those in the narrow gap, so that to good approximation $\mathcal{P}$ may be treated as a constant within each bulk fluid [see Leal (1992, pp. 387–396) for a more detailed discussion]. Thus, the lubrication problem is uncoupled from the external flow problems, where the pressure is approximately that of a stagnant fluid. This is the justification for assuming that $\mathcal{P}$ is known at each end of the gap.

We consider now the force which the fluid exerts on the upper surface in Fig. 6-10. This force (per unit width) will have a horizontal or drag component (F_x) and a vertical or lift component (F_y). The stress exerted by the fluid on the upper surface is $\mathbf{s}=\mathbf{n}\cdot\boldsymbol{\sigma}$, where

$$\mathbf{n}(x)=\frac{1}{g}\left(\frac{dh}{dx}\mathbf{e}_x - \mathbf{e}_y\right), \qquad g(x)=\left[\left(\frac{dh}{dx}\right)^2 + 1\right]^{1/2}. \qquad (6.6\text{-}42)$$

The function $g(x)$ serves to normalize $\mathbf{n}$. To focus on the forces due to fluid motion (i.e., to exclude buoyancy), we will use $\mathcal{P}$ in the normal stresses instead of P [see Eq. (5.8-4)]. (Note, however, that lubrication problems usually involve large pressure variations within thin layers of fluid, so that static pressure variations are negligible and the distinction between $\mathcal{P}$ and P is quantitatively unimportant.) Thus, the components of $\mathbf{s}$ are calculated as

$$s_x=\frac{1}{g}\left(\frac{dh}{dx}\sigma_{xx} - \sigma_{yx}\right)=\frac{1}{g}\left[\frac{dh}{dx}(-\mathcal{P}+\tau_{xx}) - \tau_{yx}\right], \qquad (6.6\text{-}43)$$

$$s_y=\frac{1}{g}\left(\frac{dh}{dx}\sigma_{xy} - \sigma_{yy}\right)=\frac{1}{g}\left[\frac{dh}{dx}\tau_{xy} - (-\mathcal{P}+\tau_{yy})\right]. \qquad (6.6\text{-}44)$$

To identify the dominant contributions to s_x and s_y, let $\varepsilon \equiv H/L \to 0$, as suggested above. It is seen from the discussion immediately following Eq. (6.6-40) that $\mathcal{P}=O(\varepsilon^{-1})$, whereas Eq. (6.6-38) indicates that $\tau_{xx}=2\mu\partial v_x/\partial x=O(\varepsilon)$ and $\tau_{yy}=2\mu\partial v_y/\partial y=O(\varepsilon)$. Thus, both of the normal viscous stresses are negligible. Again referring to the discussion following Eq. (6.6-40), $\tau_{yx}=\tau_{xy}=O(1)$. Given that the coefficient of the $O(\varepsilon^{-1})$ pressure in Eq. (6.6-43) is $dh/dx=O(\varepsilon)$, the contribution of the pressure to s_x is $O(1)$, similar to the shear stress. Accordingly, the x component of the stress vector reduces to

$$s_x=-\frac{1}{g}\left[\frac{dh}{dx}\mathcal{P} + \mu\frac{\partial v_x}{\partial y}\right]. \qquad (6.6\text{-}45)$$

In evaluating τ_{yx} in Eq. (6.6-45), the $\partial v_y/\partial x$ term was neglected. In Eq. (6.6-44) the shear stress and pressure contributions are $O(\varepsilon)$ and $O(\varepsilon^{-1})$, respectively, so that only the pressure needs to be considered in calculating the y component of the stress vector. Equation (6.6-44) reduces to

$$s_y = \frac{\mathcal{P}}{g}. \tag{6.6-46}$$

To compute the corresponding forces, the stresses must be integrated over the surface. A differential element of length along the upper surface in Fig. 6-10 is $g(x)\,dx$, so that, using Eqs. (6.6-45) and (6.6-46), the force components are

$$F_x = -\int_{x_1}^{x_2} \left(\frac{dh}{dx}\mathcal{P} + \mu \frac{\partial v_x}{\partial y}\bigg|_{y=h} \right) dx, \tag{6.6-47}$$

$$F_y = \int_{x_1}^{x_2} \mathcal{P}\,dx. \tag{6.6-48}$$

In addition to the drag and lift forces, it is of interest sometimes to compute the torque (per unit width) exerted by the fluid on the upper surface. The torque about position $\mathbf{r}_0$ is given by

$$\mathbf{G} = \int_{x_1}^{x_2} (\mathbf{r} - \mathbf{r}_0) \times s(\mathbf{n}) g \, dx. \tag{6.6-49}$$

In keeping with the lubrication approximation, it is permissible to set $g = 1$.

Example 6.6-4 Slider Bearing   The slider bearing shown in Fig. 6-11 provides a good illustration of the principles of lubrication. The main features of this problem are that the flow results entirely from the relative motion of the surfaces (i.e., $\Delta\mathcal{P} = 0$) and that the gap height decreases monotonically in the direction of flow. The inclined surface is taken to be planar, as given by Eq. (6.6-18). It is assumed that $\mathcal{P}(0) = \mathcal{P}(L) = 0$ and that the relative velocity U is known. The main objective is to calculate the drag and lift forces.

Given that $\Delta\mathcal{P} = 0$, Eq. (6.6-41) is used to evaluate the flow rate as

$$q = \frac{U}{2} \frac{\displaystyle\int_0^L h^{-2}\,dx}{\displaystyle\int_0^L h^{-3}\,dx} = \left(\frac{h_0 h_L}{h_0 + h_L} \right) U. \tag{6.6-50}$$

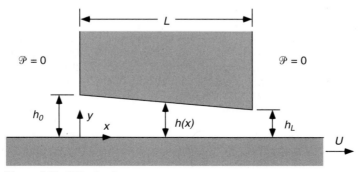

Figure 6-11. Slider bearing.

Setting $x_1 = 0$ in Eq. (6.6-40) and performing the integrations, it is found that

$$\mathcal{P}(x) = \frac{6\mu UL}{h^2} \frac{(h_0 - h)(h - h_L)}{(h_0^2 - h_L^2)}. \tag{6.6-51}$$

The pressure reaches a maximum value of

$$\mathcal{P}_{max} = \frac{3\,\mu UL}{2\,h_0 h_L} \frac{(h_0 - h_L)}{(h_0 + h_L)} \tag{6.6-52}$$

at $x = h_0 L/(h_0 + h_L)$. Given that $0 < h_L < h_0$, it is seen that the pressure maximum always occurs in the downstream half of the gap, $\frac{1}{2} < x/L < 1$. For $h_0/h_L = 2$, we obtain $\mathcal{P}_{max} = \mu UL/h_0^2$ at $x/L = \frac{2}{3}$. This is consistent with the pressure scale inferred from Eq. (6.6-40).

Using Eq. (6.6-47), the drag force on the top surface (per unit width) is found to be

$$F_x = \frac{6\mu UL}{(h_0 - h_L)} \left[\frac{2}{3} \ln \frac{h_0}{h_L} - \left(\frac{h_0 - h_L}{h_0 + h_L} \right) \right]. \tag{6.6-53}$$

Using Eq. (6.6-48), the lift force is

$$F_y = \frac{6\mu UL^2}{(h_0 - h_L)^2} \left[\ln \frac{h_0}{h_L} - 2\left(\frac{h_0 - h_L}{h_0 + h_L} \right) \right]. \tag{6.6-54}$$

The ratio F_y/F_x equals the load which can be supported by the bearing divided by the drag, and therefore it measures the effectiveness of the lubrication; the higher this number, the better. Using Eqs. (6.6-53) and (6.6-54), this force ratio is

$$\frac{F_y}{F_x} = \frac{L}{(h_0 - h_L)} \left[\frac{3(h_0 + h_L)\,\ln\,(h_0/h_L) - 6(h_0 - h_L)}{2(h_0 + h_L)\,\ln\,(h_0/h_L) - 3(h_0 - h_L)} \right] \tag{6.6-55}$$

in general, or

$$\frac{F_y}{F_x} = 0.41 \frac{L}{h_0} \tag{6.6-56}$$

for $h_0/h_L = 2$. It is seen that thin fluid-filled gaps can provide extremely favorable load-to-drag ratios.

Example 6.6-5 Sliding Cylinder   In each of the preceding examples the lubrication approximation was applied to a fluid region with a definite length. For instance, with the slider bearing of Example 6.6-4 the thin fluid-filled gap extended from $x = 0$ to $x = L$. In other situations the limits to be imposed on x are not so clear. An example is shown in Fig. 6-12, in which it is assumed

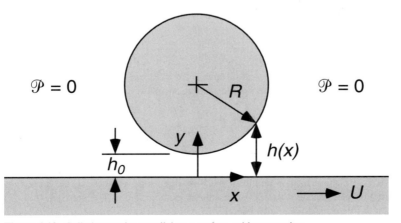

Figure 6-12. Cylinder moving parallel to a surface, without rotation.

that a long cylinder of radius R, immersed in a liquid, moves parallel to a planar surface without rotating. Adopting coordinates in which $x=0$ is always the point of minimum separation, the cylinder is fixed and the surface moves at a constant velocity U. The minimum separation distance is h_0, where $h_0 \ll R$. It is desired to determine the flow rate q in the gap between the surfaces.

The fact that $\mathcal{P}=0$ outside the gap suggests that we set $\Delta\mathcal{P}=0$ in Eq. (6.6-41) and integrate from a "far upstream" to a "far downstream" location to evaluate q. Thus,

$$q = \frac{U}{2} \frac{\int_{-x_0}^{x_0} h^{-2}\, dx}{\int_{-x_0}^{x_0} h^{-3}\, dx}, \tag{6.6-57}$$

where x_0 remains to be determined. Noting that $h(x)$ is symmetric about $x=0$, the expression for q is rewritten as

$$q = \frac{U}{2} \frac{\int_{0}^{x_0} h^{-2}\, dx}{\int_{0}^{x_0} h^{-3}\, dx}. \tag{6.6-58}$$

Although it is tempting to choose $x_0=R$, care is needed because the lubrication approximation used to derive Eq. (6.6-58) is valid only when the gap is narrow. In other words, we want x_0 to be small enough that we can use the lubrication equations, but large enough that, to good approximation, $\mathcal{P}=0$ at $x=\pm x_0$.

The distance between the surfaces is given by

$$h(x) = h_0 + R - \sqrt{R^2 - x^2}. \tag{6.6-59}$$

A simpler expression, which facilitates evaluation of the integrals in Eq. (6.6-58), is obtained by expanding Eq. (6.6-59) in a Taylor series about $x=0$. Retaining only the first two nonzero terms, the result is

$$h(x) = h_0 + \frac{x^2}{2R}, \tag{6.6-60}$$

which turns out to be an excellent approximation to $h(x)$ even for fairly large x. For example, at $|x/R|=0.5$ the gap height is underestimated by $\leqslant 7\%$ for any value of h_0/R. Using Eq. (6.6-60), the geometric requirement for the lubrication approximation is expressed as

$$\left|\frac{dh}{dx}\right| = \left|\frac{x}{R}\right| \ll 1. \tag{6.6-61}$$

Thus, the lubrication approximation becomes questionable for $|x/R|$ exceeding only 0.1 to 0.2. It is seen that the analysis will be limited more by the restriction in Eq. (6.6-61) than by errors in Eq. (6.6-60).

It is informative to rearrange Eq. (6.6-60) as

$$\frac{h(x)}{h_0} = 1 + \left(\frac{x}{\ell}\right)^2, \qquad \ell \equiv (2h_0 R)^{1/2}, \tag{6.6-62}$$

which suggests that ℓ is the characteristic length in the flow direction. In other words, h changes by an appreciable fraction of h_0 when x varies by an amount comparable to ℓ. Using Eq. (6.6-62) to evaluate the pertinent integrals, it is found that

$$\int_0^{x_0} h^{-1}\,dx = \frac{\ell}{h_0}\,a\!\left(\frac{x_0}{\ell}\right), \qquad a(X)=\tan^{-1}(X), \tag{6.6-63}$$

$$\int_0^{x_0} h^{-2}\,dx = \frac{\ell}{h_0^2}\,b\!\left(\frac{x_0}{\ell}\right), \qquad b(X)=\frac{1}{2}\!\left[\frac{X}{1+X^2}+a(X)\right], \tag{6.6-64}$$

$$\int_0^{x_0} h^{-3}\,dx = \frac{\ell}{h_0^3}\,c\!\left(\frac{x_0}{\ell}\right), \qquad c(X)=\frac{1}{4}\!\left[\frac{X}{(1+X^2)^2}+3b(X)\right]. \tag{6.6-65}$$

The functions $a(X)$, $b(X)$, and $c(X)$ range from zero at $X=0$ to asymptotic values of $\pi/2$, $\pi/4$, and $3\pi/16$, respectively, at $X=\infty$. Of particular interest is the behavior of b and c, which are the functions needed to compute $\mathscr{P}$ and q [see Eqs. (6.6-40) and (6.6-58)]. The asymptotes of b and c are closely approached already at $X=3$; the ratio $b(3)/c(3)$ is within 1% of the value of $b(\infty)/c(\infty)$. In other words, for pressure and flow calculations, $x=\pm 3\ell$ provides a practical definition of "far upstream" and "far downstream."

Using Eqs. (6.6-64) and (6.6-65) in Eq. (6.6-58), with the asymptotic value of $b/c=4/3$, the final result for the flow rate is

$$q=\frac{2}{3}\,Uh_0, \tag{6.6-66}$$

which is exact in the limit $h_0/R\to 0$. Choosing $x_0=3\ell=3(2h_0R)^{1/2}$ to ensure that $\mathscr{P}\cong 0$ at the ends of the region being modeled, and requiring that $x_0\le 0.2\,R$ for the lubrication approximation, we conclude that Eq. (6.6-66) will be accurate if $h_0/R\le 10^{-3}$. If Re exceeds unity, making the dynamic requirement of Re $|x_0/R|\ll 1$ more restrictive than Eq. (6.6-61), then h_0/R must be even smaller.

For additional information on the fluid dynamics of lubrication see Cameron (1966) or Pinkus and Sternlicht (1961). Numerous applications of the lubrication approximation in polymer processing are discussed by Middleman (1977), Pearson (1985), and Tanner (1985).

References

Atkinson, B., M. P. Brocklebank, C. C. H. Card, and J. M. Smith. Low Reynolds number developing flows. *AIChE J.* 15: 548–553, 1969.

Atkinson, B., Z. Kemblowski, and J. M. Smith. Measurements of velocity profile in developing liquid flows. *AIChE J.* 13: 17–20, 1967.

Cameron, A. *The Principles of Lubrication.* Wiley, New York, 1966.

Chandrasekhar, S. *Hydrodynamic and Hydromagnetic Stability.* Oxford University Press, London, 1961.

Christiansen, E. B. and H. E. Lemmon. Entrance region flow. *AIChE J.* 11: 995–999, 1965.

Drazin, P. G. and W. H. Reid. *Hydrodynamic Stability.* Cambridge University Press, Cambridge, 1981.

Leal, L. G. *Laminar Flow and Convective Transport Processes.* Butterworth-Heinemann, Boston, 1992.

Middleman, S. *Fundamentals of Polymer Processing.* McGraw-Hill, New York, 1977.

Pearson, J. R. A. *Mechanics of Polymer Processing.* Elsevier, London, 1985.

Pinkus, O. and B. Sternlicht. *Theory of Hydrodynamic Lubrication.* McGraw-Hill, New York, 1961.

Reynolds, O. On the theory of lubrication and its application to Mr. Beauchamp Tower's experiments, including an experimental determination of the viscosity of olive oil. *Philos. Trans. R. Soc. Lond. A* 177: 157–234, 1886.

Schlicting, H. *Boundary-Layer Theory,* sixth edition. McGraw-Hill, New York, 1968.
Sutera, S. P. and R. Skalak. The history of Poiseuille's law. *Annu. Rev. Fluid Mech.* 25: 1–19, 1993.
Tanner, R. I. *Engineering Rheology.* Clarendon, Oxford, 1985.
Taylor, G. I. Deposition of a viscous fluid on a plane surface. *J. Fluid Mech.* 9: 218–224, 1960.

Problems

6-1. Non-Newtonian Couette Flow

In Example 6.3-1 it was shown that for a Newtonian fluid in plane Couette flow the velocity profile is linear. Show that this must be true for all generalized Newtonian fluids. (*Hint:* First show that τ_{yx} is independent of y.)

6-2. Paint Film

Wet paint often exhibits a yield stress, and it can be approximated as a Bingham fluid. Consider a layer of paint on a vertical wall, with a uniform layer thickness H and a yield stress τ_0.

(a) Derive an expression for the maximum value of H which can be sustained without having the paint "run" (i.e., flow down the wall). Denote this thickness as H_0.
(b) For situations where $H > H_0$, derive an expression for the fully developed velocity profile, $v_x(y)$.

6-3. Rotating Rod

A long cylindrical rod of radius R is positioned vertically in a large container of liquid and rotated at an angular velocity ω. Far from the rod the liquid-air interface is at a level $z = h_0$.

(a) Determine the velocity and pressure in the liquid.
(b) Assuming that the immersed length of rod is L, calculate the torque which must be applied to maintain the steady rotation.
(c) Neglecting the effects of surface tension, determine the height of the liquid–air interface, $h(r)$.

6-4. Falling Film with Temperature-Dependent Viscosity

Suppose that the liquid film shown in Fig. 6-4 is not isothermal and that the temperature-dependence of the viscosity is not negligible. In particular, assume that

$$T(y) = T_0 + (T_H - T_0)\frac{y}{H}, \qquad \mu = \frac{\mu_0}{1 + a(T - T_0)},$$

where $a > 0$. That is, the temperature profile across the film is linear, and the viscosity decreases with increasing temperature (as is usual for liquids). Determine $v_x(y)$ and the mean velocity, U.

6-5. Flow in an Annulus (Power Law)

Consider steady, pressure-driven flow of a power-law fluid in the annular space between two fixed, concentric cylinders. Assuming that $d\mathcal{P}/dz$ is given, determine $v_z(r)$ for $R_1 \leq r \leq R_2$.

6-6. Parallel-Plate Channel with Cross-Flow

Assume that the walls of the parallel-plate channel of Fig. 6-2 are permeable to fluid and that the external pressures are manipulated so that there is a constant inward velocity at one wall and a constant outward velocity at the other. In other words, $v_y = v_w$ (a constant) at both walls, $y = \pm H$. Because the flows normal to the walls balance, the mean axial velocity remains constant at U.

Figure P6-8. Flow in a cavity.

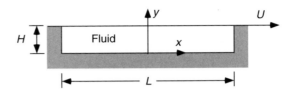

Determine the fully developed velocity and pressure fields. Is the axial pressure drop affected by the cross-flow?

6-7. Tube Flow with Slip
The possibility of unequal tangential velocities at a fluid–solid interface is sometimes modeled by assuming that the velocity difference is proportional to the shear stress at the interface. For slip at the wall of a cylindrical tube of radius R, with fully developed flow, this condition is written as

$$v_z(R) = -\beta \tau_{rz}(R),$$

where β is a positive constant.
Derive an expression for the volumetric flow rate using the slip boundary condition given above. How does this result differ from the dependence on R in Eq. (6.2-23), namely $Q \propto R^4$?

6-8. Flow in a Cavity
Consider steady flow in the closed cavity shown in Fig. P6-8. The bottom and side walls are fixed, while the top surface moves at a constant speed U. The fluid-filled space is relatively narrow, such that $H/L \ll 1$.

    (a) Determine the fully developed velocity profile, $v_x(y)$, in the central part of the cavity (i.e., neglecting the regions adjacent to the side walls).
    (b) Calculate the shear stress exerted on the fluid by the moving surface.

6-9. Free Surface in an Accelerated Tank
Suppose that a rectangular tank, partly filled with liquid and open to the atmosphere, is subjected to a constant horizontal acceleration $\mathbf{a} = a\mathbf{e}_x$, as shown in Fig. P6-9. Determine the level of the gas–liquid interface in the moving tank, $h(x)$, given that the level at rest is h_0.

6-10. Start-Up of Couette Flow
In plane Couette flow, Fig. 6-5(b), assume that the fluid is initially at rest and that the movement of the bottom surface begins suddenly at $t=0$.

    (a) Derive an expression for $v_x(y, t)$ valid for all $t>0$.
    (b) Estimate the time required to approach to within 5% of the steady-state velocity.

Figure P6-9. Rectangular tank subjected to a constant horizontal acceleration.

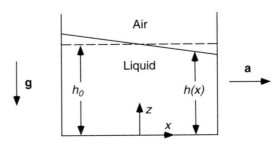

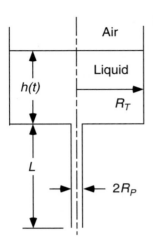

Figure P6-12. A cylindrical tank connected to an outlet pipe.

6-11. Flow in a Square Duct

Consider fully developed flow in a channel of square cross section. The coordinates shown in Fig. 6-1 may be used, with the cross section taken to be a square of side length $2H$.

(a) Assuming that $d\mathcal{P}/dx$ is given, determine $v_x(y, z)$.
(b) Relate $d\mathcal{P}/dx$ to the volumetric flow rate, Q.

6-12. Time Required to Drain a Tank

A pipe of radius R_P and length L is connected to the bottom of a cylindrical tank of radius R_T, as shown in Fig. P6-12. The top of the tank and bottom of the pipe are open to the atmosphere. At $t=0$ a valve is opened, allowing the tank to drain. Assume that the only significant flow resistance is that in the pipe, and that the pipe flow is laminar and fully developed. The instantaneous liquid level in the tank is $h(t)$ and the initial level is h_0.

(a) Assuming that flow in the pipe is pseudosteady, determine the time required to drain the tank (but not the pipe).
(b) What is required for the pseudosteady approximation to be valid?

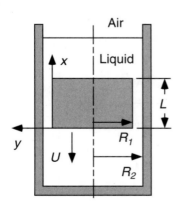

Figure P6-13. Falling cylinder viscometer.

6-13. Falling Cylinder Viscometer

A possible (although not commonly employed) approach for measuring the viscosity of a liquid is to determine the speed at which a solid cylinder falls through a closely fitting tube filled with the liquid. Such a system is shown in Fig. P6-13. The gap between the concentric cylinders is of thickness $H \equiv R_2 - R_1$. Assuming that the gap is thin ($H/R_1 \ll 1$), local rectangular coordinates can be used, as shown. The densities of the cylinder and liquid are ρ_c and ρ, respectively.

Assuming that the flow in the gap between the cylinder and tube is fully developed, derive an expression which relates the viscosity to the cylinder velocity (U) and the other parameters. State any additional geometric or dynamic assumptions which are needed.

6-14. Bubble Rising in a Tube

Consider a large gas bubble rising in a liquid-filled tube. If the bubble volume (V_B) and tube radius (R_T) are such that $V_B^{1/3} \gg R_T$, then the bubble will be highly elongated. Assume that the bulk of the bubble forms a cylinder of radius R_B and length L, as shown in Fig. P6-14(a). The liquid film between the bubble and tube wall is expected to be very thin ($H \equiv R_T - R_B \ll R_B$). Consequently, local rectangular coordinates can be used to describe the flow in the film, as shown in Fig. P6-14(b). It is convenient to fix the coordinate origin on the wall.

Assuming that the flow in the thin liquid film is fully developed and that the bubble dimensions are known, derive an expression for the steady rise velocity of the bubble, U. State any additional geometric or dynamic assumptions that are needed. (*Hint:* The bubble velocity is not uniform, so that the velocity at the gas–liquid interface differs from U, the mean value for the bubble.)

6-15. Tube with Spatially Periodic Radius

Consider steady flow in a circular tube in which the radius varies with axial position as

$$R(z) = R_0 + R_1 \sin\left(\frac{\pi z}{\ell}\right).$$

That is, the mean radius is R_0 and the amplitude and period of the variations are R_1 and 2ℓ, respectively. The volumetric flow rate, Q, is specified.

(a) Under what conditions will the lubrication approximation be applicable?

(b) Determine $\mathbf{v}(r, z)$ and $\mathcal{P}(z)$ using the lubrication approximation. For a tube of length

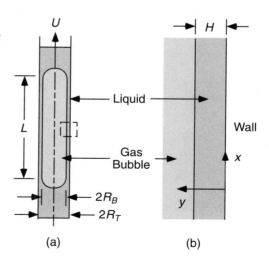

Figure P6-14. A large bubble rising in a liquid-filled tube. (a) Overall view; (b) enlargement of area shown by dashed rectangle in (a).

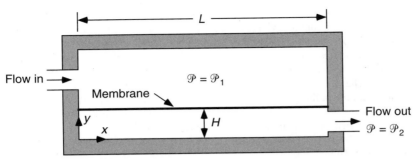

Figure P6-16. In-line particle filter.

$L \gg \ell$, how will the overall pressure drop compare with that for a tube of constant radius, $R = R_0$?

6-16. In-Line Particle Filter

Figure P6-16 shows the arrangement of a type of in-line filter designed to remove occasional particulates from a liquid. The flow resistance within the top chamber is negligible, and the dynamic pressure is constant there at $\mathcal{P}_1$. The pressure at the exit of the bottom chamber is $\mathcal{P}_2$. The local transmembrane velocity $v_w(x)$ is related to the transmembrane pressure drop $\Delta P(x)$ by $v_w = k_w \Delta P$, where k_w is the hydraulic permeability of the membrane. It is desired to relate the overall pressure drop to q, the volumetric flow rate per unit width.

(a) Assuming that the lubrication approximation is valid for flow in the bottom chamber, derive the relationship between q and $\mathcal{P}_1 - \mathcal{P}_2$.

(b) What order-of-magnitude criteria must be satisfied for the results of part (a) to be valid?

6-17. Condensation on a Vertical Wall (Constant Rate)

As shown in Fig. P6-17, steady condensation of a vapor on a vertical wall results in a condensate film of thickness $\delta(x)$ which flows down the wall at a mean velocity $u(x)$. Assume that the rate of condensation, expressed as the velocity v_c normal to the vapor–liquid interface, is constant.

(a) State the relationship between $u(x)$ and $\delta(x)$, assuming that the fluid dynamic conditions are such that $v_x(x, y)$ in the liquid is parabolic in y.

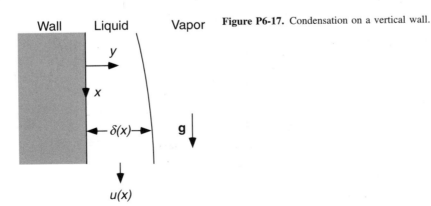

Figure P6-17. Condensation on a vertical wall.

(b) Relate $d\delta/dx$ to $u(x)$ and v_c. Determine $\delta(x)$ and $u(x)$, assuming that $\delta(0)=0$.

(c) Identify the order-of-magnitude criteria which are sufficient to justify the fluid dynamic assumptions. Express these constraints in terms of δ and u.

6-18. Sliding Cylinder

For the physical situation of Example 6.6-5, it is desired to derive expressions for the forces on the cylinder in the limit $h_0/R\rightarrow0$. The forces and torque referred to below are all expressed per unit length of cylinder.

(a) Evaluate $v_x(x,y)$ and $\mathscr{P}(x)$ in the gap between the cylinder and the flat surface.

(b) Show that there is no lift force (i.e., $F_y=0$).

(c) Evaluate the drag force, F_x. Show that there is no net contribution of the viscous stresses to F_x, and sketch several streamlines to explain why.

(d) Show that no applied torque is needed to prevent rotation of the cylinder (i.e., $G=0$).

6-19. Wire Coating

In one type of continuous coating process a wire is pulled from a reservoir of the liquid coating material through a tapered die, as shown in Fig. P6-19. In general, the final coating thickness (h_∞) will depend on the wire velocity (U), the die dimensions, the pressure difference maintained across the die $(\Delta\mathscr{P}\equiv\mathscr{P}_0-\mathscr{P}_L)$, and the liquid properties. Assume that the lubrication approximation is applicable, and that the coating is thin compared to the radius of the wire, so that rectangular coordinates can be used.

(a) Relate h_∞ to $\Delta\mathscr{P}$, U, and the die geometry.

(b) If h_0 is too large, some of the liquid will recirculate instead of being drawn out with the wire, an undesirable situation. Assuming that h_∞ is given, determine the maximum opening for which there will be no recirculation $(h_0=h_{max})$. Also, sketch several streamlines for a case where $h_0>h_{max}$. To guide your sketch, find the equation for the streamline which separates the forward-flow and recirculating regions.

6-20. Calendering

A step in the manufacture of certain polymeric films is to pass the liquid polymer between a pair of calender rolls. We consider here a simplified model of this process for a Newtonian fluid. As shown in Fig. P6-20, the counterrotation of the rolls forces the liquid through the narrow gap between them. Each roll has a radius R and is rotated at a constant angular velocity ω, giving a linear velocity ωR at its surface. The minimum half-height of the gap is such that $h_0/R\ll1$. Assume that the film separates from the rolls at some position $x=x_s$ (not known in advance), and thereafter behaves as a solid. Assume also that $\mathscr{P}=0$ in the upstream reservoir and at $x=x_s$.

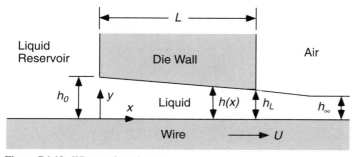

Figure P6-19. Wire coating process.

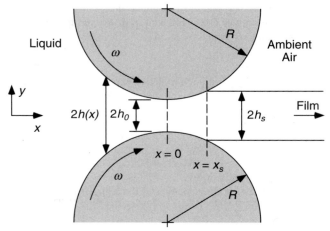

Figure P6-20. Calendering.

(a) Use the lubrication approximation to derive expressions for $v_x(x, y)$ and $\mathcal{P}(x)$ in the gap; x_s will appear as a parameter.

(b) Use conservation of mass to determine x_s and the final film thickness (h_s). Show that $h_s/h_0 = 1.226$.

6-21. Rotating Cylinders

As shown in Fig. P6-21, assume that two parallel cylinders separated by a thin film of liquid are rotated in the same direction. Each cylinder has a radius R and is rotated at a constant angular velocity ω, giving a linear velocity ωR at its surface. The minimum gap half-height is such that $h_0/R \ll 1$. Assume that the lubrication approximation is valid and that $\mathcal{P} = 0$ in the bulk fluid.

(a) Determine $v_x(x, y)$ and show that $\mathcal{P} = 0$ everywhere.

(b) Calculate the torque (per unit length) which must be applied to either cylinder to maintain the rotation, in the limit $h_0/R \to 0$.

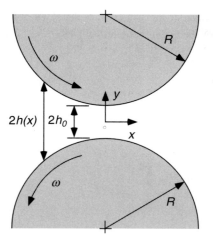

Figure P6-21. Corotating cylinders separated by a thin liquid film.

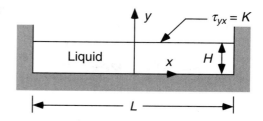

Figure P6-22. A thin liquid film with a constant shear stress at the top surface.

6-22. Flow in a Thin Liquid Film Driven by Surface Tension

Consider the two-dimensional situation shown in Fig. P6-22, in which there is a shallow layer of a Newtonian liquid in a rectangular container. The fluid is bounded at the bottom and sides by solid surfaces. The nature of the top surface is (for the moment) unspecified, except that there is a constant shear stress ($\tau_{yx} = K$) and no normal component of velocity at $y = H$. The depth-to-length ratio is such that $H/L \ll 1$.

(a) Assume that the liquid film is thin enough that the flow is fully developed (except very near the sides). Determine $d\mathcal{P}/dx$ and $v_x(y)$.

(b) It is suggested that if the top surface is a gas–liquid interface, then the mysterious stress of magnitude K could be due to a gradient in surface tension. A gradient in surface tension might result, for example, from a temperature gradient in the x direction. Show that this idea is consistent with an interfacial stress balance, and evaluate $d\gamma/dx$. If $d\gamma/dT < 0$ (as in any pure liquid) and $K > 0$, which side of the container must be hotter? (Continue to assume here that H is independent of x.)

(c) Show that if the top surface is in fact a gas–liquid interface, then H cannot be constant. [*Hint:* Show that the pressure gradient in the liquid found in part (a) is inconsistent with a constant pressure P_0 at the gas–liquid interface. Ignore wetting phenomena at the sides.]

(d) Use a lubrication analysis to determine $H(x)$, assuming that the mean film thickness is given as H_0. What is needed to ensure that the geometric and dynamic requirements for the lubrication approximation, $|dH/dx| \ll 1$ and $\text{Re}|dH/dx| \ll 1$, are satisfied?

6-23. Brush Application of a Liquid onto a Surface

In the transfer of a liquid to a flat surface by a brush, the brush bristles are parallel to the surface. Liquid is removed from the brush by a tangential force at the surface acting parallel to the axis of the bristles, and replaced by liquid flowing normal to the bristles from the interior of the brush. In other words, the brush acts as a reservoir for the liquid. A simplified model for this process, in which the cylindrical bristles are replaced by flat surfaces of indefinite height, was

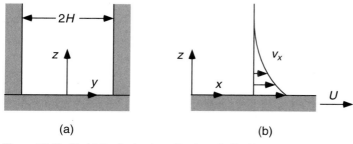

(a)   (b)

Figure P6-23. Model for the brush application of a liquid onto a flat surface.

suggested by Taylor (1960). Figure P6-23(a) is a view parallel to the axis of the model bristles, which are spaced at a distance $2H$. The "brush" is regarded as stationary and the surface is moving at a velocity $v_x = U$, as shown in Fig. P6-22(b).

(a) Determine the fully developed velocity, $v_x(y, z)$.
(b) Relate the final thickness L of the liquid film to H, U, and the fluid properties. Neglect the width of the model bristles. (*Hint:* Calculate the volumetric flow rate Q per channel, and assume that far from the brush there is no motion in the liquid relative to the surface.)

Chapter 7

CREEPING FLOW

7.1 INTRODUCTION

The theory of low Reynolds number flow, also termed *creeping flow* or *Stokes flow,* has applications to numerous technologies and natural phenomena involving small length (L) and velocity (U) scales and/or extremely viscous fluids. Examples include flow in porous media and in the microcirculation (small L and U), suspension rheology (small L, dictated by particle size), the behavior of colloidal dispersions (small L), and the processing of polymer melts (large μ). The focus of this chapter is the creeping flow of Newtonian fluids.

The chapter begins with an overview of some of the distinctive properties of creeping flow, which result from the virtual absence of inertial effects. Several approaches for the solution of creeping flow problems are then discussed, starting with examples of unidirectional and nearly unidirectional flows. The application of the stream function to two-dimensional problems is introduced, including a derivation of Stokes' law for the drag on a sphere. Singular (or point-force) solutions of Stokes' equation, which have found much use in describing the behavior of particulate systems, are discussed next. Following that are derivations of some fundamental results pertinent to suspensions, including Faxen's laws for the motion of a single sphere and Einstein's result for the viscosity of a dilute suspension of spheres. The chapter closes by focusing on a rather unexpected aspect of Stokes' equation, namely, its failure to provide an adequate approximation to the flow far from a sphere at arbitrarily small but nonzero Reynolds number. The use of singular perturbation theory to obtain corrections to Stokes' law for finite Reynolds number is described.

7.2 GENERAL FEATURES OF LOW REYNOLDS NUMBER FLOW

For vanishingly small Reynolds number the Navier–Stokes equation is approximated by *Stokes' equation,*

$$\boldsymbol{\nabla}\mathscr{P} = \mu\nabla^2\mathbf{v}. \tag{7.2-1}$$

As discussed in Section 5.10, Stokes' equation is expected to be valid for time-dependent as well as steady flow, if Re$\rightarrow 0$ and if the time scale for any applied force greatly exceeds the viscous time scale, L^2/ν. The elimination of the term $\rho D\mathbf{v}/Dt$ from the Navier–Stokes equation makes Eq. (7.2-1) pseudosteady and linear. As discussed below, the distinctive features of creeping flow are a consequence of these characteristics.

The pseudosteady character of creeping flow (i.e., the absence of a time derivative in Stokes' equation) means that, mathematically, time enters such problems only as a parameter. Physically, it implies that the velocity and pressure fields will respond almost instantaneously to changes in applied pressures or the movement of surfaces. This behavior is illustrated well by films or videos of blood flow in the microcirculation. After occlusion of a microvessel the flow (as revealed by the movement of red blood cells) does not coast to a stop, it stops immediately. The virtual absence of inertia caused by such small length and velocity scales may be difficult to accept at first, because it contradicts everyday experience. That experience with air and water mainly involves large Reynolds numbers.

Equation (7.2-1) and the continuity equation for an incompressible fluid constitute a linear set of differential equations, a fact which has enormous consequences for the analysis of creeping flow. This linearity implies that eigenfunction expansion and Green's function methods like those discussed in Chapter 4, as well as many other classical methods for solving linear partial differential equations, can be brought to bear on creeping flow problems. Uniqueness proofs apply, and the linear superposition of solutions can be exploited. The application of some of these methods is discussed later in this chapter.

An important physical implication of linearity is that creeping flow is *reversible.* Consider, for example, flow past a stationary, submerged object where the velocity is uniform far from the object. That is, $\mathbf{v}\rightarrow\mathbf{u}$ as $r\rightarrow\infty$, where $\mathbf{u}$ is a constant vector and the origin is located at the center of the object. Because the velocity field depends linearly on $\mathbf{u}$ (which is the only nonhomogeneity in the differential equations and boundary conditions), replacing $\mathbf{u}$ by $-\mathbf{u}$ will simply change $\mathbf{v}$ to $-\mathbf{v}$. That is, there is a reversal in the flow direction, but without a change in the streamline pattern. One consequence of this is that the drag on the object will be unchanged by the flow reversal. (However, the invariance to flow reversal does *not* imply that the drag is independent of the orientation of the object relative to $\mathbf{u}$; that is true only for a sphere.) Another consequence of the sign reversal in $\mathbf{v}$, together with the pseudosteady nature of creeping flow, is that if $\mathbf{u}$ were decreased to zero and then reversed in the same experiment, a fluid particle would (in principle) retrace the same streamline.

The linearity of the creeping flow equations is especially useful in analyzing the motion of suspended particles. The solutions to certain fundamental problems, such as (a) the torque on a sphere rotating in an otherwise quiescent fluid (Example 7.3-1) and (b) the drag on a sphere translating (without rotation) in such a fluid (Example 7.4-2), can be combined to describe more complex motions. This is done by formulating the relationships as a linear algebra problem and by constructing resistance matrices which relate the linear and angular velocities to the forces and torques. Resistance matrices have been derived not just for single particles in unbounded fluids, but for particles interacting with other particles or with walls (e.g., Happel and Brenner, 1965).

An interesting property of the pressure in creeping flow follows directly from continuity and Stokes' equation. Taking the divergence of both sides of Eq. (7.2-1), we obtain

$$\nabla \cdot (\nabla \mathcal{P}) = \nabla \cdot (\mu \nabla^2 \mathbf{v}) = \mu \nabla^2 (\nabla \cdot \mathbf{v}). \tag{7.2-2}$$

Recalling that $\nabla \cdot \nabla = \nabla^2$ and using continuity, it follows that

$$\nabla^2 \mathcal{P} = 0. \tag{7.2-3}$$

Because $\nabla \cdot (\rho \mathbf{g}) = 0$, Laplace's equation governs P as well as $\mathcal{P}$. Thus, if the pressure or pressure gradient is known at all boundaries, the pressure field can be determined without explicit consideration of $\mathbf{v}$. A complementary result is obtained for the velocity. Applying the Laplace operator instead of the divergence to both sides of Stokes' equation, and using Eq. (7.2-3), we find that

$$\nabla^4 \mathbf{v} = \mathbf{0}, \tag{7.2-4}$$

where $\nabla^2(\nabla^2) = \nabla^4$. This analog of Laplace's equation is called the *biharmonic equation.*

The equations governing the vorticity and stream function in creeping flow also have simple forms. Taking the curl of both sides of Stokes' equation and recalling that $\nabla \times \nabla b = \mathbf{0}$ for any scalar function b (Table A-1), we obtain

$$\nabla^2 (\nabla \times \mathbf{v}) = \nabla^2 \mathbf{w} = \mathbf{0}. \tag{7.2-5}$$

Thus, the vorticity also satisfies Laplace's equation. The equations for the stream function are obtained by recognizing that the left-hand sides of the equations given in Table 5-11 come from the inertial terms in the Navier–Stokes equation. Neglecting those terms leads to

$$\nabla^4 \psi = 0, \qquad E^4 \psi = 0, \tag{7.2-6}$$

where ∇^4 is used for planar flows and E^4 for axisymmetric flows (see Table 5-11).

Another important feature of creeping flow concerns the calculation of surface forces. If gravitational effects are neglected or, equivalently, if $\boldsymbol{\sigma}$ is written using $\mathcal{P}$ instead of P, then Eq. (7.2-1) is the same as

$$\nabla \cdot \boldsymbol{\sigma} = \mathbf{0}. \tag{7.2-7}$$

[This is seen most easily from Eq. (5.3-26).] Applying this result to a macroscopic volume V of fluid bounded by the surface S and using the divergence theorem for tensors, we obtain

$$\int_V \nabla \cdot \boldsymbol{\sigma} \, dV = \mathbf{0} = \int_S \mathbf{n} \cdot \boldsymbol{\sigma} \, dS. \tag{7.2-8}$$

Suppose now that S consists of two parts, an inner closed surface S_1 and an outer surface S_2 which completely surrounds S_1, such that V is the fluid volume between the two closed surfaces. Then, Eq. (7.2-8) leads to

$$\int_{S_1} \mathbf{n} \cdot \boldsymbol{\sigma} \, dS = -\int_{S_2} \mathbf{n} \cdot \boldsymbol{\sigma} \, dS, \tag{7.2-9}$$

which indicates that the force on S_1 is transmitted fully to S_2. One way in which Eq. (7.2-9) may be used is in calculating the fluid-dynamic force on a submerged object (fluid or solid), whose surface corresponds to S_1. If $\boldsymbol{\sigma}$ is known, but if the shape of the object is irregular, Eq. (7.2-9) provides the option of calculating the force by integrating $\mathbf{n} \cdot \boldsymbol{\sigma}$ over some more convenient surface S_2 in the fluid (e.g., a large, imaginary sphere). (This simplification is much like that discussed in Example 5.5-2, in connection with pressure forces in static fluids.)

One other general result for creeping flow which must be mentioned is the *reciprocal theorem*, reported originally by H. A. Lorentz in 1906 [see Happel and Brenner, (1965, pp. 85–87)]. Let $\mathbf{v}'$ be a velocity field which satisfies continuity and Stokes' equation throughout some volume V bounded by the surface S (which may consist of two or more distinct parts, as above), and let $\boldsymbol{\sigma}'$ be the corresponding stress tensor. As in Eq. (7.2-7), gravitational and other body forces are excluded. Now let $\mathbf{v}''$ and $\boldsymbol{\sigma}''$ be the velocity and stress for creeping flow of the same fluid in the identical geometry, but with different boundary conditions on S. Then, the Lorentz reciprocal theorem states that

$$\int_S \mathbf{n} \cdot \boldsymbol{\sigma}' \cdot \mathbf{v}'' \, dS = \int_S \mathbf{n} \cdot \boldsymbol{\sigma}'' \cdot \mathbf{v}' \, dS. \tag{7.2-10}$$

This elegant result is used in many derivations in creeping flow theory. As illustrated by Example 7.6-1, it leads to remarkable simplifications in certain force calculations.

Following Happel and Brenner (1965), the proof of Eq. (7.2-10) begins by using the constitutive equation for an incompressible, Newtonian fluid to write $\boldsymbol{\sigma}'$ as

$$\boldsymbol{\sigma}' = -\mathscr{P}' \boldsymbol{\delta} + 2\mu \boldsymbol{\Gamma}'. \tag{7.2-11}$$

Gravitational effects have been excluded, as mentioned above, so that the pressure is written as $\mathscr{P}$ instead of P. Forming the double-dot product of each term with $\boldsymbol{\Gamma}''$ gives

$$\boldsymbol{\sigma}' : \boldsymbol{\Gamma}'' = -\mathscr{P}' \boldsymbol{\delta} : \boldsymbol{\Gamma}'' + 2\mu \boldsymbol{\Gamma}' : \boldsymbol{\Gamma}''. \tag{7.2-12}$$

It can be shown that

$$\boldsymbol{\delta} : \boldsymbol{\Gamma}'' = \nabla \cdot \mathbf{v}'' = 0, \tag{7.2-13}$$

so that Eq. (7.2-12) reduces to

$$\boldsymbol{\sigma}' : \boldsymbol{\Gamma}'' = 2\mu \boldsymbol{\Gamma}' : \boldsymbol{\Gamma}''. \tag{7.2-14}$$

Interchanging the primes and double primes gives the analogous result,

$$\boldsymbol{\sigma}'' : \boldsymbol{\Gamma}' = 2\mu \boldsymbol{\Gamma}'' : \boldsymbol{\Gamma}'. \tag{7.2-15}$$

The double-dot product of any two tensors is commutative, so that $\boldsymbol{\Gamma}' : \boldsymbol{\Gamma}'' = \boldsymbol{\Gamma}'' : \boldsymbol{\Gamma}'$ and, from Eqs. (7.2-14) and (7.2-15),

$$\boldsymbol{\sigma}' : \boldsymbol{\Gamma}'' = \boldsymbol{\sigma}'' : \boldsymbol{\Gamma}'. \tag{7.2-16}$$

Using the identity

$$\boldsymbol{\sigma} : \boldsymbol{\Gamma} = \nabla \cdot (\boldsymbol{\sigma} \cdot \mathbf{v}), \tag{7.2-17}$$

which requires that Eq. (7.2-7) be valid, and integrating over V, Eq. (7.2-16) leads to

$$\int_V \boldsymbol{\nabla} \cdot (\boldsymbol{\sigma}' \cdot \mathbf{v}'') \, dV = \int_V \boldsymbol{\nabla} \cdot (\boldsymbol{\sigma}'' \cdot \mathbf{v}') \, dV. \tag{7.2-18}$$

Finally, application of the divergence theorem for tensors gives Eq. (7.2-10).

A standard source of information on many fundamental aspects of creeping flow, including those mentioned above, is the book by Happel and Brenner (1965). An informative and more recent treatment of the subject is in Leal (1992, pp. 119–274).

7.3 UNIDIRECTIONAL AND NEARLY UNIDIRECTIONAL SOLUTIONS

Presented in this section are the solutions of two well-known problems involving creeping flow. One (the rotating sphere) is a unidirectional flow, whereas the other (squeeze flow) is a lubrication-like problem. Both are solved using methods like those in Chapter 6. In neither case is it possible to obtain an analytical solution except for vanishingly small Reynolds number, which emphasizes the simplifications inherent in the pseudosteady, linear nature of Stokes' equation.

Example 7.3-1 Torque on a Rotating Sphere   Consider a solid sphere of radius R which is rotating at an angular velocity ω about the z axis, as shown in Fig. 7-1. The sphere is immersed in a large volume of fluid. The length and velocity scales in this problem are R and ωR, respectively, suggesting that Stokes' equation will be a good approximation if $Re = \omega R^2/\nu \ll 1$.

A good starting point is to identify the simplest form of velocity field which is consistent with the rotational symmetry about the z axis, the continuity equation, and the boundary conditions. The rotation ensures that v_ϕ will be important, whereas the axisymmetric geometry implies that no quantity will depend on ϕ. Notice from Table 5-6 that if v_ϕ is independent of ϕ, then the continuity equation is satisfied if $v_r = v_\theta = 0$. The boundary conditions for the velocity,

$$v_r = 0, \quad v_\theta = 0, \quad v_\phi = R\omega \sin\theta \quad \text{at } r = R, \tag{7.3-1}$$

Figure 7-1. Rotation of a sphere about the z axis at an angular velocity ω

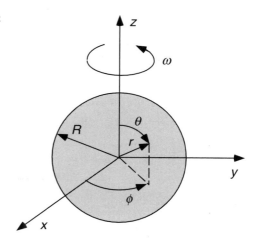

$$v_r = 0, \quad v_\theta = 0, \quad v_\phi = 0 \qquad \text{at } r = \infty, \tag{7.3-2}$$

are also consistent with the hypothesis that $v_r = v_\theta = 0$. Paying special attention to the dependence of v_ϕ on θ required by Eq. (7.3-1), the simplest form for the velocity is evidently

$$v_r = 0, \quad v_\theta = 0, \quad v_\phi = f(r) \sin \theta. \tag{7.3-3}$$

It remains to be confirmed that this form for **v** satisfies Stokes' equation.

With $v_r = v_\theta = 0$, the r and θ components of Stokes' equation reduce to $\partial \mathcal{P}/\partial r = \partial \mathcal{P}/\partial \theta = 0$, indicating (together with the symmetry) that $\mathcal{P}$ is constant throughout the fluid. The ϕ component of Stokes' equation becomes

$$0 = \frac{1}{r^2} \frac{\partial}{\partial r}\left(r^2 \frac{\partial v_\phi}{\partial r}\right) + \frac{1}{r^2 \sin \theta} \frac{\partial}{\partial \theta}\left(\sin \theta \frac{\partial v_\phi}{\partial \theta}\right) - \frac{v_\phi}{r^2 \sin^2 \theta}. \tag{7.3-4}$$

Substituting the assumed form for v_ϕ into Eq. (7.3-4) and the boundary conditions, all terms involving θ cancel or factor out, leaving

$$0 = \frac{d}{dr}\left(r^2 \frac{df}{dr}\right) - 2f, \quad f(R) = \omega R, \qquad f(\infty) = 0. \tag{7.3-5}$$

This ability to obtain a differential equation and boundary conditions for $f(r)$ confirms that the velocity is of the form given by Eq. (7.3-3).

The differential equation for $f(r)$ is equidimensional, with solutions of the form $f = r^n$. Substitution of this trial solution in Eq. (7.3-5) leads to the conclusion that $n = 1$ or -2. Thus, the general solution is

$$f(r) = ar + br^{-2}, \tag{7.3-6}$$

where a and b are constants. The boundary conditions require that $a = 0$ and $b = R^3 \omega$, so that the fluid velocity is given by

$$v_\phi(r, \theta) = \frac{\omega R^3 \sin \theta}{r^2}. \tag{7.3-7}$$

Of interest is the torque (G) exerted by the fluid on the sphere. A point on the surface is at a distance $R \sin \theta$ from the axis of rotation, so that the torque on a differential element of surface dS is given by

$$dG = \tau_{r\phi}(R, \theta)R \sin \theta \, dS, \tag{7.3-8}$$

$$dS = R^2 \sin \theta \, d\theta \, d\phi. \tag{7.3-9}$$

Using Eq. (7.3-7) to evaluate $\tau_{r\phi}$, we find that

$$\tau_{r\phi}(R, \theta) = \mu r \frac{\partial}{\partial r}\left(\frac{v_\phi}{r}\right)\bigg|_{r=R} = -3\mu\omega \sin \theta. \tag{7.3-10}$$

Finally, integrating over the sphere surface gives $dV = r^2 \cdot \sin\theta \cdot d\theta \, dr \, d\phi$

$$G = 2\pi R^3 \int_0^\pi \tau_{r\phi}(R, \theta)\sin^2 \theta \, d\theta = -8\pi\mu\omega R^3. \tag{7.3-11}$$

The surrounding fluid resists the rotation of the sphere, which is in the $+\phi$ direction, so that $G < 0$.

It is noteworthy that the use of Eq. (7.3-3) reduces not just Stokes' equation, but also the *full* Navier–Stokes equation, to Eq. (7.3-4). In other words, the assumed form of the velocity implies that the inertial terms in the ϕ-component of momentum vanish identically. Why, then, was it necessary to assume that $Re \ll 1$? The answer concerns the pressure and the other components of momentum. Note from Table 5-9 that if Re were not small, the inertial terms in the r and θ components of the Navier–Stokes equation would require that $\mathcal{P} = \mathcal{P}(r, \theta)$. In particular, using the creeping-flow velocity field in the r and θ components yields

$$\frac{\partial \mathcal{P}}{\partial r} = \frac{\rho v_\phi^2}{r} = \frac{\rho \omega^2 R^6 \sin^2 \theta}{r^5}, \tag{7.3-12}$$

$$\frac{\partial \mathcal{P}}{\partial \theta} = \rho v_\phi^2 \cot \theta = \frac{\rho \omega^2 R^6 \sin^2 \theta \cos \theta}{r^4}. \tag{7.3-13}$$

Integrating these equations gives

$$\mathcal{P}(r, \theta) = -\frac{1}{4} \left(\frac{\rho \omega^2 R^6 \sin^2 \theta}{r^4} \right) + C_1(\theta), \tag{7.3-14}$$

$$\mathcal{P}(r, \theta) = \frac{1}{2} \left(\frac{\rho \omega^2 R^6 \sin^2 \theta}{r^4} \right) + C_2(r), \tag{7.3-15}$$

Inspection of Eqs. (7.3-14) and (7.3-15) indicates that these two expressions for $\mathcal{P}$ cannot be reconciled using any choice of the functions $C_1(\theta)$ and $C_2(r)$. Because the leading terms do not match, at least one of those unknown functions would have to depend on both r and θ, thereby violating Eqs. (7.3-12) and/or (7.3-13). We conclude that Eq. (7.3-7) satisfies all three components of conservation of momentum only for vanishingly small Re, where (as noted above) $\mathcal{P}$ is constant throughout the fluid. There is no exact, analytical solution to the rotating sphere problem for moderate or large Reynolds number.

Example 7.3-2 Squeeze Flow As shown in Fig. 7-2, it is assumed that two disks of radius R are brought together, forcing out the incompressible fluid between them. A normal force $F(t)$ applied to both disks causes each to move toward the midplane at velocity $U(t)$. The disk separation is $2H(t)$, so that $U = -dH/dt$. To obtain creeping flow with negligible edge effects, we assume that $Re = H_0 U_0 / \nu \ll 1$ and $H_0/R \ll 1$, where $2H_0$ is the initial (maximum) separation and U_0 is the maximum value of $U(t)$. The flow is assumed to be axisymmetric, with $v_\theta = 0$. It is assumed further that $\mathcal{P} = 0$ at the edges of the disks, where the fluid exits.

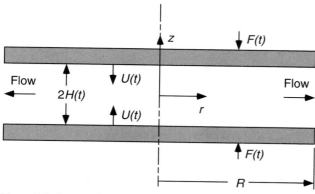

Figure 7-2. Squeeze flow. A force $F(t)$ applied to each of two disks causes them to move together at velocity $U(t)$, squeezing out the fluid between them.

The small Reynolds number makes the problem pseudosteady, and the assumption that $H_0/R \ll 1$ implies that the flow will be almost unidirectional. That is, the continuity equation requires that $v_z/v_r \sim H_0/R$ (see below). The smallness of v_z implies that $\mathscr{P} = \mathscr{P}(r, t)$ only and that we should focus on the r component of Stokes' equation. In addition, the relatively narrow gap will make $\partial^2 v_r/\partial z^2$ the dominant viscous term. Thus, by analogy with the lubrication approximation (Section 6.6), Stokes' equation reduces to

$$0 = -\frac{\partial \mathscr{P}}{\partial r} + \mu \frac{\partial^2 v_r}{\partial z^2}. \tag{7.3-16}$$

The boundary conditions for v_r are

$$\frac{\partial v_r}{\partial z} = 0 \quad \text{at} \quad z = 0, \qquad v_r = 0 \quad \text{at} \quad z = H. \tag{7.3-17}$$

Integrating Eq. (7.3-16) and applying these boundary conditions gives

$$v_r(r, z, t) = -\frac{H^2}{2\mu} \frac{\partial \mathscr{P}}{\partial r}\left[1 - \left(\frac{z}{H}\right)^2\right]. \tag{7.3-18}$$

The radial velocity depends on r through $\partial \mathscr{P}/\partial r$ (which remains to be determined) and depends on t through both $H(t)$ and $\partial \mathscr{P}/\partial r$.

The pressure gradient and v_z are evaluated by using the continuity equation,

$$\frac{1}{r}\frac{\partial}{\partial r}(rv_r) + \frac{\partial v_z}{\partial z} = 0. \tag{7.3-19}$$

Integrating Eq. (7.3-19) over z leads to an expression for v_z,

$$v_z = -\int_0^z \frac{1}{r}\frac{\partial}{\partial r}(rv_r)\, dZ. \tag{7.3-20}$$

To evaluate the integral we need to know how v_r, and therefore $\partial \mathscr{P}/\partial r$, depends on r. A clue is offered by the fact that the boundary conditions for v_z,

$$v_z = 0 \quad \text{at} \quad z = 0, \qquad v_z = -U \quad \text{at} \quad z = H, \tag{7.3-21}$$

do not require that v_z depend on r. Assuming that $v_z = v_z(z, t)$ only, Eq. (7.3-20) implies that v_r and $\partial \mathscr{P}/\partial r$ are both proportional to r. Accordingly, we assume that the pressure gradient is of the form

$$\frac{\partial \mathscr{P}}{\partial r} = rf(t), \tag{7.3-22}$$

where $f(t)$ is to be determined. Using Eqs. (7.3-18) and (7.3-22) in Eq. (7.3-20), it is found that

$$v_z(z, t) = \frac{H^3 f}{\mu}\left[\left(\frac{z}{H}\right) - \frac{1}{3}\left(\frac{z}{H}\right)^3\right]. \tag{7.3-23}$$

The function $f(t)$ is obtained now by evaluating Eq. (7.3-23) at $z = H$, where $v_z = -U$. This gives

$$f(t) = -\frac{3}{2}\frac{\mu U}{H^3}. \tag{7.3-24}$$

and, using Eq. (7.3-22),

$$\frac{\partial \mathcal{P}}{\partial r} = -\frac{3}{2}\frac{\mu U r}{H^3}. \tag{7.3-25}$$

Integrating Eq. (7.3-25) and requiring that $\mathcal{P}=0$ at $r=R$ gives

$$\mathcal{P}(r,\ t) = \frac{3}{4}\frac{\mu U R^2}{H^3}\left[1-\left(\frac{r}{R}\right)^2\right]. \tag{7.3-26}$$

The ability to evaluate $\mathcal{P}$ in this manner confirms the assumption that v_z is independent of r. Using Eqs. (7.3-25) and (7.3-24) in Eqs. (7.3-18) and (7.3-23), respectively, we obtain the final expressions for the velocity components,

$$v_r(r,\ z,\ t) = \frac{3}{4}U\left(\frac{r}{H}\right)\left[1-\left(\frac{z}{H}\right)^2\right], \tag{7.3-27}$$

$$v_z(z,\ t) = -\frac{3}{2}U\left[\left(\frac{z}{H}\right)-\frac{1}{3}\left(\frac{z}{H}\right)^3\right]. \tag{7.3-28}$$

As with the result derived for $\mathcal{P}$, the transient nature of the flow is implicit in the functions $U(t)$ and $H(t)$.

The applied force $F(t)$ can be related now to the velocity $U(t)$ and the half-spacing $H(t)$. Ignoring any gravitational or buoyancy contributions, the force applied to either plate is given by

$$F(t) = 2\pi\int_0^R \mathcal{P}(r,\ t)r\ dr = \frac{3\pi}{8}\frac{\mu U R^4}{H^3}. \tag{7.3-29}$$

In calculating $F(t)$ we have made use of the fact that the normal viscous stress for an incompressible, Newtonian fluid at any solid surface is zero (i.e., $\tau_{zz}|_{z=H}=0$). The relationship among the force, plate speed, and separation given by Eq. (7.3-29), obtained first by J. Stefan in 1874, is called the *Stefan equation*. Because H decreases with time, U must decrease (as H^3) if F is held constant, or F must increase (as H^{-3}) if U is to be held constant. Recalling that $U=-dH/dt$, Eq. (7.3-29) provides a differential equation which could be solved to determine $H(t)$ for any specified $F(t)$.

We conclude this example by commenting on the neglect of inertia. The time derivative and spatial acceleration terms in the r-momentum equation are $\sim\rho V^2/R$, and the dominant viscous term is $\sim\mu V/H_0^2$, where V is the scale for v_r. As suggested by the continuity equation and confirmed by Eq. (7.3-27), a suitable scale is $V=U_0R/H_0$. Thus, a sufficient condition for neglecting inertial effects is $U_0H_0/\nu\ll1$, as stated at the outset. In particular, it is not necessary to meet the more stringent requirement of $VH_0/\nu=U_0R/\nu\ll1$. If the applied force is constant, the requirement that $U_0H_0/\nu\ll1$ is satisfied if

$$F\ll\frac{\mu^2}{\rho}\left(\frac{R}{H_0}\right)^4. \tag{7.3-30}$$

For a constant force F, the maximum value of U is obtained at $t=0$, so that U_0 may be interpreted as the initial disk velocity.

7.4 STREAM FUNCTION SOLUTIONS

In this section are examples of two-dimensional creeping flow problems (planar or axisymmetric) in which there are two nonzero velocity components of comparable magnitude. One of these analyses leads to Stokes' law for the drag on a sphere. In such problems the stream function (Section 5.9) provides an effective approach for obtaining a solution. In addition to the examples, general solutions of the stream-function form of Stokes' equation are discussed briefly.

Example 7.4-1 Flow Near a Corner   Consider the situation depicted in Fig. 7-3, in which a fluid is bounded by two planar, solid surfaces which meet at an angle α. One surface slides past the other at a velocity U. Both surfaces are assumed to extend indefinitely. A flow like this might occur, for example, near the opening of a die in a coating or extrusion process. There is no fixed length scale in this problem, which suggests that the distance from the corner (r) be used in estimating the relative importance of inertial and viscous effects. In other words, the pertinent Reynolds number is $Re(r) = Ur/\nu$, and Stokes' equation will be valid close to the corner. It is desired to characterize the flow in this region. This problem is one of several involving sharp corners which are discussed in Moffat (1964).

The stream function, $\psi(r, \theta)$, for a planar flow in cylindrical coordinates is defined as (Table 5-11)

$$v_r = \frac{1}{r}\frac{\partial \psi}{\partial \theta}, \qquad v_\theta = -\frac{\partial \psi}{\partial r}. \tag{7.4-1}$$

From Eq. (7.2-6), the stream-function form of Stokes' equation is simply

$$\nabla^4 \psi = 0, \tag{7.4-2}$$

$$\nabla^4 \equiv \nabla^2(\nabla^2), \qquad \nabla^2 \equiv \frac{1}{r}\frac{\partial}{\partial r}\left(r\frac{\partial}{\partial r}\right) + \frac{1}{r^2}\frac{\partial^2}{\partial \theta^2}. \tag{7.4-3}$$

The no-slip and no-penetration conditions at the solid surfaces, $\theta = 0$ and $\theta = \alpha$, are expressed as

$$v_r(r, 0) = U, \qquad v_r(r, \alpha) = 0, \tag{7.4-4}$$

$$v_\theta(r, 0) = 0, \qquad v_\theta(r, \alpha) = 0. \tag{7.4-5}$$

Using Eq. (7.4-1), these boundary conditions are written in terms of ψ as

$$\frac{\partial \psi}{\partial \theta}(r, 0) = rU, \qquad \frac{\partial \psi}{\partial \theta}(r, \alpha) = 0, \tag{7.4-6}$$

$$\frac{\partial \psi}{\partial r}(r, 0) = 0, \qquad \frac{\partial \psi}{\partial r}(r, \alpha) = 0. \tag{7.4-7}$$

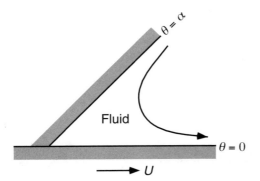

Figure 7-3. Flow near a corner due to a moving surface.

To determine the stream function we must solve Eq. (7.4-2) with these boundary conditions. Once $\psi(r, \theta)$ is known, the velocity components are calculated from Eq. (7.4-1).

Examining the boundary conditions in Eqs. (7.4-6) and (7.4-7), and in particular the no-slip condition at the moving surface, it appears that the solution may be of the form

$$\psi(r, \theta) = rf(\theta). \tag{7.4-8}$$

Substituting Eq. (7.4-8) in Eq. (7.4-2), it is found that all terms containing r factor out, leaving

$$\frac{d^4f}{d\theta^4} + 2\frac{d^2f}{d\theta^2} + f = \left(\frac{d^2}{d\theta^2} + 1\right)^2 f = 0. \tag{7.4-9}$$

Expressed in terms of f, the boundary conditions become

$$\frac{df}{d\theta}(0) = U, \qquad f(0) = f(\alpha) = \frac{df}{d\theta}(\alpha) = 0. \tag{7.4-10}$$

This ability to obtain a problem for $f(\theta)$ which is consistent with Stokes' equation and all boundary conditions confirms that the assumed form of the solution is correct.

Equation (7.4-9) is solved by rewriting it as two coupled, second-order differential equations involving $f(\theta)$ and another function $g(\theta)$, such that

$$\left(\frac{d^2}{d\theta^2} + 1\right) f = g, \tag{7.4-11}$$

$$\left(\frac{d^2}{d\theta^2} + 1\right) g = 0. \tag{7.4-12}$$

After solving Eq. (7.4-12) for g, Eq. (7.4-11) becomes

$$\frac{d^2f}{d\theta^2} + f = a \cos \theta + b \sin \theta, \tag{7.4-13}$$

where a and b are constants. The general solution for this nonhomogeneous equation is

$$f(\theta) = A \cos \theta + B \sin \theta + C\theta \cos \theta + D\theta \sin \theta, \tag{7.4-14}$$

where $C = -b/2$, $D = a/2$, and A and B are additional constants. Using Eq. (7.4-10) to evaluate the four constants in Eq. (7.4-14), the stream function is found to be

$$\psi(r, \theta) = \frac{Ur}{\alpha^2 - \sin^2 \alpha} [\alpha^2 \sin \theta - (\sin^2 \alpha)\theta \cos \theta - (\alpha - \sin \alpha \cos \alpha)\theta \sin \theta]. \tag{7.4-15}$$

The velocity components for any angle α can be obtained now, if desired, by using Eq. (7.4-15) in Eq. (7.4-1).

For perpendicular surfaces ($\alpha = \pi/2$), Eq. (7.4-15) simplifies to

$$\psi(r, \theta) = \frac{Ur}{(\pi^2/4) - 1} \left[\frac{\pi}{2}\left(\frac{\pi}{2} - \theta\right)\sin \theta - \theta \cos \theta\right]. \tag{7.4-16}$$

The velocity components for this special case

$$v_r(\theta) = \frac{U}{(\pi^2/4) - 1} \left[\left(\frac{\pi^2}{4} - \frac{\pi}{2}\theta - 1\right)\cos \theta - \left(\frac{\pi}{2} - \theta\right)\sin \theta\right], \tag{7.4-17}$$

$$v_\theta(\theta) = -\frac{U}{(\pi^2/4) - 1} \left[\frac{\pi}{2}\left(\frac{\pi}{2} - \theta\right)\sin \theta - \theta \cos \theta\right]. \tag{7.4-18}$$

Surprisingly, neither v_r nor v_θ depends on r; this is true for any angle α. Another distinctive aspect of this problem is that $\mathcal{P}$ and $\tau_{r\theta}$ both vary as r^{-1}, and therefore are singular at $r=0$. Although the stresses are singular at $r=0$, the force on an imaginary cylindrical surface of radius r remains finite in the limit $r \to 0$. That is, the surface area varies as r and the stresses as r^{-1}, so that their product is finite. This illustrates the fact that an infinite *stress* at a point is not necessarily unrealistic, although an infinite *force* is not allowed.

Example 7.4-2 Uniform Flow Past a Solid Sphere   The slow motion of a solid sphere relative to a stagnant, viscous fluid, first analyzed by Stokes (1851), is the most widely known problem in low Reynolds number hydrodynamics.[1] As shown in Fig. 7-4, it is convenient to choose a reference frame in which the sphere (of radius R) is stationary and the fluid far from the sphere is moving at a uniform velocity, $\mathbf{v} = \mathbf{U} = U\,\mathbf{e}_z$. It is desired to determine the velocity, the pressure, and the drag on the sphere, assuming that $\mathrm{Re} = UR/\nu \ll 1$.

The stream function, $\psi(r,\theta)$, for an axisymmetric flow in spherical coordinates is defined as (Table 5-11)

$$v_r = \frac{1}{r^2 \sin\theta}\frac{\partial\psi}{\partial\theta}, \qquad v_\theta = -\frac{1}{r\sin\theta}\frac{\partial\psi}{\partial r}, \tag{7.4-19}$$

and the corresponding stream-function form of Stokes' equation is

$$E^4\psi = 0, \tag{7.4-20}$$

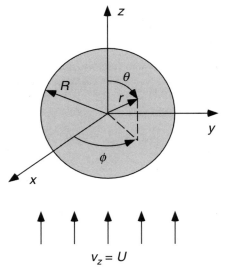

Figure 7-4. Flow around a stationary, solid sphere, with uniform velocity U far from the sphere.

$v_z = U$

---

[1] During his career at Cambridge University, George Stokes (1819–1903) made major theoretical and experimental contributions to many areas of physical science, including fluid mechanics, solid mechanics, optics, and the use of fluorescence (Parkinson, in Gillispie, Vol. 13, 1976, pp. 74–79). He is known also for his work in mathematics. Stokes shares credit (with Navier) for deriving the momentum conservation equation for Newtonian fluids and was largely responsible for the concept of the no-slip boundary condition at a solid surface. The paper in which he reported his calculation of the drag on a sphere (Stokes 1851) is one of the most famous in the history of fluid mechanics, in that it contains the first solutions to viscous flow problems. It preceded by some 10 years the first published derivations of Poiseuille's equation (Chapter 6).

$$E^4 \equiv E^2(E^2), \qquad E^2 \equiv \frac{\partial^2}{\partial r^2} + \frac{\sin\theta}{r^2}\frac{\partial}{\partial\theta}\left(\frac{1}{\sin\theta}\frac{\partial}{\partial\theta}\right). \tag{7.4-21}$$

Equation (7.4-20) is similar to the biharmonic equation, but the E^2 operator defined in Eq. (7.4-21) is not quite the same as the Laplacian (∇^2) involving r and θ.

In spherical (r, θ) coordinates, the boundary conditions of no slip and no penetration at the sphere surface, and uniform flow far from the sphere, are expressed as

$$v_r(R, \theta) = 0, \qquad\qquad v_\theta(R, \theta) = 0, \tag{7.4-22}$$

$$v_r(\infty, \theta) \to U\cos\theta, \qquad v_\theta(\infty, \theta) \to -U\sin\theta. \tag{7.4-23}$$

The boundary conditions for the stream function at the sphere surface, corresponding to Eq. (7.4-22), are

$$\frac{\partial\psi}{\partial\theta}(R, \theta) = 0, \qquad \frac{\partial\psi}{\partial r}(R, \theta) = 0. \tag{7.4-24}$$

The stream-function conditions for the uniform flow far from the sphere, corresponding to Eq. (7.4-23), are

$$\frac{\partial\psi}{\partial\theta}(\infty, \theta) \to r^2 U \sin\theta\cos\theta, \tag{7.4-25}$$

$$\frac{\partial\psi}{\partial r}(\infty, \theta) \to rU \sin^2\theta. \tag{7.4-26}$$

Integrating Eqs. (7.4-25) and (7.4-26) and comparing the results, we conclude that

$$\psi(\infty, \theta) \to \frac{r^2 U \sin^2\theta}{2}. \tag{7.4-27}$$

The absolute value of the stream function is arbitrary, so that in writing Eq. (7.4-27) the integration constant was set equal to zero. It will be seen that this choice gives $\psi = 0$ as the streamline corresponding to the sphere surface.

The required behavior of ψ far from the sphere, as given by Eq. (7.4-27), suggests that we seek a solution to Eq. (7.4-20) of the form

$$\psi(r, \theta) = f(r)\sin^2\theta. \tag{7.4-28}$$

Substitution of Eq. (7.4-28) into Eq. (7.4-20) results in

$$\left[\frac{d^2}{dr^2} - \frac{2}{r^2}\right]^2 f = 0. \tag{7.4-29}$$

From Eqs. (7.4-24) and (7.4-27), the boundary conditions for f are

$$f(R) = 0, \qquad \frac{df}{dr}(R) = 0, \qquad f(\infty) \to \frac{r^2 U}{2}. \tag{7.4-30}$$

Much like the preceding examples, the fact that all terms involving θ could be factored out to obtain Eq. (7.4-29), and that all of the original boundary conditions are satisfied using Eq. (7.4-30), confirms that Eq. (7.4-28) is correct.

Equation (7.4-29) is another example of an equidimensional equation, with solutions of the form $f = r^n$. Substitution shows that n can assume the values 4, 2, 1, and -1. Thus,

$$f(r) = Ar^4 + Br^2 + Cr + \frac{D}{r}, \tag{7.4-31}$$

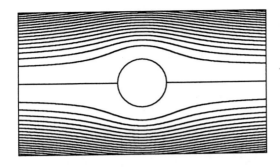

Figure 7-5. Streamlines for creeping flow past a solid sphere, assuming a uniform velocity far from the sphere. The stream function is given by Eq. (7.4-32), and there are equal increments in ψ between adjacent streamlines.

where A, B, C, and D are constants. Using the boundary conditions in Eq. (7.4-30), we find that $A = 0$, $B = U/2$, $C = -\frac{3}{4} UR$, and $D = \frac{1}{4} UR^3$. Thus, the stream function is given by

$$\psi(r, \theta) = UR^2 \sin^2 \theta \left[\frac{1}{2} \left(\frac{r}{R} \right)^2 - \frac{3}{4} \left(\frac{r}{R} \right) + \frac{1}{4} \left(\frac{R}{r} \right) \right]. \tag{7.4-32}$$

Note that $\psi = 0$ on the sphere surface, as mentioned above. The streamlines are plotted in Fig. 7-5, where it is seen that the flow has fore-and-aft symmetry.

The velocity components are evaluated from Eqs. (7.4-19) and (7.4-32) as

$$v_r(r, \theta) = U \cos \theta \left[1 - \frac{3}{2} \left(\frac{R}{r} \right) + \frac{1}{2} \left(\frac{R}{r} \right)^3 \right], \tag{7.4-33}$$

$$v_\theta(r, \theta) = -U \sin \theta \left[1 - \frac{3}{4} \left(\frac{R}{r} \right) - \frac{1}{4} \left(\frac{R}{r} \right)^3 \right]. \tag{7.4-34}$$

The pressure $\mathcal{P}$ is obtained by substituting Eqs. (7.4-33) and (7.4-34) into either the r or θ components of Stokes' equation and integrating. Letting $\mathcal{P} = 0$ far from the sphere, the result is

$$\mathcal{P}(r, \theta) = -\frac{3}{2} \frac{\mu U}{R} \left(\frac{R}{r} \right)^2 \cos \theta. \tag{7.4-35}$$

This completes the calculation of the velocity and pressure fields.

To determine the drag, we must consider both normal and tangential forces at the sphere surface. The normal viscous stress, $\tau_{rr} = 2\mu \, \partial v_r/\partial r$, vanishes at the surface, as usual. The normal stress due to $\mathcal{P}$ acts in the $-r$ direction, and the component of this stress in the z direction is given by $-\mathcal{P} \cos \theta$. The only nonvanishing shear stress is $\tau_{r\theta}$, which has a component $-\tau_{r\theta} \sin \theta$ in the z direction. The shear stress at the surface is evaluated as

$$\tau_{r\theta}(R, \theta) = \mu r \frac{\partial}{\partial r} \left(\frac{v_\theta}{r} \right) \bigg|_{r=R} = -\frac{3}{2} \frac{\mu U}{R} \sin \theta. \tag{7.4-36}$$

An element of surface area is given by $dS = R^2 \sin \theta \, d\theta \, d\phi$. Accordingly, the drag (F_D) is calculated as

$$F_D = -2\pi R^2 \int_0^\pi [\mathcal{P}(R, \theta) \cos \theta + \tau_{r\theta}(R, \theta) \sin \theta] \sin \theta \, d\theta = 6\pi\mu UR, \tag{7.4-37}$$

which is *Stokes' law.* One-third of the drag on a solid sphere is from the pressure, and two-thirds is from the shear stress. Note that by using $\mathcal{P}$ as the pressure instead of P, static pressure variations were excluded from this calculation [see Eq. (5.8-4)]. A correction to Stokes' law for small but nonzero Re is derived in Section 7.7.

The terminal velocity of a spherical particle rising or settling in a fluid is obtained by completing the overall force balance. From Eq. (5.5-6), the net force due to buoyancy (static pressure and gravity) is $\frac{4}{3}\pi R^3 g\Delta\rho$, where $\Delta\rho$ is the density difference between the sphere and the fluid. Equating this force with the drag from Eq. (7.4-37), it is found that

$$U = \frac{2}{9}\frac{R^2 g\Delta\rho}{\mu}. \tag{7.4-38}$$

Results analogous to Eqs. (7.4-37) and (7.4-38) are obtained for a fluid sphere rising or settling in another (immiscible) fluid, as outlined in Problem 7-2.

General Solutions for the Stream Function

The biharmonic equations for planar, creeping flow in rectangular or cylindrical coordinates ($\nabla^4\psi=0$) and the closely related equations for axisymmetric flow in cylindrical or spherical coordinates ($E^4\psi=0$) can be solved using eigenfunction expansions in much the same way as Laplace's equation ($\nabla^2\psi=0$). Indeed, for a given coordinate system, the solutions of these fourth-order equations are closely related to the solutions of Laplace's equation. General solutions for the stream function are available for each of the aforementioned types of flows (Happel and Brenner, 1965; Leal, 1992). For axisymmetric flow in spherical coordinates, the general solution for $\psi(r, \theta)$ is

$$\psi(r, \eta) = \sum_{n=1}^{\infty} [A_n r^{n+3} + B_n r^{n+1} + C_n r^{2-n} + D_n r^{-n}] Q_n(\eta), \tag{7.4-39}$$

$$Q_n(\eta) \equiv \int_{-1}^{\eta} P_n(x)\, dx, \tag{7.4-40}$$

where $\eta = \cos\theta$ and $P_n(x)$ are the Legendre polynomials (see Table 4-4). As described in Happel and Brenner (1965, pp. 133–138) and Leal (1992, pp. 162–164), Eq. (7.4-39) is obtained by rewriting Eq. (7.4-20) as a pair of coupled, second-order equations and using the method of separation of variables. [The conversion of the fourth-order differential equation to a pair of second-order equations is analogous to what was done to obtain $f(\theta)$ in Example 7.4-1.] The eigenfunctions defined by Eq. (7.4-40), a type of *Gegenbauer polynomial,* are orthogonal on the interval $[-1, 1]$ with respect to the weighting function $w(\eta) = (1 - \eta^2)^{-1}$. They are not normalized. As may be inferred from Eq. (7.4-40), Legendre's equation is obtained by differentiating the relevant form of Gegenbauer's equation and replacing $dQ_n/d\eta$ by P_n.

The solutions for ψ in the preceding two examples each contained only one separable term, rather than an infinite series. This is because the boundary data in each case corresponded to one of the eigenfunctions of the respective general solutions. For axisymmetric flow in spherical coordinates, the first three eigenfunctions (written in terms of θ) are

$$Q_0(\theta) = 1 + \cos\theta, \qquad Q_1(\theta) = -\frac{1}{2}\sin^2\theta, \qquad Q_2(\theta) = -\frac{1}{2}\sin^2\theta\cos\theta.$$

$$\tag{7.4-41}$$

In the case of Stokes' problem (Example 7.4-2), the boundary conditions could be satisfied using just Q_1 [see Eq. (7.4-27)]. Accordingly, only the $n=1$ term remained in the series. (An analogous simplification of a Fourier series was seen in Example 4.8-3.) In the next example we make explicit use of the general solution for spheres.

Example 7.4-3 Axisymmetric Extensional Flow Past a Solid Sphere   In this analysis, following Leal (1992, pp. 171–174), we consider flow past a stationary sphere in which the unperturbed velocity far from the sphere is given by

$$\mathbf{u} = -\gamma(x\mathbf{e}_x + y\mathbf{e}_y - 2z\mathbf{e}_z), \tag{7.4-42}$$

where $\gamma \,(= \Gamma/\sqrt{3})$ is a constant. The corresponding rate of strain is

$$\mathbf{\Gamma} = -\gamma(\mathbf{e}_x\mathbf{e}_x + \mathbf{e}_y\mathbf{e}_y - 2\mathbf{e}_z\mathbf{e}_z), \tag{7.4-43}$$

in which it is seen that there are only diagonal components. For $\gamma > 0$, in what is termed a *uniaxial extensional flow*, the streamlines far from the sphere are shown qualitatively in Fig. 7-6. The flow in any x–y plane is toward the z axis, about which there is rotational symmetry, and there is reflective symmetry about the plane $z=0$. For $\gamma<0$ the arrows are reversed and a *biaxial extensional flow* results. As shown, the sphere is centered at the origin. An important aspect of this problem is that the symmetry ensures that there is no net fluid-dynamic force on the sphere. This fact is exploited in calculating the viscosity of a dilute suspension in Section 7.6.

The stream function $\psi(r, \theta)$ is governed again by Eq. (7.4-20), which has the general solution given by Eq. (7.4-39). To identify the terms which must be present in the solution, we once again examine the boundary conditions. As in the previous example, the no-slip and no-penetration conditions at the sphere surface are represented by Eq. (7.4-24). To identify the functional form for $\psi(r, \theta)$ far from the sphere, we first convert $\mathbf{u}$ from rectangular to spherical coordinates. From Section A.7, the coordinates and base vectors in the two systems are related as

$$\mathbf{e}_x = \sin\theta\cos\phi\,\mathbf{e}_r + \cos\theta\cos\phi\mathbf{e}_\theta - \sin\phi\mathbf{e}_\phi, \qquad x = r\sin\theta\cos\phi, \tag{7.4-44a}$$

$$\mathbf{e}_y = \sin\theta\sin\phi\,\mathbf{e}_r + \cos\theta\sin\phi\mathbf{e}_\theta + \cos\phi\mathbf{e}_\phi, \qquad y = r\sin\theta\sin\phi, \tag{7.4-44b}$$

$$\mathbf{e}_z = \cos\theta\mathbf{e}_r - \sin\theta\mathbf{e}_\theta, \qquad z = r\cos\theta. \tag{7.4-44c}$$

Using Eq. (7.4-44) in Eq. (7.4-42), the unperturbed velocity far from the sphere is rewritten as

$$\mathbf{u} = \gamma\, r(3\cos^2\theta - 1)\mathbf{e}_r - 3\gamma\, r\sin\theta\cos\theta\mathbf{e}_\theta. \tag{7.4-45}$$

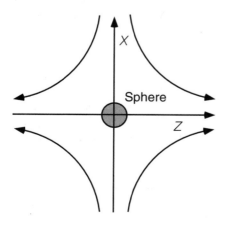

Figure 7-6. Uniaxial extensional flow past a sphere.

Sphere

Figure 7-7. Streamlines for creeping flow past a solid sphere, assuming a uniaxial extensional flow far from the sphere. The stream function is given by Eq. (7.4-48), and there are equal increments in ψ between adjacent streamlines.

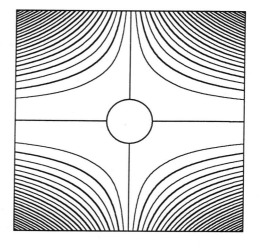

Integrating both parts of Eq. (7.4-19) using Eq. (7.4-45) and comparing the results, we conclude that

$$\psi(\infty,\theta) \to \gamma r^3 \sin^2 \theta \cos \theta. \qquad (7.4\text{-}46)$$

A comparison of Eq. (7.4-46) with Eq. (7.4-41) reveals now that the only eigenfunction needed to represent the unperturbed flow is Q_2. Because no other dependence on θ is required by the boundary conditions at the sphere surface, we conclude that Eq. (7.4-39) reduces to

$$\psi(r,\ \theta) = (A_2 r^5 + B_2 r^3 + C_2 + D_2 r^{-2})\left(-\frac{1}{2} \sin^2 \theta \cos \theta\right). \qquad (7.4\text{-}47)$$

Using Eqs. (7.4-24) and (7.4-46) to evaluate the four constants, the final expression for the stream function is found to be

$$\psi(r,\ \theta) = \gamma R^3 \left[\left(\frac{r}{R}\right)^3 - \frac{5}{2} + \frac{3}{2}\left(\frac{R}{r}\right)^2\right] \sin^2 \theta \cos \theta. \qquad (7.4\text{-}48)$$

These streamlines are plotted in Fig. 7-7, with the orientation of the plot being the same as the qualitative diagram in Fig. 7-6.

Differentiation of Eq. (7.4-48) according to Eq. (7.4-19) yields the velocity components,

$$v_r(r,\ \theta) = \gamma R\left[\left(\frac{r}{R}\right) - \frac{5}{2}\left(\frac{R}{r}\right)^2 + \frac{3}{2}\left(\frac{R}{r}\right)^4\right](3\cos^2\theta - 1), \qquad (7.4\text{-}49)$$

$$v_\theta(r,\ \theta) = -3\gamma R\left[\left(\frac{r}{R}\right) - \left(\frac{R}{r}\right)^4\right]\cos\theta \sin\theta. \qquad (7.4\text{-}50)$$

Finally, the pressure is found to be

$$\mathcal{P}(r,\ \theta) = -5\mu\gamma\left(\frac{R}{r}\right)^3(3\cos^2\theta - 1), \qquad (7.4\text{-}51)$$

where $\mathcal{P}$ has been set equal to zero far from the sphere. A direct calculation using Eqs. (7.4-50) and (7.4-51) confirms that the net viscous and pressure forces on the sphere are both zero.

7.5 POINT-FORCE SOLUTIONS

In this section we derive the point-force solution for Stokes' equation and briefly discuss how it and related singular solutions are applied in the analysis of creeping flow. The material presented here is analogous to that given for Laplace's equation in Sections 4.9 and 4.10.

Point-Force Solution

The fundamental solution for Stokes' flow corresponds to a point force $\mathbf{F}$ acting on the fluid at the origin, $\mathbf{r}=\mathbf{0}$. It satisfies

$$\nabla \mathscr{P} = \mu \nabla^2 \mathbf{v} + \delta(\mathbf{r})\mathbf{F} \tag{7.5-1}$$

together with the continuity equation and the requirement that $\mathbf{v} \to \mathbf{0}$ and $\mathscr{P} \to 0$ as $r \to \infty$. The product of the Dirac delta and $\mathbf{F}$ represents a force per unit volume on the fluid. As discussed in numerous sources (e.g., Happel and Brenner, 1965; Kim and Karrila, 1991; Ladyzhenskaya, 1969; Leal, 1992; Russel et al., 1989), the point-force solution, usually called the *stokeslet,* is given by

$$\mathbf{v}(\mathbf{r}) = \frac{1}{8\pi\mu}\left(\frac{\delta}{r} + \frac{\mathbf{rr}}{r^3}\right) \cdot \mathbf{F} = \mathbf{I}(\mathbf{r}) \cdot \mathbf{F}, \tag{7.5-2}$$

$$\mathscr{P}(\mathbf{r}) = \frac{1}{4\pi r^3}\, \mathbf{r} \cdot \mathbf{F}. \tag{7.5-3}$$

The multiplier of the force in Eq. (7.5-2),

$$\mathbf{I}(\mathbf{r}) = \frac{1}{8\pi\mu}\left(\frac{\delta}{r} + \frac{\mathbf{rr}}{r^3}\right), \qquad r = |\mathbf{r}|, \tag{7.5-4}$$

is called the *Oseen tensor.* (The use of $\mathbf{I}$ for the Oseen tensor departs from our usual practice of using boldface Greek letters for tensors.) It is important to bear in mind that the stokeslet represents only a *disturbance* to the flow associated with the force $\mathbf{F}$; ordinarily, there is a base (or unperturbed) velocity and pressure field to which such disturbances must be added.

A relatively simple way to derive the stokeslet is to examine the solution to Stokes' problem (Example 7.4-2) for a vanishingly small sphere (i.e., the limit $R \to 0$). Discarding the leading terms in Eqs. (7.4-33) and (7.4-34), which correspond to the unperturbed (uniform) velocity far from the sphere, we obtain

$$v_r(r, \theta) = -U \cos\theta \left[\frac{3}{2}\left(\frac{R}{r}\right) - \frac{1}{2}\left(\frac{R}{r}\right)^3\right], \tag{7.5-5}$$

$$v_\theta(r, \theta) = U \sin\theta \left[\frac{3}{4}\left(\frac{R}{r}\right) + \frac{1}{4}\left(\frac{R}{r}\right)^3\right] \tag{7.5-6}$$

for the velocity disturbance. Using Stokes' law to relate R to U and the drag force gives

$$R = -\frac{F_z}{6\pi\mu U}. \tag{7.5-7}$$

The minus sign is needed because $\mathbf{F}$ here is the force exerted on the fluid by the sphere. That is, $F_z = -F_D$, where F_D is the drag given by Eq. (7.4-37). Using Eq. (7.5-7) to rewrite the terms involving r in Eqs. (7.5-5) and (7.5-6), we obtain

$$\frac{R}{r} = -\frac{F_z}{6\pi\mu Ur}, \qquad \left(\frac{R}{r}\right)^3 = -\frac{F_z}{6\pi\mu Ur}\left(\frac{R}{r}\right)^2. \tag{7.5-8}$$

It is seen now that in the limit $R \to 0$ with F_z held constant, the cubic terms will vanish. Thus, combining Eqs. (7.5-5) and (7.5-6), the velocity disturbance due to a point force is

$$\mathbf{v}(r,\ \theta) = \frac{F_z}{8\pi\mu r}(2\cos\theta\,\mathbf{e}_r - \sin\theta\,\mathbf{e}_\theta). \tag{7.5-9}$$

The final form for the velocity disturbance is obtained by using the relations in Section A.7 to convert $\mathbf{e}_r$ and $\mathbf{e}_\theta$ to rectangular base vectors, and to express the spherical angles in terms of r and the rectangular coordinates. Then, recognizing that for this problem $F_z\mathbf{e}_z = \mathbf{F}$ and $zF_z\mathbf{r} = (\mathbf{r}\cdot\mathbf{F})\,\mathbf{r} = \mathbf{rr}\cdot\mathbf{F}$, we obtain Eq. (7.5-2). Similar manipulations transform Eq. (7.4-35) to Eq. (7.5-3), except that no terms are eliminated and the limit $R \to 0$ is not needed.

Alternative approaches for deriving the point-force solution include the use of Green's functions [due to C. W. Oseen, as described in Happel and Brenner (1965, pp. 79–83)], Fourier transforms (Ladyzhenskaya, 1969, pp. 50–51; Russel et al., 1989, pp. 31–34), and superposition of harmonic functions (Lamb, 1932, pp. 594–597; Leal, 1992, pp. 230–232). Other singular solutions, corresponding to force dipoles, force quadra-poles, and the like, have been derived (e.g., Lamb, 1932). The use of such solutions in problem solving was developed extensively by Burgers (1938) [see also Chwang and Wu (1975)].

Integral Representations

The solutions to problems involving complex geometries can be constructed by the su-perposition of stokeslets (or other singular solutions) in various ways. For a point force located at position $\mathbf{r} = \mathbf{r}'$ instead of the origin, the Oseen tensor becomes

$$\mathbf{I}(\mathbf{r} - \mathbf{r}') = \frac{1}{8\pi\mu}\left(\frac{\boldsymbol{\delta}}{R} + \frac{(\mathbf{r} - \mathbf{r}')(\mathbf{r} - \mathbf{r}')}{R^3}\right), \qquad R = |\mathbf{r} - \mathbf{r}'|. \tag{7.5-10}$$

As an example of an integral representation of a velocity field in creeping flow, consider a continuous distribution of forces over a surface S. If the unperturbed velocity at posi-tion $\mathbf{r}$ is $\mathbf{u}(\mathbf{r})$, then the velocity as influenced by the surface stresses is expressed as

$$\mathbf{v}(\mathbf{r}) = \mathbf{u}(\mathbf{r}) - \int_S \mathbf{I}(\mathbf{r} - \mathbf{r}')\cdot\mathbf{s}(\mathbf{r}')\,dS', \tag{7.5-11}$$

where $\mathbf{s} = \mathbf{n}\cdot\boldsymbol{\sigma}$ and $\mathbf{n}$ is the unit normal pointing into the fluid. The force exerted on the fluid by a differential element of surface is $-\mathbf{s}\,dS$, which leads to the minus sign. The integration is over all positions $\mathbf{r}'$ on the surface. An analogous relation can be written for a continuous distribution of forces over a volume V.

As in the modeling of heat transfer or other phenomena described by Laplace's equation, in the application of integral representations to creeping flow problems the distribution of sources (in this case stresses or forces) is usually unknown and must be determined numerically. Various boundary-integral techniques have been developed to solve the integral equations. A more thorough, introductory-level discussion of integral formulations in creeping flow is provided in Leal (1992, pp. 229–251). A comprehensive treatment, including the numerical implementation of such methods, is given by Kim and Karrila (1991).

This section concludes by using Eq. (7.5-11) to investigate the form of the velocity disturbance far from a submerged object of arbitrary shape.

Example 7.5-1 Multipole Expansion To approximate the velocity disturbance far from an object, we derive a multipole expansion analogous to that presented in Example 4.10-1. In this approach the integral in Eq. (7.5-11) is evaluated by expanding the Oseen tensor in a Taylor series about the origin (i.e., the center of the object). As discussed previously, the rationale for such an expansion is that, sufficiently far from the object, the distinction between the origin and the actual position $\mathbf{r}'$ of a point on the surface will become unimportant. Using the first two terms in the Taylor series of Eq. (A.6-8), we obtain

$$\mathbf{I}(\mathbf{r}-\mathbf{r}') = \mathbf{I}(\mathbf{r}) - \mathbf{r}' \cdot (\nabla \mathbf{I})|_{\mathbf{r}'=\mathbf{0}} + \cdots . \tag{7.5-12}$$

Using Eq. (7.5-12), the integral in Eq. (7.5-11) becomes

$$\int_S \mathbf{I}(\mathbf{r}-\mathbf{r}') \cdot \mathbf{s}(\mathbf{r}') \, dS' = \mathbf{I}(\mathbf{r}) \cdot \int_S \mathbf{s}(\mathbf{r}') \, dS' + \int_S \mathbf{r}' \cdot (\nabla \mathbf{I})|_{\mathbf{r}'=\mathbf{0}} \cdot \mathbf{s}(\mathbf{r}') \, dS' + \cdots . \tag{7.5-13}$$

The first integral in the expansion is just the force exerted on the object by the fluid. Letting $\mathbf{F}$ represent the force exerted on the fluid by the object, it follows that

$$-\mathbf{I}(\mathbf{r}) \cdot \int_S \mathbf{s}(\mathbf{r}') \, dS' = \mathbf{I}(\mathbf{r}) \cdot \mathbf{F}, \tag{7.5-14}$$

which is the same as the velocity disturbance given by Eq. (7.5-2). In other words, to first approximation, any distant object is perceived as a point force.

It follows from Eq. (7.5-2) that any object on which there is a net force yields a velocity disturbance which is $O(r^{-1})$ as $r \to \infty$. The behavior of the stress far from such an object is revealed by Eq. (7.2-9). As the outer test surface S_2 is moved very far from the object, its area is $O(r^2)$, so that for a finite force on the object the stress disturbance must be $O(r^{-2})$. The solution to Stokes' problem (Example 7.4-2) provides an example of such velocity and stress disturbances.

To evaluate the second integral in the expansion in Eq. (7.5-13) we employ the identity

$$\mathbf{r} \cdot \nabla \mathbf{I} \cdot \mathbf{s} = \mathbf{s}\mathbf{r} : \nabla \mathbf{I}, \tag{7.5-15}$$

which uses the fact that the Oseen tensor is symmetric. This gives

$$\int_S \mathbf{r}' \cdot (\nabla \mathbf{I})|_{\mathbf{r}'=\mathbf{0}} \cdot \mathbf{s}(\mathbf{r}') \, dS' = \mathbf{D} : (\nabla \mathbf{I})|_{\mathbf{r}'=\mathbf{0}}, \tag{7.5-16}$$

$$\mathbf{D} \equiv \int_S \mathbf{s}(\mathbf{r}') \, \mathbf{r}' \, dS' = \int_S [\mathbf{n}(\mathbf{r}') \cdot \boldsymbol{\sigma}(\mathbf{r}')] \mathbf{r}' \, dS'. \tag{7.5-17}$$

Thus, the rearrangement of terms in Eq. (7.5-15) makes it possible in Eq. (7.5-16) to separate $\nabla \mathbf{I}$ from the surface integral. At this level of approximation, all of the information on the shape of the object and the distribution of stress on its surface is contained in the *dipole tensor*, **D**.

The velocity perturbation represented by Eq. (7.5-16) is $O(r^{-2})$, as determined by $\nabla \mathbf{I}$ [see Eq. (7.5-10)]. This is the dominant, long-range effect on the velocity field for any freely suspended particle (i.e., for $\mathbf{F} = \mathbf{0}$), where the disturbance represented by Eq. (7.5-14) is absent. The disturbance in the stress due to a freely suspended particle is $O(r^{-3})$. Velocity and stress disturbances which are $O(r^{-2})$ and $O(r^{-3})$, respectively, are exemplified by the solutions for the rotating sphere (Example 7.3-1) and the sphere in axisymmetric extensional flow (Example 7.4-3); in both cases $\mathbf{F} = \mathbf{0}$.

The fact that Eq. (7.5-16) represents the main long-range effect of a freely suspended particle makes $\mathbf{D}$ a critical quantity in calculating the effective viscosity of a suspension, as discussed in Example 7.6-2.

7.6 PARTICLE MOTION AND SUSPENSION VISCOSITY

In this section we derive two fundamental results which pertain to dilute suspensions of particles. One is Faxén's first law, which provides a general relationship among the force on a spherical particle, its velocity, and the unperturbed velocity field far from the particle. Faxén's second law, which involves torque and angular velocity in the same context, is discussed briefly. The other result that is derived is Einstein's expression for the effective viscosity of a dilute suspension of solid spheres.

Example 7.6-1 Faxén's Laws   The following derivation of Faxén's first law, which follows the approach of Brenner (1964), makes use of the Lorentz reciprocal theorem. The objective is to calculate the force on a small, solid sphere in an arbitrary velocity field. In general, the flow will be three-dimensional, making the boundary-value problem for $\mathbf{v}$ and $\mathscr{P}$ exceedingly difficult to solve. Thus, the direct approach of calculating the velocity and pressure throughout the fluid, and then integrating the pressure and shear stress over the particle surface, is often impractical. However, Faxén's law provides a very simple expression for the force in an unbounded flow, in terms of the unperturbed velocity far from the sphere and the force given by Stokes' law.

The reciprocal theorem, which relates the velocities and stresses from two different creeping flow problems involving the same geometry (see Section 7.2), is stated as

$$\int_S \mathbf{n} \cdot \boldsymbol{\sigma}' \cdot \mathbf{v}'' \, dS = \int_S \mathbf{n} \cdot \boldsymbol{\sigma}'' \cdot \mathbf{v}' \, dS. \tag{7.6-1}$$

The two flows will be called the "primed" problem and the "double-primed" problem. As the primed problem we select a sphere of radius R moving at velocity $-\mathbf{U}$ in a fluid which is at rest far from the sphere. Except for a change of reference frame, this is the problem solved in Example 7.4-2 to derive Stokes' law. Thus, writing Stokes' law in vector form, the force on the sphere is

$$\mathbf{F}' = 6\pi\mu R\mathbf{U}. \tag{7.6-2}$$

As the second physical situation we consider a sphere moving at velocity $\mathbf{U}_0$ through a fluid which has an unperturbed velocity $\mathbf{u}(\mathbf{r})$ far from the sphere. The sphere is assumed not to rotate. Here we modify the reference frame by subtracting $\mathbf{u}(\mathbf{r})$ from the sphere and fluid velocities. Thus, the velocity of the sphere center in the double-primed problem is $\mathbf{U}_0 - \mathbf{u}_0$, where $\mathbf{u}_0$ is the (original) unperturbed velocity evaluated at the sphere center; in the double-primed problem the fluid velocity far from the sphere is zero. We wish to determine the force $\mathbf{F}''$ which acts on the sphere in this second situation.

It is important to note that subtracting the unperturbed velocity does not affect the force in the second problem, even though $\mathbf{u}(\mathbf{r})$ is not a constant. Because $\mathbf{u}(\mathbf{r})$ is a creeping-flow solution, the stress calculated from $\mathbf{u}(\mathbf{r})$ must satisfy Eq. (7.2-7) everywhere, including the volume presently occupied by the sphere. Accordingly, the net force at the sphere surface due to $\mathbf{u}(\mathbf{r})$ must vanish, as given by the second equality in Eq. (7.2-8).

In both problems the fluid is considered to be bounded by the sphere surface (S_0) and by an outer surface (S_∞) which encloses the sphere and is arbitrarily far from it. In other words, S consists of S_0 plus S_∞. Because we have selected the problems so that $\mathbf{v} = \mathbf{0}$ on S_∞, those parts of the integrals in Eq. (7.6-1) vanish. Evaluating the remaining part of the integral on the right-hand side of Eq. (7.6-1), we obtain

$$\int_{S_0} \mathbf{n} \cdot \boldsymbol{\sigma}'' \cdot \mathbf{v}' \, dS = -\mathbf{U} \cdot \int_{S_0} \mathbf{n} \cdot \boldsymbol{\sigma}'' \, dS = -\mathbf{U} \cdot \mathbf{F}''. \tag{7.6-3}$$

The evaluation of the left-hand side of Eq. (7.6-1) is facilitated by the fact that the stress vector in Stokes' problem, $\mathbf{n} \cdot \boldsymbol{\sigma}'$, is independent of position on the sphere surface. That is, the bracketed quantity in Eq. (7.4-37) is independent of θ as well as ϕ, as the reader may confirm. Accordingly,

$$\mathbf{n} \cdot \boldsymbol{\sigma}' = \frac{\mathbf{F}'}{4\pi R^2} = \left(\frac{3\mu}{2R}\right)\mathbf{U} \tag{7.6-4}$$

and the left-hand side of Eq. (7.6-1) becomes

$$\int_{S_0} \mathbf{n} \cdot \boldsymbol{\sigma}' \cdot \mathbf{v}'' \, dS = \left(\frac{3\mu}{2R}\right)\mathbf{U} \cdot \int_{S} [\mathbf{U}_0 - \mathbf{u}(\mathbf{r})] \, dS. \tag{7.6-5}$$

Comparing Eqs. (7.6-3) and (7.6-5), we obtain

$$\mathbf{U} \cdot \mathbf{F}'' = \left(\frac{3\mu}{2R}\right)\mathbf{U} \cdot \int_{S_0} [\mathbf{u}(\mathbf{r}) - \mathbf{U}_0] \, dS \tag{7.6-6}$$

or, factoring out the arbitrary velocity $\mathbf{U}$,

$$\mathbf{F}'' = \frac{3\mu}{2R} \int_{S_0} [\mathbf{u}(\mathbf{r}) - \mathbf{U}_0] \, dS. \tag{7.6-7}$$

Because $\mathbf{U}_0$ does not depend on surface position, Eq. (7.6-7) reduces easily to

$$\mathbf{F}'' = \frac{3\mu}{2R}\left[\int_{S_0} \mathbf{u}(\mathbf{r}) \, dS - 4\pi R^2 \mathbf{U}_0\right]. \tag{7.6-8}$$

All that is needed now to calculate the force is to integrate the unperturbed velocity over the sphere surface. In particular, Eq. (7.6-8) indicates that it is *not* necessary to solve a boundary-value problem for the second physical situation. Thus, the reciprocal theorem has provided an enormous simplification.

To evaluate the integral in Eq. (7.6-8), we expand the unperturbed velocity in a Taylor series about the sphere center,

$$\mathbf{u}(\mathbf{r}) = \mathbf{u}_0 + \mathbf{r} \cdot (\nabla \mathbf{u})_0 + \frac{1}{2}\mathbf{r}\mathbf{r} : (\nabla\nabla\mathbf{u})_0 + \cdots, \tag{7.6-9}$$

where the subscript 0 is used to denote evaluation at the center ($\mathbf{r} = \mathbf{0}$). What needs to be integrated now over the sphere surface are the terms $\mathbf{r}$, $\mathbf{rr}$, and so on. It is found that (Section A.6)

$$\int_{S_0} x_i \, dS = 0, \quad \int_{S_0} x_i x_j \, dS = \frac{4}{3} \pi R^4 \delta_{ij}, \quad i \text{ and } j = 1, 2, 3. \tag{7.6-10}$$

Thus, the symmetry of the sphere causes the integral of $\mathbf{r}$ to vanish. This is true for *all odd powers* of $\mathbf{r}$, because the integrand in each scalar component of such terms will be an odd function of one or more of the rectangular coordinates. Hence, the contributions from negative and positive positions will cancel. It follows that none of the terms in the Taylor series involving velocity derivatives of odd order will contribute to the integral. Using Eqs. (7.6-9) and (7.6-10), the first three terms of the expansion give

$$\int_{S_0} \mathbf{u}(\mathbf{r}) \, dS = 4\pi R^2 \left[\mathbf{u}_0 + \frac{R^2}{6} (\nabla^2 \mathbf{u})_0 \right]. \tag{7.6-11}$$

The terms which have not yet been discussed, and which are not shown in Eq. (7.6-11), involve even-order derivatives of the velocity of the form $\nabla^{2n}\mathbf{u}$, where $n \geq 2$. It follows from Eq. (7.2-4) that *all* of these higher-order terms are identically zero in creeping flow. Thus, remarkably, Eq. (7.6-11) is exact.

Using Eq. (7.6-11) in Eq. (7.6-8) and dropping the primes from $\mathbf{F}$, we obtain *Faxén's first law* [due to H. Faxén in 1927; see Brenner (1964)]:

$$\mathbf{F} = 6\pi\mu R \left[(\mathbf{u}_0 - \mathbf{U}_0) + \frac{R^2}{6}(\nabla^2 \mathbf{u})_0 \right]. \tag{7.6-12}$$

For a uniform unperturbed velocity, where $\nabla^2 \mathbf{u} = \mathbf{0}$, Stokes' law is recovered; the drag is seen to be proportional to the velocity difference between the unperturbed flow and the sphere, as expected. An interesting implication of Eq. (7.6-12) is that for any linear flow, the velocity of a freely suspended sphere equals that of the unperturbed flow evaluated at the sphere center. That is, for constant $\nabla\mathbf{u}$, where $\nabla^2\mathbf{u} = \mathbf{0}$, if $\mathbf{F} = \mathbf{0}$, then $\mathbf{U}_0 = \mathbf{u}_0$.

Faxén's second law, which is derived in a similar manner using the results of Example 7.3-1 (Brenner, 1964), is stated as

$$\mathbf{G} = 8\pi\mu R^3 \left[\frac{1}{2}(\nabla \times \mathbf{u})_0 - \boldsymbol{\omega} \right], \tag{7.6-13}$$

where $\mathbf{G}$ is the torque and $\boldsymbol{\omega}$ is the angular velocity of the sphere. Equation (7.6-13) shows that a freely suspended sphere ($\mathbf{G} = \mathbf{0}$) rotates at an angular velocity equal to one-half of the unperturbed vorticity. This is the same as the relationship between angular velocity and vorticity in rigid-body rotation of a fluid (see Section 5.2).

The reciprocal theorem has been used also to obtain analogs of Faxén's laws for objects other than solid spheres (Brenner, 1964; Rallison, 1978).

Example 7.6-2 Effective Viscosity of a Dilute Suspension of Spheres   The flows of suspensions are governed by at least three length scales: the size of the suspended particles, the average spacing between particles, and the characteristic dimension of the container or conduit in which the flow occurs (L). In colloidal dispersions, where electrostatic or other nonhydrodynamic forces between particles are often important, there may be additional length scales associated with those forces (Russel et al., 1989). The particle size and spacing (and the length scales for any nonhydrodynamic forces) relate to the microstructure of the suspension, whereas L is usually a macroscopic dimension. If L greatly exceeds the microstructural length scales, the suspension may be viewed as a homogeneous fluid with certain effective properties. A major objective of studies of suspension rheology is to relate the effective viscosity to the microstructure of the suspension. It is found that dilute suspensions of uncharged spheres behave as Newtonian fluids. The objective here is to compute the effective viscosity of such a suspension.

In addition to having more than one length scale, suspension flows are also characterized by more than one velocity scale. In general, there will be a microscopic scale U which character- izes relative velocities in the vicinity of a single particle, and a macroscopic scale U_L. When combined with the corresponding length scales, these velocity scales yield both particle and mac- roscopic Reynolds numbers. The analysis that follows, which applies ultimately to a suspension of spheres of radius R, assumes that $Re_R = UR/\nu \ll 1$. Although the effective viscosity of the suspension (which is governed by the microstructure) is derived using a creeping flow analysis, the result applies to macroscopic flow calculations in which $Re_L = U_L L/\nu$ is not necessarily small.

Only a few restrictions are imposed initially, allowing the first part of the analysis to be fairly general. It is assumed that the suspending fluid (continuous phase) is Newtonian with vis- cosity μ_0 and that the particles and fluid have the same density. The particles are assumed to have no preferred orientation at equilibrium, so that the fluid is isotropic. It is assumed also that there are no external couples acting on the particles, so that the particles are torque-free and the average stress in the suspension is symmetric (see Section 5.4). After deriving several general relationships involving the average stress and average rate of strain in a suspension of rigid particles, the problem is simplified by considering a very dilute suspension of spheres. The overall approach is conceptually similar to, but less general than, that employed by Batchelor (1970).

Similar to Eq. (7.2-11), the stress in the fluid phase (without particles) is governed by

$$\boldsymbol{\sigma} = -\mathscr{P}\boldsymbol{\delta} + 2\mu_0\boldsymbol{\Gamma}. \tag{7.6-14}$$

For the suspension we seek an analogous relation,

$$\langle\boldsymbol{\sigma}\rangle = -\mathscr{P}_{\text{eff}}\boldsymbol{\delta} + 2\mu_{\text{eff}}\langle\boldsymbol{\Gamma}\rangle, \tag{7.6-15}$$

where $\mathscr{P}_{\text{eff}}$ and μ_{eff} are the effective pressure and effective viscosity, respectively. The average stress, $\langle\boldsymbol{\sigma}\rangle$, and average rate of strain, $\langle\boldsymbol{\Gamma}\rangle$, are defined as

$$\langle\boldsymbol{\sigma}\rangle = \frac{1}{V}\int_V \boldsymbol{\sigma}\, dV, \qquad \langle\boldsymbol{\Gamma}\rangle = \frac{1}{V}\int_V \boldsymbol{\Gamma}\, dV, \tag{7.6-16}$$

where V is a volume of suspension which includes both particles and fluid. As in the discussion of the continuum hypothesis in Chapter 1, this averaging volume must be large enough to encom- pass a statistically representative sample of particles, but small enough that $V^{1/3} \ll L$, where L is the length scale at which the effective viscosity will be applied.

Separating the volume integrals in Eq. (7.6-16) into contributions from the volumes occu- pied by the fluid (V_f) and the particles (V_p), we obtain

$$\langle\boldsymbol{\sigma}\rangle - 2\mu_0\langle\boldsymbol{\Gamma}\rangle = \frac{1}{V}\int_{V_f} (\boldsymbol{\sigma} - 2\mu_0\boldsymbol{\Gamma})\, dV + \frac{1}{V}\int_{V_P} (\boldsymbol{\sigma} - 2\mu_0\boldsymbol{\Gamma})\, dV$$

$$= -\frac{\boldsymbol{\delta}}{V}\int_{V_f} \mathscr{P}\, dV + \frac{1}{V}\int_{V_P} \boldsymbol{\sigma}\, dV. \tag{7.6-17}$$

In the second equality we have used the fact that $\boldsymbol{\Gamma} = \mathbf{0}$ within any rigid particle. The last term is rewritten using the identity

$$\boldsymbol{\sigma} = \boldsymbol{\nabla}\cdot(\boldsymbol{\sigma}\mathbf{r}) - (\boldsymbol{\nabla}\cdot\boldsymbol{\sigma})\mathbf{r}, \tag{7.6-18}$$

where $\mathbf{r}$ is the position vector; this relation depends on $\boldsymbol{\sigma}$ being symmetric. With no inertia and negligible body forces, $\boldsymbol{\nabla}\cdot\boldsymbol{\sigma}$ vanishes in a solid particle as it does in a fluid [see Eq. (7.2-7)]. Thus, using Eq. (7.6-18) and the divergence theorem, the last integral of Eq. (7.6-17) is expressed as

$$\int_{V_P} \boldsymbol{\sigma} \, dV = \int_{V_P} \nabla \cdot (\boldsymbol{\sigma} \mathbf{r}) \, dV = \int_{S_P} (\mathbf{n} \cdot \boldsymbol{\sigma}) \mathbf{r} \, dS. \tag{7.6-19}$$

Using Eq. (7.6-19) and defining $\mathscr{P}_{\text{eff}}$ as the volume-average pressure in Eq. (7.6-17), we obtain

$$\langle \boldsymbol{\sigma} \rangle - 2\mu_0 \langle \boldsymbol{\Gamma} \rangle = -\mathscr{P}_{\text{eff}} \boldsymbol{\delta} + \frac{1}{V} \int_{S_P} (\mathbf{n} \cdot \boldsymbol{\sigma}) \mathbf{r} \, dS. \tag{7.6-20}$$

As it stands, Eq. (7.6-20) requires integration over the surface of *every particle* in the volume *V*. Even for particles of identical size and shape, different results are obtained for each particle at any given instant. This is because the stress on the surface of any given particle is influenced, in general, by velocity perturbations due to many other particles and therefore depends on their instantaneous positions. For nonspherical particles, the stress depends also on the particle orientations. Thus, the integral must, in general, represent an average over all possible particle configurations (i.e., an ensemble average). The nature of the averaging process is discussed in more detail in Batchelor (1970). For a dilute suspension of identical spheres, however, the problem is greatly simplified. When particle–particle interactions are negligible, the integral is the same for each particle. Thus, for *m* spheres per unit volume, we have

$$\frac{1}{V} \int_{S_P} (\mathbf{n} \cdot \boldsymbol{\sigma}) \mathbf{r} \, dS = m \int_{S_0} (\mathbf{n} \cdot \boldsymbol{\sigma}) \mathbf{r} \, dS = m\mathbf{D}, \tag{7.6-21}$$

where S_0 denotes the surface of a *single* sphere. The last form of Eq. (7.6-21) involves the dipole tensor **D** introduced in Example 7.5-1.

If the suspension is Newtonian, then Eq. (7.6-15) will apply to any type of flow. This suggests that to calculate μ_{eff} for a dilute suspension of identical spheres, it is sufficient to evaluate **D** for any flow involving a freely suspended sphere. Following Leal (1992, pp. 174–177), we use the results for axisymmetric extensional flow (see Example 7.4-3). Choosing the origin as the center of the sphere, so that $\mathbf{n} = \mathbf{e}_r$ and $\mathbf{r} = R\mathbf{e}_r$, the integral which gives **D** is

$$\mathbf{D} = R^3 \int_0^{2\pi} \int_0^{\pi} \left[-\mathscr{P}(R, \theta)\mathbf{e}_r\mathbf{e}_r + \tau_{r\theta}(R, \theta)\mathbf{e}_\theta\mathbf{e}_r \right] \sin \theta \, d\theta \, d\phi. \tag{7.6-22}$$

Using the relations in Section A.7, it is found that

$$\int_0^{2\pi} \mathbf{e}_r\mathbf{e}_r \, d\phi = \pi(\sin^2 \theta \mathbf{e}_x\mathbf{e}_x + \sin^2 \theta \mathbf{e}_y\mathbf{e}_y + 2 \cos^2 \theta \mathbf{e}_z\mathbf{e}_z), \tag{7.6-23}$$

$$\int_0^{2\pi} \mathbf{e}_\theta\mathbf{e}_r \, d\phi = \pi \sin \theta \cos \theta (\mathbf{e}_x\mathbf{e}_x + \mathbf{e}_y\mathbf{e}_y - 2\mathbf{e}_z\mathbf{e}_z). \tag{7.6-24}$$

Evaluating $\tau_{r\theta}(R, \theta)$ and $\mathscr{P}(R, \theta)$ using Eqs. (7.4-50) and (7.4-51), respectively, the result for **D** is

$$\mathbf{D} = -\frac{20}{3} \pi R^3 \mu_0 \gamma (\mathbf{e}_x\mathbf{e}_x + \mathbf{e}_y\mathbf{e}_y - 2\mathbf{e}_z\mathbf{e}_z). \tag{7.6-25}$$

When **D** is symmetric and has zero trace, as in Eq. (7.6-25), it is identical to the *stresslet* tensor introduced by Batchelor (1970) and used in many subsequent studies; for more information on the relationship between **D** and the stresslet see, for example, Kim and Karrila (1991, pp. 27–31).

For a dilute suspension, $\langle \boldsymbol{\Gamma} \rangle$ is very nearly the average rate of strain in the unperturbed flow. For axisymmetric extensional flow, the unperturbed rate of strain is a constant. Using Eq. (7.4-43) to evaluate the rate of strain, Eq. (7.6-25) reduces to

$$\mathbf{D} = \frac{20}{3} \pi R^3 \mu_0 \langle \mathbf{\Gamma} \rangle = 5 V_0 \mu_0 \langle \mathbf{\Gamma} \rangle, \qquad (7.6\text{-}26)$$

where V_0 is the sphere volume. Combining Eq. (7.6-26) with Eqs. (7.6-20) and (7.6-21) gives

$$\langle \mathbf{\sigma} \rangle = -\mathcal{P}_{\text{eff}} \mathbf{\delta} + 2\mu_0 \left(1 + \frac{5}{2}\phi \right) \langle \mathbf{\Gamma} \rangle, \qquad (7.6\text{-}27)$$

where $\phi = m V_0$ is the volume fraction of spheres in the suspension. Finally, comparing Eq. (7.6-27) with Eq. (7.6-15), we conclude that

$$\frac{\mu_{\text{eff}}}{\mu_0} = 1 + \frac{5}{2}\phi. \qquad (7.6\text{-}28)$$

This well-known result, due originally to Einstein in 1906 (with a correction in 1911), shows that the particles make the apparent viscosity of the suspension greater than that of the continuous phase. The viscosity ratio depends only on the volume fraction of the spheres.

The effective viscosities of relatively dilute suspensions are expressed more generally as

$$\frac{\mu_{\text{eff}}}{\mu_0} = 1 + C_1 \phi + C_2 \phi^2 + O(\phi^3). \qquad (7.6\text{-}29)$$

The result that $C_1 = 5/2$ for solid spheres was calculated originally from the increased viscous dissipation of mechanical energy in a suspension (Einstein, 1956, pp. 49–54); see Chapter 9 for information on viscous dissipation. The dissipation analysis is presented in a more complete form in Happel and Brenner (1965, pp. 431–441). Another approach, used by Burgers (1938), is based on the superposition of velocity perturbations far from a sphere in a simple shear flow. Taylor (1932) extended Einstein's analysis to immiscible fluid spheres and found that

$$C_1 = \frac{\mu_0 + \frac{5}{2}\mu_1}{\mu_0 + \mu_1}, \qquad (7.6\text{-}30)$$

where μ_1 is the viscosity of the discontinuous phase (i.e., the suspended droplets). For a low-viscosity droplet, $\mu_1/\mu_0 \rightarrow 0$ and $C_1 = 1$; Einstein's result, $C_1 = 5/2$, is recovered for $\mu_1/\mu_0 \rightarrow \infty$. The coefficient C_2 in Eq. (7.6-29) reflects hydrodynamic and other interactions between pairs of spheres and is influenced by the spatial distribution of the particles. It depends, for example, on the extent of Brownian motion of the particles. Various theoretical and experimental results for C_2 are reviewed in Russel et al. (1989, pp. 456–503).

7.7 CORRECTIONS TO STOKES' LAW

Stokes' law, derived in Example 7.4-2 by assuming that $Re = UR/\nu \ll 1$, is remarkably accurate even for Re approaching unity. For $Re = 0.5$, for example, it underestimates the drag by only about 10%. The full extent of this good fortune can be appreciated by considering the magnitude of the inertial terms, which were neglected in the analysis. Specifically, we examine what happens far from the sphere, where $r \gg R$. In that region the dominant inertial terms in either the r or θ components (as exemplified by $\rho v_\theta^2/r$) are $\sim \rho U^2/r$; the largest viscous terms in either equation (as exemplified by $2\mu v_r/r^2$) are

$\sim \mu U/r^2$. Thus, far from the sphere, the ratio of the dominant inertial to dominant viscous terms is given by

$$\frac{\text{inertial}}{\text{viscous}} \sim \frac{\rho U^2/r}{\mu U/r^2} = \left(\frac{r}{R}\right)\text{Re}. \qquad (7.7\text{-}1)$$

The rather startling conclusion is that, no matter how small Re is, there is always a region far from the sphere in which inertial effects are dominant. Thus, Stokes' equation is not a *uniformly valid* approximation to the Navier–Stokes equation in this problem for Re $\neq$ 0; it fails in the region far from the sphere. This deficiency has little effect on the drag calculation for Re $\to$ 0, as evidenced by the success of Stokes' law. However, it had an important effect on attempts made over nearly a century to derive corrections to Stokes' law for small but finite Re. The history of those efforts provides an interesting case study, and the ultimate resolution of the mathematical difficulties illustrates the power of singular perturbation analysis.

To improve upon Stokes' solution for flow around a sphere, we need to consider the full Navier–Stokes equation. It is easiest to use dimensionless quantities defined as

$$\tilde{r} \equiv \frac{r}{R}, \qquad \tilde{\psi} \equiv \frac{\psi}{UR^2}, \qquad \tilde{E}^2 \equiv R^2 E^2, \qquad \tilde{E}^4 \equiv R^4 E^4. \qquad (7.7\text{-}2)$$

The operator $\tilde{E}^2$ is given by Eq. (7.4-21), with $\tilde{r}$ substituted for r. The stream-function form of the Navier–Stokes equation becomes

$$\tilde{E}^4\tilde{\psi} = -\frac{\text{Re}}{\tilde{r}^2 \sin \theta} \frac{\partial(\tilde{\psi}, \tilde{E}^2\tilde{\psi})}{\partial(\tilde{r}, \theta)} + \frac{2\text{Re}\tilde{E}^2\tilde{\psi}}{\tilde{r}^2 \sin^2 \theta}\left(\frac{\partial\tilde{\psi}}{\partial\tilde{r}} \cos \theta - \frac{\partial\tilde{\psi}}{\partial\theta}\frac{\sin \theta}{\tilde{r}}\right). \qquad (7.7\text{-}3)$$

The boundary conditions for $\tilde{\psi}(\tilde{r}, \theta)$ for a fixed, solid sphere and a uniform unperturbed velocity, corresponding to Eqs. (7.4-24) and (7.4-27), are

$$\frac{\partial\tilde{\psi}}{\partial\theta}(1, \theta) = 0, \qquad \frac{\partial\tilde{\psi}}{\partial\tilde{r}}(1, \theta) = 0 \qquad (7.7\text{-}4)$$

$$\tilde{\psi}(\infty, \theta) \to \frac{\tilde{r}^2}{2}\sin^2 \theta. \qquad (7.7\text{-}5)$$

A seemingly reasonable approach for small but nonzero Re is to assume a regular perturbation expansion, such that

$$\tilde{\psi}(\tilde{r}, \theta) = \tilde{\psi}_0(\tilde{r}, \theta) + \text{Re } \tilde{\psi}_1(\tilde{r}, \theta) + O(\text{Re}^2). \qquad (7.7\text{-}6)$$

Substituting Eq. (7.7-6) into Eq. (7.7-3) and collecting terms involving like powers of Re, we find that $\tilde{\psi}_0$ is simply Stokes' solution, as expected. Rewriting Eq. (7.4-32) in dimensionless form, we have

$$\tilde{\psi}_0 = \left(\frac{1}{2}\tilde{r}^2 - \frac{3}{4}\tilde{r} + \frac{1}{4}\tilde{r}^{-1}\right)\sin^2 \theta. \qquad (7.7\text{-}7)$$

Considering now the $O(\text{Re})$ terms, we conclude that the first correction to Stokes' result, $\tilde{\psi}_1$, must satisfy

$$\tilde{E}^4\tilde{\psi}_1 = -\frac{1}{\tilde{r}^2 \sin \theta}\frac{\partial(\tilde{\psi}_0, \tilde{E}^2\tilde{\psi}_0)}{\partial(\tilde{r}, \theta)} + \frac{2\tilde{E}^2\tilde{\psi}_0}{\tilde{r}^2 \sin^2 \theta}\left(\frac{\partial\tilde{\psi}_0}{\partial\tilde{r}}\cos \theta - \frac{\partial\tilde{\psi}_0}{\partial\theta}\frac{\sin \theta}{\tilde{r}}\right). \qquad (7.7\text{-}8)$$

Using Eq. (7.7-7) to evaluate $\tilde{\psi}_0$, Eq. (7.7-8) becomes

$$\tilde{E}^4\tilde{\psi}_1 = -\frac{9}{4}(2\tilde{r}^{-2} - 3\tilde{r}^{-3} + \tilde{r}^{-5}) \sin^2\theta\cos\theta. \tag{7.7-9}$$

The boundary conditions to be satisfied by $\tilde{\psi}_1$ are Eq. (7.7-4) and, from Eq. (7.7-5),

$$\frac{\tilde{\psi}_1}{\tilde{r}^2} \to 0 \qquad \text{as} \qquad \tilde{r} \to \infty. \tag{7.7-10}$$

That is, $\tilde{\psi}_1$ must grow less rapidly than $\tilde{r}_2$ as $\tilde{r} \to \infty$.

The complete solution to Eq. (7.7-9) will be the sum of a particular solution to the nonhomogeneous equation ($\tilde{\psi}_{1p}$) and a solution to the homogeneous form of the equation ($\tilde{\psi}_{1h}$). The particular solution is found to be

$$\tilde{\psi}_{1p} = -\frac{3}{32}(2\tilde{r}^2 - 3\tilde{r} + 1 - \tilde{r}^{-1} + \tilde{r}^{-2}) \sin^2\theta\cos\theta. \tag{7.7-11}$$

The homogeneous solution will also consist of products of $\tilde{r}^n$ with functions of θ. For $\tilde{\psi}_{1p} + \tilde{\psi}_{1h}$ to grow less rapidly than $\tilde{r}^2$ as $\tilde{r} \to \infty$, $\tilde{\psi}_{1h}$ cannot have terms with $n > 2$. Moreover, there must be an $n = 2$ term in $\tilde{\psi}_{1h}$ that exactly cancels the term involving $\tilde{r}^2 \sin^2\theta \cos\theta$ in $\tilde{\psi}_{1p}$, if Eq. (7.7-10) is to be satisfied for all values of θ. Accordingly, we seek a homogeneous solution of the form

$$\tilde{\psi}_{1h} = f(\tilde{r}) \sin^2\theta\cos\theta. \tag{7.7-12}$$

Substituting Eq. (7.7-12) into the homogeneous form of Eq. (7.7-9), we find that $f(\tilde{r})$ must satisfy

$$\frac{d^4f}{d\tilde{r}^4} - \frac{12}{\tilde{r}^2}\frac{d^2f}{d\tilde{r}^2} + \frac{24}{\tilde{r}^3}\frac{df}{d\tilde{r}} = 0. \tag{7.7-13}$$

Assuming solutions of the form $f = \tilde{r}^n$, Eq. (7.5-13) yields a characteristic equation which has the roots $n = -2, 0, 3$, and 5. Thus, the homogeneous solution lacks the required term with $n = 2$. We are forced to conclude that the seemingly well-posed problem for $\tilde{\psi}_1$ has no solution. This conclusion, reported by A. N. Whitehead in 1889, is known as *Whitehead's paradox.*[2]

As explained first by C. W. Oseen in 1910, the reason for Whitehead's paradox is the aforementioned failure of Stokes' equation far from the sphere. Oseen's method of proceeding, which we will not describe in detail, was to approximate the inertial terms in the Navier–Stokes equation on the basis of physical intuition. The term $\mathbf{v}\cdot\nabla\mathbf{v}$, neglected entirely in Stokes' equation, was approximated as $\mathbf{U}\cdot\nabla\mathbf{v}$, where $\mathbf{U}$ is the uniform velocity vector far from the sphere. This approximation was motivated by the fact that the inertial terms only become important far from the sphere, where $\mathbf{v} \cong \mathbf{U}$. The use of $\mathbf{U}\cdot\nabla\mathbf{v}$ also made the Navier–Stokes equations linear, facilitating their solution. Although this approach yielded a first correction to Stokes' law that was in excellent agreement with data for the drag on spheres, it left certain basic questions unanswered. Why should

---

[2] Alfred North Whitehead (1861–1947) is better known for his contributions to mathematics, logic, the philosophy of science and metaphysics. It is possible that this early encounter with Stokes flow encouraged him to direct his intellectual abilities elsewhere.

the substitution of $\mathbf{U}$ for $\mathbf{v}$ in just one term work so well? How could one further refine the approximations to the velocity field and drag coefficient?

The relationship between Oseen's approximation and the Navier–Stokes equation remained unclear until the 1950s, when singular perturbation techniques were applied to low Re flow past a sphere (Proudman and Pearson, 1957). In that approach, Eq. (7.7-3) is recognized as a properly scaled form of the Navier–Stokes equation only in the region adjacent to the sphere. Near the sphere the viscous terms are dominant for small Re, and Stokes' equation ($\tilde{E}^4\tilde{\psi}=0$) is a suitable first approximation. By contrast, in the region far from the sphere the inertial terms become comparable to the viscous terms, even for very small Re. The analysis which follows is similar to that presented by Van Dyke (1964, pp. 149–161).

As discussed in Section 7.5, the velocity perturbation far from the sphere is the same as that caused by an equivalent point force, suggesting that the sphere radius R is no longer the important length scale. A more relevant characteristic length far from the sphere is the radius at which the inertial terms become important. From Eq. (7.7-1), this radius corresponds to $(r/R)\text{Re} \sim 1$, or $r \sim R/\text{Re}$. This suggests that

$$\hat{r} = \text{Re}\ \tilde{r} \qquad (7.7\text{-}14)$$

will be the properly scaled radial coordinate in the region far from the sphere.

Modifying the stream function and the differential operators for the region far from the sphere, according to Eq. (7.7-14), we obtain

$$\hat{\psi} = \text{Re}^2\tilde{\psi}, \qquad \hat{E}^2 = \text{Re}^{-2}\tilde{E}^2, \qquad \hat{E}^4 = \text{Re}^{-4}\tilde{E}^4. \qquad (7.7\text{-}15)$$

The operator $\hat{E}^2$ is given by Eq. (7.4-21), with r replaced by $\hat{r}$. With these definitions substituted into Eq. (7.7-3), the Navier–Stokes equation for the region far from the sphere becomes

$$\hat{E}^4\hat{\psi} = -\frac{1}{\hat{r}^2\sin\theta}\frac{\partial(\hat{\psi},\ \hat{E}^2\hat{\psi})}{\partial(\hat{r},\ \theta)} + \frac{2\hat{E}^2\hat{\psi}}{\hat{r}^2\sin^2\theta}\left(\frac{\partial\hat{\psi}}{\partial\hat{r}}\cos\theta - \frac{\partial\hat{\psi}}{\partial\theta}\frac{\sin\theta}{\hat{r}}\right). \qquad (7.7\text{-}16)$$

Because Re does not appear in Eq. (7.7-16), the new scaling is evidently consistent with the idea that inertial and viscous terms remain comparable far from the sphere, even for $\text{Re} \to 0$.

The original stream function $\tilde{\psi}$ must continue to satisfy the boundary conditions on the sphere surface, Eq. (7.7-4). However, the boundary condition far from the sphere must be satisfied by $\hat{\psi}(\hat{r},\ \theta)$ instead of $\tilde{\psi}(\tilde{r},\ \theta)$. Thus, instead of Eq. (7.7-5), we have

$$\hat{\psi}(\infty,\ \theta) \to \frac{1}{2}\hat{r}^2\sin^2\theta. \qquad (7.7\text{-}17)$$

The remaining constants in the two solutions, those which cannot be evaluated from the boundary conditions, must be determined by asymptotic matching of $\tilde{\psi}$ and $\hat{\psi}$ where the two regions overlap. For sufficiently small Re, large values of $\tilde{r}$ will correspond to small values of $\hat{r}$. For $\text{Re} \to 0$, the matching requirement is

$$\lim_{\tilde{r}\to\infty} \text{Re}^2\tilde{\psi} = \lim_{\hat{r}\to 0} \hat{\psi}. \qquad (7.7\text{-}18)$$

We will now determine the first few terms in the expansions for $\tilde{\psi}$ and $\hat{\psi}$. The respective expansions are written as

$$\tilde{\psi}(\tilde{r},\,\theta) = \sum_{n=0}^{\infty} \tilde{F}_n(\mathrm{Re})\,\tilde{\psi}_n(\tilde{r},\,\theta), \qquad (7.7\text{-}19)$$

$$\hat{\psi}(\hat{r},\,\theta) = \sum_{n=0}^{\infty} \hat{F}_n(\mathrm{Re})\,\hat{\psi}_n(\hat{r},\,\theta), \qquad (7.7\text{-}20)$$

where only the coefficients $\tilde{F}_n$ and $\hat{F}_n$ depend on Re. In a regular perturbation expansion we would have $\tilde{F}_n = \mathrm{Re}^n$ as in the Whitehead expansion, Eq. (7.7-6). For the present, singular case we do not know in advance the dependence of $\tilde{F}_n$ and $\hat{F}_n$ on Re, but will assume that

$$\lim_{\mathrm{Re}\to 0} \tilde{F}_{n+1}/\tilde{F}_n \to 0, \qquad \lim_{\mathrm{Re}\to 0} \hat{F}_{n+1}/\hat{F}_n \to 0. \qquad (7.7\text{-}21)$$

The scaling of Eqs. (7.7-3) and (7.7-16) ensures that $\tilde{F}_0 = \hat{F}_0 = 1$.

The leading terms in Eqs. (7.7-19) and (7.7-20) are identified by recognizing that the former represents a perturbation about Stokes' solution and that the latter represents a perturbation about the uniform, unperturbed velocity. Thus, $\tilde{\psi}_0$ is Stokes' solution [Eq. (7.7-7)], $\hat{\psi}_0$ corresponds to the unperturbed velocity, and

$$\tilde{\psi} = \left(\frac{1}{2}\,\tilde{r}^2 - \frac{3}{4}\,\tilde{r} + \frac{1}{4}\,\tilde{r}^{-1}\right)\sin^2\theta + \sum_{n=1}^{\infty} \tilde{F}_n(\mathrm{Re})\,\tilde{\psi}_n, \qquad (7.7\text{-}22)$$

$$\hat{\psi} = \frac{1}{2}\,\hat{r}^2\sin^2\theta + \sum_{n=1}^{\infty}\hat{F}_n(\mathrm{Re})\,\hat{\psi}_n. \qquad (7.7\text{-}23)$$

We have already seen that $\tilde{\psi}_0$ satisfies the $O(1)$ terms in Eq. (7.7-3). It is easily confirmed that $\hat{\psi}_0$, the leading term in Eq. (7.7-23), satisfies Eq. (7.7-16) at $O(1)$. Indeed, it is found that $\hat{E}^2\hat{\psi}_0 = 0$, so that all terms in Eq. (7.7-16) vanish. Moreover, $\tilde{\psi}_0$ and $\hat{\psi}_0$ satisfy the boundary conditions for the respective problems, Eqs. (7.7-4) and (7.7-17), and the matching condition, Eq. (7.7-18).

We can infer the form of the second term in the expansion far from the sphere by examining the behavior of $\tilde{\psi}_0$ for large $\hat{r}$. In terms of the original variables,

$$\tilde{\psi}_0 = \frac{1}{2}\,\tilde{r}^2\sin^2\theta - \frac{3}{4}\,\tilde{r}\sin^2\theta + O(r^{-1}). \qquad (7.7\text{-}24)$$

Converting Eq. (7.7-24) to the radial coordinate used in the region far from the sphere, we obtain

$$\mathrm{Re}^2\,\tilde{\psi}_0 = \frac{1}{2}\,\hat{r}^2\sin^2\theta - \frac{3}{4}\,\mathrm{Re}\,\hat{r}\sin^2\theta + O(\mathrm{Re}^2\,\hat{r}^{-1}). \qquad (7.7\text{-}25)$$

A comparison of Eq. (7.7-25) with Eq. (7.7-23) suggests that the next term of the expansion far from the sphere will be $O(\mathrm{Re})$. In other words, we conclude that $\hat{F}_1 = \mathrm{Re}$. Substituting Eq. (7.7-20) into Eq. (7.7-16) and equating the $O(\mathrm{Re})$ terms, we find that $\hat{\psi}_1$ must satisfy

$$\hat{E}^4\hat{\psi}_1 = -\frac{1}{\hat{r}^2\sin\theta}\left(\frac{\partial\hat{\psi}_0}{\partial\hat{r}}\frac{\partial\hat{E}^2\hat{\psi}_1}{\partial\theta} - \frac{\partial\hat{\psi}_0}{\partial\theta}\frac{\partial\hat{E}^2\hat{\psi}_1}{\partial\hat{r}}\right)$$

$$+\frac{2\hat{E}^2\hat{\psi}_1}{\hat{r}^2 \sin^2\theta}\left(\frac{\partial\hat{\psi}_0}{\partial\hat{r}}\cos\theta-\frac{\partial\hat{\psi}_0}{\partial\theta}\frac{\sin\theta}{\hat{r}}\right),\qquad(7.7\text{-}26)$$

where we have used the fact that $\hat{E}^2\hat{\psi}_0=0$. It is interesting to note that Eq. (7.7-26) is the same as Oseen's approximation to the Navier–Stokes equation, although it was obtained in an entirely different manner.

Using the expression for, $\hat{\psi}_0$, Eq. (7.7-26) reduces to

$$\left(\hat{E}^2+\frac{\sin\theta}{\hat{r}}\frac{\partial}{\partial\theta}-\cos\theta\frac{\partial}{\partial\hat{r}}\right)\hat{E}^2\hat{\psi}_1=0.\qquad(7.7\text{-}27)$$

The boundary condition for $\hat{\psi}_1$ [analogous to Eq. (7.7-10)] is

$$\frac{\hat{\psi}_1}{\hat{r}^2}\rightarrow 0\qquad \text{as }\hat{r}\rightarrow\infty.\qquad(7.7\text{-}28)$$

The solution to Eq. (7.7-27) which satisfies Eq. (7.7-28) is

$$\hat{\psi}_1=c_1(1+\cos\theta)\left[1-\exp\left(-\frac{\hat{r}}{2}(1-\cos\theta)\right)\right],\qquad(7.7\text{-}29)$$

where c_1 is an unknown constant. This solution was obtained by using a transformation of the form $\hat{E}^2\hat{\psi}_1=\exp(z/2)\phi_1(\hat{r},\ \theta)=\exp[(\hat{r}/2)\cos\theta]\phi_1\ (\hat{r},\ \theta)$, which is motivated by the uniform velocity in the $+z$ direction far from the sphere. A transformation like this is often useful in problems involving uniform velocities (see Problem 4-20).

The constant c_1 in Eq. (7.7-29) is found by matching the two expansions. To apply the matching condition, Eq. (7.7-18), we first examine the behavior of Eq. (7.7-29) for small $\hat{r}$. Expanding the exponential, we obtain

$$\hat{\psi}_1=\frac{c_1}{2}\hat{r}\sin^2\theta-\frac{c_1}{8}\hat{r}^2\sin^2\theta(1-\cos\theta)+O(\hat{r}^3).\qquad(7.7\text{-}30)$$

Comparing Eq. (7.7-30) with the $O(\text{Re})$ term in Eq. (7.7-25), we find that $c_1=-3/2$. The $O(\text{Re})$ part of the solution far from the sphere, evaluated for small $\hat{r}$, is now

$$\hat{\psi}_1=-\frac{3}{4}\hat{r}\sin^2\theta+\frac{3}{16}\hat{r}^2\sin^2\theta(1-\cos\theta)+O(\hat{r}^3).\qquad(7.7\text{-}31)$$

We have now found one term of the expansion for the near region and two terms of the expansion for the far region. To obtain a first correction to Stokes' law, we need the second term of the expansion for the region near the sphere. The form of this next term is suggested by the largest "unmatched" term in Eq. (7.7-31). Combining Eq. (7.7-31) with Eq. (7.7-23), and converting to the original variables, we obtain

$$\text{Re}^{-2}\hat{\psi}=\frac{1}{2}\bar{r}^2\sin^2\theta-\frac{3}{4}\bar{r}\sin^2\theta+\frac{3}{16}\text{Re }\bar{r}^2\sin^2\theta(1-\cos\theta)+O(\text{Re }\bar{r}^3).$$

$$(7.7\text{-}32)$$

Comparing Eq. (7.7-32) with Eq. (7.7-22), we see that the Re $\bar{r}^2$ term in Eq. (7.7-32) must come from $\bar{\psi}_1$ and that $\bar{F}_1=\text{Re}$. Thus, coincidentally, the first two terms of the expansion for the region near the sphere have the same dependence on Re as those in the regular perturbation expansion attempted by Whitehead. It follows that $\bar{\psi}_1$ is governed by Eq. (7.7-9), which has the particular solution given by Eq. (7.7-11). Noticing that Eq. (7.7-11) already matches the Re $\tilde{r}^2 \sin^2 \theta \cos \theta$ term in Eq. (7.7-32), we conclude that the homogeneous solution must be of the form

$$\tilde{\psi}_{1h} = f(\tilde{r}) \sin^2 \theta, \tag{7.7-33}$$

where $f(\tilde{r})$ is a different function than that appearing in Eq. (7.7-12). We have already seen that the downfall of Whitehead's approach was the requirement that $\tilde{\psi}_{1h}$ cancel $\tilde{\psi}_{1p}$ for large $\tilde{r}$. It is now clear that, instead, $\tilde{\psi}_{1h}$ should be chosen to match the appropriate part of $\hat{\psi}$, as was done in selecting Eq. (7.7-33).

To complete the solution for $\tilde{\psi}_1$, we note that $\tilde{\psi}_{1h}$ has the same form as Stokes' solution, Eq. (7.7-7). Moreover, from Eq. (7.7-9), $\tilde{E}^4\tilde{\psi}_{1h} = 0$. Thus, $f(\tilde{r})$ in Eq. (7.7-33) is given by Eq. (7.4-31) and

$$\tilde{\psi}_{1h} = (A\tilde{r}^4 + B\tilde{r}^2 + C\tilde{r} + D\tilde{r}^{-1}) \sin^2 \theta. \tag{7.7-34}$$

We conclude from a comparison with Eq. (7.7-32) that $A = 0$ and $B = 3/16$ are needed to match the next part of the expansion developed for the region far from the sphere. It follows from the no-slip condition on the sphere surface, Eq. (7.7-4), that $C = -9/32$ and $D = 3/32$. This completes the determination of $\tilde{\psi}_1$.

The expansion for the stream function in the region near the sphere is now

$$\tilde{\psi} = \frac{1}{4}(2\tilde{r}^2 - 3\tilde{r} + \tilde{r}^{-1}) \sin^2 \theta$$

$$+ \frac{3}{32} \mathrm{Re}[(2\tilde{r}^2 - 3\tilde{r} + \tilde{r}^{-1}) - (2\tilde{r}^2 - 3\tilde{r} + 1 - \tilde{r}^{-1} + \tilde{r}^{-2}) \cos \theta] \sin^2 \theta + O(\tilde{F}_2).$$

$$\tag{7.7-35}$$

The $O(1)$ term in Eq. (7.7-35) yields Stokes' law, as shown in Example 7.4-2. The first correction to the stream function, and therefore the first corrections to the velocity components, the pressure, and the drag coefficient, are all evidently proportional to Re. A straightforward but lengthy calculation reveals that the drag force on the sphere (F_D) is

$$F_D = 6\pi\mu UR \left[1 + \frac{3}{8}\mathrm{Re} + O(\tilde{F}_2) \right]. \tag{7.7-36}$$

The first correction term, $\frac{3}{8}$ Re, is the same as that derived by Oseen. This agreement stems from the coincidental correspondence between Oseen's equation and Eq. (7.7-26), as already noted.

The singular perturbation analysis for the sphere has been carried out to include additional terms. More complete results for the drag are

$$F_D = 6\pi\mu UR \left[1 + \frac{3}{8}\mathrm{Re} + \frac{9}{40}\mathrm{Re}^2\left(\ln \mathrm{Re} + \gamma + \frac{5}{3} \ln 2 - \frac{323}{360} \right) \right.$$

$$\left. + \frac{27}{80} \mathrm{Re}^3 \ln \mathrm{Re} + O(\mathrm{Re}^3) \right], \tag{7.7-37}$$

where γ is Euler's constant ($\gamma = 0.5772 \dots$). We see now that $\tilde{F}_2 = \mathrm{Re}^2 \ln \mathrm{Re}$, $\tilde{F}_3 = \mathrm{Re}^2$, and so on. The absence of a simple power series in the small parameter (Re) is typical of singular perturbation expansions. The $Re^2 \ln Re$ term was obtained by Proudman and Pearson (1957), and the higher-order terms were obtained by Chester and Breach (1969).

A noteworthy feature of Eq. (7.7-35) is that the streamlines with $O(Re)$ contributions no longer show the fore-and-aft symmetry which was present in Stokes' solution (Fig. 7-5); in particular, the $\cos \theta \; \sin^2 \theta$ term does not yield the previous symmetry about $\theta = \pi/2$. Thus, even a small amount of inertia prevents the flow from being exactly reversible. The lack of fore-and-aft symmetry is evident also in the stream-function expansion for the region far from the sphere,

$$\hat{\psi} = \frac{1}{2} \hat{r}^2 \sin^2 \theta - \frac{3}{2} Re(1 + \cos \theta)\left[1 - \exp\left(-\frac{\hat{r}}{2}(1 - \cos \theta)\right)\right] + O(Re^2 \ln Re).$$

$$(7.7\text{-}38)$$

Again, at $O(Re)$ there are angular terms other than $\sin^2 \theta$. More insight is gained by examining the velocity field far from the sphere. Using Eq. (7.7-38) to evaluate $\tilde{v}_r = v_r/U$, we obtain

$$\tilde{v}_r = -\frac{1}{\hat{r}^2 \sin \theta} \frac{\partial \hat{\psi}}{\partial \theta}$$

$$\cong \cos \theta + \frac{3}{2} Re\left[-\frac{(1 + \cos \theta)}{2\hat{r}}\exp\left(-\frac{\hat{r}}{2}(1 - \cos \theta)\right) + \frac{1}{\hat{r}^2}\left(1 - \exp\left(-\frac{\hat{r}}{2}(1 - \cos \theta)\right)\right)\right].$$

$$(7.7\text{-}39)$$

The uniform velocity far from the sphere corresponds to $\hat{v}_r = \cos \theta$; all of the $O(Re)$ terms represent deviations from the uniform flow. It is seen that when $\hat{r}$ is large, these deviations will be greatest at locations where the exponential terms do not vanish. In other words, the velocity perturbations are greatest for values of $\hat{r}$ and θ such that $(\hat{r}/2(1 - \cos \theta)$ is not large. With $\hat{r} \to \infty$, this implies that the largest perturbations are for $\cos \theta \to 1$ or $\theta \to 0$, which corresponds to positions directly downstream from the sphere. Similar conclusions are reached for $\tilde{v}_\theta$. Thus, the sphere is predicted to have a wake. Noting that $(1 - \cos \theta) \cong \theta^2/2$ for small θ, it is seen that the wake spreads at an angle θ_w which varies as $\hat{r}^{-1/2}$. The dimensionless width of the wake, $\hat{r} \sin \theta_w \cong \hat{r}\theta_w$, therefore varies as $\hat{r}^{1/2}$.

The failure of Stokes' equation to provide a uniformly valid approximation to the Navier–Stokes equation for small but nonzero Re has also influenced attempts to calculate the drag on a cylinder oriented perpendicular to the flow. With the cylinder, however, the singular character of the problem affects even the first approximation to the drag, making the failure more dramatic than for a sphere. Thus, it is found that for flow around a long cylinder (or any other two-dimensional object) there is no solution to Stokes' equation which will yield the appropriate unperturbed velocity far from the object (see Problem 7-11); this perplexing result, noted by Stokes himself, is called *Stokes' paradox*. A consequence of there being no solution to Stokes' equation is that there is no drag formula for a cylinder which is analogous to Stokes' law. Stokes' paradox was resolved in much the same way as Whitehead's: The drag on a cylinder was derived first using the Oseen approximation (Lamb, 1932, pp. 614–617) and the analysis was later refined using singular perturbation methods (Kaplun, 1957; Proudman and Pearson, 1957; Van Dyke, 1964, pp. 161–165). From Kaplun (1957), the drag force per unit length on a cylinder of radius R is given by

$$F_D = 4\pi\mu U\varepsilon[1 - 0.87\varepsilon^2 + O(\varepsilon^3)], \qquad \varepsilon = \left[\ln\left(\frac{4}{\mathrm{Re}}\right) + \frac{1}{2} - \gamma\right]^{-1}, \qquad (7.7\text{-}40)$$

where $\mathrm{Re} \equiv UR/\nu$ and γ ($= 0.5772 \ldots$) is Euler's constant.

References

Batchelor, G. K. The stress system in a suspension of force-free particles. *J. Fluid Mech.* 41:545–570, 1970.

Brenner, H. The Stokes resistance of an arbitrary particle—IV. Arbitrary fields of flow. *Chem. Eng. Sci.* 19: 703–727, 1964.

Brinkman, H. C. A calculation of the viscous force exerted by a flowing fluid on a dense swarm of particles. *Appl. Sci. Res.* A1: 27–34, 1947.

Bungay, P. M. and H. Brenner. The motion of a closely fitting sphere in a fluid-filled tube. *Int. J. Multiphase Flow* 1: 25–56, 1973.

Burgers, J. M. On the motion of small particles of elongated form, suspended in a viscous liquid. In: *Second Report on Viscosity and Plasticity.* Kon. Ned. Akad.Wet., Verhand. (Eerste Sectie), Dl. XVI, No. 4. N. V. Noord-Hollandsche, Uitgeversmaatschappij, Amsterdam, 1938, Chapter III, pp. 113–184.

Chester, W. and D. R. Breach. On the flow past a sphere at low Reynolds number. *J. Fluid Mech.* 37: 751–760, 1969.

Chwang, A. T. and T. Y. Wu. Hydromechanics of low-Reynolds-number flow. Part 2. Singularity method for Stokes flows. *J. Fluid Mech.* 67: 787–815, 1975.

Einstein, A. *Investigations on the Theory of the Brownian Movement.* Dover, New York, 1956. [This is an edited English translation of Einstein's early work on diffusion and molecular dimensions; the original papers appeared between 1905 and 1911.]

Gillispie, C. G. (Ed.). *Dictionary of Scientific Biography.* Scribner's, New York, 1970–1980.

Happel, J. and H. Brenner. *Low Reynolds Number Hydrodynamics.* Prentice-Hall, Englewood Cliffs, NJ, 1965 [reprinted by Martinus Nijhoff, The Hague, 1983].

Harper, J. F. The motion of bubbles and drops through liquids. *Adv. Appl. Mech.* 12: 59–129, 1972.

Kaplun, S. Low Reynolds number flow past a circular cylinder. *J. Math. Mech.* 6: 585–593, 1957.

Kim, S. and S. J. Karrila. *Microhydrodynamics.* Butterworth-Heinemann, Boston, 1991.

Ladyzhenskaya, O. A. *The Mathematical Theory of Viscous Incompressible Flow,* second edition. Gordon and Breach, New York, 1969.

Lamb, H. *Hydrodynamics,* sixth edition. Cambridge University Press, Cambridge, 1932.

Leal, L. G. *Laminar Flow and Convective Transport Processes.* Butterworth-Heinemann, Boston, 1992.

Moffat, H. K. Viscous and resistive eddies near a sharp corner. *J. Fluid Mech.* 18: 1–18, 1964.

Proudman. I. and J. R. A. Pearson. Expansions at small Reynolds numbers for the flow past a sphere and a circular cylinder. *J. Fluid Mech.* 2: 237–262, 1957.

Rallison, J. M. Note on the Faxén relations for a particle in Stokes flow. *J. Fluid Mech.* 88: 529–533, 1978.

Russel, W. B., D. A. Saville, and W. R. Schowalter. *Colloidal Dispersions.* Cambridge University Press, Cambridge, 1989.

Stokes, G. G. On the effect of the internal friction of fluids on the motion of pendulums. *Trans. Cambridge Philos. Soc.* 9: 8–106, 1851.

Taylor, G. I. The viscosity of a fluid containing small drops of another fluid. *Proc. R. Soc. (Lond.)* A 138: 41–48, 1932.

Van Dyke, M. *Perturbation Methods in Fluid Mechanics*. Academic Press, New York, 1964.
Young, N. O., J. S. Goldstein, and M. J. Block. The motion of bubbles in a vertical temperature
gradient. *J. Fluid Mech.* 6: 350–356, 1959.

Problems

7-1. General Formula for the Drag on a Sphere in Axisymmetric Flow

With reference to the general solution for the stream function given by Eq. (7.4-39), show
that the drag on a spherical object in any axisymmetric creeping flow is given by

$$F_D = 4\pi\mu C_1.$$

Note that C_1 has units of velocity times length. Apply this result to uniform flow (Example 7.4-
2) and axisymmetric extensional flow (Example 7.4-3) past a solid sphere. [*Hint:* Calculate the
drag by integrating the stress over a large spherical surface surrounding the object, keeping in
mind that the velocity and stress perturbations decay as r^{-1} and r^{-2}, respectively, far from the
object; see Eq. (7.2-9) and Example 7.5-1.)

7-2. Flow Past a Small Bubble or Liquid Droplet

Consider a fluid sphere of radius R moving through a large volume of stagnant liquid at a
constant velocity U. The sphere may be a gas bubble or an immiscible liquid. The internal (bubble
or droplet) and external viscosities are μ_1 and μ_2, respectively. Both fluids are Newtonian with
constant properties. It is convenient to choose a reference frame fixed at the center of the sphere,
as shown in Fig. 7-4.

(a) Reynolds numbers can be calculated using either the internal or external fluid properties.
For the special case of a gas bubble, which Reynolds number is likely to be limiting in
determining if Stokes' equation can be applied throughout the system?

(b) Assuming that Stokes' equation is valid everywhere, find the general solutions for the
internal and external velocities, $\mathbf{v}^{(1)}$ and $\mathbf{v}^{(2)}$, and identify all boundary conditons which
must be satisfied.

(c) If the viscosity ratio is defined as $\lambda \equiv \mu_1/\mu_2$, show that

$$v_r^{(1)} = -\frac{U \cos \theta}{2(1+\lambda)}\left[1 - \left(\frac{r}{R}\right)^2\right],$$

$$v_\theta^{(1)} = \frac{U \sin \theta}{2(1+\lambda)}\left[1 - 2\left(\frac{r}{R}\right)^2\right],$$

$$v_r^{(2)} = U \cos \theta\left[1 - \frac{2+3\lambda}{2(1+\lambda)}\left(\frac{R}{r}\right) + \frac{\lambda}{2(1+\lambda)}\left(\frac{R}{r}\right)^3\right],$$

$$v_\theta^{(2)} = -U \sin \theta\left[1 - \frac{2+3\lambda}{4(1+\lambda)}\left(\frac{R}{r}\right) - \frac{\lambda}{4(1+\lambda)}\left(\frac{R}{r}\right)^3\right].$$

Discuss the limiting behavior of these results for $\lambda \to 0$ and $\lambda \to \infty$.

(d) Use the general result in Problem 7-1 to show that the drag on the fluid sphere is given
by

$$F_D = 2\pi\mu_2 UR\left(\frac{2+3\lambda}{1+\lambda}\right).$$

(e) Complete the force balance on the sphere, to show that the terminal velocity is given by

$$U = \frac{2}{3}\left(\frac{1+\lambda}{2+3\lambda}\right)\frac{R^2 g(\rho_2 - \rho_1)}{\mu_2}.$$

This is called the *Hadamard–Rybczynski equation* (after J. Hadamard and D. Rybczynski, who both reported this analysis independently in 1911). Discuss the limiting behavior of the drag and terminal velocity for $\lambda \to 0$ and $\lambda \to \infty$.

The motion of small bubbles and drops often deviates from the predictions of the Hadamard–Rybczynski analysis, due to the presence of surfactants. The balance between surfactant convection and diffusion can lead to gradients in surfactant concentration at the interface and, consequently, gradients in surface tension. A nonuniform surface tension affects the tangential stress balance (see Section 5.7). A review of many aspects of the motion of bubbles and drops is given by Harper (1972).

7-3. Flow in Porous Media: Brinkman's Equation and Darcy's Law

A relation sometimes used to describe viscous flow in porous media is *Brinkman's equation* (Brinkman, 1947),

$$\mu\nabla^2\mathbf{v} - \frac{\mu}{\kappa}\mathbf{v} - \nabla\mathcal{P} = \mathbf{0}.$$

This is a semiempirical combination of Stokes' equation and *Darcy's law*, the latter being given by

$$\mathbf{v} = -\frac{\kappa}{\mu}\nabla\mathcal{P},$$

where κ is the Darcy permeability of the porous material (see Problem 2-2). Because Brinkman's equation has second-order derivatives of $\mathbf{v}$, it can satisfy no-slip conditions at solid surfaces bounding the porous material (e.g., the walls of a packed-bed reactor), whereas Darcy's law cannot. In that sense, Brinkman's equation is more exact than Darcy's law.

Consider a parallel-plane channel with plate spacing $2H$, as shown in Fig. P7-3. The channel is filled with a porous material of known κ (e.g., solid spheres of diameter $d_s \ll H$). The mean (superficial) velocity through the channel is U.

(a) Under certain conditions, Darcy's law will provide a good approximation to the flow throughout most of the channel, except in thin regions (boundary layers) near the wall. When will this be the case?

(b) Determine $\mathbf{v}$ and $\mathcal{P}$ for fully developed flow using Darcy's law.

(c) Determine $\mathbf{v}$ and $\mathcal{P}$ for fully developed flow using Brinkman's equation. How does the pressure drop for a channel of length L compare with the Darcy result from part (b)? When the boundary layers are very thin, is it important to include them in the calculation of $|\Delta\mathcal{P}|$?

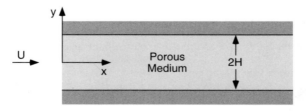

Figure P7-3. Flow through a parallel-plate channel filled with a porous material.

SIDE VIEW   TOP VIEW

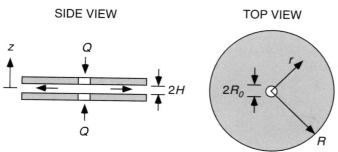

Figure P7-5. Radial flow between parallel disks.

7-4. Flow Between Rotating Disks

A viscous fluid is confined between parallel disks of radius R which are separated by a distance H. The top ($z=H$) and bottom ($z=0$) disks are rotated at angular velocities ω_H and ω_0, respectively. The system is at steady state and the dynamic conditions are such that Stokes' equation is applicable.

(a) Show that there is a solution to the θ-momentum equation of the form $v_\theta = r\, f(z)$, and determine $f(z)$.

(b) Show that $v_r = v_z = 0$ and $\mathcal{P} = $ constant satisfy all of the remaining equations. This confirms that the flow is purely rotational under these conditions.

(c) Show that a purely rotational flow does not exist if inertia is present.

(d) Calculate the external torque which must be applied to the upper disk to maintain steady rotation at given values of ω_H and ω_0.

(e) What Reynolds number must be small for Stokes' equation to be applicable? That is, how should Re be defined?

7-5. Radial Flow Between Parallel Disks

Consider steady, axisymmetric flow of a viscous fluid in the gap between two disks of radius R, as shown in Fig. P7-5. Fluid is introduced at a volumetric flow rate Q through small inlets at the center of each disk, and it flows radially outward; the total flow rate is $2Q$. The inlet radius (R_0) and disk spacing ($2H$) are such that $R_0/H \ll 1$ and $H/R \ll 1$.

(a) Relate $U(r)$, the mean radial velocity in the gap, to Q and the geometric quantities.

(b) Let $U_0 \equiv U(H)$ and assume that $v_z = v_\theta = 0$. If entrance and exit effects are neglected and if $\mathrm{Re} \equiv U_0 H\rho/\mu \ll 1$, show that the continuity and momentum equations (with appropriate boundary conditions) have a solution of the form

$$v_r = \frac{f(z)}{r}, \qquad \mathcal{P} = \mathcal{P}(r).$$

(c) Determine $v_r(r, z)$ for the conditions of part (b). Express v_r in terms of the known flow rate, Q.

7-6. Cone-and-Plate Viscometer

A standard device for measuring viscosities is the cone-and-plate viscometer, as shown in Fig. P7-6. A pool of liquid is placed on a flat surface, which is brought into contact with an inverted cone. Torque measurements are made with the top piece, of radius R, rotated at an angular velocity ω and the bottom piece stationary. The angle β between the surfaces is small. Spherical coordinates (r, θ, ϕ) are used in the analysis, such that the rotation is in the $+\phi$ direction.

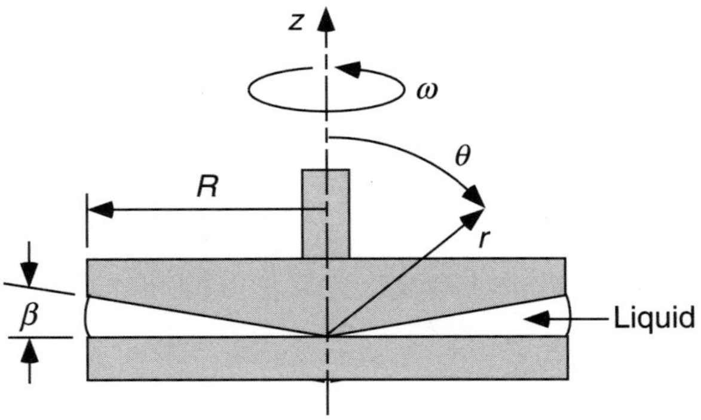

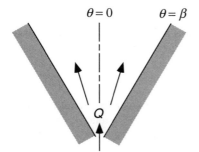

Figure P7-6. Cone-and-plate viscometer.

(a) Show that a velocity field of the form $v_\phi = v_\phi(r, \theta)$ and $v_r = v_\theta = 0$ is consistent with conservation of mass.

(b) Assuming that Stokes' equation is applicable, show that $v_\phi = rf(\theta)$ is consistent with conservation of momentum and the boundary conditions at the solid surfaces. Do this by deriving the differential equation and boundary conditions for $f(\theta)$.

(c) Assuming that $\beta \ll 1$, use the results of (b) to find $v_\phi(r, \theta)$.

(d) Calculate the torque (G) exerted on the bottom surface by the liquid. Show that

$$G = \frac{2\pi}{3} \frac{\mu\omega R^3}{\beta}.$$

7-7. Flow Through a Cone

Consider axisymmetric creeping flow in the conical region shown in Fig. P7-7, in which fluid emerges from a small hole at the apex of the cone at a volumetric flow rate Q. In spherical coordinates the solid surface corresponds to $\theta = \beta$.

(a) Show that a purely radial flow is consistent with continuity and with the boundary conditions at the solid surface.

(b) Show that a purely radial flow is consistent also with Stokes' equation, and determine the velocity and pressure fields. [*Hint:* Let $v_r(r, \eta) = f(\eta)/r^2$, where $\eta \equiv \cos \theta$.)

(c) For $\beta = \pi/2$, which corresponds to a hole in a flat wall, show that the velocity from (b) reduces to

$$v_r(r, \theta) = \frac{3Q \cos^2 \theta}{2\pi r^2}$$

as given in Happel and Brenner (1965, p. 140).

$\theta = 0 \qquad \theta = \beta$

Figure P7-7. Flow through a cone of angle β at a volumetric flow rate Q.

Figure P7-8. Liquid flow due to a moving plate, in the presence of a gas–liquid interface.

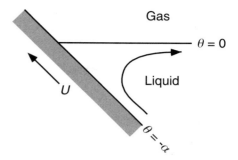

7-8. Flow of a Liquid with a Free Surface Near a Moving Plate

As shown in Fig. P7-8, a solid plate inclined at an angle α relative to horizontal is being withdrawn from a liquid at a tangential velocity U. The gas–liquid interface (corresponding to $\theta=0$ in cylindrical coordinates) is assumed to be flat. That is, wetting phenomena and the tendency for the moving plate to disturb the interface are neglected. Assume that Stokes' equation is applicable.

(a) Neglecting the shear stress at the gas–liquid interface, determine the velocity in the liquid. [*Hint:* Assume that the stream function is of the form $\psi(r,\theta)=rf(\theta)$.]

(b) Show that the velocity at the gas–liquid interface is given by

$$v_r(r,0)=U\left(\frac{\sin\alpha-\alpha\cos\alpha}{\alpha-\sin\alpha\cos\alpha}\right)$$

as reported by Moffat (1964).

7-9. Thermocapillary Motion of a Bubble

If a temperature gradient is present in a pure liquid, the variation in surface tension from one side of a small gas bubble to the other tends to make the bubble move in the direction of increasing temperature. Thus, as shown by Young et al. (1959), a vertical temperature gradient, in which temperature decreases with increasing height, can cause a bubble to move downward. To analyze this phenomenon, let the z direction be vertical (such that $\mathbf{g}=-g\mathbf{e}_z$) and assume that $\partial T/\partial z=-G$ far from the bubble, where $G>0$. Assume also that $d\gamma/dT=-K$, where $K>0$; that is, the surface tension decreases with increasing temperture. As in Fig. 7-4, it is convenient to adopt a reference frame in which the center of the bubble is stationary and $\mathbf{v}=U\mathbf{e}_z$ far from the bubble. Accordingly, $U<0$ and $U>0$ correspond to bubbles which are rising and falling, respectively. The bubble radius is R. It is assumed that Stokes' equation is valid everywhere and that all thermophysical properties except γ are constant. For simplicity, assume that the thermal conductivity, viscosity, and density of the gas are all much smaller than those in the liquid.

(a) Assuming that the bubble acts as a thermal insulator and that convective energy transport in the liquid is negligible (i.e., $\text{Pe}=UR/\alpha\ll1$, as discussed in Chapter 9), determine the interfacial temperature and show that the surface-tension gradient is given by

$$\frac{d\gamma}{d\theta}=-\frac{3}{2}KGR\sin\theta.$$

(*Hint:* Use the results of Example 4.8-3.)

(b) State the general form of the stream function, and use it to satisfy all boundary conditions for the liquid except for the tangential stress balance at the bubble surface. This will leave one constant undetermined (see below).

(c) Neglecting the shear stress exerted on the liquid by the gas, show that

$$\tau_{r\theta}(R,\ \theta)+\frac{1}{R}\frac{d\gamma}{d\theta}=0,$$

where $\tau_{r\theta}$ is evaluated in the liquid. Use this result to find the remaining constant in the stream function.

(d) Use a force balance on the bubble to show that

$$U=\frac{R}{\mu}\left(\frac{KG}{2}-\frac{\rho g R}{3}\right),$$

where the viscosity and density are those of the liquid. (*Hint:* Evaluate the drag using the general result given in Problem 7-1.) Thus, as shown both theoretically and experimentally by Young et al. (1959), the bubble will be *stationary* if

$$KG=\frac{2}{3}\rho g R.$$

7-10. Motion of a Small Sphere in a Tube

Consider a solid sphere of radius a positioned on the axis of a fluid-filled cylindrical tube of radius b. Far upstream and downstream from the sphere the fluid has the parabolic velocity profile characteristic of Poiseuille flow. The sphere is relatively small, such that $\lambda \equiv a/b \ll 1$.

(a) Use Faxén's laws to estimate the force required to keep the sphere stationary.
(b) If the sphere is suspended freely in the fluid, use Faxén's laws to estimate its translational and angular velocities.

The results using Faxén's laws are not exact here, due to hydrodynamic interactions between the sphere and the tube wall, although the wall effects are minimized for a small sphere located on the axis. Results valid for arbitrary λ (including the wall effects) are given by Bungay and Brenner (1973).

7-11. Flow Past a Long Cylinder

Consider the two-dimensional analog of Example 7.4-2, namely, Stokes' flow perpendicular to the axis of a long cylinder of radius R. Let the cylinder be aligned with the z axis, and assume that far from the cylinder the velocity is uniform with $v_x = U$.

(a) Using cylindrical coordinates, show that a solution of the form $\psi(r,\ \theta)=f(r)\sin\theta$ is consistent with all of the boundary conditions.
(b) Based on Stokes' equation, show that the general solution for $f(r)$ is

$$f(r)=Ar^{-1}+Br+Cr\ \ln r+Dr^3$$

and that this solution is incapable of satisfying the condition $v_x \to U$ as $r \to \infty$. Thus, this problem has no solution for $\mathrm{Re}=0$. As discussed in Section 7.7, the absence of such a solution for any two-dimensional, unbounded flow is known as *Stokes' paradox*.

7-12. Settling of a Cylinder on a Flat Surface

A long cylinder is moving toward a flat surface under the influence of a constant force (e.g., gravity), with the cylinder axis parallel to the surface. The cylinder radius is R and the minimum surface-to-surface distance is denoted as $h_0(t)$. (The situation is the same as that depicted in Fig. 6-12, except that $h = h(x, t)$ and $U = 0$.)

Assuming that the initial separation is much smaller than the cylinder radius, determine $h_0(t)$. How long will it take the cylinder to reach the surface? (*Hint:* This is a transient lubrication problem, made pseudosteady by the fact that $Re \ll 1$.)

CHAPTER 8

LAMINAR FLOW AT
HIGH REYNOLDS NUMBER

8.1 INTRODUCTION

High Reynolds number flows involve large velocities, large length scales, and/or small kinematic viscosities. Two prototypical examples are shown schematically in Fig. 8-1. The most studied situation, with applications ranging from particle settling to airfoil performance, is flow relative to a submerged object. As shown in Fig. 8-1(a), the typical problem involves a uniform velocity U far from the object, which is chosen to be stationary. The size of the fluid region is indefinitely large, so that the only geometric length scale is the characteristic dimension L of the object and $Re = UL/\nu$. Although *external* flows such as that have received the most analytical attention, internal flows such as the stirred tank in Fig. 8-1(b) are another important type of high Reynolds number flow. If the angular velocity of the impeller is ω and the radius of the tank is R, then ωR is a characteristic linear velocity and $Re = \omega R^2/\nu$. Whether external or internal, high Reynolds number flows contain two or more dynamically distinct regions. In most of the fluid volume the effects of viscosity are virtually absent, whereas in thin boundary layers next to the surfaces the viscous stresses are very important. For high Re flow around a submerged object the boundary layer is next to the object; for high Re flow in a stirred tank there are boundary layers at the top, bottom, side, and impeller surface. The fact that different physical factors are dominant in various regions causes these to be, in essence, singular perturbation problems.

This chapter begins with a discussion of the general features of laminar flow at high Reynolds number. The focus is on the differential equations used to provide first approximations to the flow in the "outer" or "inviscid" regions and in the "inner" or boundary layer regions. The relationship between the two types of regions is discussed, and the important phenomenon of boundary layer separation is described. The remainder

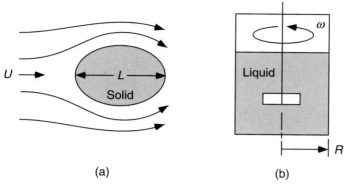

Figure 8-1. Representative examples of high Reynolds number flow. (a) Flow past a submerged object; (b) flow in a stirred tank.

of the chapter consists mainly of a series of examples intended to illustrate the physical phenomena and analytical approaches in more detail. Included are exact solutions to certain inviscid flow problems and both exact and approximate solutions of the boundary layer equations. Despite their name, boundary layers occur not just at solid surfaces or fluid–fluid interfaces. This is illustrated by examples involving wakes and jets.

8.2 GENERAL FEATURES OF HIGH REYNOLDS NUMBER FLOW

Inviscid Flow and Irrotational Flow

Given that the Reynolds number is a measure of the importance of inertial effects (or convective momentum transfer) relative to viscous effects (or diffusive momentum transfer), in analyzing flows at high Re it is logical to begin with the limiting case in which viscous effects are entirely absent. This is the way the analysis of such flows developed historically. Setting $\mu = 0$ in the dimensional Navier–Stokes equation, which is equivalent to setting $\text{Re} = \infty$ in the dimensionless form, gives the momentum equation for an *inviscid fluid*. It is

$$\rho \frac{D\mathbf{v}}{Dt} = \rho\left(\frac{\partial \mathbf{v}}{\partial t} + \mathbf{v} \cdot \nabla \mathbf{v}\right) = -\nabla \mathscr{P} \tag{8.2-1}$$

in general, or

$$\rho \mathbf{v} \cdot \nabla \mathbf{v} = -\nabla \mathscr{P} \tag{8.2-2}$$

for steady flow. In introductory courses in fluid mechanics the latter equation is encountered frequently, but mainly in another form. To obtain the more familiar equation, we use the identity

$$\mathbf{v} \cdot \nabla \mathbf{v} = \nabla\left(\frac{v^2}{2}\right) - \mathbf{v} \times (\nabla \times \mathbf{v}) \tag{8.2-3}$$

[Table A-1, (11), with $\mathbf{a} = \mathbf{b} = \mathbf{v}$] and form the dot product of $\mathbf{v}$ with each term to give

$$\mathbf{v} \cdot (\mathbf{v} \cdot \nabla) = \mathbf{v} \cdot \nabla \left(\frac{v^2}{2} \right) - \mathbf{v} \cdot [\mathbf{v} \times (\nabla \times \mathbf{v})] = \mathbf{v} \cdot \nabla \left(\frac{v^2}{2} \right). \tag{8.2-4}$$

Accordingly, the dot product of $\mathbf{v}$ with Eq. (8.2-2) yields, after rearrangement,

$$\mathbf{v} \cdot \nabla \left(\frac{v^2}{2} + \frac{P}{\rho} + gh \right) = 0. \tag{8.2-5}$$

Here we have set $\mathcal{P} = P + \rho gh$, where h is the height above some horizontal reference plane. Recalling now that in steady flow $\mathbf{v} \cdot \nabla b$ gives the rate of change of the scalar b along a streamline, Eq. (8.2-5) is seen to be equivalent to

$$\Delta \left(\frac{v^2}{2} + \frac{P}{\rho} + gh \right) = 0, \tag{8.2-6}$$

where the difference operator (Δ) refers to any two points along the same streamline. This is *Bernoulli's equation*, one of the cornerstones of classical fluid mechanics. It describes the interconversions of kinetic energy, pressure, and potential energy which occur in an inviscid fluid. In such a fluid there is no internal friction, and thus there is no dissipative conversion of mechanical energy to heat (see Chapter 9).

To obtain boundary-value problems which will permit the calculation of $\mathbf{v}$ and $\mathcal{P}$ throughout an inviscid fluid, we return to Eqs. (8.2-1) and (8.2-2). These are nonlinear partial differential equations, and their solution (together with the continuity equation) is generally very difficult. However, placing a restriction on the vorticity leads to enormous simplifications. The simplifications are described first as motivation, and then the assumption concerning the vorticity is shown to be valid for a large class of inviscid flows.

Assume now that the flow is *irrotational*; that is, $\mathbf{w} \equiv \nabla \times \mathbf{v} = \mathbf{0}$ everywhere. From the relationship between the stream function and vorticity given in Eq. (5.9-13), it follows that for either steady or unsteady flow we have

$$\nabla^2 \psi = 0 \qquad \text{(planar flow, } w = 0) \tag{8.2-7a}$$

$$E^2 \psi = 0 \qquad \text{(axisymmetric flow, } w = 0). \tag{8.2-7b}$$

Thus, for two-dimensional irrotational flow the momentum and continuity equations are satisfied by solving Laplace's equation (or a closely related linear equation) for the stream function, instead of Eqs. (8.2-1) or (8.2-2). Time is not involved in Eq. (8.2-7), indicating that irrotational flow is pseudosteady. Recall that for creeping flow, the analogous stream-function equations are fourth order rather than second order [see Eq. (7.2-6)].

Irrotational flow problems can be formulated also in terms of the *velocity potential*, $\phi(\mathbf{r}, t)$. By definition, the velocity potential is a scalar function whose gradient is such that

$$\mathbf{v} \equiv \nabla \phi. \tag{8.2-8}$$

(Some authors introduce a minus sign.) In general, there is no guarantee that a function $\phi(\mathbf{r}, t)$ with this property will exist. However, if it does exist, taking the divergence of both sides of Eq. (8.2-8) and using continuity shows that

$$\nabla^2 \phi = \mathbf{0}. \tag{8.2-9}$$

Thus, if there is a velocity potential, it will satisfy Laplace's equation. Any flow for which Eqs. (8.2-8) and (8.2-9) are valid is called a *potential flow*. It is seen that potential flows are also pseudosteady. To establish when a potential flow will exist, we take the curl of both sides of Eq. (8.2-8) and use the fact that $\nabla \times \nabla b = \mathbf{0}$ for any scalar function b. This gives

$$\nabla \times \mathbf{v} = \nabla \times \nabla \phi = \mathbf{0}, \tag{8.2-10}$$

which implies that a necessary condition for potential flow is that the velocity field be irrotational. Thus, Eq. (8.2-7) for the stream function and Eq. (8.2-9) for the velocity potential have a similar basis; either can be used as the governing equation for an irrotational flow. In principle, the velocity potential can be used in three-dimensional as well as two-dimensional problems (unlike the stream function), although its usual application is to two-dimensional flows. Whether the stream function or the velocity potential is employed, the main point is that the many powerful methods which can be applied to Laplace's equation, including those described in Chapter 4, make irrotational flow problems relatively easy to solve.

It should be clear now that irrotationality simplifies flow analyses considerably. The relevance of this to inviscid flow follows from the *vorticity transport equation*,

$$\frac{D\mathbf{w}}{Dt} = \mathbf{w} \cdot \nabla \mathbf{v} + \nu \nabla^2 \mathbf{w}, \tag{8.2-11}$$

which is derived by taking the curl of each term in the full Navier–Stokes equation (see Problem 5-2). For steady flow of an inviscid fluid, Eq. (8.2-11) reduces to

$$\mathbf{v} \cdot \nabla \mathbf{w} = \mathbf{w} \cdot \nabla \mathbf{v}. \tag{8.2-12}$$

Consider now flow around a submerged object, and suppose that $\mathbf{w} = \mathbf{0}$ at some location on each streamline. This will occur, for example, if the velocity is uniform far upstream from the object. It follows from Eq. (8.2-12) that $\mathbf{v} \cdot \nabla \mathbf{w} = \mathbf{0}$ at the reference location, or that the rate of change of $\mathbf{w}$ there along the streamline is zero. A sequence of small displacements along the streamline then yields the conclusion that $\mathbf{w}$ will not change from its reference value of zero. This result is summarized as follows: *In an inviscid fluid, an irrotational flow will remain irrotational.* Thus, for a uniform unperturbed velocity, which is the situation of most interest in flow around submerged objects, the flow of an inviscid fluid will be irrotational everywhere. This contrasts, for example, with creeping flow around a sphere (Example 7.4-2), where it can be shown that the flow is irrotational only at great distances from the sphere.

There is a special form of Bernoulli's equation for irrotational flow. From Eqs. (8.2-2) and (8.2-3) and the assumption that $\nabla \times \mathbf{v} = \mathbf{0}$, we obtain

$$\nabla \left(\frac{v^2}{2} + \frac{P}{\rho} + gh \right) = \mathbf{0} \qquad (w = 0). \tag{8.2-13}$$

Whereas Eq. (8.2-6) indicates that the sum in parentheses is constant only along a given *streamline*, Eq. (8.2-13) implies that it is constant *everywhere* in the fluid (i.e., the same for all streamlines). This emphasizes that the assumption of irrotational flow, which led to Eq. (8.2-13), is stronger than the assumption of inviscid flow, which is all that was needed to obtain the normal form of Bernoulli's equation.

The relationships given above were well known to physical scientists and mathematicians by the early 1800s.[1] Indeed, the major emphasis of theoretical fluid dynamics in the nineteenth century was the study of potential flow. Ignoring viscosity leads to fairly accurate results in certain high Reynolds number applications, such as in wave mechanics, but there was early evidence that this approach is erroneous in other situations. Perhaps the best-known symptom of trouble, called *d'Alembert's paradox*, was the finding that for potential flow past any solid object the drag is zero (see Example 8.3-1). As this prediction of zero drag is counter even to everyday experience, it is remarkable that the focus on potential theory persisted as long as it did. The rather unfortunate result, as described by Lighthill (1963) and Schlicting (1968), is that by the late 1800s there was little connection between theoretical and experimental fluid dynamics. The brilliant concept of the boundary layer, introduced by L. Prandtl in 1904, resolved d'Alembert's paradox and (eventually) helped reunify the theoretical and experimental branches of the subject. Incidentally, it also breathed new life into potential theory by suggesting that it might adequately approximate the flow in the outer region outside the boundary layer.[2]

Laminar Boundary Layers

The essential limitation of inviscid flow theory and the mathematical basis for boundary layers are seen by examining a dimensionless form of the Navier–Stokes equation. At steady state, and using the inertial pressure scale ρU^2 to nondimensionalize $\mathcal{P}$, it is found that

$$\tilde{\mathbf{v}} \cdot \tilde{\boldsymbol{\nabla}} \tilde{\mathbf{v}} + \tilde{\boldsymbol{\nabla}} \tilde{\mathcal{P}} = \mathrm{Re}^{-1} \tilde{\nabla}^2 \tilde{\mathbf{v}} \tag{8.2-14}$$

as discussed in Section 5.10. If Re$\gg$1 it is tempting to ignore the viscous terms, leading to

$$\tilde{\mathbf{v}} \cdot \tilde{\boldsymbol{\nabla}} \tilde{\mathbf{v}} + \tilde{\boldsymbol{\nabla}} \tilde{\mathcal{P}} = 0, \tag{8.2-15}$$

which is the dimensionless form of Eq. (8.2-2). That this inviscid momentum equation cannot be a uniformly valid approximation throughout the fluid, even for Re$\rightarrow\infty$, is indicated by the fact that it is a first-order differential equation, whereas the full Navier–Stokes equation is second order. Because Eq. (8.2-15) can satisfy only one boundary condition in each coordinate, it cannot satisfy both a no-slip condition at the surface of an object and a specified velocity far away. As is shown by the examples in Section 8.3, it is the no-slip condition which must be discarded. What retarded progress in the theory of high Reynolds number flow was the failure of analysts to appreciate the importance of this boundary condition.

---

[1] The equations of motion for an inviscid fluid (continuity and momentum) were established in a paper by the Swiss mathematician Leonhard Euler (1707–1783) in 1755. In the opinion of Dugas (1988, pp. 301–304), "so perfect is this paper that not a line has aged." Euler, the most prolific mathematician in history, also made key contributions to several areas of mechanics.

[2] Ludwig Prandtl (1875–1953) was the originator of the German school of aerodynamics. He was responsible not only for the theory of laminar boundary layers, but also for the first solutions to turbulent flow problems, using his "mixing length" concept (Chapter 13). His research (and that of his students at Göttingen) on aerodynamic drag and lift established the main principles of modern wing design. The history of boundary layer theory is reviewed in Tani (1977), and an engaging essay on Prandtl's life is that of Lienhard (in Gillispie, Vol. 11, 1975, pp. 123–125).

The conceptual advance made by Prandtl was to realize that Eq. (8.2-15) is based on scaling assumptions which are incorrect in the vicinity of a solid object. In passing from Eq. (8.2-14) to Eq. (8.2-15) the implicit assumption is that $\tilde{\nabla}^2\tilde{\mathbf{v}} = O(1)$ everywhere as Re→∞, allowing the viscous terms to be neglected. However, the need for a second-order equation indicates that, at least in part of the flow, $\mathrm{Re}^{-1}\tilde{\nabla}^2\tilde{\mathbf{v}} = O(1)$ and $\tilde{\nabla}^2\tilde{\mathbf{v}} = O(\mathrm{Re})$. The region where viscous and inertial effects are comparable, and where the velocity gradients are improperly scaled, is the boundary layer. The fact that some of the terms in $\tilde{\nabla}^2\tilde{\mathbf{v}}$ are unbounded as Re→∞ reflects a problem not with the velocity scale (i.e., U, the velocity far from the object) but with the assumed length scale (i.e., L, the overall size of the object). The key conclusion is that there is a boundary layer with a thickness (δ) which is small compared with L and which decreases with increasing Re.

This chapter is concerned only with steady, two-dimensional flows (planar or axisymmetric). In preparation for deriving the governing equations for the boundary layer we observe that, to first approximation, the thinness of this region will make a surface appear flat. Assuming that at every location the boundary layer thickness is much smaller than the curvature of the surface, it is permissible in planar flows to use local rectangular coordinates. As shown in Fig. 8-2, x is interpreted as the arc length from the leading edge of an object and the surface corresponds to $y = 0$; the boundary layer thickness is a function of x only. This is the situation considered in the discussion which follows.

An order-of-magnitude analysis is used now to relate the dimensionless boundary layer thickness, $\tilde{\delta} \equiv \delta/L$, to Re, and to reveal the dominant terms in the momentum equations. The dimensionless continuity equation is

$$\frac{\partial \tilde{v}_x}{\partial \tilde{x}} + \frac{\partial \tilde{v}_y}{\partial \tilde{y}} = 0.$$

1     $\tilde{V}/\tilde{\delta}$

(8.2-16)

Under each term is its order-of-magnitude estimate. The choice of L and U as length and velocity scales ensures that $\Delta\tilde{v}_x \sim 1$ and $\Delta\tilde{x} \sim 1$, from which it follows that $\partial\tilde{v}_x/\partial\tilde{x} \sim 1$, independent of Re. Whereas velocity variations in the x direction occur over an actual distance which is $\sim L$ or a dimensionless distance which is ~ 1, those in the y direction occur over the boundary layer thickness, δ or $\tilde{\delta}$. Thus, $\partial\tilde{v}_y/\partial\tilde{y} \sim \tilde{V}/\tilde{\delta}$, where

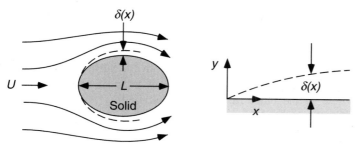

Figure 8-2. (Left) Two-dimensional flow around a submerged object with characteristic dimension L and approach velocity U. (Right) An enlargement of the boundary layer showing the local rectangular coordinates. The boundary layer thickness is denoted by $\delta(x)$.

$\tilde{V} \equiv V/U$ and V is the scale for v_y. Equation (8.2-16) implies that the linkage between the unknown length and velocity scales, V and $\tilde{\delta}$, is such that $\tilde{V} \sim \tilde{\delta}$. The velocity component normal to the surface arises from the fact that, as the approaching stream reaches the leading edge of the object, the no-slip condition forces part of the fluid to decelerate. Accordingly, to conserve mass, there must be a small velocity component directed away from the surface. The thicker the boundary layer, the larger must be the velocity normal to the surface.

The orders of magnitude of the terms in the x and y components of the Navier–Stokes equation are given by

$$\tilde{v}_x \frac{\partial \tilde{v}_x}{\partial \tilde{x}} + \tilde{v}_y \frac{\partial \tilde{v}_x}{\partial \tilde{y}} + \frac{\partial \tilde{\mathcal{P}}}{\partial \tilde{x}} = \mathrm{Re}^{-1}\left(\frac{\partial^2 \tilde{v}_x}{\partial \tilde{x}^2} + \frac{\partial^2 \tilde{v}_x}{\partial \tilde{y}^2}\right),$$

$\quad\quad (1)(1) \quad (\tilde{\delta})(\tilde{\delta}^{-1}) \quad\quad\quad\quad \mathrm{Re}^{-1}(1) \;\; \mathrm{Re}^{-1}(\tilde{\delta}^{-2})$ $\quad\quad$ (8.2-17)

$\dfrac{1}{Re}\cdot\dfrac{1}{\delta^2}\sim1$

$\delta^2\sim Re^{-1}$

$\delta\sim Re^{-\frac{1}{2}}$

$$\tilde{v}_x \frac{\partial \tilde{v}_y}{\partial \tilde{x}} + \tilde{v}_y \frac{\partial \tilde{v}_y}{\partial \tilde{y}} + \frac{\partial \tilde{\mathcal{P}}}{\partial \tilde{y}} = \mathrm{Re}^{-1}\left(\frac{\partial^2 \tilde{v}_y}{\partial \tilde{x}^2} + \frac{\partial^2 \tilde{v}_y}{\partial \tilde{y}^2}\right).$$

$\quad\quad (1)(\tilde{\delta}) \quad (\tilde{\delta})(1) \quad\quad\quad\quad \mathrm{Re}^{-1}(\tilde{\delta}) \;\; \mathrm{Re}^{-1}(\tilde{\delta}^{-1})$ $\quad\quad$ (8.2-18)

In estimating the terms involving $\tilde{v}_y$ we have used the information that $\tilde{v}_y \sim \tilde{V} \sim \tilde{\delta}$. Because $\tilde{\delta} \ll 1$, it is evident from Eq. (8.2-17) that the dominant viscous term in the x-momentum equation is $\partial^2 \tilde{v}_x/\partial \tilde{y}^2$. If this viscous term is to be comparable to the inertial terms, which are both of similar magnitude, then the boundary layer thickness must be such that

$$\tilde{\delta} \sim \mathrm{Re}^{-1/2}. \quad\quad \delta \sim \frac{1}{\sqrt{Re}} \quad\quad (8.2\text{-}19)$$

The same conclusion follows from Eq. (8.2-18). Thus, scaling considerations require that the boundary layer thickness vary as the inverse square root of the Reynolds number. This is an extremely powerful result, because it does not depend on the precise shape of the object.

The other important information which is inferred from Eqs. (8.2-17) and (8.2-18) concerns the pressure. Because $\partial \tilde{\mathcal{P}}/\partial \tilde{y}$ is bounded by the other terms in Eq. (8.2-18), $\partial \tilde{\mathcal{P}}/\partial \tilde{y} \sim \tilde{\delta}$; the same reasoning with Eq. (8.2-17) indicates that $\partial \tilde{\mathcal{P}}/\partial \tilde{x} \sim 1$. These results indicate that $\tilde{\mathcal{P}}$ depends mainly on $\tilde{x}$. To show this more clearly, we expand $\tilde{\mathcal{P}}$ in a Taylor series about a given position on the surface to give

$$\tilde{\mathcal{P}}(\tilde{x}, \tilde{y}) = \tilde{\mathcal{P}}(\tilde{x}, 0) + \left.\frac{\partial \tilde{\mathcal{P}}}{\partial \tilde{y}}\right|_{\tilde{y}=0} \tilde{y} + \cdots = \tilde{\mathcal{P}}(\tilde{x}, 0) + O(\tilde{\delta}^2). \quad\quad (8.2\text{-}20)$$

Thus, the error in assuming that $\tilde{\mathcal{P}} = \tilde{\mathcal{P}}(\tilde{x})$ is $O(\tilde{\delta}^2)$ or, from Eq. (8.2-19), $O(\mathrm{Re}^{-1})$. The statement that $\tilde{\mathcal{P}} \cong \tilde{\mathcal{P}}(\tilde{x})$ implies that pressure variations from the surface to the outer edge of the boundary layer are negligible. In other words, the pressure in the boundary layer is dictated by the flow in the outer (inviscid) region. Letting $\tilde{\mathcal{P}}^0$ represent the pressure obtained by solving the *inviscid* flow equations, Eqs. (8.2-15) and (8.2-16), the boundary layer pressure is given by

$$\tilde{\mathcal{P}}(\tilde{x}) = \lim_{\tilde{y} \to 0} \tilde{\mathcal{P}}^0(\tilde{x}, \tilde{y}). \quad\quad (8.2\text{-}21)$$

The right-hand side of Eq. (8.2-21) is written as a limit, rather than as $\tilde{\mathcal{P}}^0(\tilde{x}, 0)$, because the outer solution is not valid exactly at the surface. The importance of Eq. (8.2-21) stems from the fact that, to first approximation, the outer flow problem is independent of that in the boundary layer. This is illustrated by the examples in Section 8.3. Thus, unless there is flow separation (see below), the outer problem can be solved first and used to evaluate the pressure in the boundary layer.

Using what has been found so far, the x-momentum equation for the boundary layer is

$$\tilde{v}_x \frac{\partial \tilde{v}_x}{\partial \tilde{x}} + \tilde{v}_y \frac{\partial \tilde{v}_x}{\partial \tilde{y}} + \frac{d\tilde{\mathcal{P}}}{d\tilde{x}} = \mathrm{Re}^{-1} \frac{\partial^2 \tilde{v}_x}{\partial \tilde{y}^2} \tag{8.2-22}$$

or, in dimensional form,

$$v_x \frac{\partial v_x}{\partial x} + v_y \frac{\partial v_x}{\partial y} + \frac{1}{\rho} \frac{d\mathcal{P}}{dx} = \nu \frac{\partial^2 v_x}{\partial y^2}. \tag{8.2-23}$$

It is informative to compare Eq. (8.2-23) with the lubrication approximation, Eq. (6.6-1). Although the two equations differ in the presence or absence of inertial terms, they are otherwise the same. Specifically, in both cases it is found that the $\partial^2 v_x/\partial x^2$ term is negligible and that $\mathcal{P} = \mathcal{P}(x)$ only. Both of those simplifications are a consequence of the fact that the equations apply to thin layers of fluid. In lubrication problems the small length scale in the cross-flow direction is due to the geometry, whereas the boundary layer thickness is a dynamic length scale dictated by the balance between inertial and viscous effects at large Re.

It is convenient to rewrite the pressure gradient in the boundary layer in terms of the outer velocity, $\tilde{v}^0$. Approaching an impermeable solid surface, the limiting behavior of the velocity components is given by

$$\lim_{\tilde{y} \to 0} \tilde{v}_x^0(\tilde{x}, \tilde{y}) \equiv \tilde{u}(\tilde{x}), \tag{8.2-24}$$

$$\lim_{\tilde{y} \to 0} \tilde{v}_y^0(\tilde{x}, \tilde{y}) = 0. \tag{8.2-25}$$

The reason that $\tilde{u}(\tilde{x})$ is not zero is that the outer solution is incapable of satisfying the no-slip condition, as already mentioned. Using Eqs. (8.2-24) and (8.2-25) in the x component of Eq. (8.2-15) gives

$$\frac{d\tilde{\mathcal{P}}}{d\tilde{x}} = \lim_{\tilde{y} \to 0} \frac{d\tilde{\mathcal{P}}^0}{dx} = -\tilde{u} \frac{d\tilde{u}}{d\tilde{x}}, \tag{8.2-26}$$

so that Eq. (8.2-22) becomes

$$\tilde{v}_x \frac{\partial \tilde{v}_x}{\partial \tilde{x}} + \tilde{v}_y \frac{\partial \tilde{v}_x}{\partial \tilde{y}} - \tilde{u} \frac{d\tilde{u}}{d\tilde{x}} = \mathrm{Re}^{-1} \frac{\partial^2 \tilde{v}_x}{\partial \tilde{y}^2}. \tag{8.2-27}$$

What remains is to identify the properly scaled y-coordinate and y-component of velocity for the boundary layer. Equations (8.2-16) and (8.2-19) imply that these are

$$\hat{y} = \tilde{y} \, \mathrm{Re}^{1/2}, \qquad \hat{v}_y = \tilde{v}_y \, \mathrm{Re}^{1/2}. \tag{8.2-28}$$

Whereas $\tilde{y}$ and $\tilde{v}_y$ were seen to be small throughout the boundary layer, the new boundary layer variables $\hat{y}$ and $\hat{v}_y$ are both $O(1)$ as Re$\rightarrow\infty$. Using these new variables, the continuity and x-momentum equations for the boundary layer are written as

$$\frac{\partial \tilde{v}_x}{\partial \tilde{x}} + \frac{\partial \hat{v}_y}{\partial \hat{y}} = 0,$$ (8.2-29)

$$\tilde{v}_x \frac{\partial \tilde{v}_x}{\partial \tilde{x}} + \hat{v}_y \frac{\partial \tilde{v}_x}{\partial \hat{y}} - \tilde{u}\frac{\partial \tilde{u}}{\partial \tilde{x}} = \frac{\partial^2 \tilde{v}_x}{\partial \hat{y}^2}.$$ (8.2-30)

Equations (8.2-29) and (8.2-30) provide two equations to determine the two remaining unknowns, $\tilde{v}_x$ and $\hat{v}_y$. The y-momentum equation, which was used in evaluating the pressure, is no longer needed. In dimensional form, the boundary layer equations are

$$\frac{\partial v_x}{\partial x} + \frac{\partial v_y}{\partial y} = 0,$$ (8.2-31)

$$v_x \frac{\partial v_x}{\partial x} + v_y \frac{\partial v_x}{\partial y} - u\frac{du}{dx} = \nu \frac{\partial^2 v_x}{\partial y^2}.$$ (8.2-32)

An alternative way to obtain Eqs. (8.2-29) and (8.2-30) is to recognize the need to rescale the small variables $\tilde{y}$ and $\tilde{v}_y$ by "stretching" them using the large parameter Re. Thus, one could start by assuming that

$$\hat{y} = \tilde{y}\,\text{Re}^a, \qquad \hat{v}_y = \tilde{v}_y\,\text{Re}^b.$$ (8.2-33)

With $\partial \hat{v}_y/\partial \hat{y} = O(1)$ by definition, we need $a = b$ to eliminate Re from Eq. (8.2-16). Then, eliminating Re from Eq. (8.2-17), while maintaining a balance between the dominant viscous and inertial terms, requires that $a = b = 1/2$. In this manner we arrive again at Eq. (8.2-28) and the other results.

The mathematical structure of boundary layer problems is best understood from the perspective of singular perturbation analysis. The boundary layer equations as stated above govern the leading term in a singular perturbation expansion for the velocity and pressure in the boundary layer. Similarly, the inviscid flow equations determine the first term in an expansion for the region outside the boundary layer. We will be concerned only with these leading terms. The solutions for the two regions are linked by asymptotic matching (see Section 3.7). The matching of the pressures was built into the boundary layer equation by the way the pressure term was evaluated using the outer velocity, and the normal components of the velocities are too small to be of concern at leading order. Thus, of greatest interest is the asymptotic matching of the tangential velocity components. This is expressed as

$$\lim_{\tilde{y}\rightarrow 0} \tilde{v}_x^0(\tilde{x},\,\tilde{y}) = \tilde{u}(\tilde{x}) = \lim_{\hat{y}\rightarrow\infty} \tilde{v}_x(\tilde{x},\,\hat{y}), \qquad \text{Re}\rightarrow\infty.$$ (8.2-34)

There are two important conceptual points here. One is that, for large enough Re, small values of $\tilde{y}$ correspond to large values of $\hat{y}$ [see Eq. (8.2-28)]. The other concerns the dual role of $u(x)$. As shown also in Fig. 8-3, it is both the outer velocity at "zero" and the boundary layer velocity at "infinity." It is evident from the figure that increases in Re, which decrease δ, will result in a progressively smoother transition between the leading terms of the boundary layer and outer solutions.

Figure 8-3. First-order matching of the boundary layer and outer solutions.

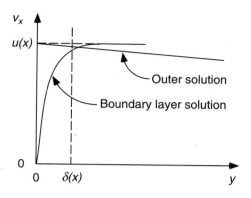

Flow Separation

It was mentioned earlier that a difficulty with potential (irrotational flow) theory is its inability to provide a reasonable approximation to the drag on a solid object. Even more serious is that it does not give even a qualitatively correct picture of the flow pattern near a blunt object. For example, for flow perpendicular to the axis of a long cylinder, potential theory yields streamlines like those shown in Fig. 8-4(a), which have fore-and-aft symmetry. (The stream function for this case is derived in Example 8.3-1.) The actual flow patterns, one of which is shown qualitatively in Fig. 8-4(b), tend to be much more complicated. For $Re_d > 6$, where Re_d is the Reynolds number based on the cylinder diameter, these involve *flow separation*. In a separated flow there is a point where the surface streamline divides, one branch departing from the surface; in Fig. 8-4(b) this is shown as occurring at about 120° from the most upstream point on the cylinder. Physically, the separation point is a location where fluid that had been flowing along the surface turns away from it. Immediately downstream from the separation point is a region where the flow near the surface is reversed. The separated streamline will reattach to the surface at some point, thereby enclosing an eddy.

As Re_d for a circular cylinder is increased from 6 to about 44, the separation point moves forward and the eddies, which are initially symmetrical, grow in size. The fairly large eddies depicted in Fig. 8-4(b) are representative of the flow pattern for $Re_d \cong 30$. For $Re_d > 44$ the upper and lower eddies (or vortices) begin to be shed and reformed alternately in a time-periodic manner, creating an extended wake. Thereafter, the flow is no longer symmetrical or steady. The flow in the wake passes through a series of other,

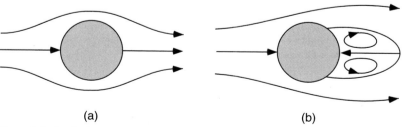

(a)   (b)

Figure 8-4. Qualitative depiction of streamlines for flow past a cylinder. (a) The prediction of potential theory; (b) one of several observed patterns involving flow separation.

increasingly complex regimes until the boundary layer becomes turbulent at $Re_d \cong 2 \times 10^5$. Theoretical and experimental information on the flow regimes for cylinders is reviewed by Churchill (1988, pp. 317–357); see also the discussion of flow separation given in Batchelor (1967, pp. 337–343). A qualitatively similar series of events is observed with spheres (e.g., Churchill, 1988, pp. 359–409).

The pressure distribution for a separated flow like that in Fig. 8-4(b) is very different than that in Fig. 8-4(a). In particular, in the separated flow the average pressure at the downstream side of the surface is found to be significantly lower than at the upstream side, so that the pressure force contributes to the drag. In any real flow there is also a contribution of the shear stress to the drag. Potential theory fails to capture either of these contributions, whereas boundary layer theory explains both. It is found that flow separation is the result of an "adverse" pressure gradient along the surface, by which is meant a pressure increase in the direction of flow. The combination of the adverse pressure gradient and the shear stress at the surface tends to decelerate the adjacent fluid. When the kinetic energy of the fluid is no longer sufficient to overcome these forces, separation and flow reversal result. The use of boundary layer theory to predict when and where separation will occur is discussed further in Section 8.4.

In all of fluid mechanics, high Reynolds number flow with separation is one of the most difficult types of problems to solve. Eddies in a separated region, especially large ones such as those shown in Fig. 8-4(b), can distort the streamlines in much of the fluid and affect even the upstream pressure distribution. Consequently, the approach outlined earlier for streamlined objects, in which one first solves the irrotational flow problem and then uses that information in the boundary layer equations, does not work for blunt objects. Without knowing in advance the extent of the eddies, which are obviously not irrotational, one cannot define the region of the fluid in which potential theory will be valid. The eddies are also not thin, precluding any analytical simplifications based on that characteristic. These difficulties have been overcome to a certain extent by obtaining numerical solutions for the entire flow field. However, the typical instability of steady flow in a wake at moderate Reynolds numbers, as occurs with a cylinder for $Re_d > 44$, makes it difficult to compute the correct flow regime.

8.3 IRROTATIONAL FLOW

The following examples illustrate the calculation of velocity and pressure fields in irrotational flow. Some of these results are used in the boundary layer problems discussed in Section 8.4.

Example 8.3-1 Flow Past a Cylinder   Consider irrotational flow past a long cylinder of radius R. As shown in Fig. 8-5, the cylinder is assumed to be aligned with the z axis, and far away there is a uniform velocity U in the x direction.

The stream function $\psi(r, \theta)$ for this planar flow is defined by

$$v_r = \frac{1}{r}\frac{\partial \psi}{\partial \theta}, \qquad v_\theta = -\frac{\partial \psi}{\partial r}. \tag{8.3-1}$$

As mentioned in Section 8.2, the stream function for a planar, irrotational flow is governed by Laplace's equation, or

Figure 8-5. Flow past a long, circular cylinder.

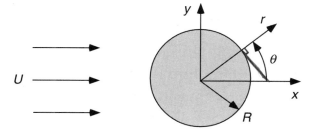

$$\nabla^2 \psi = \left[\frac{1}{r} \frac{\partial}{\partial r} \left(r \frac{\partial}{\partial r} \right) + \frac{1}{r^2} \frac{\partial^2}{\partial \theta^2} \right] \psi = 0. \tag{8.3-2}$$

The applicable boundary conditions for the velocity components are

$$v_r(R, \ \theta) = 0, \tag{8.3-3}$$

$$v_r(\infty, \ \theta) = U \cos \theta, \qquad v_\theta(\infty, \ \theta) = -U \sin \theta. \tag{8.3-4}$$

The no-slip condition at the cylinder surface has been omitted because it cannot be satisfied, as will be shown. In terms of the stream function, the no-penetration condition becomes

$$\frac{\partial \psi}{\partial \theta}(R, \ \theta) = 0 \tag{8.3-5}$$

and the unperturbed velocity field is consistent with

$$\psi(\infty, \ \theta) \rightarrow Ur \sin \theta. \tag{8.3-6}$$

Equation (8.3-6) was obtained by using Eq. (8.3-1) in Eq. (8.3-4), integrating, and setting the respective integration constants equal to zero. The problem for the stream function is now completely specified.

Based on the boundary conditions, we assume a solution of the form

$$\psi(r, \ \theta) = f(r) \sin \theta. \tag{8.3-7}$$

Substituting this expression into Eqs. (8.3-2), (8.3-5), and (8.3-6) gives

$$f'' + \frac{f'}{r} - \frac{f}{r^2} = 0, \tag{8.3-8}$$

$$f(R) = 0, \qquad f(\infty) \rightarrow Ur. \tag{8.3-9}$$

Equation (8.3-8) is an equidimensional equation, with solutions of the form $f = r^n$. The characteristic equation obtained using this trial solution has the roots $n = 1$ and $n = -1$. Using Eq. (8.3-9) to determine the two constants, the solution for the stream function is found to be

$$\psi(r, \ \theta) = UR \sin \theta \left(\frac{r}{R} - \frac{R}{r} \right). \tag{8.3-10}$$

The streamlines for this flow are plotted in Fig. 8-6, in which the fore-and-aft symmetry is evident. The velocity components calculated from Eq. (8.3-10) are

$$v_r(r, \ \theta) = U \cos \theta \left[1 - \left(\frac{R}{r} \right)^2 \right], \tag{8.3-11}$$

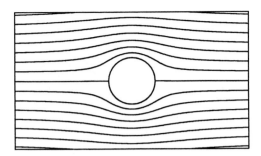

Figure 8-6. Streamlines for potential flow past a solid cylinder, assuming a uniform unperturbed velocity which is normal to the cylinder axis. The stream function is given by Eq. (8.3-10), and there are equal increments in ψ between adjacent streamlines.

$$v_\theta(r,\ \theta) = -U \sin\ \theta\left[1 + \left(\frac{R}{r}\right)^2\right]. \tag{8.3-12}$$

The tangential velocity evaluated at the cylinder surface is

$$v_\theta(R,\ \theta) = -2U \sin\ \theta, \tag{8.3-13}$$

which confirms that the no-slip condition is not satisfied. Because Eq. (8.3-2) is second order, there were no degrees of freedom to satisfy boundary conditons beyond Eqs. (8.3-5) and (8.3-6). The two surface locations at which both v_r and v_θ vanish, $\theta = \pi$ and $\theta = 0$, are called the upstream and downstream *stagnation points,* respectively.

A relatively simple way to evaluate the pressure is to use the special form of Bernoulli's equation derived for irrotational flow, Eq. (8.2-13). Setting $\mathcal{P} = 0$ far from the cylinder, that equation reduces to

$$\frac{v^2}{2} + \frac{\mathcal{P}}{\rho} = \frac{U^2}{2}. \tag{8.3-14}$$

Using Eqs. (8.3-11) and (8.3-12) in Eq. (8.3-14), we obtain

$$\mathcal{P}(r,\ \theta) = \frac{\rho U^2}{2}\left(\frac{R}{r}\right)^2\left[2 - 4\sin^2\ \theta - \left(\frac{R}{r}\right)^2\right] \tag{8.3-15}$$

for the full pressure field, and

$$\mathcal{P}(R,\ \theta) = \frac{\rho U^2}{2}[1 - 4\sin^2\ \theta] \tag{8.3-16}$$

for the pressure on the surface. The maximum pressure, equal to $\rho U^2/2$, is seen to occur at the stagnation points; this follows also by inspection of Eq. (8.3-14). The symmetry of $\sin^2\ \theta$ about $\theta = \pi/2$ and $\theta = 3\pi/2$ indicates that the pressure has fore-and-aft symmetry, which implies that there is no net pressure force on the cylinder. With neither a pressure force nor a viscous force in this model, there is no drag. This is an example of *d'Alembert's paradox,* mentioned in Section 8.2.

The pressure gradient along the surface is

$$\frac{\partial \mathcal{P}}{\partial \theta}(R,\ \theta) = -4\rho U^2 \sin\ \theta \cos\ \theta. \tag{8.3-17}$$

It is seen that the pressure gradient favors forward flow on the upstream side of the cylinder (i.e., $\partial \mathcal{P}/\partial \theta > 0$ for $\pi > \theta > \pi/2$) but opposes it on the downstream side. The development of a significant adverse pressure gradient on the trailing side, which is the cause of boundary layer separation, is a general feature of flow past blunt objects. Boundary layer separation for cylinders is discussed further at the end of Example 8.4-2.

Example 8.3-2 Flow Between Intersecting Planes We consider now irrotational flow near the intersection of two stationary, flat surfaces. As shown in Fig. 8-7, the surfaces correspond to $\theta=0$ and $\theta=\gamma\pi$ in cylindrical coordinates. The planes intersect at $r=0$, and the fluid motion in the region shown is assumed to result from some unspecified flow at $r=\infty$.

The stream function and differential equation are given again by Eqs. (8.3-1) and (8.3-2). Noting that the two planes form a continuous surface across which there is no flow, we conclude that the stream function must have the same constant value on both; that constant is taken to be zero. Thus, the no-penetration condition is expressed as

$$\psi(r,\,0)=0, \qquad \psi(r,\,\gamma\pi)=0, \tag{8.3-18}$$

$$\psi(0,\,\theta)=0. \tag{8.3-19}$$

Equation (8.3-19) is deduced from the fact that the line where the two planes intersect is part of the surface and therefore must have the same value of ψ. As in the preceding example, it will not be possible to satisfy the no-slip condition. Because the nature of the flow at $r=\infty$ is unspecified, the solution will contain an undetermined constant(s).

The boundary conditions do not suggest a particular functional form for $\psi(r,\,\theta)$, so that the solution is constructed as an eigenfunction expansion. The expansion is written as

$$\psi(r,\,\theta)=\sum_{n=1}^{\infty}\psi_n(r)\Phi_n(\theta), \tag{8.3-20}$$

$$\Phi_n(\theta)=\sqrt{\frac{2}{\gamma\pi}}\,\sin\frac{n\theta}{\gamma}, \tag{8.3-21}$$

$$\psi_n(r)=\langle\psi,\,\Phi_n\rangle=\int_0^{\gamma\pi}\psi\Phi_n\,d\theta, \tag{8.3-22}$$

where the basis functions have been obtained from case I of Table 4-1; see Section 4.7. Applying the finite Fourier transform, the problem for $\psi_n(r)$ is found to be

$$r^2\frac{d^2\psi_n}{dr^2}+r\frac{d\psi_n}{dr}-\left(\frac{n}{\gamma}\right)^2\psi_n=0, \tag{8.3-23}$$

$$\psi_n(0)=0. \tag{8.3-24}$$

Once again, the differential equation is equidimensional. The solution to Eq. (8.3-23) which satisfies Eq. (8.3-24) is

$$t(t-1)+t-s^2=0$$
$$t^2-A \quad t+\infty s^2=0$$
$$(t+s)(t-s)$$

Figure 8-7. Flow between intersecting planes.

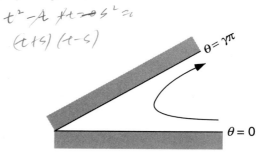

$$\psi_n(r) = a_n r^{n/\gamma}, \tag{8.3-25}$$

where a_n is undetermined. The overall solution is given by

$$\psi(r,\ \theta) = \sum_{n=1}^{\infty} A_n r^{n/\gamma} \sin \frac{n\theta}{\gamma}, \tag{8.3-26}$$

where $A_n = (2/\gamma\pi)^{1/2} a_n$. To determine a_n or A_n, we would need to specify the velocity profile at some radial position other than the origin (e.g., at $r=\infty$). In the next example, where this solution is applied to a slightly different geometry, certain assumptions are made about the velocity at $r=\infty$.

Example 8.3-3 Symmetric Flow in a Wedge-Shaped Region   The analysis in Example 8.3-2 applies without change to the symmetric flow shown in Fig. 8-8. In this case $\theta=0$ represents a symmetry plane rather than a solid surface. The reason that Eq. (8.3-26) continues to apply is that the no-slip condition is not satisfied. For irrotational flow there is no distinction between the solid surface at $\theta=0$ in Fig. 8-7 and the symmetry plane at $\theta=0$ in Fig. 8-8.

Assume now that the flow at $r=\infty$ is such that only the first term in Eq. (8.3-26) is needed. The stream function is then

$$\psi(r,\ \theta) = A_1 r^{1/\gamma} \sin \frac{\theta}{\gamma} \tag{8.3-27}$$

and the velocity components are given by

$$v_r(r,\ \theta) = \frac{A_1}{\gamma} r^{(1-\gamma)/\gamma} \cos \frac{\theta}{\gamma}, \tag{8.3-28}$$

$$v_\theta(r,\ \theta) = -\frac{A_1}{\gamma} r^{(1-\gamma)/\gamma} \sin \frac{\theta}{\gamma}. \tag{8.3-29}$$

The physical interpretation of this family of solutions is simplest for $\gamma=1$, where the two planes reduce to a single flat plate of zero thickness. The stream function for the flat plate is given by

$$\psi(r,\ \theta) = A_1 r \sin \theta. \tag{8.3-30}$$

Comparing this stream function with Eq. (8.3-6), which represents the unperturbed flow in Fig. 8-5, we see that Eq. (8.3-30) corresponds to a uniform velocity A_1 parallel to the plate. The impor-

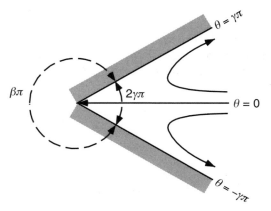

Figure 8-8. Symmetric flow in a wedge-shaped region.

tance of the general wedge-flow results given by Eqs. (8.3-27)–(8.3-29) is that this is one of the few irrotational flows which permit a similarity solution to the boundary layer equations. Consequently, this class of flows has been studied in great detail, as discussed in Section 8.4.

Other Methods

Among the many other methods for solving irrotational flow problems, one that has been used extensively is *conformal mapping*. This method is based on the properties of analytic (i.e., differentiable) functions of a complex variable. From the definitions of the stream function and velocity potential for a two-dimensional flow in rectangular coordinates, it is found that ψ and ϕ are related by

$$\frac{\partial \phi}{\partial x} = \frac{\partial \psi}{\partial y}, \qquad \frac{\partial \phi}{\partial y} = -\frac{\partial \psi}{\partial x}. \tag{8.3-31}$$

These relations are equivalent to the *Cauchy–Riemann equations* involving the real (ϕ) and imaginary (ψ) parts of an analytic function; they are necessary and sufficient conditions for the function to be analytic. It follows that the real and imaginary parts of *any* analytic function correspond to the velocity potential and stream function, respectively, for *some* irrotational flow. In most instances, that flow will be physically uninteresting. Conformal mapping is a technique by which a known solution for one geometric situation (i.e., uniform flow parallel to a flat plate) is transformed into the solution for another geometry (i.e., flow past a wedge of arbitrary angle). An introduction to conformal mapping may be found in any book on complex analysis and in most general texts on applied mathematics (e.g., Hildebrand, 1976, pp. 628–638).

A more thorough introduction to irrotational flow theory is given in Batchelor (1967, pp. 378–506). The books by Lamb (1932) and Milne-Thomson (1960) are devoted largely to inviscid flow; they include applications to subjects not considered here, such as wave motion and the lift on airfoils.

8.4 BOUNDARY LAYERS NEAR SOLID SURFACES

The examples in this section involve both exact and approximate solutions of the boundary layer equations. Flow past a flat plate oriented parallel to the approaching fluid, which is the simplest boundary layer problem, is discussed first. This is followed by an analysis of flow past wedge-shaped objects, which illustrates the important influence of the pressure gradient on the boundary layer and provides insight into the phenomenon of boundary layer separation. Finally, an approximate integral method for solving the boundary layer equations is discussed.

Example 8.4-1 Flow Past a Flat Plate The problem of parallel flow past a flat plate of negligible thickness is unique in that the pressure gradient in the boundary layer is zero. Because an adverse pressure gradient cannot develop, boundary layer separation does not occur; a thin flat plate is the ultimate streamlined object. Let x and y be the tangential and normal coordinates, respectively, with the origin at the upstream edge of the plate. The irrotational flow solution in this case is just $v_x = U$ everywhere, where U is the (constant) unperturbed velocity. Thus, $u(x) = U$ and the boundary layer momentum equation, Eq. (8.2-32), reduces to

$$v_x \frac{\partial v_x}{\partial x} + v_y \frac{\partial v_x}{\partial y} = \nu \frac{\partial^2 v_x}{\partial y^2}. \tag{8.4-1}$$

With two velocity components involved it is advantageous to employ the stream function. Given that $v_x = \partial \psi / \partial y$ and that u and δ are the scales for v_x and y, respectively, in the boundary layer, we infer that

$$\psi \sim u(x)\delta(x). \tag{8.4-2}$$

Although u is constant for a flat plate, the variation in the boundary layer thickness δ (which is qualitatively like that shown in Fig. 8-2) indicates that there is an inherent dependence of ψ on x. To obtain a similarity solution for the stream function, as described below, it is therefore necessary to divide ψ by a function which is proportional to the boundary layer thickness (i.e., a position-dependent scale factor).

The solution is obtained using the *scaled* variables introduced in Section 8.2. The dimensionless quantities are then

$$\tilde{x} = \frac{x}{L}, \qquad \hat{y} = \frac{y}{L} \mathrm{Re}^{1/2}, \qquad \mathrm{Re} = \frac{UL}{\nu}, \tag{8.4-3}$$

$$\tilde{v}_x = \frac{v_x}{U}, \qquad \hat{v}_y = \frac{v_y}{U} \mathrm{Re}^{1/2}, \qquad \tilde{\mathcal{P}} = \frac{\mathcal{P}}{\rho U^2}, \tag{8.4-4}$$

where L is the length of the plate. The dimensionless stream function is defined as

$$\tilde{v}_x = \frac{\partial \hat{\psi}}{\partial \hat{y}}, \qquad \hat{v}_y = -\frac{\partial \hat{\psi}}{\partial \tilde{x}}, \qquad \hat{\psi} = \frac{\psi}{UL} \mathrm{Re}^{1/2}. \tag{8.4-5}$$

Rewriting Eq. (8.4-1) in terms of the dimensionless stream function, we obtain

$$\frac{\partial \hat{\psi}}{\partial \hat{y}} \frac{\partial^2 \hat{\psi}}{\partial \tilde{x} \partial \hat{y}} - \frac{\partial \hat{\psi}}{\partial \tilde{x}} \frac{\partial^2 \hat{\psi}}{\partial \hat{y}^2} = \frac{\partial^3 \hat{\psi}}{\partial \hat{y}^3}. \tag{8.4-6}$$

As mentioned above, to obtain a similarity solution the stream function is modified by dividing it by a position-dependent scale factor. The modified stream function f is assumed to be a function only of the variable η, such that

$$f(\eta) = \frac{\hat{\psi}(\tilde{x}, \hat{y})}{g(\tilde{x})}, \qquad \eta = \frac{\hat{y}}{g(\tilde{x})}, \tag{8.4-7}$$

where the scale factor g is proportional to the boundary layer thickness. Employing the chain rule for differentiation, the required derivatives of the original stream function are rewritten as

$$\frac{\partial \hat{\psi}}{\partial \hat{y}} = f', \qquad \frac{\partial^2 \hat{\psi}}{\partial \hat{y}^2} = \frac{f''}{g}, \qquad \frac{\partial^3 \hat{\psi}}{\partial \hat{y}^3} = \frac{f'''}{g^2}, \tag{8.4-8}$$

$$\frac{\partial \hat{\psi}}{\partial \tilde{x}} = g'(-\eta f' + f), \qquad \frac{\partial^2 \hat{\psi}}{\partial \tilde{x} \partial \hat{y}} = -\frac{g'}{g} \eta f'', \tag{8.4-9}$$

where the primes indicate differentiation with respect to η for f and $\tilde{x}$ for g. Using these expressions, Eq. (8.4-6) reduces to

$$f''' + (gg')ff'' = 0. \tag{8.4-10}$$

The quantity in parentheses in Eq. (8.4-10) must be a constant for a similarity solution to exist, and that constant is chosen here as 1. Because the boundary layer thickness at the leading edge of the plate is zero, we require that $g(0) = 0$. Solving for g (as in Section 3.5) gives

$$g(\tilde{x}) = (2\tilde{x})^{1/2}, \qquad \eta = \frac{\hat{y}}{(2\tilde{x})^{1/2}}, \tag{8.4-11}$$

which shows that the boundary layer thickness varies along a flat plate as $x^{1/2}$. Equation (8.4-10), a third-order differential equation, requires three boundary conditions. The boundary conditions for f are derived from the no-penetration condition ($\hat{v}_y = 0$ at $\hat{y} = 0$), the no-slip condition ($\tilde{v}_x = 0$ at $\hat{y} = 0$), and the asymptotic matching condition [$\tilde{v}_x = 1$ at $\hat{y} = \infty$, equivalent to Eq. (8.2-34)]. The complete problem for $f(\eta)$ is

$$f''' + ff'' = 0, \tag{8.4-12}$$

$$f(0) = 0, \qquad f'(0) = 0, \qquad f'(\infty) = 1. \tag{8.4-13}$$

Equation (8.4-12) is called the *Blasius equation* after H. Blasius (a student of Prandtl), who reported this analysis in 1908 [see Schlicting (1968, pp. 126–127)]. (The original form of the Blasius equation corresponds to $gg' = 1/2$ rather than 1.) Although Eq. (8.4-12) is nonlinear and must be solved numerically, the simplification in reducing the original partial differential equation to an ordinary differential equation is considerable. The solution for the velocity profile is discussed in Example 8.4-2.

An interesting aspect of this problem is seen by rewriting η as

$$\eta = \frac{1}{\sqrt{2}}\frac{\hat{y}}{\tilde{x}^{1/2}} = \frac{1}{\sqrt{2}}\frac{y}{L}\left(\frac{UL}{\nu}\right)^{1/2}\left(\frac{L}{x}\right)^{1/2} = \frac{1}{\sqrt{2}}\frac{y}{x}\mathrm{Re}_x^{1/2}, \qquad \mathrm{Re}_x \equiv \frac{Ux}{\nu}, \tag{8.4-14}$$

which shows that the solution does not depend on L. Thus, the local velocity is independent of the plate length; it is governed by a Reynolds number (Re_x) based on the distance from the leading edge. The reason that L does not affect the solution is that the $\partial^2 v_x / \partial x^2$ term is neglected in the boundary layer approximation. Without this term, neither momentum nor vorticity can diffuse upstream. One may think of this as a situation in which fluid dynamic information is transmitted only in the direction of flow, leaving the fluid unaware of what is happening downstream. Near the leading edge, where Re_x is small, the boundary layer approximation breaks down and the flow behavior is more complicated than that described here. Nonetheless, it has been confirmed experimentally that the solution of the Blasius equation gives very accurate predictions of velocity profiles and shear stresses for flat plates [see Schlicting (1968, pp. 132–133)].

To calculate the drag we need the local shear stress, which is given by

$$\tau_0 \equiv \mu \frac{\partial v_x}{\partial y}\bigg|_{y=0} = \frac{\mu U}{L}\mathrm{Re}^{1/2}\frac{f''(0)}{g} = 0.332\frac{\mu U}{L}\mathrm{Re}^{1/2}\tilde{x}^{-1/2}. \tag{8.4-15}$$

Here we have made use of the the numerical solution of Eq. (8.4-12), which gives $f''(0) = 0.46960$ (Howarth, 1938; Jones and Watson, 1963). The drag on both sides of a flat plate of width W is then

$$F_D = 2WL\int_0^1 \tau_0 \, d\tilde{x} = 1.328\,\mu UW\,\mathrm{Re}^{1/2}. \tag{8.4-16}$$

Drag results for various submerged objects are expressed in dimensionless form using the *drag coefficient* C_D, defined as

$$C_D = \frac{2F_D}{\rho U^2 A}, \tag{8.4-17}$$

where A is the projected or wetted area, depending on the shape. Setting $A = 2WL$ for the flat plate, Eq. (8.4-16) gives

$$C_D = \frac{1.328}{\mathrm{Re}^{1/2}} \tag{8.4-18}$$

as a first approximation to the drag coefficient for $\mathrm{Re} \to \infty$. Imai (1957) showed that, surprisingly, a second term could also be obtained from the leading-order boundary layer solution, giving

$$C_D = \frac{1.328}{\mathrm{Re}^{1/2}} + \frac{2.326}{\mathrm{Re}} + \cdots . \tag{8.4-19}$$

The calculation of higher-order approximations for boundary layer flow along a flat plate is discussed in Van Dyke (1964, pp. 121–146).

Example 8.4-2 Flow Past a Wedge   The shape of an object affects the boundary layer mainly by determining the pressure gradient along the surface. Flow past a wedge of angle $\beta\pi$, as shown in Fig. 8-9(a), yields a different pressure profile for each value of β, thereby offering insight into boundary layer behavior in a number of situations. The flow configurations corresponding to various wedge angles are shown in the other panels of Fig. 8-9. Two of the special cases are planar stagnation flow ($\beta = 1$) and parallel flow past a flat plate ($\beta = 0$). As reported first by Falkner and Skan (1931) and extended by Hartree (1937), a similarity transformation can be found for an arbitrary wedge angle, thereby simplifying the solution.

The velocity field for irrotational flow past a wedge is obtained from Example 8.3-3. Setting $\theta = \gamma\pi$ in Eq. (8.3-28) yields the tangential velocity at the surface. Letting $x = r$, $C = -A_1/\gamma$, and $m = (1 - \gamma)/\gamma$ and using the relationship between β and γ indicated in Fig. 8-8, we find that

$$u(x) = Cx^m, \tag{8.4-20}$$

$$m = \frac{\beta}{2 - \beta}. \tag{8.4-21}$$

To define dimensionless quantities we regard L as some representative distance along the surface from the stagnation point and choose $U = u(L)$, so that Eq. (8.4-20) becomes

$$\tilde{u}(x) = \tilde{x}^m. \tag{8.4-22}$$

It will be shown that, in actuality, the solution is independent of the choices of L and U.

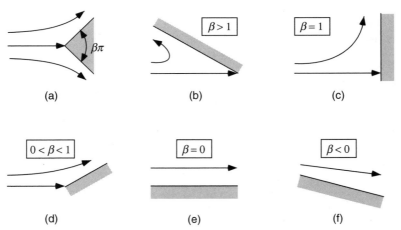

Figure 8-9. Flow past a wedge. (a) Definition of the wedge angle, β; (b)–(f) special cases.

The remainder of the analysis closely parallels that for the flat plate in Example 8.4-1. The governing equation for the stream function is

$$\frac{\partial \hat{\psi}}{\partial \hat{y}} \frac{\partial^2 \hat{\psi}}{\partial \tilde{x} \partial \hat{y}} - \frac{\partial \hat{\psi}}{\partial \tilde{x}} \frac{\partial^2 \hat{\psi}}{\partial \hat{y}^2} - \tilde{u} \frac{d\tilde{u}}{d\tilde{x}} = \frac{\partial^3 \hat{\psi}}{\partial \hat{y}^3}. \tag{8.4-23}$$

Taking into account the variations in u, the modified stream function and similarity variable are defined as

$$f(\eta) = \frac{\hat{\psi}(\tilde{x}, \hat{y})}{\tilde{u}(\tilde{x})g(\tilde{x})}, \qquad \eta = \frac{\hat{y}}{g(\tilde{x})}. \tag{8.4-24}$$

Changing to the new dependent and independent variables, Eq. (8.4-23) becomes

$$f''' + ff'' \left[g \frac{d}{d\tilde{x}}(\tilde{u}g) \right] + g^2 \frac{d\tilde{u}}{d\tilde{x}}[1 - (f')^2] = 0. \tag{8.4-25}$$

For a similarity solution to exist, the terms in Eq. (8.4-25) which involve $\tilde{x}$ must be constant. Accordingly, we assume that

$$g \frac{d}{d\tilde{x}}(\tilde{u}g) = C_1, \tag{8.4-26}$$

$$g^2 \frac{d\tilde{u}}{d\tilde{x}} = C_2, \tag{8.4-27}$$

where C_1 and C_2 are constants. Only one of these constants may be chosen independently. Letting $C_2 = \beta$ and using Eq. (8.4-22) in Eq. (8.4-27), we obtain

$$g(\tilde{x}) = \left(\frac{2}{1+m} \right)^{1/2} \tilde{x}^{(1-m)/2}. \tag{8.4-28}$$

Thus, in this family of flows the boundary layer thickness varies as a power of $\tilde{x}$, the exponent being determined by the wedge angle. It follows from Eqs. (8.4-24) and (8.4-28) that

$$\eta = \left(\frac{1+m}{2} \right)^{1/2} \frac{\hat{y}}{\tilde{x}^{(1-m)/2}} = \left(\frac{1+m}{2} \right)^{1/2} \frac{y}{x} \mathrm{Re}_x^{1/2}, \qquad \mathrm{Re}_x \equiv \frac{u(x)x}{\nu}. \tag{8.4-29}$$

The second expression for η confirms that the choices of L and U do not affect the solution; that is, the wedge-flow problem has no fixed length or velocity scale. Assuming that the edge of the boundary layer corresponds to $\eta \sim 1$, Eq. (8.4-29) indicates that $\delta/x \sim \mathrm{Re}_x^{-1/2}$, analogous with Eq. (8.2-19). Substituting Eqs. (8.4-22) and (8.4-28) into Eq. (8.4-26), we find that $C_1 = 1$. With $C_1 = 1$ and $C_2 = \beta$, Eq. (8.4-25) becomes

$$f''' + ff'' + \beta[1 - (f')^2] = 0, \tag{8.4-30}$$

which is called the *Falkner-Skan equation*. The boundary conditions for $f(\eta)$ are the same as in Eq. (8.4-13). The Blasius equation is recovered by setting $\beta = 0$.

The Falkner–Skan equation has no known analytical solution. Numerical results from Hartree (1937) are plotted in Fig. 8-10 as $v_x/u(x)$ versus η for various wedge angles. For $\beta \geq 0$, the position where $v_x/u = 0.95$ corresponds to values of η ranging from about 1.5 to 3. These values provide a fairly precise measure of the boundary layer thickness, refining the order-of-magnitude estimate given earlier. The boundary layer thickness is greater at the smaller values of β, including the one case shown where $\beta < 0$.

Cases where $\beta < 0$ resemble the situation on the trailing side of a blunt object, where the

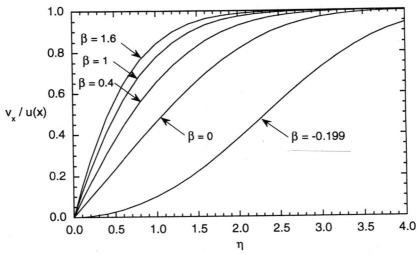

Figure 8-10. Solutions to the Falkner–Skan equation for various wedge angles, based on numerical results tabulated by Hartree (1937).

surface is inclined away from the main flow [see Fig. 8-9(f)]. For $\beta<0$, there is an inflection point in the velocity profile. Of special interest is that the angle $\beta=-0.199$ represents a limiting case; no solutions to the Falkner–Skan equation corresponding to forward flow exist for $\beta<-0.199$. The results for this particular angle exemplify what happens at a separation point, as shown qualitatively in Fig. 8-11. To the left of the separation point, as at location 1, $\partial v_x/\partial y>0$ at $y=0$ and all flow is from left to right. To the right of the separation point, as at location 3 in Fig. 8-11, $\partial v_x/\partial y<0$ at $y=0$ and there is flow reversal in the vicinity of the surface. At the separation point, which is location 2, $\partial v_x/\partial y=0$ at $y=0$ and fluid which had been flowing along the surface

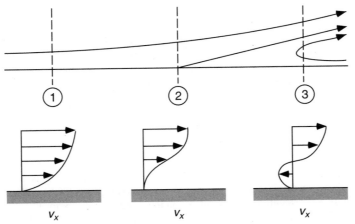

Figure 8-11. Boundary layer separation. Qualitative velocity profiles are shown for three positions along a surface, position 2 being the point of separation.

departs from it. The wedge angle $\beta = -0.199$ corresponds to the critical condition of $\partial v_x/\partial y = 0$ at $y = 0$, as shown by the computed velocity profile in Fig. 8-10.

It was mentioned in Section 8.2 that the underlying cause of boundary layer separation is an adverse pressure gradient near the surface. Using Eqs. (8.2-26) and (8.4-22) we find that for wedge flows we have

$$\frac{d\tilde{\mathscr{P}}}{d\tilde{x}} = -\tilde{u}\,\frac{d\tilde{u}}{d\tilde{x}} = -m\tilde{x}^{2m-1}. \tag{8.4-31}$$

Thus, $d\tilde{\mathscr{P}}/d\tilde{x} < 0$ for $m > 0$, favoring forward flow. For $m < 0$, however, the pressure increases in the direction of flow. If the rate of increase is large enough (as happens for $\beta = -0.199$ or at an actual separation point), forward flow along the surface is no longer possible.

Near the upstream stagnation point on a circular cylinder or any similar blunt object, the velocity profile in the boundary layer closely resembles that for planar stagnation flow ($\beta = 1$, Fig. 8-10). Moving downstream, the pressure gradient at the surface eventually becomes less and less favorable, and the velocity profile undergoes changes which are qualitatively like those shown in Fig. 8-10 for progressive decreases in β. The pressure at the surface of a cylinder in irrotational flow was derived in Example 8.3-1, where it was pointed out that an adverse pressure gradient exists for angles greater than 90° from the upstream stagnation point (i.e., on the entire downstream side). If the pressure gradient in the boundary layer equations is evaluated using that analysis, separation is predicted to occur at an angle of 105° (see Problem 8-4). For large Reynolds numbers ($1 \times 10^4 < \mathrm{Re}_d < 1 \times 10^5$) the actual separation point is at about 81°. The reason for this discrepancy is that, as mentioned in Section 8.2, the actual pressure profile in the separated flow differs considerably from that predicted by potential theory. If a measured pressure is used in the boundary layer equations, the predicted separation point is within about 1° of the measured value. As already mentioned, the boundary layer approximation is no longer valid beyond the point of separation; few theories predict so accurately their own limitations. The theoretical and experimental results for cylinders summarized above are reviewed in more detail by Schlicting (1968, pp. 20–21, 158–162, 201–203) and Churchill (1988, pp. 317–357).

Example 8.4-3 Kármán–Pohlhausen Method   An integral form of the boundary layer momentum equation is obtained by integrating Eq. (8.2-32) over the thickness of the boundary layer; this was done already in Example 6.5-1. The result is the *integral momentum equation* or *Kármán integral equation*,

$$\frac{\tau_0}{\rho} = \frac{d}{dx}\int_0^\delta v_x(u - v_x)\,dy + \frac{du}{dx}\int_0^\delta (u - v_x)\,dy, \tag{8.4-32}$$

where $\tau_0(x)$ is the shear stress exerted on the surface and $\delta(x)$ is the boundary layer thickness. This relation assumes that $v_y = 0$ at $y = 0$ but does not require that $v_x = 0$. Beyond the basic boundary layer assumptions, the only approximation made in deriving Eq. (8.4-32) is that the asymptotic matching requirement for the tangential velocities is replaced by the assumption that $v_x = u$ at a finite distance from the surface. In other words, the boundary layer is assigned a definite thickness, δ.

The *Kármán–Pohlhausen method* is an approach for obtaining approximate solutions to boundary layer problems using Eq. (8.4-32). The major approximation is that the dependence of v_x on y is assumed to have a simple functional form, allowing τ_0 and the integrals to be evaluated readily in terms of $\delta(x)$ and the known function $u(x)$. This leads to an ordinary differential equation for $\delta(x)$, which is solved to complete the analysis. In this approach momentum is conserved in the overall sense expressed by Eq. (8.4-32), but not at each point in the fluid. The Kármán–Pohlhausen method is similar in spirit to the integral approximations used in Section 3.8 and

Example 6.5-1, although more sophisticated in selecting a functional form for the dependent variable (i.e., v_x). It is designed specifically for boundary layers at stationary solid surfaces.

It is assumed that the tangential velocity is of the form

$$\frac{v_x(x, y)}{u(x)} = f(x, \eta) \equiv a(x)\eta + b(x)\eta^2 + c(x)\eta^3 + d(x)\eta^4, \qquad \eta \equiv \frac{y}{\delta(x)}. \qquad (8.4\text{-}33)$$

Although η is analogous to the similarity variables used in the preceding examples, the velocity profiles here are not, in general, self-similar because the coefficients of the fourth-order polynomial in η depend on x. This dependence on x is very important, in that it permits the velocity profile to evolve from a shape which accompanies a favorable pressure gradient (i.e., the $\beta = 1$ curve in Fig. 8-10) to a shape corresponding to separation (the $\beta = -0.199$ curve). Indeed, one of the main attractions of the Kármán–Pohlhausen method is that it can be used in situations where similarity solutions cannot be found.

The no-slip condition at the surface is automatically satisfied by Eq. (8.4-33), so that four additional pointwise conditions involving the dimensionless function $f(x, \eta)$ are needed to relate the coefficient functions to $\delta(x)$. For $v_x = u$ at the outer edge of the boundary layer it is necessary that

$$f(x, 1) = 1. \qquad (8.4\text{-}34)$$

To make τ_{yx} continuous at the outer edge of the boundary layer we also require that

$$\frac{\partial f}{\partial \eta}(x, 1) = 0. \qquad (8.4\text{-}35)$$

The remaining two conditions are obtained by examining the differential form of the boundary layer equation,

$$v_x \frac{\partial v_x}{\partial x} + v_y \frac{\partial v_x}{\partial y} - u \frac{du}{dx} = v \frac{\partial^2 v_x}{\partial y^2}, \qquad (8.4\text{-}36)$$

at the outer edge of the boundary layer and at the surface. Noting that (by definition) the inertial terms at $y = \delta$ reduce to $u\, du/dx$, we find that

$$\frac{\partial^2 f}{\partial \eta^2}(x, 1) = 0. \qquad (8.4\text{-}37)$$

In other words, the viscous term is zero at the outer edge of the boundary layer. At $y = 0$, the no-slip and no-penetration conditions cause the inertial terms in Eq. (8.4-36) to vanish, leaving

$$\frac{\partial^2 v_x}{\partial y^2}(x, 0) = -\frac{u}{v}\frac{du}{dx} \qquad (8.4\text{-}38)$$

or, in dimensionless form,

$$\frac{\partial^2 f}{\partial \eta^2}(x, 0) = -\Lambda(x), \qquad \Lambda(x) \equiv \frac{\delta^2}{v}\frac{du}{dx}. \qquad (8.4\text{-}39)$$

Because the shape of an object enters a boundary layer problem by determining the value of du/dx, the function $\Lambda(x)$ is called the *shape factor.* Inasmuch as the differential formulation of the same boundary layer problem would require boundary conditions only for the velocity, and not its first or second derivatives, Eqs. (8.4-35), (8.4-37), and (8.4-39) may seem superfluous. Each is physically meaningful, however; an exact solution would automatically satisfy all of these conditions (provided, of course, that $y = \delta$ is interpreted as $\hat{y} \to \infty$).

The four requirements stated above for f lead to four algebraic equations for the coefficients, which are solved to give

$$a = 2 + \frac{\Lambda}{6}, \qquad b = -\frac{\Lambda}{2}, \qquad c = -2 + \frac{\Lambda}{2}, \qquad d = 1 - \frac{\Lambda}{6}. \tag{8.4-40}$$

The velocity profile is rewritten now as

$$\frac{v_x}{u(x)} = f(x, \eta) = F(\eta) + \Lambda(x)G(\eta), \tag{8.4-41}$$

$$F(\eta) \equiv 2\eta - 2\eta^3 + \eta^4, \qquad G(\eta) \equiv \frac{1}{6}(\eta - 3\eta^2 + 3\eta^3 - \eta^4). \tag{8.4-42}$$

Because $u(x)$ is known and $\Lambda(x)$ is determined by u and δ, the only independent unknown that remains is $\delta(x)$.

The qualitative behavior of the shape factor and its relation to boundary layer separation are worth noting. At the front of a rounded object, $du/dx > 0$ and, from Eq. (8.4-39), $\Lambda > 0$. The shape factor changes sign along with the pressure gradient, and it tends to become increasingly negative as one moves toward the rear of the object. As discussed in connection with Fig. 8-11, the critical condition for separation is that $\partial v_x/\partial y = 0$ at $y = 0$. That is equivalent to

$$\frac{\tau_0}{\rho} = \nu \frac{\partial v_x}{\partial y}\bigg|_{y=0} = \frac{\nu u}{\delta} \frac{\partial f}{\partial \eta}(x, 0) = \frac{\nu u a}{\delta} = \frac{\nu u}{\delta}\left(2 + \frac{\Lambda}{6}\right) = 0. \tag{8.4-43}$$

Accordingly, separation is predicted to occur at the point where $\Lambda = -12$. The use of the Kármán–Pohlhausen method to estimate the location of that point is outlined in Problem 8-4.

To derive the differential equation for $\delta(x)$ we use $v_x = uf$ in Eq. (8.4-32) to obtain

$$\frac{\tau_0}{\rho} = \frac{d}{dx}\int_0^1 u^2 \delta\, f(1-f)d\eta + \frac{du}{dx}\int_0^1 u\delta(1-f)\,d\eta, \tag{8.4-44}$$

where y has been converted to η. Evaluating f and τ_0 as in Eqs. (8.4-41) and (8.4-43), respectively, this becomes

$$\frac{\nu u}{\delta}\left(2 + \frac{\Lambda}{6}\right) = I_1 \frac{d}{dx}(u^2\delta) + I_2 \frac{d}{dx}(u^2\delta\Lambda) - I_3 \frac{d}{dx}(u^2\delta\Lambda^2) + (I_4 - I_5\Lambda)u\delta\frac{du}{dx}, \tag{8.4-45}$$

where the constants I_n are

$$I_1 \equiv \int_0^1 F(1-F)\,d\eta \qquad = \frac{37}{315}, \tag{8.4-46}$$

$$I_2 \equiv \int_0^1 G(1-2F)\,d\eta \qquad = -\frac{1}{945}, \tag{8.4-47}$$

$$I_3 \equiv \int_0^1 G^2 d\eta \qquad = \frac{1}{9072}, \tag{8.4-48}$$

$$I_4 \equiv \int_0^1 (1-F)\,d\eta \qquad = \frac{3}{10}, \tag{8.4-49}$$

$$I_5 \equiv \int_0^1 G \, d\eta \qquad\qquad = \frac{1}{120}. \tag{8.4-50}$$

The final form of the differential equation is obtained by expanding the various derivatives in Eq. (8.4-45) and employing the dimensionless quantities

$$h(\tilde{x}) \equiv \left(\frac{\delta}{L}\right)^2 \mathrm{Re}, \qquad \Lambda(\tilde{x}) \equiv h\frac{d\tilde{u}}{d\tilde{x}}, \qquad \Omega(\tilde{x}) \equiv h^2\tilde{u}\frac{d^2\tilde{u}}{d\tilde{x}^2}, \tag{8.4-51}$$

where $\tilde{x} = x/L$, $\tilde{u} = u/U$, and $\mathrm{Re} = UL/\nu$, as used previously. The result is

$$\frac{dh}{d\tilde{x}} = \frac{2}{\tilde{u}}\frac{\left[2 + \left(\frac{1}{6} - 2I_1 - I_4\right)\Lambda - (2I_2 - I_5)\Lambda^2 + 2I_3\Lambda^3 - (I_2 - 2I_3\Lambda)\Omega\right]}{\left[I_1 + 3I_2\Lambda - 5I_3\Lambda^2\right]}. \tag{8.4-52}$$

For an object with a sharp leading edge (i.e., a wedge with $0 < \beta < 1$) the initial condition is $h(0) = 0$; for an object with a rounded front, which locally resembles a plane oriented normal to the approaching flow, $dh/d\tilde{x} = 0$ at $\tilde{x} = 0$. It should be remarked that Eq. (8.4-52) differs from the differential equations usually used to implement the Kármán–Pohlhausen method, although they yield equivalent results. The traditional approaches involve several additional transformations, designed in part to make manual calculations more tractable [see Schlicting (1968, pp. 192–206)].

As a specific example, we again consider a thin flat plate oriented parallel to the approaching flow. In this case $\tilde{u} = 1$, $\Lambda = 0$, $\Omega = 0$, and $h(0) = 0$. The solution to Eq. (8.4-52) is

$$h(\tilde{x}) = \frac{4\tilde{x}}{I_1} \tag{8.4-53}$$

and the dimensional boundary layer thickness is

$$\delta(x) = \left(\frac{4}{I_1}\frac{\nu x}{U}\right)^{1/2}. \tag{8.4-54}$$

Thus, it is found that the boundary layer thickness on a flat plate varies as $x^{1/2}$, in agreement with the similarity solution in Example 8.4-1. The local shear stress is given by

$$\frac{\tau_0(x)}{\mu U/L} = K\,\mathrm{Re}^{1/2}\left(\frac{x}{L}\right)^{-1/2}, \qquad \mathrm{Re} \equiv \frac{UL}{\nu} \tag{8.4-55}$$

or, equivalently,

$$\frac{\tau_0(x)}{\rho U^2} = K\,\mathrm{Re}_x^{-1/2}, \qquad \mathrm{Re}_x \equiv \frac{Ux}{\nu}, \tag{8.4-56}$$

where K is a constant. By comparing Eq. (8.4-55) with Eq. (8.4-15), it is seen that the exact result is $K = 0.332$. To evaluate K from the Kármán–Pohlhausen solution we use Eq. (8.4-54) in Eq. (8.4-43); the result is $K = I_1^{1/2} = 0.343$, which is only 3% too large. Similar accuracy is achieved for planar stagnation flow, where $\Lambda \neq 0$ (see Problem 8-3).

8.5 INTERNAL BOUNDARY LAYERS

The boundary layer approximation is not restricted to fluid layers near solid surfaces. It applies also at interfaces where two fluids with different velocities are brought into contact, as well as at locations within a single fluid where the dynamic conditions make inertial and viscous effects comparable in a thin region. In the latter cases the boundary layers are "internal" rather than at an interface. Presented here are two examples of internal boundary layers, one involving the wake of a streamlined object (a flat plate) and the other a jet within a stagnant volume of the same fluid.

Example 8.5-1 Flow in the Wake of a Flat Plate   At the trailing edge of a flat plate the boundary layers on the two sides detach and merge to form a wake, as shown in Fig. 8-12. In the wake the no-slip condition at the solid surface is replaced by the condition of zero shear stress at the plane of symmetry ($y=0$). Moving downstream, the wake grows in thickness and the velocity differences decay until, eventually, the entire fluid is again at the uniform velocity which characterizes the unperturbed flow. This example focuses on the region relatively far downstream from the plate, where the velocity differences are small relative to the velocity in the free stream. As shown by Tollmien [see Schlicting (1968, pp. 166-170)], a fairly simple expression for the velocity can be derived for this part of the wake.

It is advantageous here to work with the *velocity difference*,

$$\Lambda(x,y) \equiv U - v_x(x, y). \tag{8.5-1}$$

Assuming that $\Lambda \ll U$, the inertial terms in the boundary layer momentum equation are rewritten as

$$v_x \frac{\partial v_x}{\partial x} = (U - \Lambda)\left(-\frac{\partial \Lambda}{\partial x}\right) \cong -U\frac{\partial \Lambda}{\partial x} \sim \frac{UV}{x}, \tag{8.5-2}$$

$$v_y \frac{\partial v_x}{\partial y} = \left(\int_0^y \frac{\partial \Lambda}{\partial x} \, dY\right)\left(-\frac{\partial \Lambda}{\partial y}\right) \sim \left(\frac{V\delta}{x}\right)\left(\frac{V}{\delta}\right) = \frac{V^2}{x}, \tag{8.5-3}$$

where V is the scale for Λ and 2δ is the thickness of the wake. It is seen that the $\partial v_x/\partial y$ term is negligible. There is also no pressure gradient in this flow, so that the momentum equation reduces to

$$U\frac{\partial \Lambda}{\partial x} = \nu \frac{\partial^2 \Lambda}{\partial y^2}. \tag{8.5-4}$$

This linear differential equation is analogous to that for transient diffusion, with x/U the time variable and ν the diffusivity. The boundary conditions for $\Lambda(x, y)$ are

$$\frac{\partial \Lambda}{\partial y}(x, 0) = 0, \qquad \Lambda(x, \infty) = 0. \tag{8.5-5}$$

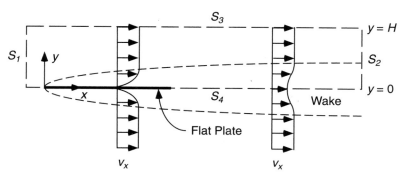

Figure 8-12. Flow in the wake of a flat plate.

No boundary condition can be applied at the downstream edge of the plate ($x=L$), because the analysis is valid only for much larger x.

What takes the place of the missing boundary condition in x is an overall momentum balance, which is derived now using the rectangular control volume depicted in Fig. 8-12. Integrating the x component of the Navier–Stokes equation over the control volume (with $\partial\mathcal{P}/\partial x=0$) gives

$$\rho\int_V \mathbf{v}\cdot\nabla v_x\, dV = \mathbf{e}_x\cdot\int_V \nabla\cdot\boldsymbol{\tau}\, dV = \mathbf{e}_x\cdot\int_S \mathbf{n}\cdot\boldsymbol{\tau}\, dS, \qquad (8.5\text{-}6)$$

where the divergence theorem for tensors has been used with the viscous term. The only place on the control surface where the viscous stresses are not negligible is the part of S_4 which coincides with the plate (see Fig. 8-12). Denoting the top surface of the plate as S_p, we obtain

$$\mathbf{e}_x\cdot\int_S \mathbf{n}\cdot\boldsymbol{\tau}\, dS = \mathbf{e}_x\cdot\int_{S_p}(-\mathbf{e}_y)\cdot\boldsymbol{\tau}\, dS = -\int_{S_p}\tau_{yx}\, dS = -\frac{F_D}{2}, \qquad (8.5\text{-}7)$$

where F_D is the total drag on the plate (both sides). With the help of continuity and the divergence theorem, the inertial or convective term in Eq. (8.5-6) is rewritten as

$$\rho\int_V \mathbf{v}\cdot\nabla v_x\, dV = \rho\int_V \nabla\cdot(\mathbf{v}v_x)\, dV = \rho\int_S v_x(\mathbf{n}\cdot\mathbf{v})\, dS. \qquad (8.5\text{-}8)$$

Evaluating the contributions from the three surfaces where there is a nonzero normal component of velocity ($S_1,\ S_2,\ S_3$), we obtain

$$\rho\int_S v_x(\mathbf{n}\cdot\mathbf{v})\, dS = -\rho WHU^2 + \rho W\left(HU^2 - 2U\int_0^H \Lambda\, dy\right) + \rho U\int_{S_3} v_y\, dS, \qquad (8.5\text{-}9)$$

where W is the plate width and H is the height of the control volume. In evaluating the middle (S_2) term we have again used the assumption that $\Lambda \ll U$.

The last integral in Eq. (8.5-9), which gives the volumetric flow rate across the top surface, is evaluated from overall conservation of mass. Conservation of mass for the control volume is expressed as

$$\int_S (\mathbf{n}\cdot\mathbf{v})\, dS = 0 = -WHU + W\int_0^H v_x\, dy + \int_{S_3} v_y\, dS, \qquad (8.5\text{-}10)$$

which leads to

$$\int_{S_3} v_y\, dS = W\int_0^H (U - v_x)\, dy = W\int_0^H \Lambda\, dy. \qquad (8.5\text{-}11)$$

Evaluating the terms in Eq. (8.5-6) and letting $H\to\infty$, overall conservation of momentum reduces to

$$F_D = 2\rho WU\int_0^\infty \Lambda\, dy. \qquad (8.5\text{-}12)$$

The absence of any fixed length scale in the far-downstream region suggests that the similarity method may work for determining $\Lambda(x, y)$. Because Λ decreases as one moves downstream, it will be necessary to introduce a scale factor which depends on x. The solution is assumed to be of the form

$$\frac{\Lambda(x,\, y)}{U} = \left(\frac{x}{L}\right)^p f(\eta), \qquad \eta \equiv \frac{y}{g(x)}, \tag{8.5-13}$$

where the constant p and the functions $f(\eta)$ and $g(x)$ are to be determined. Changing variables in the usual manner, Eq. (8.5-4) becomes

$$f'' + \left(\frac{Ugg'}{\nu}\right)\eta f' - \left(\frac{Upg^2}{\nu x}\right)f = 0. \tag{8.5-14}$$

If the assumed form of solution is valid, then both terms in parentheses must be constant. Accordingly, we set

$$\frac{Ugg'}{\nu} = C_1 \tag{8.5-15}$$

$$\frac{Upg^2}{\nu x} = C_2. \tag{8.5-16}$$

Choosing $C_1 = 2$ and integrating Eq. (8.5-15) gives

$$g(x) = 2\left(\frac{\nu x}{U}\right)^{1/2}, \qquad \eta = \frac{y}{2}\left(\frac{U}{\nu x}\right)^{1/2}, \tag{8.5-17}$$

where the integration constant in g has been set equal to zero. Using this expression for g in Eq. (8.5-16), we find that $C_2 = 4p$. This confirms that Eq. (8.5-13) is consistent with the differential form of the momentum equation. Notice that if the integration constant in g had not been set equal to zero, we would have obtained the contradictory result that C_2 is a function of x.

The constant p is evaluated using the overall momentum balance. Substituting the assumed form of solution into Eq. (8.5-12) and using $dy = g \, d\eta$ gives

$$F_D = 4\rho W U^2 \left(\frac{x}{L}\right)^p \left(\frac{\nu x}{U}\right)^{1/2} \int_0^\infty f \, d\eta. \tag{8.5-18}$$

Because all other terms in Eq. (8.5-18) are constant (including the integral), the powers of x must cancel. Accordingly, $p = -1/2$ and (from above) $C_2 = 4p = -2$.

The complete problem for $f(\eta)$ is now

$$f'' + 2(\eta f' + f) = 0, \tag{8.5-19}$$

$$f'(0) = 0, \qquad f(\infty) = 0, \tag{8.5-20}$$

where Eq. (8.5-20) follows directly from the boundary conditions for Λ in Eq. (8.5-5). The first integration of Eq. (8.5-19) is simplified by noticing that $\eta f' + f = (\eta f)'$. Integrating twice gives

$$f(\eta) = a e^{-\eta^2} + b, \tag{8.5-21}$$

where a and b are constants. The boundary condition at infinity requires that $b = 0$, whereas the boundary condition at zero is satisfied for any choice of a. The expression for the velocity difference is now

$$\frac{\Lambda(x,\, y)}{U} = a\left(\frac{x}{L}\right)^{-1/2} e^{-\eta^2}. \tag{8.5-22}$$

To evaluate a we return to the overall momentum balance, Eq. (8.5-18). Using Eq. (8.4-16) for F_D, Eq. (8.5-21) for f, and $p = -1/2$, it is found that

$$a = \frac{1.328}{2\sqrt{\pi}} = 0.211. \tag{8.5-23}$$

The final result is

$$\frac{\Lambda(x, y)}{U} = 0.211 \left(\frac{x}{L}\right)^{-1/2} \exp\left(-\frac{Uy^2}{4\nu x}\right). \tag{8.5-24}$$

It is seen that the velocity difference at the center of the wake varies as $x^{-1/2}$, whereas the width of the wake (as reflected by g) varies as $x^{1/2}$.

As discussed in Schlicting (1968, p. 169), it has been found that Eq. (8.5-24) is satisfactory for $x/L > 3$. An analysis which includes the region immediately downstream from the plate is given in Goldstein (1930).

Example 8.5-2 Planar Jet   An internal boundary layer is created also by a narrow jet of fluid entering a large, stagnant volume of the same fluid. Some of the initial momentum of the jet is transferred to the outer fluid, causing it to be entrained. As this happens the jet gradually slows and spreads out. Given that the pressure in the outer region (stagnant fluid) is imposed on the boundary layer (jet), $\mathscr{P}$ is constant in the jet. We focus here on the planar geometry depicted in Fig. 8-13, in which the jet is assumed to emerge from a narrow slit in a wall. The key characteristic of the jet is its initial momentum (assumed to be known), rather than its initial thickness (negligible) or flow rate. The analysis which follows is due to Schlicting (1968, pp. 170–174).

Both velocity components are important here, so that the stream function $\psi(x, y)$ is employed. Converting Eq. (8.2-23) (with $d\mathscr{P}/dx = 0$) to the stream function gives

$$\frac{\partial \psi}{\partial y} \frac{\partial^2 \psi}{\partial x \partial y} - \frac{\partial \psi}{\partial x} \frac{\partial^2 \psi}{\partial y^2} = \nu \frac{\partial^3 \psi}{\partial y^3}. \tag{8.5-25}$$

The absence of a fixed length scale suggests that a similarity solution is possible. Introducing a scale factor to account for the slowing of the jet as it moves away from the wall, we assume a solution of the form

$$\psi(x, y) = x^p f(\eta), \qquad \eta \equiv \frac{y}{g(x)}, \tag{8.5-26}$$

where the constant p and the functions $f(\eta)$ and $g(x)$ must be determined. Changing variables, Eq. (8.5-25) becomes

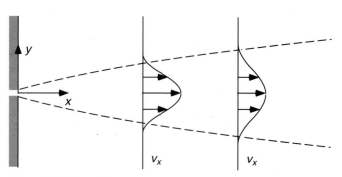

Figure 8-13. Planar jet.

$$f''' + C_1 ff'' + C_2(f')^2 = 0, \qquad (8.5\text{-}27)$$

$$C_1 = \frac{px^{p-1}g}{\nu} \qquad (8.5\text{-}28)$$

$$C_2 = \frac{1}{\nu}(x^p g' - px^{p-1}g), \qquad (8.5\text{-}29)$$

where it is assumed that C_1 and C_2 are constants. Choosing $C_1 = 2$ for later convenience, we obtain

$$g(x) = \frac{2\nu x^{1-p}}{p}, \qquad g'(x) = \frac{2\nu(1-p)x^{-p}}{p}. \qquad (8.5\text{-}30)$$

Equations (8.5-29) and (8.5-30) imply that $C_2 = 2(1-2p)/p$, a constant, confirming that the assumed form of solution is consistent with the momentum equation.

An integral form of conservation of momentum is used to determine p. The desired form of the momentum equation is obtained by recalling that the derivation of Eq. (8.4-32) did not require that $v_x = 0$ at $y = 0$ and by observing that the equation remains valid for $\delta \to \infty$. Setting $u = 0$ (stagnant outer fluid) and $\tau_0 = 0$ (no stress at the symmetry plane) then gives

$$\frac{d}{dx} \int_0^\infty v_x^2 \, dy = 0. \qquad (8.5\text{-}31)$$

From Eq. (8.5-31) and the symmetry of the jet we conclude that

$$K = \int_{-\infty}^\infty v_x^2 \, dy, \qquad (8.5\text{-}32)$$

where K is a constant; Schlicting termed this quantity the *kinematic momentum.* It is assumed that the kinematic momentum, which is independent of distance from the wall, is given. Noting that $v_x = \partial\psi/\partial y = x^p f'/g$ and using Eq. (8.5-30) to evaluate g, the overall momentum balance gives

$$K = \frac{x^{2p}}{g} \int_{-\infty}^\infty (f')^2 \, d\eta = \frac{px^{3p-1}}{\nu} \int_0^\infty (f')^2 \, d\eta, \qquad (8.5\text{-}33)$$

which implies that $p = \frac{1}{3}$. It follows that

$$g(x) = 6\nu x^{2/3}, \qquad \eta = \frac{y}{6\nu x^{2/3}}, \qquad (8.5\text{-}34)$$

and $C_2 = 2$.

The boundary conditions for f are obtained from the requirements that $v_y = 0$ and $\partial v_x/\partial y = 0$ at $y = 0$, and that $v_x = 0$ at $y = \infty$. The complete problem for $f(\eta)$ is now

$$f''' + 2[ff'' + (f')^2] = 0, \qquad (8.5\text{-}35)$$

$$f(0) = f''(0) = 0, \qquad f'(\infty) = 0. \qquad (8.5\text{-}36)$$

Amazingly, this nonlinear, third-order differential equation has an analytical solution. The first integration of Eq. (8.5-35) is facilitated by noticing that $(ff')' = ff'' + (f')^2$. Together with the boundary conditions at $\eta = 0$, this yields

$$f'' + 2 ff' = f'' + (f^2)' = 0. \qquad (8.5\text{-}37)$$

A second integration then gives

$$f' + f^2 = b^2, \tag{8.5-38}$$

where b is a constant. The integration constant can be written in this manner, with b a real number, because $f(0) = 0$ and (from $v_x > 0$) $f'(0) > 0$. Equation (8.5-38) is separable and can be integrated once more to obtain

$$\eta = \frac{1}{2b} \ln\left[\frac{1 + (f/b)}{1 - (f/b)}\right] = \frac{1}{b} \tanh^{-1}\left(\frac{f}{b}\right), \tag{8.5-39}$$

which is inverted to give

$$f(\eta) = b \tanh b\eta. \tag{8.5-40}$$

The solution is complete now except for the value of b.

One of the boundary conditions has not been used, namely $f'(\infty) = 0$. However, differentiation of Eq. (8.5-40) gives

$$f'(\eta) = b^2(1 - \tanh^2 b\eta), \tag{8.5-41}$$

which, together with $\tanh(\infty) = 1$, indicates that the remaining boundary condition is satisfied for any value of b. To determine b we return to the overall momentum balance. Using Eq. (8.5-41) in Eq. (8.5-33) and letting t be a dummy variable, we obtain

$$K = \frac{b^3}{3\nu} \int_0^\infty (1 - \tanh^2 t)^2 \, dt = \frac{2b^3}{9\nu}. \tag{8.5-42}$$

Accordingly, the velocity normal to the wall is given by

$$v_x(x, y) = \frac{1}{6}\left(\frac{9}{2}\right)^{2/3}\left(\frac{K^2}{\nu x}\right)^{1/3}(1 - \tanh^2 b\eta) = 0.4543\left(\frac{K^2}{\nu x}\right)^{1/3}(1 - \tanh^2 b\eta), \tag{8.5-43}$$

$$b\eta = \frac{1}{6}\left(\frac{9}{2}\right)^{1/3}\left(\frac{K}{\nu^2}\right)^{1/3}\frac{y}{x^{2/3}} = 0.2752\left(\frac{K}{\nu^2}\right)^{1/3}\frac{y}{x^{2/3}}. \tag{8.5-44}$$

Finally, the volumetric flow rate per unit width is

$$q(x) = 2\int_0^\infty v_x \, dy = \left[2\left(\frac{9}{2}\right)^{1/3}\int_0^\infty (1 - \tanh^2 t) \, dt\right](K\nu x)^{1/3} = 3.3019(K\nu x)^{1/3}. \tag{8.5-45}$$

In summary, it has been shown that the width [Eq. (8.5-34)], velocity [Eq. (8.5-43)], and volumetric flow rate of the jet [Eq. (8.5-45)] vary as $x^{2/3}$, $x^{-1/3}$, and $x^{1/3}$, respectively. The increase in q reflects the entrainment of the surrounding fluid. Experimental evidence in support of the predicted behavior is cited in Schlicting (1968, p. 174).

References

Batchelor, G. K. *An Introduction to Fluid Dynamics.* Cambridge University Press, Cambridge, 1967.

Churchill, S. W. *Viscous Flows.* Butterworths, Boston, 1988.

Cochran, W. G. The flow due to a rotating disk. *Proc. Cambridge Philos. Soc.* 30: 365–375, 1934.

Denn, M. M. *Process Fluid Mechanics.* Prentice-Hall, Englewood Cliffs, NJ, 1980.

Dugas, R. *A History of Mechanics.* Dover, New York, 1988. [This is a reprint of a book originally published in 1955.]

Falkner, V. M. and S. M. Skan. Solutions of the boundary-layer equations. *Philos. Mag.* 12: 865–896, 1931.

Gillispie, C. G. (Ed.). *Dictionary of Scientific Biography.* Scribner's, New York, 1970–1980.

Goldstein, S. Concerning some solutions of the boundary layer equations in hydrodynamics. *Proc. Cambr. Philos. Soc.* 26: 1–30, 1930.

Hartree, D. R. On an equation occurring in Falkner and Skan's approximate treatment of the equations of the boundary layer. *Proc. Cambr. Philos. Soc.* 33: 223–239, 1937.

Hildebrand, F. B. *Advanced Calculus for Applications,* second edition. Prentice-Hall, Englewood Cliffs, NJ, 1976.

Howarth, L. On the solution of the laminar boundary layer equations. *Proc. R. Soc. Lond.* A 164: 547–579, 1938.

Imai, I. Second approximation to the laminar boundary layer flow over a flat plate. *J. Aeronaut. Sci.* 24: 155–156, 1957.

Jones, C. W. and E. J. Watson. Two-dimensional boundary layers. In *Laminar Boundary Layers,* L. Rosenhead, Ed. Clarendon Press, Oxford, 1963, pp. 198–257.

Lamb, H. *Hydrodynamics,* sixth edition. Cambridge University Press, Cambridge, 1932.

Lighthill, M. J. Introduction. Real and ideal fluids. In *Laminar Boundary Layers,* L. Rosenhead, Ed. Clarendon Press, Oxford, 1963, pp. 1–45.

Milne-Thomson, L. M. *Theoretical Hydrodynamics,* fourth edition. Macmillan, New York, 1960.

Rogers, M. H. and G. N. Lance. The rotationally symmetric flow of a viscous fluid in the presence of an infinite rotating disk. *J. Fluid Mech.* 7: 617–631, 1960.

Schlicting, H. *Boundary-Layer Theory,* sixth edition. McGraw-Hill, New York, 1968.

Tani, I. History of boundary layer theory. *Annu. Rev. Fluid Mech.* 9: 87–111, 1977.

Van Dyke, M. *Perturbation Methods in Fluid Mechanics.* Academic Press, New York, 1964.

Van Dyke, M. *An Album of Fluid Motion.* Parabolic Press, Stanford, CA, 1982.

Problems

8-1. Irrotational and Viscous Flows Past a Sphere

   This problem concerns flow past a sphere of radius R with a uniform approach velocity U.

   (a) Determine the velocity and pressure for irrotational flow.

   (b) As a contrast to part (a), where the vorticity is zero everywhere, calculate the vorticity distribution for creeping flow (see Example 7.4-2). Show that this flow is irrotational only for $r \rightarrow \infty$. Is the vorticity field spherically symmetric?

8-2. Hele-Shaw Flow

   Consider viscous flow past an object confined between closely spaced, parallel plates, as shown in Fig. P8-2. It is assumed that $H \ll L$, where L is the characteristic dimension of the object and $2H$ is the plate spacing. Far from the object the flow is unidirectional with a maximum velocity v_0, such that

$$v_x = v_0 \left[1 - \left(\frac{z}{H} \right)^2 \right], \qquad r = (x^2 + y^2)^{1/2} \rightarrow \infty.$$

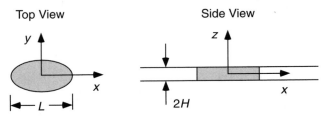

Figure P8-2. Arrangement for Hele-Shaw flow. An object of characteristic dimension L is confined between parallel plates with spacing $2H$.

(a) Show that a solution of the continuity and Stokes' equations which is consistent with the unperturbed velocity given above is

$$v_x(x, y, z) = u_x(x, y)\left[1 - \left(\frac{z}{H}\right)^2\right], \qquad v_y(x, y, z) = u_y(x, y)\left[1 - \left(\frac{z}{H}\right)^2\right],$$

and $v_z = 0$, where **u** is the velocity for *irrotational* flow past a *long* object of the same cross section. Furthermore, show that the streamlines in any plane of constant z are identical to those for the corresponding irrotational flow.

(b) The solution in part (a) satisfies the no-slip condition at the plates but not at the surface of the object. At approximately what distance from the object will it be valid?

(c) Show that to neglect inertia in this situation it is sufficient that

$$\left(\frac{v_0 H}{\nu}\right)\frac{H}{L} \ll 1, \qquad \frac{H}{L} \ll 1,$$

as with the lubrication approximation.

This type of flow is called *Hele-Shaw flow* after H. S. Hele-Shaw, who described it first in 1898 [see Schlicting (1968, p. 116)]. Its significance is that it provides an experimental arrangement for visualizing the streamlines of any planar potential flow (except very near an object). This includes flows which, because of separation, do not otherwise exist in reality! Several excellent examples are pictured in Van Dyke (1982, pp. 8–10).

8-3. Planar Stagnation Flow

In planar stagnation flow, shown in Fig. 8-9(c), the outer velocity near the surface is given by Eq. (8.4-20) with $m = 1$:

$$u(x) = Cx.$$

(a) State the differential equation obtained for this geometry using the Kármán–Pohlhausen method.

(b) Solve the equation from part (a) to determine $\delta(x)$ and $\tau_0(x)$. Specifically, show that if the shear stress is written as

$$\frac{\tau_0}{\rho u^2} = K \, Re_x^{-1/2},$$

then $K = 1.196$. For comparison, the exact result from the numerical solution of the Falkner–Skan equation is $K = 1.233$ (Jones and Watson, 1963, p. 232).

8-4. Boundary Layer Separation on a Cylinder

Consider flow past a circular cylinder, as shown in Fig. 8-5. Assume that the pressure is given by the solution for irrotational flow.

(a) Use the results of Example 8.3-1 to determine $u(x)$, where $x=0$ at the leading stagnation point.

(b) State the differential equation obtained for this geometry using the Kármán–Pohlhausen method.

(c) Use the equation from part (b) to compute $\delta(0)$.

(d) Solve the differential equation for $h(x)$ numerically and determine the location of the separation point. You should obtain an angle of 107°; the exact value for this idealized pressure field is reported to be 104.5° (Schlicting, 1968, p. 203). (*Hint:* The fourth-order Runge–Kutta method is satisfactory for the integration. To avoid the singularity at $x=0$, start the integration at a small, but nonzero, value of x.)

8-5. Solution of the Blasius Equation

A numerical solution for the Blasius analysis of flow past a flat plate is obtained relatively simply, as outlined below. The problem to be solved consists of Eqs. (8.4-12) and (8.4-13).

(a) Assume that $F(\zeta)$ is the solution to the closely related problem,

$$F''' + FF'' = 0; \qquad F(0)=F'(0)=0, \quad F''(0)=1,$$

where $F'=dF/d\zeta$, and so on. This differs from the Blasius problem only in the last boundary condition. Show that $f(\eta)=a\,F(a\eta)$, where a is a constant, satisfies Eq. (8.4-12).

(b) Show that

$$f''(0)=[F'(\infty)]^{-3/2}.$$

(c) Solve the problem in part (a) numerically to determine $F(\zeta)$ and $f''(0)$; the latter quantity determines the drag on a flat plate, as shown in Eq. (8.4-15). Finally, compute the velocity profile for the boundary layer next to a flat plate (the curve for $\beta=0$ in Fig. 8-10). (*Hint:* Rewrite the third-order equation as a system of three first-order equations involving the dependent variables f, f', and f''. The fourth-order Runge–Kutta method is a good choice for the integration.)

The approach just described exploits a special property of the Blasius equation; it does not work, for example, with the Falkner–Skan equation. Whereas the solution of a boundary-value problem such as this ordinarily requires numerous iterations to match the data at both boundaries, the ability to compute $f''(0)$ as in part (b) reduces the effort to just two integrations.

8-6. Circular Jet

Consider a jet of fluid which enters a large, stagnant volume of the same fluid via a small, circular hole in a wall. This problem is the axisymmetric analog of Example 8.5-2, and it too has an exact, analytical solution. The boundary layer momentum equation in this case is

$$v_r\frac{\partial v_z}{\partial r}+v_z\frac{\partial v_z}{\partial z}=\frac{\nu}{r}\frac{\partial}{\partial r}\left(r\frac{\partial v_z}{\partial r}\right)$$

and the kinematic momentum, which again is a known constant, is defined as

$$K\equiv 2\pi\int_0^\infty v_z^2 r\,dr.$$

(a) To obtain a similarity solution, assume that the stream function is of the form

$$\psi(r, z) = -\nu z^p F(\eta), \qquad \eta \equiv \frac{r}{z^n},$$

where the factor ν has been introduced for later convenience. Show that $p=n=1$.

(b) After one integration of the differential equation for $F(\eta)$, show that the problem becomes

$$FF' = F' - \eta F'', \qquad F(0) = F'(0) = 0.$$

(c) If $F(\eta)$ is a solution of the differential equation in part (b), show that $F(\xi)$ is also a solution, where $\xi = a\eta$ and a is a constant. (The Blasius equation also has this property; see Problem 8-5.) Moreover, show that if ξ replaces η as the independent variable, the solution to the problem in part (b) is

$$F(\xi) = \frac{\xi^2}{1 + (\xi^2/4)}.$$

This completes the solution for the jet, except for the value of a.

(d) Evaluate a in terms of the kinematic momentum and other known constants, and show that the volumetric flow rate in the jet is

$$Q = 8\pi\nu z$$

as given in Schlicting (1968, p. 221). Interestingly, this result indicates that Q is independent of K.

8-7. Flow in a Convergent Channel (Hamel Flow)

Two-dimensional flow in channels bounded by nonparallel, planar walls was investigated by G. Hamel in 1916 [see Schlicting (1968, p. 102)]. We consider here the convergent case shown in Fig. P8-7(a), in which the flow is directed toward the narrow end of the channel, the fluid being withdrawn though a narrow slit at the intersection of the walls. The dynamic conditions are assumed to result in irrotational flow in the center of the channel and a boundary layer at each wall. This is one of the few problems in which analytical solutions can be obtained for both the irrotational and boundary layer regions.

(a) Show that in the irrotational region, where the flow is purely radial, the velocity is given by

$$v_r(r) = -\frac{q}{\alpha\pi r},$$

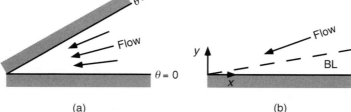

| (a) | (b) |

Figure P8-7. Flow in a converging channel. (a) Overall view; (b) enlargement showing boundary layer (BL) near one wall.

where q (>0) is the rate of fluid withdrawal per unit width. It is assumed here that the thickness of the boundary layers is negligible.

(b) Referring now to the boundary layer coordinates in Fig. P8-7(b), show that

$$\bar{u}(\tilde{x}) = -\frac{1}{\tilde{x}}.$$

How must the velocity scale, U, be defined?

(c) Use the Falkner–Skan analysis to show that

$$f''' - (f')^2 + 1 = 0; \qquad f'(0) = 0, \quad f'(\infty) = 1, \quad f''(\infty) = 0,$$

where $f(\eta)$ is as defined in Eq. (8.4-24). Notice that $f(0) = 0$, which corresponds to the no-penetration condition, is replaced by $f''(\infty) = 0$ (see Example 8.4-3). Determine $g(\tilde{x})$.

(d) A first integration of the differential equation in part (c) is accomplished by multiplying by f'' and noting that $[(f'')^2]' = 2f''f'''$ and $[(f')^3]' = 3(f')^2 f''$. Show that

$$f'' = \sqrt{\frac{2}{3}}(1 - f')\sqrt{f' + 2}.$$

(e) Noticing that the differential equation in part (d) is separable, show that the final expression for the tangential velocity is

$$\frac{v_x}{u} = f'(\eta) = 3 \tanh^2\left[\frac{\eta}{\sqrt{2}} + \tanh^{-1}\sqrt{\frac{2}{3}}\right] - 2$$

as given in Schlicting (1968, p. 153).

It is interesting that, for the opposite case of flow in a divergent channel, there is no boundary layer regime. A detailed discussion of Hamel flow solutions, not only in the context of boundary layers but also in relation to creeping flow and the lubrication approximation, is given in Denn (1980, pp. 211–229).

8-8. Rotating Disk

Consider a solid disk of indefinite size, rotating at angular velocity ω, which is in contact with a large volume of fluid. The surface of the disk is at $z = 0$ and the liquid occupies the space $z > 0$. The effect of the rotation is to draw fluid toward the disk axially and to throw it outward radially as it approaches the surface. An integral approximate solution for this problem was reported first by T. von Kármán in 1921. A numerical solution of the equations given below was obtained originally by Cochran (1934), and the results were refined and extended by Rogers and Lance (1960). Rotating disks are used extensively in experimental research because of their special mass transfer characteristics (see Problem 10-2).

(a) Show that the continuity and Navier–Stokes equations are satisfied by velocity and pressure fields of the form

$$v_r = r\omega F(\zeta), \qquad v_\theta = r\omega G(\zeta), \qquad v_z = \sqrt{\nu\omega}\,H(\zeta), \qquad \mathcal{P} = \rho\nu\omega\,P(\zeta),$$

where $\zeta \equiv z\sqrt{\omega/\nu}$. Specifically, show that the functions F, G, H, and P are governed by

$$H' + 2F = 0,$$

$$F'' - F^2 + G^2 - HF' = 0,$$

$$G'' - 2FG - HG' = 0,$$

$$H'' - HH' - P' = 0$$

with the boundary conditions

$$F(0) = 0, \qquad G(0) = 1, \qquad H(0) = 0, \qquad P(0) = 0, \qquad F(\infty) = 0, \qquad G(\infty) = 0.$$

From their numerical solution, Rogers and Lance (1960) found that $F'(0) = 0.5102$ and $G'(0) = 0.6159$.

(b) Calculate the torque on a disk of radius R wetted on one side, assuming that edge effects are negligible.

(c) It is the behavior of v_z near the surface which influences the mass transfer coefficient for a rotating disk, as shown in Problem 10-2. Show that for $\zeta \to 0$ we have

$$H(\zeta) = -0.5102\zeta^2 + O(\zeta^3).$$

8-9. Rotating Fluid Near a Surface (Bödewadt Flow)

In this problem a large volume of rotating fluid is assumed to be in contact with a stationary, planar surface. The fluid is in rigid-body rotation at angular velocity ω, except in the boundary layer at the surface. This is the opposite of the rotating disk flow considered in Problem 8-8, and the secondary flow (that superimposed on the rotation) is opposite in sign: The surface creates a flow which is radially inward and axially upward. This problem was solved first by U. Bödewadt in 1940. The numerical results were refined by Rogers and Lance (1960) as part of a more comprehensive investigation of various combinations of rotating disks and rotating fluids.

(a) According to the boundary layer approximation, the radial pressure gradient in the outer fluid (the part undergoing rigid-body rotation) is imposed on the boundary layer. Determine that pressure gradient.

(b) Using a velocity field of the same form as that in Problem 8-8, show that the functions F, G, and H are governed by

$$H' + 2F = 0,$$

$$F'' - F^2 + G^2 - HF' = 1,$$

$$G'' - 2FG - HG' = 0$$

with the boundary conditions

$$F(0) = 0, \qquad G(0) = 0, \qquad H(0) = 0, \qquad F(\infty) = 0, \qquad G(\infty) = 1.$$

From their numerical solution, Rogers and Lance (1960) found that $F'(0) = -0.9420$ and $G'(0) = 0.7729$.

(c) Calculate the torque on an area of surface of radius R.

(d) The behavior of v_z near the surface is important in determining the mass transfer coefficient. Show that for $\zeta \to 0$ we have

$$H(\zeta) = 0.9420\zeta^2 + O(\zeta^3).$$

8-10. Axisymmetric Stagnation Flow

Suppose that a fluid moving perpendicular to a flat surface is forced to flow radially as it approaches the surface; this is the axisymmetric analog of planar stagnation flow.

(a) By analogy with the planar case, state the continuity and momentum equations for the boundary layer. Assume that the solution for irrotational flow gives

$$u(r) = ar,$$

where a is a constant.

(b) Show that the transformation

$$v_r = u(r)\phi'(\zeta), \qquad v_z = -2\sqrt{a\nu}\,\phi(\zeta), \qquad \zeta = z\sqrt{a/\nu}$$

satisfies the continuity equation and reduces the momentum equation and boundary conditions to

$$\phi''' + 2\phi\phi'' + 1 - (\phi')^2 = 0; \qquad \phi(0) = \phi'(0) = 0, \quad \phi'(\infty) = 1.$$

(c) Show that near the surface (i.e., for small ζ) the velocity components are given by

$$\frac{v_r}{u(r)} = \phi''(0)\zeta + O(\zeta^2), \qquad \frac{v_z}{u(r)}\,\text{Re}_r^{1/2} = -\phi''(0)\zeta^2 + O(\zeta^3),$$

where $\text{Re}_r = ur/\nu$. A numerical solution to this problem yields $\phi''(0) = 1.3120$ (Schlicting, 1968, p. 90).

8-11. Boundary Layer Equation for a Power-Law Fluid

Derive the boundary layer momentum equation for a power-law fluid, in dimensionless form. How should the Reynolds number be defined, and how does the boundary layer thickness vary with Re?

Chapter 9

FORCED-CONVECTION HEAT AND
MASS TRANSFER IN CONFINED
LAMINAR FLOWS

9.1 INTRODUCTION

This chapter is concerned with forced-convection heat and mass transfer in tubes, liquid films, and other confined geometries. In *forced convection* the flow is caused by an applied pressure or moving surface, as in Chapters 6–8, and is more or less independent of the heat or mass transfer process. Thus, the velocity can be determined first, and it is a known quantity in the energy or species conservation equation. This is in contrast to *free convection* or *buoyancy-driven flow*, where it is the effect of temperature or solute concentration on the fluid density which causes the flow. In free convection the fluid mechanics and heat and/or mass transfer problems must be considered simultaneously, as discussed in Chapter 12. Although several of the derivations in this chapter apply to forced convection in general, the examples involve only confined laminar flows of the types considered in Chapter 6. Heat and mass transfer in unconfined laminar flows is the subject of Chapter 10, and transport in turbulent flows is discussed in Chapter 13.

A key parameter in any forced-convection heat or mass transfer problem is the Péclet number (Pe), which indicates the importance of convection relative to conduction or diffusion. The chapter begins with a discussion of certain approximations which are justified when $Pe \gg 1$, which is the situation of interest here. Attention is given then to heat and mass transfer coefficients in confined flows. The dimensionless heat transfer coefficient (Nusselt number, Nu) and mass transfer coefficient (Sherwood number, Sh) are defined, and examples are given to illustrate their evaluation for laminar flow in tubes. The dependence of Nu and Sh on axial position and on the type of boundary condition at the wall is discussed. After that, the energy equation for a pure fluid is generalized to include terms associated with the compressibility of the fluid and with viscous dissipation of energy, effects which were neglected in the approximate form of the energy equation presented in Chapter 2. The chapter concludes with an analysis of solute dispersion in tube flow, called *Taylor dispersion*.

9.2 PÉCLET NUMBER

The significance of the Péclet number is clearest when one examines the dimensionless energy or species conservation equation for steady-state problems without volumetric source terms. In such cases no other parameters are present in the differential equation. Under those conditions both equations have the form

$$\text{Pe } \tilde{\mathbf{v}} \cdot \tilde{\nabla} \Theta = \tilde{\nabla}^2 \Theta, \qquad \text{Pe} \equiv \frac{UL}{\alpha} \quad \text{or} \quad \frac{UL}{D_i}, \tag{9.2-1}$$

where Θ is the dimensionless temperature or concentration, and the velocity and differential operators have been made dimensionless using the characteristic velocity U and characteristic length L. Because Pe is the coefficient of the convective terms, it evidently indicates the importance of convection relative to the molecular transport processes, as already mentioned. Indeed, noticing that α/L and D_i/L both have units of velocity, Pe can be interpreted as the ratio of a characteristic velocity for convection to a characteristic velocity for conduction or diffusion. In mass transfer this ratio will differ, in general, for each species in a mixture, and when distinctions are needed we will denote the Péclet number for species i as Pe_i. A comparison of Eq. (9.2-1) with Eqs. (5.10-14) or (5.10-15) reveals that Pe plays the same role in the energy or species equations as Re does in the Navier–Stokes equation. The identification of the velocity and length scales which are appropriate for a given problem is crucial, of course, in attempting to draw meaningful conclusions from the magnitudes of either Pe or Re.

Certain general features of heat and mass transfer in confined flows are illustrated by a simple physical situation. The problem to be analyzed involves steady heat transfer to a moving solid, as depicted in Fig. 9-1. A plate of thickness L is assumed to be moving at a speed U in the x direction. In the region $0 \le x \le K$ the top surface receives an incident heat flux q_0, whereas the rest of the plate is effectively insulated (i.e., the heat fluxes at the surfaces are negligible). Far "upstream" and "downstream" from the heated section the plate is assumed to approach constant temperatures; the upstream value is specified as T_0 and the downstream value is determined by the plate velocity and heating rate. It is assumed that $T = T(x, y)$ only. In the first of the examples which

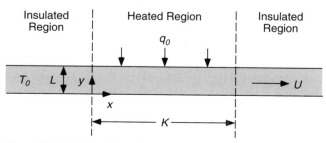

Figure 9-1. Steady heating of a solid plate moving at a constant speed.

follow, an exact solution is obtained for any value of Pe. The simplified analysis in the second example illustrates the approximations which apply when Pe is large.

Example 9.2-1 Heating of a Moving Plate at Arbitrary Pe   The energy equation and boundary conditions for $T(x, y)$ in the two-dimensional problem of Fig. 9-1 are

$$U\frac{\partial T}{\partial x} = \alpha\left(\frac{\partial^2 T}{\partial x^2} + \frac{\partial^2 T}{\partial y^2}\right), \tag{9.2-2}$$

$$T(-\infty, y) = T_0, \tag{9.2-3}$$

$$\frac{\partial T}{\partial x}(\infty, y) = 0, \tag{9.2-4}$$

$$\frac{\partial T}{\partial y}(x, 0) = 0, \tag{9.2-5}$$

$$\frac{\partial T}{\partial y}(x, L) = \begin{cases} 0, & x<0 \text{ or } x>K, \\ q_0/k, & 0\le x\le K. \end{cases} \tag{9.2-6}$$

Although this linear boundary-value problem can be solved using the methods of Chapter 4, a less detailed solution will suffice here. Suppose that it is desired to know only the cross-sectional average temperature at a given x, defined as

$$\bar{T}(x) = \frac{1}{L}\int_0^L T(x, y)\, dy. \tag{9.2-7}$$

The ordinary differential equation governing $\bar{T}(x)$ is obtained by averaging each term in Eq. (9.2-2). The result is

$$U\frac{d\bar{T}}{dx} = \alpha\frac{d^2\bar{T}}{dx^2} + \frac{\alpha}{L}\frac{\partial T}{\partial y}\bigg|_{y=0}^{y=L} \tag{9.2-8}$$

or, using the boundary conditions at the top and bottom surfaces,

$$\frac{d^2\bar{T}}{dx^2} - \frac{U}{\alpha}\frac{d\bar{T}}{dx} = 0, \qquad x<0 \quad \text{or} \quad x>K, \tag{9.2-9a}$$

$$\frac{d^2\bar{T}}{dx^2} - \frac{U}{\alpha}\frac{d\bar{T}}{dx} = -\frac{q_0}{kL}, \qquad 0\le x\le K. \tag{9.2-9b}$$

The integration employed here to obtain a one-dimensional problem is similar to that used in the fin analysis in Example 3.3-2, but it is unnecessary in the present problem to assume that temperature variations in the y direction are small.

The form of the governing equations is simplified by using the dimensionless quantities

$$\zeta \equiv \frac{x}{L}, \qquad \Theta \equiv \frac{\bar{T}-T_0}{q_0 L/k}, \qquad \text{Pe} \equiv \frac{UL}{\alpha}, \qquad \lambda \equiv \frac{K}{L}. \tag{9.2-10}$$

The dimensionless average temperature is governed by

$$\frac{d^2\Theta}{d\zeta^2} - \text{Pe}\frac{d\Theta}{d\zeta} = 0, \qquad \zeta<0 \quad \text{or} \quad \zeta>\lambda, \tag{9.2-11a}$$

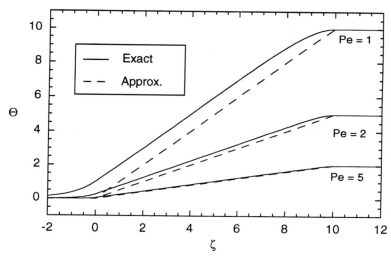

Figure 9-2. Temperature profiles for a moving plate heated for $0 \leq \zeta \leq 10$. The "Exact" curves are based on Eqs. (9.2-14)–(9.2-16) with $\lambda = 10$; the approximate results ("Approx.") are from Eq. (9.2-20).

$$\frac{d^2\Theta}{d\zeta^2} - \text{Pe}\,\frac{d\Theta}{d\zeta} = -1, \qquad 0 \leq \zeta \leq \lambda, \tag{9.2-11b}$$

$$\Theta(-\infty) = 0, \tag{9.2-12}$$

$$\frac{d\Theta}{d\zeta}(\infty) = 0. \tag{9.2-13}$$

In Eq. (9.2-10) we selected as the characteristic dimension the plate thickness (L) rather than the length of the heated region (K). Choosing L instead of K will enable us to examine situations where $K/L \gg 1$ and the exact value of K is unimportant. For confined flows in general, a cross-sectional dimension is usually the appropriate length scale.

An exact solution for $\Theta(\zeta)$ is obtained by dividing the plate into three regions. The general solutions for the upstream and downstream regions are obtained by integrating Eq. (9.2-11a), and that for the heated region is found from Eq. (9.2-11b). The integration constants are evaluated by requiring that the upstream and downstream solutions satisfy Eqs. (9.2-12) and (9.2-13), respectively, and that the temperature and axial heat flux be continuous where the regions join ($\zeta = 0$ and $\zeta = \lambda$). The resulting solution is

$$\Theta_1(\zeta) = \frac{e^{\text{Pe}\,\zeta}}{\text{Pe}^2}\,(1 - e^{-\text{Pe}\,\lambda}), \qquad \zeta < 0, \tag{9.2-14}$$

$$\Theta_2(\zeta) = \frac{1}{\text{Pe}^2}\,(1 - e^{\text{Pe}\,(\zeta - \lambda)}) + \frac{\zeta}{\text{Pe}}, \qquad 0 < \zeta < \lambda, \tag{9.2-15}$$

$$\Theta_3(\zeta) = \frac{\lambda}{\text{Pe}}, \qquad \zeta > \lambda. \tag{9.2-16}$$

The exact temperature profiles for three values of Pe, as computed from Eqs. (9.2-14)–(9.2-16), are shown by the solid curves in Fig. 9-2.

Example 9.2-2 Heating of a Moving Plate at Large Pe The analysis just completed did not place any restrictions on Pe. We focus now on the behavior of the temperature in the heated region when Pe is large. Using Eq. (9.2-15) to evaluate the terms in Eq. (9.2-11b) arising from convection and axial conduction, their ratio is found to be

$$\left| \frac{\text{convection}}{\text{conduction}} \right| = \frac{\text{Pe } d\Theta/d\zeta}{d^2\Theta/d\zeta^2} = e^{\text{Pe}(\lambda - \zeta)} - 1. \tag{9.2-17}$$

As expected, Eq. (9.2-17) indicates that convection is the dominant mode of axial energy transport in the heated region when Pe is large (except for $\zeta \cong \lambda$). Indeed, if Pe $\lambda > 50$, the axial conduction term is $<10\%$ of the convection term over $>90\%$ of the heated region. If the length of the heated region is 10 times the plate thickness (i.e., $\lambda = 10$), this criterion for neglecting axial conduction requires only that Pe>5. It is important to keep in mind that Eq. (9.2-17) only compares *axial* conduction with convection. In the transverse (y) direction, conduction is the only mechanism for energy transport, and thus it is essential at all values of Pe. For the two insulated regions, the convection and axial conduction terms must be equal for all Pe, because there are no other non-zero terms in Eq. (9.2-11a).

Anticipating that conduction in the heated (middle) region would be negligible for large Pe, approximate expressions for Θ_1 and Θ_2 could have been obtained somewhat more easily by first neglecting $d^2\Theta/d\zeta^2$ in Eq. (9.2-11b). The solutions which result are

$$\Theta_1(\zeta) = \frac{e^{\text{Pe }\zeta}}{\text{Pe}^2}, \qquad\qquad \zeta < 0, \tag{9.2-18}$$

$$\Theta_2(\zeta) = \frac{1}{\text{Pe}}\left(\frac{1}{\text{Pe}} + \zeta\right), \qquad 0 \le \zeta \le \lambda. \tag{9.2-19}$$

Note that because the actual second-order differential equation was replaced by a first-order equation in deriving Eq. (9.2-19), only the upstream boundary condition could be satisfied. A comparison of Eq. (9.2-19) with the corresponding exact solution for the heated region, Eq. (9.2-15), reveals good agreement when Pe$(\lambda - \zeta) \gg 1$. That is, Eq. (9.2-19) loses accuracy near the downstream end of the heated section $(\zeta \to \lambda)$, but the region where there is significant error becomes smaller as Pe is increased. Agreement between Eqs. (9.2-18) and (9.2-14) for the upstream insulated region requires only that Pe$\lambda \gg 1$. We conclude that the existence of a downstream insulated region has little effect on the temperature profile elsewhere, provided that Pe is large and that the heated section is not too short.

At high Pe, conduction upstream from the heated section should be rendered ineffective by the motion of the plate. Thus, for sufficiently large Pe, we expect not only that axial conduction will be negligible in the heated section, but that the temperature in the upstream region will be virtually unperturbed by the heating. In other words, $\Theta_1 = 0$ to good approximation. Under these conditions,

$$\Theta_2(\zeta) = \frac{\zeta}{\text{Pe}}, \qquad 0 < \zeta < \lambda. \tag{9.2-20}$$

This result is obtained from the original problem by considering only the heated region, neglecting axial conduction, and applying the upstream boundary condition at $\zeta = 0$ rather than at $\zeta = -\infty$. Thus, Eq. (9.2-20) is the solution to a problem involving one region, in contrast to the three-region problem solved in Example 9.2-1. This temperature profile is an excellent approximation provided that Pe$(\lambda - \zeta) \gg 1$ and Pe$\zeta \gg 1$. That is, accuracy is lost near both ends of the heated section, but [as with Eq. (9.2-19)] the regions where the errors occur vanish as Pe $\to \infty$. As shown by the dashed curves in Fig. 9-2, the temperature profile from Eq. (9.2-20) is almost indistinguishable from that of Eq. (9.2-15) for Pe$=5$.

The simplifications made in Example 9.2-2 are valid for many heat and mass transfer problems involving confined flows. In summary, for large Pe:

(1) Axial conduction (or diffusion) may be neglected relative to axial convection (provided there are other nonzero terms in the differential equations).

(2) The "inlet" temperature or concentration may be specified at the beginning of the "active" section, rather than at the true inlet.

(3) The conditions at the outlet may be ignored, provided that, in addition to large Pe, the length of the active section greatly exceeds the cross-sectional dimension.

What constitutes a large enough Péclet number to apply any or all of these approximations depends on the physical situation and the degree of precision required. Based on several analyses of heat or mass transfer in circular tubes or parallel-plate channels with axial conduction or diffusion (Ash and Heinbockel, 1970; Hsu, 1971; Papoutsakis et al., 1980a,b; Shah and London, 1978; Tan and Hsu, 1972), we suggest that Pe > 10 is a reasonable guideline for engineering work, where Pe is based on the mean fluid velocity and the full cross-sectional dimension (diameter or plate spacing). Péclet numbers much larger than this are common in applications, and the above three simplifications have been employed in the majority of published studies of heat and mass transfer in confined flows. Unless noted otherwise, subsequent examples of convective heat and mass transfer in this chapter assume that Pe is large enough for these approximations to be valid.

9.3 NUSSELT AND SHERWOOD NUMBERS

Heat and Mass Transfer Coefficients

As introduced in Chapter 2, heat and mass transfer coefficients are frequently used to describe transport between an interface and the bulk of a given fluid. The driving force used in the expression for the energy or species flux is the difference in temperature or concentration between the interface and the bulk fluid. When the fluid extends an indefinite distance from the interface and eventually reaches a constant temperature or concentration, the bulk temperature (T_b) or bulk concentration of species i (C_{ib}) in the fluid are simply those constant values far from the interface. In confined flows, however, where none of the fluid is "at infinity," it is necessary to employ some sort of cross-sectional average value for the bulk temperature or bulk concentration. For reasons which will be made clear, the preferred choice is a velocity-weighted average. Accordingly,

$$T_b(z) = \frac{\int_A T v_z \, dA}{\int_A v_z \, dA}, \qquad C_{ib}(z) = \frac{\int_A C_i v_z \, dA}{\int_A v_z \, dA}, \qquad (9.3\text{-}1)$$

where the flow is assumed to be in the z direction and A is the cross-sectional area. Thus, the denominator equals the volumetric flow rate, Q. It follows that, for example, if axial diffusion of species i is negligible, then QC_{ib} is the molar flow rate of i. If all of the fluid passing a given axial position were collected and mixed, then C_{ib} would be the resulting concentration. Accordingly, T_b and C_{ib} are sometimes called "mixing cup" quantities.

Example 9.3-1 Hollow-Fiber Dialyzer To illustrate why bulk quantities in confined flows are defined as in Eq. (9.3-1), we examine steady mass transfer in a dialysis device consisting of many hollow fibers arranged in parallel. A single fiber of internal radius R and length L is depicted in Fig. 9-3. The fiber wall is a membrane which permits diffusional transport of solutes between the liquid solution flowing inside the fiber and an external solution, the dialysate. By adjusting the composition of the dialysate, solutes can be added to or removed from the internal solution as desired. In such a system there are three mass transfer resistances in series, associated with (1) the fluid inside the fiber, (2) the membrane, and (3) the dialysate. Only the first two resistances are considered here. The concentration of solute i in the internal solution is $C_i(r, z)$. The flux from the internal fluid to the wall is given by

$$N_{ir}(R, z) = k_{ci}(z)[C_{ib}(z) - C_i(R, z)], \qquad (9.3\text{-}2)$$

where it is assumed that the mass transfer coefficient is a known function of position, $k_{ci}(z)$. The flux of solute i through the membrane (based on the inner radius) is

$$N_{ir}(R, z) = k_{mi}[C_i(R, z) - C_{id}], \qquad (9.3\text{-}3)$$

where k_{mi} is the permeability of the membrane to i (analogous to a mass transfer coefficient), and C_{id} is the concentration in the dialysate at the outer membrane surface. For simplicity, it is assumed that the dialysate flow rate is sufficiently large that C_{id} is independent of position; thus, C_{id} is treated as a known constant, as is k_{mi}. The internal flow is assumed to be fully developed.

The differential form of the species transport equation for the fluid inside the fiber is

$$v_z \frac{\partial C_i}{\partial z} = \frac{D_i}{r} \frac{\partial}{\partial r}\left(r \frac{\partial C_i}{\partial r}\right), \qquad (9.3\text{-}4)$$

where axial diffusion has been neglected. Rather than attempting to solve Eq. (9.3-4), we wish to take advantage of the available information on k_{ci} by using a macroscopic (partially lumped) model. Integrating Eq. (9.3-4) over r gives

$$\int_0^R v_z \frac{\partial C_i}{\partial z} r\, dr = RD_i \frac{\partial C_i}{\partial r}(R, z) = -RN_{ir}(R, z). \qquad (9.3\text{-}5)$$

Using Eq. (9.3-1), the left-hand side is rewritten as

$$\int_0^R v_z \frac{\partial C_i}{\partial z} r\, dr = \frac{d}{dz}\int_0^R C_i v_z r\, dr = \frac{R^2 U}{2} \frac{dC_{ib}}{dz}, \qquad \text{from page 255} \qquad (9.3\text{-}6)$$

$\rightarrow \int_A v_z\, dA$

$(6.2\text{-}19)$

where U is the mean velocity. The key point to notice is that the integration of the convective term leads naturally to the velocity-weighted average used to define the bulk concentration. The

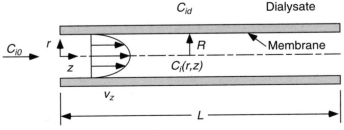

Figure 9-3. Dialysis in a hollow fiber.

flux on the right-hand side of Eq. (9.3-5) can be evaluated using either Eq. (9.3-2) or Eq. (9.3-3). Choosing the former and combining Eqs. (9.3-5) and (9.3-6), we obtain

$$\frac{dC_{ib}}{dz} = -\frac{2k_{ci}}{RU}[C_{ib}(z) - C_i(R, z)]. \tag{9.3-7}$$

The concentration at the inner membrane surface is eliminated now by equating the two expressions for the flux, Eqs. (9.3-2) and (9.3-3). After rearrangement, this gives

$$C_{ib}(z) - C_i(R, z) = \left(\frac{k_{mi}}{k_{ci} + k_{mi}}\right)[C_{ib}(z) - C_{id}]. \tag{9.3-8}$$

The final differential equation for the bulk concentration is

$$\frac{dC_{ib}}{dz} = -\frac{2}{RU}\left(\frac{k_{ci}k_{mi}}{k_{ci} + k_{mi}}\right)(C_{ib} - C_{id}), \qquad C_{ib}(0) = C_{io}, \tag{9.3-9}$$

where C_{io} is the inlet concentration. Thus, we have reduced Eq. (9.3-4) to an ordinary differential equation in which C_{ib} is the only unknown. Equation (9.3-9) can be integrated using elementary analytical or numerical methods, depending on the functional form for $k_{ci}(z)$.

If k_{ci} is independent of z, then Eq. (9.3-9) gives

$$\frac{C_{ib}(z) - C_{id}}{C_{io} - C_{id}} = \exp\left[-\frac{2z}{RU}\left(\frac{k_{ci}k_{mi}}{k_{ci} + k_{mi}}\right)\right], \tag{9.3-10}$$

from which one can readily calculate the overall rate of removal of solute i from the fiber, equal to $Q[C_{io} - C_{ib}(L)]$. The quantity $k_{ci}k_{mi}/(k_{ci} + k_{mi})$ is the *overall mass transfer coefficient*. The derivation is easily generalized to include a mass transfer resistance in the external fluid, by distinguishing between the surface and bulk concentrations of i in the dialysate. As discussed later, the assumption of a constant heat or mass transfer coefficient in a tube is accurate only for very long tubes. Design methods for hollow-fiber dialyzers are detailed in Colton and Lowrie (1981).

In summary, we have shown that the bulk concentration defined in Eq. (9.3-1) is the one which arises naturally in a partially lumped model for solute transport in a tube. If the mass transfer coefficient is a known function of position, calculations of overall mass transfer rates become elementary. Obviously, this strategy depends on having experimental or theoretical values for k_{ci}. Because many design calculations (e.g., for heat exchangers) are based on the approach illustrated here, much of the literature on convective heat and mass transfer concerns the evaluation of transfer coefficients. For *laminar* flow in tubes, heat and mass transfer coefficients can be computed from first principles, as shown in this chapter; for *turbulent* flow in tubes, one must rely on empirical correlations, as discussed in Chapter 13.

Dimensional Analysis: Nusselt and Sherwood Numbers

Information on heat and mass transfer coefficients is correlated most efficiently using dimensionless groups. The dimensionless forms of the heat transfer coefficient (h) and mass transfer coeffieient (k_{ci}) are the *Nusselt number* (Nu) and *Sherwood number* (Sh$_i$), respectively. They are given by

$$Nu = \frac{hL}{k}, \qquad Sh_i = \frac{k_{ci}L}{D_i}, \tag{9.3-11}$$

where L is the characteristic length (e.g., tube diameter). Although Nu and Sh$_i$ are used throughout this book, another dimensionless transfer coefficient in the engineering literature is the *Stanton number,* $St = Nu/Pe$ or $St_i = Sh_i/Pe_i$. The Nusselt and Sherwood numbers should not be confused with Biot numbers for heat or mass transfer (Section 3.3), although they look the same. The key distinction is that the thermal conductivity and diffusivity in Nu and Sh_i are those of the *fluid* to which h and k_{ci} apply, whereas the thermal conductivity and diffusivity in the respective Biot numbers are for a *solid* in contact with a fluid. In other words, Nu and Sh_i each pertain to a single (fluid) phase, whereas Bi expresses the relative resistances of two phases (fluid and solid) arranged in series.

In correlating information on Nu or Sh_i, either experimental or theoretical, we are concerned with the minimum set of dimensionless variables on which they depend. The following discussion applies to steady flows of incompressible Newtonian fluids. The conclusions from the dimensional analysis are valid for laminar internal flows (tubes or channels) or external flows (flow around objects). If the velocity and the temperature (or concentration) are interpreted as time-smoothed quantities, then the conclusions concerning Nu and Sh_i apply also to turbulent flow (see Chapter 13 for a discussion of time-smoothing in turbulence). All physical properties of the fluid are assumed to be constant. Viscous dissipation, neglected here, is considered in Section 9.6. The dimensionless variables for buoyancy-driven flow are discussed in Chapter 12.

The pertinent dimensionless quantities are identified by inspecting the governing equations. We start with the velocity field, because it is needed for the convective terms in Eq. (9.2-1). From the dimensionless Navier–Stokes equation (Section 5.10) we infer that, for a fluid bounded by stationary solid surfaces,

$$\tilde{\mathbf{v}} = \tilde{\mathbf{v}}(\tilde{\mathbf{r}}, \text{Re, geometric ratios}), \qquad (9.3\text{-}12)$$

where $\tilde{\mathbf{r}}$ is dimensionless position. "Geometric ratios" refers here to all ratios of lengths needed to define the system geometry. For example, choosing the diameter of a circular tube as the characteristic dimension, the ratio of tube length to tube diameter completes the geometric information needed. For a tube of constant, rectangular cross section there are three dimensions (two cross-sectional dimensions and a tube length) and therefore two geometric ratios. The important point is that because no other independent variables or parameters appear in the continuity equation, Navier–Stokes equation, and boundary conditions, $\tilde{\mathbf{v}}$ must depend at most on the quantities listed in Eq. (9.3-12), even if we are unable to predict the precise functional form of the relationship. The pressure is regarded as simply another dependent variable governed by the same quantities as $\tilde{\mathbf{v}}$, and therefore it is not listed in Eq. (9.3-12).

Equation (9.2-1) implies that the dimensionless temperature or concentration depends on the same quantities as the dimensionless velocity, with the addition of the Péclet number. In other words,

$$\Theta = \Theta(\tilde{\mathbf{r}}, \text{Re, Pe, geometric ratios}). \qquad (9.3\text{-}13)$$

It is assumed here that no additional parameters are introduced by the thermal or concentration boundary conditions. For example, suppose that the fluid temperature is T_0 at the inlet of a tube and is T_w at the tube walls. Choosing $\Theta = (T - T_w)/(T_0 - T_w)$ gives dimensionless temperatures of unity and zero, respectively, at these boundaries. For a constant heat flux q_w at the tube wall, choosing $\Theta = (T - T_0)/(q_w L/k)$ again yields boundary conditions which involve only pure numbers, so that Θ still depends on just the quantities listed in Eq. (9.3-13). Situations where additional parameters appear include mass transfer problems with reactions and heat or mass transfer problems with wall resistances (e.g., Example 9.3-1).

The Reynolds and Péclet numbers are natural parameters, in that they appear directly in the Navier–Stokes and the energy or species equations, respectively. However, any other independent pair of groups derived from Re and Pe is equally valid from the standpoint of dimensional analysis. In particular, we note that for heat transfer we have $\mathrm{Pe} = \mathrm{Re}\,\mathrm{Pr}$, where $\mathrm{Pr} = \nu/\alpha$ is the Prandtl number; for mass transfer, $\mathrm{Pe}_i = \mathrm{Re}\,\mathrm{Sc}_i$, where $\mathrm{Sc}_i = \nu/D_i$ is the Schmidt number. Thus, Eq. (9.3-13) can be written also as

$$\Theta = \Theta(\tilde{\mathbf{r}},\ \mathrm{Re},\ \mathrm{Pr},\ \text{geometric ratios}) \qquad \text{(temperature)}, \qquad (9.3\text{-}14a)$$

$$\Theta = \Theta(\tilde{\mathbf{r}},\ \mathrm{Re},\ \mathrm{Sc}_i,\ \text{geometric ratios}) \qquad \text{(concentration).} \qquad (9.3\text{-}14b)$$

Equation (9.3-13) is advantageous when the dimensionless velocity is independent of Re (i.e., in creeping flow or fully developed, laminar flow), whereas Eq. (9.3-14) is used otherwise.

Having identified the quantities which determine the dimensionless temperature or concentration fields, we can draw similar conclusions concerning functions of Θ. For example, consider the heat transfer coefficient in a tube. By definition, the local heat transfer coefficient is given by

$$h = \frac{-k(\partial T/\partial n)_w}{T_b - T_w} = \frac{-k(\partial \Theta/\partial n)_w}{\Theta_b - \Theta_w}, \qquad (9.3\text{-}15)$$

where n is a coordinate normal to the tube wall (positive direction outward) and the subscript w denotes evaluation at the wall. It follows from Eq. (9.3-15) that

$$\mathrm{Nu} = \frac{hL}{k} = \frac{-(\partial \Theta/\partial \tilde{n})_w}{\Theta_b - \Theta_w}, \qquad (9.3\text{-}16)$$

where $\tilde{n}$ is a dimensionless coordinate. It is seen that the Nusselt number is simply the dimensionless temperature gradient at the tube wall divided by the dimensionless temperature difference. It follows from Eqs. (9.3-14) and (9.3-16) that the local Nusselt number has the functional dependence

$$\mathrm{Nu} = \mathrm{Nu}\,(\tilde{\mathbf{r}}_w,\ \mathrm{Re},\ \mathrm{Pr},\ \text{geometric ratios}), \qquad (9.3\text{-}17)$$

where $\tilde{\mathbf{r}}_w$ is dimensionless position on the wall, and that Nu averaged over the tube surface is

$$\overline{\mathrm{Nu}} = \overline{\mathrm{Nu}}\,(\mathrm{Re},\ \mathrm{Pr},\ \text{geometric ratios}). \qquad (9.3\text{-}18)$$

The conclusions expressed by Eqs. (9.3-17) and (9.3-18) apply quite generally to steady forced convection, with either constant temperature or constant heat-flux boundary conditions. The flow may be laminar or turbulent, developing or fully developed, and confined or unconfined. For cases where $\tilde{\mathbf{v}}$ does not depend explicitly on Re, the result corresponding to Eq. (9.3-18) is

$$\overline{\mathrm{Nu}} = \overline{\mathrm{Nu}}\,(\mathrm{Pe},\ \text{geometric ratios}) \qquad [\tilde{\mathbf{v}} \neq \tilde{\mathbf{v}}(\mathrm{Re})]. \qquad (9.3\text{-}19)$$

In other words, for velocity fields of this type Re and Pr will appear only as the product Pe, and the individual values of Re and Pr are of no significance for heat transfer.

The analogous results for mass transfer are

$$Sh_i = Sh_i(\tilde{\mathbf{r}}_w, \text{ Re, } Sc_i, \text{ geometric ratios}),\qquad\qquad\qquad (9.3\text{-}20)$$

$$\overline{Sh_i} = \overline{Sh_i} \text{ (Re, } Sc_i, \text{ geometric ratios)},\qquad\qquad\qquad\quad (9.3\text{-}21)$$

$$\overline{Sh_i} = \overline{Sh_i} \text{ (Pe}_i, \text{ geometric ratios)} \qquad [\tilde{\mathbf{v}} \neq \tilde{\mathbf{v}}(\text{Re})].\qquad (9.3\text{-}22)$$

Additional parameters are involved if there are chemical reactions (Damköhler numbers) or permeable walls (dimensionless permeabilities). The parameters for these and other, more complex situations in heat or mass transfer can be identified by writing all of the differential equations and boundary conditions in dimensionless form and proceeding as above.

Qualitative Behavior of the Nusselt and Sherwood Numbers

Before analyzing any examples in detail, we give an overview of the typical behavior of Nu and Sh_i for confined laminar flows. These remarks apply to steady transport in tubes of any cross-sectional shape and in liquid films, provided that the tube cross section or film thickness is constant. In such flows the velocity profile becomes fully developed, given enough length (see Section 6.5). It is assumed that Pe is large and that the temperature (or concentration) boundary conditions undergo a step change at $z=0$. For example, the wall temperature changes from the inlet value to some new value at $z=0$, and it remains at the new value for all $z>0$. As shown qualitatively in Fig. 9-4, in these situations Nu is very large near $z=0$ and declines with increasing z. Typically, Nu initially varies as some inverse power of z, so that a log–log plot of Nu(z) is linear at small z, as indicated. For long enough tubes or films, Nu approaches a constant, even though the temperature may continue to depend on z. The position at which Nu becomes essentially constant separates the *thermal entrance region* from the *thermally fully developed* region. With a suitable cross-sectional dimension as the characteristic length, Nu ~ 1 (or $Sh_i \sim 1$) for large z. In general, the length of the thermal or concentration entrance region differs from that of the velocity entrance region.

The trends in Nu (or Sh_i) can be understood in terms of the distance $\delta(z)$ that temperature (or concentration) disturbances at the wall penetrate into the fluid. Considering heat transfer in a tube, an order-of-magnitude estimate of the temperature gradient at the wall is $(\partial T/\partial n)_w \sim (T_b - T_w)/\delta$. It follows from Eq. (9.3-16) that

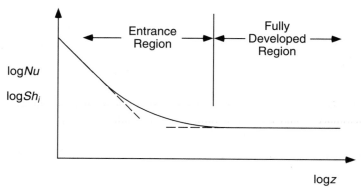

Figure 9-4. Qualitative behavior of Nu or Sh_i in confined flows with step-change boundary conditions at $z=0$.

$$\text{Nu} \sim \frac{L}{\delta(z)}. \tag{9.3-23}$$

Thus, Nu is very large near $z=0$, where only a thin layer of fluid near the wall has been affected. Moving downstream, conduction normal to the wall causes progressive increases in δ. Eventually, when the temperature changes have reached the center of the tube, $\delta \sim L$ and Nu ~ 1. Similar statements apply to mass transfer and Sh_i.

Entrance regions and fully developed regions are considered in turn in Sections 9.4 and 9.5.

9.4 ENTRANCE REGION

Basic Problem Formulation

Here and in Section 9.5 are four examples which illustrate various aspects of convective transport in confined flows, all involving heat transfer in circular tubes. The focus of this section is the thermal entrance region, whereas that of Section 9.5 is the thermally fully developed region. We begin by stating the problem formulation which serves as a common point of departure. The tube wall is assumed to be at temperature T_0 for $z<0$ and T_w for $z>0$, where T_0 and T_w are constants. The fluid enters at $T=T_0$, and the flow is assumed to be laminar and fully developed for $z>0$, the region modeled. Neglecting axial conduction (i.e., assuming large Pe), the dimensional problem involving $T(r, z)$ is

$$V_z = 2U\left[1-\left(\frac{r}{R}\right)^2\right]\frac{\partial T}{\partial z} = \frac{\alpha}{r}\frac{\partial}{\partial r}\left(r\frac{\partial T}{\partial r}\right), \tag{9.4-1}$$

$$T(r, 0) = T_0, \tag{9.4-2}$$

$$\frac{\partial T}{\partial r}(0, z) = 0, \tag{9.4-3}$$

$$T(R, z) = T_w, \tag{9.4-4}$$

where R is the tube radius and U is the mean velocity. This is called the *Graetz problem*, after L. Graetz, whose pioneering analysis was published in 1883 [see, for example, Brown (1960)].

A useful set of dimensionless variables is

$$\eta \equiv \frac{r}{R}, \quad \zeta \equiv \frac{z}{R\,\text{Pe}}, \quad \Theta \equiv \frac{T-T_0}{T_w-T_0}, \quad \text{Pe} \equiv \frac{2UR}{\alpha}. \tag{9.4-5}$$

The noteworthy aspect of the independent variables is that Pe has been embedded in the axial coordinate. This choice of ζ is motivated by the fact that when all terms in Eq. (9.4-1) are made dimensionless, z and Pe appear only as a ratio; this is not true if axial conduction is included. The Péclet number is based on the mean velocity and the diameter, which is the usual convention for circular tubes. The equations which govern $\Theta(\eta, \zeta)$ are

$$(1-\eta^2)\frac{\partial\Theta}{\partial\zeta} = \frac{1}{\eta}\frac{\partial}{\partial\eta}\left(\eta\frac{\partial\Theta}{\partial\eta}\right), \tag{9.4-6}$$

$$\Theta(\eta, 0) = 0, \tag{9.4-7}$$

$$\frac{\partial \Theta}{\partial \eta}(0, \zeta) = 0, \tag{9.4-8}$$

$$\Theta(1, \zeta) = 1. \tag{9.4-9}$$

Notice that all parameters have been eliminated from this set of equations, so that, in principle, a single solution for $\Theta(\eta, \zeta)$ will apply to all conditions. In actuality, it is advantageous to use different solution strategies for the entrance region and fully developed region, as will be shown.

Example 9.4-1 Nusselt Number in the Entrance Region of a Tube with Specified Temperature   The objective is to obtain an asymptotic approximation for Nu which is valid at small ζ, for the heat transfer problem just stated. The temperature can be expressed as an eigenfunction expansion, as described in Section 9.5, but that type of solution is not very helpful for the early part of the entrance region. The main reason another approach is needed is that the resulting Fourier series converges very slowly for small ζ, just as the analogous series representations for transient conduction problems converge very slowly for small t. A secondary reason is that the basis functions which arise in this problem, called *Graetz functions*, are less convenient for calculations than any of the functions encountered in Chapter 4. Thus, instead of using the finite Fourier transform method, we will take advantage of the thinness of the nonisothermal region and develop a similarity solution.

For small ζ the temperature changes will be confined to a region near the tube wall, so that it is useful to define a new radial coordinate which is based at the wall,

$$\chi \equiv 1 - \eta. \tag{9.4-10}$$

Using this new coordinate, Eq. (9.4-6) becomes

$$(2\chi - \chi^2)\frac{\partial \Theta}{\partial \zeta} = \frac{\partial^2 \Theta}{\partial \chi^2} - \frac{1}{1-\chi}\frac{\partial \Theta}{\partial \chi}. \tag{9.4-11}$$

The boundary conditions follow readily from Eqs. (9.4-7)–(9.4-9).

An order-of-magnitude analysis will be used to identify the dominant terms in Eq. (9.4-11) and to infer how the thermal penetration depth grows with distance along the tube. We are concerned with axial positions such that $\tilde{\delta} \ll 1$, where $\tilde{\delta}$ is the penetration depth divided by the tube radius. At these axial positions the fluid remains at the inlet temperature, except near the wall. Thus, Θ varies from zero (the inlet temperature) to one (the wall temperature) as χ varies by the amount $\tilde{\delta}$. Comparable variations in Θ occur over the axial distance ζ. It follows that for the nonisothermal region near the wall,

$$(2\chi - \chi^2)\frac{\partial \Theta}{\partial \zeta} \sim (\tilde{\delta})\left(\frac{1}{\zeta}\right), \tag{9.4-12}$$

$$\frac{\partial^2 \Theta}{\partial \chi^2} \sim \frac{1}{\tilde{\delta}^2}, \tag{9.4-13}$$

$$\frac{1}{1-\chi}\frac{\partial \Theta}{\partial \chi} \sim (1)\left(\frac{1}{\tilde{\delta}}\right). \tag{9.4-14}$$

In Eq. (9.4-12) we have used the fact that $\chi^2 \ll \chi$. A comparison of Eqs. (9.4-13) and (9.4-14) indicates that the $\partial\Theta/\partial\chi$ term, which is associated with the curvature of the tube wall, is negligible.

The remaining two terms in Eq. (9.4-11), one representing axial convection and the other representing radial conduction, are evidently the dominant ones. From Eqs. (9.4-12) and (9.4-13) we
infer that

$$\tilde{\delta} \sim \zeta^{1/3}. \tag{9.4-15}$$

It follows from Eq. (9.3-23) that the Nusselt number varies as $\zeta^{-1/3}$, or that

$$Nu = C\left(\frac{z}{R}\right)^{-1/3} Pe^{1/3}, \tag{9.4-16}$$

where C is a constant which is of order of magnitude unity.

Equation (9.4-16), which required little mathematical effort, represents most of what there
is to know about heat transfer in the early part of the thermal entrance region. What remains is to
evaluate the constant C, for which we need to determine the temperature profile. Retaining only
the dominant terms in Eq. (9.4-11), the energy equation is

$$2\chi \frac{\partial \Theta}{\partial \zeta} = \frac{\partial^2 \Theta}{\partial \chi^2}. \tag{9.4-17}$$

There is only a χ term in the coefficient of $\partial \Theta / \partial \zeta$, reflecting the fact that the curvature in the
parabolic velocity profile is not evident near the tube wall. This linearization of the velocity profile
near the wall is referred to as the *Lévêque approximation* after M. Lévêque, who published the
similarity solution for the entrance region in 1928. The boundary conditions for $\Theta(\chi, \zeta)$ are

$$\Theta(\chi, 0) = 0, \tag{9.4-18}$$

$$\Theta(0, \zeta) = 1, \tag{9.4-19}$$

$$\Theta(\infty, \zeta) = 0. \tag{9.4-20}$$

Equation (9.4-20), which replaces the centerline condition, is based on the fact that the tube center
is many characteristic lengths from the wall.

We assume now that $\Theta = \Theta(s)$ only, where

$$s \equiv \frac{\chi}{g(\zeta)}. \tag{9.4-21}$$

The position-dependent length scale, $g(\zeta)$, must be proportional to the thermal penetration depth,
so that Eq. (9.4-15) implies that $g \propto \zeta^{1/3}$. Converting Eq. (9.4-17) to the similarity variable, we
obtain

$$\frac{d^2 \Theta}{ds^2} + 2s^2 (g^2 g') \frac{d\Theta}{ds} = 0. \tag{9.4-22}$$

For s to be the only independent variable, the term in parentheses must be constant; for later
convenience, the constant is chosen as 3/2. Accordingly, recognizing that $3g^2 g' = (g^3)'$ and requiring that $g(0) = 0$, we find that

$$g(\zeta) = \left(\frac{9}{2}\zeta\right)^{1/3}, \qquad s = \left(\frac{2}{9}\right)^{1/3} \frac{\chi}{\zeta^{1/3}}. \tag{9.4-23}$$

The equations governing $\Theta(s)$ are now

$$\left. \frac{d^2 \Theta}{ds^2} + 3s^2 \frac{d\Theta}{ds} = 0 \right. \tag{9.4-24}$$

$$\left. \Theta(0) = 1, \qquad \Theta(\infty) = 0, \right. \tag{9.4-25}$$

which have the solution

$$\Theta(s) = \frac{3}{\Gamma(\frac{1}{3})} \int_s^\infty e^{-t^3} dt, \tag{9.4-26}$$

where $\Gamma(x)$ is the gamma function and $\Gamma(\frac{1}{3}) = 2.67894$ (Abramowitz and Stegun, 1970, p. 253). The Nusselt number (based on the tube diameter) is calculated as

$$\mathrm{Nu} \equiv \frac{2hR}{k} = \frac{2(-\partial\Theta/\partial\chi)|_{\chi=0}}{\Theta_w - \Theta_b} = -2 \left.\frac{\partial\Theta}{\partial\chi}\right|_{\chi=0} = \frac{6}{\Gamma(\frac{1}{3})g(\zeta)}. \tag{9.4-27}$$

Because the nonisothermal region is very thin and almost all of the fluid remains at the inlet temperature of zero, we have set $\Theta_b = 0$. The final result is

$$\mathrm{Nu} = 1.357 \left(\frac{R}{z}\right)^{1/3} \mathrm{Pe}^{1/3} \qquad (T_w \text{ constant}). \tag{9.4-28}$$

This result confirms the form for Nu given by Eq. (9.4-16), and it shows that $C = 1.357$.

For a circular tube with a specified heat flux at the wall, the Nusselt number in the early entrance region is

$$\mathrm{Nu} = 1.640 \left(\frac{R}{z}\right)^{1/3} \mathrm{Pe}^{1/3} \qquad (q_w \text{ constant}). \tag{9.4-29}$$

Thus, the functional form of Nu is the same as for a specified temperature, but the value of C is somewhat larger (see Problem 9-3).

The Lévêque approximation is highly accurate only for the early part of the thermal entrance region, but it has been extended by constructing solutions as power-series expansions in g (Newman, 1969; Worsøe-Schmidt, 1967).

Example 9.4-2 Thermal Entrance Length   The thermal entrance length (L_T) for a tube is the axial position at which Nu approaches its constant, far-downstream value to within some specified tolerance. The dimensionless form of the Graetz problem, which shows that Nu is a function of ζ only, implies that $z = L_T$ corresponds to a specific value of ζ, which we denote as ζ_T. Thus, the entrance length must be of the form

$$\frac{L_T}{R} = K\,\mathrm{Pe}, \tag{9.4-30}$$

where $K(=\zeta_T)$ is a constant. This expression is valid only for $\mathrm{Pe} \gg 1$, which was an assumption in the Graetz formulation. In this example we use an integral approximation to estimate K.

The required integral equation is obtained by first integrating Eq. (9.4-6) over the radial coordinate to give

$$\frac{d}{d\zeta} \int_0^1 \Theta(1 - \eta^2)\eta \, d\eta = \left.\frac{\partial\Theta}{\partial\eta}\right|_{\eta=1}. \tag{9.4-31}$$

The radial temperature profile is approximated now by assuming that all temperature changes occur within a dimensional distance $\delta(\zeta)$ from the tube wall; the corresponding dimensionless distance is $F(\zeta) = \delta(\zeta)/R$. By definition, $F(0) = 0$ and $F(\zeta_T) = 1$. That is, in this analysis the entrance region is assumed to end at the axial position where the temperature changes just reach the tube centerline. The assumed form of the temperature profile makes it convenient to employ a new radial coordinate,

$$s \equiv \frac{1-\eta}{F(\zeta)}, \tag{9.4-32}$$

which is analogous to the similarity variable used in Example 9.4-1. It is assumed that for $s \geq 1$, which corresponds to the "core" region in the center of the tube, the temperature remains at the inlet value, $\Theta = 0$. After changing variables from η to s, Eq. (9.4-31) becomes

$$\frac{d}{d\zeta} \int_0^1 (2F^2 s - 3F^3 s^2 + F^4 s^3) \Theta \, ds = -\frac{1}{F} \frac{\partial \Theta}{\partial s}\Big|_{s=0}. \tag{9.4-33}$$

Similar to the Kármán–Pohlhausen method for momentum boundary layers (Example 8.4-3), it is assumed now that $\Theta(s, \zeta)$ is of the form

$$\Theta(s, \zeta) = a(\zeta) + b(\zeta)s + c(\zeta)s^2 + d(\zeta)s^3 + e(\zeta)s^4. \tag{9.4-34}$$

Five conditions must be imposed to determine the coefficient functions. The first three, which follow from the wall and core temperatures and the continuity of the radial heat flux, are

$$\Theta(0, \zeta) = 1, \tag{9.4-35}$$

$$\Theta(1, \zeta) = 0, \tag{9.4-36}$$

$$\frac{\partial \Theta}{\partial s}(1, \zeta) = 0. \tag{9.4-37}$$

The fourth and fifth conditions are obtained by noticing that the left-hand side of Eq. (9.4-6), and also its η derivative, must vanish at the tube wall; this is a consequence of the no-slip condition and the constancy of the temperature along the wall. These last two conditions for Θ are written as

$$\frac{\partial^2 \Theta}{\partial s^2}(0, \zeta) - F \frac{\partial \Theta}{\partial s}(0, \zeta) = 0, \tag{9.4-38}$$

$$\frac{\partial^3 \Theta}{\partial s^3}(0, \zeta) - F \frac{\partial^2 \Theta}{\partial s^2}(0, \zeta) + F^2 \frac{\partial \Theta}{\partial s}(0, \zeta) = 0. \tag{9.4-39}$$

Using Eq. (9.4-34) in Eqs. (9.4-35)–(9.4-39) and solving for the coefficients, it is found that

$$a(\zeta) = 1, \quad b(\zeta) = -\frac{4}{3+F}, \quad c(\zeta) = -\frac{2F}{3+F}, \quad d(\zeta) = 0, \quad e(\zeta) = \frac{1+F}{3+F}. \tag{9.4-40}$$

Substituting the resulting expression for Θ in Eq. (9.4-33) and (with the help of symbolic manipulation software) performing the differentiation and integration involving s, we obtain

$$\frac{d}{d\zeta}\left[\frac{F^2}{840}\left(\frac{560 - 80F - 129F^2 + 35F^3}{3+F}\right)\right] = \frac{4}{F(3+F)}, \quad F(0) = 0, \tag{9.4-41}$$

which is a nonlinear, first-order differential equation for the one remaining unknown, $F(\zeta)$. Integrating this separable equation from the initial condition to $\zeta = \zeta_T$ and $F = 1$ gives, finally,

$$\zeta_T = K = 0.0628. \tag{9.4-42}$$

This value of ζ_T or K is in reasonable agreement with the exact results from the eigenfunction solution of the Graetz problem, where Nu is within 6% of the fully developed value for $\zeta = 0.06$ and within 1% for $\zeta = 0.10$ (Shah and London, 1978).

The thermal entrance length for a constant flux at the wall is similar to that for a constant temperature. Likewise, the results for parallel-plate channels are similar to those for tubes, if R is replaced by the channel half-height (H). Thus, the thermal entrance lengths for these laminar flows can be summarized as

$$\frac{L_T}{L_0} \cong 1 + 0.1 \, \text{Pe}, \qquad \text{Pe} = \frac{2UL_0}{\alpha}, \qquad L_0 = R \text{ or } H \tag{9.4-43}$$

which is analogous to Eq. (6.5-3) for the velocity entrance length. The additive term "1," which is a correction for small Pe, is based on the typical extent of edge effects in conduction (Example 4.5-2).

9.5 FULLY DEVELOPED REGION

The calculation of Nu for $z \rightarrow \infty$ usually requires the evaluation of at least one term of the eigenfunction expansion for the temperature field. This is illustrated in Example 9.5-1 for a tube with a specified wall temperature. In some problems, however, including those with a specified heat flux, the mathematical problem can be made much simpler. The constant-flux case for a circular tube is examined in Example 9.5-2.

Example 9.5-1 Nusselt Number in the Fully Developed Region of a Tube with Specified Temperature   It is desired to determine the fully developed value of Nu for the Graetz problem. The only change from Eqs. (9.4-6)–(9.4-9) is that the dimensionless temperature is redefined as $\Theta = (T - T_w)/(T_0 - T_w)$, so that $\Theta = 1$ at the inlet and $\Theta = 0$ at the wall. (This yields a slight simplification in the form of the solution.) The differential equation and boundary conditions are linear, and the finite Fourier transform method can be used to derive an eigenfunction expansion for the temperature. Rewriting Eq. (9.4-6) as

$$\frac{\partial \Theta}{\partial \zeta} = \frac{1}{(1 - \eta^2)\eta} \frac{\partial}{\partial \eta}\left(\eta \frac{\partial \Theta}{\partial \eta}\right) \tag{9.5-1}$$

and using the approach of Chapter 4, it is found that the eigenvalue problem is

$$\frac{1}{(1 - \eta^2)\eta} \frac{d}{d\eta}\left(\eta \frac{d\Phi}{d\eta}\right) = -\lambda^2 \Phi, \qquad \Phi(1) = 0. \tag{9.5-2}$$

As the second boundary condition, it is sufficient to require that $\Phi(0)$ be finite. Equation (9.5-2) is a Sturm–Liouville problem with $w(\eta) = (1 - \eta^2)\eta$, $p(\eta) = \eta$, and $q(\eta) = 0$. Except for the factor $(1 - \eta^2)$, the differential equation is identical to Bessel's equation. The eigenfunctions defined by Eq. (9.5-2) are called *cylindrical Graetz functions*. As with Bessel's equation, only one of the independent solutions is finite at $\eta = 0$. The basis functions are written as $\Phi_n(\eta) = a_n \, G(\lambda_n \eta)$, where $G(x)$ denotes the Graetz function that is finite at $x = 0$ and a_n is a normalization constant.

The temperature is expanded as

$$\Theta(\eta, \zeta) = \sum_{n=1}^{\infty} \Theta_n(\zeta)\Phi_n(\eta), \tag{9.5-3}$$

$$\Theta_n(\zeta) = \langle \Theta, \, \Phi_n \rangle \equiv \int_0^1 \Theta \Phi_n (1 - \eta^2)\eta \, d\eta. \tag{9.5-4}$$

Using the inner product defined by Eq. (9.5-4), the energy equation is transformed to

$$\frac{d\Theta_n}{d\zeta} + \lambda_n^2\,\Theta_n = 0 \tag{9.5-5}$$

and the inlet condition becomes

$$\Theta_n(0) = \langle 1,\, \Phi_n \rangle \equiv b_n. \tag{9.5-6}$$

Solving for Θ_n, Eq. (9.5-3) is written now as

$$\Theta(\eta,\, \zeta) = \sum_{n=1}^{\infty} c_n e^{-\lambda_n^2 \zeta} G(\lambda_n\, \eta), \tag{9.5-7}$$

where $c_n = a_n b_n$.

Before returning to the eigenvalue problem, we examine what is needed to compute the Nusselt number for $\zeta \to \infty$. With a wall temperature of zero, Nu is given by

$$\mathrm{Nu}(\zeta) \equiv \frac{2hR}{k} = -\frac{2}{\Theta_b(\zeta)}\frac{\partial\Theta}{\partial\eta}(1,\, \zeta). \tag{9.5-8}$$

The calculation of Nu is simplified by using an overall energy balance to relate the bulk temperature to the temperature gradient at the wall. It is readily shown that the left-hand side of Eq. (9.4-31) equals $(1/4)\,d\Theta_b/d\zeta$, so that we obtain

$$\frac{d\Theta_b}{d\zeta} = 4\frac{\partial\Theta}{\partial\eta}(1,\, \zeta). \tag{9.5-9}$$

Integrating from $\zeta = \infty$ (where $\Theta_b = 0$) to a finite value of ζ gives

$$\Theta_b(\zeta) = -4\int_{\zeta}^{\infty}\frac{\partial\Theta}{\partial\eta}(1,\, t)\, dt, \tag{9.5-10}$$

where t is a dummy variable. This result is valid for any axial position. Now, from Eqs. (9.5-7) and (9.5-10) we see that, for $\zeta \to \infty$,

$$\Theta(\eta,\, \zeta) \to c_1 e^{-\lambda_1^2 \zeta} G(\lambda_1 \eta), \tag{9.5-11}$$

$$\frac{\partial\Theta}{\partial\eta}(1,\, \zeta) \to c_1 e^{-\lambda_1^2 \zeta}\lambda_1\frac{dG}{d\eta}(\lambda_1), \tag{9.5-12}$$

$$\Theta_b(\zeta) = -\frac{4c_1}{\lambda_1}e^{-\lambda_1^2 \zeta}\frac{dG}{d\eta}(\lambda_1). \tag{9.5-13}$$

The most slowly decaying term eventually dominates, so that only the smallest eigenvalue (λ_1) is needed. Using these results to evaluate the temperature gradient and bulk temperature in Eq. (9.5-8), it is found that c_1, the exponentials, and the Graetz functions all cancel. The expression for the fully developed Nusselt number is simply

$$\mathrm{Nu}(\infty) = \frac{\lambda_1^2}{2}. \tag{9.5-14}$$

Thus, Nu becomes constant at large ζ, even though the temperature continues to depend on axial position. All that is needed to calculate $\mathrm{Nu}(\infty)$ is the first eigenvalue.

Graetz functions are not widely available in computationally convenient form. However, a transformation used by a number of authors (Ash and Heinbockel, 1970; Papoutsakis et al.,

1980a,b) facilitates calculations for the cylindrical case that is of interest here. In particular, the variable changes

$$s = \lambda \eta^2, \qquad G(\lambda \eta) = Y(s)e^{-s/2} \tag{9.5-15}$$

transform the differential equation in the eigenvalue problem for G to

$$s \frac{d^2 Y}{ds^2} + (1 - s) \frac{dY}{ds} - \left(\frac{1}{2} - \frac{\lambda}{4} \right) Y = 0. \tag{9.5-16}$$

This is of the same form as *Kummer's equation,*

$$x \frac{d^2 y}{dx^2} + (b - x) \frac{dy}{dx} - ay = 0, \tag{9.5-17}$$

the solutions to which are the *confluent hypergeometric functions* $M(a, b, x)$ and $U(a, b, x)$ (Abramowitz and Stegun, 1970, p. 504). Because $U(a, b, 0)$ is unbounded, only M (also called *Kummer's function*) is needed. Comparing Eqs. (9.5-16) and (9.5-17), we find that

$$G(\lambda \eta) = e^{-\lambda \eta^2/2} M\left(\frac{1}{2} - \frac{\lambda}{4}, 1, \lambda \eta^2 \right) \tag{9.5-18}$$

and the characteristic equation for the Graetz problem, $G(\lambda) = 0$, becomes

$$M\left(\frac{1}{2} - \frac{\lambda}{4}, 1, \lambda \right) = 0. \tag{9.5-19}$$

Kummer's function is evaluated from its power-series representation (Abramowitz and Stegun, 1970, p. 504) as

$$M(a, b, x) = \sum_{n=0}^{\infty} \frac{(a)_n x^n}{(b)_n n!}, \tag{9.5-20}$$

$$(a)_n = a(a+1)(a+2) \ldots (a+n-1), \qquad (a)_0 = 1, \tag{9.5-21}$$

$$(b)_n = b(b+1)(b+2) \ldots (b+n-1), \qquad (b)_0 = 1. \tag{9.5-22}$$

Finding the smallest root of Eq. (9.5-19) using 20 terms of the series for M, we obtain $\lambda_1 = 2.7044$. Finally, the Nusselt number is calculated as

$$\text{Nu} = 3.657 \qquad (T_w \text{ constant}). \tag{9.5-23}$$

As discussed in Brown (1960) and Shah and London (1978), L. Graetz determined the first two eigenvalues in 1883 and W. Nusselt independently calculated the first three in 1910; both of those early workers obtained results sufficiently accurate to give $\text{Nu} = 3.66$. When digital computers became available it was possible at long last to calculate large numbers of eigenvalues, thereby allowing the eigenfunction expansion to be extended to relatively early parts of the entrance region. Sellars et al. (1956) were pioneers in this regard. The extensive review of Shah and London (1978) is a good source of tabulated results for this and related problems.

Example 9.5-2 Nusselt Number in the Fully Developed Region of a Tube with Specified Flux
It is desired now to calculate the far-downstream value of Nu for a circular tube with a specified heat flux at the wall, q_w. The dimensionless problem formulation is as given by Eqs. (9.4-5)–(9.4-9), except that the boundary condition at the wall is replaced by

$$\frac{\partial \Theta}{\partial \eta}(1, \zeta) = -1. \tag{9.5-24}$$

To make the temperature gradient at the wall unity, the dimensionless temperature is defined here as $\Theta = (T - T_0)/(q_w R/k)$.

As with the constant-temperature boundary condition considered in Example 9.5-1, a complete solution to the constant-flux problem can be obtained by using the finite Fourier transform method and Graetz (or Kummer) functions. However, the temperature field for $\zeta \to \infty$ is obtained more easily by a linear superposition approach, in which the temperature is written as the sum of the solutions of three problems. Explicit solutions are needed for only two of the problems, both of which involve only polynomials of a single variable. Adopting the superposition approach, we introduce the functions $\Omega(\eta, \zeta)$, $\psi(\eta)$, and $\phi(\zeta)$, such that

$$\Theta(\eta, \zeta) = \Omega(\eta, \zeta) + \psi(\eta) + \phi(\zeta). \tag{9.5-25}$$

Clearly, the *sum* of the three new functions must satisfy the partial differential equation and boundary conditions for Θ. Provided that is the case, we are free to choose the conditions satisfied by the individual functions in any way we find advantageous. Our strategy is to select the subsidiary problems in such a way that only $\psi(\eta)$ and $\phi(\zeta)$ need to be evaluated to determine Nu.

Suppose now that we define Ω such that

$$(1 - \eta^2) \frac{\partial \Omega}{\partial \zeta} = \frac{1}{\eta} \frac{\partial}{\partial \eta} \left(\eta \frac{\partial \Omega}{\partial \eta} \right), \tag{9.5-26}$$

$$\Omega(\eta, 0) = -\psi(\eta) - \phi(0), \tag{9.5-27}$$

$$\frac{\partial \Omega}{\partial \eta}(0, \zeta) = 0, \tag{9.5-28}$$

$$\frac{\partial \Omega}{\partial \eta}(1, \zeta) = 0. \tag{9.5-29}$$

Thus, Ω satisfies the original partial differential equation, and Eq. (9.5-27) ensures that the inlet condition is satisfied. Because both of the boundary conditions in η have been chosen so that they are homogeneous, the solution for $\Omega(\eta, \zeta)$ will be of the same form as that given for $\Theta(\eta, \zeta)$ in Eq. (9.5-7), although the eigenvalues will be different. We conclude that $\Omega \to 0$ as $\zeta \to \infty$. It will be shown that the other functions in Eq. (9.5-25) do not vanish, so that no further consideration of Ω is needed in determining the temperature far from the inlet.

Because $\Omega(\eta, \zeta)$ satisfies the original partial differential equation, so must the sum of $\psi(\eta)$ and $\phi(\zeta)$. Substituting that sum into Eq. (9.4-6) and rearranging gives

$$\frac{d\phi}{d\zeta} = \frac{1}{\eta(1 - \eta^2)} \frac{d}{d\eta} \left(\eta \frac{d\psi}{d\eta} \right). \tag{9.5-30}$$

The function of ζ on the left-hand side can match the function of η on the right-hand side only if both equal a constant. Denoting that constant as C_0, the problems for $\psi(\eta)$ and $\phi(\zeta)$ become

$$\frac{1}{\eta(1 - \eta^2)} \frac{d}{d\eta} \left(\eta \frac{d\psi}{d\eta} \right) = C_0; \quad \frac{d\psi}{d\eta}(0) = 0, \quad \frac{d\psi}{d\eta}(1) = -1, \tag{9.5-31}$$

$$\frac{d\phi}{d\zeta} = C_0. \tag{9.5-32}$$

Determining $\psi(\eta)$ and $\phi(\zeta)$ is straightforward. Integrating the differential equation for ψ and applying the boundary condition at $\eta = 1$ indicates that $C_0 = -4$. It is found that for $\zeta \to \infty$,

$$\Theta(\eta, \zeta) \to \psi(\eta) + \phi(\zeta) = -\eta^2 + \frac{\eta^4}{4} - 4\zeta + C_1, \tag{9.5-33}$$

where C_1 is another constant.

It is not necessary to evaluate C_1 to calculate Nu, but this will be done to complete the solution for the temperature field far from the inlet. We cannot apply the inlet condition to Eq. (9.5-33), which is an asymptotic expression for large ζ. However, we can use the fact that axial variations in the *bulk* temperature result only from the known heat flux at the tube wall. Evaluating the bulk temperature using Eq. (9.5-33), we find that

$$\Theta_b(\zeta) \to -\frac{7}{24} - 4\zeta + C_1 \tag{9.5-34}$$

for $\zeta \to \infty$. Another expression for the bulk temperature is obtained from the overall energy balance, Eq. (9.5-9), which is valid for all ζ and any type of boundary condition at the wall. Using the constant-flux boundary condition in that relation gives

$$\frac{d\Theta_b}{d\zeta} = -4, \qquad \Theta_b(0) = 0, \tag{9.5-35}$$

which has the solution

$$\Theta_b = -4\zeta. \tag{9.5-36}$$

Comparing Eqs. (9.5-34) and (9.5-36), we find that $C_1 = 7/24$. Accordingly, for $\zeta \to \infty$,

$$\Theta(\eta, \zeta) \to -\eta^2 + \frac{\eta^4}{4} - 4\zeta + \frac{7}{24}. \tag{9.5-37}$$

The Nusselt number for large ζ is calculated as

$$\mathrm{Nu} = \frac{2}{\Theta_b(\zeta) - \Theta(1, \zeta)} = \frac{48}{11} = 4.364 \qquad (q_w \text{ constant}). \tag{9.5-38}$$

As in Example 9.5-1, it is seen that Nu becomes constant far from the inlet, even though the temperature continues to vary with ζ.

Figure 9-5 shows plots of Nu(ζ) for both the constant-temperature and constant-flux cases.

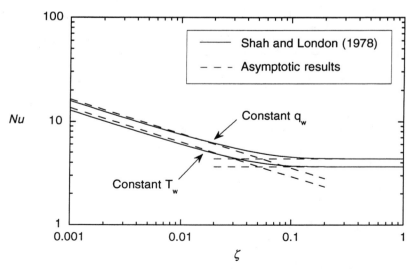

Figure 9-5. Nusselt number for fully developed laminar flow in a circular tube with a constant heat flux (q_w) or constant temperature (T_w) at the wall. The asymptotic results are based on Eqs. (9.4-28), (9.4-29), (9.5-23), and (9.5-38). The other results were obtained by calculating the first 121 terms in the eigenfunction expansions for the temperature (Shah and London, 1978).

In addition to the asymptotic results for large and small ζ, included are intermediate values taken from the review by Shah and London (1978, pp. 103 and 127). The entrance-region results given by Eqs. (9.4-28) and (9.4-29) are quite accurate for $\zeta < 0.03$, whereas the fully developed values in Eqs. (9.5-23) and (9.5-38) are satisfactory for most purposes when $\zeta > 0.08$. At any given value of ζ, the constant-flux results are about 20% larger than those for constant temperature.

For mass transfer between a dilute solution and a tube wall, the results for the Sherwood number are entirely analogous to those shown in Fig. 9-5 for the Nusselt number.

Mixed Boundary Conditions and Other Conduit Shapes

Thus far, the discussion of Nu has focused entirely on circular tubes with constant-temperature or constant-flux boundary conditions. A more general form of boundary condition in heat or mass transfer involves a linear combination of the field variable and its gradient at the wall, the mixed or Robin boundary condition. Three situations where that type of condition arises are: (1) heat transfer, if the resistance of the wall is not negligible; (2) mass transfer, if the wall is a membrane with a finite permeability (Example 9.3-1); and (3) mass transfer, if the wall catalyzes a first-order, irreversible reaction with a moderate rate constant. The more general boundary condition is written in dimensionless form as

$$\frac{\partial \Theta}{\partial \eta}(1, \zeta) + B\Theta(1, \zeta) = 0, \tag{9.5-39}$$

where B is the *wall conductance parameter.* The definitions of B for the three situations just mentioned are

$$B = \frac{h_w R}{k} \quad \text{or} \quad B = \frac{k_{mi} R}{D_i} \quad \text{or} \quad B = \frac{k_r R}{D_i}, \tag{9.5-40}$$

where h_w is the heat transfer coefficient for the wall, k_{mi} is the permeability of the wall (membrane) to solute i, and k_r is the rate constant for the heterogeneous reaction. In the first two cases, B is similar to the inverse of a Biot number, the difference being that the heat and mass transfer coefficients for the fluid are replaced by k/R and D_i/R, respectively. The "wall thermal resistance parameter" used by Shah and London (1978) equals $(2B)^{-1}$ from the first expression, whereas the "wall Sherwood number" employed by Colton et al. (1971) and Colton and Lowrie (1981) is equivalent to $2B$ from the second expression. The wall conductance parameter for the reaction case equals the Damköhler number.

The effect of B on the fully developed value of Nu is shown in Fig. 9-6. Results are shown both for circular tubes and parallel-plate channels; for a channel with plate spacing $2H$, R in Nu and B is replaced by H. The results for the two geometries are very similar, with Nu $\cong 4$ for all conditions. The constant-flux and constant-temperature cases correspond to the limits $B \to 0$ and $B \to \infty$, respectively, and serve to bracket the other results. That is, $\text{Nu}_T \leq \text{Nu} \leq \text{Nu}_q$ for both geometries, where Nu_T and Nu_q are the respective constant-temperature and constant-flux values. The continuous curves were calculated from

$$\text{Nu} = \frac{2B + \text{Nu}_q}{(2B/\text{Nu}_T) + 1}, \tag{9.5-41}$$

which differs from the exact results computed from the eigenvalues of Sideman et al. (1964) by < 0.3% over the entire range of B (see Problem 9-12). This simple interpola-

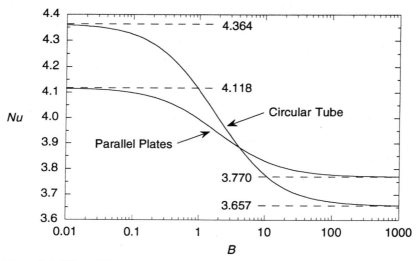

Figure 9-6. Effect of the wall conductance parameter (B) on the fully developed values of Nu for circular tubes and parallel-plate channels.

tion formula is inaccurate for the early entrance region, where it is found that the values of Nu for any finite B approach those for the constant-flux case [see Colton et al. (1971)]. For length-averaged values of Nu or Sh_i which are useful in design calculations, see Colton et al. (1971), Colton and Lowrie (1981), and Shah and London (1978). Shah and London (1978) also summarize many results for conduits other than circular tubes or parallel-plate channels. Values of Nu are available for ducts with rectangular, triangular, annular, and numerous other cross sections.

Axial Conduction

Analyses of Graetz-type problems which include axial conduction are complicated in most cases by the fact that the eigenvalue problem that results is not self-adjoint. Accordingly, the eigenfunctions, while linearly independent, are not orthogonal. Hsu (1971), Tan and Hsu (1972), and Ash and Heinbockel (1970) used the Gram–Schmidt procedure to construct a set of orthogonal functions, whereas Papoutsakis et al. (1980a,b) created a self-adjoint problem by decomposing the second-order partial differential equation into a pair of first-order partial differential equations. Acrivos (1980) obtained regular perturbation solutions to the constant-flux and constant-temperature problems for small Pe.

Velocity and Thermal Entrance Regions

All of the analyses discussed above assumed that the velocity profile was fully developed. Given that the hydrodynamic inlet is often where the new thermal conditions begin, we need to examine what happens when the velocity and temperature fields develop simultaneously. The approximations which are applicable (if any) depend on the velocity and thermal entrance lengths. There are four basic possibilities, as shown in Fig. 9-7. Far from the inlet, the velocity and temperature profiles are both fully developed, as

Figure 9-7. Development of the velocity and temperature fields in a tube (see text).

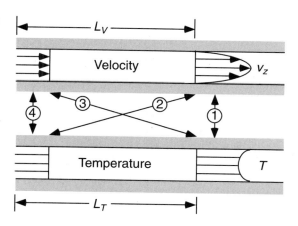

depicted by case 1 and as assumed in Examples 9.5-1 and 9.5-2. Closer to the inlet, one needs to consider the relative magnitudes of L_V and L_T. From Eqs. (9.4-43) and (6.5-3), we see that $L_T/L_V \cong \text{Pr}$ when Pe and Re are large. Thus, for $\text{Pr} \gg 1$, the velocity changes much faster than the temperature. This results in the situation shown in case 2 and assumed in Examples 9.4-1 and 9.4-2. Case 2 is particularly important for mass transfer, because $\text{Sc} \gg 1$ for all liquids. Case 3 in Fig. 9-7, the opposite of case 2, assumes that $\text{Pr} \ll 1$; it is relatively rare, applying mainly to liquid metals. Here, the temperature may enter a kind of fully developed regime while the velocity still approximates plug flow. As the velocity profile changes, the temperature will eventually leave this regime. Case 4, which is typical of heat or mass transfer in gases ($\text{Pr} \sim 1$ and $\text{Sc} \sim 1$), is a situation where the profiles develop at about the same rate. Numerical solutions are generally needed, as reviewed in Shah and London (1978).

9.6 CONSERVATION OF ENERGY: MECHANICAL EFFECTS

The approximate form of the energy equation which we have been using, Eq. (2.4-3), is restricted to pure, incompressible fluids and considers only thermal forms of energy. It is adequate for many applications, but for a more general description of energy transfer in fluids we need to consider mechanical effects as well. In the derivation which follows, the total energy is regarded as the sum of the internal energy and kinetic energy. It is the internal energy (rather than the enthalpy) which is conserved in an isolated system, making it the appropriate thermodynamic function with which to begin. Gravitational effects can be incorporated into an energy balance in either of two ways, by including a potential energy term in the total energy or by including an energy input term due to gravitational work. We adopt the latter approach, because it is easily generalized to include nongravitational body forces in multicomponent systems (see Chapter 11). Thus, in what follows, the only mechanical term in the "total energy" is the kinetic energy.

The internal energy per unit mass is denoted as $\hat{U}$ and the kinetic energy per unit mass is given by $v^2/2$, where $v^2 = \mathbf{v} \cdot \mathbf{v}$. The conservation equation written for the sum of internal and kinetic energy, for a fixed control volume, is

$$\frac{d}{dt}\int_V \rho\left(\hat{U}+\frac{v^2}{2}\right) dV = -\int_S \rho\left(\hat{U}+\frac{v^2}{2}\right)\mathbf{v}\cdot\mathbf{n}\, dS + \dot{Q} - \dot{W}, \tag{9.6-1}$$

where $\dot{Q}$ is the rate of heat transfer from the surroundings to the fluid inside the control volume, and $\dot{W}$ is the rate of work done by the fluid on its surroundings. This is just a statement of the first law of thermodynamics for a control volume. Whereas in elementary texts on thermodynamics the first law is generally written for a system of fixed mass, the first term on the right-hand side of Eq. (9.6-1) allows for energy transfer due to bulk flow through the control surface.

The surface integral in Eq. (9.6-1) accounts for convective energy transport, so that $\dot{Q}$ represents conduction only. The rate of heat transfer into the control volume by conduction is

$$\dot{Q} = -\int_S \mathbf{q}\cdot\mathbf{n}\, dS, \tag{9.6-2}$$

where $\mathbf{q}$ is evaluated using Fourier's law.

The rate of work done by the fluid on its surroundings has two parts, one due to surface forces and the other due to body forces. The contribution to the work from forces exerted on the surface S is evaluated in terms of the stress tensor ($\boldsymbol{\sigma}$), decomposed as usual into pressure and viscous contributions:

$$\boldsymbol{\sigma} = -P\boldsymbol{\delta} + \boldsymbol{\tau}. \tag{9.6-3}$$

According to the definition of $\boldsymbol{\sigma}$, the force exerted by the surroundings on a surface element dS is $(\mathbf{n}\cdot\boldsymbol{\sigma})\, dS$. Thus, the surface force exerted by the fluid on the surroundings is $-(\mathbf{n}\cdot\boldsymbol{\sigma})\, dS$. The rate of work done on the fluid just outside this differential surface element is $-\mathbf{v}\cdot(\mathbf{n}\cdot\boldsymbol{\sigma})\, dS$. The total rate of surface work on the surroundings ($\dot{W}_S$) is then

$$\dot{W}_S = -\int_S \mathbf{v}\cdot(\mathbf{n}\cdot\boldsymbol{\sigma})\, dS = \int_S P\mathbf{v}\cdot\mathbf{n}\, dS - \int_S \mathbf{v}\cdot(\mathbf{n}\cdot\boldsymbol{\tau})\, dS. \tag{9.6-4}$$

Assuming that gravity is the only body force, the rate of work done by the surroundings on a volume element dV is $\rho\,(\mathbf{v}\cdot\mathbf{g})\, dV$. Accordingly, the gravitational work done by the system on its surroundings is

$$\dot{W}_G = -\int_V \rho(\mathbf{v}\cdot\mathbf{g})\, dV. \tag{9.6-5}$$

The total rate of work done on the surroundings is $\dot{W}_S + \dot{W}_G$.

Certain other types of work, including that due to an electrical power source, can also be included. In previous chapters we used H_V to represent such external sources, which are encountered more frequently in heat transfer problems in solids than in fluids. We omit the H_V term here, with the understanding that (if needed) it may simply be added to the other volumetric source terms in the final, differential form of the energy equation.

Evaluating the heat transfer and work terms as just described, Eq. (9.6-1) becomes

$$\frac{d}{dt}\int_V \rho\left(\hat{U}+\frac{v^2}{2}\right) dV + \int_S \rho\left(\hat{U}+\frac{v^2}{2}\right)\mathbf{v}\cdot\mathbf{n}\, dS$$

$$= -\int_S \mathbf{q} \cdot \mathbf{n} \, dS - \int_S P\mathbf{v} \cdot \mathbf{n} \, dS + \int_S (\boldsymbol{\tau} \cdot \mathbf{v}) \cdot \mathbf{n} \, dS + \int_V \rho(\mathbf{v} \cdot \mathbf{g}) \, dV, \qquad (9.6\text{-}6)$$

where we have used the identity $\mathbf{v} \cdot (\mathbf{n} \cdot \boldsymbol{\tau}) = (\boldsymbol{\tau} \cdot \mathbf{v}) \cdot \mathbf{n}$. To obtain a differential form of Eq. (9.6-6), we convert the surface integrals to volume integrals and let $V \to 0$, as in Section 2.2. The left-hand side of the resulting differential equation is simplified by using the continuity equation and introducing the material derivative, as in the derivation of Eq. (2.3-7). The result of these manipulations is

$$\rho \frac{D}{Dt}\left(\hat{U} + \frac{v^2}{2}\right) = -\boldsymbol{\nabla} \cdot \mathbf{q} - \boldsymbol{\nabla} \cdot (P\mathbf{v}) + \boldsymbol{\nabla} \cdot (\boldsymbol{\tau} \cdot \mathbf{v}) + \rho(\mathbf{v} \cdot \mathbf{g}), \qquad (9.6\text{-}7)$$

which is a general expression for conservation of energy at any point in a pure fluid.

It will be more useful to have an energy equation that involves the temperature. The first steps in obtaining such a relation are to evaluate $\rho D(v^2/2)/Dt$ and to subtract that expression from Eq. (9.6-7), thereby obtaining an equation for the internal energy alone. Once this is done, there are two basic options. One is to express $\hat{U}$ as a function of T and ρ, which is the approach taken in Bird et al. (1960, pp. 314–315). The other is to convert $\hat{U}$ to the enthalpy per unit mass, $\hat{H}$, and then to express $\hat{H}$ as a function of T and P. That is the approach used here.

A relationship involving mechanical forms of energy is obtained from the Cauchy momentum equation,

$$\rho \frac{D\mathbf{v}}{Dt} = -\boldsymbol{\nabla} P + \boldsymbol{\nabla} \cdot \boldsymbol{\tau} + \rho \mathbf{g}. \qquad (9.6\text{-}8)$$

By forming the dot product of $\mathbf{v}$ with Eq. (9.6-8), we obtain a scalar equation with terms having the same dimensions as those in Eq. (9.6-7). The result is

$$\rho \mathbf{v} \cdot \frac{D\mathbf{v}}{Dt} = -\mathbf{v} \cdot \boldsymbol{\nabla} P + \mathbf{v} \cdot (\boldsymbol{\nabla} \cdot \boldsymbol{\tau}) + \rho(\mathbf{v} \cdot \mathbf{g}). \qquad (9.6\text{-}9)$$

Using Eq. (8.2-4) to evaluate the left-hand side, Eq. (9.6-9) is rearranged to

$$\rho \frac{D}{Dt}\left(\frac{v^2}{2}\right) = -\mathbf{v} \cdot \boldsymbol{\nabla} P + \mathbf{v} \cdot (\boldsymbol{\nabla} \cdot \boldsymbol{\tau}) + \rho(\mathbf{v} \cdot \mathbf{g}). \qquad (9.6\text{-}10)$$

This is called the *mechanical energy equation*.[1] Subtracting Eq. (9.6-10) from Eq. (9.6-7) yields the desired equation involving only the internal energy, which is

$$\rho \frac{D\hat{U}}{Dt} = -\boldsymbol{\nabla} \cdot \mathbf{q} - P(\boldsymbol{\nabla} \cdot \mathbf{v}) + \boldsymbol{\tau}:(\boldsymbol{\nabla}\mathbf{v}). \qquad (9.6\text{-}11)$$

---

[1] As explained in Bird et al. (1960, pp. 314–315), an alternate form of the mechanical (or total) energy equation includes potential energy on the left-hand side. Defining the gravitational potential energy ϕ by $\mathbf{g} = -\boldsymbol{\nabla}\phi$ and recognizing that in terrestrial applications ϕ is independent of time, it follows from Eq. (9.6-10) that

$$\rho \frac{D}{Dt}\left(\frac{v^2}{2} + \phi\right) = -\mathbf{v} \cdot \boldsymbol{\nabla} P + \mathbf{v} \cdot (\boldsymbol{\nabla} \cdot \boldsymbol{\tau}).$$

A comparison of this with Eq. (9.6-10) confirms that gravity may be included either as a potential energy term on the left-hand side or as an energy source term on the right-hand side. It cannot appear in both places.

In manipulating the viscous terms we have made use of Eq. (A.4-14), which requires that τ be symmetric.

The specific internal energy and specific enthalpy are related as

$$\hat{U} = \hat{H} - \frac{P}{\rho},$$ (9.6-12)

where $1/\rho$ is the specific volume. The left-hand side of Eq. (9.6-11) is rewritten now as

$$\rho \frac{D\hat{U}}{Dt} = \rho \frac{D\hat{H}}{Dt} - \rho \frac{D}{Dt}\left(\frac{P}{\rho}\right) = \rho \frac{D\hat{H}}{Dt} - \frac{DP}{Dt} + \frac{P}{\rho}\frac{D\rho}{Dt}.$$ (9.6-13)

The material derivative of the density is evaluated by rearranging the continuity equation, Eq. (2.3-1), to obtain

$$\frac{D\rho}{Dt} = -\rho(\nabla \cdot \mathbf{v}).$$ (9.6-14)

Combining Eqs. (9.6-13) and (9.6-14) with Eq. (9.6-11) gives

$$\rho \frac{D\hat{H}}{Dt} = -\nabla \cdot \mathbf{q} + \frac{DP}{Dt} + \tau:(\nabla \mathbf{v}).$$ (9.6-15)

This is the general equation for conservation of energy written in terms of the specific enthalpy.

The dependence of the enthalpy on temperature and pressure is expressed by (e.g., Denbigh, 1966, p. 98)

$$d\hat{H} = \hat{C}_p \, dT + \left[\frac{1}{\rho} - T\left(\frac{\partial(1/\rho)}{\partial T}\right)_P\right] dP,$$ (9.6-16)

which permits the left-hand side of Eq. (9.6-15) to be rewritten as

$$\rho \frac{D\hat{H}}{Dt} = \rho \hat{C}_p \frac{DT}{Dt} + \left[1 - \rho T\left(\frac{\partial(1/\rho)}{\partial T}\right)_P\right]\frac{DP}{Dt}.$$ (9.6-17)

Using Eq. (9.6-17) in Eq. (9.6-15), we obtain

$$\rho \hat{C}_p \frac{DT}{Dt} = -\nabla \cdot \mathbf{q} - \frac{T}{\rho}\left(\frac{\partial\rho}{\partial T}\right)_P \frac{DP}{Dt} + \tau:(\nabla \mathbf{v}).$$ (9.6-18)

This form of the energy equation is valid for any pure (single component) fluid with a symmetric stress tensor. A comparison with Eq. (2.4-1) (without H_V) reveals that there are now two additional terms on the right-hand side. The one involving the density and pressure is related to the work required to compress the fluid and is referred to below as the *compressibility* term. The one involving the viscous stress represents the conversion of kinetic energy to heat, due to the internal friction in the fluid. This irreversible process is called *viscous dissipation*.

For a *Newtonian fluid,* the rate of viscous dissipation is proportional to the viscosity, and is evaluated as

$$\tau:\nabla \mathbf{v} = \mu \Phi.$$ (9.6-19)

The *viscous dissipation function,* Φ, is related to the various spatial derivatives of the velocity as shown in Table 5-10. It is seen that all contributions to Φ involve *squares* of velocity derivatives, indicating that $\Phi \geq 0$. That is, viscous dissipation always acts as a *heat source*, which reflects the irreversibility of the frictional losses of mechanical energy. For Newtonian fluids, including those with variable density, Eq. (9.6-18) becomes

$$\rho \hat{C}_p \frac{DT}{Dt} = -\nabla \cdot \mathbf{q} - \frac{T}{\rho}\left(\frac{\partial \rho}{\partial T}\right)_P \frac{DP}{Dt} + \mu\Phi. \tag{9.6-20}$$

The compressibility term has a simple form for ideal gases, where $(\partial \rho/\partial T)_P = -\rho/T$. Thus, the energy equation for an *ideal gas* is

$$\rho \hat{C}_p \frac{DT}{Dt} = -\nabla \cdot \mathbf{q} + \frac{DP}{Dt} + \mu\Phi. \tag{9.6-21}$$

For a *Newtonian fluid* which is *incompressible* in the sense that $(\partial \rho/\partial T)_P = 0$, the energy equation simplifies further to

$$\rho \hat{C}_p \frac{DT}{Dt} = -\nabla \cdot \mathbf{q} + \mu\Phi. \tag{9.6-22}$$

The principal forms of the energy equation derived here are summarized in Table 9-1. Many other forms of conservation of energy are given in Bird et al. (1960).

The different uses of the term "incompressible" are potentially confusing, especially in connection with gases. From the viewpoint of the continuity and momentum equations, gases often behave as if their density is constant. As discussed in Section 2.3, this is generally true for velocities much smaller than the speed of sound. Nonetheless, the equation of state for a gas is such that $(\partial \rho/\partial T)_P \neq 0$. Whether or not Eq. (9.6-21) for an ideal gas can be approximated by Eq. (9.6-22) depends on the relative magnitudes of the pressure and temperature variations; this is illustrated by the numerical values for air given in Section 2.3. Thus, there is a distinction between fluid dynamic and thermodynamic incompressibility.

Viscous dissipation introduces a new dimensionless parameter. For steady flow of a Newtonian, incompressible fluid, Eq. (9.2-1) becomes

$$\text{Pe } \tilde{\mathbf{v}} \cdot \tilde{\nabla}\Theta = \tilde{\nabla}^2\Theta + \text{Br } \tilde{\Phi}, \tag{9.6-23}$$

TABLE 9-1
Conservation of Energy for a Pure Fluid in Terms of Temperature and Pressure

General:	$\rho \hat{C}_p \dfrac{DT}{Dt} = -\nabla \cdot \mathbf{q} - \dfrac{T}{\rho}\left(\dfrac{\partial \rho}{\partial T}\right)_P \dfrac{DP}{Dt} + \tau:(\nabla\mathbf{v})$	(A)
Newtonian[a]:	$\rho \hat{C}_p \dfrac{DT}{Dt} = -\nabla \cdot \mathbf{q} - \dfrac{T}{\rho}\left(\dfrac{\partial \rho}{\partial T}\right)_P \dfrac{DP}{Dt} + \mu\Phi$	(B)
Ideal Gas:	$\rho \hat{C}_p \dfrac{DT}{Dt} = -\nabla \cdot \mathbf{q} + \dfrac{DP}{Dt} + \mu\Phi$	(C)
Incompressible Newtonian[b]:	$\rho \hat{C}_p \dfrac{DT}{Dt} = -\nabla \cdot \mathbf{q} + \mu\Phi$	(D)

[a] The dissipation function, Φ, is given in Table 5-10.
[b] As used here, "incompressible" means that $(\partial \rho/\partial T)_P = 0$.

$$\tilde{\Phi} = \left(\frac{L}{U}\right)^2 \Phi, \tag{9.6-24}$$

$$Br = \frac{\mu U^2}{k \Delta T}, \tag{9.6-25}$$

where ΔT is the temperature scale used in defining Θ. The *Brinkman number,* Br, expresses the relative importance of viscous dissipation and heat conduction. (Some authors employ the *Eckert number,* Ec = Br/Pr.) Reasoning as in Section 9.3, we infer that

$$Nu = Nu \ (\tilde{r}_w, \ Re, \ Pr, \ Br, \ geometric \ ratios) \tag{9.6-26}$$

for steady flow. The results presented for Nu in Sections 9.4 and 9.5 all correspond to Br→0, in which case viscous dissipation is negligible. Viscous dissipation tends to be important mainly for polymeric fluids (large μ) or high-speed flows (large Φ).

9.7 TAYLOR DISPERSION

If a soluble substance is injected rapidly into a fluid flowing in a tube, a sharp peak in solute concentration is created. As the peak travels along the tube it gradually broadens. The axial dispersion of solute which causes the peak broadening resembles molecular diffusion, but is actually a combined effect of diffusion and convection. This phenomenon, called *Taylor dispersion,*[2] has applications as diverse as the analysis of peak broadening in chromatographic separations, the prediction of the degree of mixing of petroleum products pumped in sequence in a pipeline, and the measurement of molecular diffusivities.

Taylor dispersion concerns the evolution of a solute concentration peak as would be seen by an observer moving with the fluid. This makes it advantageous to employ a coordinate origin that moves at the mean velocity of the fluid. Let z represent axial position relative to such an origin. For sufficiently long observation times, it will be shown that the cross-sectional average solute concentration $\overline{C}(z, t)$ is governed by an equation of the form

$$\frac{\partial \overline{C}}{\partial t} = K \frac{\partial^2 \overline{C}}{\partial z^2}, \tag{9.7-1}$$

$$\overline{C} = \frac{1}{A} \int_A C \, dA, \tag{9.7-2}$$

where A denotes the cross-sectional area. The *dispersivity, K,* is a constant whose value depends on the shape of the tube cross section and the form of the fully developed velocity profile; it does not in general equal the molecular diffusivity, D. Given the

---

[2] During his career at Cambridge University, G. I. Taylor (1886–1975) made major contributions to several areas of applied mechanics and transport theory. He had a remarkable ability to examine a phenomenon using a relatively simple experiment and was able to build on the experimental observations a thorough and imaginative theoretical analysis. Among his best-known contributions to fluid mechanics are his concepts of isotropic turbulence and turbulent energy spectra, as well as his experimental and theoretical investigation of instabilities in Couette flows.

nonuniform axial velocity in a tube, it is remarkable that Eq. (9.7-1) has the same form as the conservation equation for one-dimensional diffusion in a stagnant fluid.

The objective of the following analysis is to evaluate K for the laminar flow of an incompressible, Newtonian fluid in a circular tube. The approach will be to calculate the first few *moments* of the concentration distribution, while avoiding the more difficult task of solving the full species conservation equation for $C(r, z, t)$. This approach, de-scribed by Aris (1956), involves fewer restrictive assumptions than the original analysis given by Taylor (1953).

To show how moments may be used to evaluate K, we first examine the relation-ship between the diffusivity of a solute and certain moments of the concentration distri-bution. The conclusions are applicable to any one-dimensional, transient diffusion pro-cess in an unbounded domain; to obtain results which will be useful for the Taylor dispersion analysis, we use Eq. (9.7-1) as the starting point. Because we will be applying these results to dispersion in a long tube of radius R, we introduce the dimensionless variables $\zeta = z/R$ and $\tau = Dt/R^2$. The resulting differential equation for $\overline{C}(\zeta, \tau)$ is

$$\frac{\partial \overline{C}}{\partial \tau} = \left(\frac{K}{D}\right) \frac{\partial^2 \overline{C}}{\partial \zeta^2}. \tag{9.7-3}$$

The dimensionless diffusivity in this equation is the coefficient on the right-hand side, namely K/D. The pth moment of $\overline{C}$ about $\zeta = 0$ is defined as

$$m_p(\tau) \equiv \int_{-\infty}^{\infty} \overline{C}(\zeta, \tau) \zeta^p \, d\zeta, \tag{9.7-4}$$

where p is a non-negative integer. It is seen that the various moments are functions only of time. If the initial concentration is given by

$$\overline{C}(\zeta, 0) = f(\zeta), \tag{9.7-5}$$

where $f(\zeta) \to 0$ as $\zeta \to \pm \infty$, then the initial values of all m_p will be finite. Because diffusion limits the rate at which the solute can move to distant locations, it follows that all m_p will remain finite for finite values of τ. Other than the restriction just stated, the precise form of $f(\zeta)$ is unimportant for our purposes.

A differential equation for $m_p(\tau)$ is obtained by multiplying both sides of Eq. (9.7-3) by ζ^p and then integrating as in Eq. (9.7-4). The left-hand side gives

$$\int_{-\infty}^{\infty} \frac{\partial \overline{C}}{\partial \tau} \zeta^p \, d\zeta = \frac{d}{d\tau} \int_{-\infty}^{\infty} \overline{C} \zeta^p \, d\zeta = \frac{dm_p}{d\tau}. \tag{9.7-6}$$

Integrating by parts twice to evaluate the right-hand side, we obtain

$$\int_{-\infty}^{\infty} \frac{\partial^2 \overline{C}}{\partial \zeta^2} \zeta^p \, d\zeta = p(p-1) \int_{-\infty}^{\infty} \overline{C} \zeta^{p-2} \, d\zeta = p(p-1)m_{p-2}. \tag{9.7-7}$$

There are no boundary terms remaining from the integrations by parts, because the assumed form of the initial concentration ensures that $\overline{C}$ and $\partial \overline{C}/\partial \zeta$ vanish as $\zeta \to \pm \infty$. Using Eqs. (9.7-6) and (9.7-7), we find that the moments must satisfy

$$\frac{dm_p}{d\tau} = \left(\frac{K}{D}\right)p(p-1)m_{p-2}, \qquad m_p(0) = M_p. \tag{9.7-8}$$

The initial values, M_p, may be computed if desired by replacing by $\overline{C}$ by $f(\zeta)$ in Eq. (9.7-4). It will not be necessary to evaluate these constants.

We will be concerned only with the moments corresponding to $p=0$, 1, and 2. Inspection of Eq. (9.7-8) indicates that the zeroth and first moments remain constant for all τ. In physical terms, m_0 is constant because it is proportional to the total amount of solute present, which does not change. Likewise m_1, which is related to the solute center of mass, remains constant because there is no directional bias in a diffusion equation such as Eq. (9.7-3). For $p=2$, integration of Eq. (9.7-8) gives

$$m_2 = 2(K/D)m_0\tau + M_2. \tag{9.7-9}$$

Thus, m_2 grows linearly with time. Eventually, for large τ, the constant (M_2) term becomes negligible, and we find that

$$\frac{K}{D} = \lim_{\tau \to \infty} \frac{m_2}{2m_0\tau}. \tag{9.7-10}$$

It is seen that the diffusivity (in this case, K/D) may be computed from the long-time behavior of the ratio m_2/m_0. The analysis which will be presented for tube flow confirms that m_2 grows linearly in time for that situation, as expected if Eq. (9.7-1) is correct. Thus, the ability to determine a constant value of K from Eq. (9.7-10) ultimately validates Eq. (9.7-1).

We turn now to the problem of evaluating the necessary moments in terms of the known quantities R, D, and U, where U is the mean velocity of the fluid relative to the tube. In the moving reference frame the conservation equation for the solute (assumed to be dilute) is

$$\frac{\partial C}{\partial \tau} + \frac{\text{Pe}}{2}(1-2\eta^2)\frac{\partial C}{\partial \zeta} = \frac{1}{\eta}\frac{\partial}{\partial \eta}\left(\eta \frac{\partial C}{\partial \eta}\right) + \frac{\partial^2 C}{\partial \zeta^2}, \tag{9.7-11}$$

where $\eta = r/R$ and $\text{Pe} = 2UR/D$. The axial velocity used in Eq. (9.7-11) was obtained by subtracting U from the usual parabolic velocity profile for a tube; in the moving reference frame the mean velocity is zero. Because the axial diffusion term has been retained, Eq. (9.7-11) is valid for any value of Pe. This equation is subject to the radial boundary conditions

$$\frac{\partial C}{\partial \eta} = 0 \qquad \text{at } \eta = 0, \tag{9.7-12}$$

$$\frac{\partial C}{\partial \eta} = 0 \qquad \text{at } \eta = 1. \tag{9.7-13}$$

As in the preceding discussion, the exact form of the initial concentration distribution turns out to be unimportant. It is only necessary to assume that the solute is initially confined to a finite length of tube, from which it follows that $C \to 0$ as $\zeta \to \pm\infty$ for all τ and that the moments $m_p(\tau)$ are finite. For simplicity, we assume also that the initial distribution is axisymmetric, so that C is always independent of the cylindrical angle θ, as implied by Eq. (9.7-11). The more general analysis given by Aris (1956) shows that the value obtained for K does not depend on the initial concentration being axisymmetric.

The moments m_p defined by Eq. (9.7-4) are based on the cross-sectional average concentration. The analysis requires that we work also with analogous moments of the local (radially dependent) concentration. These moments are defined by

$$C_p(\eta, \tau) \equiv \int_{-\infty}^{\infty} C(\eta, \zeta, \tau) \zeta^p \, d\zeta. \tag{9.7-14}$$

The two types of moments are related according to

$$m_p(\tau) = \overline{C}_p(\tau) = \frac{1}{A} \int_A C_p \, dA = 2 \int_0^1 C_p(\eta, \tau) \eta \, d\eta. \tag{9.7-15}$$

A partial differential equation governing $C_p(\eta, \tau)$ is obtained by integrating Eq. (9.7-11) over ζ, as was done in deriving Eq. (9.7-8). The result is

$$\frac{\partial C_p}{\partial \tau} = \frac{1}{\eta} \frac{\partial}{\partial \eta} \left(\eta \frac{\partial C_p}{\partial \eta} \right) + p(p-1)C_{p-2} + \frac{\text{Pe}}{2}(1-2\eta^2)pC_{p-1}. \tag{9.7-16}$$

The ordinary differential equation which governs $m_p(\tau)$ is found by integrating Eq. (9.7-16) over η, leading to

$$\frac{dm_p}{d\tau} = p(p-1)m_{p-2} + p \, \text{Pe} \int_0^1 C_{p-1}(1-2\eta^2)\eta \, d\eta. \tag{9.7-17}$$

Equations (9.7-16) and (9.7-17) constitute a sequence of differential equations which can be used to compute progressively higher moments. The utility of this approach stems from the fact that to calculate K/D, we need not proceed beyond m_2. To obtain m_2 we will need to find C_0, C_1, and m_1, although the complete details of these lower moments are unnecessary. As already noted, m_0 is simply a constant, proportional to the total amount of solute present.

Setting $p = 0$ in Eq. (9.7-16), the governing equation for $C_0(\eta, \tau)$ is seen to be

$$\frac{\partial C_0}{\partial \tau} = \frac{1}{\eta} \frac{\partial}{\partial \eta} \left(\eta \frac{\partial C_0}{\partial \eta} \right). \tag{9.7-18}$$

The boundary conditions are analogous to Eqs. (9.7-12) and (9.7-13) and the initial condition is of the form

$$C_0(\eta, 0) = f(\eta), \tag{9.7-19}$$

where $f(\eta)$ is computed from the initial concentration distribution. Using the methods of Chapter 4, the solution is found to be

$$C_0(\eta, \tau) = m_0 + \sum_{n=1}^{\infty} a_n J_0(\lambda_n \eta) e^{-\lambda_n^2 \tau}, \tag{9.7-20}$$

$$a_n = \frac{2}{J_0^2(\lambda_n)} \int_0^1 f(\eta) J_0(\lambda_n \eta) \eta \, d\eta, \tag{9.7-21}$$

where $J_1(\lambda_n) = 0$. That the leading constant in Eq. (9.7-20) must equal m_0 is seen by using Eq. (9.7-20) in Eq. (9.7-15) and evaluating the integral.

The differential equation for $m_1(\tau)$, obtained by setting $p = 1$ in Eq. (9.7-17), is

$$\frac{dm_1}{d\tau} = \text{Pe} \int_0^1 C_0(1 - 2\eta^2)\eta \, d\eta. \tag{9.7-22}$$

The constant (m_0) term in C_0 does not contribute to the integral in Eq. (9.7-22), as is easily confirmed. From Eq. (9.7-20), the remainder of the integral is $O(e^{-\lambda_1^2 \tau})$ as $\tau \to \infty$, where λ_1 $(= 3.83)$ is the smallest positive root of $J_1(\lambda)$. We expect the $O(e^{-\lambda_1^2 \tau})$ term to be negligible for $\lambda_1^2 \tau > 3$, or $\tau > 0.2$. Accordingly, $dm_1/d\tau \to 0$ fairly quickly, and $m_1 \to m_{1\infty}$, a constant. As already noted, m_1 is proportional to the center of mass for the solute. Thus, after a short period of adjustment (depending on the initial concentration distribution), the solute center of mass moves at the mean velocity of the fluid. If the initial pulse is symmetric about $\zeta = 0$, it turns out that $m_1 = 0$ for all τ.

The differential equation for $C_1(\eta, \tau)$ is

$$\frac{\partial C_1}{\partial \tau} = \frac{1}{\eta} \frac{\partial}{\partial \eta}\left(\eta \frac{\partial C_1}{\partial \eta}\right) + \frac{\text{Pe}}{2}(1 - 2\eta^2)C_0, \tag{9.7-23}$$

which is seen to be nonhomogeneous because of the C_0 term. The boundary conditions and initial condition for C_1 are similar to those for C_0. Because it is a linear, nonhomogeneous differential equation, the general solution to Eq. (9.7-23) is the solution to the homogeneous equation added to a particular solution of the nonhomogeneous equation. The particular solution of Eq. (9.7-23) consists of a sum of terms corresponding to those in Eq. (9.7-20); the nth term of the particular solution is denoted as $\tilde{C}_{1,n}$. It is readily shown that $\tilde{C}_{1,0}$ (the part of the particular solution which corresponds to the constant, m_0) is a function of η only, such that

$$\tilde{C}_{1,0}(\eta) = -\frac{\text{Pe} \, m_0}{8}\left(\eta^2 - \frac{\eta^4}{2}\right). \tag{9.7-24}$$

The remaining terms all have the form

$$\tilde{C}_{1,n}(\eta, \tau) = \text{Pe} \, b_n g_n(\eta) e^{-\lambda_n^2 \tau}, \tag{9.7-25}$$

where b_n is a constant and

$$\frac{1}{\eta} \frac{d}{d\eta}\left(\eta \frac{dg_n}{d\eta}\right) + \lambda_n^2 g_n = -(1 - 2\eta^2) J_0(\lambda_n \eta). \tag{9.7-26}$$

Each of the functions $g_n(\eta)$ must satisfy the homogeneous boundary conditions.

The homogeneous form of Eq. (9.7-23) is the same as Eq. (9.7-18) and has solutions

$$\hat{C}_{1,j}(\eta, \tau) = c_j J_0(\lambda_j \eta) e^{-\lambda_j^2 \tau}, \tag{9.7-27}$$

where c_j is a constant that is determined by the initial condition. Adding the particular and homogeneous solutions, the complete solution to Eq. (9.7-23) is of the form

$$C_1(\eta, \tau) = \sum_{n=1}^{\infty} \tilde{C}_{1,n}(\eta, \tau) + \sum_{j=1}^{\infty} \hat{C}_{1,j}(\eta, \tau)$$

$$= -\frac{\mathrm{Pe}\, m_0}{8}\left(\eta^2 - \frac{\eta^4}{2}\right) + \mathrm{Pe} \sum_{n=1}^{\infty} b_n g_n(\eta) e^{-\lambda_n^2 \tau} + c_0 + \sum_{j=1}^{\infty} c_j J_0(\lambda_j\, \eta) e^{-\lambda_n^2 \tau}.$$

(9.7-28)

The behavior of $C_1(\eta,\, \tau)$ for large τ is seen to be

$$C_1(\eta,\, \tau) = c_0 - \frac{\mathrm{Pe}\, m_0}{8}\left(\eta^2 - \frac{\eta^4}{2}\right) + O(e^{-\lambda_1^2 \tau}).$$

(9.7-29)

Any more detailed consideration of the time-dependent terms in $C_1(\eta,\, \tau)$ will be unnecessary.

The governing equation for $m_2(\tau)$ is

$$\frac{dm_2}{d\tau} = 2m_0 + 2\,\mathrm{Pe} \int_0^1 C_1(1 - 2\eta^2)\eta\, d\eta.$$

(9.7-30)

Considering the behavior of m_2 only for large τ, Eqs. (9.7-29) and (9.7-30) indicate that

$$\frac{dm_2}{d\tau} = 2m_0 + \frac{\mathrm{Pe}^2\, m_0}{96} + O(e^{-\lambda_1^2 \tau}).$$

(9.7-31)

Neglecting the terms which decay in time and then integrating, we arrive at the desired ratio of moments:

$$\frac{m_2}{2m_0} = \left(1 + \frac{\mathrm{Pe}^2}{192}\right)\tau \qquad (\tau \to \infty).$$

(9.7-32)

Finally, using Eq. (9.7-32) in Eq. (9.7-10), we evaluate the dispersivity as

$$K = D\left(1 + \frac{\mathrm{Pe}^2}{192}\right) = D + \frac{U^2 R^2}{48D}.$$

(9.7-33)

Thus, the dispersivity K appearing in Eq. (9.7-1) is the sum of the molecular diffusivity and a term which, surprisingly, varies as D^{-1}. As discussed in detail by Taylor (1953), the second term arises from the combined effects of axial convection and radial diffusion. The numerical coefficient (1/48) depends on the shape of the velocity profile, which in this case was parabolic. For a uniform velocity (plug flow), the convective contribution to dispersion vanishes and $K = D$.

The elegant result given by Eq. (9.7-33) is valid for conditions even more general than those assumed in the above derivation. As already mentioned, Aris (1956) showed that the same result is obtained if the initial concentration distribution is not axisymmetric. Although our derivation focused on a pulse input of solute, Taylor (1953) showed that the convective dispersion result applies equally to a step change in solute concentration; he also provided several comparisons between theory and experiment. While giving results for the special case of laminar flow in a cylindrical tube, Aris (1956) also showed how to compute K for any other cross-sectional geometry and any fully developed flow.

The analysis of Taylor dispersion in tube flow is the prototype for a general class of convective-diffusion problems, in which it is desired to calculate effective transport properties through spatial or temporal averaging. In this case the problem was to find the "effective diffusivity" (K) which could be used with the cross-sectional average concentration. In porous media, a basic problem is to determine coefficients which can be used to describe convection and diffusion using, say, the superficial velocity and the volume-average solute concentration. A discussion of averaging methods for many such problems is given in Brenner and Edwards (1993).

References

Abramowitz, M. and I. A. Stegun. *Handbook of Mathematical Functions.* U.S. Department of Commerce, National Bureau of Standards, Washington, DC, 1970.

Acrivos, A. The extended Graetz problem at low Peclet numbers. *Appl. Sci. Res.* 36: 35–40, 1980.

Aris, R. On the dispersion of a solute in a fluid flowing through a tube. *Proc. R. Soc. Lond. A* 235: 67–77, 1956.

Ash, R. L. and J. H. Heinbockel. Note on heat transfer in laminar, fully developed pipe flow with axial conduction. *Z. Angew. Math. Phys.* 21: 266–269, 1970.

Bird, R. B., W. E. Stewart, and E. N. Lightfoot. *Transport Phenomena.* Wiley, New York, 1960.

Brenner, H. and D. A. Edwards. *Macrotransport Processes.* Butterworth-Heinemann, Boston, 1993.

Brown, G. M. Heat or mass transfer in a fluid in laminar flow in a circular or flat conduit. *AIChE J.* 6: 179–183, 1960.

Colton, C. K. and E. G. Lowrie. Hemodialysis: physical principles and technical considerations. In *The Kidney,* Vol. II, B. M. Brenner and F. C. Rector, Jr., Eds., second edition, WB Saunders, Philadelphia, 1981, pp. 2425–2489.

Colton, C. K., K. A. Smith, P. Stroeve, and E. W. Merrill. Laminar flow mass transfer in a flat duct with permeable walls. *AIChE J.* 17: 773–780, 1971.

Denbigh, K. *The Principles of Chemical Equilibrium,* second edition. Cambridge University Press, Cambridge, 1968.

Hsu, C.-J. An exact analysis of low Peclet number thermal entry region heat transfer in transversely nonuniform velocity fields. *AIChE J.* 17: 732–740, 1971.

Newman, J. Extension of the Lévêque solution. *J. Heat Transfer* 91: 177–178, 1969.

Olbrich, W. E. and J. D. Wild. Diffusion from the free surface into a liquid film in laminar flow over defined shapes. *Chem. Eng. Sci.* 24: 25–32, 1969.

Papoutsakis, E., D. Ramkrishna, and H. C. Lim. The extended Graetz problem with Dirichlet wall boundary conditions. *Appl. Sci. Res.* 36: 13–34, 1980a.

Papoutsakis, E., D. Ramkrishna, and H. C. Lim. The extended Graetz problem with prescribed wall flux. *AIChE J.* 26: 779–787, 1980b.

Sellars, J. R., M. Tribus, and J. S. Klein. Heat transfer to laminar flow in a round tube or flat conduit—the Graetz problem extended. *Trans. ASME* 78: 441–448, 1956.

Shah, R. K. and A. L. London. *Laminar Flow Forced Convection in Ducts.* Academic Press, New York, 1978.

Sideman, S., D. Luss, and R. E. Peck. Heat transfer in laminar flow in circular and flat conduits with (constant) surface resistance. *Appl. Sci. Res. A* 14:157–171, 1964.

Tan, C. W. and C.-J. Hsu. Low Péclét number mass transfer in laminar flow through circular tubes. *Int. J. Heat Mass Transfer* 15: 2187–2201, 1972.

Taylor, G. Dispersion of soluble matter in solvent flowing slowly through a tube. *Proc. R. Soc. Lond. A* 219: 186–203, 1953.

Worsøe-Schmidt, P. M. Heat transfer in the thermal entrance region of circular tubes and annular passages with fully developed laminar flow. *Int. J. Heat Mass Transfer* 10: 541–551, 1967.

Problems

9-1. Heat Transfer in a Parallel-Plate Channel with Constant Wall Flux: Fully Developed Region

Consider fully developed flow in a parallel-plate channel in which there is a constant heat flux q_w at both surfaces for $x>0$. The plate spacing is $2H$. Show that the Nusselt number for large x is given by

$$\mathrm{Nu} = \frac{2hH}{k} = \frac{70}{17} = 4.118$$

as indicated in Fig. 9-6.

9-2. Heat Transfer in a Parallel-Plate Channel with Constant Wall Flux: Entrance Region

For the parallel-plate channel of Problem 9-1, derive an asymptotic expression for the Nusselt number valid for small x. (*Hint:* To obtain a similarity solution, define a new dependent variable as $Q \equiv \partial\Theta/\partial Y$, where Y is a dimensionless coordinate normal to the wall. This variable is constant at the walls for $x>0$, whereas Θ is not.)

9-3. Heat Transfer in a Tube with Constant Wall Flux: Entrance Region

For the situation described in Example 9.5-2, derive an asymptotic expression for Nu valid for small z. That is, derive Eq. (9.4-29). (*Hint:* Change the dependent variable as described in Problem 9-2.)

9-4. Absorption in a Falling Film with a Homogeneous Reaction

Consider a liquid film of thickness L flowing down a vertical surface, as shown in Fig. P9-4. The flow is laminar and fully developed. Beginning at $x=0$, the liquid contacts a gas containing species A, which dissolves in the liquid and undergoes an irreversible reaction. The liquid-phase concentration of species A at the gas–liquid interface ($y=0$) is a constant, C_{A0}. The solid wall is impermeable and unreactive.

(a) Assuming that the reaction is first order with rate constant k_1, derive an expression for the liquid-phase Sherwood number for species A, valid for large x. Examine the limiting behavior of Sh_A for slow and fast reactions.

(b) Repeat part (a) for a zeroth-order reaction with rate k_0.

Figure P9-4. Falling liquid film, with absorption of species A from the gas.

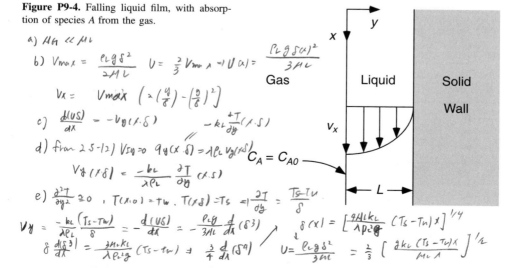

9-5. Heat Transfer in a Hydrodynamic Entrance Region for Pr≪1

Consider heat transfer in a tube in which the temperature and velocity profiles develop simultaneously and Pr≪1. The tube of radius R has a constant wall temperature T_w for $z>0$. Assume that the temperature and velocity are both uniform at $z=0$, such that $T=T_0$ and $v_z=U$. As suggested in the discussion of case 3 of Fig. 9-7, a good approximation is to set $v_z=U$ in the energy equation.

(a) Derive an expression for Nu which is valid when the temperature changes are confined near the tube wall. Approximately how small must z be for this result to be accurate?
(b) Repeat part (a) for z large enough that temperature changes occur over the entire tube cross section, but still small enough that $v_z \cong U$. Obtain an order-of-magnitude estimate of the range of z for which this result will be valid.

9-6. Absorption in a Falling Film without Reaction

Consider absorption in the falling liquid film described in Problem 9-4, but without any chemical reaction.

(a) Determine Sh_A in the liquid for $x \to 0$.
(b) Determine Sh_A in the liquid for $x \to \infty$, using the following information. For $\Theta(\eta, \zeta)$ governed by

$$(1-\eta^2)\frac{\partial \Theta}{\partial \zeta} = \frac{\partial^2 \Theta}{\partial \eta^2},$$

$$\Theta(\eta, 0)=0, \qquad \Theta(0, \zeta)=1, \qquad \frac{\partial \Theta}{\partial \eta}(1, \zeta)=0$$

the expansion for the dimensionless concentration is written as

$$\Theta(\eta, \zeta)=1-\sum_{n=1}^{\infty} b_n e^{-\lambda_n^2 \zeta} H(\lambda_n, \eta),$$

where $H(\lambda_n \eta)$ are rectangular Graetz functions. From Olbrich and Wild (1969), the first eigenvalue (squared) is given by

$$\lambda_1^2 = 5.1217.$$

[The answer for (b) is $Sh_A = k_c L/D_A = 3.414$.]

9-7. Effect of Axial Conduction on the Nusselt Number

The effect of the Péclet number on the temperature field and Nusselt number can be calculated most easily for the idealized case of plug flow in a parallel-plate channel. As shown in Fig. P9-7, assume that $v_x = U$ in a channel with plate spacing $2H$. At $x=0$ the temperature of both

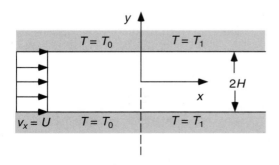

Figure P9-7. Plug flow in a parallel-plate channel with a step change in the wall temperature.

walls changes from T_0 to T_1. Assume that the channel extends long distances upstream and downstream from $x=0$. Let $\Theta = (T-T_1)/(T_0-T_1)$, $X=x/H$, $Y=y/H$, and $Pe=2HU/\alpha$.

(a) Assuming that Pe is not necessarily large, use the finite Fourier transform method to determine $\Theta(X, Y)$ for $X<0$. The variable change $\Theta_n(X)=\exp(Pe\ X/4)\Psi_n(X)$ will help in solving for the transformed temperature. Some of the constants remain undetermined at this stage (see below). Approximately how far upstream from $X=0$ is $\Theta(X, Y)$ affected?

(b) Repeat part (a) for $X>0$, and evaluate the remaining constants by matching the two solutions at $X=0$.

(c) Determine $\Theta(X, Y)$ for $Pe\to\infty$.

(d) Derive expressions for Nu for finite Pe and for $Pe\to\infty$. Show that $Nu\to\pi^2/2$ as $X\to\infty$ for any value of Pe. Compare plots of Nu versus X/Pe for $Pe=1$ and $Pe\to\infty$. (*Hint:* You should find that, as X/Pe becomes smaller, Nu is increasingly enhanced by axial conduction. It is sufficient to consider only $X>0$.)

9-8. Viscous Dissipation in a Couette Viscometer *

Consider the effects of viscous dissipation on the shear stress (τ_{yx}) measured in a Couette viscometer, when the dependence of the viscosity on temperature is taken into account. Let b be the gap width, U the speed of the surface at $y=b$, and T_0 the temperature of both surfaces. Assume that over some range of temperatures the viscosity is related to T according to

$$\mu = \frac{\mu_0}{1+a(T-T_0)}.$$

Note that for $a>0$, μ decreases with increasing T, as is usual for liquids. Assume that all other fluid properties are constant.

(a) Determine the temperature profile in terms of the shear stress and Brinkman number. It will be convenient to employ the following dimensionless temperature, dimensionless shear stress, and Brinkman number:

$$\Theta = \frac{T-T_0}{\mu_0 U^2/k}, \quad S = \frac{\tau_{yx}b}{\mu_0 U}, \quad Br = \frac{\mu_0 U^2 a}{k}.$$

(b) Use the result from (a) to determine the velocity profile in terms of S and Br.

(c) Find a relationship between S and Br that is valid for any Br, and then reduce this to a simpler expression for $Br\ll1$. As $Br\to0$, what is the limiting value of S? How large must Br be to cause S to deviate by 1% from the value it would have under perfectly isothermal conditions?

*This problem was suggested by R. C. Armstrong.

9-9. Temperature Rise for Adiabatic Flow in a Tube

Consider fully developed flow of a Newtonian fluid in a thermally insulated tube of radius R and length L. The mean velocity is U and all physical properties are constant. Use the viscous dissipation function to calculate the increase in bulk temperature between the inlet and outlet.

9-10. Taylor Dispersion in a Parallel-Plate Channel

Show that the Taylor dispersivity for laminar flow in a parallel-plate channel with plate spacing $2H$ is

$$K=D+\frac{2U^2H^2}{105D}.$$

9-11. Dispersion in a Tube Following a Step Change in Concentration

The results of the Taylor dispersion analysis for laminar flow in a tube can be applied to a variety of initial conditions, not just pulse inputs of solute. Consider a step change in solute concentration in a circular tube, such that at $t=0$, $C=C_0$ for $z<0$ and $C=0$ for $z>0$.

(a) Use the similarity method to determine $\overline{C}(z, t)$. [*Hint:* Based on the expected symmetry of the concentration profile, what is the value of $\overline{C}(0, t)$?]

(b) Calculate the length $L(t)$ over which 95% of the concentration change occurs at time t. An approach like this has been used to determine the extent of mixing in pipelines, where fluids of different composition are pumped in sequence and it is desired to predict how much fluid will be contaminated by mixing.

9-12. Effect of the Wall Conductance on the Nusselt Number in a Tube

Consider heat transfer in a circular tube with a finite wall conductance; the boundary condition at the wall is given by Eq. (9.5-39). The objective is to derive the results presented for tubes in Fig. 9-6.

(a) Derive an expression which relates Nu(∞) to B and λ_1, analogous to Eq. (9.5-14).

(b) Based on the results of Sideman et al. (1964), the smallest eigenvalues for selected values of B are:

B:	0.5	1	2	5	10	20	∞
λ_1:	1.27163	1.64125	1.99999	2.35665	2.51675	2.60689	2.70436

Use these results to calculate Nu(∞).

(c) Show that for any finite wall conductance, Nu will approach the *constant flux* result for $\zeta \rightarrow 0$.

9-13. Heat Transfer for a Power-Law Fluid in a Tube with a Constant Wall Flux

As shown in Example 6.2-2, for fully developed, laminar flow of a power-law fluid in a tube, the velocity profile is

$$v_z = \left(\frac{3n+1}{n+1}\right)U\left[1-\left(\frac{r}{R}\right)^{(n+1)/n}\right].$$

(a) Calculate Nu for large z, assuming a constant heat flux at the wall. Your result should reduce to Eq. (9.5-38) for $n=1$, the Newtonian case.

(b) Derive Nu(z) for small z, again for a constant-flux boundary condition. Here the result for $n=1$ is given by Eq. (9.4-29).

9-14. Melting of a Vertical Sheet of Ice

As shown in Fig. P9-14, assume that as a vertical sheet of ice melts it forms a film of water of thickness $L(x)$, which runs down the surface. Heat transfer from the water–air interface to the atmosphere is characterized by a heat transfer coefficient h (assumed constant). The ambient air temperature, T_A, is only slightly above the freezing temperature of water, T_F. The ice temperature is constant at T_F.

(a) Determine the mean water velocity, $U(x)$, in terms of $L(x)$, the density (ρ_w) and viscosity (μ_w) of water, and the gravitational acceleration. Assume that the fluid dynamic conditions are such that $v_x(x, y)$ is parabolic in y.

(b) Relate dL/dx to the local rate of melting, expressed as the water velocity $v_M(x)$ normal to the ice–water interface.

(c) Relate the local rate of melting to h, the given temperatures, the latent heat of fusion

$1 = ice$
$2 = water$
$n = v$

$\frac{q}{A}\bigg|_1 - \frac{q}{A}\bigg|_2 = \hat{\lambda}\,\rho_{\frac{1}{2}}^{\prime}\left(v_{n}\big|_2 - v_{n}\big|_1\right) \quad \rho_w\,\overline{v}_m\,\hat{\lambda} = k_w \frac{\partial T}{\partial \theta}\bigg|_{y=0} \quad \varkappa\left(\frac{d^2T}{dy^2}=0 \to T(y)=T_F + T(x,y)\right)$

$-k_w \frac{\partial T}{\partial \theta}\bigg|_{y=0} \quad \rho_w \quad v_m \quad D \qquad \rho_w\,v_m\,\hat{\lambda} = \frac{k_w}{L}\left(T(x,L)-T_F\right) = h\left(T_A - T_F\right)$

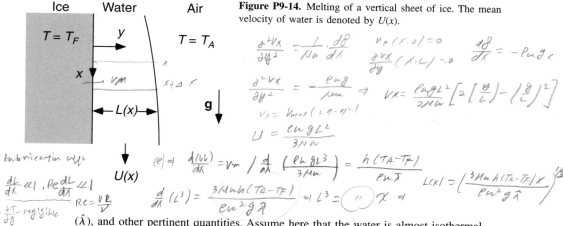

$V_m(x) \Delta x + U(\tau) L(x) = U(x+\Delta x) V(x+\Delta x)$

$= \frac{d(UL)}{dx} = V_m(x)$

Figure P9-14. Melting of a vertical sheet of ice. The mean velocity of water is denoted by $U(x)$.

Ice — $T = T_F$

Water — y, v_M, $\leftarrow L(x) \rightarrow$

Air — $T = T_A$, $g \downarrow$

$\dfrac{\partial^2 v_x}{\partial y^2} = -\dfrac{1}{\mu_w}\dfrac{d\delta}{dx}$ $\quad u_x(x,0)=0$ $\quad \dfrac{d\delta}{dx} = -\rho_w g c$

$\dfrac{\partial v_x}{\partial y}(x,L) = 0$

$\dfrac{\partial^2 v_x}{\partial y^2} = -\dfrac{\rho_w g}{\mu_w}$ $\Rightarrow$ $v_x = \dfrac{\rho_w g L^2}{2\mu_w}\left[2\left(\dfrac{y}{L}\right) - \left(\dfrac{y}{L}\right)^2\right]$

$v_x = v_{max}(2\eta - \eta^2)$

$U = \dfrac{\rho_w g L^2}{3\mu_w}$

$(\rho) \dfrac{d(UL)}{dx} = V_m \left/ \dfrac{d}{dx}\left(\dfrac{\rho_w g L^3}{3\mu_w}\right) = \dfrac{h(T_A - T_F)}{\rho_w \lambda}\right.$

$\dfrac{d}{dx}(L^3) = \dfrac{3\mu_w h(T_A-T_F)}{\rho_w^2 g \lambda} \Rightarrow L^3 = (\,\cdots\,) x \Rightarrow$ $L(x) = \left(\dfrac{3\mu_w h(T_A-T_F)/x}{\rho_w^2 g \lambda}\right)^{1/3}$

Re $= \dfrac{v L}{\nu}$

($\hat{\lambda}$), and other pertinent quantities. Assume here that the water is almost isothermal.
(d) For the conditions of parts (a) and (c), determine $U(x)$ and $L(x)$.
(e) What quantitative criteria must be satisfied for the approximations in parts (a) and (c) to be valid? Specifically, when will the velocity profile be parabolic in y? When can one neglect $\partial T/\partial y$ in the water? When can one neglect $\partial T/\partial x$ in the water, even if $\partial T/\partial y$ is not negligible?

$U(x) = \dfrac{\rho_w g}{3\mu_w}\left(\dfrac{3\mu_w h(T_A-T_F)x}{\rho_w^2 g \lambda}\right)^{2/3}$

9-15. Nusselt Number for a Parallel-Plate Channel with Viscous Dissipation

Consider a parallel-plate channel with the walls maintained at different temperatures. The inlet temperature of the fluid (at $x=0$) is T_0, the temperature at the bottom surface ($y=-H$) is T_1, and the temperature at the top surface ($y=H$) is T_2 ($>T_1$). Assume that the flow is fully developed and that the fluid is Newtonian with constant physical properties. The Brinkman number is not necessarily small.

(a) Determine the temperature profile for large x, including the effects of viscous dissipation.
(b) Evaluate Nu for the top surface as a function of Br. Do the same for the bottom surface.
(c) Show that under certain conditions Nu <0 at the hotter (top) surface. Explain why this happens, using a sketch of $T(y)$.

9-16. Condensation on a Vertical Wall

As shown in Fig. P9-16, a pure vapor at its saturation temperature T_s is brought into contact with a vertical wall maintained at some cooler temperature T_w. As a consequence, a film of condensate of thickness $\delta(x)$ is formed, which flows down the wall. The mean downward velocity of the liquid is denoted by $U(x)$. Far from the wall, the vapor is stagnant. The system is at steady state.

(a) Explain briefly why the rate of condensation will be essentially independent of the density (ρ_G), viscosity (μ_G), and thermal conductivity (k_G) of the gas.
(b) State the relationship between $U(x)$ and $\delta(x)$, assuming that the fluid-dynamic conditions are such that $v_x(x, y)$ in the liquid is parabolic in y.
(c) Relate $d\delta/dx$ to $U(x)$ and the local rate of condensation. The condensation rate is described by the liquid velocity $v_c(x)$ normal to the vapor–liquid interface.
(d) Use an energy balance to relate the local rate of condensation to the latent heat and other pertinent quantities.

air/water interface $-\dfrac{k_w}{L}[T(x,L) - T_F] = h[T(x,L) - T_A]$

$T(x,L) \cong T_F$

$$Nu_B = L\left(-\frac{\partial \theta}{\partial y}\Big|_{\theta=0}\right) \Big/ \frac{\theta/y=0-\theta_0}{1} = \left(\frac{3}{2\pi}\right)^{1/2} Pe_B^{1/2}\left(\frac{L}{x}\right)^{1/2}, \quad Pe_B = \frac{UL}{\alpha_B}$$

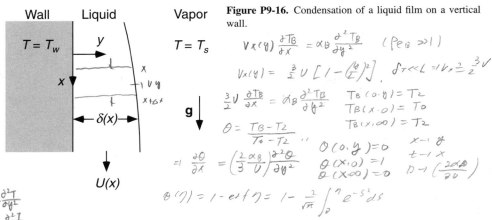

Figure P9-16. Condensation of a liquid film on a vertical wall.

$$V_x(y)\frac{\partial T_B}{\partial x} = \alpha_B \frac{\partial^2 T_B}{\partial y^2} \quad (Pe_B \gg 1)$$

$$V_x(y) = \frac{3}{2}U\left[1 - \left(\frac{y}{L}\right)^2\right], \quad \delta_T \ll L, \quad U_x = \frac{3}{2}V$$

$$\frac{3}{2}U\frac{\partial T_B}{\partial x} = \alpha_B \frac{\partial^2 T_B}{\partial y^2} \qquad T_B(0,y) = T_2$$
$$T_B(x,0) = T_0$$
$$T_B(x,\infty) = T_2$$

$$\theta = \frac{T_B - T_2}{T_0 - T_2}$$

$$=1 \quad \frac{\partial \theta}{\partial x} = \left(\frac{2\alpha_B}{3U}\right)\frac{\partial^2\theta}{\partial y^2}$$

$$\theta(0,y) = 0 \qquad x \to y$$
$$\theta(x,0) = 1 \qquad t \to x$$
$$\theta(x,\infty) = 0 \qquad D \to \left(\frac{2\alpha_B}{3U}\right)$$

$$\theta(\eta) = 1 - erf\,\eta = 1 - \frac{2}{\sqrt{\pi}}\int_0^\eta e^{-s^2}ds$$

(left margin handwritten:)
$$V_x\frac{\partial T}{\partial x} \ll \alpha_L \frac{\partial^2 T}{\partial y^2}$$
$$V_y\frac{\partial T}{\partial y} \ll \alpha_L \frac{\partial^2 T}{\partial y^2}$$
$$V_y \sim \frac{U\delta}{\lambda}$$
$$\frac{U\delta\,\Delta T}{x\,\delta} = \frac{U_0 T}{\lambda} \sim V_x\frac{\partial T}{\partial x}$$
$$\frac{V_0 T}{x} \ll \alpha_L\frac{\Delta T}{\delta^2}$$
$$\left(\frac{U\delta}{\alpha_L}\right)\left(\frac{\delta}{x}\right) \ll 1$$
$$Pe_{(L)}\frac{d\delta}{dx} \ll 1$$

(e) Assuming that the dominant type of energy transfer within the liquid is heat conduction across the film (i.e., in the y direction), evaluate $\delta(x)$.

(f) State order-of-magnitude criteria which are sufficient to justify the assumptions made in parts (b) and (e), concerning the parabolic velocity profile and the dominance of y conduction. Express these constraints in terms of δ and U. $\frac{d\delta}{dx} \ll 1$, $Pe_{(L)}\frac{d\delta}{dx} \ll 1$, $Pe_{(L)} = \frac{U_{(L)}\,\delta(x)}{V_L}$

9-17. Heat Transfer in Two-Phase Channel Flow

Immiscible fluids A and B, which are initially at different temperatures, are brought into contact in a parallel-plate channel beginning at $x=0$, as shown in Fig. P9-17. The channel walls are maintained at constant temperatures T_1 and T_2, and the initial (inlet) temperature of each fluid equals that of the adjacent wall. The fluids have the same density ($\rho_A = \rho_B = \rho$) and viscosity ($\mu_A = \mu_B = \mu$) but have different heat capacities ($\hat{C}_{pA} \neq \hat{C}_{pB}$), different thermal conductivities ($k_A \neq k_B$), and different thermal diffusivities ($\alpha_A \neq \alpha_B$). Assume that the system is at steady state, that the flow is fully developed for $x>0$, and that all of the thermophysical properties are constant. The mean velocity over the entire channel (U) is given and viscous dissipation is negligible. The temperature at the fluid–fluid interface ($y=0$) is denoted as $T_0(x)$.

(a) Determine the Nusselt number for heat transfer between the fluid–fluid interface and fluid B, for small x. Let $Nu_B \equiv h_B L/k_B$. (*Hint:* Assume that T_0 is constant at small x.)

(b) Evaluate T_0 for small x, confirming that T_0 is constant.

(c) Evaluate T_0 for large x. How does the result compare with that for small x?

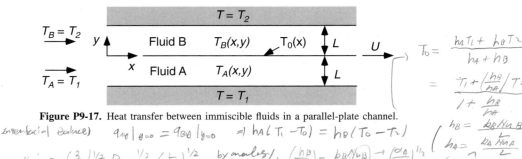

Figure P9-17. Heat transfer between immiscible fluids in a parallel-plate channel.

(handwritten below figure:)

$$\text{Interfacial balance)} \quad q_{Ay}\Big|_{\theta=0} = q_{B\theta}\Big|_{y=0} \quad =) \; h_A(T_1 - T_0) = h_B(T_0 - T_2)$$

$$To = \frac{h_A T_1 + h_B T_2}{h_A + h_B}$$

$$= \frac{T_1 + \left(\frac{h_B}{h_A}\right)T_2}{1 + \frac{h_B}{h_A}}$$

$$Nu_A = \left(\frac{3}{2\pi}\right)^{1/2} Pe_A^{1/2}\left(\frac{L}{x}\right)^{1/2} \quad \text{by analogy}, \quad \left(\frac{h_B}{h_A}\right) = \frac{k_B/Nu_B}{k_A/Nu_A} = \left(\frac{\alpha_A}{\alpha_B}\right)^{1/2}$$

$$\left(\frac{h_B}{h_A} = \frac{k_B Nu_B}{L}\right)$$
$$\left(h_A = \frac{k_A Nu_A}{L}\right)$$

$$To = \frac{T_1 + \beta\,T_2}{1 + \beta}, \quad \beta = \left(\frac{k_B\,\hat{C}_{pB}}{k_A\,\hat{C}_{pA}}\right)^{1/2} \qquad = \left(\frac{k_B\,\hat{C}_{pB}}{k_A\,\hat{C}_{pA}}\right)^{1/2}$$

$$\text{for large } x =) \quad T_A = T_A(\theta) \qquad T_A = T_0 + (T_0 - T_1)\frac{y}{L}$$
$$T_B = T_B(y) \qquad T_B = T_0 + (T_2 - T_0)\frac{y}{L}$$

$$q_{Ay} = \frac{k_B(T_B - T_0)}{L} = \frac{k_B(T_0 - T_2)}{L} = q_{By}$$

$$To = \frac{k_A\,T_1 + k_B\,T_2}{k_A + k_B}$$

Chapter 10

FORCED-CONVECTION HEAT AND MASS TRANSFER IN UNCONFINED LAMINAR FLOWS

10.1 INTRODUCTION

The analysis of forced-convection heat and mass transfer begun in Chapter 9 is extended here to unconfined flows. Several examples of unconfined flows were discussed in Chapters 7 and 8. The prototypical situation involves a uniform fluid stream passing a stationary object, and the objective is to characterize the rate of heat or mass transfer from the object to the surrounding fluid. The analysis of heat and mass transfer in this type of situation is inherently more complicated than in confined flows, because there is no analog to the fully developed flow which occurs in tubes. That is, the velocity fields in unconfined flows are always at least two-dimensional. The analytical problems are mitigated somewhat if there are thermal or concentration boundary layers. These occur for $Pe \gg 1$, just as momentum boundary layers occur for $Re \gg 1$. Fortunately, heat and mass transfer at large Péclet number is also the situation of greatest practical importance. Thus, this chapter is concerned almost exclusively with boundary layers.

To obtain a thermal or concentration boundary layer it is not necessary that there be a momentum boundary layer. That is, $Pe = Re \, Pr$ (or $Pe = Re \, Sc$) can be large even if Re is small or moderate. To emphasize this independence, thermal boundary layers are introduced in the context of creeping flow past spheres. Of course, momentum boundary layers will often coexist with thermal (or concentration) boundary layers, and these situations are discussed next. The remainder of the chapter is devoted to certain generalizations which can be made concerning the functional forms of the Nusselt and Sherwood numbers, based on scaling concepts. These "scaling laws" provide useful guidance even for situations where analytical or numerical solutions of the differential equations are impractical.

411

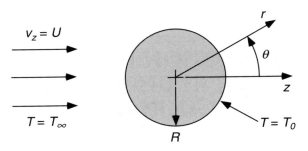

Figure 10-1. Flow past a solid sphere maintained at constant temperature, with uniform fluid temperature and velocity far from the sphere.

10.2 HEAT AND MASS TRANSFER IN CREEPING FLOW

Heat or mass transfer from a solid sphere to a fluid at low Reynolds number is pertinent to systems involving small particles, such as suspensions and aerosols, and also provides an introduction to the key features of thermal or concentration boundary layers. The two examples in this section both involve the situation illustrated in Fig. 10-1. Fluid with an approach velocity U and temperature T_∞ flows past a stationary sphere of radius R and constant temperature T_0. Assuming that $\mathrm{Re} = UR/\nu \ll 1$, we will calculate the temperature field and average Nusselt number for both small and large Péclet number. Given that $\mathrm{Pe} = \mathrm{Re}\,\mathrm{Pr}$, for many fluids a small Péclet number will result automatically from the small Reynolds number. However, a large Péclet number can also be achieved in creeping flow, provided that the Prandtl number ($\mathrm{Pr} = \nu/\alpha$) is sufficiently large. The analogous situation in mass transfer is quite common, in that the Schmidt number ($\mathrm{Sc} = \nu/D$) in liquids is always large.

We begin with the basic problem formulation that is common to both examples. The initial set of dimensionless quantities is defined as

$$\xi \equiv \frac{r}{R}, \qquad \Theta \equiv \frac{T - T_\infty}{T_0 - T_\infty}, \qquad \mathrm{Pe} \equiv \frac{UR}{\alpha}. \tag{10.2-1}$$

For this axisymmetric situation $\Theta = \Theta(\xi, \theta)$, and the energy equation and boundary conditions are

$$\mathrm{Pe}\ \tilde{\mathbf{v}} \cdot \tilde{\nabla}\Theta = \tilde{\nabla}^2\Theta, \tag{10.2-2}$$

$$\Theta(1, \theta) = 1, \qquad \Theta(\infty, \theta) = 0, \tag{10.2-3}$$

where $\tilde{\nabla}$ and $\tilde{\nabla}^2$ are operators involving only ξ and θ. As shown in Example 7.4-2, the velocity components for creeping flow are given by

$$\tilde{v}_r = \frac{v_r}{U} = \cos\theta \left[1 - \frac{3}{2\xi} + \frac{1}{2\xi^3}\right], \tag{10.2-4}$$

$$\tilde{v}_\theta = \frac{v_\theta}{U} = -\sin\theta \left[1 - \frac{3}{4\xi} - \frac{1}{4\xi^3}\right]. \tag{10.2-5}$$

Example 10.2-1 Nusselt Number for a Sphere with Pe≪1 When Pe is small we expect convection to have little effect on the temperature profile in the fluid, conduction being the dominant mechanism for heat transfer. Equation (10.2-2) suggests that

$$\tilde{\nabla}^2 \Theta = O(\text{Pe}) \qquad (\text{Pe} \to 0). \tag{10.2-6}$$

The θ-dependent velocity terms are absent from this equation at $O(1)$, and the boundary conditions involve only ξ. Thus, we conclude that for $\text{Pe} \to 0$ the temperature field is spherically symmetric, to first approximation. With $\Theta = \Theta(\xi)$ only, the $O(1)$ part of Eq. (10.2-6) gives

$$\frac{d}{d\xi}\left(\xi^2 \frac{d\Theta}{d\xi}\right) = 0. \tag{10.2-7}$$

The solution that satisfies the boundary conditions is simply

$$\Theta = \frac{1}{\xi}. \tag{10.2-8}$$

The Nusselt number based on the sphere diameter is then

$$\text{Nu} = \frac{2hR}{k} = -2 \left.\frac{\partial \Theta}{\partial \xi}\right|_{\xi=1} = 2 \qquad (\text{Pe} \to 0), \tag{10.2-9}$$

where the temperature driving force used with the heat transfer coefficient is $T_0 - T_\infty$. This result, along with its analog for mass transfer ($\text{Sh} = 2$), is widely used in transport calculations involving small particles. It is easily confirmed that the same result is obtained for a constant-flux boundary condition. For $\text{Pe} \to 0$ the heat or mass transfer coefficient is uniform over the surface, so that the average Nusselt number ($\overline{\text{Nu}}$) for either boundary condition is the same as the local value.

Results for $\overline{\text{Nu}}$ or $\overline{\text{Sh}}$ as $\text{Pe} \to 0$ are obtainable, in principle, for any geometric shape which allows solution of the steady conduction or diffusion problem in an unbounded fluid. Values for several shapes, using constant-temperature boundary conditions, are shown in Table 10-1. It is seen that $\overline{\text{Nu}}$ or $\overline{\text{Sh}}$ for various compact objects is similar to that for spheres.

When Pe is small but nonzero, Eq. (10.2-9) will not be exact. It appears from Eq. (10.2-6) that we could obtain corrections to this limiting expression by expanding the temperature field as a regular perturbation series in Pe. However, as shown by Acrivos and Taylor (1962), the perturbation problem turns out to be singular. The difficulty which arises is analogous to that encountered in seeking corrections to Stokes' law, and the singular perturbation analysis parallels that discussed for the hydrodynamic problem in Section 7.7. The heat transfer problem is made singular by the fact that for arbitrarily small but nonzero values of Pe, there is always a region far from the sphere where the convection terms in the energy equation are comparable to the conduction terms. Thus, the scaling assumed in Eq. (10.2-6) is valid only in a region not too far from the sphere. As with the derivation of Stokes' law, the singularity does not affect the first term in the solution near the sphere, so that Eq. (10.2-9) is correct as a first approximation. The result for $\overline{\text{Nu}}$ obtained by Acrivos and Taylor (1962) is

TABLE 10-1
Nusselt and Sherwood Numbers for Various Shapes as $\text{Pe} \to 0$

Shape	L	$\overline{\text{Nu}} = \overline{h}L/k$ or $\overline{\text{Sh}} = \overline{k}_c L/D$	Reference
Sphere	$2R$	2	
Oblate spheroid[a] ($a/b = 1/2$)	$2b$	2.40	Skelland and Cornish (1963)
Circular disk	$2R$	$8/\pi = 2.55$	Skelland and Cornish (1963)
Square plate[b]	a	$\pi/\ln 4 = 2.27$	Kutateladze (1963)

[a] The minor axis is $2a$ and the major axis is $2b$. The reference gives an expression for $0 \le a/b \le 1$.
[b] The length of a side is a. The reference gives a general expression for a rectangular plate.

$$\overline{Nu} = 2\left[1 + \frac{1}{4}Pe_d + \frac{1}{8}Pe_d^2 \ln Pe_d + 0.01702 \ Pe_d^2 + \frac{1}{32}Pe_d^3 \ln Pe_d + O(Pe_d^3)\right], \qquad (10.2\text{-}10)$$

where $Pe_d = 2UR/\alpha$ is the Péclet number based on the *diameter* of the sphere. Interestingly, Acrivos and Taylor (1962) showed that the first two terms in Eq. (10.2-10) remain the same for any value of Re, provided that the flow is laminar.

Example 10.2-2 Nusselt Number for a Sphere with Pe→∞   We turn now to the other limiting case, Pe≫1. If we try to neglect the right-hand side of Eq. (10.2-2) for large Pe, we obtain

$$\tilde{\mathbf{v}} \cdot \tilde{\nabla}\Theta = 0. \qquad (10.2\text{-}11)$$

This may seem at first to be an attractive simplification, but its shortcomings soon become evident. Mathematically, by omitting the conduction terms we have reduced the order of the differential equation and have thereby lost the ability to satisfy one of the temperature boundary conditions. Physically, there is no longer any mechanism for heat transfer from the sphere to the fluid. Indeed, recalling that for steady conditions the quantity $\tilde{\mathbf{v}} \cdot \tilde{\nabla}\Theta$ represents the rate of change of Θ along a streamline, we see that Eq. (10.2-11) implies that all fluid which is at $\Theta = 0$ (i.e., $T = T_\infty$) far upstream must remain at that temperature as it passes the sphere. In other words, the entire body of fluid is predicted to remain at the approach temperature.

The types of difficulties encountered with Eq. (10.2-11) are familiar. Attempts to neglect heat conduction (or species diffusion) throughout a fluid at large Pe are analogous to attempts to neglect viscous stresses throughout a fluid at large Re (Chapter 8). Both situations lead to singular perturbation problems, in which separate approximations must be developed for boundary layer and outer regions. For the present heat transfer problem, we conclude that both conduction and convection are important in a boundary layer next to the sphere. The boundary layer becomes thinner as Pe increases, such that $\tilde{\nabla}^2\Theta = O(Pe)$ in this region as Pe→∞ [see Eq. (10.2-2)]. The bulk of the fluid constitutes the outer region, in which convection is dominant and Eq. (10.2-11) is a good approximation. Thus, $\Theta = 0$ in the outer region, to first approximation.

For the analysis of the thermal boundary layer, we write the operators in Eq. (10.2-2) in spherical coordinates to obtain

$$\tilde{v}_r \frac{\partial \Theta}{\partial \xi} + \frac{\tilde{v}_\theta}{\xi}\frac{\partial \Theta}{\partial \theta} = Pe^{-1}\left\{\frac{\partial^2 \Theta}{\partial \xi^2} + \frac{2}{\xi}\frac{\partial \Theta}{\partial \xi} + \frac{1}{\xi^2 \sin \theta}\frac{\partial}{\partial \theta}\left(\sin \theta \frac{\partial \Theta}{\partial \theta}\right)\right\}. \qquad (10.2\text{-}12)$$

As was the case for certain conduction problems in spherical coordinates in Chapter 4, it is advantageous to define a new angular coordinate as

$$\eta \equiv \cos \theta, \qquad (10.2\text{-}13)$$

which varies from -1 to 1. Evaluating $\tilde{v}_r$ and $\tilde{v}_\theta$ using Eqs. (10.2-4) and (10.2-5) and converting θ to η, Eq. (10.2-12) becomes

$$\eta\left(1 - \frac{3}{2\xi} + \frac{1}{2\xi^3}\right)\frac{\partial \Theta}{\partial \xi} + \frac{(1 - \eta^2)}{\xi}\left(1 - \frac{3}{4\xi} - \frac{1}{4\xi^3}\right)\frac{\partial \Theta}{\partial \eta} = Pe^{-1}\left\{\frac{\partial^2 \Theta}{\partial \xi^2} + \frac{2}{\xi}\frac{\partial \Theta}{\partial \xi}\right.$$
$$\left. + \frac{1}{\xi^2}\frac{\partial}{\partial \eta}\left[(1 - \eta^2)\frac{\partial \Theta}{\partial \eta}\right]\right\}. \qquad (10.2\text{-}14)$$

Major simplifications in Eq. (10.2-14) are achieved next by exploiting the fact that the boundary layer is thin. To resolve the scaling problem discussed above, we define a new radial coordinate for the boundary layer as

$$Y \equiv (\xi - 1)Pe^b, \qquad (10.2\text{-}15)$$

where the constant b is to be determined. Bearing in mind that Pe is large and that it will be necessary to stretch the original coordinate, it must turn out that $b > 0$. The velocity terms in Eq. (10.2-14) are simplified by expanding the coefficients involving ξ as polynomials in x, where $x \equiv \xi - 1$. To do this, we first notice that

$$\xi^{-n} = [1 + (\xi - 1)]^{-n} = (1 + x)^{-n}. \tag{10.2-16}$$

A standard formula for binomial series is

$$(1 + x)^{-n} = 1 - nx + \frac{n(n + 1)}{2!} x^2 + O(x^3). \tag{10.2-17}$$

Because $x \ll 1$ in the boundary layer, only a few terms are needed in any such series. Taking the radial velocity term as an example, we obtain

$$1 - \frac{3}{2\xi} + \frac{1}{2\xi^3} = 1 - \frac{3}{2}[1 - x + x^2] + \frac{1}{2}[1 - 3x + 6x^2] + O(x^3) = \frac{3}{2}x^2 + O(x^3). \tag{10.2-18}$$

Doing the same for the angular velocity term and converting ξ to Y, Eq. (10.2-14) becomes

$$\left[\frac{3}{2}Y^2 \text{Pe}^{-2b} + O(\text{Pe}^{-3b})\right] \eta \text{ Pe}^b \frac{\partial \Theta}{\partial Y} + \frac{(1 - \eta^2)}{(1 + Y\text{Pe}^{-b})}\left[\frac{3}{2}Y\text{Pe}^{-b} + O(\text{Pe}^{-2b})\right]\frac{\partial \Theta}{\partial \eta}$$

$$= \text{Pe}^{-1}\left[\text{Pe}^{2b} \frac{\partial^2 \Theta}{\partial Y^2} + \frac{2\text{Pe}^b}{(1 + Y \text{ Pe}^{-b})} \frac{\partial \Theta}{\partial Y}\right] + \text{Pe}^{-1}\left[\frac{1}{(1 + Y \text{ Pe}^{-b})^2} \frac{\partial}{\partial \eta}\left((1 - \eta^2) \frac{\partial \Theta}{\partial \eta}\right)\right]. \tag{10.2-19}$$

It is seen in the top line that both convection terms are $O(\text{Pe}^{-b})$. The dominant conduction term, that involving $\partial^2\Theta/\partial Y^2$, is $O(\text{Pe}^{2b-1})$. Maintaining a balance between convection and conduction as $\text{Pe} \to \infty$ therefore requires that $-b = 2b - 1$, or $b = \frac{1}{3}$. The required stretching factor of $\text{Pe}^{1/3}$ for the radial coordinate implies that the thickness of the thermal boundary layer varies as $\text{Pe}^{-1/3}$.

Setting $b = \frac{1}{3}$ and retaining only the dominant terms, Eq. (10.2-19) simplifies to

$$\frac{3}{2}Y^2\eta \frac{\partial \Theta}{\partial Y} + \frac{3}{2}Y(1 - \eta^2) \frac{\partial \Theta}{\partial \eta} = \frac{\partial^2 \Theta}{\partial Y^2} + O(\text{Pe}^{-1/3}). \tag{10.2-20}$$

In going from Eq. (10.2-19) to Eq. (10.2-20), conduction parallel to the surface was neglected, as was the effect of the surface curvature on conduction normal to the surface. Those effects are $O(\text{Pe}^{-1/3})$, as indicated, so they will not influence the first term in the expansion for the boundary layer. Neglecting surface curvature and conduction parallel to the surface is valid for thermal boundary layers in general, when seeking only an $O(1)$ approximation. The boundary conditions for $\Theta(Y, \eta)$ are

$$\Theta(0, \eta) = 1, \qquad \Theta(\infty, \eta) = 0. \tag{10.2-21}$$

We can deduce the dependence of Nu on Pe before solving for Θ. Of greatest interest is the average Nusselt number, which is given by

$$\overline{\text{Nu}} = \frac{\int_0^\pi \text{Nu}(\theta)\sin\theta \, d\theta}{\int_0^\pi \sin\theta \, d\theta} = \frac{1}{2}\int_{-1}^1 \left(-2\frac{\partial \Theta}{\partial \xi}\right)_{\xi=1} d\eta \tag{10.2-22}$$

$$= \left\{\int_{-1}^1 \left(-\frac{\partial \Theta}{\partial Y}\right)_{Y=0} d\eta\right\}\text{Pe}^{1/3} \equiv C \text{ Pe}^{1/3}$$

Thus, $\overline{\text{Nu}}$ varies as $\text{Pe}^{1/3}$, because the boundary layer thickness varies as $\text{Pe}^{-1/3}$. To evaluate the proportionality constant C we must solve for the temperature field.

To determine the temperature, a similarity variable is defined as

$$s \equiv \frac{Y}{g(\eta)}.$$ (10.2-23)

Making the variable changes in Eq. (10.2-20), the energy equation becomes

$$\frac{d^2\Theta}{ds^2} + s^2 \left\{ \frac{3}{2}(1-\eta^2)g^2 \frac{dg}{d\eta} - \frac{3}{2}g^3\eta \right\} \frac{d\Theta}{ds} = 0.$$ (10.2-24)

To eliminate η from Eq. (10.2-24), it is necessary that the term in brackets be a constant. Choosing that constant as 3, the differential equation for $\Theta(s)$ is

$$\frac{d^2\Theta}{ds^2} + 3s^2 \frac{d\Theta}{ds} = 0.$$ (10.2-25)

The boundary conditions in Eq. (10.2-21) are transformed to

$$\Theta(0) = 1, \qquad \Theta(\infty) = 0.$$ (10.2-26)

The solution for the dimensionless temperature is

$$\Theta(s) = \frac{\int_s^\infty e^{-t^3} dt}{\int_0^\infty e^{-t^3} dt} = \frac{1}{\Gamma(4/3)} \int_s^\infty e^{-t^3} dt,$$ (10.2-27)

where $\Gamma(4/3) = 0.8930$.

The next step is to determine the scale factor, $g(\eta)$. Noting that $3g^2 \, dg/d\eta = d(g^3)/d\eta$, Eqs. (10.2-24) and (10.2-25) imply that g^3 satisfies

$$(1-\eta^2) \frac{d}{d\eta}(g^3) - 3\eta g^3 = 6.$$ (10.2-28)

The initial condition which g^3 (or g) must satisfy is not obvious. What we will require is that the boundary layer thickness be *finite* at the upstream stagnation point ($\theta = \pi$ or $\eta = -1$). This is based on the reasoning that temperature disturbances should not be able to propagate long distances directly upstream for $\text{Pe} \to \infty$. Because the boundary layer thickness is proportional to $g(\eta)$, this implies that $g(-1)$ is finite. The solution to Eq. (10.2-28) that satisfies this requirement is

$$g(\eta) = \frac{6^{1/3}}{(1-\eta^2)^{1/2}} \left[\int_{-1}^{\eta} (1-t^2)^{1/2} \, dt \right]^{1/3}.$$ (10.2-29)

This completes the solution for the temperature. It is interesting to note that although $g(-1)$ is finite, there is no restriction on the magnitude of $g(1)$. Indeed, Eq. (10.2-29) indicates that $g(1) = \infty$, so that for large Pe there is a thermal wake of indefinite length.

Having determined $\Theta(s)$, the constant C in Eq. (10.2-22) is evaluated as

$$C = \int_{-1}^{1} \left(-\frac{d\Theta}{ds} \right)_{s=0} \frac{1}{g} \, d\eta \doteq 1.249.$$ (10.2-30)

The resulting expression for $\overline{\text{Nu}}$, using the Péclet number based on the sphere radius (Pe), is

$$\overline{\text{Nu}} = 1.249 \text{Pe}^{1/3} \qquad (\text{Pe} \to \infty).$$ (10.2-31)

Using the more conventional Péclet number based on the sphere diameter (Pe_d), the result is

$$\overline{\text{Nu}} = 0.991 \text{Pe}_d^{1/3} \qquad (\text{Pe}_d \to \infty),$$ (10.2-32)

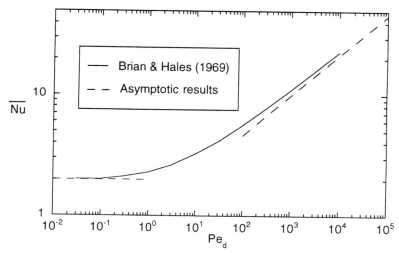

Figure 10-2. Average Nusselt number for heat transfer to a sphere in creeping flow. The dashed lines are from Eqs. (10.2-9) and (10.2-32), and the solid curve is based on the finite-difference results of Brian and Hales (1969).

which was derived first by Levich (1962, pp. 80–85) (although with a coefficient of 1.01). An additional term in the expansion for $\overline{Nu}$ at large Pe was obtained by Acrivos and Goddard (1965). Again using quantities based on the sphere diameter, their result is written as

$$\overline{Nu} = Pe_d^{1/3}[0.991 + 0.922\,Pe_d^{-1/3} + O(Re) + O(Pe_d^{-1/3})].\tag{10.2-33}$$

The Nusselt numbers calculated from Eqs. (10.2-9) and (10.2-32), together with numerical results obtained by Brian and Hales (1969) for intermediate values of Pe_d, are shown in Fig. 10-2. It is seen that the overall behavior of $\overline{Nu}$ is revealed fairly well by the asymptotic results alone.

One of the features of creeping flow mentioned in Chapter 7 is that the drag coefficient for an object of arbitrary shape is unchanged if the flow direction is reversed. The same is true for the average Nusselt or Sherwood number (Brenner, 1970).

It should be noted that measured values of $\overline{Nu}$ or $\overline{Sh}$ for spheres are often larger than those predicted by the creeping-flow results discussed here, as a consequence of buoyancy-induced flow. As discussed in Chapter 12, either temperature or concentration gradients may cause variations in fluid density which lead to such flow. At low Reynolds number, especially, this additional source of convection can increase the heat or mass transfer coefficient significantly. A key parameter in buoyancy-driven flow (free convection) is the Grashof number (Gr) for heat transfer or mass transfer (Section 12.2). Data for solid spheres in liquids (where $Sc \gg 1$) suggest that free convection is negligible when $Gr/(Re^2Sc^{1/3}) < 6$ (Garner and Keey, 1958).

10.3 HEAT AND MASS TRANSFER IN LAMINAR BOUNDARY LAYERS

For heat or mass transfer to or from a submerged object, a thermal or concentration boundary layer will result if the Péclet number is large. As shown in Section 10.2, this is true even if the Reynolds number is small. In this section we focus on situations in which Re and Pe are *both* large, so that momentum and thermal (or momentum and concentration) boundary layers coexist. It will be seen that the relative thicknesses of the two boundary layers, which are determined by the Prandtl number (or Schmidt number), are crucial for understanding how the boundary layers interact. The discussion here is restricted to steady, two-dimensional flows.

If the density and viscosity of a fluid are assumed to be constant, then the velocity in the momentum boundary layer is governed by the equations derived in Section 8.2. To briefly restate what was found there, the (unscaled) dimensionless quantities are defined as

$$\tilde{x} \equiv \frac{x}{L}, \quad \tilde{y} \equiv \frac{y}{L}, \quad \tilde{v}_x \equiv \frac{v_x}{U}, \quad \tilde{v}_y \equiv \frac{v_y}{U}, \quad \tilde{u} \equiv \frac{u}{U}, \quad \tilde{\nabla} \equiv L\nabla, \quad \tilde{\nabla}^2 \equiv L^2\nabla^2,$$

$$(10.3\text{-}1)$$

where L is the characteristic length (e.g., a representative dimension of a submerged object), U is the characteristic velocity (e.g., the approach velocity of the fluid), and $u(x)$ is the velocity from the "outer" or inviscid flow problem evaluated at the surface. With Re $\gg 1$, the equations for the momentum boundary layer are

$$\frac{\partial \tilde{v}_x}{\partial \tilde{x}} + \frac{\partial \tilde{v}_y}{\partial \tilde{y}} = 0, \tag{10.3-2}$$

$$\left(\text{from } 8.2\text{-}27\right) \qquad \tilde{v}_x \frac{\partial \tilde{v}_x}{\partial \tilde{x}} + \tilde{v}_y \frac{\partial \tilde{v}_x}{\partial \tilde{y}} = \tilde{u}\frac{d\tilde{u}}{d\tilde{x}} + \text{Re}^{-1} \frac{\partial^2 \tilde{v}_x}{\partial \tilde{y}^2}. \tag{10.3-3}$$

Local rectangular coordinates are employed, and in Eq. (10.3-3) the second derivative in the x direction has been neglected. Those approximations, as well as the way in which the pressure gradient is related to u, are justified by the fact that the boundary layer is thin.

As discussed in Example 10.2-2, "thinness" approximations analogous to those in Eq. (10.3-3) apply to the energy equation when Pe $\gg 1$. Thus, the temperature in the thermal boundary layer is governed by

$$\tilde{v}_x \frac{\partial \Theta}{\partial \tilde{x}} + \tilde{v}_y \frac{\partial \Theta}{\partial \tilde{y}} = \text{Pe}^{-1} \frac{\partial^2 \Theta}{\partial \tilde{y}^2} = \text{Re}^{-1} \text{Pr}^{-1} \frac{\partial^2 \Theta}{\partial \tilde{y}^2}. \tag{10.3-4}$$

All terms which act as volumetric energy sources, including viscous dissipation, are assumed here to be negligible; such terms may be added to the right-hand side, if necessary. The dimensionless form of the species conservation equation for a concentration boundary layer is identical to Eq. (10.3-4), except that Sc replaces Pr. In that case, a source term will be needed if there is a homogeneous chemical reaction.

The presence in Eq. (10.3-3) of Re, a large parameter, indicates that the variables defined in Eq. (10.3-1) are not properly scaled for a momentum boundary layer. It was shown in Section 8.2 that the *scaled* coordinate and velocity in the y direction are

$$\hat{y} \equiv \tilde{y}\,\text{Re}^{1/2}, \quad \hat{v}_y \equiv \tilde{v}_y\,\text{Re}^{1/2}. \tag{10.3-5}$$

In terms of these variables, the governing equations for the boundary layers become

$$\frac{\partial \tilde{v}_x}{\partial \tilde{x}} + \frac{\partial \hat{v}_y}{\partial \hat{y}} = 0, \tag{10.3-6}$$

$$\tilde{v}_x \frac{\partial \tilde{v}_x}{\partial \tilde{x}} + \hat{v}_y \frac{\partial \tilde{v}_x}{\partial \hat{y}} = \bar{u} \frac{d\bar{u}}{d\tilde{x}} + \frac{\partial^2 \tilde{v}_x}{\partial \hat{y}^2},$$ (10.3-7)

$$\tilde{v}_x \frac{\partial \Theta}{\partial \tilde{x}} + \hat{v}_y \frac{\partial \Theta}{\partial \hat{y}} = \mathrm{Pr}^{-1} \frac{\partial^2 \Theta}{\partial \hat{y}^2}.$$ (10.3-8)

It is seen that with this scaling, the Reynolds number is eliminated from the momentum and energy equations. The only parameter which remains is the Prandtl number. If $\mathrm{Pr} \sim 1$, then the variables are adequately scaled not just for the momentum boundary layer, but also for the thermal boundary layer. Put another way, the thicknesses of the thermal and momentum boundary layers are roughly the same when $\mathrm{Pr} \sim 1$.

To illustrate the extent to which the two types of boundary layers may be similar, assume that $\mathrm{Pr} = 1$ exactly, in which case

$$\tilde{v}_x \frac{\partial \Theta}{\partial \tilde{x}} + \hat{v}_y \frac{\partial \Theta}{\partial \hat{y}} = \frac{\partial^2 \Theta}{\partial \hat{y}^2}.$$ (10.3-9)

If $\Theta = (T - T_0)/(T_\infty - T_0)$, where T_0 is the (constant) temperature at the surface of some object and T_∞ is the fluid temperature far away, the boundary conditions for $\Theta(\tilde{x}, \hat{y})$ become

$$\Theta(0, \hat{y}) = 1, \qquad \Theta(\tilde{x}, 0) = 0, \qquad \Theta(\tilde{x}, \infty) = 1, \qquad (10.3\text{-}10)$$

where $\tilde{x} = 0$ corresponds to the leading edge (or upstream stagnation point) of the object. Notice now that if we replace Θ in Eq. (10.3-9) by $\tilde{v}_x$, we obtain the differential equation which governs $\tilde{v}_x$ for flow past a flat plate (see Example 8.4-1). Likewise, replacing Θ by $\tilde{v}_x$ in Eq. (10.3-10) yields the fluid-dynamic boundary conditions for a flat plate. Because Θ and $\tilde{v}_x$ satisfy the same differential equation and boundary conditions in this circumstance, we conclude that

$$\Theta = \tilde{v}_x \qquad \text{(isothermal flat plate, } \mathrm{Pr} = 1\text{)}. \qquad (10.3\text{-}11)$$

Obviously, then, the thermal and momentum boundary layer thicknesses are identical for this special case. The analogy between the temperature and velocity fields is inexact for other geometries. That is because the pressure gradient is not zero (i.e., $u \, du/dx \neq 0$), and there is no corresponding term in the energy equation. Nonetheless, the thermal and momentum boundary layer thicknesses will be similar in any system where $\mathrm{Pr} \sim 1$.

When Pr is very large or small, the thicknesses of the momentum boundary layer (δ_M) and thermal boundary layer (δ_T) are no longer comparable. In Chapter 8 it was shown that at a given position on a surface, δ_M increases with the kinematic viscosity; more precisely, $\delta_M \propto \nu^{1/2}$ for Newtonian fluids. Similarly, at a given position, δ_T should increase with α. Suppose now that it were possible to vary α without affecting ν. This would make δ_T larger or smaller, without affecting δ_M. As we have just seen, $\delta_M \sim \delta_T$ when $\mathrm{Pr} \sim 1$ or $\alpha \sim \nu$. Thus, we expect that

$$\left| \frac{\delta_M}{\delta_T} \ll 1 \qquad \text{for } \mathrm{Pr} \ll 1, \right. \qquad \left(\mathrm{Pr} = \frac{\nu}{\alpha} \right) \qquad (10.3\text{-}12)$$

$$\left| \frac{\delta_M}{\delta_T} \gg 1 \qquad \text{for } \mathrm{Pr} \gg 1. \right. \qquad (10.3\text{-}13)$$

Temperature and velocity profiles corresponding to these two situations are shown qualitatively in Fig. 10-3. For very small Pr (e.g., a liquid metal), most of the thermal bound-

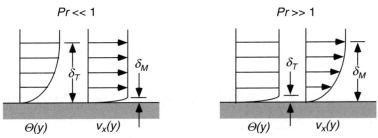

Figure 10-3. Temperature and velocity profiles in boundary layers next to a solid surface, for small or large values of the Prandtl number.

ary layer is in the momentum outer region; for very large Pr (e.g., a heavy lubricating oil), the thermal boundary layer occupies only a small fraction of the momentum boundary layer.

For mass transfer in boundary layers, the relationship analogous to Eq. (10.3-13) is

$$\frac{\delta_M}{\delta_C} \gg 1 \qquad \text{for } \text{Sc} \gg 1, \tag{10.3-14}$$

where δ_C is the thickness of the concentration boundary layer. Because Sc is always large in liquids, the concentration boundary layer is always much thinner than the momentum boundary layer. There is no mass-transfer analog of Eq. (10.3-12) because there are no fluids with $\text{Sc} \ll 1$. The smallest Schmidt numbers are for gases, where $\text{Sc} \sim 1$ (Chapter 1).

For extreme values of Pr it is possible to simplify the velocity terms in the energy equation. Suppose first that $\text{Pr} \ll 1$, in which case most of the thermal boundary layer is in the momentum outer region. This suggests that the details of the velocity profile in the momentum *boundary layer* will have little effect on the temperature and that we can approximate the convective terms using the velocity in the momentum *outer region*. Accordingly, we set

$$\tilde{v}_x = \tilde{u}(x), \tag{10.3-15}$$

$$\hat{v}_y = -\int_0^{\hat{y}} \frac{\partial \tilde{v}_x}{\partial \tilde{x}} \, d\hat{y}' = -\frac{d\tilde{u}}{d\tilde{x}} \hat{y}. \tag{10.3-16}$$

Equation (10.3-16), which is based on the continuity equation, assumes that $\hat{v}_y = 0$ at $\hat{y} = 0$. Using these expressions to evaluate the velocity components in Eq. (10.3-8), we obtain

$$\tilde{u} \frac{\partial \Theta}{\partial \tilde{x}} - \frac{d\tilde{u}}{d\tilde{x}} \hat{y} \frac{\partial \Theta}{\partial \hat{y}} = \text{Pr}^{-1} \frac{\partial^2 \Theta}{\partial \hat{y}^2} \qquad (\text{Pr} \ll 1). \tag{10.3-17}$$

This is a suitable approximation for the energy equation, except that the very different thicknesses of the momentum and thermal boundary layers require that the scaling be adjusted.

To rescale the y coordinate for the *thermal* boundary layer, we need a transformation which will eliminate Pr from Eq. (10.3-17). The left-hand side of that equation is $O(1)$ as Pr$\rightarrow 0$. Thus, by analogy with Eq. (10.3-5), we incorporate Pr into the coordinate by letting

$$Y \equiv \hat{y} \, \mathrm{Pr}^{1/2} = \tilde{y} \, \mathrm{Re}^{1/2} \, \mathrm{Pr}^{1/2} \qquad (\mathrm{Pr} \ll 1). \tag{10.3-18}$$

The stretching factor used here to obtain Y from $\tilde{y}$ implies that δ_T varies as $\mathrm{Re}^{-1/2} \, \mathrm{Pr}^{-1/2}$, whereas we knew before that δ_M varies as $\mathrm{Re}^{-1/2}$. Thus, δ_M / δ_T scales as $\mathrm{Pr}^{1/2}$, a small number, consistent with Eq. (10.3-12). Using the new coordinate in Eq. (10.3-17), the energy equation becomes

$$\tilde{u} \frac{\partial \Theta}{\partial \tilde{x}} - \frac{d\tilde{u}}{d\tilde{x}} Y \frac{\partial \Theta}{\partial Y} = \frac{\partial^2 \Theta}{\partial Y^2} \qquad (\mathrm{Pr} \ll 1). \tag{10.3-19}$$

Aside from having the proper scaling, the advantage of Eq. (10.3-19) is that the velocities are evaluated explicitly in terms of $\tilde{u}(\tilde{x})$. Thus, to determine Θ, it is unnecessary to solve the momentum boundary layer problem. Only the momentum outer solution is needed.

For Pr$\gg 1$, the entire thermal boundary layer is embedded within a small part of the momentum boundary layer; see Fig. 10-3. This suggests that for the convective terms in the energy equation, one may approximate the velocities using the momentum boundary layer results for small $\hat{y}$. Approximations to the velocities which are valid very close to a stationary solid surface are

$$\hat{v}_x = \hat{\tau}_0(\tilde{x})\hat{y}, \tag{10.3-20}$$

$$\hat{v}_y = -\int_0^{\hat{y}} \left(\frac{d\hat{\tau}_0}{d\tilde{x}} \, \hat{y}' \right) d\hat{y}' = -\frac{d\hat{\tau}_0}{d\tilde{x}} \frac{\hat{y}^2}{2}, \tag{10-3.21}$$

$$\hat{\tau}_0(\tilde{x}) \equiv \left. \frac{\partial \hat{v}_x}{\partial \hat{y}} \right|_{\hat{y}=0} = \mathrm{Re}^{-1/2} \left. \frac{\partial \tilde{v}_x}{\partial \tilde{y}} \right|_{\tilde{y}=0} = \mathrm{Re}^{-1/2} \frac{L}{\mu U} \left. \tau_{yx} \right|_{y=0}. \tag{10.3-22}$$

Equation (10.3-20) is analogous to the linearization of the velocity profile used to analyze thermal entrance regions in tubes (i.e., the Lévêque approximation in Section 9.4). Equation (10.3-21), which is based on continuity, assumes that $\hat{v}_y = 0$ at $\hat{y} = 0$. As defined in Eq. (10.3-22), $\hat{\tau}_0$ is the dimensionless velocity gradient (or shear stress) at the surface, based on the variables scaled for the momentum boundary layer. Substituting Eqs. (10.3-20) and (10.3-21) into Eq. (10.3-8), the energy equation is now

$$\hat{\tau}_0 \hat{y} \frac{\partial \Theta}{\partial \tilde{x}} - \frac{d\hat{\tau}_0}{d\tilde{x}} \frac{\hat{y}^2}{2} \frac{\partial \Theta}{\partial \hat{y}} = \mathrm{Pr}^{-1} \frac{\partial^2 \Theta}{\partial \hat{y}^2} \qquad (\mathrm{Pr} \gg 1). \tag{10.3-23}$$

The presence of the large parameter (Pr) is evidence that the scaling must be adjusted again for the thermal boundary layer.

The fact that $\hat{\tau}_0$ is a velocity gradient based on properly scaled variables indicates that it is $O(1)$ as Re$\rightarrow\infty$. Moreover, the fact that the velocity in forced convection (with constant fluid properties) is independent of the temperature field implies that $\hat{\tau}_0$ is also $O(1)$ as Pr$\rightarrow\infty$. Thus, as was true for small Pr, the only variable which needs adjustment for large Pr is the y coordinate.

We now let

$$Y = \hat{y} \, \mathrm{Pr}^b, \tag{10.3-24}$$

where b is a constant. Equation (10.3-23) becomes

$$\mathrm{Pr}^{-b}\left(\hat{\tau}_0 Y \frac{\partial \Theta}{\partial \tilde{x}} - \frac{d\hat{\tau}_0}{d\tilde{x}} \frac{Y^2}{2} \frac{\partial \Theta}{\partial Y}\right) = \mathrm{Pr}^{2b-1} \frac{\partial^2 \Theta}{\partial Y^2} \qquad (\mathrm{Pr} \gg 1). \qquad (10.3\text{-}25)$$

In the thermal boundary layer, the conduction and convection terms must balance (by definition), implying that $-b = 2b - 1$ or $b = \frac{1}{3}$. We conclude that the scaled y coordinate and energy equation for this case are

$$Y \equiv \hat{y}\mathrm{Pr}^{1/3} = \tilde{y}\,\mathrm{Re}^{1/2}\mathrm{Pr}^{1/3} \qquad (\mathrm{Pr} \gg 1), \qquad (10.3\text{-}26)$$

$$\hat{\tau}_0 Y \frac{\partial \Theta}{\partial \tilde{x}} - \frac{d\hat{\tau}_0}{d\tilde{x}} \frac{Y^2}{2} \frac{\partial \Theta}{\partial Y} = \frac{\partial^2 \Theta}{\partial Y^2} \qquad (\mathrm{Pr} \gg 1). \qquad (10.3\text{-}27)$$

The stretching factor used to obtain Y from $\tilde{y}$ implies that δ_T varies as $\mathrm{Re}^{-1/2}\mathrm{Pr}^{-1/3}$, and that δ_M/δ_T scales as $\mathrm{Pr}^{1/3}.$ This is now a large number, consistent with Eq. (10.3-13). Aside from the scaling, the advantage of Eq. (10.3-27) is that the y-dependence of the velocities is given explicitly. The only unknown in the velocities is $\hat{\tau}_0$, which must be found by solving the momentum boundary layer equations.

For mass transfer at large Schmidt number, Sc replaces Pr in Eq. (10.3-26). Equation (10.3-27) applies as is, with Θ now representing a dimensionless concentration.

Boundary layer equations scaled as described above for extreme values of Pr were derived by Morgan and Warner (1956) and Morgan et al. (1958). Those authors appear to have been the first to recognize the implications of the scaling for the Nusselt number, a subject which is explored in Section 10.4. Similarity solutions for Eqs. (10.3-19) and (10.3-27) and for related forced-convection equations are given, for example, in Acrivos et al. (1960) and Acrivos (1962).

10.4 SCALING LAWS FOR NUSSELT AND SHERWOOD NUMBERS

The key quantity for calculating the rate of heat or mass transfer from an object to the surrounding fluid is the average Nusselt or Sherwood number. As discussed in Section 9.3, dimensional analysis for forced convection indicates that $\overline{\mathrm{Nu}}$ depends on Re and Pr and that $\overline{\mathrm{Sh}}$ depends on Re and Sc. (Other parameters, such as Br and Da, may be present also.) The relationships, either empirical or theoretical, are often of the form

$$\overline{\mathrm{Nu}} = C\,\mathrm{Re}^a\mathrm{Pr}^b, \qquad \overline{\mathrm{Sh}} = C\,\mathrm{Re}^a\mathrm{Sc}^b, \qquad (10.4\text{-}1)$$

where a, b, and C are constants determined by the geometry and flow regime. Equations (10.2-9) and (10.2-32) are examples of such relationships. In this section we discuss how the values of a and b can be predicted for various situations. These predictions, which are limited to laminar flows with thermal or concentration boundary layers, are obtained by using scaling concepts. Accordingly, we refer to the results as "scaling laws." Being able to predict the exponents in Eq. (10.4-1) without doing detailed calculations is of considerable value, both for anticipating the form of theoretical results and for extrapolating from limited experimental data.

For an object with a thermal boundary layer (i.e., for Pe $\gg$ 1), the local Nusselt number is expressed as

$$\mathrm{Nu} = \frac{hL}{k} = -\frac{1}{\Delta\Theta}\frac{\partial\Theta}{\partial\tilde{y}}\bigg|_{\tilde{y}=0}, \tag{10.4-2}$$

where $\Delta\Theta$ is the dimensionless temperature difference between the surface and the bulk fluid, and $\tilde{y}=y/L$. The y coordinate scaled for the thermal boundary layer is assumed to be given by

$$Y = \tilde{y}\,\mathrm{Re}^a\,\mathrm{Pr}^b, \tag{10.4-3}$$

where a and b depend on the dynamic conditions. Equation (10.4-2) is rewritten now as

$$\mathrm{Nu} = \left(-\frac{1}{\Delta\Theta}\frac{\partial\Theta}{\partial Y}\bigg|_{Y=0}\right)\frac{\partial Y}{\partial\tilde{y}} \equiv f(\tilde{\mathbf{r}}_s)\,\mathrm{Re}^a\,\mathrm{Pr}^b, \tag{10.4-4}$$

where $f(\tilde{\mathbf{r}}_s)$ is a dimensionless function of position on the surface. Averaging Eq. (10.4-4) over the surface S yields the expression for $\overline{\mathrm{Nu}}$ in Eq. (10.4-1), where C is the $O(1)$ constant given by

$$C = \frac{1}{S}\int_S\left(-\frac{1}{\Delta\Theta}\frac{\partial\Theta}{\partial Y}\bigg|_{Y=0}\right)dS = \frac{1}{S}\int_S f(\mathbf{r}_s)\,dS. \tag{10.4-5}$$

The calculation of this constant for creeping flow past a solid sphere is shown in Eq. (10.2-30). The important point in Eqs. (10.4-4) and (10.4-5) is that the careful scaling of all variables makes f and C independent of Re and Pr, so that the dependence of Nu or $\overline{\mathrm{Nu}}$ on those parameters is determined entirely by the relationship between Y and $\tilde{y}$. Thus, the problem of predicting the exponents in Eq. (10.4-1) is reduced to that of scaling the y coordinate. Scaling the y coordinate [i.e., determining the stretching factor in Eq. (10.4-3)] is equivalent, in turn, to finding the thickness of the thermal boundary layer.

The scaling of y is guided mainly by the fact that a thermal boundary layer is a region in which conduction and convection are both important, even for arbitrarily large values of Pe. For a steady, two-dimensional flow, the energy equation for the boundary layer indicates that

$$\tilde{v}_x\frac{\partial\Theta}{\partial\tilde{x}} \sim \mathrm{Re}^{-1}\mathrm{Pr}^{-1}\frac{\partial^2\Theta}{\partial\tilde{y}^2} = \mathrm{Re}^{2a-1}\mathrm{Pr}^{2b-1}\frac{\partial^2\Theta}{\partial Y^2}. \tag{10.4-6}$$

The convection term involving $\tilde{v}_y$ is not considered here, because the continuity equation ordinarily ensures that it is no larger than the one involving $\tilde{v}_x$. Examples of this were seen in Sections 10.2 and 10.3. Assuming in Eq. (10.4-6) that the derivatives involving $\tilde{x}$ and Y are $O(1)$, we infer that

$$\tilde{v}_x = O(\mathrm{Re}^{2a-1}\mathrm{Pr}^{2b-1}). \tag{10.4-7}$$

This shows how the magnitude of $\tilde{v}_x$ in the thermal boundary layer is related to Re and Pr. It is worth noting that this relationship is obtained even if the wrong temperature scale is employed, because Θ appears in all terms of the energy equation. That is, in forced convection the temperature scale cancels out. More careful scaling of the temperature is required in free convection, as discussed in Chapter 12.

What is needed now is independent information on the magnitude of $\tilde{v}_x$ in various flow regimes. Combined with Eq. (10.4-7), this will allow us to infer the values of a

and *b*. To estimate $\tilde{v}_x$ in the thermal boundary layer, we expand the velocity about the surface as

$$\tilde{v}_x = \tilde{v}_x \bigg|_{\tilde{y}=0} + \frac{\partial \tilde{v}_x}{\partial \tilde{y}} \bigg|_{\tilde{y}=0} \tilde{y} + \cdots. \tag{10.4-8}$$

As described below for several types of flow, the values of the scaling exponents are inferred from the first nonzero term in this expansion. The key fluid-dynamic considerations are the type of interface and the range of Re.

Fluid–Fluid Interface

If the fluid of interest is in contact with another fluid, there will usually be motion at the interface. That is, the first term in the expansion in Eq. (10.4-8) will not vanish. For example, we may wish to determine the resistance to heat or mass transfer in the fluid surrounding a gas bubble or liquid drop. The "fluid of interest" is then the surrounding fluid. If the coordinates are fixed at the center of the bubble or drop, the tangential velocity at the interface will ordinarily be comparable to the approach velocity of the surrounding fluid. An example is provided by the results given in Problem 7-2 for creeping flow around a spherical bubble or drop. Assuming that the interfacial velocity is similar to the imposed velocity scale, $\tilde{v}_x \sim 1$ in the thermal boundary layer near the interface. With $\tilde{v}_x = O(1)$, Eq. (10.4-7) implies that $a = b = \frac{1}{2}$. Provided that Pe >> 1 and the flow is laminar, no restrictions are placed here on Re and Pr (or Sc).

Fluid–Solid Interface for Small or Moderate Re and Large Pr

If the object in question is a stationary solid, the no-slip condition ensures that the first nonzero term in Eq. (10.4-8) will be the one proportional to $\tilde{y}$. If Re is not large (i.e., if there is not a momentum boundary layer), then $\partial \tilde{v}_x / \partial \tilde{y} = O(1)$. In other words, $\tilde{v}_x$ and $\tilde{y}$ are both correctly scaled for the fluid-dynamic problem. Thus, the velocity near the surface will scale as $\tilde{y}$, which in the thermal boundary layer is $O(\mathrm{Re}^{-a}\mathrm{Pr}^{-b})$ [see Eq. (10.4-3)]. Equating these exponents with those in Eq. (10.4-7), we conclude that $a = b = \frac{1}{3}$. This type of scaling was seen in the analysis of heat transfer in creeping flow past a solid sphere (Example 10.2-2).

Fluid–Solid Interface for Large Re and Large Pr

Again, the first nonvanishing term in the expansion for the velocity is that proportional to $\tilde{y}$. What differs from the previous case is that with a momentum boundary layer the velocity gradient at the surface is large. Specifically, laminar boundary layer theory predicts that $\partial \tilde{v}_x / \partial \tilde{y} = O(\mathrm{Re}^{1/2})$. It follows that the velocity in the thermal boundary layer is $O(\mathrm{Re}^{1/2-a}\mathrm{Pr}^{-b})$ and that $a = \frac{1}{2}$ and $b = \frac{1}{3}$. This scaling was derived in Section 10.3 [see Eq. (10.3-26)].

Fluid–Solid Interface for Large Re and Small Pr

In boundary layer flow at small Pr the thermal boundary layer resides mainly in the momentum outer region, as discussed in Section 10.3. Thus, we conclude directly that $\tilde{v}_x = O(1)$ in the thermal boundary layer; Eq. (10.4-8) is not used. Coincidentally, the

TABLE 10-2
Forms of the Average Nusselt Number for Laminar Flow with Pe $\gg$ 1

Case	Other phase	Re	Pr	$\overline{Nu}$
1	Fluid or solid	$\gg 1$	Any[a]	$B(Pr)\,Re^{1/2}$
2	Fluid	Any[a]	Any[a]	$C\,Re^{1/2}\,Pr^{1/2}$
3	Solid	$\ll 1$ or ~ 1	$\gg 1$	$C\,Re^{1/3}\,Pr^{1/3}$
4	Solid	$\gg 1$	$\gg 1$	$C\,Re^{1/2}\,Pr^{1/3}$
5	Solid	$\gg 1$	$\ll 1$	$C\,Re^{1/2}\,Pr^{1/2}$

[a]This parameter is not restricted individually, but it must be such that $Pe = Re\,Pr \gg 1$.

scaling is the same as for a fluid–fluid interface, or $a = b = \frac{1}{2}$. This result has been given already as Eq. (10.3-18).

Summary of Scaling Laws

The results for the average Nusselt number in the situations just discussed are summarized in Table 10-2. The analogous predictions for the average Sherwood number are given in Table 10-3. Case 1 in each table, which is for any situation with large Re, follows jointly from the other results (excluding case 3). Thus, it was not discussed separately. The coefficient B in case 1 is a function of Pr or Sc and the geometry, whereas the coefficient C in the other entries depends on the geometry only. These scaling laws apply to laminar flow with Pe $\gg$ 1. As discussed at the end of this section, they are not necessarily accurate for flows with boundary layer separation. Following are two examples which illustrate their application.

Example 10.4-1 Heat Transfer from a Bubble or Drop Assuming that Pe $\gg$ 1, heat transfer from a gas bubble or liquid drop rising or falling in an immiscible liquid is an example of case 2 of Table 10-2. The characteristic velocity (U) for a single bubble or drop is its terminal velocity, and the customary choice for the characteristic length (L) is $2R$, where R is the radius of a sphere of equivalent volume. Thus, the average Nusselt number for the external fluid is predicted to be of the form

$$\overline{Nu} = \frac{2hR}{k} = C\,Pe_d^{1/2},\qquad(10.4\text{-}9)$$

where $C \sim 1$.

TABLE 10-3
Forms of the Average Sherwood Number for Laminar Flow with Pe $\gg$ 1

Case	Other phase	Re	Sc	$\overline{Sh}$
1	Fluid or solid	$\gg 1$	Any[a]	$B(Sc)\,Re^{1/2}$
2	Fluid	Any[a]	Any[a]	$C\,Re^{1/2}\,Sc^{1/2}$
3	Solid	$\ll 1$ or ~ 1	$\gg 1$	$C\,Re^{1/3}\,Sc^{1/3}$
4	Solid	$\gg 1$	$\gg 1$	$C\,Re^{1/2}\,Sc^{1/3}$

[a]This parameter is not restricted individually, but it must be such that $Pe = Re\,Sc \gg 1$.

For bubbles or drops moving at low Reynolds number, deformation from the equilibrium (spherical) shape is minimal. The problem of diffusion from a spherical bubble or drop in creeping flow was solved first by Levich (1962, pp. 404–408). For the special case of a gas bubble, where the internal viscosity is negligible, his result corresponds to Eq. (10.4-9) with $C = 0.651$ (see Problem 10-1).

The mechanics of small bubbles and drops is complicated by gradients in surface tension. In water, especially, it is very difficult to exclude trace contaminants which act as surfactants. Flow-induced variations in surfactant concentration create gradients in surface tension which oppose tangential motion at the interface. For example, if the interfacial motion carries surfactant toward the south pole of a bubble, the surface tension there will be reduced. The higher surface tension near the north pole will tend to cause flow from south to north. In other words, the original motion is resisted. These effects become more prominent as R decreases, all other factors being equal. In very small bubbles or drops the interfacial motion is arrested completely, and the drag coefficient becomes the same as for a solid sphere. For air bubbles in water, the transition from fluid to solid behavior commonly occurs at $R \sim 1$ mm. If a bubble or drop is small enough to act as a solid, the heat transfer regime will correspond to case 3 of Table 10-2 instead of case 2. Thus, in applying the scaling laws, caution is needed to ensure that the actual flow is qualitatively similar to what was assumed.

For bubbles or drops moving at large Reynolds number, the form of the Nusselt number may be affected by flow separation and/or turbulence. Comments on heat and mass transfer in flows with boundary layer separation are given at the end of this section.

The effects of surfactants on the mechanics of bubbles and drops depend on the kinetics of adsorption, the diffusivity of surfactant within the interface, and the equation of state which relates the surface tension to the surfactant concentration. Models to describe these effects are discussed in Harper (1972) and Edwards et al. (1991). A recent review of heat and mass transfer results for liquid droplets is Ayyaswamy (1995).

Example 10.4-2 Mass Transfer at a Gas–Liquid interface in a Stirred Reactor   Consider the laboratory-scale reactor shown in Fig. 10-4. It is desired to determine the overall mass transfer coefficients for various dissolved gases at the gas–liquid interface, under nonreactive conditions, when the liquid is stirred at a given rate. In this type of apparatus the flow patterns are complex enough to make it very difficult to predict the mass transfer coefficients by solving the governing differential equations. An experimental approach is needed. One such approach is to flow a selected gas mixture through the head space until equilibrium is reached, thereby achieving a desired initial concentration of some component in the liquid. Rapidly changing the gas mixture allows

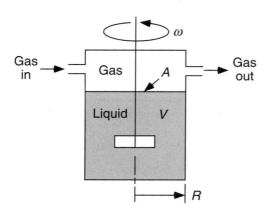

Figure 10-4. Schematic of a stirred two-phase reactor. A fixed volume of liquid is stirred at an angular velocity ω, while gas flows continuously through the head space. The radius of the cylindrical vessel is R, the liquid volume is V, and the area of the gas–liquid interface is A.

one to reduce the head-space concentration of that component to zero, after which its disappearance from the liquid is monitored. Assume that a particular vessel is fitted with a detector which allows the liquid-phase concentration of nitric oxide (NO) to be monitored as a function of time. It is desired to extrapolate the overall mass transfer coefficient measured for NO at a given temperature to other temperatures and/or other chemical species.

We will be analyzing data obtained under conditions where the flow is laminar and Pe is large (see below). In these experiments, the NO concentration in the gas phase was reduced to zero at $t = 0$. For $t > 0$ the concentration of NO in most of the liquid was approximately uniform at the bulk value, $C_{NO}(t)$. Choosing the entire liquid as the control volume, a macroscopic mass balance for NO gives

$$V\frac{dC_{NO}}{dt} = -k_{NO}^{(0)} A C_{NO}, \qquad (10.4\text{-}10)$$

where V is the liquid volume, A is the area of the gas–liquid interface, and $k_{NO}^{(O)}$ is the overall mass transfer coefficient for NO at that interface. The solution to Eq. (10.4-10) is

$$\frac{C_{NO}(t)}{C_{NO}(0)} = \exp\left[-\left(\frac{k_{NO}^{(0)}A}{V}\right)t\right] \equiv \exp(-b_{NO}t). \qquad (10.4\text{-}11)$$

Thus, with A and V known, $k_{NO}^{(O)}$ can be computed from the slope (b_{NO}) of a semilogarithmic plot of the concentration data.

The liquid and gas provide resistances in series near the gas–liquid interface, so that the overall mass transfer coefficient is related to those in the liquid ($k_{NO}^{(L)}$) and gas ($k_{NO}^{(G)}$) as

$$\frac{1}{k_{NO}^{(O)}} = \frac{1}{k_{NO}^{(L)}} + \frac{K_{NO}}{k_{NO}^{(G)}} \qquad (10.4\text{-}12)$$

where K_{NO} is the liquid-to-gas concentration ratio at equilibrium (0.047 at 23°C). Because the Péclet number is large and the interface is between two fluids, case 2 of Table 10-3 can be applied to *both* phases. Thus, we expect the average Sherwood numbers to be of the form

$$\overline{Sh}_{NO}^{(j)} \equiv \frac{k_{NO}^{(j)}R}{D_{NO}^{(j)}} = B_j(Pe_{NO}^{(j)})^{1/2}, \qquad (10.4\text{-}13)$$

$$Pe_{NO}^{(j)} \equiv \frac{\omega R^2}{D_{NO}^{(j)}}, \qquad (10.4\text{-}14)$$

where R is the radius of the vessel, ω is the angular velocity of the stirrer, and $j = L$ or G. The coefficient B_j in Eq. (10.4-13) is an unknown function of the reactor geometry (height-to-diameter ratio, impeller shape, relative position of the interface, etc.), but it is expected that $B_j \sim 1$ for both phases. Using Eq. (10.4-13) and recalling from Chapter 1 that a typical ratio of gas to liquid diffusivities is $\sim 10^4$, we estimate that

$$\frac{k_{NO}^{(G)}}{k_{NO}^{(L)}} \sim \left(\frac{D_{NO}^{(G)}}{D_{NO}^{(L)}}\right)^{1/2} \sim 10^2. \qquad (10.4\text{-}15)$$

Together with the small value of K_{NO}, this indicates that the mass transfer resistance in the gas is negligible. Thus, the overall mass transfer coefficient essentially equals that in the liquid.

Having found that the controlling resistance is in the liquid, we can use Eq. (10.4-13) to extrapolate a measured value of the overall mass transfer coefficient to other conditions. For example, the values for NO at temperatures T_1 and T_2 are predicted to be related as

$$\frac{k_{NO}^{(O)}(T_2)}{k_{NO}^{(O)}(T_1)} = \left[\frac{D_{NO}^{(L)}(T_2)}{D_{NO}^{(L)}(T_1)}\right]^{1/2} \qquad (10.4\text{-}16)$$

An analogous relation would be used to extrapolate the results for NO to another dissolved gas, the correction factor once again being the square root of the ratio of the liquid-phase diffusivities.

A reactor of the type shown in Fig. 10-4 was used by Lewis and Deen (1994). With $R = 3.1$ cm, $\omega = 100$ rpm $= 10.5$ s^{-1}, and water at 23°C, for which $\nu = 9.3 \times 10^{-3}$ cm^2/s, the liquid-phase Reynolds number was Re $= \omega R^2/\nu = 1 \times 10^4$. At this value of Re the flow is expected to be laminar, with momentum boundary layers at the side, bottom, and gas–liquid interface. Boundary layer flow in a stirred cylindrical container was studied by Colton and Smith (1972), who found that flow was laminar for Re $< 1.5 \times 10^4$. (That study focused on mass transfer at the base of a stirred cylindrical container, rather than at a gas–liquid interface.) The aqueous diffusivity for NO at 23°C is 2.67×10^{-5} cm^2/s, yielding $Sc_{NO} = 3 \times 10^2$ and $Pe_{NO} = 3 \times 10^6$. Thus, as already mentioned, the Péclet number was very large. Experiments with NO at 23°C and 37°C yielded $b_{NO} = 0.75 \pm 0.01$ and 1.02 ± 0.02, respectively, in units of 10^{-3} s^{-1}. Using the measured value for b_{NO} at 23°C, along with the aqueous diffusivities for NO at 23°C (see above) and 37°C (5.08×10^{-5} cm^2/s), Eq. (10.4-16) predicts that $b_{NO} = 1.03 \times 10^{-3}$ s^{-1} at 37°C. This is identical to the observed value, within experimental error.

It is worth noting that if we had employed a stagnant-film model (Chapter 2), the prediction for the mass transfer coefficient would have been much less accurate. In a stagnant-film model the mass transfer coefficient at any type of interface is proportional to D_{NO}, as compared with $D_{NO}^{1/2}$ for a fluid–fluid interface using boundary layer theory. Using the stagnant-film approach, we would have predicted that $b_{NO} = 1.43 \times 10^{-3}$ s^{-1} at 37°C, a value which is 40% too large.

Heat and Mass Transfer with Boundary Layer Separation

As discussed in Chapter 8, a hallmark of high Reynolds number flow past blunt objects is boundary layer separation. For laminar flow past solid bodies, the typical finding is that most or all of the facing (or upstream) side of the object has a boundary layer with a thickness which varies as Re$^{-1/2}$, as predicted from the scaling of the momentum equation. However, the adverse pressure gradient on the trailing (or downstream) side of a blunt object forces the boundary layer to depart from the surface. Beyond the separation point, the boundary layer scaling fails and there is no simple theory to take its place. Because an implicit assumption in the scaling laws derived here for high Reynolds number is that there is a momentum boundary layer on the entire surface, their application to blunt objects is questionable. To the extent that most of the transport occurs in the boundary layer on the facing side, where the heat or mass transfer coefficient should be largest, the scaling laws will be satisfactory. There is no guarantee that transport at the facing side will be dominant, however. To provide examples of heat and mass transfer from blunt objects at large Re, a brief discussion follows of experimental results for cylinders and spheres. In this discussion, Re is based on the diameter of the cylinder or sphere.

Extensive reviews of the many results for heat transfer from cylinders in crossflow are found in Morgan (1975) and in Zukauskas and Ziugzda (1985). The discussion here is restricted to data for $40 < Re < 2 \times 10^5$, the upper limit being the approximate point at which the flow becomes fully turbulent. The results in Morgan (1975) are for air only

(Pr $= 0.7$), so that comparisons are limited to case 1 of Table 10-2 (i.e., $a = \frac{1}{2}$). In addition to air, Zukauskas and Ziugzda (1985) employed a range of liquids with Pr as large as 10^4. This makes case 4 of Table 10-2 applicable (i.e., $a = \frac{1}{2}$ and $b = \frac{1}{3}$). The summary correlation given in Zukauskas and Ziugzda (1985, p. 162) indicates that $b = 0.37$, in reasonable agreement with the theory. In both studies it is concluded that $a = 0.47$–0.50 for values of Re up to about 10^3, again consistent with the scaling laws. At higher Re, however, the recommended value of a increases to 0.6–0.8, depending on which correlation is used. Thus, a stronger-than-predicted dependence on the Reynolds number is evident well below the critical value of Re.

Many empirical correlations have been proposed for mass transfer from solid spheres, as summarized in Skelland (1974, pp. 276–277). An experimental study of heat transfer from spheres is reported in Raithby and Eckert (1968). In general, the pattern which emerges is similar to that for cylinders. That is, the mass transfer data are reported to be consistent with $b = \frac{1}{3}$. The mass transfer data for Re up to values of several thousand also indicate that $a \cong 0.5$. These exponents match those in case 4 of Table 10-3. However, the mass transfer correlations intended for larger values of Re tend to have $a \cong 0.6$. Likewise, the heat transfer correlation of Raithby and Eckert (1968), which is for $3.6 \times 10^3 \leq \mathrm{Re} \leq 5.2 \times 10^4$, gives $a = 0.58$.

In empirical correlations for the Nusselt or Sherwood numbers, Reynolds number exponents slightly below unity are typical of turbulent flow; some examples are given in Chapter 13. Based on dimensional analysis and a few physical assumptions, Ruckenstein (1986) argued that for turbulent flow one should always expect $\frac{3}{4} \leq a \leq 1$. Thus, the results for cylinders and spheres indicate that the Nusselt and Sherwood numbers become increasingly "turbulent" as Re is increased above about 10^3. The change is evidently gradual, and it begins well before the boundary layer becomes completely turbulent. Local measurements on cylinders show that heat transfer from the trailing side grows in relative importance as Re is elevated (Zukauskas and Ziugzda (1985, pp. 97–101). This is not surprising, given the increasingly turbulent character of the wake and separation regions (see Chapter 8).

References

Acrivos, A. On the solution of the convection equation in laminar boundary layer flows. *Chem. Eng. Sci.* 17: 457–465, 1962.

Acrivos, A. and J. D. Goddard. Asymptotic expansions for laminar forced-convection heat and mass transfer. Part 1. Low speed flows. *J. Fluid Mech.* 23: 273–291, 1965.

Acrivos, A. and T. D. Taylor. Heat and mass transfer from single spheres in Stokes flow. *Phys. Fluids* 5: 387–394, 1962.

Acrivos, A., M. J. Shah, and E. E. Petersen. Momentum and heat transfer in laminar boundary-layer flows of non-Newtonian fluids past external surfaces. *AIChE J.* 6: 312–317, 1960.

Ayyaswamy, P. S. Direct-contact transfer processes with moving liquid droplets. *Adv. Heat Transfer* 26: 1–104, 1995.

Brenner, H. Invariance of the overall mass transfer coefficient to flow reversal during Stokes flow past one or more particles of arbitrary shape. *Chem. Eng. Progr. Symp. Ser.,* no. 105, pp. 123–126, 1970.

Brian, P. L. T. and H. B. Hales. Effects of transpiration and changing diameter on heat and mass transfer to spheres. *AIChE J.* 15: 419–425, 1969.

Colton, C. K. and K. A. Smith. Mass transfer to a rotating fluid. Part II. Transport from the base of an agitated cylindrical tank. *AIChE J.* 18: 958–967, 1972.

Edwards, D. A., H. Brenner, and D. T. Wasan. *Interfacial Transport Processes and Rheology.* Butterworth-Heinemann, Boston, 1991.

Garner, F. H. and R. B. Keey. Mass transfer from single solid spheres—I. Transfer at low Reynolds numbers. *Chem. Eng. Sci.* 9: 119–129, 1958.

Harper, J. F. The motion of bubbles and drops through liquids. *Adv. Appl. Mech.* 12: 59–129, 1972.

Kutateladze, S. S. *Fundamentals of Heat Transfer.* Edward Arnold, London, 1963.

Levich, V. G. *Physicochemical Hydrodynamics.* Prentice-Hall, Englewood Cliffs, NJ, 1962.

Lewis, R. S., and W. M. Deen. Kinetics of the reaction of nitric oxide with oxygen in aqueous solutions. *Chem. Res. Toxicol.* 7: 568–574, 1994.

Morgan, G. W., A. C. Pipkin, and W. H. Warner. On heat transfer in laminar boundary-layer flows of liquids having a very small Prandtl number. *J. Aeronaut. Sci.* 25: 173–180, 1958.

Morgan, G. W. and W. H. Warner. On heat transfer in laminar boundary layers at high Prandtl number. *J. Aeronaut. Sci.* 23: 937–948, 1956.

Morgan, V. T. The overall convective heat transfer from smooth circular cylinders. *Adv. Heat Transfer* 11: 199–264, 1975.

Raithby, G. D. and E. R. G. Eckert. The effect of turbulence parameters and support position on the heat transfer from spheres. *Int. J. Heat Mass Transfer* 11: 1233–1252, 1968.

Ruckenstein, E. Scaling in laminar and turbulent heat and mass transfer. In *Handbook of Heat and Mass Transfer,* Vol. 1, N. P. Cheremisinoff, Ed. Gulf, Houston, 1986.

Schlicting, H. *Boundary Layer Theory,* sixth edition. McGraw-Hill, New York, 1968, p. 90.

Skelland, A. H. P. *Diffusional Mass Transfer.* Wiley, New York, 1974.

Skelland, A. H. P. and A. R. H. Cornish. Mass transfer from spheroids to an air stream. *AIChE J.* 9: 73–76, 1963.

Zukauskas, A. and J. Ziugzda. *Heat Transfer of a Cylinder in Crossflow.* Hemisphere, Washington, 1985.

Problems

10-1. Absorption from a Small Bubble

Consider a small gas bubble of radius R rising in a stagnant liquid at a velocity U. Assume that the bubble consists of pure gas A, which dissolves in the liquid. The liquid concentrations of A in equilibrium with the gas and far from the bubble are C_{Ao} and $C_{A\infty}$, respectively. Assume that $Re = UR/\nu \ll 1$ in both phases and that the liquid viscosity greatly exceeds that of the gas. Also assume that Sc in the liquid is sufficiently large that $Pe = Re\, Sc \gg 1$. Finally, assume that mass transfer is slow enough that a pseudosteady analysis is appropriate (i.e., R remains nearly constant).

(a) Use the results of Problem 7-2 to show that the liquid velocity is well approximated by

$$v_r = U \cos\theta \left[1 - \left(\frac{R}{r}\right) \right],$$

$$v_\theta = -U \sin\theta \left[1 - \frac{1}{2}\left(\frac{R}{r}\right) \right].$$

(b) Determine the local and average Sherwood numbers for the liquid phase, for $Pe \to \infty$. Show that

(handwritten at top) $2ff' + f'^2 = 2?$ $(f^2)' + \frac{\cdot}{x} f^2 = \frac{4}{x}$ $=1$ $f^2 = \frac{9}{x^2} + 2.$

$$\overline{Sh} = 0.651 Pe_d^{1/2}$$

as found by Levich (1962, pp. 404–408).

10-2. Mass Transfer from a Rotating Disk

Rotating disk electrodes are used extensively in studies of electrochemical kinetics, because of their special mass transfer characteristics. Consider a solid disk, rotating at angular velocity ω, which is in contact with a Newtonian liquid with constant physical properties. The surface of the disk is at $z=0$, and the liquid occupies the space $z>0$. Assume that solute i reacts rapidly and irreversibly at the surface and that the concentration of i far from the disk is $C_{i\infty}$.

The fluid dynamics of rotating disks is discussed in Problem 8-8. It is found that the velocity and pressure can be expressed as functions of ζ only, where $\zeta \equiv z\sqrt{\omega/\nu}$. In particular, it is found that $v_z = \sqrt{\nu\omega}\,H(\zeta)$. An expansion for $H(\zeta)$ which is valid near the disk surface is

$$H(\zeta) = -a\zeta^2 + \frac{1}{3}\zeta^3 + \cdots, \qquad a = 0.51023.$$

(a) Assuming that the reactant concentration C_i approaches a constant as $r\to\infty$, show that it is sufficient to assume that $C_i = C_i(z)$ only. Consequently, C_i is governed by

$$v_z \frac{dC_i}{dz} = D_i \frac{d^2 C_i}{dz^2}.$$

(b) Derive an expression for Sh which is valid in the limit $\mathrm{Sc}\to\infty$. Show that

$$\mathrm{Sh} \equiv \frac{k_{ci}\sqrt{\nu/\omega}}{D_i} = 0.6204\ \mathrm{Sc}^{1/3}$$

as obtained by Levich (1962, pp. 60–72).

(c) If the reaction on the disk is confined to a circular area of radius R, an alternative definition of the Sherwood number is $\mathrm{Sh} \equiv k_{ci} R/D_i$. If this definition is adopted, which is the corresponding situation in Table 10-3? Explain.

10-3. Heat Transfer in Two-Dimensional Stagnation Flow

A stream of fluid at temperature T_∞ is directed perpendicular to a flat, solid surface. The surface is located in the plane $y=0$ and the approaching flow is in the $-y$ direction. The surface is maintained at constant temperature T_0 for $-L \le x \le L$. Outside this "stripe," which is much narrower than the overall dimensions, the surface is thermally insulated. For this two-dimensional stagnation flow, v_x from the momentum outer solution, evaluated at the surface, is given by

$$u(x) = Cx,$$

where C is a constant. The Reynolds number based on the half-width of the active stripe is $\mathrm{Re} = CL^2/\nu$. The shear stress at the surface is given by

(handwritten left margin) $\left(\frac{2}{x}\right)^{1/2} \cdot Re^{1/2}\ P_L^{1/2}$

$$\tau_0(x) = \mu \left(\frac{C^3}{\nu}\right)^{1/2} Bx,$$

(handwritten right) $\bar{x}\frac{\partial\theta}{\partial\bar{y}} - Y\frac{\partial\theta}{\partial Y} = \frac{\partial^2\theta}{\partial Y^2}$, $\frac{\partial\theta}{\partial x}(0,y)=0$. $\theta(\bar{x},0)=1$ $\theta(\bar{x},\infty)=0$

$s = \frac{Y}{f(\bar{x})}$ 4 $\frac{d^2\theta}{ds^2} + \left(\frac{3}{2}ff' + f\cdot\right)\frac{d\theta}{ds} = 0$

where $B = 1.2326$ (Schlicting, 1968). Assume that viscous dissipation is negligible.

(a) Derive an expression for Nu_L ($= hL/k$) which is valid for $\mathrm{Pr}\ll 1$.

(b) Derive an expression for Nu_L which is valid for $\mathrm{Pr}\gg 1$.

(handwritten right) $Pr \gg 1$ $v(1) = u(x) = Cx.$ $v_y = -Cy.$ $\bar{x} = \frac{x}{L}$ $\hat{y} = \frac{y}{L}$, $\theta = \frac{T - T_\infty}{T_0 - T_\infty}$ $Y = \hat{y}\,Re^{1/2}\,Pr^{1/2}$

The following definite integrals may be helpful: *(handwritten)* $\sim Re^{1/2}Pr^{1/3}$.

$$\int_0^\infty \exp(-x^n)\,dx = \frac{\Gamma(1/n)}{n}, \qquad n = 1,2,3,\cdots,$$

(handwritten lower left) $\frac{\hat{v}_x}{L} = \frac{?}{?}\hat{\theta}$ $Pr \gg 1$ $Y = \frac{\hat{y}}{2}Re^{1/2}Pr^{1/3}$

$$\Gamma(1)=1, \quad \Gamma\left(\tfrac{1}{2}\right)=\sqrt{\pi}, \quad \Gamma\left(\tfrac{1}{3}\right)=2.6789, \quad \Gamma\left(\tfrac{1}{4}\right)=3.6256.$$

10-4. Heat Transfer from a Flat Plate

A stream of fluid at temperature T_∞ is directed parallel to a thin, flat plate which is at constant temperature T_0. The plate is located in the plane $y=0$ and extends from $x=0$ to $x=L$. The velocity far from the plate is U. For $\mathrm{Re} \equiv UL/\nu \gg 1$, the shear stress at the surface is given by

$$\tau_0(x)=\mu B\left(\frac{U^3}{\nu x}\right)^{1/2},$$

where $B=0.332$ (Chapter 8). Assume that viscous dissipation is negligible.

(a) Derive an expression for Nu_L ($=hL/k$) which is valid for $\mathrm{Pr}\ll 1$.
(b) Derive an expression for Nu_L which is valid for $\mathrm{Pr}\gg 1$.

The integrals given at the end of Problem 10-3 may be helpful.

10-5. Heat Transfer to a Power-Law Fluid

Consider steady, two-dimensional flow of a polymer solution parallel to an isothermal flat plate, as described in Problem 10-4. Far from the plate the velocity is uniform at U. Assume that near the plate the rheology is approximated well by the power-law model,

$$\mu=m(2\Gamma)^{n-1},$$

where m and n are constants (Section 5.6). The conditions are such that, based on representative values of μ near the plate, $\mathrm{Re}_L \equiv UL\rho/\mu \gg 1$ and $\mathrm{Pr}\gg 1$. Assume that viscous dissipation is negligible.

(a) Derive an order-of-magnitude estimate for the momentum boundary layer thickness, $\delta_M(x)$.
(b) Derive an order-of-magnitude estimate for the thermal boundary layer thickness, $\delta_T(x)$.
(c) Determine the dependence of Nu_L on U, where $\mathrm{Nu}_L \equiv hL/k$. In particular, if U is doubled, how much will Nu_L change for the power-law fluid? Compare this with the effect of doubling U for a Newtonian fluid.

A detailed analysis of this problem is given in Acrivos et al. (1960).

10-6. Nusselt Numbers for Wedge Flows

As discussed in Chapter 8, two-dimensional flows past wedge-shaped objects yield limiting "outer" velocities which are of the form

$$u(x)=Cx^m,$$

where C and m are constants. The parameter m is related to the wedge angle. Assume that $m\ge 0$ and that the surface extends a distance L from the stagnation point, so that L is the characteristic dimension of the object. Assume also that $\mathrm{Re} \equiv CL^{m+1}/\nu \gg 1$. The fluid temperature far from the surface is T_∞ and the surface temperature is T_0, both constant. Viscous dissipation is negligible.

(a) Derive a general expression for $\mathrm{Nu}_L \equiv hL/k$ for wedge flows, valid for $\mathrm{Pr}\ll 1$. The result will involve m.
(b) Derive a general expression for Nu_L for wedge flows, valid for $\mathrm{Pr}\gg 1$. The result will involve the dimensionless shear stress at the wall, $\hat{\tau}_0$.

Chapter 11

MULTICOMPONENT ENERGY AND MASS TRANSFER

11.1 INTRODUCTION

Thus far in this book we have treated energy and mass transfer as separate, but mathematically analogous, phenomena. This was advantageous because the equations describing energy transfer in a single-component fluid and mass transfer in an isothermal mixture are often identical, and those equations have widespread application. In this chapter we depart from that approach, discussing a variety of situations in which Fourier's law and Fick's law are inadequate for describing the molecular contributions to energy and mass transfer. More complex constitutive equations may be needed for either or both of two reasons. One is that the various fluxes in a system may directly influence one another. For example, molecular energy transport in a mixture may result from diffusion of species having different specific enthalpies, as well as from heat conduction. As another example, the diffusional flux of any species in a multicomponent gas mixture is capable of influencing the diffusional fluxes of all other species, an effect distinct from any influence on the mixture-average velocity. Fourier's law does not include diffusional transport of enthalpy, and Fick's law has no provision for multicomponent diffusional interactions. A second reason that we need more general constitutive equations is that mass transfer may result from "driving forces" which we have not yet considered, including electric fields and gradients in temperature or pressure. These forces act differently on the various chemical species in a mixture, and therefore they cause fluxes relative to the average velocity.

The objective of this chapter is to present a much more general discussion of the constitutive equations for energy and mass transfer than that given in Chapter 1, by describing the various couplings among the fluxes and by including a more comprehensive set of driving forces. The new phenomena are introduced in several steps, gradually building toward the most general description of multicomponent energy and mass transfer. In Section 11.2 we extend the energy conservation equation to multicomponent mixtures. This provides the basis for solving a variety of problems involving simultaneous heat and mass transfer, as described in Section 11.3. The main new element in such problems is the diffusional transport of enthalpy. Section 11.4 introduces the notion of coupled fluxes, including certain concepts from nonequilibrium thermodynamics. The specific new elements there are species fluxes due to thermal diffusion and energy transfer due to the diffusion–thermo effect. The remainder of the chapter is devoted mainly to mass transfer. Section 11.5 is concerned with the simplest type of multicomponent diffusion, due to concentration gradients in an isothermal, isobaric gas mixture. Section 11.6 broadens the treatment of mass transfer in dilute mixtures, where multicomponent diffusional effects (direct couplings of species fluxes) are negligible. The main consideration there is that diffusion may result from gradients in pressure or electrical potential, as well as from gradients in concentration; there are effects also of thermodynamic nonidealities. Section 11.7 is devoted to the important special case of transport of ions in electrolyte solutions. Finally, Section 11.8 contains the general description of multicomponent diffusion, including all of the aforementioned driving forces. Certain results are derived there which are presented without proof in the earlier sections.

11.2 CONSERVATION OF ENERGY: MULTICOMPONENT SYSTEMS

A number of general expressions for conservation of energy in single-component fluids were derived in Section 9.6. The objective here is to extend those results to multicomponent mixtures. Before proceeding, the reader may find it helpful to review the notation for species velocities and fluxes given in Section 1.2.

The conservation equation for internal and kinetic energy in a single-component fluid, Eq. (9.6-7), is

$$\rho \frac{D}{Dt}\left(\hat{U}+\frac{v^2}{2}\right) = -\nabla\cdot\mathbf{q}-\nabla\cdot(P\mathbf{v})+\nabla\cdot(\boldsymbol{\tau}\cdot\mathbf{v})+\rho(\mathbf{v}\cdot\mathbf{g}). \qquad (11.2\text{-}1)$$

To extend this equation to a mixture, we need to reconsider the term $\rho(\mathbf{v}\cdot\mathbf{g})$, which represents the energy input from the gravitational force. A distinctive feature of a mixture is that the body force per unit mass on chemical species i, denoted by $\mathbf{g}_i$, may differ for each species. The most common example is the action of an electric field on an ion, for which $\mathbf{g}_i$ will depend on molecular charge. Thus, $\mathbf{g}_i \neq \mathbf{g}$ in general. For each chemical species the energy input will be given by $\rho_i\,(\mathbf{v}_i\cdot\mathbf{g}_i)=\mathbf{n}_i\cdot\mathbf{g}_i$. Replacing $\rho(\mathbf{v}\cdot\mathbf{g})$ in Eq. (11.2-1) by the sum of these contributions for all n chemical species in the mixture, we obtain

$$\rho \frac{D}{Dt}\left(\hat{U}+\frac{v^2}{2}\right) = -\nabla\cdot\mathbf{q}-\nabla\cdot(P\mathbf{v})+\nabla\cdot(\boldsymbol{\tau}\cdot\mathbf{v})+\sum_{i=1}^{n}\mathbf{n}_i\cdot\mathbf{g}_i. \qquad (11.2\text{-}2)$$

For the special case of $\mathbf{g}_i=\mathbf{g}$, Eq. (11.2-2) reduces to Eq. (11.2-1).

For Eq. (11.2-2) to properly describe energy transfer in a mixture, it is necessary also to reconsider the meaning of the "heat flux," $\mathbf{q}$. This flux represents all energy transfer relative to the mass-average velocity of the fluid. For pure fluids this flux is given simply by Fourier's law, $\mathbf{q} = -k\nabla T$, but for mixtures there are additional contributions to $\mathbf{q}$. In general, as discussed in Hirschfelder et al. (1954),

$$\mathbf{q} = -k\nabla T + \sum_{i=1}^{n} \mathbf{J}_i \overline{H}_i + \mathbf{q}^{(x)}. \tag{11.2-3}$$

The first new term in Eq. (11.2-3) represents energy transfer due to the diffusion of species having different partial molar enthalpies $(\overline{H}_i)$. The second new term, $\mathbf{q}^{(x)}$, represents the Dufour or "diffusion–thermo" effect, which is discussed further in Section 11.4. Whereas the diffusional transfer of enthalpy is often very important, the Dufour flux is usually negligible.

   To obtain a conservation equation involving only internal energy, we proceed as in Section 9.6 to derive a comparable equation for kinetic energy, which is then subtracted from Eq. (11.2-2). Conservation of momentum for a multicomponent fluid is given by

$$\rho \frac{D\mathbf{v}}{Dt} = -\nabla P + \nabla \cdot \boldsymbol{\tau} + \sum_{i=1}^{n} \rho_i \mathbf{g}_i. \tag{11.2-4}$$

Equation (11.2-4) differs from the momentum equation for a pure fluid only in that it allows for the possibility of body forces other than gravity. The corresponding kinetic energy equation is

$$\rho \frac{D}{Dt} \left(\frac{v^2}{2} \right) = -\mathbf{v} \cdot \nabla P + \mathbf{v} \cdot (\nabla \cdot \boldsymbol{\tau}) + \sum_{i=1}^{n} \rho_i (\mathbf{v} \cdot \mathbf{g}_i). \tag{11.2-5}$$

Subtracting Eq. (11.2-5) from Eq. (11.2-2), we obtain

$$\rho \frac{D\hat{U}}{Dt} = -\nabla \cdot \mathbf{q} - P(\nabla \cdot \mathbf{v}) + \boldsymbol{\tau} : \nabla\mathbf{v} + \sum_{i=1}^{n} \mathbf{j}_i \cdot \mathbf{g}_i. \tag{11.2-6}$$

Comparing Eq. (11.2-6) with Eq. (9.6-11), we see that the one new term in the internal energy equation is $\sum \mathbf{j}_i \cdot \mathbf{g}_i$. This summation vanishes for a pure fluid $(\mathbf{j}_i = \mathbf{0})$ or when the body force is the same for all species in a mixture $(\mathbf{g}_i = \mathbf{g}$ for all $i)$.

   In multicomponent energy calculations it is usually more convenient to work with partial molar enthalpies than with internal energy. The total enthalpy per unit mass is given by

$$\hat{H} = \hat{U} + \frac{P}{\rho}. \tag{11.2-7}$$

Introducing Eq. (11.2-7) into Eq. (11.2-6), we obtain

$$\rho \frac{D\hat{H}}{Dt} = -\nabla \cdot \mathbf{q} + \frac{DP}{Dt} + \boldsymbol{\tau} : \nabla\mathbf{v} + \sum_{i=1}^{n} \mathbf{j}_i \cdot \mathbf{g}_i, \tag{11.2-8}$$

which is analogous to Eq. (9.6-15). The total enthalpy and partial molar enthalpies are related by

$$\rho\hat{H} = \sum_{i=1}^{n} C_i \overline{H}_i. \tag{11.2-9}$$

Using Eq. (11.2-9) and continuity, one finds that

$$\rho \frac{D\hat{H}}{Dt} + \nabla \cdot \mathbf{q} = \frac{\partial}{\partial t} \left(\sum_{i=1}^{n} C_i \overline{H}_i \right) + \nabla \cdot \left(\mathbf{v} \sum_{i=1}^{n} C_i \overline{H}_i + \mathbf{q} \right). \tag{11.2-10}$$

The last term on the right-hand side of Eq. (11.2-10) is simplified by defining a *multi-component energy flux* relative to fixed coordinates (**e**):

$$\mathbf{e} \equiv \mathbf{v} \sum_{i=1}^{n} C_i \overline{H}_i + \mathbf{q} = -k \nabla T + \sum_{i=1}^{n} \mathbf{N}_i \overline{H}_i + \mathbf{q}^{(x)}. \tag{11.2-11}$$

Substituting Eq. (11.2-10) into Eq. (11.2-8) and using the definition of **e**, we obtain

$$\frac{\partial}{\partial t} \left(\sum_{i=1}^{n} C_i \overline{H}_i \right) = -\nabla \cdot \mathbf{e} + \frac{DP}{Dt} + \boldsymbol{\tau} : \nabla \mathbf{v} + \sum_{i=1}^{n} \mathbf{j}_i \cdot \mathbf{g}_i. \tag{11.2-12}$$

This is the general form of the multicomponent energy equation in terms of partial molar enthalpies. It contains no special assumptions about the nature of the mixture. To apply it one needs constitutive equations to describe the various fluxes in the mixture, including the dependence of the transport coefficients on composition. In addition, one needs thermodynamic data which describe the dependence of all $\overline{H}_i$ on temperature, pressure, and composition.

The steady-state form of Eq. (11.2-12) is

$$0 = -\nabla \cdot \mathbf{e} + \mathbf{v} \cdot \nabla P + \boldsymbol{\tau} : \nabla \mathbf{v} + \sum_{i=1}^{n} \mathbf{j}_i \cdot \mathbf{g}_i. \tag{11.2-13}$$

The energetic contributions of the pressure ($\mathbf{v} \cdot \nabla P$) and viscous dissipation ($\boldsymbol{\tau} : \nabla \mathbf{v}$) terms are often negligible, as discussed in Sections 2.4 and 9.6. Moreover, in most chemical engineering applications it is unnecessary to consider body forces other than gravity. Under these conditions the energy equation reduces to

$$\nabla \cdot \mathbf{e} = 0. \tag{11.2-14}$$

This simple form of the multicomponent energy equation is the starting point for many engineering calculations. Examples of its application are given in Section 11.3.

11.3 SIMULTANEOUS HEAT AND MASS TRANSFER

The most common examples of simultaneous heat and mass transfer involve phase changes (e.g., condensation or evaporation of mixtures) or heat effects due to chemical reactions. In these problems the latent heats or heats of reaction cause energy transfer to be dependent on mass transfer, so that the temperature and concentration fields are coupled. (Mass and energy fluxes may also be coupled more directly, as discussed in Section 11.4.) In such problems thermal effects usually predominate, and mechanical forms of energy can be neglected. Thus, at steady state, Eq. (11.2-14) ordinarily provides a good approximation to the energy conservation equation. If the Dufour flux is neglected, the energy flux defined by Eq. (11.2-11) simplifies to

$$\mathbf{e} = -k\nabla T + \sum_{i=1}^{n} \mathbf{N}_i \overline{H}_i. \tag{11.3-1}$$

Thus, Eqs. (11.2-14) and (11.3-1) provide the usual starting point for steady-state analyses.

The interfacial energy balance for a multicomponent system is obtained from Eq. (B) of Table 2-1 by setting $\mathbf{F} = \mathbf{e}$ and $b = \Sigma C_i \overline{H}_i$. For any given point on the interface between phases A and B, this yields

$$\left[\left(\mathbf{e} - \mathbf{v}_I \sum_{i=1}^{n} C_i \overline{H}_i \right)_B - \left(\mathbf{e} - \mathbf{v}_I \sum_{i=1}^{n} C_i \overline{H}_i \right)_A \right] \cdot \mathbf{n}_I = 0. \tag{11.3-2}$$

As used before, $\mathbf{v}_I$ is the interfacial velocity and $\mathbf{n}_I$ is a unit vector normal to the interface; both of these vectors may vary with time and/or position, thereby describing a deforming, translating, and/or nonplanar interface. By setting the right-hand side of Eq. (11.3-2) equal to zero, we are assuming no heat generation at the interface due to an external power source. It is seen that for *stationary interfaces* ($\mathbf{v}_I = \mathbf{0}$), the normal component of $\mathbf{e}$ (i.e., $e_n = \mathbf{n}_I \cdot \mathbf{e}$) is continuous across the interface. The effects of latent heats and heats of heterogeneous reactions are embodied in e_n, as will be seen.

Condensation and Evaporation

The interfacial conditions needed to describe phase changes in multicomponent systems will be illustrated using condensation and evaporation. In either process, if the coordinates are chosen so that the interface is stationary, Eq. (11.3-2) requires that

$$(-k\nabla T)_n^{(L)} + \sum_{i=1}^{n} (N_{in} \overline{H}_i)^{(L)} = (-k\nabla T)_n^{(G)} + \sum_{i=1}^{n} (N_{in} \overline{H}_i)^{(G)}, \tag{11.3-3}$$

where the subscript n denotes vector components normal to the interface, the superscripts L and G refer to the liquid and gas, respectively, and all quantities are evaluated at the interface. Assuming that no chemical reactions occur at the interface, conservation of species i requires that

$$N_{in}^{(L)} = N_{in}^{(G)}. \tag{11.3-4}$$

Combining Eqs. (11.3-3) and (11.3-4), we find that

$$(-k\nabla T)_n^{(L)} = (-k\nabla T)_n^{(G)} + \sum_{i=1}^{n} N_{in} \lambda_i, \tag{11.3-5}$$

where $\lambda_i \equiv \overline{H}_i^{(G)} - \overline{H}_i^{(L)}$, the molar latent heat for species i. The assumption of equal temperatures in the two phases, along with the thermodynamic relationships between the gas and liquid mole fractions (as functions of temperature and pressure), completes the specification of conditions at the gas–liquid interface.

Heterogeneous Reactions

A representative situation with a heterogeneous reaction involves gaseous species reacting at the surface of an impermeable solid catalyst. With $N_{in} = 0$ at the solid side of the interface, the interfacial energy balance for a stationary particle is

$$(-k\nabla T)_n^{(S)} = (-k\nabla T)_n^{(G)} + \sum_{i=1}^{n}(N_{in}\overline{H}_i)^{(G)}. \qquad (11.3\text{-}6)$$

As discussed in Section 2.7, the interfacial species balances require that

$$\frac{N_{in}^{(G)}}{\xi_i} = \frac{N_{jn}^{(G)}}{\xi_j}, \qquad (11.3\text{-}7)$$

where ξ_i are the stoichiometric coefficients of the reaction. (Recall that $\xi_i<0$ for reactants, $\xi_i>0$ for products, and $\xi_i=0$ for inert species.) The molar heat of reaction (ΔH_R) is defined as

$$\Delta H_R \equiv \sum_{i=1}^{n}\xi_i\overline{H}_i^{(G)}, \qquad (11.3\text{-}8)$$

so that $\Delta H_R<0$ for exothermic reactions and $\Delta H_R>0$ for endothermic reactions. As with the quantities appearing in the interfacial balances, the partial molar enthalpies in Eq. (11.3-8) are understood to be evaluated at the interface.

The enthalpy term in Eq. (11.3-6) will now be expressed in terms of ΔH_R and the flux of an arbitrarily chosen reactant or product. Choosing A as the reference species, we obtain

$$\sum_{i=1}^{n}(N_{in}\overline{H}_i)^{(G)} = \sum_{i=1}^{n}\frac{\xi_i}{\xi_A}(N_{An}\overline{H}_i)^{(G)} = \frac{N_{An}^{(G)}}{\xi_A}\Delta H_R. \qquad (11.3\text{-}9)$$

The interfacial energy balance becomes

$$(-k\nabla T)_n^{(S)} = (-k\nabla T)_n^{(G)} + \frac{N_{An}^{(G)}}{\xi_A}\Delta H_R. \qquad (11.3\text{-}10)$$

Comparing Eqs. (11.3-5) and (11.3-10), we see that latent heats and heats of heterogeneous reactions have analogous effects. That is, both cause an imbalance (or discontinuity) in the energy fluxes at the interface due to conduction.

Homogeneous Reactions

No new interfacial conditions are required to describe transport in nonisothermal systems with homogeneous reactions. As shown by Example 11.3-2, heat effects in homogeneous reactions are embodied in the energy conservation equation itself, Eq. (11.2-14), rather than in the boundary conditions. This is analogous to how chemical kinetic data are used: The rate of a homogeneous reaction appears in the species conservation equation, whereas the rate of a heterogeneous reaction appears in a boundary condition.

Example 11.3-1 Evaporation of a Water Droplet   We consider now the evaporation of a small water droplet suspended in a stream of nitrogen. The objective is to determine the rate of evaporation, given the droplet radius and the temperature and composition far from the droplet. For simplicity, we assume that the droplet radius $R(t)$ changes slowly with time, so that a pseudo-steady model can be used. We assume also that Pe for mass and heat transfer is small, which reduces the problem to one involving steady diffusion and conduction with spherical symmetry. The coordinate origin is chosen as the center of the droplet. The mole fraction of species i and the temperature far from the droplet are denoted by $x_{i\infty}$ and T_∞, respectively, with $i = N$ for nitrogen and $i = W$ for water.

The conservation equation for species i in the gas reduces to

$$\frac{d}{dr}(r^2 N_{ir}) = 0, \tag{11.3-11}$$

from which it follows that

$$N_{ir}(r) = \frac{R^2}{r^2} N_{ir}(R). \tag{11.3-12}$$

To proceed further in evaluating the species fluxes, we need to consider the mass-transfer conditions at the gas–liquid interface. It will be assumed that the droplet is pure water (i.e., there is negligible nitrogen in the liquid). Applying Eq. (B) of Table 2-1 to species i at the gas–liquid interface and noting that the interfacial velocity is dR/dt, we obtain

$$\left(N_{ir} - \frac{dR}{dt}C_i\right)^{(G)} = \left(N_{ir} - \frac{dR}{dt}C_i\right)^{(L)}. \tag{11.3-13}$$

If the water concentration in the droplet is denoted as C_L, then this interfacial balance implies that

$$N_{Wr}(R) = \frac{dR}{dt}(C_W(R) - C_L), \tag{11.3-14}$$

$$N_{Nr}(R) = \frac{dR}{dt}C_N(R), \tag{11.3-15}$$

where $N_{ir}(R)$ and $C_i(R)$ represent quantities evaluated in the *gas*. Within the *droplet*, the fluxes of both species are zero. The low density of a gas at ambient conditions, relative to liquid water, ensures that $C_L \gg C_W$ or C_N. Accordingly, Eqs. (11.3-14) and (11.3-15) imply that, to good approximation,

$$N_{Wr}(R) \cong -C_L \frac{dR}{dt}. \tag{11.3-16}$$

Moreover, it is seen that the nitrogen flux at the interface is negligible relative to the water flux. It follows then from Eq. (11.3-12) that the nitrogen flux is negligible *throughout the gas*.

The molar-average velocity is chosen as the diffusional reference frame. Setting $N_{Nr} = 0$ in the corresponding form of Fick's law (Tables 1-2 and 1-3), the flux of water in the gas becomes

$$N_{Wr}(r) = x_W(N_{Wr} + N_{Nr}) - CD_{WN}\frac{dx_W}{dr} = -\frac{CD_{WN}}{1 - x_W}\frac{dx_W}{dr}. \tag{11.3-17}$$

Combining Eqs. (11.3-12) and (11.3-17) and integrating from $r = R$ to $r = \infty$, we obtain

$$N_{Wr}(R) = \frac{CD_{WN}}{R}\ln\left(\frac{1 - x_{W\infty}}{1 - x_W(R)}\right). \tag{11.3-18}$$

It was assumed in the integration that CD_{WN} is constant, which is a good approximation for an isobaric gas with moderate temperature differences.

Equation (11.3-18) relates two unknown quantities, the water flux at the droplet surface, $N_{Wr}(R)$, and the water mole fraction in the gas next to the surface, $x_W(R)$. If the system were isothermal, then the surface temperature would simply be T_∞, and the analysis would be completed by calculating $x_W(R)$ from the vapor pressure of water at $T = T_\infty$. For the nonisothermal system considered here, it is necessary to use the energy equation to calculate the surface temperature.

The applicable energy conservation equation is Eq. (11.2-14), which for this system is analogous to Eq. (11.3-11). Integrating over r we obtain

$$e_r(r) = \frac{I_1}{r^2}, \tag{11.3-19}$$

where I_1 is an integration constant. From Eq. (11.3-1) we obtain

$$e_r(r) = -k_G \frac{dT}{dr} + N_{Wr}\left[\overline{C}_{pW}(T - T_0) + \overline{H}_W{}^0\right], \tag{11.3-20}$$

where k_G is the thermal conductivity of the gas. Here we have expressed the partial molar enthalpy of water vapor in terms of the molar heat capacity $\overline{C}_{pW}$ and the enthalpy $\overline{H}_W{}^0$ at some reference temperature T_0. This ignores the enthalpy of mixing and therefore requires that the mixture be ideal. Equating e_r from Eqs. (11.3-19) and (11.3-20) and using Eq. (11.3-12), we obtain

$$I_1 = -k_G r^2 \frac{dT}{dr} + R^2 N_{Wr}(R)\left[\overline{C}_{pW}(T - T_0) + \overline{H}_W{}^0\right] \tag{11.3-21}$$

or

$$I_2 = -k_G r^2 \frac{dT}{dr} + R^2 N_{Wr}(R)\overline{C}_{pW}T, \tag{11.3-22}$$

where I_2 is another constant (replacing I_1).

The constant I_2 is evaluated using an interfacial energy balance. Because the interface is moving, we return to Eq. (11.3-2) instead of using Eq. (11.3-5). Under the assumed pseudosteady conditions with spherical symmetry, it is readily shown that the droplet must be isothermal. That is, the droplet temperature is governed by Laplace's equation, with a constant temperature at the surface, implying that T is constant throughout the droplet (Example 4.2-1). Moreover, using the results of the interfacial nitrogen balance, it is found that the enthalpy terms involving nitrogen exactly cancel. Collecting the remaining terms, the interfacial energy balance is written as

$$k_G \frac{dT}{dr}(R) = -\overline{\lambda}C_L \frac{dR}{dt} = \overline{\lambda}N_{Wr}(R), \tag{11.3-23}$$

where $\overline{\lambda} \equiv \overline{H}_W{}^{(G)} - \overline{H}_W{}^{(L)}$ is the molar latent heat for evaporation of water. Evaluating Eq. (11.3-22) at $r = R$ and using Eq. (11.3-23) to determine the temperature gradient, it is found that

$$I_2 = -R^2 N_{Wr}(R)\left[\overline{\lambda} - \overline{C}_{pW}T(R)\right]. \tag{11.3-24}$$

Still remaining to be determined are the water flux and temperature at the surface, $N_{Wr}(R)$ and $T(R)$, respectively.

Using Eq. (11.3-24) in Eq. (11.3-22), we complete the integration of the energy equation. Integrating from $r = R$ to $r = \infty$ (with k_G and $\overline{C}_{pW}$ assumed constant), we find that

$$N_{Wr}(R) = \frac{k_G}{\overline{C}_{pW}R} \ln\left\{1 + \frac{\overline{C}_{pW}[T_\infty - T(R)]}{\overline{\lambda}}\right\}. \tag{11.3-25}$$

Equating the expressions for the water flux obtained from the analyses of species transport, Eq. (11.3-18), and energy transport, Eq. (11.3-25), we arrive at a relationship between the surface temperature and surface composition. The result is

$$\left(\frac{1 - x_{W\infty}}{1 - x_W(R)}\right)^{D_{WN}/\alpha^*} = 1 + \frac{\overline{C}_{pW}[T_\infty - T(R)]}{\overline{\lambda}}, \tag{11.3-26}$$

where $\alpha^* = k_G/(C\overline{C}_{pW})$ is a modified thermal diffusivity.[1] A second relationship between $T(R)$ and $x_W(R)$ is obtained from vapor–liquid equilibrium data. For example, assuming ideal gas behavior, we obtain

$$x_W(R) = \frac{P_W[T(R)]}{P}, \tag{11.3-27}$$

where $P_W(T)$, the vapor pressure of water, has been equated with the partial pressure of water at the interface. The analysis may be completed by solving Eqs. (11.3-26) and (11.3-27) simultaneously to determine $T(R)$ and $x_W(R)$. Once those quantities are known, $N_{Wr}(R)$ can be calculated from either Eq. (11.3-18) or Eq. (11.3-25). Finally, dR/dt can be determined from Eq. (11.3-16).

Example 11.3-2 Nonisothermal Reaction in a Porous Catalyst Consider an irreversible chemical reaction involving a binary gas mixture. The reaction is of the form $A \rightarrow B$ and takes place within a porous catalyst pellet. As mentioned in Example 3.7-2, the usual approach for modeling diffusion in such a system is to employ an effective diffusivity, D_E, which is measured for the materials of interest. It is assumed that the flux of A is given by

$$\mathbf{N}_A = -D_E \nabla C_A. \tag{11.3-28}$$

We also employ an effective thermal conductivity for the pellet, k_E. Because some of the pellet volume is occupied by solids and because the pores are not straight, D_E will tend to be less than D_{AB}, the binary gas diffusivity. Ordinarily, k_E for a catalyst pellet mainly reflects heat conduction through the solid. The poorly conducting, gas-filled pores will cause k_E to be lower than that for a nonporous but otherwise similar solid. In the analysis which follows, D_E and k_E are both treated as constants. The effective homogeneous reaction rate is based on the total rate of reaction within any small, representative volume (including solids and gas-filled pores). Assuming that D_E, k_E, and the function $R_{VA} = R_{VA}(C_A, T)$ are known, we wish to assess the temperature variations at steady state, allowing for a nonzero heat of reaction. The analysis which follows is valid for a pellet of any shape.

Making use of the stoichiometry, the species conservation equations are

$$\nabla \cdot \mathbf{N}_A = R_{VA} = -\nabla \cdot \mathbf{N}_B. \tag{11.3-29}$$

Employing Eq. (11.3-28) to evaluate the flux, the equation for A becomes

$$0 = D_E \nabla^2 C_A + R_{VA}. \tag{11.3-30}$$

Because R_{VA} depends on T, integration of Eq. (11.3-30) requires information on the temperature profile within the pellet. Once again, conservation of energy is expressed by Eq. (11.2-14). As in the preceding example, we treat the gas mixture as ideal and write the partial molar enthalpies as

$$\overline{H}_i = \overline{H}_i^0 + \overline{C}_{pi}(T - T_0), \tag{11.3-31}$$

where $\overline{H}_i^0$, $\overline{C}_{pi}$, and T_0 are constants. The heat of reaction is expressed as

$$\Delta H_R = \overline{H}_B - \overline{H}_A = \Delta H_R^0 + (\overline{C}_{pB} - \overline{C}_{pA})(T - T_0), \tag{11.3-32}$$

---

[1] The usual thermal diffusivity, α, is related to α^* by

$$\alpha = \frac{k}{\rho \hat{C}_p} = \frac{k}{C\overline{C}_p} = \left(\frac{\overline{C}_{pW}}{\overline{C}_p}\right)\alpha^*,$$

where $\overline{C}_p$ is the mean molar heat capacity for the mixture. The ratio D_{WN}/α^* is a modified Lewis number.

where $\Delta H_R^0 \equiv \overline{H}_B^0 - \overline{H}_A^0$ is the heat of reaction at temperature T_0. To simplify the analysis, we assume now that $\overline{C}_{pi}\Delta T \ll \Delta H_R^0$, where $i = A$ or B and ΔT is the maximum temperature difference within the pellet. Accordingly, the individual enthalpies and the heat of reaction are all regarded as constant. The energy equation then becomes

$$0 = \nabla \cdot \mathbf{e} = \nabla \cdot (-k_E \nabla T + \mathbf{N}_A \overline{H}_A + \mathbf{N}_B \overline{H}_B) = -k_E \nabla^2 T - R_{VA}\Delta H_R^0. \tag{11.3-33}$$

To obtain the last equality, we used the assumed constancy of the partial molar enthalpies and employed Eq. (11.3-29) to relate the divergence of each species flux to the reaction rate. In this example and in other problems involving homogeneous reactions, the heat of reaction enters the analysis when one evaluates the divergence of the total enthalpy flux (i.e., $\nabla \cdot \sum \mathbf{N}_i \overline{H}_i$).

A very useful relationship between the reactant concentration and the temperature within the pellet is derived by multiplying Eq. (11.3-30) by $\Delta H_R^0/k_E$ and Eq. (11.3-33) by $1/k_E$. Adding the results, we obtain

$$\nabla^2(\gamma C_A - T) = 0, \tag{11.3-34}$$

$$\gamma \equiv \frac{D_E \Delta H_R^0}{k_E}. \tag{11.3-35}$$

Equation (11.3-34) indicates that a function consisting of a certain linear combination of C_A and T satisfies Laplace's equation. According to the results in Example 4.2-1, if that function is constant on the boundary, then it must equal the same constant everywhere. We conclude that if the concentration and temperature on the pellet surface are uniform at C_{AS} and T_S, respectively, then

$$T - T_S = \gamma(C_A - C_{AS}) \tag{11.3-36}$$

throughout the pellet. This implies that for any interior points where the reactant is largely consumed, such that $C_A \ll C_{AS}$, there will be a *maximum* temperature difference of

$$(T - T_S)_{\max} = -\frac{D_E \Delta H_R^0}{k_E} C_{AS}. \tag{11.3-37}$$

The algebraic sign of $(T - T_S)_{\max}$ depends, of course, on the sign of ΔH_R^0. This result, obtained originally by Prater (1958), provides a simple way to estimate an upper bound for temperature variations with a given reaction and catalyst. It is independent of the reaction rate law and is also independent of the pellet size and shape. Moreover, the analysis is readily extended to other stoichiometries.

To proceed further, we would need to specify the geometry and the dependence of R_{VA} on T and C_A. Given the relationship between T and C_A in Eq. (11.3-36), we could then solve Eq. (11.3-30) for C_A. Unless the pellet is nearly isothermal, numerical methods are needed. Given C_A, the temperature distribution and the overall reaction rate are calculated in a straightforward manner.

11.4 INTRODUCTION TO COUPLED FLUXES

In the elementary forms of the constitutive equations each of the fluxes is associated with a single "driving force," the driving forces being gradients in temperature or composition. Thus, Fourier's law relates the energy flux to the temperature gradient, and the various forms of Fick's law relate the diffusional flux of species i to a gradient in the concentration, mole fraction, or mass fraction of i. In nonisothermal and/or multicomponent mixtures, however, it has been found that any of the driving forces may give rise to any of the fluxes. That is, in addition to the direct effects given by Fourier's and Fick's laws and also in addition to the diffusional transport of enthalpy in a mixture, there are coupling effects. Species fluxes result from a temperature gradient, even in the absence of concentration gradients or bulk flow; this phenomenon is called "thermal diffusion" or the "Soret effect." Likewise, there is an energy flux caused by concentration gradients, even if there is no temperature gradient; this is the "diffusion–thermo" or Dufour flux, $\mathbf{q}^{(x)}$, already mentioned in Section 11.2. In addition, a flux of species i in a multicomponent mixture may result from a concentration gradient for any other species j. These coupling effects, which are predicted by the kinetic theory of gases as well as being observed experimentally, require that we seek more general constitutive equations.

To describe the coupling of fluxes in a more precise yet general way, we let $\mathbf{f}^{(1)}$, $\mathbf{f}^{(2)}, \ldots, \mathbf{f}^{(N)}$ represent a set of fluxes and let $\mathbf{X}^{(1)}$, $\mathbf{X}^{(2)}, \ldots \mathbf{X}^{(N)}$ be a set of driving forces. For example, $\mathbf{f}^{(1)}$ might be the energy flux, $\mathbf{f}^{(2)}$ the flux of species A, $\mathbf{f}^{(3)}$ the flux of species B, and so on. The driving forces are related to gradients in temperature and in the variables used to describe composition (concentrations, mole fractions, or mass fractions). It turns out that for any given system the number of independent fluxes is equal to the number of independent driving forces. Moreover, the number of independent energy and mass fluxes is equal to the number of chemical species, n. The possible force–flux relations may be represented as a set of linear, homogeneous, algebraic equations, written as

$$\mathbf{f}^{(i)} = \sum_{j=1}^{n} L_{ij} \mathbf{X}^{(j)}, \qquad (11.4\text{-}1)$$

where L_{ij} is the coefficient relating flux i to force j. For these coefficients to be scalars, as indicated, the material must be isotropic. The fluxes and forces are numbered so that in the $n \times n$ matrix of coefficients the diagonal elements (L_{ii}) correspond to the more familiar direct effects, whereas the off-diagonal elements (L_{ij}, $i \neq j$) account for the various couplings which occur.

The linearity of the relations between $\mathbf{f}^{(i)}$ and $\mathbf{X}^{(j)}$, as expressed by Eq. (11.4-1), is one of the postulates of nonequilibrium thermodynamics, a body of theory which deals with dynamic processes in systems not too far removed from equilibrium [see de Groot and Mazur (1962) for an extensive discussion of this subject]. While in principle these linear force–flux relations may fail if the gradients represented by the $\mathbf{X}^{(j)}$ become too large, there appear to be no examples of such a failure in heat or mass transfer. It is worth noting that the elementary constitutive equations are themselves linear force–flux relations.

The existence of coupled fluxes implies that, in a multicomponent mixture, there are many more transport coefficients than the ones which appear in the elementary laws. The attendant experimental difficulties are mitigated somewhat by the fact that not all of the coefficients L_{ij} are independent. A fundamental theorem of nonequilibrium thermodynamics, due to Onsager, is that for a "proper choice" of fluxes and forces, the coefficient matrix is symmetric:

$$L_{ij} = L_{ji}. \qquad (11.4\text{-}2)$$

Thus, with n forces and fluxes, the number of independent transport coefficients is reduced from n^2 to $n(n+1)/2$. It follows that the measurement of a particular flux provides information also on the "reciprocal" flux whose coefficient is positioned on the other

TABLE 11-1
Transport Coefficients for a Nonisothermal
Binary Mixture

Flux[a]	Associated Gradient	
	Temperature	Mass fraction
Energy	k (Fourier)	$D_A^{(T)}$ (Dufour)
Species A	$D_A^{(T)}$ (Soret)	D_{AB} (Fick)

[a] The energy flux here is the sum of the Fourier and Dufour fluxes, or $\mathbf{q} - \sum_{i=1}^{n} \mathbf{J}_i \bar{H}_i$. The species flux is $\mathbf{j}_A$.

side of the matrix diagonal. The relationships among coefficients represented by Eq. (11.4-2) are often referred to as the "Onsager reciprocal relations." Equation (11.4-2) must be modified if an external magnetic field is present [see de Groot and Mazur (1962, p. 39)]. What constitutes a thermodynamically proper choice of fluxes and forces for energy and mass transfer is discussed in Section 11.8.

The coefficients which govern energy and mass transfer in a nonisothermal, binary mixture are shown in Table 11-1. In a binary mixture with fluxes expressed relative to any mixture-average velocity, only one of the species fluxes is mathematically independent (see Table 1-2). Accordingly, it is necessary to consider only one species flux in addition to the energy flux. In the 2×2 matrix relating the fluxes and driving forces, the diagonal coefficients correspond to k and D_{AB}, while the off-diagonal couplings are described by one additional coefficient, $D_A^{(T)}$, the *thermal diffusion coefficient*.

For a binary mixture of A and B (gas, liquid, or solid), the contribution of thermal diffusion to the mass flux of A relative to the mass-average velocity may be written as

$$\mathbf{j}_A^{(T)} = -D_A^{(T)} \nabla \ln T. \tag{11.4-3}$$

The coefficient $D_A^{(T)}$ has units of kg m^{-1} s^{-1}. It may be positive or negative, depending on the relative molecular weights of A and B. If A has the larger molecular weight (i.e., $M_A > M_B$), then generally $D_A^{(T)} > 0$, so that thermal diffusion of A will be from hotter to colder temperatures. As is readily shown,[2] $D_B^{(T)} = -D_A^{(T)}$. Thus, there is only one independent thermal diffusivity for a binary mixture, just as there is only one independent mass diffusivity (i.e., $D_{AB} = D_{BA}$). The thermal diffusivity depends strongly on the concentrations of A and B (see Example 11.4-1).

The Dufour energy flux is considerably more complicated than the thermal diffusion flux, and there appears to be no general expression which is suitable for all fluids. For a binary gas mixture, it is expressed as (Hirschfelder et al., 1954)

$$\mathbf{q}^{(x)} = \frac{RTD_A^{(T)}}{D_{AB}} \left(\frac{x_A}{M_B} + \frac{x_B}{M_A} \right) (\mathbf{v}_A - \mathbf{v}_B). \tag{11.4-4}$$

(Throughout this chapter, R is used only for the gas constant.)

---

[2] By definition, $\mathbf{j}_A + \mathbf{j}_B = 0$. For this to hold for any combination of temperature and concentration gradients, it must be true that $\mathbf{j}_A^{(T)} + \mathbf{j}_B^{(T)} = -(D_A^{(T)} + D_B^{(T)}) \nabla \ln T = 0$. Accordingly, $D_B^{(T)} = -D_A^{(T)}$.

In a mixture containing n components, there are $n-1$ independent diffusional fluxes. As discussed for ideal gases in Section 11.5 and more generally in Section 11.8, these involve $n(n-1)/2$ multicomponent mass diffusivities. In addition, there are $n-1$ multicomponent thermal diffusivities and one thermal conductivity. Thus, there is a total of $n(n+1)/2$ transport coefficients for energy and mass transfer, as indicated earlier.

The following examples, which involve binary gas mixtures, illustrate the magnitudes of the thermal diffusion and Dufour fluxes.

Example 11.4-1 Thermal Diffusion in a Binary Gas Mixture   Assume that gaseous species A and B are confined between parallel surfaces maintained at different temperatures, as shown in Fig. 11-1. Assume further that the surfaces are impermeable, that there are no chemical reactions, and that convection (e.g., buoyancy-induced flow) is absent. With no convection, the steady-state temperature and mole fractions will depend only on y. The objective is to determine the mole fraction difference, $x_A(L) - x_A(0)$, induced by the temperature difference.

For this one-dimensional configuration, the continuity equation reduces to $d(\rho v_y)/dy = 0$. With constant ρv_y and impermeable walls, it follows that $v_y = 0$ for all y. Similarly, from the conservation equation for species A, $N_{Ay} = J_{Ay} = 0$ for all y. Considering both the Fick's law and thermal diffusion contributions to the flux of A, we have

$$J_{Ay} = -\frac{\rho D_{AB}}{M_A}\frac{d\omega_A}{dy} - \frac{D_A^{(T)}}{M_A}\frac{d\ln T}{dy} = 0. \tag{11.4-5}$$

Differential changes in the mass fraction and mole fraction are related by

$$d\omega_A = \frac{M_A M_B C^2}{\rho^2}\,dx_A. \tag{11.4-6}$$

Combining Eqs. (11.4-5) and (11.4-6), we obtain

$$dx_A = -x_A x_B \alpha_A d\ln T, \tag{11.4-7}$$

$$\alpha_A \equiv \frac{\rho D_A^{(T)}}{C^2 M_A M_B x_A x_B D_{AB}}. \tag{11.4-8}$$

Whereas $D_A^{(T)}$ depends strongly on the mole fractions of A and B, the *thermal diffusion factor* α_A defined by Eq. (11.4-8) is nearly independent of the mole fractions. In other words, it is approximately true that $D_A^{(T)} \propto x_A x_B$. Although α_A depends on temperature, we will treat it here as a constant.

Composition changes are expected to be small in this system, so that the average mole

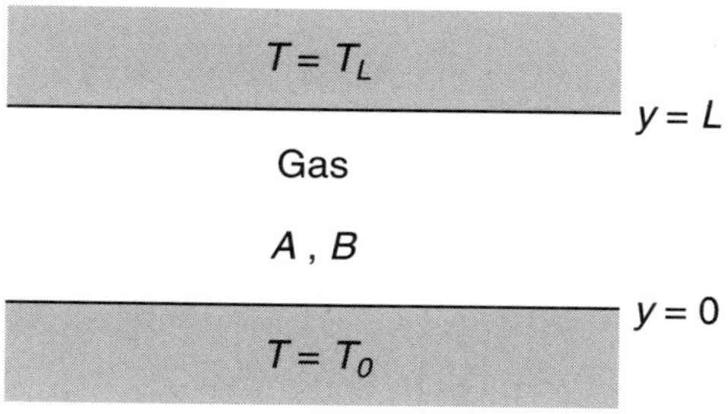

Figure 11-1. A gas mixture of species A and B confined between parallel surfaces. Constant temperatures are maintained at the top and bottom, such that $T_L > T_0$.

$T = T_L$

$y = L$

Gas

A , B

$y = 0$

$T = T_0$

fractions, $\langle x_A \rangle$ and $\langle x_B \rangle$, can be substituted for x_A and x_B on the right-hand side of Eq. (11.4-7). Making those substitutions and completing the integration, we obtain

$$x_A(L) - x_A(0) \cong -\langle x_A \rangle\langle x_B \rangle\alpha_A \ln(T_L/T_0). \tag{11.4-9}$$

Equation (11.4-9) can be used to judge whether or not thermal diffusion is negligible for a binary gas subjected to a given range of temperatures. For gases at 200–400 K and atmospheric pressure the value of $|\alpha_A|$ usually does not exceed 0.3 and is often as low as 0.01 (Grew and Ibbs, 1952; Vasaru et al., 1969). For $|\alpha_A| = 0.3$, $T_L = 300$ K, $T_0 = 200$ K, and $\langle x_A \rangle = \langle x_B \rangle = 0.5$, we obtain $|x_A(L) - x_A(0)| = 0.03$. Thus, the thermally induced change in x_A under these conditions is a small fraction of $\langle x_A \rangle$. Nonetheless, thermal diffusion has been used effectively in gas separations, especially of isotopes. A steep temperature gradient between two *vertical* surfaces, combined with buoyancy-induced flow, can produce large separation factors in a tall column. An overview of thermal diffusion is provided by Grew and Ibbs (1952), and the analysis and use of thermal diffusion columns is discussed by Vasaru et al. (1969).

Example 11.4-2 Dufour Flux in a Binary Gas Mixture   Consider a stagnant-film model for energy and mass transfer in a binary gas next to a catalytic surface, as shown in Fig. 11-2. At the solid surface ($y = L$) there is a fast, irreversible reaction with the stoichiometry $A \rightarrow 2B$. The temperature and composition of the bulk gas (at $y = 0$) are given, and there is no heat conduction or diffusion within the solid. The system is at steady state and convection is assumed to be absent, so that temperature and composition depend only on y. The objective is to assess the relative importance of the Dufour flux in energy transfer across the gas film.

As in the previous example, the impermeable solid surface dictates that $v_y = 0$ for all y. Moreover, in the absence of a homogeneous reaction, it is again true that $N_{iy} = J_{iy} = $ constant for all y, where $i = A$ or B; in this case, because of the heterogeneous reaction, the species fluxes are not zero. Neglecting thermal diffusion and assuming that ρ and D_{AB} are approximately constant,

$$J_{Ay} = \frac{D_{AB} C_{A0}}{L} = -\frac{J_{By}}{2}, \tag{11.4-10}$$

where we have used the fast-reaction boundary condition, $C_A(L) = 0$, and the reaction stoichiometry.

The energy conservation equation reduces in this case to $de_y/dy = 0$ or $e_y = $ constant. With no energy transfer in the solid, the interfacial energy balance requires that $e_y(L) = 0$, so that $e_y = 0$ for all y. Accordingly,

$$e_y = -k\frac{dT}{dy} + J_{Ay}\overline{H}_A + J_{By}\overline{H}_B + q_y^{(x)} = 0, \tag{11.4-11}$$

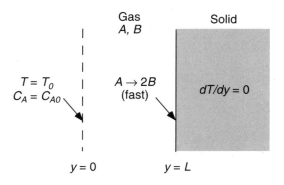

Gas
A, B

Solid

$T = T_0$
$C_A = C_{A0}$

$A \rightarrow 2B$
(fast)

$dT/dy = 0$

$y = 0$ $y = L$

Figure 11-2. A gas mixture of A and B in contact with a catalytic surface where there is a fast, irreversible reaction.

where k is the thermal conductivity of the gas mixture. Using the stoichiometry and introducing the heat of reaction, Eq. (11.4-11) becomes

$$e_y = -k\frac{dT}{dy} - J_{Ay}\Delta H_R + q_y^{(x)} = 0. \tag{11.4-12}$$

Equation (11.4-4) is used now to evaluate the Dufour flux. Recognizing that with $v_y = 0$ the species fluxes are related to the species velocities by $J_{iy} = C_i v_{iy}$, and noting that the stoichiometry requires that $M_B = M_A/2$, the Dufour flux is given by

$$q_y^{(x)} = \frac{RTD_A^{(T)}J_{Ay}}{CD_{AB}M_A}\frac{(1+x_A)^2}{x_A x_B}. \tag{11.4-13}$$

Introducing the thermal diffusion factor defined by Eq. (11.4-8), the Dufour flux simplifies to

$$q_y^{(x)} = \alpha_A RT(1+x_A)J_{Ay}. \tag{11.4-14}$$

Using Eqs. (11.4-14) and (11.4-10) in Eq. (11.4-12), the overall energy flux becomes

$$e_y = -k\frac{dT}{dy} - \frac{D_{AB}C_{A0}}{L}[\Delta H_R - \alpha_A RT(1+x_A)] = 0. \tag{11.4-15}$$

If the Dufour flux is neglected and if k and ΔH_R are assumed constant, then it follows from Eq. (11.4-15) that the temperature difference between the surface and the bulk gas is given by

$$T_L - T_0 = -\frac{D_{AB}C_{A0}\Delta H_R}{k}, \tag{11.4-16}$$

where $T_L = T(L)$. Thus, the temperature difference is set by the requirement that there be a certain rate of heat conduction through the gas, to or from the surface depending on the sign of ΔH_R. Recalling now that $\alpha_A > 0$ for $M_A > M_B$, in the present example the Dufour energy flux is directed toward the surface (i.e., $q_y^{(x)} > 0$). Accordingly, the effect of the Dufour flux will be to increase $|T_L - T_0|$ for an exothermic reaction ($\Delta H_R < 0$) and to decrease $|T_L - T_0|$ for an endothermic reaction ($\Delta H_R > 0$). However, Eq. (11.4-15) indicates that the Dufour effect will be important only if α_A is comparable to $|\Delta H_R/RT|$. For a typical heat of reaction of 30 kcal/mol, $|\Delta H_R/RT| = 50$ at $T = 300$ K. With $\alpha_A = 0.3$, the Dufour flux will be only about 1% of the conduction (Fourier) flux.

11.5 STEFAN–MAXWELL EQUATIONS

The Stefan–Maxwell equations describe multicomponent diffusion in ideal gases. Thermodynamic ideality requires that the gas density be low, and the Stefan–Maxwell equations assume also that the effects of any temperature and/or pressure gradients on diffusion are negligible. Those conditions are often met, so that the Stefan–Maxwell equations have proven to be very useful, as discussed in Cussler (1976). They have been derived from the kinetic theory of gases, and they are obtained also as a special case from the general treatment of multicomponent diffusion given in Section 11.8. Because they ignore diffusional driving forces other than gradients in mole fraction and are therefore relatively simple, the Stefan–Maxwell equations provide a good introduction to the modeling of multicomponent diffusion.

The gradient in the mole fraction of any given species is related to the velocities (or fluxes) of *all* species in a mixture. For an ideal gas without large temperature or pressure gradients, the specific relationship is

$$\nabla x_i = \sum_{\substack{j=1 \\ j \neq i}}^{n} \frac{x_i x_j}{D_{ij}} (\mathbf{v}_j - \mathbf{v}_i), \tag{11.5-1}$$

where n is the number of components in the gas mixture, and D_{ij} are the usual binary diffusivities. Writing Eq. (11.5-1) for the various species in the mixture generates a set of relations which is one form of the *Stefan–Maxwell equations*. Because the mole fractions sum to unity, the gradients of the mole fractions sum to zero. Accordingly, only $n-1$ of these equations provide independent information.

The driving force for diffusion in Eq. (11.5-1) is ∇x_i, and the fluxes are related to the species velocities (i.e., $\mathbf{v}_i = \mathbf{N}_i / C_i$). Inasmuch as each driving force is expressed as a linear combination of fluxes, Eq. (11.5-1) represents an inverted form of Eq. (11.4-1). While seemingly more awkward than a set of equations modeled after Eq. (11.4-1), the Stefan–Maxwell equations have a major advantage: The coefficients D_{ij} can be determined individually from binary diffusion data, and they are independent of the mole fractions. The coefficients which result from inverting the Stefan–Maxwell equations, thereby solving for the fluxes in terms of the ∇x_i, are not nearly as simple. Thus, the "backward" or Stefan–Maxwell form of equation is much more useful in multicomponent diffusion calculations than is a "forward," or generalized Fick's law, form of equation.

A mechanical analogy is helpful in understanding the form of Eq. (11.5-1). It is well known that if two solid surfaces are in close contact and if they slide relative to one another, then a frictional force results. Likewise, Eq. (11.5-1) suggests that each relative velocity between pairs of chemical species, $\mathbf{v}_j - \mathbf{v}_i$, results in a kind of frictional force acting on a molecular level. The magnitude of this frictional force is proportional to the frequency of encounters between i and j (and thus to $x_i x_j$) and is inversely proportional to D_{ij}. The mole fraction gradient which causes the relative motion between i and the other species must match the sum of the opposing frictions. Note that because only *differences* in species velocities appear in Eq. (11.5-1), adding the same vector to $\mathbf{v}_i$ and $\mathbf{v}_j$ does not change the equations. Accordingly, the Stefan–Maxwell equations remain the same for any diffusional reference frame (i.e., any choice of mixture-average velocity).

A more practical form of the Stefan–Maxwell equations is obtained by replacing the velocities in Eq. (11.5-1) by the molar fluxes, using $\mathbf{N}_i = \mathbf{v}_i C_i$. The result is

$$\nabla x_i = \sum_{\substack{j=1 \\ j \neq i}}^{n} \frac{1}{CD_{ij}} (x_i \mathbf{N}_j - x_j \mathbf{N}_i). \tag{11.5-2}$$

Example 11.5-1 Relationship Between Stefan–Maxwell Equations and Fick's Law To show that the Stefan–Maxwell equations reduce to Fick's law for a binary mixture, we apply Eq. (11.5-2) to a mixture of species A and B:

$$\nabla x_A = \frac{1}{CD_{AB}} (x_A \mathbf{N}_B - x_B \mathbf{N}_A). \tag{11.5-3}$$

As already mentioned, only $n-1$ of the Stefan–Maxwell equations are independent, so the corresponding equation involving ∇x_B is redundant. Rearranging Eq. (11.5-3) using $x_B = 1 - x_A$, we obtain

$$\mathbf{N}_A = x_A(\mathbf{N}_A + \mathbf{N}_B) - CD_{AB}\nabla x_A. \tag{11.5-4}$$

This is the same as Fick's law, using the molar-average velocity (see Tables 1-2 and 1-3).

Example 11.5-2 Ternary Gas Diffusion with a Heterogeneous Reaction   Consider nitrogen
and hydrogen diffusing to a catalytic surface, where they react to form ammonia. The overall
reaction is written as

$$\frac{1}{2} N_2 + \frac{3}{2} H_2 \rightarrow NH_3.$$

For simplicity, we consider again a one-dimensional, stagnant-film geometry, at steady state. Temperature and pressure are assumed to be constant. The gas composition is assumed to be known
at the plane $y = 0$ (mole fractions x_{i0}), and the solid surface is located at $y = L$. The objective is to
determine the rate of formation of NH_3. For convenience, we number the components as follows:
$1 = N_2$, $2 = H_2$, and $3 = NH_3$.

Two independent Stefan–Maxwell equations, based on Eq. (11.5-2), are

$$C\frac{dx_1}{dy} = \frac{1}{D_{12}}(x_1 N_{2y} - x_2 N_{1y}) + \frac{1}{D_{13}}(x_1 N_{3y} - x_3 N_{1y}), \tag{11.5-5}$$

$$C\frac{dx_2}{dy} = \frac{1}{D_{12}}(x_2 N_{1y} - x_1 N_{2y}) + \frac{1}{D_{23}}(x_2 N_{3y} - x_3 N_{2y}). \tag{11.5-6}$$

For steady, one-dimensional diffusion with no homogeneous reactions, the species conservation
equation reduces to

$$\frac{dN_{iy}}{dy} = 0. \tag{11.5-7}$$

Thus, the fluxes of all three species are constant.

For a specified temperature and pressure, the total molar concentration C is a known constant. The unknowns in Eqs. (11.5-5) and (11.5-6) are then $x_i(y)$ and N_{iy}, where $i = 1$, 2, or 3.
One of the mole fractions can always be expressed in terms of the others, thereby reducing the
number of unknowns. Choosing x_3 as the mole fraction to be eliminated,

$$x_3 = 1 - x_1 - x_2. \tag{11.5-8}$$

Additionally, all but one of the fluxes can be eliminated by using the reaction stoichiometry. Using
Eq. (11.3-7), with $\xi_1 = -\frac{1}{2}$, $\xi_2 = -\frac{3}{2}$, and $\xi_3 = 1$, we find that

$$N_{1y} = \frac{\xi_1}{\xi_3} N_{3y} = -\frac{1}{2} N_{3y}, \tag{11.5-9}$$

$$N_{2y} = \frac{\xi_2}{\xi_3} N_{3y} = -\frac{3}{2} N_{3y}. \tag{11.5-10}$$

Using Eqs. (11.5-8)–(11.5-10) in Eqs. (11.5-5) and (11.5-6), we obtain

$$-\frac{2C}{N_{3y}}\frac{dx_1}{dy} = \frac{1}{D_{12}}(3x_1 - x_2) + \frac{1}{D_{13}}(-1 - x_1 + x_2), \tag{11.5-11}$$

$$-\frac{2C}{N_{3y}}\frac{dx_2}{dy} = \frac{1}{D_{12}}(x_2 - 3x_1) + \frac{1}{D_{23}}(-3 + 3x_1 + x_2). \tag{11.5-12}$$

Equations (11.5-11) and (11.5-12), with the conditions $x_1(0) = x_{10}$ and $x_2(0) = x_{20}$, can be solved
simultaneously to determine $x_1(y)$ and $x_2(y)$. Both of these functions will depend parametrically

on the ammonia flux, N_{3y}, which is still unknown. To complete the solution, the reaction rate must be related to the concentrations of hydrogen and nitrogen at the catalytic surface. This kinetic information can be expressed generally as

$$-N_{3y} = R_{S3} = f[x_1(L), x_2(L), T, P].$$ (11.5-13)

Equation (11.5-13) provides the additional relation needed to determine N_{3y}.

The solution for $x_1(y)$ and $x_2(y)$ is simplified by using the experimental observation that for this particular system, $D_{12} \cong D_{23}$. Equation (11.5-12) then simplifies to

$$-\frac{2C}{N_{3y}} \frac{dx_2}{dy} \cong \frac{1}{D_{12}} (-3 + 2x_2),$$ (11.5-14)

which no longer contains $x_1(y)$. Solving Eq. (11.5-14), we obtain

$$x_2 = \frac{3}{2} + \left(x_{20} - \frac{3}{2}\right) \exp[-yN_{3y}/(CD_{12})].$$ (11.5-15)

Substituting this result into Eq. (11.5-11) and solving again, we find that

$$x_1 = \frac{1}{2} + \left(x_{20} - \frac{3}{2}\right) \exp[-yN_{3y}/(CD_{12})]$$
$$+ (1 + x_{10} - x_{20}) \exp[-(3 - \gamma)yN_{3y}/(2CD_{12})],$$ (11.5-16)

where $\gamma = D_{12}/D_{13}$.

For the special case of fast reaction kinetics, the diffusion of either N_2 or H_2 (but usually not both) will be rate-limiting. Under these conditions, Eq. (11.5-13) is replaced by

$$x_i(L) = 0,$$ (11.5-17)

where $i = 1$ or $i = 2$, depending on the values of x_{10}, x_{20}, and the other parameters. The identity of the rate-limiting reactant [i.e., the value of i in Eq. (11.5-17)] is determined by trial and error. If $i = 1$ is correct, the solution should give $x_2(L) \geq 0$, and vice versa.

Several other examples of the application of the Stefan–Maxwell equations to steady diffusion in ternary gas mixtures are given in Toor (1957). A more extensive and general treatment of multicomponent diffusion (including liquids and solids as well as gases) is provided by Cussler (1976). We return to the subject of multicomponent diffusion in Section 11.8.

11.6 GENERALIZED DIFFUSION IN DILUTE MIXTURES

In Section 11.4 it was described how diffusion may result from gradients in temperature as well as from gradients in concentration. The objective here is to generalize the concept of diffusion further to include other driving forces (e.g., pressure gradients). The present discussion is restricted to dilute mixtures, which are characterized by the fact that the mole fraction of one component (e.g., the solvent in a liquid solution) is near unity. Consequently, encounters between the other species are relatively rare and multicomponent diffusional effects are negligible.

In a *dilute* mixture the diffusional flux of solute i is written generally as

$$\mathbf{J}_i = -D_i C \mathbf{d}_i - \frac{D_i^{(T)}}{M_i} \nabla \ln T.$$ (11.6-1)

The second term on the right-hand side represents thermal diffusion (Section 11.4), whereas the first term describes all other types of diffusion. As usual for diffusivities in dilute solutions, the subscripts referring to the abundant species have been omitted. The new quantity is the diffusional driving force $\mathbf{d}_i$, defined as

$$CRT\mathbf{d}_i = C_i \nabla_{T,P}\mu_i + (C_i \overline{V}_i - \omega_i)\nabla P - \rho_i\left(\mathbf{g}_i - \sum_{k=1}^{n}\omega_k\mathbf{g}_k\right), \tag{11.6-2}$$

$$C_i\nabla_{T,P}\mu_i = \sum_{j=1}^{n}\left(\frac{\partial\mu_i}{\partial x_j}\right)_{T,P,x_k}\nabla x_j, \tag{11.6-3}$$

where μ_i and $\overline{V}_i$ are the chemical potential and partial molar volume of species i, respectively. As described in Section 11.8, where this quantity is derived, $CRT\mathbf{d}_i$ is the force per unit volume tending to move species i relative to an isothermal mixture.

The term in Eq. (11.6-2) involving $\nabla_{T,P}\mu_i$, the gradient in chemical potential at constant temperature and pressure, describes diffusion in response to *concentration* gradients. Although the mixture is assumed here to be dilute, it may be thermodynamically nonideal. Thus, Eq. (11.6-3) allows for the possibility that the chemical potential of species i is influenced by the mole fractions of *all* components. It is often convenient to rewrite Eq. (11.6-3) as

$$\nabla_{T,P}\mu_i = RT\nabla \ln(\gamma_i x_i), \tag{11.6-4}$$

where γ_i is the activity coefficient of species i. The effects of the thermodynamic nonidealities are embedded now in γ_i, which depends on the local composition; $\gamma_i = 1$ for an ideal solution. The second term on the right-hand side of Eq. (11.6-2) represents diffusion due to *pressure* gradients, whereas the third term describes diffusion in response to *body forces,* such as gradients in electrostatic potential. If there is no body force other than gravity, such that $\mathbf{g}_i = \mathbf{g}$, then

$$\rho_i\left(\mathbf{g}_i - \sum_{k=1}^{n}\omega_k\mathbf{g}_k\right) = \rho_i\left(\mathbf{g} - \mathbf{g}\sum_{k=1}^{n}\omega_k\right) = \mathbf{0}. \tag{11.6-5}$$

Thus, gravity does not contribute to the diffusional force, $\mathbf{d}_i$.

The effects of solution nonidealities and of pressure gradients are illustrated by the examples which follow. The relationship between electrostatic potentials and the transport of ions in electrolyte solutions, which requires more extensive discussion, is the subject of Section 11.7.

Example 11.6-1 Concentration Diffusion in a Nonideal Solution We assume here that the effects of pressure and temperature gradients are negligible and that there are no body forces other than gravity. Accordingly, the only diffusional force is that due to $\nabla_{T,P}\mu_i$. Using Eq. (11.6-4) in Eq. (11.6-1), the expression for the flux becomes

$$\mathbf{J}_i = -\frac{D_i}{RT}(CRT\mathbf{d}_i) = -\frac{D_i}{RT}(C_i\nabla_{T,P}\mu_i) = -D_i C_i\nabla \ln(\gamma_i x_i). \tag{11.6-6}$$

Thus, the rate of diffusion is influenced by the gradient in both the activity coefficient and the mole fraction.

For a *binary* solution at constant temperature and pressure, it is informative to rearrange the right-hand side of Eq. (11.6-6) as

$$C_i \nabla \ln(\gamma_i x_i) = \frac{C_i}{\gamma_i x_i} \nabla(\gamma_i x_i) = \frac{C}{\gamma_i} (\gamma_i \nabla x_i + x_i \nabla \gamma_i) = C\left(1 + \frac{d \ln \gamma_i}{d \ln x_i}\right)\nabla x_i. \tag{11.6-7}$$

The last equality makes use of the fact that, under these conditions, $\gamma_i = \gamma_i(x_i)$ only. Substituting Eq. (11.6-7) into Eq. (11.6-6), we obtain

$$\mathbf{J}_i \equiv -D_i^* \nabla C_i, \tag{11.6-8}$$

$$D_i^* = \left(1 + \frac{d \ln \gamma_i}{d \ln x_i}\right) D_i. \tag{11.6-9}$$

Thus, if we relate the flux to the concentration gradient in the usual way, as in Eq. (11.6-8), then the apparent diffusivity D_i^* is given by Eq. (11.6-9). In general, D_i^* is more sensitive to x_i than is the "true" diffusivity, D_i. For an ideal solution, the two diffusivities are equal.

Example 11.6-2 Pressure Diffusion in an Ultracentrifuge An ultracentrifuge offers a means to achieve separations based on very large pressure gradients. As shown in Fig. 11-3, the liquid in a closed sample cell is assumed to occupy radial positions $a - (L/2) \leq r \leq a + (L/2)$, and the centrifuge is assumed to spin at a constant angular velocity Ω. The objective is to determine the concentration profile $C_i(r)$ for an uncharged solute under isothermal conditions. In this closed system at steady state there will be no relative motion among the species, so that $\mathbf{J}_i = \mathbf{0}$ for all i. Thus, for constant T, Eq. (11.6-1) implies that $\mathbf{d}_i = \mathbf{0}$ for all i. This will be true even if the solution is not dilute. That is, when all diffusional fluxes vanish, multicomponent diffusional effects are eliminated (see Section 11.8).

We consider now the various contributions to $\mathbf{d}_i$. With uncharged solutes the only body force is gravity, and as shown by Eq. (11.6-5), gravity does not affect diffusion. Thus, the separation achieved in the ultracentrifuge is evidently due to a balance between concentration diffusion and pressure diffusion. Using Eq. (11.6-4) to evaluate the radial component of the chemical potential gradient, we obtain

$$C_i\left(\frac{\partial \mu_i}{\partial r}\right)_{T,P} = CRT\left(1 + \frac{\partial \ln \gamma_i}{\partial \ln x_i}\right)\frac{dx_i}{dr}. \tag{11.6-10}$$

The radial pressure gradient is evaluated by noting that $v_r = v_z = 0$ and $v_\theta = r\Omega$ for rigid-body rotation at angular velocity Ω. The r-momentum equation then reduces to

$$\frac{\partial P}{\partial r} = \frac{\rho v_\theta^2}{r} = \rho \Omega^2 r. \tag{11.6-11}$$

Using Eqs. (11.6-10) and (11.6-11) in Eq. (11.6-2) and recalling that $\mathbf{d}_i = \mathbf{0},$ we obtain

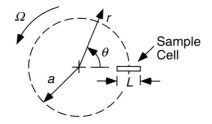

Figure 11-3. Schematic representation of a sample cell of length L in an ultracentrifuge. The mean radius of rotation is a and the angular velocity is Ω

$$\left(1+\frac{\partial \ln \gamma_i}{\partial \ln x_i}\right)\frac{dx_i}{dr}=(\omega_i-C_i\overline{V}_i)\frac{\rho\Omega^2 r}{CRT}.$$

(11.6-12)

This result is valid for ideal or nonideal solutions containing any number of components. The physical significance of Eq. (11.6-12) is seen by noting that $C_i\overline{V}_i$ is the volume fraction of species i in the mixture. Thus, the "denser species," those for which the mass fraction (ω_i) exceeds the volume fraction, will tend to migrate away from the axis of rotation (i.e., $dx_i/dr>0$); the less dense species will migrate toward the axis $(dx_i/dr<0)$.

Further insight is gained by integrating Eq. (11.6-12), after introducing certain approximations. Assume now that the solution is both ideal and dilute, so that no correction is needed for variations in the activity coefficient and C is constant. Assume also that the length of the sample cell is much smaller than the axis of rotation (i.e., $L\ll a$), so that r on the right-hand side of Eq. (11.6-12) can be replaced by a. Equation (11.6-12) then becomes

$$\frac{dC_i}{dr}=C_iM_i\left(1-\frac{\overline{V}_i\rho}{M_i}\right)\frac{\Omega^2 a}{RT}.$$

(11.6-13)

We now introduce a dimensionless radial coordinate,

$$\eta\equiv\frac{r-[a-(L/2)]}{L},$$

(11.6-14)

such that $\eta=0$ at the inner end of the sample cell and $\eta=1$ at the outer end. Changing from r to η in Eq. (11.6-13) and integrating gives

$$C_i=C_i(0)\exp(\Lambda_i\,\eta),$$

(11.6-15)

$$\Lambda_i\equiv M_i\left(1-\frac{\overline{V}_i\rho}{M_i}\right)\frac{\Omega^2 aL}{RT}.$$

(11.6-16)

The concentration ratio from the outer to the inner surface of the sample cell is given simply by $\exp(\Lambda_i)$.

From Tanford (1961), some representative parameter values are $a=6.5$ cm, $L=1.0$ cm, $\Omega=6.3\times10^3$ s^{-1} (60,000 rpm), and $T=293$ K, for which $\Omega^2 aL/(RT)\cong1.1\times10^{-2}$ mol/g. For a typical protein in water, $M_i=1.0\times10^5$ g/mol and $\overline{V}_i/M_i=0.75$ cm^3/g, yielding $\Lambda_i\cong260$. With this large value of Λ_i, $>90\%$ of the protein is calculated to reside within the outer 1% (or 0.1 mm) of the sample cell. By contrast, for a typical inorganic salt with $M_i=65$ g/mol and $\overline{V}_i/M_i=0.3$ cm^3/g, $\Lambda_i\cong0.5$. These calculations illustrate the much greater effectiveness of ultracentrifugation for polymeric solutes. That is, a large value of M_i is much more important for the separation than is a small value of $\overline{V}_i/M_i$.

11.7 TRANSPORT IN ELECTROLYTE SOLUTIONS

The most important body force in mass transfer is that exerted by an electric field on an ion. In an electrolyte solution the electric current is carried by the dissolved ions (rather than by free electrons), and the ion fluxes are influenced by the electric field. Transport calculations for electrolyte solutions are complicated by the fact that the field strength is dependent on the concentrations of the various charged species. Thus, the electric field cannot usually be specified independently, and it is an additional unknown to be calculated. The interdependence of the ion concentrations and the electric field causes the fluxes of all ions to be coupled to one another, even for dilute solutions.

Ion Fluxes

The electric force on an ion is the product of the electric field ($\mathbf{E}$) and the ionic charge. The electrostatic potential ϕ, customarily expressed in volts, is defined by $\mathbf{E} = -\nabla\phi$. Expressing the charge in coulombs (C), the charge per mole of ion i is $z_i F$, where z_i is the valence (e.g., $+1$ for Na^+, -2 for SO_4^{-2}) and F is Faraday's constant (9.652×10^4 C/mol). Thus, the electric force per mole of i is $-z_i F\nabla\phi$. The total body force on ion i (per unit mass) is then

$$\mathbf{g}_i = \mathbf{g} + \mathbf{g}_i^{(e)}, \tag{11.7-1}$$

$$\mathbf{g}_i^{(e)} = -\frac{z_i F}{M_i}\nabla\phi, \tag{11.7-2}$$

where $\mathbf{g}_i^{(e)}$ is the electrostatic contribution.

Using Eqs. (11.7-1) and (11.7-2) to evaluate the body force term in the expression for $\mathbf{d}_i$, Eq. (11.6-2), we obtain

$$-\rho_i\left(\mathbf{g}_i - \sum_{k=1}^{n}\omega_k\mathbf{g}_k\right) = z_i C_i F\nabla\phi\left(1 - \frac{M_i}{z_i M}\sum_{k=1}^{n}z_k x_k\right), \tag{11.7-3}$$

where M is the mean molecular weight for the solution. The last term on the right-hand side of Eq. (11.7-3), involving the summation, is negligible in almost all applications. For a solution which is locally electroneutral, the summation vanishes identically [see Eq. (11.7-9)]. For a dilute but not necessarily electroneutral solution, that summation is also very small, because the sum of all ion mole fractions is much less than unity. Neglecting that term, Eq. (11.6-2) becomes

$$CRT\mathbf{d}_i = C_i\nabla_{T,P}\mu_i + (C_i\bar{V}_i - \omega_i)\nabla P + z_i C_i F\nabla\phi. \tag{11.7-4}$$

This expression for the diffusional force remains fairly general. When combined with the generalized Stefan–Maxwell equations (Section 11.8), it provides a basis for multicomponent diffusion calculations involving ions. It also leads to the pseudobinary flux equation which is at the heart of most descriptions of transport in electrolyte solutions, as described below.

We assume now that the electrolyte solution is ideal and dilute and that pressure diffusion is negligible. Under these conditions Eq. (11.7-4) reduces to

$$C\mathbf{d}_i = \nabla C_i + z_i C_i\left(\frac{F}{RT}\right)\nabla\phi. \tag{11.7-5}$$

Combining this result with Eq. (11.6-1) and neglecting thermal diffusion, we obtain

$$\mathbf{J}_i = -D_i\left[\nabla C_i + z_i C_i\left(\frac{F}{RT}\right)\nabla\phi\right], \tag{11.7-6}$$

which is one form of the *Nernst–Planck equation*. In addition to the usual term describing concentration diffusion, Eq. (11.7-6) includes a contribution to the flux from the movement of species i in an electric field. Inspection of the latter term reveals that

RT/F has the same units as ϕ. Accordingly, RT/F is a natural scale for the electrostatic potential in ion diffusion problems; at 25°C, $RT/F = 25.7$ mV. Including the convective part of the flux, an alternative form of the Nernst–Planck equation is

$$\mathbf{N}_i = C_i \mathbf{v} - D_i \left[\nabla C_i + z_i C_i \left(\frac{F}{RT} \right) \nabla \phi \right]. \tag{11.7-7}$$

In either form, the Nernst–Planck equation may be viewed as a generalization of Fick's law for dilute electrolyte solutions. This pseudobinary flux equation is used widely in electrochemical engineering, colloid science, and biophysics, just as Fick's law is employed widely for systems with uncharged solutes.

Electroneutrality and Poisson's Equation

In a mass transfer problem involving a dilute electrolyte solution with n components (including the solvent), there are $n-1$ unknown solute concentrations (C_i) and an unknown electrostatic potential (ϕ). Conservation equations written for each of the $n-1$ solutes leave us one short of the number of equations needed to determine the n unknowns. The additional equation comes from either of two relations involving the charge density. The net charge density (electrical charge concentration) at any point in a solution (ρ_e) is given by

$$\rho_e = F \sum_{i=1}^{n} z_i C_i. \tag{11.7-8}$$

The most common approach is to assume that the solution is locally electroneutral, or that

$$\sum_{i=1}^{n} z_i C_i = 0. \tag{11.7-9}$$

This is an excellent approximation for most situations, failing only in the vicinity of charged interfaces. The other approach, based on Poisson's equation (see below), is used in systems where interfacial effects are prominent.

To understand the limitations of Eq. (11.7-9), we need to consider how the charge density in an electrolyte solution varies near a charged surface. To preserve overall (macroscopic) electroneutrality, any net charge residing at an interface must be balanced by an opposite charge in the solution. The electrostatic potential due to the surface charge tends to attract oppositely charged ions toward the surface, and repel ions of like charge. Opposing this is the tendency of diffusion to make the solution uniform. The net effect, as illustrated in Fig. 11-4, is that $|\rho_e|$ decays gradually to zero with increasing distance from the surface. The combination of a charged surface and an oppositely charged layer of solution is called an *electrical double layer*. The decay of ρ_e with distance is roughly exponential, with a space constant equal to the *Debye length*, λ. In other words, λ is the characteristic thickness of the *diffuse* part of the double layer, the part residing in the solution. For a univalent–univalent salt in water at 25°C, $\lambda = 3.0$ $C_s^{-1/2}$, where λ is in Å and the salt concentration C_s is in molar units (see Example 11.7-3). Thus, the distances over which ρ_e differs appreciably from zero are ordinarily very small, only ~ 10 Å for a 0.1 M salt solution. When the dimensions of the system greatly exceed λ, the use of Eq. (11.7-9) usually results in negligible error.

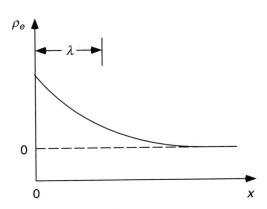

Figure 11-4. Charge density in solution (ρ_e) as a function of distance (x) from a charged interface. Here the interface is assumed to be negatively charged, so that $\rho_e \geq 0$. This figure corresponds to the simplest model, in which all ions are assumed to behave as point charges. The Debye length is denoted as λ.

To analyze transport in systems where interfacial charge effects are prominent, such as colloidal suspensions and porous media, we need an exact relationship between ρ_e and ϕ. Such a relationship is derived from *Gauss' law*, one of Maxwell's equations of electricity and magnetism. It is written as

$$\int_S \varepsilon \mathbf{E} \cdot \mathbf{n} \, dS = Q_e, \qquad (11.7\text{-}10)$$

where ε is the dielectric permittivity ($\varepsilon = \varepsilon_o = 8.854 \times 10^{-12}$ C V^{-1} m^{-1} in vacuum), and Q_e is the total charge enclosed by the surface S. If we assume that the volume enclosed by S is occupied by an electrolyte solution, then Q_e is the integral of ρ_e over that volume. Evaluating Q_e in that manner and using $\mathbf{E} = -\nabla\phi$, Eq. (11.7-10) becomes

$$-\int_S \varepsilon \nabla\phi \cdot \mathbf{n} \, dS = \int_V \rho_e \, dV. \qquad (11.7\text{-}11)$$

Applying the divergence theorem to Eq. (11.7-11) and letting $V \to 0$, we obtain

$$\nabla \cdot (\varepsilon \nabla\phi) = -\rho_e. \qquad (11.7\text{-}12)$$

This fundamental result from electrostatics is called *Poisson's equation*. It is usually simplified by assuming that ε is constant, in which case

$$\nabla^2\phi = -\frac{\rho_e}{\varepsilon}. \qquad (11.7\text{-}13)$$

Poisson's equation (in either form) is used in place of Eq. (11.7-9) when the system to be modeled has significant deviations from electroneutrality.

Current Density

A key feature of electrolyte solutions, already mentioned, is that the electrical current is carried by the dissolved ions. The *electric current density* (or flux of charge) $\mathbf{i}$ is given by

$$\mathbf{i} = F \sum_{i=1}^{n} z_i \mathbf{N}_i. \qquad (11.7\text{-}14)$$

Thus, the current density is a weighted sum of the species fluxes, the weighting factor being the valence. Using the Nernst–Planck equation to evaluate $\mathbf{N}_i$, the current density is expressed as

$$\mathbf{i} = F\mathbf{v}\sum_{i=1}^{n} z_i C_i - F\sum_{i=1}^{n} z_i D_i \nabla C_i - \frac{F^2}{RT}\nabla\phi\sum_{i=1}^{n} z_i^2 D_i C_i. \tag{11.7-15}$$

As shown, three factors may contribute to the current: convection, diffusion, and ion movement induced by the electric field. Assuming local electroneutrality, as in Eq. (11.7-9), there will be no convective contribution to the current. If the composition is uniform ($\nabla C_i = \mathbf{0}$), there will also be no contribution from diffusion. Under these conditions Eq. (11.7-15) reduces to

$$\mathbf{i} = -\kappa_e \nabla\phi \quad \text{(electroneutral, } \nabla C_i = \mathbf{0}), \tag{11.7-16}$$

$$\kappa_e \equiv \frac{F^2}{RT}\sum_{i=1}^{n} z_i^2 D_i C_i, \tag{11.7-17}$$

where κ_e is the *electrical conductance*. Equation (11.7-16) is analogous to Ohm's law. Notice in Eq. (11.7-17) that all ions make a positive contribution to the conductance, irrespective of the sign of z_i. The contribution of each ion to κ_e is proportional to C_i, as one might expect.

To summarize, for the analysis of transport in a dilute electrolyte solution containing n species, there will be at least n unknowns, consisting of $n-1$ solute concentrations and ϕ. To determine these unknowns there are $n-1$ solute conservation equations and either electroneutrality [Eq. (11.7-9)] or Poisson's equation [Eq. (11.7-13)]. For the simplest situation, where $\mathbf{v}$ is known and the solution contains only a *binary electrolyte* (a salt consisting of a single type of cation and single type of anion), there are three unknowns. Even with the dilute solution and electroneutrality approximations, the governing set of equations (two differential and one algebraic) is nonlinear and strongly coupled, due to the interdependence of the ion fluxes and potential gradient. Nonetheless, useful analytical results can be obtained for a variety of situations. A sampling of these is provided by the examples which follow.

Example 11.7-1 Effective Diffusivity of a Binary Electrolyte   The objective here is to determine how the individual ion diffusivities contribute to the apparent diffusivity of a dissolved salt. Consider a salt whose stoichiometric formula has ν_+ cations (positively charged ions) of charge number z_+ and has ν_- anions (negatively charged ions) of charge number z_-. For example, for $CuCl_2$, $\nu_+ = 1$, $\nu_- = 2$, $z_+ = 2$, and $z_- = -1$. Neutrality of the salt *formula* requires that

$$z_+\nu_+ + z_-\nu_- = 0. \tag{11.7-18}$$

If the salt *solution* is assumed to be locally electroneutral, Eq. (11.7-9) becomes

$$z_+C_+ + z_-C_- = 0. \tag{11.7-19}$$

Combining Eqs. (11.7-18) and (11.7-19), we find that

$$C_s = \frac{C_+}{\nu_+} = \frac{C_-}{\nu_-}. \tag{11.7-20}$$

Thus, it is sufficient to work with a single concentration, C_s, the "salt concentration."

Assuming the solution to be dilute, we combine the Nernst–Planck equation [Eq. (11.7-7)] with the species conservation equation for a constant-property fluid, which results in

$$\frac{DC_i}{Dt}=D_i\left[\nabla^2C_i+z_i\left(\frac{F}{RT}\right)\nabla\cdot(C_i\nabla\phi)\right]+R_{Vi}. \tag{11.7-21}$$

Assuming no homogeneous reactions and using Eq. (11.7-20), Eq. (11.7-21) becomes the following for the *cation:*

$$\frac{DC_s}{Dt}=D_+\left[\nabla^2C_s+z_+\left(\frac{F}{RT}\right)\nabla\cdot(C_s\nabla\phi)\right]. \tag{11.7-22}$$

Doing the same for the *anion* we obtain

$$\frac{DC_s}{Dt}=D_-\left[\nabla^2C_s+z_-\left(\frac{F}{RT}\right)\nabla\cdot(C_s\nabla\phi)\right]. \tag{11.7-23}$$

Subtracting Eq. (11.7-23) from Eq. (11.7-22) gives

$$0=(D_+-D_-)\nabla^2C_s+(z_+D_+-z_-D_-)\left(\frac{F}{RT}\right)\nabla\cdot(C_s\nabla\phi). \tag{11.7-24}$$

Using this last result to eliminate $\nabla\cdot(C_s\nabla\phi)$ from either Eq. (11.7-22) or Eq. (11.7-23), we obtain finally

$$\frac{DC_s}{Dt}=D_s\nabla^2C_s, \tag{11.7-25}$$

$$D_s\equiv\frac{(z_+-z_-)D_+D_-}{z_+D_+-z_-D_-}. \tag{11.7-26}$$

Thus, the salt concentration C_s is governed by the usual conservation equation for a dilute solution of a neutral solute, with an effective diffusivity D_s given by Eq. (11.7-26). Accordingly, $C_s\,(\mathbf{r},\,t)$ can be determined without having to evaluate ϕ.

Example 11.7-2 Diffusion Potential and Electroneutrality   Concentration gradients in an electrolyte solution tend to create gradients in the electrostatic potential, ϕ. From Eq. (11.7-15), for a solution with local electroneutrality but with $\nabla C_i\neq\mathbf{0}$, we obtain

$$\nabla\phi=-\frac{\mathbf{i}}{\kappa_e}-\frac{F}{\kappa_e}\sum_{i=1}^{n}z_iD_i\nabla C_i. \tag{11.7-27}$$

The first term on the right-hand side is the ohmic contribution to the potential gradient, whereas the second term creates what is called a *diffusion potential.* It is evident that even if $\mathbf{i}=\mathbf{0}$, a potential gradient will tend to arise if $\nabla C_i\neq\mathbf{0}$ for any of the ions. The exception is if D_i is the same for all ions, in which case electroneutrality guarantees that the diffusion potential will vanish. Thus, diffusion potentials arise from a combination of ion concentration gradients and unequal ion diffusivities.

A specific example will illustrate the typical magnitudes of diffusion potentials and also provide an assessment of the accuracy of the electroneutrality approximation. Assume that NaCl, KCl, or HCl diffuses along a channel of length L, where $C_s=C_s(x)$ only, $C_s(0)=C_0$, and $C_s(L)=C_L$. The solution is dilute, the system is at steady state, there is no convection, and there is no current flow. Assuming electroneutrality, $C_s=C_+=C_-$. With $\mathbf{i}=\mathbf{0}$, it follows from Eq. (11.7-14) that $\mathbf{N}_+=\mathbf{N}_-$. Equating the ion fluxes in the x direction, we find that

$$D_+\left[\frac{dC_s}{dx}+C_s\left(\frac{F}{RT}\right)\frac{d\phi}{dx}\right]=D_-\left[\frac{dC_s}{dx}-C_s\left(\frac{F}{RT}\right)\frac{d\phi}{dx}\right].$$ (11.7-28)

Solving for the potential gradient, we obtain

$$\frac{d\phi}{dx}=\left(\frac{D_- -D_+}{D_- +D_+}\right)\left(\frac{RT}{F}\right)\frac{d\ln C_s}{dx}.$$ (11.7-29)

Integrating Eq. (11.7-29) from $x=0$ to $x=L$ results in

$$\phi(0)-\phi(L)=\left(\frac{D_- -D_+}{D_- +D_+}\right)\left(\frac{RT}{F}\right)\ln\left(\frac{C_0}{C_L}\right).$$ (11.7-30)

The infinite-dilution diffusivities in water at 25°C are 2.032×10^{-5} cm²/s (Cl⁻), 1.334×10^{-5} cm²/s (Na⁺), 1.957×10^{-5} cm²/s (K⁺), and 9.312×10^{-5} cm²/s (H⁺) (Newman, 1973, p.230). Assuming that $C_0/C_L=10$, the values of $\phi(0)-\phi(L)$ are $+12.3$, $+1.1$, and -37.9 mV for NaCl, KCl, and HCl, respectively. The diffusion potential is nearest zero when the cation and anion diffusivities are most nearly matched, as with KCl.

Assuming that the expression for the potential gradient in Eq. (11.7-29) is accurate, we can estimate the deviation from electroneutrality. Evaluating $d^2\phi/dx^2$ by differentiating Eq. (11.7-29) and substituting the result into Poisson's equation, Eq. (11.7-13), we obtain

$$\frac{d^2\phi}{dx^2}=\left(\frac{D_- -D_+}{D_- +D_+}\right)\left(\frac{RT}{F}\right)\left[\frac{1}{C_s}\frac{d^2C_s}{dx^2}-\frac{1}{C_s^2}\left(\frac{dC_s}{dx}\right)^2\right]=\frac{-F(C_+ -C_-)}{\varepsilon}.$$ (11.7-31)

To estimate $C_+ -C_-$ from this relation, we need to determine $C_s(x)$. Using the conservation equation for a binary electrolyte [Eq. (11.7-25)], we find that

$$\frac{d^2C_s}{dx^2}=0,$$ (11.7-32)

$$\frac{dC_s}{dx}=\frac{C_L-C_0}{L},$$ (11.7-33)

$$C_s=C_0+(C_L-C_0)\frac{x}{L}.$$ (11.7-34)

Substituting these results into Eq. (11.7-31) and solving for $C_+ -C_-$ gives

$$C_+ -C_-=\frac{\varepsilon RT}{F^2}\left(\frac{D_- -D_+}{D_- +D_+}\right)\left[\frac{(C_L-C_0)}{L[C_0+(C_L-C_0)(x/L)]}\right]^2.$$ (11.7-35)

The greatest deviations from electroneutrality will evidently occur when the values of D_- and D_+ are very different and when the values of C_0 and C_L are very different. Assuming that $C_0\gg C_L$, Eq. (11.7-35) indicates that

$$|C_+ -C_-|\sim\frac{\varepsilon RT}{F^2L^2}\left(\frac{C_0}{C_L}\right)^2.$$ (11.7-36)

Using $\varepsilon/\varepsilon_0=78$ (relative dielectric constant for water), $L=1$ cm, and $C_0/C_L=10$, we obtain $|C_+ -C_-|\sim10^{-13}$ M. This exceedingly small deviation from electroneutrality supports the use of Eq. (11.7-9) in modeling electrolyte transport in macroscopic systems.

Example 11.7-3 Poisson–Boltzmann Equation and Double-Layer Potential   A common starting point for the analysis of transport phenomena in double layers is the assumption that all ion concentrations obey a *Boltzmann distribution:*

$$C_i = C_{i\infty} \exp(-z_i F\phi/RT). \tag{11.7-37}$$

The concentration far from the interface, where ϕ is taken to be zero, is denoted by $C_{i\infty}$. The assumption underlying Eq. (11.7-37) is that any transport processes which may be occurring perturb C_i at most slightly from its equilibrium distribution.[3] Using Eq. (11.7-37) to evaluate ρ_e in Poisson's equation and assuming that ε is constant, we obtain

$$\nabla^2\phi = -\frac{F}{\varepsilon} \sum_{i=1}^{n} z_i C_{i\infty} \exp\left(-\frac{z_i F\phi}{RT}\right), \tag{11.7-38}$$

which is the *Poisson–Boltzmann equation.* Together with suitable boundary conditions (see below), it may be used to calculate ϕ within the diffuse double layer.

When the electrostatic potential is sufficiently small, Eq. (11.7-38) can be simplified by expanding the exponential in a Taylor series. The first two terms give

$$\nabla^2\phi = -\frac{F}{\varepsilon} \sum_{i=1}^{n} z_i C_{i\infty}\left(1 - \frac{z_i F\phi}{RT} + \cdots\right). \tag{11.7-39}$$

With the bulk solution taken to be electroneutral, the first term of the expansion vanishes. Retaining only the next term, we find that

$$\nabla^2\phi = \frac{\phi}{\lambda^2}, \tag{11.7-40}$$

$$\lambda \equiv \left(\frac{F^2}{\varepsilon RT} \sum_{i=1}^{n} z_i^2 C_{i\infty}\right)^{-1/2}. \tag{11.7-41}$$

Equation (11.7-40) is called the *linearized Poisson–Boltzmann equation;* unlike Eq. (11.7-38), it is linear in ϕ. The Debye length, mentioned earlier, is given explicitly by Eq. (11.7-41).

The boundary condition relating the electrostatic potential to the charge density at an interface is derived from Gauss' law. As shown in Fig. 11-5, suppose that the interface between two materials having dielectric constants ε_1 and ε_2 has a local surface charge density q_e (net charge per unit area, positive or negative). Applying Eq. (11.7-10) to a control volume of infinitesimal dimensions centered on the interface, we find that

$$q_e = \varepsilon_1\left(\frac{\partial\phi}{\partial n}\right)_1 - \varepsilon_2\left(\frac{\partial\phi}{\partial n}\right)_2, \tag{11.7-42}$$

where the potential gradients are evaluated at the interface, and the normal coordinate n is directed from phase 1 to phase 2.

As a specific example of calculating ϕ in a double layer, consider a planar interface between a solid and an electrolyte solution, where the solution occupies the space $y>0$. Assuming that Eq. (11.7-40) is applicable and that $\phi = \phi(y)$ only, we have

$$\frac{d^2\phi}{dy^2} = \frac{\phi}{\lambda^2}. \tag{11.7-43}$$

---

[3] The Boltzmann distribution is the equilibrium condition for a dilute solution at constant T and P, as may be demonstrated by purely thermodynamic arguments. This equilibrium condition follows also from the Nernst–Planck equation. That is, from Eq. (11.7-6), $\mathbf{J}_i = 0$ when $\ln C_i + z_i F\phi/(RT)$ is a constant. This leads to Eq. (11.7-37).

Figure 11-5. Schematic of a charged interface.

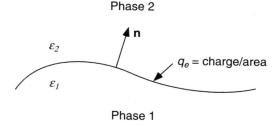

Assuming that $C_i \to C_{i\infty}$ as $y \to \infty$, the boundary condition required for consistency with Eq. (11.7-37) is

$$\phi(\infty) = 0. \tag{11.7-44}$$

The other boundary condition is obtained from Eq. (11.7-42). Assuming that there is no potential gradient within the solid and setting $n = y$ and $\varepsilon_2 = \varepsilon$, we obtain

$$\frac{d\phi}{dy}(0) = -\frac{q_e}{\varepsilon}, \tag{(11.7-45)}$$

where ε is the dielectric permittivity of the liquid. Integrating Eq. (11.7-43) and applying these boundary conditions, the potential is found to be

$$\phi = \frac{\lambda q_e}{\varepsilon} e^{-y/\lambda}. \tag{11.7-46}$$

Accordingly, ϕ decays exponentially from an interfacial value of $\lambda q_e / \varepsilon$ to zero, with a space constant λ. This result is used in Eq. (11.7-13) to determine the solution charge density, $\rho_e(y)$, as

$$\rho_e = -\varepsilon \frac{d^2\phi}{dy^2} = -\frac{q_e}{\lambda} e^{-y/\lambda}. \tag{11.7-47}$$

Thus, ρ_e also decays exponentially with distance from the interface, as was depicted qualitatively in Fig. 11-4. It is easily confirmed that the overall system, interface plus solution, is electrically neutral.

Example 11.7-4 Electroosmotic Flow *Electroosmosis* is the term for fluid flow across a membrane or other porous medium resulting from an applied electric field. Such flow results from the action of the applied field on the diffuse part of the electric double layer in the pores (i.e., the fluid where $\rho_e \neq 0$). The following analysis is based on that in Newman (1973, pp. 193–196).

Consider a cylindrical pore of radius a. We will assume that the pore wall has a uniform charge density (q_e) and that the axial component of the electric field (E_z) is constant; both q_e and E_z are assumed to be known. We assume also that there are no axial variations in composition and that the potentials are sufficiently small to expand the exponential in the Boltzmann distribution as

$$C_i = C_{i0} \exp\left[\frac{-z_i F(\phi - \phi_0)}{RT}\right] \cong C_{i0}\left[1 - \frac{z_i F(\phi - \phi_0)}{RT}\right]. \tag{11.7-48}$$

The concentration and potential at the pore centerline are denoted by C_{i0} and $\phi_0(z)$, respectively. Implicit in Eq. (11.7-48) is the assumption that the diffuse double layer maintains its equilibrium structure, even for $E_z \neq 0$. The objective is to determine the velocity, $v_z(r)$, resulting from a given applied field, E_z.

Considering both gravitational and electrical body forces, the momentum equation for a multicomponent fluid, Eq. (11.2-4), becomes

$$\rho \frac{D\mathbf{v}}{Dt} = -\nabla \mathcal{P} + \nabla \cdot \tau + \rho_e \mathbf{E}. \tag{11.7-49}$$

For fully developed flow with $v_\theta = 0$ and axial symmetry, we conclude as usual from continuity that $v_r = 0$. To examine "pure" electroosmosis, with no Poiseuille flow contribution, we assume that $\partial \mathcal{P}/\partial z = 0$. The z component of Eq. (11.7-49), for a Newtonian fluid, is then

$$0 = \frac{\mu}{r} \frac{d}{dr} \left(r \frac{dv_z}{dr} \right) + \rho_e E_z. \tag{11.7-50}$$

From Poisson's equation, we find that

$$\rho_e = -\varepsilon \nabla^2 \phi = -\frac{\varepsilon}{r} \frac{\partial}{\partial r}\left(r \frac{\partial \phi}{\partial r} \right). \tag{11.7-51}$$

The term $\partial^2 \phi/\partial z^2$ does not appear because $E_z = -\partial \phi/\partial z$ has been assumed to be constant. Using Eq. (11.7-51) to eliminate ρ_e from Eq. (11.7-50), and rearranging, we obtain

$$\frac{d}{dr} \left(r \frac{dv_z}{dr} \right) = \frac{\varepsilon E_z}{\mu} \frac{\partial}{\partial r} \left(r \frac{\partial \phi}{\partial r} \right). \tag{11.7-52}$$

Integrating once and requiring that dv_z/dr and $\partial \phi/\partial r$ both vanish at $r = 0$, we find that

$$\frac{dv_z}{dr} = \frac{\varepsilon E_z}{\mu} \frac{\partial \phi}{\partial r}. \tag{11.7-53}$$

Integrating again gives

$$v_z = \frac{\varepsilon E_z}{\mu} (\phi - \phi_a), \tag{11.7-54}$$

where $\phi_a \equiv \phi(a, z)$, the unknown potential at the pore wall, has been introduced to satisfy the no-slip condition.

To complete the solution for v_z it is necessary to evaluate $\phi - \phi_a$. Using Eq. (11.7-48) in Eq. (11.7-51), we obtain

$$\frac{1}{r} \frac{\partial}{\partial r} \left(r \frac{\partial \phi}{\partial r} \right) = -\frac{F}{\varepsilon} \left[\sum_{i=1}^{n} z_i C_{i0} - \sum_{i=1}^{n} \frac{z_i^2 C_{i0} F(\phi - \phi_0)}{RT} \right]. \tag{11.7-55}$$

This result is analogous to Eq. (11.7-40), the linearized Poisson–Boltzmann equation derived in Example 11.4-3. However, in deriving Eq. (11.7-40) the term corresponding to the first summation in Eq. (11.7-55) vanished. In electroosmosis the fluid at the pore centerline is not necessarily electroneutral, so that this term must be retained. In solving Eq. (11.7-55) it is convenient to introduce the dimensionless variables

$$\tilde{r} = \frac{r}{\lambda}, \qquad \tilde{\phi} = \frac{(\phi - \phi_0)F}{RT}, \tag{11.7-56}$$

where λ is given by Eq. (11.7-41), with $C_{i\infty}$ replaced by C_{i0}. The modified Poisson–Boltzmann equation is now

$$\frac{1}{\tilde{r}} \frac{\partial}{\partial \tilde{r}} \left(\tilde{r} \frac{\partial \tilde{\phi}}{\partial \tilde{r}} \right) = -\Lambda + \tilde{\phi}, \tag{11.7-57}$$

$$\Lambda \equiv \sum_{i=1}^{n} z_i C_{i0} \Big/ \sum_{i=1}^{n} z_i^2 C_{i0}. \tag{11.7-58}$$

The particular solution to Eq. (11.7-57) is $\tilde{\phi} = \Lambda$. The homogeneous form of Eq. (11.7-57) is the modified Bessel equation of order zero, which has the independent solutions $I_0(\tilde{r})$ and $K_0(\tilde{r})$; the latter is excluded because it is unbounded at the centerline. The solution is then

$$\tilde{\phi} = \Lambda[1 - I_0(\tilde{r})], \tag{11.7-59}$$

where the coefficient of $I_0(\tilde{r})$ has been chosen to satisfy the condition $\tilde{\phi}(0) = 0$.

It remains now to determine the constant Λ. Assuming no potential variations in the solid surrounding the pore, Eq. (11.7-42) indicates that

$$q_e = \varepsilon \left. \frac{\partial \phi}{\partial r} \right|_{r=a} = \frac{\varepsilon RT}{\lambda F} \left. \frac{\partial \tilde{\phi}}{\partial \tilde{r}} \right|_{\tilde{r}=\tilde{a}}, \tag{11.7-60}$$

where $\tilde{a} = a/\lambda$. Using Eq. (11.7-59) in Eq. (11.7-60) and solving for Λ, we obtain

$$\Lambda = -\frac{\lambda F q_e}{\varepsilon RT I_1(\tilde{a})}. \tag{11.7-61}$$

Finally, using Eqs. (11.7-59) and (11.7-61) to evaluate $\phi - \phi_a$ in Eq. (11.7-54), we find that

$$v_z = -\frac{\lambda q_e E_z}{\mu} \left[\frac{I_0(\tilde{a}) - I_0(\tilde{r})}{I_1(\tilde{a})} \right]. \tag{11.7-62}$$

It is seen that for a negatively charged pore wall ($q_e < 0$), where the solution carries a net positive charge ($\rho_e > 0$), v_z will have the same sign as E_z. That is, for $q_e < 0$ the flow will be in the direction of decreasing potential.

When the pore radius greatly exceeds the Debye length, $\tilde{a} \gg 1$ and $\tilde{r} \to \tilde{a}$ in the double layer. Under these conditions, we have

$$\frac{I_0(\tilde{a}) - I_0(\tilde{r})}{I_1(\tilde{a})} \to 1 - e^{-y/\lambda}, \tag{11.7-63}$$

where $y \equiv a - r$ is the distance from the pore wall. The velocity profile simplifies to

$$v_z = -\frac{\lambda q_e E_z}{\mu}(1 - e^{-y/\lambda}) \qquad (a/\lambda \gg 1). \tag{11.7-64}$$

Because $y/\lambda \gg 1$ except very near the pore wall, $v_z \cong -\lambda q_e E_z/\mu$ (a constant) throughout almost the entire pore cross section. If the thin double-layer region is ignored, the velocity profile approximates that for plug flow, with "slip" at the wall.

Electroosmosis is only one of several *electrokinetic phenomena* associated with transport in diffuse double layers. Others include: *electrophoresis,* the migration of a charged particle or macromolecule in an applied field; *streaming potential,* the potential difference due to an imposed pressure or imposed volume flow, at zero current; *streaming current,* the current due to an imposed pressure or imposed volume flow, at zero potential difference; and *sedimentation potential,* the potential difference due to sedimentation of charged particles.

A good general source of information on transport in electrolyte solutions is Newman (1973). For information on electrokinetic phenomena in particular, see Newman (1973) or Dukhin and Derjaguin (1974).

11.8 GENERALIZED STEFAN–MAXWELL EQUATIONS

The Stefan–Maxwell equations introduced in Section 11.5 are strictly applicable only to isothermal, isobaric, ideal gases. The objective of this section is to develop an analogous but much more general set of equations to describe multicomponent diffusion in gases, liquids, or solids, allowing for gradients in temperature and pressure, for electric or other body forces, and for mixture nonidealities. The approach to be taken builds on the concepts of nonequilibrium thermodynamics introduced in Section 11.4 and involves specification of a complete set of fluxes f_i and forces X_j. The development here closely parallels that given by Lightfoot (1974). For additional information on the application of nonequilibrium thermodynamics to transport problems, see Beris and Edwards (1994).

As noted in Section 11.4, the Onsager reciprocal relations, Eq. (11.4-2), are valid only for certain choices of fluxes and forces. Within the framework of nonequilibrium thermodynamics, the key requirement for a proper set of forces and fluxes is that they must account for the rate of production of entropy. Based on Eq. (2.3-7), a conservation equation involving entropy per unit mass $(\hat{S})$ is written as

$$\rho \frac{D\hat{S}}{Dt} = -\nabla \cdot \mathbf{j}_s + \sigma, \tag{11.8-1}$$

where $\mathbf{j}_s$ is the entropy flux relative to $\mathbf{v}$, and σ is the volumetric rate of entropy production. It follows from the second law of thermodynamics that $\sigma > 0$ for any spontaneous process; the quantity $T\sigma$ represents the *irreversible rate of energy dissipation,* per unit volume. A suitable set of fluxes and forces is one for which

$$T\sigma = \sum_i \mathbf{f}^{(i)} \cdot \mathbf{X}^{(i)}, \tag{11.8-2}$$

where the summation is over all "conjugate pairs" of fluxes and forces. Thus, the set of fluxes and forces must be complete in the sense that it must account for the magnitude of $T\sigma$. Obviously, the units for $\mathbf{f}^{(i)}$ and $\mathbf{X}^{(i)}$ are also constrained by Eq. (11.8-2).

For nonisothermal flow of chemically reactive mixtures, there are four processes contributing to the rate of entropy production: (1) heat conduction, (2) diffusion of chemical species, (3) viscous dissipation of energy, and (4) chemical reactions. Heat conduction and species diffusion involve fluxes and forces which are vectors (first-order tensors); viscous dissipation involves second-order tensors (τ and $\nabla\mathbf{v}$, Section 9.6); rates of reaction and the associated chemical affinities (differences in chemical potential between reactants and products) are scalars (zero-order tensors). [Accordingly, not all of the flux and force terms in Eq. (11.8-2) are vectors.] It has been shown that for an isotropic material, there can be no direct coupling between processes described by tensors differing in order by an odd number; this is called the *Curie–Prigogine principle.* Thus, coupling may occur between the heat flux and species flux vectors, but neither of those vectors may be coupled to either the viscous stress tensor or to reaction rates. It must be emphasized that in the present context, "coupling" means the existence of a coefficient L_{ij} in a relation like Eq. (11.4-1). It follows from these restrictions that, to describe energy and mass transfer in a viscous, reactive, binary mixture, no other "new" coupling coefficients are needed beyond $D_A^{(T)}$.

A postulate of nonequilibrium thermodynamics is that the various relations among intensive variables obtained from classical thermodynamics are valid locally. This assumption of small departures from local equilibrium appears to be suitable for a very wide range of applications. It is much like the assumption that an equilibrium equation of state can be used to describe local pressure–temperature–density relationships in the flow of a compressible fluid. Application of the concept of local equilibrium will allow us to evaluate the volumetric rate of energy dissipation $(T\sigma)$. Together with Eq. (11.8-2), this will suggest a suitable set of forces and fluxes for multicomponent mass transfer under nonisothermal, nonisobaric conditions.

We begin with a form of the Gibbs equation,

$$T d\hat{S} = d\hat{U} + P d\hat{V} - \sum_{i=1}^{n} \frac{\mu_i}{M_i} d\omega_i, \tag{11.8-3}$$

where $\hat{V}$ is the specific volume (i.e., $\hat{V} = 1/\rho$), μ_i is the chemical potential of species i, and n is the number of components. Assuming local equilibrium in a reactive, flowing mixture, Eq. (11.8-3) becomes

$$\rho T \frac{D\hat{S}}{Dt} = \rho \frac{D\hat{U}}{Dt} + \rho P \frac{D\hat{V}}{Dt} - \rho \sum_{i=1}^{n} \frac{\mu_i}{M_i} \frac{D\omega_i}{Dt}. \tag{11.8-4}$$

Using Eq. (11.2-6) to evaluate $D\hat{U}/Dt$ in Eq. (11.8-4) and using the continuity and species conservation equations to evaluate $D\hat{V}/Dt$ and $D\omega_i/Dt$, respectively, we obtain, after some manipulation,

$$\rho T \frac{D\hat{S}}{Dt} = -T \nabla \cdot \left[(\mathbf{q}/T) - \sum_{i=1}^{n} (\mathbf{J}_i \mu_i / T) \right] - \mathbf{q} \cdot \nabla \ln T$$

$$- \sum_{i=1}^{n} \mathbf{j}_i \cdot \left[T \nabla \left(\frac{\mu_i}{TM_i} \right) - \mathbf{g}_i \right] + \boldsymbol{\tau} : \nabla \mathbf{v} - \sum_{i=1}^{n} \mu_i R_{Vi}. \tag{11.8-5}$$

This result is compared with Eq. (11.8-1), multiplied by T:

$$\rho T \frac{D\hat{S}}{Dt} = -T \nabla \cdot \mathbf{j}_s + T\sigma. \tag{11.8-6}$$

A correspondence between Eqs. (11.8-5) and (11.8-6) is established by defining the entropy flux as

$$\mathbf{j}_s \equiv \frac{1}{T} \left[\mathbf{q} - \sum_{i=1}^{n} \mathbf{J}_i \mu_i \right]. \tag{11.8-7}$$

With this definition, which is consistent with the expectation that $\mathbf{j}_s = \mathbf{0}$ at equilibrium, the first term on the right-hand side of Eq. (11.8-5) is the same as the first term on the right of Eq. (11.8-6). Equating the remaining terms, we find that the volumetric rate of energy dissipation is

$$T\sigma = -\mathbf{q} \cdot \nabla \ln T - \sum_{i=1}^{n} \mathbf{j}_i \cdot \left[T \nabla \left(\frac{\mu_i}{TM_i} \right) - \mathbf{g}_i \right] + \boldsymbol{\tau} : \nabla \mathbf{v} - \sum_{i=1}^{n} \mu_i R_{Vi}. \tag{11.8-8}$$

It is seen that additive contributions to energy dissipation arise from each of the rate processes: energy transfer, mass fluxes, viscous flow, and chemical reactions.

It is convenient for applications to rearrange the energy and mass flux terms in Eq. (11.8-8). Instead of employing the full, multicomponent energy flux $\mathbf{q}$ defined by Eq. (11.2-3), we will subtract from $\mathbf{q}$ the term representing diffusional transport of enthalpy. This leaves the sum of the Fourier and Dufour fluxes, or

$$\mathbf{q} - \sum_{i=1}^{n} \mathbf{J}_i \overline{H}_i = \mathbf{q}^{(c)} + \mathbf{q}^{(x)} \equiv -k\nabla T + \mathbf{q}^{(x)}, \tag{11.8-9}$$

where the Fourier or conduction flux is denoted by $\mathbf{q}^{(c)}$. Equation (11.8-8) is rewritten now as

$$T\sigma = -\left(\mathbf{q}^{(c)} + \mathbf{q}^{(x)}\right) \cdot \nabla \ln T - CRT \sum_{i=1}^{n} \frac{\mathbf{j}_i \cdot \mathbf{d}_i}{\rho_i} + \tau : \nabla \mathbf{v} - \sum_{i=1}^{n} \mu_i R_{Vi}, \tag{11.8-10}$$

where $\mathbf{d}_i$ is the diffusional driving force defined by Eq. (11.6-2). In obtaining Eq. (11.8-10) from Eq. (11.8-8), we used the thermodynamic identities

$$\overline{V}_i = (\partial \mu_i / \partial P)_{T,x_j}, \tag{11.8-11}$$

$$\mu_i = \overline{H}_i + T(\partial \mu_i / \partial T)_{P,x_j}. \tag{11.8-12}$$

The utility of the quantity $\mathbf{d}_i$ comes from the fact that $CRT\mathbf{d}_i$ is a force per unit volume tending to move species i relative to the mixture. This interpretation of $\mathbf{d}_i$ is perhaps most evident in the body force term in Eq. (11.6-2), which involves the difference between the body force on species i and the total body force on the mixture. The physical interpretation of $\mathbf{d}_i$ suggests that the sum of all $\mathbf{d}_i$ should vanish. Using Eq. (11.6-2) and the Gibbs–Duhem equation,

$$-S\,dT + V\left(dP - \sum_{i=1}^{n} C_i\,d\mu_i\right) = 0, \tag{11.8-13}$$

one confirms that, indeed,

$$\sum_{i=1}^{n} \mathbf{d}_i = \mathbf{0}. \tag{11.8-14}$$

We now have a basis for identifying all of the force–flux couplings which arise in energy and mass transfer. As discussed in connection with Eq. (11.4-1), one way to express those relationships is to write the fluxes as a linear combination of force terms. An alternative approach is to write the forces as a linear combination of flux terms, or

$$\mathbf{X}^{(i)} = \sum_{j=1}^{n} B_{ij} \mathbf{f}^{(j)}, \tag{11.8-15}$$

where the coefficients B_{ij} are related to the coefficients L_{ij} in Eq. (11.4-1) and are independent of $\mathbf{X}^{(i)}$ and $\mathbf{f}^{(j)}$. Specifically, the B_{ij} matrix is the inverse of the L_{ij} matrix. Equation (11.8-15), which is like the Stefan–Maxwell equations (Eq. [11.5-1)], is the preferred form. A comparison of Eq. (11.8-2) with Eq. (11.8-10) allows us to identify forces and fluxes which are consistent with Eq. (11.8-15). According to the Curie–Prigogine principle, the viscous dissipation and chemical reaction terms in Eq. (11.8-10) cannot be directly coupled with energy or mass fluxes. Based on the remaining terms in Eq. (11.8-10), if the forces are chosen as $-\nabla \ln T$ and $-CRT\mathbf{d}_i$, then the corresponding fluxes must be $(\mathbf{q}^{(c)} + \mathbf{q}^{(x)})$ and $\mathbf{j}_i / \rho_i$, respectively. Recalling that $\mathbf{j}_i / \rho_i = \mathbf{v}_i - \mathbf{v}$, Eq. (11.8-15) becomes

$$- \nabla \ln T = B_{00} (\mathbf{q}^{(c)} + \mathbf{q}^{(x)}) + \sum_{j=1}^{n} B_{0j} (\mathbf{v}_j - \mathbf{v}), \tag{11.8-16}$$

$$- CRT \mathbf{d}_i = B_{i0} (\mathbf{q}^{(c)} + \mathbf{q}^{(x)}) + \sum_{j=1}^{n} B_{ij} (\mathbf{v}_j - \mathbf{v}), \qquad i = 1, 2, \ldots, n. \tag{11.8-17}$$

The subscript 0 has been chosen to refer to the thermal terms, so that the coefficient relating the temperature gradient to the energy flux is written as B_{00}. These equations imply that the B_{ij} matrix is $(n+1) \times (n+1)$, which seems inconsistent with Eq. (11.8-15). However, as shown below, the matrix of *independent* coefficients is indeed $n \times n$. Equation (11.8-17) is a generalization of the Stefan–Maxwell equations, although not yet in the most convenient form.

The transport properties of a multicomponent fluid are embodied in the *phenomenological coefficients*, B_{ij}. Before arriving at the final form of the generalized Stefan–Maxwell equations, it is useful to establish certain properties of these coefficients and to relate them to more familiar quantities. From Eq. (11.4-2) and the relationship between B_{ij} and L_{ij}, the B_{ij} matrix must be symmetric, or

$$B_{ij} = B_{ji}. \tag{11.8-18}$$

This immediately reduces the apparent number of independent coefficients in Eqs. (11.8-16) and (11.8-17) from $(n+1)^2$ to $(n+1)(n+2)/2$. As already mentioned, however, not all of the remaining coefficients are independent. An additional set of relationships among the coefficients follows from Eq. (11.8-14). Summing Eq. (11.8-17) over all components, we find that

$$- CRT \sum_{i=1}^{n} \mathbf{d}_i = (\mathbf{q}^{(c)} + \mathbf{q}^{(x)}) \sum_{i=1}^{n} B_{io} + \sum_{j=1}^{n} (\mathbf{v}_j - \mathbf{v}) \sum_{i=1}^{n} B_{ij} = 0. \tag{11.8-19}$$

As suggested by the form of Eq. (11.8-15), the fluxes may be viewed as a set of mathematically independent variables which determine the forces. For Eq. (11.8-19) to hold for an arbitrary set of values for the fluxes $(\mathbf{q}^{(c)} + \mathbf{q}^{(x)})$ and $(\mathbf{v}_j - \mathbf{v})$, it is necessary that

$$\sum_{i=1}^{n} B_{ij} = 0, \qquad j = 0, 1, 2, \ldots, n. \tag{11.8-20}$$

This additional set of constraints reduces the number of independent coefficients to $n(n+1)/2$. Thus, three coefficients are needed to completely describe heat and mass transfer in a binary mixture (Table 11-1), and six are needed for a ternary mixture.

The relationship between B_{00} and the thermal conductivity is derived by applying Eq. (11.8-16) to a fluid where diffusional fluxes are absent (i.e., $\mathbf{v}_j - \mathbf{v} = \mathbf{0}$). In this case, as shown by Eq. (11.4-4) for a binary gas and as discussed below, the Dufour energy flux also vanishes (i.e., $\mathbf{q}^{(x)} = \mathbf{0}$). Using Fourier's law in Eq. (11.8-16), it is seen then that $B_{00} = (kT)^{-1}$. Because all of the phenomenological coefficients, including B_{00}, are independent of the fluxes and forces, this interpretation of B_{00} remains valid when diffusion is present. Using $B_{00} = (kT)^{-1}$ and $B_{0j} = B_{j0}$ in Eq. (11.8-16), we find that the multicomponent Dufour flux is given by

$$\mathbf{q}^{(x)} = -kT \sum_{j=1}^{n} B_{j0}(\mathbf{v}_j - \mathbf{v}).$$ (11.8-21)

This confirms that the Dufour flux vanishes when diffusion is absent.

The coefficients B_{i0} ($i \neq 0$) in Eq. (11.8-17) describe multicomponent thermal diffusion effects. Of greater interest in mass transfer are the coefficients B_{ij} ($i \neq 0$, $j \neq 0$). To put Eq. (11.8-17) in a form more closely resembling the Stefan–Maxwell equations, we define *multicomponent diffusivities*, $\mathcal{D}_{ij}$, by

$$\mathcal{D}_{ij} \equiv -\frac{x_i x_j CRT}{B_{ij}}, \qquad i \text{ and } j = 1, 2, \ldots, n.$$ (11.8-22)

Properties of $\mathcal{D}_{ij}$ which follow from Eqs. (11.8-18), (11.8-20), and (11.8-22) are that

$$\mathcal{D}_{ij} = \mathcal{D}_{ji},$$ (11.8-23)

$$\sum_{i=1}^{n} \frac{x_i}{\mathcal{D}_{ij}} = 0, \qquad j = 1, 2, \ldots, n.$$ (11.8-24)

The advantage of the coefficients $\mathcal{D}_{ij}$ is that they are much less dependent on composition than are the corresponding B_{ij}. Moreover, for gas mixtures, $\mathcal{D}_{ij} = D_{ij}$; that is, the multicomponent diffusivity in a gas is the same as the usual binary diffusivity of species i and j.

The interpretation of $\mathcal{D}_{ij}$ for liquids and solids is not as straightforward. In a liquid solution, for example, let the solvent be denoted as component 1, so that the solutes are identified by the indices $i = 2, 3, \ldots, n$. Multicomponent diffusivities for the various solute–solvent pairs are similar to the corresponding binary diffusivities. Although diffusivities depend to some extent on the solution composition, it is true nonetheless that $\mathcal{D}_{i1} \cong D_{i1}$. For solute–solute pairs, however, it is commonly the case that there is no measurable binary diffusivity which corresponds even approximately to $\mathcal{D}_{ij}$. This will be true if i and j are both solids or immiscible fluids under the conditions of interest, such that they cannot form a binary solution.

Using Eq. (11.8-22), Eq. (11.8-17) is rearranged to

$$\mathbf{d}_i = \sum_{j=1}^{n} \frac{x_i x_j}{\mathcal{D}_{ij}} (\mathbf{v}_j - \mathbf{v}) - \frac{B_{i0}}{CRT} \left(\mathbf{q}^{(c)} + \mathbf{q}^{(x)} \right).$$ (11.8-25)

The summation term in Eq. (11.8-25) now resembles that in the Stefan–Maxwell equations, except that it has $\mathbf{v}$ in place of $\mathbf{v}_i$. In fact, any velocity other than $\mathbf{v}_j$ may be substituted here for $\mathbf{v}$. This is demonstrated by noting, from Eq. (11.8-24), that

$$\sum_{j=1}^{n} \frac{x_i x_j}{\mathcal{D}_{ij}} \mathbf{v} = \sum_{j=1}^{n} \frac{x_i x_j}{\mathcal{D}_{ij}} \mathbf{v}_k = 0, \qquad k \neq j.$$ (11.8-26)

Thus, Eq. (11.8-25) is rewritten more generally as

$$\mathbf{d}_i = \sum_{\substack{j=1 \\ j \neq k}}^{n} \frac{x_i x_j}{\mathcal{D}_{ij}} (\mathbf{v}_j - \mathbf{v}_k) - \frac{B_{i0}}{CRT} \left(\mathbf{q}^{(c)} + \mathbf{q}^{(x)} \right).$$ (11.8-27)

Finally, evaluating $\mathbf{q}^{(c)}$ using Fourier's law, neglecting $\mathbf{q}^{(x)}$, and setting $b_i = B_{i0}$, we obtain

$$\mathbf{d}_i = \sum_{\substack{j=1 \\ j \neq k}}^{n} \frac{x_i x_j}{\mathcal{D}_{ij}} (\mathbf{v}_j - \mathbf{v}_k) + \frac{k b_i}{CR} \boldsymbol{\nabla} \ln T. \tag{11.8-28}$$

The relations given by Eq. (11.8-28) are called the *generalized Stefan–Maxwell equations*. It follows from Eq. (11.8-14) that only $n-1$ of these equations are independent. Although $\mathbf{v}_k$ in Eq. (11.8-28) may be chosen at will, according to what is convenient for a particular problem, it is usually equated with either $\mathbf{v}$ or $\mathbf{v}_i$.

In summary, multicomponent mass transfer is described by the generalized Stefan–Maxwell equations, together with the driving force constraint [Eq. (11.8-14)] and the two sets of relationships among the multicomponent diffusivities [(Eqs. (11.8-23) and (11.8-24)]. The following examples demonstrate the relationship between the generalized Stefan–Maxwell equations and the equations usually used to describe multicomponent diffusion in gases and in dilute mixtures, respectively.

Example 11.8-1 Multicomponent Diffusion in Ideal Gases   We consider first the diffusional driving force, $\mathbf{d}_i$, defined by Eq. (11.6-2). For an ideal mixture, $\boldsymbol{\nabla}_{T,P} \mu_i = RT \boldsymbol{\nabla} \ln x_i$. Assuming also constant pressure and no body forces other than gravity, we find that

$$\mathbf{d}_i = \boldsymbol{\nabla} x_i. \tag{11.8-29}$$

Setting $\mathbf{v}_k = \mathbf{v}_i$ in Eq. (11.8-28), using Eq. (11.8-29), and neglecting thermal diffusion, we obtain

$$\boldsymbol{\nabla} x_i = \sum_{\substack{j=1 \\ j \neq i}}^{n} \frac{x_i x_j}{\mathcal{D}_{ij}} (\mathbf{v}_j - \mathbf{v}_i). \tag{11.8-30}$$

As already mentioned, for gases we have $\mathcal{D}_{ij} = D_{ij}$. Thus, Eq. (11.8-30) is identical to Eq. (11.5-1), the (original) Stefan–Maxwell equations.

There is a subtle implication in choosing $\mathbf{v}_k = \mathbf{v}_i$, as just done in obtaining Eq. (11.8-30). With this choice, the diagonal elements $\mathcal{D}_{ii}$ (or D_{ii}) of the multicomponent diffusivity matrix are not needed. Thus, there was no need in Section 11.5 for a relation analogous to Eq. (11.8-24), which provides a way to compute each of the $\mathcal{D}_{ii}$ (or D_{ii}) from the off-diagonal coefficients. It must be emphasized that the $\mathcal{D}_{ii}$ (or D_{ii}) in the multicomponent diffusion equations are not the self-diffusivities which might be measured by studying the diffusion of a tracer form of species i relative to unlabeled i. Indeed, applying Eq. (11.8-24) to gases, by replacing $\mathcal{D}_{ij}$ with D_{ij}, we obtain

$$D_{ij} = -x_i \left[\sum_{\substack{j=1 \\ j \neq i}}^{n} \frac{x_j}{D_{ij}} \right]^{-1}, \qquad i = 1, 2, \ldots, n. \tag{11.8-31}$$

Because the mole fractions and the off-diagonal D_{ij} are all positive numbers, it is evident that $D_{ii} < 0$! This makes it clear that the D_{ii} are not ordinary diffusivities.

Example 11.8-2 Multicomponent Diffusion in Dilute Mixtures   A dilute mixture is one which consists mainly of a single abundant species, denoted by 1, such that

$$\sum_{j=2}^{n} x_j << x_1. \tag{11.8-32}$$

Dilute mixtures may exist as gases, liquids, or solids; when the mixture is a liquid solution, the abundant component is the solvent. The objective here is to derive Eq. (11.6-1), the flux equation for a dilute mixture, from the generalized Stefan–Maxwell equations. We begin by choosing $\mathbf{v}_k = \mathbf{v}_i$ and using $\mathbf{v}_i = \mathbf{N}_i / C_i$ in Eq. (11.8-28) to give

$$\mathbf{d}_i = \sum_{\substack{j=1 \\ j \neq i}}^{n} \frac{x_j}{\mathcal{D}_{ij}} \left(x_i \mathbf{v}_j - \frac{\mathbf{N}_i}{C} \right) + \frac{kb_i}{CR} \, \nabla \ln T. \tag{11.8-33}$$

Assuming for the moment that all of the fluxes are collinear, so that $\mathbf{v}_j = (|\mathbf{v}_j|/|\mathbf{v}_1|)\mathbf{v}_1$, the summation is expanded as

$$\sum_{\substack{j=1 \\ j \neq i}}^{n} \frac{x_j}{\mathcal{D}_{ij}} \left(x_i \mathbf{v}_j - \frac{\mathbf{N}_i}{C} \right) = \frac{x_i \mathbf{v}_1}{\mathcal{D}_{i1}} \left[x_1 + \sum_{\substack{j=2 \\ j \neq i}}^{n} x_j \frac{\mathcal{D}_{i1}}{\mathcal{D}_{ij}} \frac{|\mathbf{v}_j|}{|\mathbf{v}_1|} \right] - \frac{\mathbf{N}_i}{C\mathcal{D}_{i1}} \left[x_1 + \sum_{\substack{j=2 \\ j \neq i}}^{n} x_j \frac{\mathcal{D}_{i1}}{\mathcal{D}_{ij}} \right], \tag{11.8-34}$$

where the diffusional interactions between species i and the abundant component have been separated from those between i and the other minor species (e.g., the other solutes in a liquid). It follows from Eq. (11.8-32) that the summations on the right-hand side of Eq. (11.8-34) will be negligible, provided that the ratios $\mathcal{D}_{i1}/\mathcal{D}_{ij}$ and $|\mathbf{v}_j|/|\mathbf{v}_1|$ are not large. Under these conditions, Eq. (11.8-33) reduces to

$$C\mathbf{d}_i = \frac{x_1}{\mathcal{D}_{i1}} (C_i \mathbf{v}_1 - \mathbf{N}_i) + \frac{kb_i}{R} \, \nabla \ln T. \tag{11.8-35}$$

For this pseudobinary situation, $\mathcal{D}_{i1} = D_{i1} \equiv D_i$. Recalling also that $\mathbf{v}_1 \cong \mathbf{v}$ and $x_1 \cong 1$ for a dilute mixture and that $\mathbf{J}_i = \mathbf{N}_i - C_i \mathbf{v}$, Eq. (11.8-35) becomes

$$\mathbf{J}_i = -D_i C\mathbf{d}_i + \frac{kb_i D_i}{R} \, \nabla \ln T. \tag{11.8-36}$$

This relation is equivalent to Eq. (11.6-1). A comparison of the thermal diffusion terms indicates that

$$b_i = -\frac{R}{kM_i} \frac{D_i^{(T)}}{D_i} \tag{11.8-37}$$

for a dilute mixture.

References

Beris, A. N. and B. J. Edwards. *Thermodynamics of Flowing Systems.* Oxford University Press, New York, 1994.

Cussler, E. L. *Multicomponent Diffusion.* Elsevier, Amsterdam, 1976.

de Groot, S. R. and P. Mazur. *Non-Equilibrium Thermodynamics.* North-Holland, Amsterdam, 1962.

Dukhin, S. S. and D. V. Derjaguin. Equilibrium double layer and electrokinetic phenomena. In *Surface and Colloid Science,* Vol. 7, E. Matijevic, Ed. Wiley, New York, 1974, pp. 49–272.

Grew, K. E. and T. L. Ibbs. *Thermal Diffusion in Gases.* Cambridge University Press, Cambridge, 1952.

Hirschfelder, J. O., C. F. Curtiss, and R. B. Bird. *Molecular Theory of Gases and Liquids.* Wiley, New York, 1954, pp. 522–523.

Lightfoot, E. N. *Transport Phenomena and Living Systems.* Wiley, New York, 1974, pp. 157–165.

Newman, J. S. *Electrochemical Systems.* Prentice-Hall, Englewood Cliffs, NJ, 1973.

Prater, C. D. The temperature produced by heat of reaction in the interior of porous particles. *Chem. Eng. Sci.* 8: 284–286, 1958.

Tanford, C. *Physical Chemistry of Macromolecules.* Wiley, New York, 1961, pp. 260 and 269.

Toor, H. L. Diffusion in three-component gas mixtures. *AIChE J.* 3: 198–207, 1957.

Vasaru, G., G. Müller, G. Reinhold, and T. Fodor. *The Thermal Diffusion Column.* VEB Deutscher Verlag der Wissenschaften, Berlin, 1969.

Problems

11-1. Combustion of a Carbon Particle

Within a certain range of temperatures the combustion of carbon occurs primarily through the reaction

$$C(s) + \frac{1}{2} O_2(g) \rightarrow CO(g)$$

at the carbon surface. Consider a spherical carbon particle of radius $R(t)$ suspended in a gas mixture, under pseudosteady conditions. Assume that the gas contains only oxygen (species A) and carbon monoxide (species B) and that the mole fractions ($x_{i\infty}$) and the temperature far from the particle (T_∞) are known. The reaction is very fast, and the particle is small enough that $Pe \ll 1$ for both heat and mass transfer.

(a) Use interfacial mass balances to relate the fluxes of A and B to the reaction rate at the surface. Also derive the relationship between the reaction rate and dR/dt. Let C_C be the molar concentration of carbon in the solid.

(b) Determine the oxygen flux at the particle surface, N_{A0}. For simplicity, you may assume that the product CD_{AB} is independent of position.

(c) Use an interfacial energy balance to derive the relationship between the heat flux at the particle, the reaction rate, and the heat of reaction. Assume here that radiation heat transfer is negligible.

(d) Determine the particle temperature, T_0, for the conditions of part (c).

(e) Explain how to include radiation heat transfer in the analysis, assuming that the particle and surroundings act as blackbodies.

(f) Describe how the analysis would be changed if the gas contained an additional, inert species (e.g., N_2). State the new governing equations and outline the approach needed to calculate the reaction rate and particle temperature.

11-2. Effect of a Reversible Reaction on Energy Transfer in a Gas

A binary gas mixture consisting of species A and B is confined between parallel plates maintained at different temperatures, as shown in Fig. P11-2. The fast, reversible reaction $A \leftrightarrow B$ occurs at the *surfaces,* so that A and B are in equilibrium there; there is no homogeneous reaction. The equilibrium constant, $K \equiv (C_B/C_A)|_{eq}$, depends on temperature. From van't Hoff's equation, the equilibrium constants at the top and bottom surfaces are related by

$$\ln \frac{K_L}{K_0} = \frac{\Delta H_R}{R} \left(\frac{1}{T_0} - \frac{1}{T_L} \right),$$

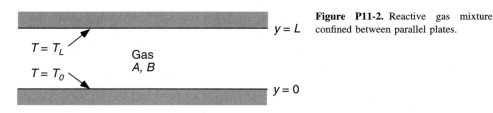

Figure P11-2. Reactive gas mixture confined between parallel plates.

where ΔH_R is the heat of reaction, assumed constant. The temperaure difference is small enough that the total molar concentration, thermal conductivity, and diffusivity are all constant to good approximation. The system is at steady state, and temperature and composition depend on y only.

(a) Determine the species fluxes in terms of K_0, K_L, and the other parameters.
(b) Derive an expression for the energy flux, again in terms of K_0, K_L, and the other parameters.
(c) In the limit $T_L \to T_0$, compare the actual energy flux with that which would occur in the absence of chemical reaction. Use the van't Hoff relationship between K_L and K_0. How (if at all) is the ratio of the energy fluxes affected by the algebraic sign of ΔH_R?

11-3. Absorption with a Temperature-Dependent Reaction
A stagnant liquid film of thickness L separates a gas from an impermeable and thermally insulated solid surface, as shown in Fig. P11-3. Species A diffuses from the gas into the liquid, where it undergoes an irreversible, first-order reaction. Species B, the product, diffuses back to the gas. The first-order, homogeneous, rate constant depends on temperature according to

$$k_v = k_{vo} \exp\left[\frac{E}{R}\left(\frac{1}{T_0} - \frac{1}{T}\right)\right] \cong k_{vo} \exp\left[\frac{E}{RT_0}\left(\frac{T-T_0}{T_0}\right)\right],$$

where k_{vo} is the rate constant at temperature T_0, E is the activation energy, and R is the gas constant. The second equality requires that $|T-T_0|/T_0 \ll 1$, which you may regard as a good approximation here. Assume that k_{vo}, E, and the heat of reaction ($\Delta H_R \equiv \overline{H}_B - \overline{H}_A$) are all known constants. Assume that the temperature (T_0) and liquid-phase concentrations (C_{A0}, C_{B0}) at the gas–liquid interface are also known and that the system is at steady state.

(a) Use conservation of energy to obtain an explicit relationship between $T(y)$ and $C_A(y)$.
(b) Derive the differential equation and boundary conditions which govern the dimensionless reactant concentration $\Theta(\eta)$, where

$$\Theta \equiv \frac{C_A}{C_{A0}}, \qquad \eta \equiv \frac{y}{L}.$$

Show that the only parameters involved are

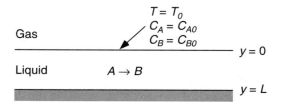

$T = T_0$
$C_A = C_{A0}$
$C_B = C_{B0}$

Figure P11-3. Absorption with a temperature-dependent reaction rate.

$$\mathrm{Da} \equiv \frac{k_{v0}L^2}{D_A},$$

$$\varepsilon \equiv \left(\frac{D_A \Delta H_R C_{A0}}{kT_0}\right)\left(\frac{E}{RT_0}\right).$$

(c) Assume now that $\varepsilon \ll 1$. Reformulate the result of part (b) as a perturbation problem, with $\Theta(\eta)$ expressed as an expansion in ε. State the differential equations and boundary conditions which govern the first two terms in the series, $\Theta_0(\eta)$ and $\Theta_1(\eta)$.

(d) Evaluate $\Theta_0(\eta)$ and $\Theta_1(\eta)$, and determine the flux of A at the gas–liquid interface at this level of approximation. Discuss the magnitude and direction of the error in the flux calculation which would result from assuming that $k_v = k_{v0}$ throughout the liquid. (It is sufficient to restrict your solution to $\mathrm{Da} \gg 1$, which is algebraically simpler than the general case.)

11-4. Burning of a Liquid Fuel Droplet

The burning of a small droplet of liquid fuel is to be modeled as a pseudosteady, spherically symmetric process, as shown in Fig. P11-4. A droplet of radius $R_0(t)$, consisting of pure fuel (species A), is surrounded by a stagnant gas which contains oxygen (species B). The combustion reaction is written as $A + mB \rightarrow nP$, where P represents all combustion products. The reaction is assumed to be extremely fast, so that it may be modeled as occurring only at a position $r = R_f(t)$, the "flame front." The fast reaction ensures that $x_A = x_B = 0$ at $r = R_f$. Thus, for $r < R_f$, only A and P are present; for $r > R_f$, there is only B and P. Assume that the pressure is constant, the gas mixture is ideal, and the combustion products do not enter the droplet.

In addition to all thermophysical properties, assume that the following quantities are given: the mole fractions in the bulk gas ($x_A = 0$, $x_B = x_{B\infty}$, $x_P = x_{P\infty}$); the bulk temperature ($T = T_\infty$); and the droplet radius at a given instant (R_0). The function $x_{A0}(T_0)$, which describes the mole fraction next to the droplet, is also given; however, the value of x_{A0} is not known until T_0, the droplet temperature, is found. The quantities which are to be determined in the analysis are T_0, T_f, R_f, and the flux of fuel at the droplet surface, $N_{Ar}(R_0) \equiv N_{A0}$.

(a) Sketch qualitatively the expected temperature and concentration profiles.

(b) Use an interfacial mass balance at the droplet surface to relate dR_0/dt to N_{A0} and N_{P0}, the fluxes evaluated on the gas side. Show that $|N_{P0}| \ll |N_{A0}|$. (Hint: Note that $C_{A0} \ll C_L$, where C_{A0} is the gas-phase concentration of A next to the droplet and C_L is the molar concentration of A in the pure liquid.)

(c) Use an interfacial mass balance at the flame front to relate the fluxes of the three species at $r = R_f$. Let $N_{if}^{(-)}$ and $N_{if}^{(+)}$ be the radial fluxes of i evaluated just inside and outside the front, respectively.

(d) By considering mass transfer in the gas inside the flame front ($R_0 < r < R_f$), relate N_{A0} to x_{A0}, R_f, and known quantities. For simplicity, you may assume that the product CD_{ij} is independent of position for all i and j.

Figure P11-4. Burning of a liquid fuel droplet.

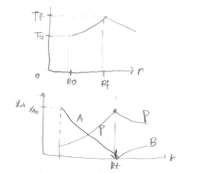

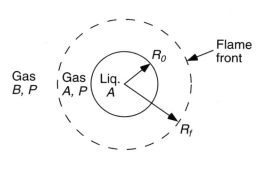

(e) By considering mass transfer outside the flame front ($r > R_f$), obtain another relationship between N_{A0} and R_f.

(f) Use an interfacial energy balance at the droplet to relate dR_0/dt to the energy flux in the gas, q_0. The molar latent heat for species A is $\overline{\lambda} \equiv \overline{H}_A(T_0) - \overline{H}_L(T_0)$.

(g) Use an interfacial energy balance at the flame front to relate the temperature gradients at $r = R_f^{(-)}$ and $R_f^{(+)}$ to the flux of A and the heat of reaction, $\Delta H_R(T_f)$.

(h) State the energy conservation equations for the two gas regions and describe how to determine the remaining unknowns.

11-5. Effect of Mass Transfer on the Heat Transfer Coefficient

The effect of mass transfer on the heat transfer coefficient at a given interface can be approximated using a stagnant film model. Assume that there is a stagnant film of fluid of thickness δ adjacent to an interface located at $x=0$. The interfacial temperature is T_0 and the bulk fluid temperature (at $x=\delta$) is T_∞. In general, there will be nonzero fluxes of one or more species normal to the interface; assume that these fluxes (N_{ix}) are known constants throughout the film (i.e., there are no chemical reactions in the fluid). Recall that the heat transfer coefficient is defined so that it represents only the *conduction* (Fourier) energy flux at $x=0$. That is,

$$h \equiv \frac{-k(\partial T/\partial x)_{x=0}}{T_0 - T_\infty}.$$

(a) Show that in the absence of mass transfer, the heat transfer coefficient is given by

$$h = \frac{k}{\delta} \equiv h_0.$$

Thus, a measured or theoretical value of h_0 determines the apparent film thickness to use in this type of model.

(b) Derive an expression for h/h_0 for the general case, including mass transfer. Is the result different for a gas than for a liquid? Discuss the conditions which will make h/h_0 either greater than or less than unity.

11-6. Film Model for Condensation

As shown in Fig. P11-6, consider a binary gas in contact with a cold, vertical surface, such that one component (species A) condenses at the surface but the other (species B) does not. Changes in the gas composition and temperature are assumed to be confined to a stagnant film of thickness δ; the thickness of the condensate layer is L. Assume that these thicknesses are known, along with the wall temperature (T_W) and the temperature and mole fraction of A in the bulk gas (T_∞ and $x_{A\infty}$). The mole fraction of A in the vapor contacting the condensate (x_{A0}) is a known function of the temperature at that interface (T_0). The temperature and composition of the gas are functions only of y.

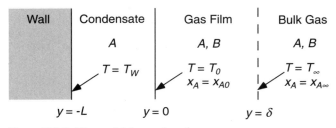

Figure P11-6. Film model for condensation.

(a) Assuming that T_0 (and thus x_{A0}) are known, determine the rate of condensation. That is, find N_{Ay}.

(b) Determine $T(y)$ in the gas film, assuming again that T_0 is known.

(c) Suppose now that T_0 is not known in advance. Use an interfacial energy balance to derive the additional equation which is needed, and describe how to compute T_0. You may assume that $T(y)$ in the liquid is approximately linear.

11-7. Nonisothermal Diffusion and Reaction in a Ternary Gas

A small, impermeable, spherical particle of radius a is suspended in a gas mixture containing components A, B, and C. The reaction $A+B \rightarrow C$ takes place at the particle surface, with a nonzero heat of reaction ΔH_R. Far from the particle the mole fractions and temperature are given as $x_{i\infty}$ ($i=A$, B, C) and T_∞, respectively, and the gas is stagnant. The mole fractions (x_{iS}) and temperature (T_S) at the particle surface are unknown. The pressure in the gas is constant (atmospheric) and the gas mixture is ideal. Assume that $x_i = x_i(r)$ and $T = T(r)$ only.

(a) Relate each of the position-dependent fluxes, $N_{ir}(r)$, to the rate of product formation at the surface, R_{SC}.

(b) Show that the mass-average velocity is zero throughout the gas (i.e., $v_r = 0$ for all r). Note that, due to variations in the mole fractions and temperature, the mass density (ρ) is not necessarily constant.

(c) Supposing that R_{SC} is known, determine $T(r)$ and T_S. For simplicity, assume that the thermal conductivity in the gas mixture (k_G) is constant and that variations in ΔH_R with temperature are negligible.

(d) Again assuming that R_{SC} is known, show how to calculate $x_i(r)$. State the differential equations and boundary conditions which must be solved, and assume that you have software to solve sets of equations of the form

$$\frac{dy_i}{dt} = f_i(t, y_1, y_2, \ldots, y_n), \qquad y_i(t_0) = Y_i, \qquad i = 1, 2, \ldots, n$$

for $t \geq t_0$, where the f_i are user-specified functions and the Y_i are constants. Explain how to proceed. You may assume that CD_{ij} is approximately constant.

(e) Suppose now that you are given the function $R_{SC}(x_A, x_B, T)$ which describes the reaction kinetics for this system at atmospheric pressure. Explain how to determine R_{SC}.

11-8. Diffusion-Limited Reaction at an Electrode

Consider a stagnant-film model for steady reaction and diffusion near a silver (Ag) cathode, as shown in Fig. P11-8. Assume that the electrolyte solution contains AgCl and KCl at bulk concentrations of C_S and C_P, respectively. The reaction at the electrode surface, which is driven by an external power source, is

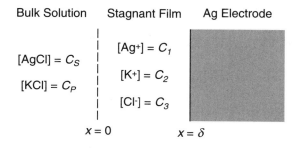

Bulk Solution   Stagnant Film   Ag Electrode

$[AgCl] = C_S$

$[KCl] = C_P$

$[Ag^+] = C_1$

$[K^+] = C_2$

$[Cl^-] = C_3$

$x = 0$ $x = \delta$

Figure P11-8. Diffusion to a silver electrode. Solid Ag is formed at the surface by reduction of Ag^+, whereas K^+ and Cl^- do not react.

$$Ag^+(aq) + e^- \rightarrow Ag(s).$$

Assume that the solution is sufficiently dilute that the Nernst–Planck equation is applicable. The film thickness (δ) greatly exceeds the Debye length.

(a) Assume that the current density at the electrode surface is $i_x(\delta) = i_0$, where i_0 is known. State the governing equations for the ion concentrations, $C_j(x)$ ($j = 1, 2, 3$), and the dimensionless electrostatic potential, $\psi(x) \equiv \phi(x)F/RT$. In other words, formulate the diffusion problem.

(b) Evaluate $d\psi/dx$ in terms of $C_3(x)$ (the Cl^- concentration) and known constants, and use the result to determine $C_3(x)$. Then, rewrite the governing equation for $C_1(x)$ (the Ag^+ concentration) in terms of the dimensionless quantities,

$$\Theta = \frac{C_1}{C_S}, \qquad \eta = \frac{x}{\delta}, \qquad \gamma = \frac{i_0\delta}{FD_1 C_S}, \qquad \beta = \frac{C_S}{C_P}.$$

(c) Assuming that AgCl is the predominant salt (i.e., $\beta \gg 1$), determine the first term in a perturbation expansion for Θ, and calculate the maximum current density (i_{max}). The maximum current is obtained when the applied voltage is sufficient to make the kinetics of the plating reaction very fast.

(d) Repeat part (c), assuming now that KCl is predominant (i.e., $\beta \ll 1$). Show that the potential gradient in this case is negligible, so that the value of i_{max} is the same as if Ag^+ diffused as an uncharged species. The suppression of potential gradients by a large concentration of an inert or "supporting" electrolyte is a very general effect (see also Problem 11-9).

11-9. Determination of Mass Transfer Coefficients by Polarography

An effective method for measuring the average mass transfer coefficient over part of a solid surface is to replace that part of the surface with an electrode and to measure the current obtained when a suitable reaction is fast enough to be limited entirely by diffusion. In practice, a voltage applied by an external power source is increased until the measured current reaches a plateau, the *limiting current*. What makes this approach especially useful is that, if the conditions are properly chosen, there is an exact analogy between diffusion of the reactive ion to the electrode and diffusion of an uncharged solute to a corresponding surface. In other words, the results are applicable not just to ions but also to nonelectrolytes. The technique is called *polarography*.

To examine the basis for the polarographic method, consider a stagnant-film model for steady diffusion and reaction at a platinum (Pt) electrode, as shown in Fig. P11-9. Assume that

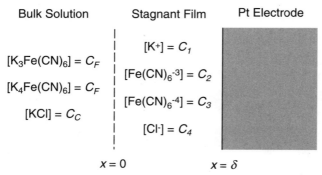

Figure P11-9. Diffusion to a platinum electrode. Ferricyanide, $Fe(CN)_6^{3-}$, is reduced to ferrocyanide, $Fe(CN)_6^{4-}$, at the electrode surface, whereas K^+ and Cl^- do not react.

the bulk solution contains equimolar amounts of $K_3Fe(CN)_6$ and $K_4Fe(CN)_6$ (both at concentration C_F), together with a large excess of KCl (at concentration C_C). The applied potential drives the reduction of ferricyanide to ferrocyanide at the electrode surface, such that

$$Fe(CN)_6^{3-}(aq) + e^- \rightarrow Fe(CN)_6^{4-}(aq).$$

The other ions, K^+ and Cl^-, do not react. Assume that the solution is sufficiently dilute that the Nernst–Planck equation is applicable. The film thickness (δ) greatly exceeds the Debye length.

(a) Assume that the current density at the electrode surface is $i_x(\delta) = i_0$, where i_0 is known. State the governing equations for the ion concentrations, $C_j(x)$ ($j = 1, 2, 3, 4$), and the dimensionless electrostatic potential, $\psi(x) \equiv \phi(x)F/RT$. In other words, formulate the diffusion problem.

(b) Rewrite the equation for the reactant (ferricyanide) in terms of the dimensionless quantities

$$\Theta = \frac{C_2}{C_F}, \qquad \eta = \frac{x}{\delta}, \qquad \gamma = \frac{i_0\delta}{FD_2C_F},$$

and $d\psi/d\eta$. Assuming that $\varepsilon = C_F/C_C \ll 1$, use the other information from (a) to show that $d\psi/d\eta = O(\varepsilon)$.

(c) Evaluate the first term in a perturbation expansion for Θ and calculate the maximum current density (i_{max}), corresponding to a diffusion-limited reaction. Relate the mass transfer coefficient to i_{max}. Show that the ferricyanide flux equals that which would be exhibited by a nonelectrolyte with the same diffusivity and bulk concentration.

11-10. Donnan Model for Diffusion Across a Charged Membrane

Membranes and other porous materials, such as particles used in liquid chromatography, are sometimes made with large concentrations of charged groups covalently bound to a polymeric network. When the polymer has a loosely crosslinked structure, so that the porous material consists mostly of water, an informative limiting model is obtained by assuming that the polymer occupies negligible volume. In this model the material is treated simply as a collection of immobilized charges. Describing the material in terms of a concentration of charge smeared uniformly throughout its volume, as done in this problem, is most appropriate when the Debye length greatly exceeds the characteristic spacing between the polymer chains.

A model for a charged membrane of thickness L is shown in Fig. P11-10. The membrane is characterized by a uniform concentration C_M of fixed, negative charges. It is desired to relate the transmembrane flux of a univalent-univalent salt to the external salt concentrations (C_0 and C_L), assuming that there is no current flow and no fluid movement. The dimensionless electrostatic potential ($\psi = \phi F/RT$) is taken to be zero in both external solutions.

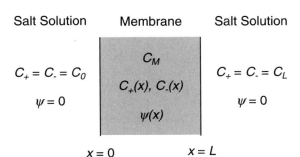

Salt Solution   Membrane   Salt Solution

C_M

$C_+ = C_- = C_0$ $C_+ = C_- = C_L$

$C_+(x), C_-(x)$

$\psi = 0$ $\psi = 0$

$\psi(x)$

$x = 0$ $x = L$

Figure P11-10. Diffusion of a univalent–univalent salt through a membrane characterized by a concentration C_M of fixed, negative charges.

(a) Derive a differential equation involving the anion concentration, $C_-(x)$, the salt flux, N_S, and known constants.

(b) Show that the Nernst–Planck equation implies that the equilibrium condition for species i is

$$\ln C_i + z_i \psi = \text{constant},$$

which is the same as the thermodynamic condition of uniform *electrochemical potential* in ideal solutions. Use this to relate $C_-(0)$ to C_0 and $C_-(L)$ to C_L.

(c) Assume now that $\gamma \equiv C_L / C_0 < 1$ and that the membrane is highly charged, such that $\varepsilon \equiv C_0 / C_M \ll 1$. Show that under these conditions the scale for $C_-(x)$ is C_0^2 / C_M, which in turn suggests that $N_S \sim (D_- / L)(C_0^2 / C_M)$. Accordingly, suitable choices for the dimensionless variables and flux are

$$\eta \equiv \frac{x}{L}, \qquad \Theta \equiv \frac{C_- C_M}{C_0^2}, \qquad \alpha \equiv \frac{N_S L C_M}{D_- C_0^2}.$$

Rewrite the differential equation and boundary conditions for the anion in dimensionless form. The addtional parameter which will appear is the diffusivity ratio, $\beta \equiv D_- / D_+$.

(d) Derive the first term in a perturbation expansion for $\Theta(\eta)$, and use this result to compute the flux. Notice that the usual proportionality between the flux and the overall concentration difference does not hold.

(e) The analysis could have been done for the cation instead of the anion. What is the scale for $C_+(x)$? What was the advantage in choosing the anion?

11-11. Diffusion Potential in Protein Ultrafiltration

Suppose that a dilute protein solution is to be concentrated using ultrafiltration. A membrane is available which allows water and salts to pass freely into the filtrate, while keeping all protein in the retentate solution. As discussed in Problem 2-8, the tendency for the protein concentration at the upstream surface of the membrane to exceed that in the bulk retentate, termed "concentration polarization," can impede the separation process. The net protein charge can be manipulated by varying the pH, and it is desired to know whether a high protein charge will lessen or increase concentration polarization.

Consider the stagnant film model depicted in Fig. P11-11. Assume that the bulk retentate has known concentrations of K^+, Cl^-, and an anionic protein denoted as Pr^{-m} ($m > 0$). (The three

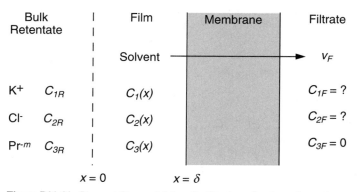

Figure P11-11. Stagnant-film model for ultrafiltration of a charged protein.

solutes are numbered as shown.) The effective film thickness (δ) and the filtrate velocity (v_F) are also given, but the filtrate concentrations of the small ions (C_{1F}, C_{2F}) are not known in advance. A good approximation is that $D_1 = D_2$. The key feature of this problem is that the large size of the protein molecule makes $D_3 \ll D_2$ or D_1 and leads to a diffusion potential.

(a) Using the Nernst–Planck approach, state the equations which govern $C_i(x)$ and $\psi(x) \equiv \phi(x)F/RT$. (Regard C_{1F} and C_{2F} as parameters, to be determined later.)

(b) Show that

$$\frac{d\psi}{dx} = \frac{m}{D_2(C_1 + C_2)}\left(v_F C_3 - D_2 \frac{dC_3}{dx}\right).$$

(c) Using the result from part (b), show that the conservation equation for the protein takes the form

$$0 = v_F C_3 - fD_3 \frac{dC_3}{dx},$$

where f is a function of the solute concentratons. Show that $f = 1$ for $m = 0$, and that $f > 1$ for $m > 0$. Thus, the apparent diffusivity of the protein is fD_3, and the protein charge *lessens* concentration polarization.

(d) Numerical methods are required to determine $C_3(x)$ and the other concentrations. Outline how to complete the solution, including how to determine C_{1F} and C_{2F}. [*Hint:* Assume diffusional equilibrium of K^+ and Cl^- across the membrane; see part (b) of Problem 11-10.]

11-12. Isothermal Diffusion and Reaction in a Ternary Gas

Consider steady diffusion and reaction of a gas in the stagnant-film geometry shown in Fig. P11-12. The gas contains species A, B, and C. The rapid and irreversible reaction $A \rightarrow mB$ (with $m \neq 1$) occurs at the catalytic surface ($y = L$), whereas species C is inert. The mole fractions of all species are assumed to be known at $y = 0$. The gas mixture is ideal, isothermal, and isobaric.

(a) Formulate the diffusion problem. That is, state the differential equations and boundary conditions which could be used to determine $x_i(y)$ and N_{iy}, where $i = A$, B, and C. Include any stoichiometric relations or auxiliary conditions which are needed.

(b) Solve for $x_C(y)$ in terms of N_{Ay} and known constants.

(c) Show that the differential equation for $x_A(y)$ can be put in the form

$$\frac{dx_A}{dy} + kx_A = p + qe^{sy},$$

where k, p, q, and s are constants. Identify each of those constants.

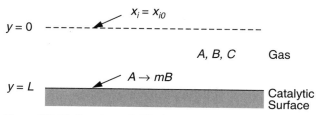

Figure P11-12. Reaction and diffusion in a ternary gas.

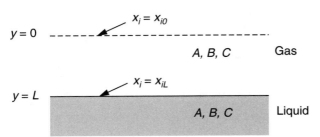

Figure P11-13. Ternary gas diffusion with evaporation and condensation.

(d) Solve for $x_A(y)$ in terms of N_{Ay} and known constants, and complete the solution by showing how to calculate N_{Ay}. Compare your results with those from Example 2.8-2.

11-13. Ternary Gas Diffusion with Evaporation and Condensation

Consider the stagnant-film arrangement shown in Fig. P11-13, in which a liquid containing components A, B, and C is in contact with a gas consisting of the same three species. The mole fractions in the bulk gas (x_{i0}) differ from those in equilibrium with the liquid (x_{iL}), causing one or more species to condense and one or more to evaporate. Assume that the total molar flux is zero, as is sometimes true in distillation. Assume also that the gas is essentially isothermal and isobaric.

Solve for the steady-state fluxes of all species in terms of the given mole fractions at $y=0$ and $y=L$.

11-14. Freezing of Salt Water*

This problem involves a simplified model for the growth of an ice crystal in salt water. The crystal, which is pure H_2O, is idealized as a sphere of radius $R(t)$. The main effect of the salt is that it lowers the freezing temperature (T_F). The freezing-point depression is described by

$$T_F = T_0 - \beta C_S,$$

where T_0 is the freezing point for pure water, β is a known (positive) constant, and C_S is the salt concentration. Far from the crystal, $T = T_\infty$, $C_S = C_\infty$, and the solution is stagnant. It is assumed that the temperature and concentration fields are both pseudosteady and spherically symmetric. For simplicity, it is assumed also that ice, pure water, and the salt solution all have the same density. With regard to notation, it is suggested that the subscript I be used for ice, W for pure water, S for salt (NaCl), and L for the salt solution.

(a) Use the continuity equation to show that $v_r = 0$ throughout the liquid.

(b) Assuming for the moment that no salt is present, determine dR/dt.

(c) Returning to the case of salt water, note that the temperature and salt concentration in the liquid at $r = R$ are both unknown, as is dR/dt. Use conservation of salt to obtain one relationship among these quantities.

(d) Use conservation of energy to obtain a second relationship among the quantities mentioned in (c). Assume that the partial molar enthalpies in the liquid are constant. That is, assume that the temperature variations in the liquid produce enthalpy changes which are negligible compared to the latent heat. (*Hint:* The temperature profile and the expression for dR/dt are very similar to those for pure water. Why?)

(e) Briefly describe how to complete the calculation of dR/dt for salt water.

*This problem was suggested by K. A. Smith.

11-15. Streaming Potential

Consider pressure-driven flow across a membrane with pore radius a and surface charge density q_e. In the absence of any transmembrane current flow, a transmembrane potential difference will develop, which is the streaming potential. Using assumptions similar to those in Example 11.7-4, relate the streaming potential (expressed as the axial potential gradient) to the volumetric flow rate Q. (Note that in electroosmosis there is a nonzero current but no pressure gradient; with the streaming potential there is a pressure gradient but no current.)

a) $\dfrac{\partial c_i}{\partial t}^0 = -\nabla \cdot N_i + \not{R_i}^0$, $N_{Ax} = const$ $\qquad N_{1X} = C_1 F V_F = ?$

$N_{2A} = C_2 F V_F = ?$

Nernst-Planck $\quad N_{ix} = C_i V_F = -D_i\left(\dfrac{dC_i}{dx} + z_i C_i \dfrac{d\psi}{dx}\right)$ $\qquad N_{3x} = 0$.

Electroneutrality $\quad C_1 = C_2 + m C_3$.

zero current $\quad \displaystyle\sum_{i=1} z_i N_{Ax} = 0 \quad \Rightarrow N_{1X} = N_{2X}$ $\qquad \dfrac{d\psi}{dx} = \dfrac{m}{D_2(C_1+C_2)}\left(C_3 V_F - D_2\dfrac{dC_3}{dx}\right)$

b) $C_1 V_F - \underset{D_2}{\not{D_1}}\left(\dfrac{dC_1}{dx} + C_1\dfrac{d\psi}{dx}\right) = C_2 V_F - D_2\left(\dfrac{dC_2}{dx} - \underset{mC_3}{C_2}\dfrac{d\psi}{dx}\right)$

$D_2(C_1+C_2)\cdot\dfrac{d\psi}{dx} = (C_1 - C_2) V_F - D_2\dfrac{d}{dx}(C_1 - C_2)$

c) $0 = C_3 V_F - D_3\left[\dfrac{dC_3}{dx} - mC_3\dfrac{d\psi}{dx}\right]$

$= C_3 V_F - D_3\dfrac{dC_3}{dx} + D_3 m C_3\left[\dfrac{m}{D_2(C_1+C_2)}\right]\left[C_3 V_F - D_2\dfrac{dC_3}{dx}\right]$

$= C_3 V_F\left[1 + \dfrac{D_3}{D_2}\left(\dfrac{m^2 C_3}{C_1+C_2}\right)\right] - D_3\dfrac{dC_3}{dx}\left[1 + \left(\dfrac{m^2 C_3}{C_1+C_2}\right)\right]$

$0 = C_3 V_F - f D_3\dfrac{dC_3}{dx} \quad\Rightarrow\quad f(C_1,C_2,C_3) = \dfrac{\left[1 + \left(\dfrac{m^2 C_3}{C_1+C_2}\right)\right]}{1 + \dfrac{D_3}{D_2}\left(\dfrac{m^2 C_3}{C_1+C_2}\right)}$

[5] $C_i(r) = C_{io}\exp\left[\dfrac{-z_i F(\phi-\phi_0)}{RT}\right] \cong C_{io}\left[1 - \dfrac{z_i F(\phi-\phi_0)}{RT}\right]$

$C_{io} = C_i(r=0)$, $\phi_0 = \phi(r=0.2)$.

$V_\theta = 0. \ V_r = 0. \ V_z(r) \quad 0 = -\dfrac{\partial p}{\partial z} + \mu\cdot\dfrac{1}{r}\dfrac{d}{dr}\left(r\dfrac{dV_z}{dr}\right) - \rho_e\dfrac{\partial\phi}{\partial z}$ ⓪ ⟨momentum⟩

Poisson $\quad \rho_e = -\epsilon\nabla^2\phi = -\epsilon\dfrac{1}{r}\dfrac{\partial}{\partial r}\left(r\dfrac{\partial\phi}{\partial r}\right) - \epsilon\dfrac{\partial}{\partial z}\left(\dfrac{\partial\phi}{\partial z}\right)$ ⑩ ⑨

$\Rightarrow \dfrac{d}{dr}\left(r\cdot\dfrac{dV_z}{dr}\right) = \dfrac{r}{\mu}\dfrac{\partial p}{\partial z} - \dfrac{\epsilon}{\mu}\dfrac{\partial\phi}{\partial z}\dfrac{\partial}{\partial r}\left(r\dfrac{\partial\phi}{\partial r}\right)$

integrate twice over r with BC's $\quad \dfrac{dV_z}{dr}\bigg|_{r=0} = \dfrac{d\phi}{dr}\bigg|_{r=0} = 0$ ⟨symmetric⟩

$V_z(a,2)=0, \ \phi(a,3)=\phi_0$.

$V_z(r) = \dfrac{1}{4\mu}\dfrac{\partial p}{\partial z}(r^2-a^2) + \dfrac{\epsilon}{\mu}\dfrac{\partial\phi}{\partial z}(\phi_0-\phi(r))$

ex 11-7-4/

$\rho_e = \sum z_i F C_i \Rightarrow \rho(r)-\phi_0 = \dfrac{-\lambda q_e}{\epsilon I_1(a/\lambda)}\left[1 - I_0(r/\lambda)\right]/\rho_A - \phi(r) = \dfrac{\lambda q_e}{\epsilon}\left[\dfrac{I_1(\varphi/\lambda) - I_1(r/\lambda)}{I_1(a/\lambda)}\right]$

$V(r) = \dfrac{1}{4\mu}\dfrac{\partial p}{\partial z}(r^2-a^2) + \dfrac{\lambda q_e}{\mu}\dfrac{\partial\phi}{\partial z}\left[1 - e^{-(a-r)/\lambda}\right]$ $a\gg\lambda \cong \dfrac{\lambda q_e}{\epsilon}\left[1 - e^{-(a-r)/\lambda}\right]I_1(a/\lambda)$

$Q = 2\pi\displaystyle\int_0^a V_z(r) r\,dr = \dfrac{2\pi}{4\mu}\dfrac{\partial p}{\partial z}\left(\dfrac{a^4}{4} - \dfrac{a^4}{2}\right) + \dfrac{2\pi\lambda q_e}{\mu}\dfrac{\partial\phi}{\partial z}\left[\dfrac{a^2}{2} - \lambda^2 e^{-a/\lambda} + \lambda^2 - a\lambda\right]$

$Q = \dfrac{\pi a^4 \rho}{8\mu L} - \dfrac{\pi a^2 \lambda q_e \phi}{4L}$ $\qquad$ ⟨negligible a$\gg\lambda$⟩

to remove ϕ dependence

$0 = \displaystyle\int_A i\cdot z\,dA = 2\pi\int_0^a \sum F z_i N_{iz}\,r\,dr = 0$

$q_{streaming}$ $\qquad C_i(r) = C_{io}\left[1 - \dfrac{z_i F}{RT}\phi(r,z)\right]$ $\qquad N_{iz} = C_i(r) V_z(r) - D_i\dfrac{\partial\phi}{\partial z} - \dfrac{F}{RT}\left(\dfrac{\partial\phi}{\partial z}\right)z_i D_i C_i(z)$

Chapter 12

TRANSPORT IN BUOYANCY-DRIVEN FLOW

12.1 INTRODUCTION

Much of this book has been concerned with *forced convection,* in which the flow is caused by an applied pressure or a moving surface. Not yet considered is flow due to density variations combined with gravity. This is known variously as *buoyancy-driven flow, free convection,* or *natural convection.* The first two terms are used here interchangeably, the first being the most descriptive and the second being the shortest. A typical example of free convection involves a vertical heated surface. The fluid near the surface is less dense than the cooler fluid far away, and therefore tends to rise. As the warm fluid rises and is replaced by fluid from below, heat transfer from the surface to the arriving fluid continues the process. Density variations and buoyancy-driven flows can result also from solute concentration gradients. Atmospheric winds are perhaps the most familiar and spectacular example of free convection, but buoyancy-driven flows exist also on small length scales. Free convection is crucial for some processes (e.g., many forms of combustion), but in others it is an unwanted byproduct of temperature or concentration gradients.

As may be apparent already, a key feature of free convection is that the velocity and temperature (or concentration) fields are closely coupled. The implication for analysis is that more differential equations must be solved simultaneously than in forced convection, making free convection problems more difficult. This chapter provides an introduction to buoyancy-driven transport, mainly in the context of heat transfer. The basic equations are derived and are then applied to confined flows. One of the examples is an analysis of hydrodynamic stability. The dimensionless parameters for free convection are identified and the boundary layer equations are given. The chapter closes with a discussion of unconfined (boundary layer) flows, including the behavior of the Nusselt number in various situations.

12.2 BUOYANCY AND THE BOUSSINESQ APPROXIMATION

Buoyancy Force

The variations in fluid density which underlie free convection preclude the use of the Navier–Stokes equation. We begin instead with the Cauchy form of conservation of momentum, written as

$$\rho \frac{D\mathbf{v}}{Dt} = -\nabla P + \rho \mathbf{g} + \nabla \cdot \boldsymbol{\tau}. \tag{12.2-1}$$

As with forced convection, it is often desirable to work with the dynamic pressure ($\mathscr{P}$) instead of the actual pressure (P). However, maintaining a simple relationship between P and $\mathscr{P}$ requires that ρ be constant (see Section 5.8). To be able to change easily from one pressure variable to the other, we redefine $\mathscr{P}$ as

$$\nabla \mathscr{P} \equiv \nabla P - \rho_0 \mathbf{g}, \tag{12.2-2}$$

where ρ_0 is the (constant) density in some arbitrarily chosen reference state (e.g., at $T = T_0$). The pressure and gravitational terms in the momentum equation are rewritten now as

$$\nabla P - \rho \mathbf{g} = (\nabla P - \rho_0 \mathbf{g}) + (\rho_0 - \rho)\mathbf{g} = \nabla \mathscr{P} + (\rho_0 - \rho)\mathbf{g}. \tag{12.2-3}$$

What this does is decompose the pressure and gravitational terms into a contribution due to the dynamic pressure gradient and a contribution due to density variations. Using Eq. (12.2-3) in Eq. (12.2-1), the momentum equation becomes

$$\rho \frac{D\mathbf{v}}{Dt} = -\nabla \mathscr{P} + (\rho - \rho_0)\mathbf{g} + \nabla \cdot \boldsymbol{\tau}. \tag{12.2-4}$$

The term $(\rho - \rho_0)\mathbf{g}$ may be viewed as a force per unit volume due to buoyancy [compare with Eq. (5.5-6)]. In essence, it is the force responsible for *free* convection. If $\rho = \rho_0$ everywhere, then the pressure and gravitational terms reduce to $-\nabla \mathscr{P}$, as in the previous chapters. This suggests that we associate the gradient in dynamic pressure with *forced* convection. This distinction between free and forced convection is appealing and at times useful, but it is made somewhat artificial by the arbitrary nature of the reference state. In other words, different choices of ρ_0 will change the relative contributions of the two terms, although their sum is unaffected. The effects of the choice of reference state are discussed further in Example 12.3-1.

Boussinesq Approximation

In a free convection problem involving a pure, nonisothermal fluid, the unknowns are the velocity, pressure, temperature, and density. The basic equations are continuity, conservation of momentum, and conservation of energy (all with variable ρ), together with an equation of state for the fluid. The equation of state, which is of the form $\rho = \rho(P, T)$, makes the fluid dynamic problem dependent on the heat transfer problem, even if all other fluid properties are constant. For an isothermal mixture in which the density depends on the local composition, the energy equation is replaced by one or more species conservation equations. Temperature and concentration variations can, of course, affect ρ simultaneously, so that both energy and species conservation equations may be needed.

Clearly, the analysis of buoyancy-driven flows may be extremely complicated. Even if one focuses on the simplest situations (e.g., a pure fluid in a parallel-plate channel), the solution to the general set of equations is so difficult that almost all published work has employed what is called the *Boussinesq approximation*.[1] Based largely on the assumption that $\Delta\rho/\rho_0 \ll 1$, where $\Delta\rho$ is the maximum change in density, this approach yields what is still a difficult set of coupled differential equations, but one which is much more amenable to analysis.

There are two parts to the Boussinesq approximation. The first is the assumption that ρ varies linearly with temperature and/or concentration. The effects of pressure variations on density are assumed to be negligible. Allowing for variations in temperature and the concentration of species i, the density is expressed as

$$\rho = \rho_0 + \left(\frac{\partial\rho}{\partial T}\right)_{T_0, C_{i0}} (T-T_0) + \left(\frac{\partial\rho}{\partial C_i}\right)_{T_0, C_{i0}} (C_i - C_{i0}), \qquad (12.2\text{-}5)$$

where the reference condition has been chosen as $T=T_0$ and $C_i=C_{i0}$. Equation (12.2-5) is just the leading part of a Taylor series expansion about the reference state. The concentration term is absent for a pure fluid, whereas additional concentration terms can be added for multicomponent mixtures. It is convenient to rewrite the density as

$$\frac{\rho_0}{\rho} = 1 + \beta(T-T_0) + \beta_i(C_i - C_{i0}), \qquad (12.2\text{-}6)$$

$$\beta = -\frac{1}{\rho}\left(\frac{\partial\rho}{\partial T}\right)\Bigg|_{T_0, C_{i0}}, \qquad \beta_i = -\frac{1}{\rho}\left(\frac{\partial\rho}{\partial C_i}\right)\Bigg|_{T_0, C_{i0}}, \qquad (12.2\text{-}7)$$

where β and β_i are the *thermal expansion coefficient* and *solutal expansion coefficient*, respectively.

Ordinarily $\beta > 0$, so that ρ decreases with increasing T. For an ideal gas $\beta = 1/T_0$, which gives $\beta = 3.4 \times 10^{-3}$ K^{-1} at 20°C. For water at 20°C, $\beta = 2.066 \times 10^{-4}$ K^{-1} (Gebhart et al., 1988, p. 946). Data for other common liquids are given in Bolz and Tuve (1970, p. 69). Water exhibits unusual behavior near its freezing point, where it is found that $\beta < 0$ for $0 < T < 4$°C and $\beta > 0$ for $T > 4$°C. When β changes rapidly with temperature, as in this case, the linearization of ρ_0/ρ in Eq. (12.2-6) will be unsatisfactory. The solutal coefficient β_i is negative in some systems and positive in others. Hereafter, we consider only temperature variations.

The second part of the Boussinesq approximation is the assumption that the variable density ρ can be replaced everywhere by the constant value ρ_0, *except* in the term $(\rho - \rho_0)\mathbf{g}$. This is crucial, because it leads to the continuity equation for a constant density fluid and allows the viscous stress to be evaluated as in the Navier–Stokes equation.

The basis for the Boussinesq approximation is examined in detail in Mihaljan (1962), where it is concluded that two dimensionless parameters must be small. The necessary conditions are given as

$$\varepsilon_1 \equiv \beta\Delta T \ll 1, \qquad (12.2\text{-}8)$$

$$\varepsilon_2 \equiv \frac{\alpha^2}{L^2 \hat{C}_p \Delta T} \ll 1, \qquad (12.2\text{-}9)$$

---

[1] Named after the French physicist J. V. Boussinesq (1842–1929).

TABLE 12-1
Values for the Parameters which Limit the Boussinesq Approximation[a]

Fluid	ε_1	ε_2
Water	2.1×10^{-3}	5.0×10^{-19}
Air	3.4×10^{-2}	4.6×10^{-14}

[a] Based on Eqs. (12.2-8) and (12.2-9) with $\Delta T = 10°C$, $L = 0.1$ m, and $T_0 = 20°C$.

where ΔT and L are the characteristic temperature difference and length, respectively. Values of ε_1 and ε_2 for air and water with $\Delta T = 10°C$ and $L = 0.1$ m are shown in Table 12-1. It is seen that Eq. (12.2-8) is by far the more restrictive requirement under these conditions, for either fluid. For liquids, this is equivalent to requiring that $\Delta \rho / \rho_0 \ll 1$. The analysis of Mihaljan (1962) assumed that density variations are linear in temperature, as in Eqs. (12.2-5) and (12.2-6), so that the requirement for linearity is in addition to Eqs. (12.2-8) and (12.2-9). The Boussinesq approximation is examined also in Spiegel and Veronis (1960), using a different approach.

Using the Boussinesq approximation, the continuity and momentum equations for a pure Newtonian fluid with constant μ become

$$\nabla \cdot \mathbf{v} = 0, \tag{12.2-10}$$

$$\frac{D\mathbf{v}}{Dt} = -\frac{\nabla \mathcal{P}}{\rho_0} - \mathbf{g}\beta(T - T_0) + \nu \nabla^2 \mathbf{v}. \tag{12.2-11}$$

Equation (12.2-10) is the usual continuity equation for constant density, and Eq. (12.2-11) differs from the Navier–Stokes equation only by the addition of the temperature term. Neglecting the viscous dissipation and compressibility effects, the energy equation is the same as used previously, namely

$$\frac{DT}{Dt} = \alpha \nabla^2 T. \tag{12.2-12}$$

Within the limitations of the Boussinesq approximation, Eqs. (12.2-10)–(12.2-12) describe any combination of forced and free convection in a pure fluid. The equations for an isothermal mixture are analogous.

12.3 CONFINED FLOWS

In this section we consider two well-known examples of confined (or internal) flows which are due to buoyancy. The first, which involves fully developed flow in a vertical channel, is one of the few free convection problems which has a simple analytical solution. The second, which concerns the stability of a layer of fluid heated from below, examines the conditions under which buoyancy effects are strong enough to cause motion in an otherwise static fluid.

Example 12.3-1 Flow in a Vertical Channel We first consider a vertical parallel-plate channel of indefinite length, with walls at different temperatures. As shown in Fig. 12-1, the channel width is $2H$ and the wall temperatures are T_1 and T_2; it is assumed that $T_2 > T_1$. In this system the heat fluxes in and out of the fluid will come into balance, yielding a thermally fully developed region with no streamwise temperature variations. With the temperature (and buoyancy force) independent of x, it is feasible to have fully developed flow. Accordingly, we assume that $v_x = v_x(y)$ and $v_y = 0$.

With temperature a function of y only, the energy equation reduces to

$$\frac{d^2T}{dy^2} = 0, \tag{12.3-1}$$

which indicates that the temperature profile is linear. Introducing a dimensionless coordinate η and the arithmetic mean temperature T_m, the linear profile is expressed as

$$T = T_m + (T_2 - T_m)\eta, \tag{12.3-2}$$

$$\eta \equiv \frac{y}{H}, \qquad T_m \equiv \frac{T_1 + T_2}{2}. \tag{12.3-3}$$

Turning now to the momentum equation, we note from the y component of Eq. (12.2-11) that, with $v_y = 0$ and $g_y = 0$, $\mathcal{P}$ must be independent of y. With $g_x = -g$, the x component becomes

$$\frac{d^2v_x}{d\eta^2} = -\frac{g\beta H^2}{\nu} \left[(T_m - T_0) + (T_2 - T_m)\eta \right] + \frac{H^2}{\mu} \frac{d\mathcal{P}}{dx}. \tag{12.3-4}$$

Integrating and using the no-slip conditions, $v_x(\pm 1) = 0$, the velocity is found to be

$$v_x = \frac{g\beta(T_2 - T_m)H^2}{6\nu}(\eta - \eta^3) - \frac{H^2}{2}\left[\frac{1}{\mu}\frac{d\mathcal{P}}{dx} - \frac{g\beta(T_m - T_0)}{\nu} \right](1 - \eta^2). \tag{12.3-5}$$

Notice that the velocity is the sum of an odd function and an even function. If $d\mathcal{P}/dx$ is assumed to be known, all that is needed to complete the solution is to specify the reference temperature, T_0.

Before choosing T_0, it is informative to derive a general relationship involving the pressure gradient, reference temperature, and mean velocity. Using Eq. (12.3-5), the mean velocity is evaluated as

$$U = \frac{1}{2}\int_{-1}^{1} v_x \, d\eta = -\frac{H^2}{3}\left[\frac{1}{\mu}\frac{d\mathcal{P}}{dx} - \frac{g\beta(T_m - T_0)}{\nu} \right]. \tag{12.3-6}$$

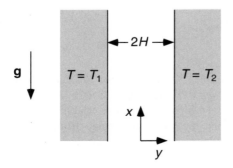

Figure 12-1. Flow in a vertical channel with walls at different temperatures.

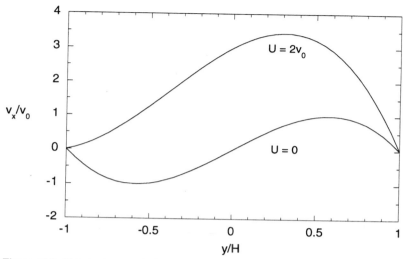

Figure 12-2. Velocity in a vertical channel with free convection only ($U=0$) or combined free and forced convection ($U=2v_0$), based on Eq. (12.3-9).

Only the even part of v_x contributes to U. This relationship shows that $d\mathcal{P}/dx$, T_0, and U cannot all be specified independently. For example, assume that the channel is closed at the ends, so that there is no net flow and $U=0$. Equation (12.3-6) reduces to

$$\frac{d\mathcal{P}}{dx} = \rho_0 g \beta (T_m - T_0). \qquad (12.3-7)$$

This illustrates the dependence of $d\mathcal{P}/dx$ on the reference state that was mentioned in Section 12.2. In particular, the pressure gradient for $U=0$ must vanish if $T_0=T_m$.

Returning to Eq. (12.3-5), we see that the simplest expression for the velocity is obtained by choosing $T_0=T_m$. In modeling a real system, another motivation for choosing an intermediate temperature is that it will minimize the errors made in linearizing the function $\rho(T)$. With $T_0=T_m$, the result for a fully enclosed channel is

$$v_x = \frac{g\beta(T_2-T_1)H^2}{12\nu}(\eta-\eta^3) \equiv \frac{3\sqrt{3}}{2} v_0(\eta-\eta^3) \qquad (U=0), \qquad (12.3-8)$$

where v_0 is the maximum velocity under these conditions. This velocity profile is antisymmetric, as shown in Fig. 12-2. As one might expect, the flow is upward in the warm half of the channel ($y>0$) and downward in the cool half ($y<0$).

If there is a net flow (i.e., $U\neq0$) the pressure gradient will not vanish. We evaluate $d\mathcal{P}/dx$ again by setting $T_0=T_m$ in Eq. (12.3-6), and use the result in Eq. (12.3-5) to obtain

$$v_x = \frac{3}{2}[\sqrt{3}\,v_0(\eta-\eta^3) + U(1-\eta^2)]. \qquad (12.3-9)$$

Equations (12.3-8) and (12.3-9) represent pure free convection and mixed free and forced convection, respectively. An example of the mixed case (with $U=2v_0$) is shown in Fig. 12-2.

Example 12.3-2. Stability of a Layer of Fluid Heated from Below The system to be analyzed, which is shown in Fig. 12-3, consists of a static layer of fluid between horizontal surfaces main-

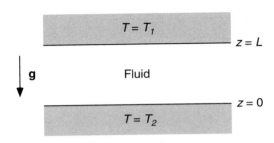

tained at different temperatures. The thickness of the fluid layer is L, and the z axis is directed upward. The temperature at the bottom (T_2) exceeds that at the top (T_1), so that the vertical temperature gradient causes the density to increase with height. This "top-heavy" arrangement tends to be unstable, but the viscosity of the fluid opposes any motion. If the temperature difference $(\Delta T = T_2 - T_1)$ is increased gradually in an experiment, it is found that flow begins only when ΔT exceeds a critical value. As viewed from above, the ensuing flow consists of a regular pattern of hexagonal cells, whose width is close to $2L$. A photograph of this pattern, from a classic study by H. Bénard in 1900, is reproduced in Chandrasekhar (1961, p. 10). The objective of the analysis is to predict the critical condition when flow begins. This much-studied physical situation is called the *Bénard problem*.

We will first determine the temperature and pressure in the static fluid and then examine what happens when the fluid experiences small but otherwise arbitrary disturbances in temperature, pressure, and velocity. The underlying idea is that any real system is subjected almost continuously to small vibrations or other upsets; how it responds is what characterizes its stability. In the Bénard problem it turns out that a given disturbance will either decay in time or grow exponentially. For conditions in which *all* possible disturbances can be shown to decay, we conclude that the base case (the static fluid) is stable. On the other hand, if at least *one* disturbance can be shown to grow, the base case is evidently unstable. The growth of disturbances leads eventually to a new flow pattern, in this case the hexagonal cells described above; the approach used here does not predict the nature of the new flow. The assumption that the disturbances are small allows us to linearize the governing equations by neglecting all terms which involve products of disturbances. This linearization is crucial to the methodology, which is called a *linear stability analysis*. The development here largely follows that in Chandrasekhar (1961).

In the static fluid the temperature and pressure depend only on z. With $\mathbf{v} = \mathbf{0}$, the energy equation reduces to

$$\frac{d^2T}{dz^2} = 0. \tag{12.3-10}$$

The solution which satisfies the boundary conditions is

$$T(z) = T_2 - Gz \equiv b_T(z), \tag{12.3-11}$$

where $G = \Delta T/L$ is the magnitude of the static temperature gradient. As in the previous example, we will choose as the reference temperature the arithmetic mean of the surface temperatures; that is, $T_0 = T_m$, where T_m is given by Eq. (12.3-3). With this choice, the z-momentum equation reduces to

$$0 = -\frac{1}{\rho_0}\frac{d\mathcal{P}}{dz} + g\beta(T - T_m). \tag{12.3-12}$$

Using Eq. (12.3-11) for T, integrating, and letting $\mathcal{P} = 0$ at $z = 0$, the dynamic pressure is evaluated as

$$\mathcal{P}(z) = \frac{\rho_0 g L^2 \beta G}{2}\left[\left(\frac{z}{L}\right) - \left(\frac{z}{L}\right)^2\right] \equiv b_P(z).$$

(12.3-13)

We are used to $\mathcal{P}$ being constant in a static fluid. Here, in order to balance the buoyancy force, the dynamic pressure must vary with height.

In the presence of disturbances, the temperature and pressure are expressed as

$$T = b_T + \theta,$$

(12.3-14)

$$\mathcal{P} = b_P + p,$$

(12.3-15)

where θ and p represent small perturbations to the base case. No special symbol is used for the velocity disturbance, because in this problem any nonzero $\mathbf{v}$ represents a perturbation. Expanding the various terms in the energy equation using Eq. (12.3-14) gives

$$\frac{\partial T}{\partial t} = \frac{\partial b_T}{\partial t} + \frac{\partial \theta}{\partial t} = \frac{\partial \theta}{\partial t},$$

(12.3-16)

$$\mathbf{v}\cdot\nabla T = \mathbf{v}\cdot\nabla b_T + \mathbf{v}\cdot\nabla\theta \cong -Gv_z,$$

(12.3-17)

$$\nabla^2 T = \nabla^2 b_T + \nabla^2 \theta = \nabla^2 \theta.$$

(12.3-18)

In Eq. (12.3-17) we have neglected $\mathbf{v}\cdot\nabla\theta$, because it involves a product of two perturbations. The other simplifications follow directly from the form of the static temperature profile, $b_T(z)$. Using these expressions in Eq. (12.2-12), the energy equation becomes

$$\frac{\partial \theta}{\partial t} = Gv_z + \alpha\nabla^2\theta.$$

(12.3-19)

Expanding Eq. (12.2-11) in a similar manner gives

$$\frac{\partial \mathbf{v}}{\partial t} = -\nabla\left(\frac{p}{\rho_0}\right) + g\beta\theta\mathbf{e}_z + \nu\nabla^2\mathbf{v}.$$

(12.3-20)

In obtaining this form of the momentum equation we neglected $\mathbf{v}\cdot\nabla\mathbf{v}$. The third basic relationship is the continuity equation, Eq. (12.2-10).

The unknowns are θ, p, and the three components of $\mathbf{v}$. The number of dependent variables is reduced by taking the curl of Eq. (12.3-20), which eliminates p. Using identity (7) of Table A-1, the result is

$$\frac{\partial \mathbf{w}}{\partial t} = g\beta\left(\frac{\partial \theta}{\partial y}\mathbf{e}_x - \frac{\partial \theta}{\partial x}\mathbf{e}_y\right) + \nu\nabla^2\mathbf{w},$$

(12.3-21)

where $\mathbf{w} \equiv \nabla\times\mathbf{v}$ is the vorticity. Taking the curl again, and using continuity and identity (10) of Table A-1, we obtain

$$\frac{\partial}{\partial t}(\nabla^2\mathbf{v}) = g\beta\left[-\frac{\partial^2\theta}{\partial x\,\partial z}\mathbf{e}_x - \frac{\partial^2\theta}{\partial y\,\partial z}\mathbf{e}_y + \left(\frac{\partial^2\theta}{\partial x^2} + \frac{\partial^2\theta}{\partial y^2}\right)\mathbf{e}_z\right] + \nu\nabla^4\mathbf{v}.$$

(12.3-22)

Because Eq. (12.3-19) does not involve v_x or v_y, it is sufficient to consider only the z component of Eq. (12.3-22), which is

$$\frac{\partial}{\partial t}(\nabla^2 v_z) = g\beta\left(\frac{\partial^2\theta}{\partial x^2} + \frac{\partial^2\theta}{\partial y^2}\right) + \nu\nabla^4 v_z.$$

(12.3-23)

The original set of dependent variables has been reduced now to the primary unknowns, θ and v_z, which are governed by Eqs. (12.3-19) and (12.3-23).

With both surfaces assumed to be maintained at constant temperature, the boundary conditions for the temperature perturbation are

$$\theta = 0 \qquad \text{at } z=0 \text{ and } z=L. \tag{12.3-24}$$

The usual no-penetration condition implies that

$$v_z = 0 \qquad \text{at } z=0 \text{ and } z=L. \tag{12.3-25}$$

Because Eq. (12.3-23) is fourth order, two more boundary conditions are needed for the velocity. Two alternative situations are considered. The more realistic one assumes that the fluid layer is bounded by solid surfaces at which the no-slip condition applies. Rearranging the continuity equation gives

$$\frac{\partial v_z}{\partial z} = -\frac{\partial v_x}{\partial x} - \frac{\partial v_y}{\partial y}. \tag{12.3-26}$$

With v_x and v_y constant on both surfaces, it follows that

$$\frac{\partial v_z}{\partial z} = 0 \qquad \text{at } z=0 \text{ and } z=L \quad \text{(solid surfaces)}. \tag{12.3-27}$$

In the other situation it is assumed that there is no shear stress at either surface. This is difficult or impossible to achieve experimentally, but it yields a problem which is much easier to solve. Thus, the solution is completed only for this case, although the results for both situations are discussed at the end of the analysis. Differentiating Eq. (12.3-26) with respect to z and then reversing the order of differentiation, we obtain

$$\frac{\partial^2 v_z}{\partial z^2} = -\frac{\partial}{\partial x}\left(\frac{\partial v_x}{\partial z}\right) - \frac{\partial}{\partial y}\left(\frac{\partial v_y}{\partial z}\right). \tag{12.3-28}$$

With $\tau_{zx} = \tau_{zy} = 0$, it follows that

$$\frac{\partial^2 v_z}{\partial z^2} = 0 \qquad \text{at } z=0 \text{ and } z=L \quad \text{(free surfaces)}. \tag{12.3-29}$$

Thus, the six boundary condtions for Eqs. (12.3-19) and (12.3-23) are provided by Eqs. (12.3-24), (12.3-25), and either (12.3-27) or (12.3-29).

The disturbance(s) to a real system might have almost any functional form. Accordingly, what is needed is a general way to represent the possible disturbances in temperature and velocity. Here we take advantage of the fact that an arbitrary function can be represented as

$$a(x,\ y,\ z,\ t) = \int_{-\infty}^{\infty} \int_{-\infty}^{\infty} a_c(z,\ t) \exp[i(c_x x + c_y y)]\ dc_x\ dc_y. \tag{12.3-30}$$

This is one form of the *Fourier integral,* which is to infinite intervals what the Fourier series is to finite intervals. The only requirements are that the function $a(x,\ y,\ z,\ t)$ be piecewise differentiable with respect to x and y in any finite interval and that the integral of $|a|$ over x and y exist [see, for example, Hildebrand (1976, pp. 234–240)]. In Eq. (12.3-30), z and t are treated as parameters. In this representation the function is decomposed into fundamental contributions corresponding to a continuous range of spatial frequencies in x and y. Each contribution, or *mode,* is associated with a particular wave number, given by

$$c = (c_x^2 + c_y^2)^{1/2}. \tag{12.3-31}$$

The corresponding wavelength is $2\pi/c$. The function $a_c(z, t)$ differs, in general, for each wave number. In applying Eq. (12.3-30) to stability problems, it is assumed that the disturbances are exponential in time, such that

$$a_c(z, t) = A(z)e^{st}. \tag{12.3-32}$$

In general, the constant s for a given wave number will be complex, but it can be shown that in the Bénard problem the imaginary part of s is zero (Chandrasekhar, 1961, pp. 24–26). Thus, disturbances will grow or decay with time, as indicated earlier, rather than oscillate. The change from stable to unstable behavior occurs when s changes sign, so that the objective of our analysis is to define the conditions under which $s = 0$.

Based on the foregoing, the amplitude of any disturbance may be viewed as a linear super-position of functions of the form

$$f(x, y, z, t) = F(z) \exp[i(c_x x + c_y y) + st], \tag{12.3-33}$$

which correspond to particular modes. The key point is that we are able to examine the stability of the system with respect to *all* disturbances by letting c vary from 0 to ∞. It is not necesary to know the contributions of individual modes to any particular disturbance. The various derivatives of f are evaluated as

$$\frac{\partial f}{\partial t} = sf, \qquad \frac{\partial^2 f}{\partial x^2} + \frac{\partial^2 f}{\partial y^2} = -c^2 f, \qquad \nabla^2 f = -c^2 f + \frac{\partial^2 f}{\partial z^2}. \tag{12.3-34}$$

The decomposition into modes is applied to the Bénard problem by letting $f = \theta$ and $F = \Theta$ for temperature, and $f = v_z$ and $F = V$ for velocity. Using Eqs. (12.3-33) and (12.3-34) in Eqs. (12.3-19) and (12.3-23), we obtain

$$s\Theta = GV + \alpha\left(\frac{d^2}{dz^2} - c^2\right)\Theta, \tag{12.3-35}$$

$$s\left(\frac{d^2}{dz^2} - c^2\right)V = -g\beta c^2\Theta + \nu\left(\frac{d^2}{dz^2} - c^2\right)^2 V. \tag{12.3-36}$$

Thus, the new temperature and velocity variables are $\Theta(z)$ and $V(z)$, respectively. Before proceeding, it is convenient to introduce the dimensionless quantities

$$\zeta \equiv \frac{z}{L}, \qquad \tau \equiv \frac{t\nu}{L^2}, \qquad \lambda \equiv cL, \qquad \sigma \equiv \frac{sL^2}{\nu}. \tag{12.3-37}$$

The temperature and velocity are left in dimensional form. Using Eq. (12.3-37) in Eqs. (12.3-35) and (12.3-36) gives

$$\left(\frac{d^2}{d\zeta^2} - \lambda^2 - \mathrm{Pr}\ \sigma\right)\Theta = -\frac{GL^2}{\alpha}V, \tag{12.3-38}$$

$$\left(\frac{d^2}{d\zeta^2} - \lambda^2\right)\left(\frac{d^2}{d\zeta^2} - \lambda^2 - \sigma\right)V = \left(\frac{g\beta L^2}{\nu}\right)\lambda^2\Theta. \tag{12.3-39}$$

Combining these two equations so as to eliminate the temperature variable, we obtain

$$\left(\frac{d^2}{d\zeta^2} - \lambda^2\right)\left(\frac{d^2}{d\zeta^2} - \lambda^2 - \sigma\right)\left(\frac{d^2}{d\zeta^2} - \lambda^2 - \mathrm{Pr}\ \sigma\right)V = -\lambda^2\mathrm{Ra}\ V, \tag{12.3-40}$$

$$\mathrm{Ra} \equiv \frac{gL^4\beta G}{\alpha\nu} = \frac{gL^3\beta\Delta T}{\alpha\nu} \tag{12.3-41}$$

where Ra is the *Rayleigh number.* Notice that the density change which destabilizes the static fluid (i.e., $\beta\Delta T$) is in the numerator of Ra, and the viscosity, which is expected to be stabilizing, is in the denominator. This suggests that stability will be lost when Ra exceeds some minimum value, which is in fact the case.

To determine the conditions corresponding to neutral stability (i.e., $s=0$), we set $\sigma=0$ in Eq. (12.3-40). The result is

$$\left(\frac{d^2}{d\zeta^2}-\lambda^2\right)^3 V = -\lambda^2 \text{ Ra } V, \tag{12.3-42}$$

which is the final form of the differential equation which governs stability. It is clearer now that the key physical parameter is Ra. Using the velocity conditions stated earlier, four of the six boundary conditions for this equation are given by

$$V=0 \qquad \text{at } \zeta=0 \text{ and } \zeta=1, \tag{12.3-43}$$

$$\frac{dV}{d\zeta}=0 \qquad \text{at } \zeta=0 \text{ and } \zeta=1 \qquad \text{(solid surfaces)}, \tag{12.3-44}$$

$$\frac{d^2V}{d\zeta^2}=0 \qquad \text{at } \zeta=0 \text{ and } \zeta=1 \qquad \text{(free surfaces)}. \tag{12.3-45}$$

Combining the temperature conditions with Eq. (12.3-39) (with $\sigma=0$), the remaining two boundary conditions are found to be

$$\left(\frac{d^2}{d\zeta^2}-\lambda^2\right)^2 V=0 \qquad \text{at } \zeta=0 \text{ and } \zeta=1. \tag{12.3-46}$$

Because the differential equation and all of the boundary conditions are homogeneous, it appears that the solid-surface and free-surface cases each constitute an eigenvalue problem. Indeed, for a given wave number (λ), there are nontrivial solutions only for specific values of Ra.

The rest of our analysis is limited to the special case of free surfaces. The eigenvalue problem is simplified by expanding Eq. (12.3-46) as

$$\frac{d^4V}{d\zeta^4}=2\lambda^2 \frac{d^2V}{d\zeta^2}-\lambda^4 V. \tag{12.3-47}$$

Notice, from Eqs. (12.3-43) and (12.3-45), that the right-hand side vanishes. Accordingly, the boundary conditions for the free-surface case are just

$$V=\frac{d^2V}{d\zeta^2}=\frac{d^4V}{d\zeta^4}=0 \qquad \text{at } \zeta=0 \text{ and } \zeta=1 \quad \text{(free surfaces)}. \tag{12.3-48}$$

It is straightforward to show that *all* even derivatives must vanish at the boundaries. The solution to Eq. (12.3-42) which has this property is

$$V=a_n \sin n\pi\zeta, \qquad n=1,2, \ldots \tag{12.3-49}$$

Substituting this into Eq. (12.3-42), we obtain

$$\left(\frac{d^2}{d\zeta^2}-\lambda^2\right)^3 \sin n\pi\zeta = -\left(n^2\pi^2+\lambda^2\right)^3 \sin n\pi\zeta = -\lambda^2 \text{ Ra } \sin n\pi\zeta, \tag{12.3-50}$$

which implies that the Rayleigh number at neutral stability is given by

$$\text{Ra} = \frac{\left(n^2\pi^2+\lambda^2\right)^3}{\lambda^2}. \tag{12.3-51}$$

TABLE 12-2
Critical Values of the Rayleigh Number for a Layer of Fluid Heated from Below

Types of surfaces	Ra_c
Both free	657.5
One solid and one free	1101
Both solid	1708

What we wish to determine is the *smallest* value of Ra at which the limits of stability are reached. For fixed λ, Eq. (12.3-51) indicates that the minimum Rayleigh number is

$$\text{Ra}_1 = \frac{\left(\pi^2 + \lambda^2\right)^3}{\lambda^2}, \tag{12.3-52}$$

which corresponds to $n = 1$. Given that disturbances of any wave number might be present, we now minimize Ra_1 with respect to λ. Using

$$\left. \frac{d\,\text{Ra}_1}{d\lambda^2} \right|_{\lambda=\lambda_c} = 0 \tag{12.3-53}$$

it is found that the critical value of λ is $\lambda_c = \pi/\sqrt{2}$. Finally, the critical Rayleigh number is calculated to be

$$\text{Ra}_c = \frac{27}{4}\pi^4 = 657.5. \tag{12.3-54}$$

For $\text{Ra} < \text{Ra}_c$, the static layer of fluid is predicted to be stable with respect to any small disturbance; for $\text{Ra} > \text{Ra}_c$, it is unstable.

The solutions of the eigenvalue problems which arise when one or both of the bounding surfaces are solid are discussed in Chandrasekhar (1961). The results for the three cases are summarized in Table 12-2. For two solid surfaces, Ra_c is increased to 1708. The trend in Ra_c is consistent with the expectation that a stationary boundary with no slip will resist flow more effectively than a boundary with zero shear stress. An experimental value for solid surfaces is 1700 ± 51 [Silveston (1958), as quoted in Chandrasekhar (1961, p. 69)], in excellent agreement with the theory. For discussions of this and other experimental work with the Bénard system, as well as discussions of theoretical analyses of the cellular flow patterns, see Chandrasekhar (1961) and Drazin and Reid (1981). Either of those books is a good general source of information on hydrodynamic stability. In addition to linear stability analysis, Drazin and Reid (1981) has an introduction to nonlinear stability.

12.4 DIMENSIONAL ANALYSIS AND BOUNDARY LAYER EQUATIONS

This section begins with a discussion of the dimensionless forms of the differential equations for free convection, along with a discussion of the dimensionless parameters involved. Proceeding largely by analogy with the derivations in Chapters 8 and 10, this information is then used to identify the equations which govern boundary layers.

Dimensional Analysis

In defining dimensionless variables we denote the characteristic velocity for a buoyancy-driven flow as U_b, the characteristic length as L, and the characteristic temperature difference as ΔT. The time scale is assumed to be L/U_b and the inertial pressure scale, $\rho_0 U_b^2$, is employed. Thus, the dimensionless variables and differential operators are

$$\tilde{\mathbf{r}} \equiv \frac{\mathbf{r}}{L}, \quad \tilde{t} \equiv \frac{tU_b}{L}, \quad \tilde{\mathbf{v}} \equiv \frac{\mathbf{v}}{U_b}, \quad \tilde{\mathcal{P}} \equiv \frac{\mathcal{P}}{\rho_0 U_b^2}, \quad \Theta \equiv \frac{T - T_0}{\Delta T}, \quad \tilde{\boldsymbol{\nabla}} \equiv L\boldsymbol{\nabla}, \quad \tilde{\nabla}^2 \equiv L^2 \nabla^2.$$

$$(12.4\text{-}1)$$

The major distinction between forced and free convection lies in the velocity scale. Whereas in forced convection the velocity scale is imposed and is therefore known in advance, in free convection U_b must be inferred by balancing the appropriate terms in the governing equations. Introducing the dimensionless quantities into Eqs. (12.2-11) and (12.2-12), the momentum and energy equations are written as

$$\frac{D\tilde{\mathbf{v}}}{D\tilde{t}} = -\tilde{\boldsymbol{\nabla}}\tilde{\mathcal{P}} - \left(\frac{gL\beta\Delta T}{U_b^2}\right)\Theta\mathbf{e}_g + \left(\frac{\nu}{U_b L}\right)\tilde{\nabla}^2\tilde{\mathbf{v}}, \qquad (12.4\text{-}2)$$

$$\frac{D\Theta}{D\tilde{t}} = \left(\frac{\alpha}{U_b L}\right)\tilde{\nabla}^2\Theta, \qquad (12.4\text{-}3)$$

where $\mathbf{e}_g$ is a unit vector which points in the direction of gravity. Our main interest here is in the dimensionless parameters, which are the three terms in parentheses. The dimensionless continuity equation, which has the same form as Eq. (12.2-10), contains no parameters and therefore adds no information at present.

To identify the velocity scale, we focus on the buoyancy term in Eq. (12.4-2). In flows where buoyancy effects are prominent, we would expect this term to be of the same order of magnitude as the inertial and pressure terms. Assuming that Θ is a properly scaled temperature, this implies that the coefficient of the buoyancy term is neither large nor small. Setting that coefficient equal to one, the characteristic velocity is defined as

$$U_b \equiv (gL\beta\Delta T)^{1/2}. \qquad (12.4\text{-}4)$$

It is conventional to express the coefficient of the viscous term in Eq. (12.4-2) as $\mathrm{Gr}^{-1/2}$, where Gr is the *Grashof number*, defined as

$$\mathrm{Gr} \equiv \frac{gL^3\beta\Delta T}{\nu^2} = \left(\frac{U_b L}{\nu}\right)^2. \qquad (12.4\text{-}5)$$

As indicated by the second equality, the Grashof number is equivalent to the square of a Reynolds number based on U_b. What in forced convection would be the Péclet number is written in free convection as

$$\frac{U_b L}{\alpha} = \left(\frac{U_b L}{\nu}\right)\left(\frac{\nu}{\alpha}\right) = \mathrm{Gr}^{1/2}\,\mathrm{Pr}. \qquad (12.4\text{-}6)$$

Accordingly, the coefficient of the conduction term in Eq. (12.4-3) is expressed as $\mathrm{Gr}^{-1/2}\mathrm{Pr}^{-1}$. Another group that is commonly encountered in free convection is the *Ray-*

leigh number, which arose in Example 12.3-2. It is related to the Grashof and Prandtl numbers as

$$\text{Ra} \equiv \frac{gL^3 \beta \Delta T}{\alpha \nu} = \left(\frac{gL^3 \beta \Delta T}{\nu^2}\right)\left(\frac{\nu}{\alpha}\right) = \text{Gr Pr}. \qquad (12.4\text{-}7)$$

Rewriting Eqs. (12.4-2) and (12.4-3) in terms of the Grashof and Prandtl numbers gives

$$\frac{D\tilde{\mathbf{v}}}{D\tilde{t}} = -\tilde{\nabla}\tilde{\mathscr{P}} - \Theta \mathbf{e}_g + \text{Gr}^{-1/2}\tilde{\nabla}^2 \tilde{\mathbf{v}}, \qquad (12.4\text{-}8)$$

$$\frac{D\Theta}{D\tilde{t}} = \text{Gr}^{-1/2}\,\text{Pr}^{-1}\,\tilde{\nabla}^2 \Theta. \qquad (12.4\text{-}9)$$

Within the restrictions of the Boussinesq approximation, these are the most general, dimensionless forms of the momentum and energy equations for buoyancy-driven flow.

We wish now to identify the factors which determine the Nusselt number in free convection. This discussion, which is restricted to steady states, parallels that in Section 9.3. Equations (12.4-8) and (12.4-9) indicate that the velocity and temperature in free convection are functions of the form

$$\tilde{\mathbf{v}} = \tilde{\mathbf{v}}(\tilde{\mathbf{r}},\,\text{Gr, Pr, geometric ratios}), \qquad (12.4\text{-}10)$$

$$\Theta = \Theta(\tilde{\mathbf{r}},\,\text{Gr, Pr, geometric ratios}). \qquad (12.4\text{-}11)$$

As a specific example, consider the velocity in a vertical parallel-plate channel, as derived in Example 12.3-1. Rewriting Eq. (12.3-8) in dimensionless form gives

$$\tilde{v}_x = \frac{v_x}{U_b} = \frac{\text{Gr}^{1/2}}{12}(\eta - \eta^3), \qquad U_b = [gH\beta(T_2 - T_1)]^{1/2}. \qquad (12.4\text{-}12)$$

Only the Grashof number is involved here, because of the special nature of the temperature field in this flow. It follows from Eqs. (12.4-10) and (12.4-11) that the local and average Nusselt numbers are of the form

$$\text{Nu} = \text{Nu}(\tilde{\mathbf{r}}_s,\,\text{Gr, Pr, geometric ratios}), \qquad (12.4\text{-}13)$$

$$\overline{\text{Nu}} = \overline{\text{Nu}}\,(\text{Gr, Pr, geometric ratios}), \qquad (12.4\text{-}14)$$

where $\mathbf{r}_s$ denotes position on a surface.

Boundary Layer Equations

It was mentioned above that the Grashof number is like the square of a Reynolds number. A comparison of Eq. (12.4-8) with Eq. (5.10-20), the corresponding momentum equation for forced convection, emphasizes that $\text{Gr}^{1/2}$ plays the same role in free convection that Re plays in forced convection. That is, both quantities indicate the relative importance of the inertial and viscous terms. Accordingly, by analogy with the findings for forced convection in Chapter 8, we conclude that there will be a momentum boundary layer in free convection when $\text{Gr}^{1/2} \gg 1$. Outside the boundary layer the inertial and buoyancy effects are important; within the boundary layer the inertial, buoyancy, and viscous terms are all comparable. The dynamic pressure term, which may or may not be important (depending on the geometry), is retained in the momentum equations for both regions. It was noted also that $\mathrm{Gr}^{1/2}\,\mathrm{Pr}$ in free convection is like $\mathrm{Pe}=\mathrm{Re}\,\mathrm{Pr}$ in forced convection. Thus, by analogy with the discussion in Chapter 10, a thermal boundary layer will exist if $\mathrm{Gr}^{1/2}\mathrm{Pr}\gg1$. Conduction is important in the thermal boundary layer, but not in the thermal outer region.

With the aforementioned analogies in mind, we now develop the equations which govern steady free convection in two-dimensional boundary layers. As usual, we only consider boundary layers which are thin enough to justify the use of local rectangular coordinates. Although the other effects of surface curvature are neglected, to compute the buoyancy force we must account for the local orientation of the gravitational vector relative to the surface. As shown in Fig. 12-4, the shape and orientation of the surface are described in terms of the angle $\phi(x)$ between $\mathbf{g}$ and the surface normal, $\mathbf{e}_y$. Thus, the unit downward normal is related to $\mathbf{e}_x$ and $\mathbf{e}_y$ by

$$\mathbf{e}_g = -\sin\,\phi(x)\mathbf{e}_x + \cos\,\phi(x)\mathbf{e}_y. \qquad (12.4\text{-}15)$$

As with boundary layers in forced convection, the thinness of the region has two basic consequences. One is that the second derivatives of velocity and temperature with respect to x are negligible. The other is that the pressure is ordinarily a function of x only, to good approximation. (This conclusion concerning the pressure is not valid for free-convection boundary layers at horizontal or nearly horizontal surfaces, where $\phi\cong0$.) Accordingly, the governing equations for free-convection boundary layers at vertical or inclined surfaces are

$$\frac{\partial\tilde{v}_x}{\partial\tilde{x}}+\frac{\partial\tilde{v}_y}{\partial\tilde{y}}=0, \qquad (12.4\text{-}16)$$

$$\tilde{v}_x\frac{\partial\tilde{v}_x}{\partial\tilde{x}}+\tilde{v}_y\frac{\partial\tilde{v}_x}{\partial\tilde{y}}=-\frac{d\tilde{\mathcal{P}}}{d\tilde{x}}+\Theta\,\sin\,\phi+\mathrm{Gr}^{-1/2}\frac{\partial^2\tilde{v}_x}{\partial\tilde{y}^2}, \qquad (12.4\text{-}17)$$

$$\tilde{v}_x\frac{\partial\Theta}{\partial\tilde{x}}+\tilde{v}_y\frac{\partial\Theta}{\partial\tilde{y}}=\mathrm{Gr}^{-1/2}\,\mathrm{Pr}^{-1}\frac{\partial^2\Theta}{\partial\tilde{y}^2}. \qquad (12.4\text{-}18)$$

As usual for flow in thin regions, the y-momentum equation has been eliminated.

The presence in Eqs. (12.4-17) and (12.4-18) of the Grashof number, which must be large to justify the boundary layer approximation, is evidence that one or more variables is improperly scaled. For momentum boundary layers in forced convection, it was shown in Chapter 8 that the variables needing attention were $\tilde{y}$ and $\tilde{v}_y$. The scaling problem was resolved by multiplying these small quantities by $\mathrm{Re}^{1/2}$ to obtain $\hat{y}$ and $\hat{v}_y$, respectively. This removed the large parameter Re from the governing equations. Based

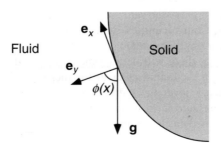

Figure 12-4. Local coordinates for boundary layers in buoyancy-driven flow.

on the analogy between Re and $Gr^{1/2}$, it is tempting to assume that the scaling in free convection will be corrected by using

$$\hat{y} = \bar{y}\, Gr^{1/4}, \qquad \hat{v}_y = \bar{v}_y Gr^{1/4}. \qquad (12.4\text{-}19)$$

With these new variables, Eqs. (12.4-16)-(12.4-18) become

$$\frac{\partial \bar{v}_x}{\partial \tilde{x}} + \frac{\partial \hat{v}_y}{\partial \hat{y}} = 0, \qquad (12.4\text{-}20)$$

$$\bar{v}_x \frac{\partial \bar{v}_x}{\partial \tilde{x}} + \hat{v}_y \frac{\partial \bar{v}_x}{\partial \hat{y}} = -\frac{d\tilde{\mathcal{P}}}{d\tilde{x}} + \Theta \sin \phi + \frac{\partial^2 \bar{v}_x}{\partial \hat{y}^2}, \qquad (12.4\text{-}21)$$

$$\bar{v}_x \frac{\partial \Theta}{\partial \tilde{x}} + \hat{v}_y \frac{\partial \Theta}{\partial \hat{y}} = Pr^{-1} \frac{\partial^2 \Theta}{\partial \hat{y}^2}. \qquad (12.4\text{-}22)$$

It is seen that Gr no longer appears in the momentum or energy equations, suggesting that (for moderate values of Pr) all variables are now properly scaled. This conclusion is correct for constant-temperature boundary conditions, where the temperature scale is predetermined, but is not correct in general, as explained below.

Implicit in the reasoning leading to Eqs. (12.4-20)–(12.4-22) is the assumption that the temperature scale is independent of the Grashof number, or (symbolically) that $\Theta = O(1)$ as $Gr \to \infty$. This is true, say, for an isothermal surface at $T = T_w$ in contact with a fluid at bulk temperature T_∞; the temperature scale is then fixed at $|T_w - T_\infty|$. However, for a surface with a constant heat flux q_w, the temperature difference across the boundary layer is not fixed. Instead, it will vary as the temperature gradient at the wall times the thermal boundary layer thickness (δ_T), or $q_w \delta_T / k$. Because the boundary layer thickness is dependent on Gr, the assumption that $\Theta = O(1)$ is no longer true. As discussed in Problem 12-2, the y coordinate, both velocity components, and the temperature must all be rescaled for the constant-flux case. For ($\bar{y}, \bar{v}_x, \hat{v}_y, \Theta$), the corresponding "stretching factors" are found to be ($Gr^{1/5}, Gr^{1/10}, Gr^{3/10}, Gr^{1/5}$). In other words, the boundary layer thickness and temperature difference both vary as $Gr^{-1/5}$ for a constant flux. A rescaling of the four quantities is required also for extreme values of Pr. This is discussed for the isothermal case in Example 12.5-2, and for the constant-flux case in Problem 12-2.

In retrospect, the scaling intricacies for free convection should not be surprising, given the interdependence of the temperature and velocity fields. In forced convection, one scaling for the momentum boundary layer suffices, because the velocity is independent of the temperature (or nearly so). In free convection, the dependence of the velocity on the temperature field leads to effects of Pr and the thermal boundary conditions, thereby creating a number of special cases.

12.5 UNCONFINED FLOWS

Example 12.5-1 Free Convection Past a Vertical Flat Plate The prototypical boundary layer problem in free convection involves a flat, vertical plate or wall maintained at constant temperature, T_w. The fluid temperature far from the wall is T_∞. The analysis is the same for hot or cold surfaces, requiring only a reversal of the x coordinate. For purposes of discussion, however, we will assume that $T_w > T_\infty$. Thus, buoyancy will cause the fluid to rise near the wall. A convenient reference temperature is $T_0 = T_\infty$, and the dimensionless temperature is chosen as

$$\Theta \equiv \frac{T - T_\infty}{T_w - T_\infty}. \tag{12.5-1}$$

The other dimensionless variables are defined as in Section 12.4, with L taken to be the vertical dimension of the plate. At the end of the analysis it is shown that the local velocity and temperature fields are independent of L. That is, it turns out that the only pertinent length scale is the distance x from the leading edge (bottom) of the plate. It was shown in Section 8.4 that this is also a feature of forced convection past a flat plate, as well as a feature of other wedge flows. Indeed, the similarity approach used here is much like that in Examples 8.4-1 and 8.4-2, and the reader may find it helpful to review those examples before proceeding. The similarity solution for the vertical plate was obtained first by E. Pohlhausen in 1930 (see Schlicting, 1968, pp. 300–302).

The fluid far from the surface is at the reference temperature. Thus, a satisfactory solution for the outer region is $\tilde{\mathbf{v}} = \mathbf{0}$, $\Theta = 0$, and $\tilde{\mathcal{P}} = $ constant; it is confirmed by inspection that this satisfies the general continuity, momentum, and energy equations. As the starting point for the boundary layer we choose Eqs. (12.4-20)–(12.4-22), which contain variables that are appropriately scaled for an isothermal surface. The trivial nature of the outer solution greatly simplifies the boundary layer problem. In Eq. (12.4-21) we set $d\mathcal{P}/dx = 0$ (as determined by the outer solution) and $\phi = \pi/2$ (see Fig. 12-4) to obtain

$$\tilde{v}_x \frac{\partial \tilde{v}_x}{\partial \tilde{x}} + \hat{v}_y \frac{\partial \tilde{v}_x}{\partial \hat{y}} = \Theta + \frac{\partial^2 \tilde{v}_x}{\partial \hat{y}^2}. \tag{12.5-2}$$

As usual with two-dimensional flows, it is advantageous to employ the stream function. The dimensionless stream function is defined here as

$$\tilde{v}_x \equiv \frac{\partial \hat{\psi}}{\partial \hat{y}}, \qquad \hat{v}_y \equiv -\frac{\partial \hat{\psi}}{\partial \tilde{x}}. \tag{12.5-3}$$

Using this in Eq. (12.5-3) gives

$$\frac{\partial \hat{\psi}}{\partial \hat{y}} \left(\frac{\partial^2 \hat{\psi}}{\partial \tilde{x} \partial \hat{y}} \right) - \frac{\partial \hat{\psi}}{\partial \tilde{x}} \left(\frac{\partial^2 \hat{\psi}}{\partial \hat{y}^2} \right) = \Theta + \frac{\partial^3 \hat{\psi}}{\partial \hat{y}^3}, \tag{12.5-4}$$

which replaces the continuity and momentum equations.

A similarity variable (η) and modified stream function (F) are defined now as

$$\eta \equiv \frac{\hat{y}}{g(\tilde{x})}, \qquad F(\eta) \equiv \frac{\hat{\psi}(\tilde{x}, \hat{y})}{c(\tilde{x})}, \tag{12.5-5}$$

where the scale factor g varies as the boundary layer thickness. The motivation for the modified stream function is explained in Section 8.4. Briefly, the stream function in a boundary layer next to a solid surface scales as the boundary layer thickness (or g) times the velocity component parallel to the surface. The local velocity scale for the present problem is unknown, so the product of the two scale factors is denoted simply as the function c. For the change of variables, the stream-function derivatives are calculated as

$$\frac{\partial \hat{\psi}}{\partial \hat{y}} = \frac{cF'}{g}, \qquad \frac{\partial^2 \hat{\psi}}{\partial \hat{y}^2} = \frac{cF''}{g^2}, \qquad \frac{\partial^3 \hat{\psi}}{\partial \hat{y}^3} = \frac{cF'''}{g^3}, \tag{12.5-6}$$

$$\frac{\partial \hat{\psi}}{\partial \tilde{x}} = -\frac{cg'}{g} \eta F' + c'F, \qquad \frac{\partial^2 \hat{\psi}}{\partial \tilde{x} \partial \hat{y}} = -\left(\frac{cg'}{g^2} - \frac{c'}{g} \right) F' - \frac{cg'}{g^2} \eta F'', \tag{12.5-7}$$

where primes are used to denote the derivatives of any function of a single variable. Using these relationships, Eq. (12.5-4) becomes

$$F''' + (c'g)FF'' + (cg' - c'g)F'^2 + \left(\frac{g^3}{c}\right)\Theta = 0. \tag{12.5-8}$$

The boundary layer energy equation, Eq. (12.4-22), is transformed in a similar manner to give

$$\Theta'' + (c'g)\text{Pr } F\Theta' = 0. \tag{12.5-9}$$

It is assumed here that $\Theta = \Theta(\eta)$ only, so that the primes on Θ mean derivatives with respect to η. With constant-temperature boundary conditions at the wall and in the bulk fluid, Θ does not have to be modified with a position-dependent scale factor.

Each quantity in parentheses in Eqs. (12.5-8) and (12.5-9) contains functions of x. If the similarity hypothesis is correct, all of those quantities must be constant. The two unknown scale factors give us two degrees of freedom in choosing those constants. As the first choice we set

$$c = g^3 \tag{12.5-10}$$

to make the coefficient of Θ unity in Eq. (12.5-8). As the second choice we let

$$c'g = 3 = 3g^3g', \tag{12.5-11}$$

where Eq. (12.5-10) has been used to obtain the second equality. Noticing that $c'g$ is proportional to $(g^4)'$, Eq. (12.5-11) indicates that

$$(g^4)' = 4, \qquad g(0) = 0. \tag{12.5-12}$$

The necessity for the condition $g(0) = 0$ is discussed later. Equations (12.5-10) and (12.5-12) imply that the scale factors are

$$g(\tilde{x}) = (4\tilde{x})^{1/4}, \qquad c(\tilde{x}) = (4\tilde{x})^{3/4}. \tag{12.5-13}$$

The functional form of g indicates that the boundary layer thickness varies as $x^{1/4}$. It is easily shown using Eq. (12.5-13) that $cg' = 1$. Thus, all of the terms involving functions of x in Eqs. (12.5-8) and (12.5-9) are indeed constants, as required.

Evaluating the scale factors as just discussed, the differential equations for the boundary layer become

$$F''' + 3FF'' - 2(F')^2 + \Theta = 0, \tag{12.5-14}$$

$$\Theta'' + 3\text{Pr } F\Theta' = 0. \tag{12.5-15}$$

The boundary conditions for F and Θ are

$$F(0) = F'(0) = F'(\infty) = \Theta(\infty) = 0, \qquad \Theta(0) = 1. \tag{12.5-16}$$

The boundary conditions for the stream function are obtained from the no-penetration and no-slip conditions at the surface ($\eta = 0$) and the stagnant conditions in the outer fluid ($\eta = \infty$). The thermal boundary conditions involve the specified temperatures at the surface, in the outer fluid, and in the fluid approaching the leading edge at $x = 0$. Our need to make the latter two equivalent to $\Theta(\infty) = 0$ is what requires that $g(0) = 0$, as given in Eq. (12.5-12).

Equations (12.5-14) and (12.5-15) are coupled, nonlinear differential equations which must be solved numerically. Before discussing the results, we mention certain alternative ways to express the similarity variable and the velocity. Using Eqs. (12.5-5) and (12.5-13), the similarity variable is given by

$$\eta = \frac{\hat{y}}{(4\tilde{x})^{1/4}} = \left(\frac{\text{Gr}_x}{4}\right)^{1/4}\frac{y}{x}, \tag{12.5-17}$$

$$\mathrm{Gr}_x \equiv \frac{gx^3\beta\Delta T}{\nu^2}, \tag{12.5-18}$$

where Gr_x is the Grashof number based on x, the distance from the leading edge; compare with Eq. (12.4-5). As is apparent from the second equality in Eq. (12.5-17), η is independent of the overall length of the plate, L. From the definition of the modified stream function, the velocity parallel to the surface is

$$\tilde{v}_x = 2\tilde{x}^{1/2} F'(\eta). \tag{12.5-19}$$

This velocity has been made dimensionless using U_b, which depends on L [see Eq. (12.4-4)]. However, Eq. (12.5-19) can be rearranged to

$$\frac{v_x}{(gx\beta\Delta T)^{1/2}} = 2F'(\eta), \tag{12.5-20}$$

which is independent of the plate length. The characteristic velocity here is similar to U_b, but it involves x instead of L. We conclude that the local velocity and temperature are both independent of the overall plate length.

An extensive set of velocity and temperature calculations for the isothermal vertical plate, for $0.01 \le \mathrm{Pr} \le 1000$, is given in Ostrach (1953). Representative profiles for four values of Pr are plotted in Figs. 12-5 and 12-6 as $F'(\eta)$ and $\Theta(\eta)$, respectively. It is seen that the position of the maximum velocity (i.e., the maximum in F') moves farther from the surface as Pr is decreased and that the peak velocity also increases in magnitude. Likewise, the temperature variations extend farther into the fluid as Pr is reduced. A comparison of the two figures indicates that for large Pr the nonzero velocities extend much farther from the surface than do the temperature changes. This suggests that the thickness of the thermal boundary layer for $\mathrm{Pr} \to \infty$ will be only a small fraction of that of the momentum boundary layer. The implications of the relative boundary layer thicknesses are discussed in Example 12.5-2, in connection with the factors which control Nu.

Measured velocity and temperature profiles near vertical flat plates are generally in very good agreement with the above theory when the flow is laminar (Ostrach, 1953; Schlicting, 1968,

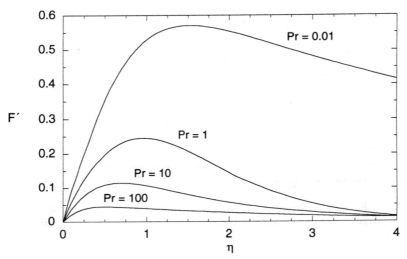

Figure 12-5. Velocity profiles for free convection past an isothermal vertical plate, from results tabulated in Ostrach (1953). The abscissa is defined by Eq. (12.5-17), and F' is related to the velocity by Eq. (12.5-20).

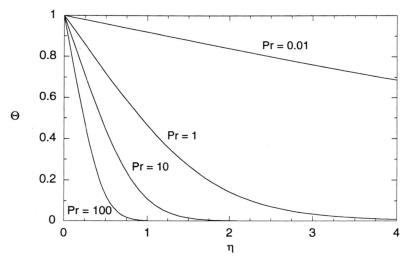

Figure 12-6. Temperature profiles for free convection past an isothermal vertical plate, from results tabulated in Ostrach (1953). The abscissa is defined by Eq. (12.5-17), and Θ is defined by Eq. (12.5-1).

p. 302). As with forced convection past flat plates, the flow becomes turbulent at a certain distance from the leading edge. For free convection at isothermal vertical plates the transition from laminar to turbulent flow occurs at values of $Ra_x = Gr_x\, Pr$ of about 10^9. Most data are for air or other fluids where Pr is neither large nor small, and the dependence on Pr implied by the use of Ra_x may not be precisely correct. A detailed discussion of transition in free convection is found in Gebhart et al. (1988, Chapter 11).

Similarity solutions for free-convection boundary layers have been found for a variety of other boundary conditions on vertical surfaces (Sparrow and Gregg, 1956; Yang, 1960; Gebhart et al., 1988, Chapter 3). The constant heat-flux condition on a vertical plate, studied by Sparrow and Gregg (1956), is important because it is often easier to achieve experimentally than constant temperature. Like the isothermal theory, the constant-flux analysis gives results which are in good ageement with experiments. That analysis is discussed in Problem 12-3. Results for many geometries are reviewed in Gebhart et al. (1988). In addition to the results for Newtonian fluids, similarity solutions for free-convection boundary layers have been obtained for power-law fluids (Acrivos, 1960). Information on the Nusselt number for Newtonian fluids is presented in the next example.

Example 12.5-2 Effects of Pr on Nu in Free Convection The rescaled coordinate and velocity defined by Eq. (12.4-19), $\hat{y}$ and $\hat{v}_y$, are appropriate for the *momentum* boundary layer near an *isothermal* surface. If $Pr \sim 1$, the thicknesses of the momentum boundary layer (δ_M) and thermal boundary layer (δ_T) will be similar, and there is no need for further adjustment of the governing equations, Eqs. (12.4-20)–(12.4-22). If Pr is very large or small, however, δ_T will not be comparable to δ_M, and the scaling will be incorrect. As shown for forced convection in Chapter 10, examining the scaling of the energy equation for extreme values of Pr reveals how Pr influences Nu. The objective here is to perform a similar analysis for free convection at an isothermal surface. The scaling analysis for a constant-flux boundary condition is discussed in Problem 12-2.

To identify the terms which must be balanced in the scaling procedure, we first review the characteristics of the various regions, as summarized in Fig. 12-7. In the momentum outer region,

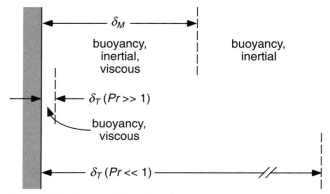

Figure 12-7. Characteristics of boundary layer and outer regions in free convection. Indicated are the relative thicknesses of the momentum and thermal boundary layers, along with the terms in the momentum equation which are important in each region.

the inertial and buoyancy terms are dominant and the viscous terms are negligible. In the momentum boundary layer, the viscous terms are important (by definition), and they are balanced by some combination of inertial and buoyancy terms. The analysis assumes that the pressure term is no larger than the buoyancy term and does not consider the pressure gradient explicitly. Based on what we know about forced-convection boundary layers, we expect δ_T/δ_M to become very large as $\mathrm{Pr}\to 0$. Thus, the thermal boundary layer will reside largely in the momentum outer region. Conversely, we expect δ_T/δ_M to become very small as $\mathrm{Pr}\to\infty$, in which case the thermal boundary layer will occupy only a small fraction of the momentum boundary layer. In this region very near the surface, it will be shown that the inertial terms become unimportant.

For extreme values of Pr, the new variables for the thermal boundary layer are defined as

$$Y=\hat{y}\,\mathrm{Pr}^m, \qquad V_x=\hat{v}_x\,\mathrm{Pr}^n, \qquad V_y=\hat{v}_y\,\mathrm{Pr}^p. \tag{12.5-21}$$

where m, n, and p are constants which are to be determined. For the isothermal surface considered here, the temperature need not be rescaled. Converting the continuity, momentum, and energy equations to the new variables, we obtain

$$\mathrm{Pr}^{-n}\frac{\partial V_x}{\partial \tilde{x}}+\mathrm{Pr}^{m-p}\frac{\partial V_y}{\partial Y}=0, \tag{12.5-22}$$

$$\mathrm{Pr}^{-2n}\,V_x\frac{\partial V_x}{\partial \tilde{x}}+\mathrm{Pr}^{m-p-n}\,V_y\frac{\partial V_x}{\partial Y}=-\frac{d\tilde{\mathscr{P}}}{d\tilde{x}}+\Theta\,\sin\,\phi+\mathrm{Pr}^{2m-n}\frac{\partial^2 V_x}{\partial Y^2}, \tag{12.5-23}$$

$$\mathrm{Pr}^{-n}\,V_x\frac{\partial\Theta}{\partial \tilde{x}}+\mathrm{Pr}^{m-p}\,V_y\frac{\partial\Theta}{\partial Y}=\mathrm{Pr}^{2m-1}\frac{\partial^2\Theta}{\partial Y^2}. \tag{12.5-24}$$

The unknown constants will be chosen so as to eliminate Pr while preserving the proper balance of terms in each equation.

Certain relationships must hold for either small or large Pr. From Eq. (12.5-22) we infer that $m-p=-n$. This implies that the two inertial terms in Eq. (12.5-23) are comparable to one another, as are the two convection terms in Eq. (12.5-24). Similar conclusions were reached in the analysis of forced convection in Chapter 10. The other general statement is that, in the thermal boundary layer, conduction and convection must be comparable. Equation (12.5-24) indicates then that $2m-1=-n$. Thus, two of the three relationships needed to determine the constants are valid for either extreme of Pr.

Consider now what happens for Pr→0. In this case, buoyancy and inertial effects are dominant in virtually the entire thermal boundary layer, which is mainly in the momentum outer region (see Fig. 12-7). For the inertial and buoyancy terms to balance, Eq. (12.5-23) requires that $n=0$. Combined with the two general relationships, this indicates that

$$m=\tfrac{1}{2}, \quad n=0, \quad p=\tfrac{1}{2} \qquad (\text{Pr}\to 0). \tag{12.5-25}$$

For Pr→∞, buoyancy and viscous effects are expected to be important within the relatively thin thermal boundary layer; we cannot yet tell about the inertial effects. For the buoyancy and viscous terms to be comparable, Eq. (12.5-23) requires that $2m-n=0$. Solving again for the three constants gives

$$m=\tfrac{1}{4}, \quad n=\tfrac{1}{2}, \quad p=\tfrac{3}{4} \qquad (\text{Pr}\to \infty). \tag{12.5-26}$$

As a consistency check, we examine what this implies about the inertial terms. Using these constants in Eq. (12.5-23), we find that the inertial terms in the thermal boundary layer are $O(\text{Pr}^{-1})$ as Pr→∞. Thus, the inertial terms are indeed negligible, as indicated in Fig. 12-7.

To infer the functional form of the Nusselt number for free-convection boundary layers, we first recall from Section 10.4 that

$$\text{Nu} \sim \frac{1}{\Delta\Theta}\left(\frac{\partial\Theta}{\partial\tilde{y}}\right)\Bigg|_{\tilde{y}=0} = \frac{1}{\Delta\Theta}\left(\frac{\partial\Theta}{\partial Y}\right)\Bigg|_{Y=0} \frac{\partial Y}{\partial\tilde{y}} \sim \frac{\partial Y}{\partial\tilde{y}}, \tag{12.5-27}$$

where $\Delta\Theta$ is the dimensionless temperature difference between the surface and the bulk fluid. Accordingly, the key piece of information is the relationship between the scaled coordinate Y and the original dimensionless coordinate $\tilde{y}$. Based on Eq. (12.4-19) and the values of m just determined, those relationships for the two extremes of Pr are

$$Y=\hat{y}\,\text{Pr}^{1/2}=\tilde{y}\,\text{Gr}^{1/4}\,\text{Pr}^{1/2} \qquad (\text{Pr}\to 0), \tag{12.5-28}$$

$$Y=\hat{y}\,\text{Pr}^{1/4}=\tilde{y}\,\text{Gr}^{1/4}\,\text{Pr}^{1/4} \qquad (\text{Pr}\to \infty). \tag{12.5-29}$$

What this means, for example, is that the thermal boundary thickness for large Pr varies as $\text{Gr}^{-1/4}\text{Pr}^{-1/4}$. Accordingly, the local Nusselt numbers for the two cases must be of the form

$$\text{Nu}(\tilde{x})=f_1(\tilde{x})\,\text{Gr}^{1/4}\,\text{Pr}^{1/2} \qquad (\text{Pr}\to 0), \tag{12.5-30}$$

$$\text{Nu}(\tilde{x})=f_2(\tilde{x})\,\text{Gr}^{1/4}\,\text{Pr}^{1/4} \qquad (\text{Pr}\to \infty), \tag{12.5-31}$$

where the functions $f_1(\tilde{x})$ and $f_2(\tilde{x})$ are independent of Gr and Pr. The functional forms of the average Nusselt numbers are then

$$\overline{\text{Nu}} = C_1\,\text{Gr}^{1/4}\,\text{Pr}^{1/2} \qquad (\text{Pr}\to 0), \tag{12.5-32}$$

$$\overline{\text{Nu}} = C_2\,\text{Gr}^{1/4}\,\text{Pr}^{1/4} \qquad (\text{Pr}\to \infty), \tag{12.5-33}$$

where C_1 and C_2 are constants.

The asymptotic behavior of the average Nusselt number is illustrated by the results of Le Fevre (1956) for the isothermal, vertical flat plate:

$$\overline{\text{Nu}} = 0.8005\,\text{Gr}^{1/4}\,\text{Pr}^{1/2} \qquad (\text{Pr}\to 0), \tag{12.5-34}$$

$$\overline{\text{Nu}} = 0.6703\,\text{Gr}^{1/4}\,\text{Pr}^{1/4} \qquad (\text{Pr}\to \infty). \tag{12.5-35}$$

It is seen that the constants in Eqs. (12.5-32) and (12.5-33) for this case are $C_1 = 0.8005$ and $C_2 = 0.6703$. Kuiken (1968, 1969) confirmed these expressions and used matched asymptotic expansions to obtain additional terms in the respective series involving Pr. An expression used by Le Fevre (1956) to fit theoretical results for all Pr is

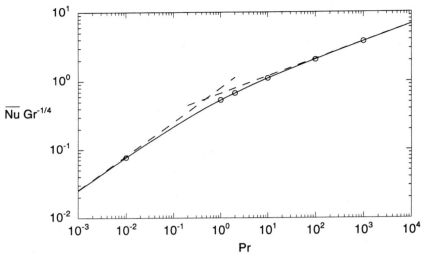

Figure 12-8. Average Nusselt number for free convection past an isothermal, vertical plate. The dashed lines are from Eqs. (12.5-34) and (12.5-35), the solid curve is from Eq. (12.5-36), and the individual symbols are based on the results of Ostrach (1953).

$$\overline{Nu} = \left(\frac{Gr\,Pr^2}{2.435 + 4.884\,Pr^{1/2} + 4.953\,Pr} \right)^{1/4}.$$ (12.5-36)

As shown in Fig. 12-8, this agrees with the two asymptotic formulas and also matches the results of Ostrach (1953) computed at discrete values of Pr.

Local values of the Nusselt number for vertical plates are usually expressed as $Nu_x = hx/k$. As an example, for the isothermal, vertical plate at large Pr, Kuiken (1968) obtained

$$Nu_x = 0.5028\,Gr_x^{1/4}\,Pr^{1/4} \qquad (Pr \to \infty,\ T_w\ constant).$$ (12.5-37)

The corresponding result for a constant-flux boundary condition (Gebhart et al., 1988, p. 89) is

$$Nu_x = 0.6316\,Gr_x^{*1/5}\,Pr^{1/5} \qquad (Pr \to \infty,\ q_w\ constant),$$ (12.5-38)

$$Gr_x^* = \frac{gx^4 \beta q_w}{\nu^2 k},$$ (12.5-39)

where Gr_x^* is the *modified Grashof number* based on x. Essentially, Gr^* is Gr with $\Delta T = q_w L/k$, and Gr_x^* is Gr_x with $\Delta T = q_w\,x/k$. The differences in exponents between Eqs. (12.5-37) and (12.5-38) are indicative of differences in the scaling of the thermal boundary layer thickness. However, the two expressions can be put in more comparable form by noting that $Gr_x^* = Gr_x\,Nu_x$. Thus, Eq. (12.5-38) is rewritten as

$$Nu_x = 0.5630\,Gr_x^{1/4}\,Pr^{1/4} \qquad (Pr \to \infty,\ q_w\ constant).$$ (12.5-40)

This differs from Eq. (12.5-37) only in that the coefficient is 12% larger. Equation (12.5-40) is impractical for calculations because the value of ΔT, which is needed to determine Gr_x, is not known in advance for the constant flux case. Equation (12.5-38) has only known constants on the right-hand side, making it the more appropriate expression. Numerous other results for vertical plates, vertical axisymmetric flows, inclined surfaces, horizontal surfaces, curved surfaces (e.g., horizontal cylinders), spheres, and other shapes are found in Gebhart et al. (1988).

Mixed Convection

The presentation of convective heat and mass transfer in largely separate discussions for forced convection (Chapters 9, 10, and 13) and free convection (the present chapter) suggests that only one mechanism is important in any given situation. This is true for many, but not all, applications. Useful guidance on when to expect forced, free, or "mixed convection" regimes in laminar or turbulent tube flow, based on the values of Re and Ra, is provided by Metais and Eckert (1964). A general analysis of combined forced and free convection in laminar boundary layers is given in Acrivos (1966), in which it is shown that the controlling parameter is Gr/Re^2 for $Pr \ll 1$ and $Gr/(Re^2 Pr^{1/3})$ for $Pr \gg 1$. Numerous other findings for mixed convection are reviewed in Gebhart et al. (1988, Chapter 10).

A good source for additional information on buoyancy effects in fluids is Gebhart et al. (1988), which has already been cited a number of times. This is a very comprehensive text with an emphasis on engineering applications. Another general source, with more of a geophysical flavor, is Turner (1973).

References

Acrivos, A. A theoretical analysis of laminar natural convection heat transfer to non-Newtonian fluids. *AIChE J.* 6: 584–590, 1960.

Acrivos, A. On the combined effect of forced and free convection heat transfer in laminar boundary layer flows. *Chem. Eng. Sci.* 21: 343–352, 1966.

Bolz, R. E. and G. L. Tuve (Eds.) *Handbook of Tables for Applied Engineering Science.* Chemical Rubber Co., Cleveland, OH, 1970.

Chandrasekhar, S. *Hydrodynamic and Hydromagnetic Stability.* Clarendon Press, Oxford, 1961.

Cormack, D. E., L. G. Leal, and J. Imberger. Natural convection in a shallow cavity with differentially heated end walls. Part 1. Asymptotic theory. *J. Fluid Mech.* 65: 209–229, 1974.

Drazin, P. G. and W. H. Reid. *Hydrodynamic Stability.* Cambridge University Press, Cambridge, 1981.

Gebhart, B., Y. Jaluria, R. L. Mahajan, and B. Sammakia. *Buoyancy-Induced Flows and Transport.* Hemisphere, New York, 1988.

Goldstein, S. *Modern Developments in Fluid Dynamics.* Clarendon Press, Oxford, 1938 [reprinted by Dover, New York, 1965].

Hildebrand, F. B. *Advanced Calculus for Applications,* second edition. Prentice-Hall, Englewood Cliffs, NJ, 1976.

Kuiken, H. K. An asymptotic solution for large Prandtl number free convection. *J. Eng. Math.* 2: 355–371, 1968.

Kuiken, H. K. Free convection at low Prandtl numbers. *J. Fluid Mech.* 37: 785–798, 1969.

Le Fevre, E. J. Laminar free convection from a vertical plane surface. *Ninth International Congress of Applied Mechanics,* Brussels, 1956, Vol. 4, pp. 168–174.

Metais, B. and E. R. G. Eckert. Forced, mixed, and free convection regimes. *J. Heat Transfer* 86: 295–296, 1964.

Mihaljan, J. M. A rigorous exposition of the Boussinesq approximations applicable to a thin layer of fluid. *Astrophys. J.* 136: 1126–1133, 1962.

Ostrach, S. An analysis of laminar free-convection flow and heat transfer about a flat plate parallel to the direction of the generating body force. *NACA Tech. Rept.* No. 1111, 1953.

Schlicting, H. *Boundary-Layer Theory,* sixth edition. McGraw-Hill, New York, 1968.

Sparrow, E. M. and J. L. Gregg. Laminar free convection from a vertical plate with uniform surface heat flux. *Trans. ASME* 78: 435–440, 1956.

Spiegel, E. A. and G. Veronis. On the Boussinesq approximation for a compressible fluid. *Astrophys. J.* 131: 442–447, 1960.

Turner, J. S. *Buoyancy Effects in Fluids.* Cambridge University Press, Cambridge, 1973.

Wooding, R. A. Convection in a saturated porous medium at large Rayleigh number or Péclet number. *J. Fluid Mech.* 15: 527–544, 1963.

Yang, K.-T. Possible similarity solutions for laminar free convection on vertical plates and cylinders. *J. Appl. Mech.* 27: 230–236, 1960.

$$ v_x \underset{neglect}{\frac{\partial T}{\partial x}} + v_y \underset{0}{\frac{\partial T}{\partial y}} = \alpha \left(\frac{\partial^2 T}{\partial x^2} + \underset{neglect}{\frac{\partial^2 T}{\partial y^2}} \right) + A \underset{0}{\frac{\partial}{\partial x}} $$

$$ \frac{d^2 T}{dx^2} = 0 \qquad \begin{pmatrix} T(o) = T_1 \\ T(L) = T_2 \end{pmatrix} \quad T = T_1 + (T_2 - T_1)\frac{x}{L} $$

$$ T_o - T_o = T - T_m $$
$$ = T_1 + (T_2 - T_1)\frac{x}{L} - \frac{T_1 + T_2}{2} $$

Problems

$$ = \Delta T \left(\frac{x}{L} - \frac{1}{2} \right) $$

12-1. Flow in a Closed Horizontal Channel Heated at One End

Consider the situation depicted in Fig. P12-1, in which the fluid in a parallel-plate channel with closed ends is subjected to a horizontal temperature gradient. The temperatures at the ends are such that $T_2 > T_1$, and the top and bottom are perfectly insulated. A buoyancy-driven flow results. If the channel is sufficiently thin, the flow will be fully developed in a core region which extends over most of the channel length; the ends of this region are indicated approximately by the dashed lines.

$$ \frac{Dv}{Dt} = -\frac{1}{\rho_o}\nabla p - g\,\beta(T - T_0) + \nu\nabla^2 v, \qquad T_0 = T_m = \frac{T_1 + T_2}{2}, \qquad \Delta T = T_2 - T_1 $$

(a) Determine $v_x(y)$ in the core region. Assume that the end regions, where $v_y \neq 0$ and the streamlines are curved, are a negligible fraction of the channel length. Assume also that convective heat transfer is negligible.

(b) What dimensionless criteria must be satisfied to make this analysis valid?

A more complete analysis of this problem, including asymptotically matched solutions for the core and end regions, is given in Cormack et al. (1974).

12-2. Scaling for a Free-Convection Boundary Layer with a Constant Heat Flux

The scaling for a free-convection boundary layer at a surface with a constant heat flux differs from that for an isothermal surface, in that the surface temperature depends on the boundary layer thickness. Thus, the temperature scale is not known in advance. As an example, consider a flat plate with vertical dimension L and constant heat flux q_w at the surface. Setting $\Delta T = q_w L/k$, the dimensionless temperature and modified Grashof number are defined as

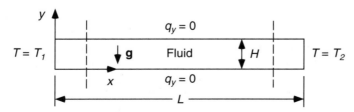

Figure P12-1. Flow in a closed horizontal channel with ends at different temperatures.

$$\Theta \equiv \frac{T-T_\infty}{q_w L/k}, \qquad Gr^* \equiv \frac{g L^4 \beta q_w}{\nu^2 k},$$

where T_∞ is the fluid temperature far from the surface, assumed constant. For large Gr* and moderate Pr, the y coordinate, velocity components, and temperature are rescaled as

$$\hat{y} \equiv \bar{y} \, Gr^{*a}, \qquad \hat{v}_x \equiv \bar{v}_x \, Gr^{*b}, \qquad \hat{v}_y \equiv \bar{v}_y \, Gr^{*c}, \qquad \hat{\Theta} \equiv \Theta \, Gr^{*d}$$

where a, b, c, and d are constants. For extreme values of Pr, these variables are adjusted further as

$$Y \equiv \hat{y} \, Pr^m, \qquad V_x \equiv \hat{v}_x \, Pr^n, \qquad V_y \equiv \hat{v}_y \, Pr^p, \qquad H \equiv \hat{\Theta} \, Pr^q,$$

where m, n, p, and q are additional constants. The objective of this problem is to determine the various constants and to infer the functional form of $\overline{Nu}$.

(a) Use the heat-flux boundary condition to show that $a = d$ and (for extreme Pr) $m = q$.

(b) Using the continuity and momentum equations, show that $a = \frac{1}{5}$, $b = \frac{1}{10}$, $c = \frac{3}{10}$, and $d = \frac{1}{5}$.

(c) For Pr$\rightarrow 0$, show that $m = \frac{2}{5}$, $n = \frac{1}{5}$, $p = \frac{3}{5}$, $q = \frac{2}{5}$, and

$$\overline{Nu} = C_1 \, Gr^{*1/5} \, Pr^{2/5} \qquad (Pr \rightarrow 0).$$

(d) For Pr$\rightarrow \infty$, show that $m = \frac{1}{5}$, $n = \frac{3}{5}$, $p = \frac{4}{5}$, $q = \frac{1}{5}$, and

$$\overline{Nu} = C_2 \, Gr^{*1/5} \, Pr^{1/5} \qquad (Pr \rightarrow \infty).$$

12-3. Free Convection Past a Vertical Plate with a Constant Heat Flux

Consider a boundary layer next to a vertical plate of height L with a constant heat flux q_w at the surface. The objective is to derive a similarity transformation which is suitable for moderate values of Pr. The basic dimensionless quantities are as defined in Section 12.4, except that with $\Delta T \equiv q_w L/k$ the modified Grashof number Gr* replaces Gr. Using the results of Problem 12-2, the appropriate variables for large Gr* and moderate Pr are

$$\hat{y} \equiv \bar{y} \, Gr^{*1/5}, \qquad \hat{v}_x \equiv \bar{v}_x \, Gr^{*1/10}, \qquad \hat{v}_y \equiv \bar{v}_y \, Gr^{*3/10}, \qquad \hat{\Theta} \equiv \Theta \, Gr^{*1/5}.$$

The similarity variable (η), modified stream function (F), and modified temperature (G) are defined as

$$\eta \equiv \frac{\hat{y}}{g(\bar{x})}, \qquad F(\eta) \equiv \frac{\hat{\psi}(\bar{x}, \hat{y})}{c(\bar{x})}, \qquad G(\eta) \equiv \frac{\hat{\Theta}(\bar{x}, \hat{y})}{b(\bar{x})}.$$

What is to be determined are the scale factors, $g(\bar{x})$, $c(\bar{x})$, and $b(\bar{x})$.

(a) Show that the ratio $b(\bar{x})/g(\bar{x})$ must be constant. What is the advantage in choosing $b(\bar{x})/g(\bar{x}) = -1$? (*Hint:* Examine the constant-flux boundary condition.)

(b) Determine the set of scale factors which gives

$$F''' + 4FF'' - 3F'^2 - G = 0,$$

$$G'' + Pr(4FG' - F'G) = 0,$$

$$F(0) = F'(0) = F'(\infty) = G(\infty) = 0, \qquad G'(0) = 1.$$

This is the set of equations solved numerically by Sparrow and Gregg (1956) in their analysis of this problem. (The function G used here is equivalent to their θ.)

(c) Confirm that the local Nusselt number is given by

$$\text{Nu}_x \equiv \frac{hx}{k} = -\frac{\text{Gr}_x^{*1/5}}{5^{1/5}G(0)}$$

as stated in Sparrow and Gregg (1956).

12-4. Temperature Variations in a Static Fluid
Assume that the temperature in a fluid is described by

$$T(x, y) = T_0 + ax + by,$$

where a and b are constants and the y axis points upward. For which combinations of a and b is it possible for the fluid to be static? Consider the possibility that either constant may be negative, zero, or positive.

12-5. Integral Boundary Layer Equations for Free Convection
Show that the integral forms of the momentum and energy equations for steady, two-dimensional, free-convection boundary layers are

$$\frac{d}{d\tilde{x}} \int_0^{\tilde{\delta}_M} \tilde{v}_x^2 \, d\tilde{y} = \sin\phi \int_0^{\tilde{\delta}_M} \Theta \, d\tilde{y} - \text{Gr}^{-1/2} \left. \frac{\partial \tilde{v}_x}{\partial \tilde{y}} \right|_{\tilde{y}=0},$$

$$\frac{d}{d\tilde{x}} \int_0^{\tilde{\delta}_T} \tilde{v}_x \Theta \, d\tilde{y} = -\text{Gr}^{-1/2} \, \text{Pr}^{-1} \left. \frac{\partial \Theta}{\partial \tilde{y}} \right|_{\tilde{y}=0},$$

where $\tilde{\delta}_M = \delta_M/L$ and $\tilde{\delta}_T = \delta_T/L$ are the dimensionless boundary layer thicknesses. Start with Eqs. (12.4-16)–(12.4-18), and assume that the outer fluid is stagnant and isothermal.

12-6. Integral Analysis for Free Convection Past a Vertical Plate
The objective is to obtain an approximate solution for the boundary layer next to an isothermal vertical plate, using the integral equations given in Problem 12-5. As in the original analysis by Squire in 1938 [see Goldstein (1938, pp. 641–643)], the functional forms suggested for the velocity and temperature are

$$\tilde{v}_x = f(\tilde{x})\eta(1-\eta)^2, \qquad \Theta = (1-\eta)^2, \qquad \eta \equiv \frac{\hat{y}}{\delta(\tilde{x})},$$

where $\hat{y}$ is defined in Eq. (12.4-19) and f and δ are functions which are to be determined. No distinction is made here between the momentum and thermal boundary layers.

(a) Discuss the qualitative behavior of the functional forms suggested for $\tilde{v}_x$ and Θ, and explain why these are reasonable choices.
(b) Show that the system of equations which must be solved is

$$\frac{1}{105} \frac{d}{d\tilde{x}} (f^2 \delta) = \frac{\delta}{3} - \frac{f}{\delta},$$

$$\frac{1}{30} \frac{d}{d\tilde{x}} (f\delta) = \frac{2}{\text{Pr } \delta},$$

$$f(0) = \delta(0) = 0.$$

(c) Determine δ and f. (Hint: Assume that $\delta = A\tilde{x}^m$ and $f = B\tilde{x}^n$.)

(d) Show that, according to this approach, $\overline{Nu}$ is given by

$$\overline{Nu} = \frac{8}{3(240)^{1/4}} \, Gr^{1/4} \, \frac{Pr^{1/2}}{[Pr + (20/21)]^{1/4}}.$$

Compare this with the exact results for the isothermal plate in Section 12.5.

12-7. Buoyancy-Driven Flow in Porous Media

Buoyancy-induced flows may occur in saturated (i.e., liquid-filled) porous media, thereby affecting, for example, heat transfer from underground pipes or storage tanks. According to Darcy's law (Problem 2-2), the velocity of a constant-density liquid in a porous solid is proportional to $\nabla P - \rho \mathbf{g}$. Using Eq. (12.2-3), the generalization to include buoyancy effects is

$$\mathbf{v} = -\frac{\kappa}{\mu} [\nabla \mathscr{P} + \rho_0 g \beta (T - T_0)],$$

where κ is the Darcy permeability. When using Darcy's law to replace the momentum equation, the ability to satisfy no-slip boundary conditions is lost. Accurate results are obtained nonetheless, provided that the system dimensions greatly exceed the microstructural length scale of the porous material. For steady conditions, the energy equation for porous media is

$$\mathbf{v} \cdot \nabla T = \alpha_{\text{eff}} \nabla^2 T,$$

where α_{eff} is the effective thermal diffusivity of the liquid–solid system.

(a) Define the dimensionless variables which convert Darcy's law and the energy equation to

$$\tilde{\mathbf{v}} = -\tilde{\nabla}\tilde{\mathscr{P}} - \Theta \mathbf{e}_g,$$

$$\tilde{\mathbf{v}} \cdot \tilde{\nabla}\Theta = Ra^{-1} \, \tilde{\nabla}^2 \Theta,$$

$$Ra = \frac{\kappa g L \beta \Delta T}{\nu \alpha_{\text{eff}}},$$

where Ra is the Rayleigh number for porous media. What velocity and pressure scales are implied?

(b) If Ra is large, a boundary layer will exist. Consider the two-dimensional configuration shown in Fig. 12-4, but with the fluid replaced by a saturated porous material. In terms of the dimensionless stream function defined as

$$\tilde{v}_x \equiv \frac{\partial \tilde{\psi}}{\partial \tilde{y}}, \qquad \tilde{v}_y \equiv -\frac{\partial \tilde{\psi}}{\partial \tilde{x}},$$

show that the boundary layer equations are

$$\frac{\partial^2 \tilde{\psi}}{\partial \tilde{y}^2} = \frac{\partial \Theta}{\partial \tilde{y}} \sin \phi,$$

$$\frac{\partial \tilde{\psi}}{\partial \tilde{y}} \frac{\partial \Theta}{\partial \tilde{x}} - \frac{\partial \tilde{\psi}}{\partial \tilde{x}} \frac{\partial \Theta}{\partial \tilde{y}} = Ra^{-1} \frac{\partial^2 \Theta}{\partial \tilde{y}^2}.$$

(c) For specified temperatures at the boundaries, the scaling of Θ is unaffected by Ra. In such cases, the equations of part (b) are rescaled by adjusting only the y coordinate and stream function as

$$\hat{y} = \tilde{y} \, Ra^a, \qquad \hat{\psi} = \tilde{\psi} \, Ra^b.$$

Show that $a=b=\frac{1}{2}$. What is the functional form of the average Nusselt number?

Several problems involving boundary layers in porous media were analyzed by Wooding (1963); see also Gebhart et al. (1988, Chapter 15).

12-8. Boundary Layer in a Porous Material Near a Heated Vertical Surface*

The basic equations governing buoyancy-induced flow in porous media are given in Problem 12-7. Consider a saturated porous material in contact with an isothermal, vertical plate of height L, at large Rayleigh number. From Problem 12-7, the scaled equations for the dimensionless stream function and temperature in the boundary layer are

$$\frac{\partial^2 \hat{\psi}}{\partial \hat{y}^2} = \frac{\partial \Theta}{\partial \hat{y}},$$

$$\frac{\partial^2 \Theta}{\partial \hat{y}^2} = \frac{\partial \hat{\psi}}{\partial \hat{y}}\frac{\partial \Theta}{\partial \hat{x}} - \frac{\partial \hat{\psi}}{\partial \hat{x}}\frac{\partial \Theta}{\partial \hat{y}}.$$

(a) Derive the boundary conditions for $\hat{\psi}$ and Θ.

(b) Show that the differential equations given above can be combined to yield a single equation for the stream function,

$$\frac{\partial^3 \hat{\psi}}{\partial \hat{y}^3} = \frac{\partial \hat{\psi}}{\partial \hat{y}}\frac{\partial^2 \hat{\psi}}{\partial \hat{x}\,\partial \hat{y}} - \frac{\partial \hat{\psi}}{\partial \hat{x}}\frac{\partial^2 \hat{\psi}}{\partial \hat{y}^2}.$$

(*Hint:* Eliminate Θ by integrating the first equation and using the boundary conditions.)

(c) A solution can be obtained by defining a similarity variable and modified stream function as

$$\eta \equiv \frac{\hat{y}}{g(\hat{x})}, \qquad F(\eta) \equiv \frac{\hat{\psi}(\hat{x}, \hat{y})}{g(\hat{x})}.$$

What is it about the boundary conditions which motivates the use of the same scale factor g in both variables?

(d) Determine g, and indicate the differential equation and boundary conditions which must be solved (numerically) to find F. Assuming that $F(\eta)$ has been evaluated, show how to calculate the local Nusselt number. What is the dependence of Nu on $\hat{x}$ and Ra?

* This problem was suggested by R. A. Brown.

[Handwritten notes follow:]

$x-$momentum $(g_x=0)$ $0 = \frac{1}{\rho_0}\frac{\partial P}{\partial x} + \nu\frac{d^2 V_x}{dy^2}$ ⇒ $\frac{\partial P}{\partial x} = P(y)$ only

$y - \| \quad (g_y = -g)$

$0 = -\frac{1}{\rho_0}\frac{\partial P}{\partial y} + g\rho\Delta T\left(\frac{x}{L}-\frac{1}{2}\right)$ + $P = \rho_0 g\rho\Delta T\left(\frac{x}{L}-\frac{1}{2}\right)y + f(x)$

⇒ $\frac{\partial P}{\partial x} = \frac{\rho_0 g\rho\Delta T}{L}y + \frac{df}{dx} = g(y)$ ⇒ $\frac{df}{dx} = C_0$ (const)

$\frac{\partial P}{\partial x} = \frac{\rho_0 g\rho\Delta T}{L}y + C_0$ $\left| + \frac{d^2 V_x}{dy^2} = \frac{1}{N}\frac{\partial P}{\partial x} = \frac{\rho_0 g\rho\Delta T}{\mu L}y + C$ ⇒ $V_x = \frac{g\rho\Delta T}{\nu L}\frac{y^3}{6} + \frac{Cy^2}{2} + ay + b$

$V(x)=0 \rightarrow | \; b=0$

$V(a)= y \rightarrow a = -\frac{g\rho\Delta T y^2}{6\nu L} - \frac{Cy}{2}$ $V_x = \frac{g y^2\rho\Delta T}{6\nu L}[\eta^3 - \eta] + \frac{C y^2}{2}(\eta^2 - \eta)$

$0 = \int_0^1 V_x \, d\eta = \frac{g y^3\rho\Delta T}{6\nu L}\left(-\frac{1}{4}\right) + \frac{C y^2}{2}\left(-\frac{1}{6}\right)$ ⇒ $\frac{C y^2}{2} = -\frac{g y^3\rho\Delta T}{4\nu L}$ If $N/\nu \ll 0$

$\boxed{V_x = \frac{g y^3\rho\Delta T}{12\nu L}(2\eta^3 - 3\eta^2 + \eta)}$ $\frac{L\nu}{H/2} = 1 + 6.1\left(\frac{V_{xy}}{\nu}\right)$ $= \frac{2L\nu}{\nu}$ cc/ $\left(=\frac{\mu}{\nu}cc\right)$

$0 \; V_x\frac{\partial T}{\partial x} \ll \alpha\frac{d^2 T}{dx^2}$ $\left|\frac{V_0 L}{\alpha} = \frac{g y^3\rho\Delta T}{12\nu\alpha} = \frac{Ra}{12}cc\right|$ $I+N/\nu \gg 10 \quad \frac{\nu}{\nu}\left(\frac{H}{2}\right)^2 cc/00$

$V_0\frac{\Delta T}{L} \ll \alpha\frac{\Delta T}{L^2}$ $\boxed{Ra \ll 10}$

Chapter 13

TRANSPORT IN TURBULENT FLOW

13.1 INTRODUCTION

The discussion of fluid mechanics and convective heat and mass transfer in the preceding chapters focused on laminar flows. Laminar-flow theory has provided much of the conceptual framework for the subject of transport phenomena, and it gives accurate predictions in applications which involve small to moderate length and velocity scales. However, the usefulness of that body of theory is limited by the instability of laminar flows at high Reynolds numbers and the emergence of turbulent flow. All turbulent flows are irregular, time-dependent, and three-dimensional, making fundamental analysis exceedingly difficult. Consequently, turbulent flow is often described as the major unresolved problem of classical physics. The difficulties notwithstanding, the importance of turbulent flow in engineering and in the natural world has engaged the interest of numerous investigators, who have developed an impressive variety of experimental and theoretical approaches. The study of turbulent flow has relied heavily on empirical relationships, supplemented where possible by analytical theory and, increasingly, direct computer simulations.

This chapter provides a brief introduction to the large and specialized subject of turbulence. The discusion here is limited to some of the basic features of momentum, heat, and mass transfer in turbulent flow, mainly in the context of flow in pipes. The focus is on the engineering description of these transport processes, and the main objective is to convey an appreciation for the differences in transport behavior between laminar and turbulent flows.

13.2 BASIC FEATURES OF TURBULENCE

Instability, Intermittency, and Transition

The various aspects of the breakdown of laminar flow and the emergence of turbulence are summarized by Schlicting (1968, pp. 431–466) and McComb (1990, pp. 110–112 and 413–425). In general, laminar flow is found to be stable when the Reynolds number is below some critical value, Re_c. The critical Reynolds number depends on the geometry: for pipe flow, $Re_c \cong 2.1 \times 10^3$ (based on diameter and mean velocity); for a flat plate parallel to the approaching stream, $Re_c \cong 6 \times 10^4$ (based on distance from the leading edge; Kays and Crawford, 1993, p.193); for a sphere or for a long cylinder with its axis perpendicular to the approaching flow, $Re_c \cong 2 \times 10^5$ (based on diameter). The onset of turbulence is evident from macroscopic pressure–flow or force measurements. For pipe flow and flow past streamlined objects, turbulence is associated with distinct increases in the friction factor and drag coefficient, respectively. Conversely, there is a sharp drop in the drag coefficient for spheres and cylinders when the boundary layer becomes fully turbulent. This is due to the movement of the separation point toward the rear and the consequent reduction in the size of the wake.

The Reynolds number at which a given type of laminar flow will actually become turbulent is only approximate, for two reasons. The first concerns the magnitude of flow disturbances. Whereas Re_c is the maximum Reynolds number at which a flow is stable relative to *large* disturbances, care taken to minimize disturbances has been found to increase the transitional value of Re by orders of magnitude. In pipe flow, examples of disturbances are vibrations and areas of flow separation upstream of the test section. In Reynolds' original pipe-flow experiments of 1883, transition occurred at $Re = 1.3 \times 10^4$ (Rott, 1990). For even more carefully controlled conditions, laminar pipe flow has been maintained for Re as large as 1.0×10^5 (Pfenniger, 1952, as cited by Bhatti and Shah, 1987); the absolute upper limit of Re (if any) remains a matter of conjecture. The second reason that the transitional value of the Reynolds number is imprecise is that transition from laminar to turbulent flow is a gradual process, rather than a "catastrophic" event. At Reynolds numbers just exceeding the critical value, laminar and turbulent flow alternate in time, a phenomenon called *intermittency*. For pipe flow, the fraction of time that the flow is turbulent grows with increasing Re, the flow usually becoming fully turbulent by $Re \cong 3000$. Thus, some degree of intermittency is typically present for $2100 < Re < 3000$.

The manner in which turbulence is initiated and spreads in a flow has received considerable attention. A number of "coherent" (i.e., recognizable) spatial patterns have been identified. In shear flows near a wall, flow visualization studies suggest that turbulence originates with "lumps" of fluid which first move parallel to the wall, then in a wavy or oscillatory manner, finally breaking up in distinct *bursts*. The radial and axial propagation of turbulence in a pipe appears to be the net effect of many such events occurring over time at different locations.

The remainder of this chapter is concerned only with fully turbulent flow. That is, intermittency and other aspects of the laminar-turbulent transition are not considered further.

Fundamental Characteristics of Fully Turbulent Flow

As already mentioned, turbulent flows share three basic characteristics: They are irregular, time-dependent, and three-dimensional. The irregularity is illustrated in Fig. 13-1, which contrasts hypothetical particle paths in laminar and turbulent pipe flow. In fully developed laminar flow, all motion is parallel to the tube axis, and the path of a given fluid particle (visualized, for example, by injection of a line of dye) is a straight line [see Fig. 13-1(a)]. If the velocity is increased so that the flow is turbulent, then the path of any particle becomes irregular [see Fig. 13-1(b)]. As indicated by the dashed and solid curves, the flow is irregular also in time; path lines begun at the same point tend to vary randomly from moment to moment. Actual photographs of dye traces in turbulent flow show that the streaks quickly become faint, a consequence largely of enhanced radial dispersion.

The variation in path lines from moment to moment, even in a flow that is macroscopically steady, reflects the fact that turbulent velocity fields are inherently time-dependent. "Macroscopically steady" refers to a situation in which average or overall quantities such as the volumetric flow rate are independent of time. The time-dependence of turbulent flow is revealed more directly by continuous measurements of one or more velocity components at a fixed location. This has been done using hot-wire anemometry or laser-Doppler velocimetry; for descriptions of these and other measurement techniques in turbulence see Hinze (1975, pp. 83–174) and McComb (1990, pp. 88–99). Such velocity measurements yield tracings like the idealized plot for tube flow shown in Fig. 13-2. Superimposed on the time-averaged or time-smoothed velocity is a series of more-or-less random fluctuations of relatively high frequency and small amplitude. The local velocity at any instant is expressed as

$$v_z = \langle v_z \rangle + u_z, \tag{13.2-1}$$

where $\langle v_z \rangle$ is the time-smoothed axial velocity at a given point in the fluid and u_z is the instantaneous deviation from that value. The fluctuation, u_z, may be either positive or negative at a given moment.

In Fig. 13-2 it is assumed that the flow is macroscopically steady for $t<t_0$, so that $\langle v_z \rangle$ during this period is just the average value of v_z over all times prior to t_0. The averaging interval t_a used in actual calculations must be long enough to encompass many fluctuations, if the calculated value of $\langle v_z \rangle$ is to converge to the true average. It is

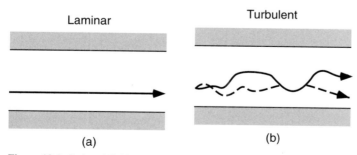

Laminar

Turbulent

(a)

(b)

Figure 13-1. Paths of fluid particles for fully developed laminar or turbulent flow in a tube. The two curves in panel (b) correspond to different times.

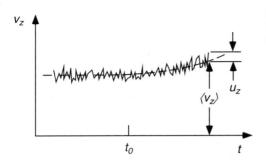

assumed in the figure that for $t > t_0$ the axial flow rate increases gradually with time. Thus, the smoothed velocity increases, as shown by the dashed curve. The concept of time-smoothing is valid for a macroscopically unsteady flow if the characteristic time for the system or process (t_p) is much larger than that of the fluctuations (t_f). That is, there must be a value of t_a such that $t_f \ll t_a \ll t_p$. The mathematical definition of time-smoothing and the manipulation of smoothed functions are discussed in Section 13.3.

The irregularity of the path lines for turbulent flow in a tube suggests that the velocity field is three-dimensional. Indeed, in any turbulent flow there are fluctuations in all three velocity components, which we express as $\mathbf{v} = \langle \mathbf{v} \rangle + \mathbf{u}$. The three-dimensional fluctuations are shown schematically in Fig. 13-3. It is important to note that fluctuations are present in each direction, even if the time-smoothed velocity component in that direction is zero. Turbulent flow is inherently rotational, and the velocity fluctuations are usually described in terms of transient eddies of various sizes and lifetimes which are superimposed on the main flow.

The velocity fluctuations are obviously a key aspect of any turbulent flow, and some measure of their average magnitude is needed. In this connection, note that $\langle u_i \rangle = 0$ (i.e., the time-averaged fluctuation must vanish) but $\langle u_i^2 \rangle > 0$. This motivates the use of the root-mean-square (rms) fluctuation, $\langle u_i^2 \rangle^{1/2}$. If the rms values of the three components of $\mathbf{u}$ are equal, the turbulence is said to be *isotropic*. Shear flows are generally not isotropic, because the cross-stream rms fluctuations at a given point typically are somewhat smaller than that in the main flow direction. A useful guideline for order-of-magnitude estimates in such flows is that the rms fluctuations of the main velocity component are about 10% of the overall mean velocity. Thus, at a representative position in a tube (i.e., not too close to the wall) we have

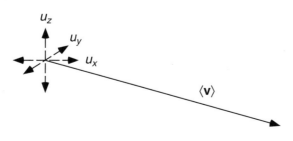

Figure 13-3. Three-dimensional nature of velocity fluctuations in turbulent flow.

$$\frac{\langle u_z^2 \rangle^{1/2}}{U} \sim 0.1, \qquad (13.2\text{-}2)$$

where $U = Q/(\pi R^2)$, Q is the volumetric flow rate, and R is the radius. [It is somewhat more accurate to replace 0.1 by $\sqrt{f}$, where f is the friction factor; see Eq. (13.2-3).] The velocity ratio in Eq. (13.2-2) (or its average for the three components) is often called the *turbulence intensity*. In another definition of turbulence intensity, U is replaced by the magnitude of the *local* time-smoothed velocity. Turbulence intensities measured in pipes are given in Laufer (1954) for most of the cross section, and in Durst et al. (1993) for positions very near the wall.

Enhanced Cross-Stream Transport

In fully developed laminar flow, transport normal to the flow direction is by the molecular or diffusive mechanisms only. Thus, in fully developed tube flow, heat is transferred in the radial direction only by conduction, and momentum is transferred only by viscous stresses. In turbulent flow, the velocity fluctuations result in cross-stream velocity components (i.e., velocity components normal to the main flow direction) which tend to be nonzero at any given instant. These lead to convective transport normal to the main flow. Although the average cross-stream velocities are zero, the net effect of the turbulent eddies is to promote cross-stream mixing. Increased radial mixing in a pipe may be either desirable or undesirable from an engineering viewpoint. Heat transfer from the fluid to the wall is more rapid, but the faster momentum transfer has the effect of increasing the pressure drop. If $\Delta \mathcal{P}$ is the difference in dynamic pressure between the ends of the pipe, the pumping power is given by $Q|\Delta \mathcal{P}|$. Thus, any increase in $\Delta \mathcal{P}$ leads to a proportionate increase in pumping costs.

It is instructive to consider some numerical examples of enhanced momentum and heat transfer in turbulent pipe flow. The *Fanning friction factor* is defined as

$$f \equiv \frac{|\Delta \langle \mathcal{P} \rangle|}{\rho U^2} \frac{R}{L}, \qquad (13.2\text{-}3)$$

where L is the pipe length and $\langle \mathcal{P} \rangle$ is the time-smoothed pressure. (Some authors employ the *Darcy friction factor*, which equals $4f$.) In essence, f is a dimensionless pressure drop. For pipes with smooth walls, it is found that f is a function of the Reynolds number only. For laminar flow, the Poiseuille equation [Eq. (6.2-23)] takes the form

$$f = \frac{16}{\text{Re}} \qquad (\text{Re} \leq 2.1 \times 10^3), \qquad (13.2\text{-}4)$$

where $\text{Re} = 2UR/\nu$. For turbulent flow, the accepted semiempirical expression is the *Kármán–Nikuradse equation*,

$$\frac{1}{\sqrt{f}} = 4 \log (\text{Re} \sqrt{f}) - 0.4 \qquad (\text{Re} \geq 3 \times 10^3), \qquad (13.2\text{-}5)$$

which provides an implicit relationship between f and Re that is highly accurate for smooth tubes. A convenient expression that is valid over a more limited range of Re is the *Blasius equation*,

$$f = 0.0791 \, \text{Re}^{-1/4} \qquad (3 \times 10^3 \leq \text{Re} \leq 1 \times 10^5). \qquad (13.2\text{-}6)$$

The effect on f of wall roughness, which is not considered here, is negligible for laminar flow but important sometimes for turbulent flow in commercial pipe. These and other results for f are reviewed in Bhatti and Shah (1987). Of the original papers dealing with both smooth and rough pipes, the two most influential are Nikuradse (1933) and Moody (1944).

The friction factor is plotted as a function of Re in Fig. 13-4. It is seen that Eqs. (13.2-4)–(13.2-6) are in excellent agreement with representative experimental data. The most noteworthy point for the present discussion is that turbulence causes a substantial increase in f. Choosing, for example, Re $= 1 \times 10^5$ and comparing the turbulent value of f with that which would result if laminar flow could be maintained, we find that the ratio of friction factors is 28. In other words, for a given set of fluid properties and pipe dimensions and a fixed flow rate, turbulence causes a nearly 30-fold increase in the pressure drop (at this Reynolds number).

Another measure of cross-stream mixing in turbulent flow is the heat transfer coefficient or (in dimensionless form) the Nusselt number. As shown in Section 9.5, for laminar flow in a tube Nu $\cong 4$ in the thermally fully developed region. A widely used empirical expression for turbulent flow in a tube is the *Colburn equation*,

$$\text{Nu} = 0.023 \ \text{Re}^{0.8} \ \text{Pr}^{1/3}. \tag{13.2-7}$$

This correlation is said to be accurate to within about $\pm 20\%$ for $1 \times 10^4 \le \text{Re} \le 1 \times 10^5$ and moderate values of Pr (Bhatti and Shah, 1987). For Re $= 1 \times 10^5$ and Pr $= 0.73$ (air at 20°C), Nu $= 210$; for Re $= 1 \times 10^5$ and Pr $= 6.9$ (water at 20°C), Nu $= 440$. Thus, the heat transfer coefficients for these two cases are 50–100 times those for laminar flow.

In laminar flow the velocity and thermal entrance lengths are determined by the

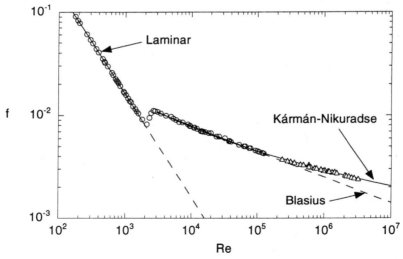

Figure 13-4. Friction factor for cylindrical tubes with smooth walls. The curves labeled "Laminar," "Kármán-Nikuradse," and "Blasius" are from Eqs. (13.2-4), (13.2-5), and (13.2-6), respectively. The circles are data from Koury (1995) for water flow in tubes of 1-cm diameter; the triangles are data from Nikuradse (1932) for 10-cm diameter.

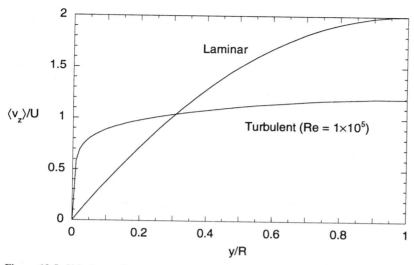

Figure 13-5. Velocity profiles in laminar and turbulent pipe flow. The radial coordinate has been reversed, such that $y = R - r$. The curve for turbulent flow was calculated from the Van Driest and Reichardt models, as described in Example 13.4-2.

rates at which momentum and heat, respectively, are able to diffuse from the tube wall to the interior of the fluid (Sections 6.5 and 9.4). This suggests that the enhanced rates of radial transport in turbulent flow might shorten the entrance lengths, which is indeed the case. Typical entrance lengths for turbulent flow are $L_V/R \leq 80$ and $L_T/R \leq 30$ (Nikuradse, 1933; Bhatti and Shah, 1987). For a hypothetical laminar flow with $Re = 1 \times 10^5$ and $Pr = 1$, the corresponding length ratios ($\cong 0.1\, Re$ or $\cong 0.1\, Pe$) would both be about 10^4, much larger than the turbulent values.

For turbulent flow in the z direction, the shear stress at a solid surface located at $y = 0$ is given by

$$\langle \tau_{yz} \rangle|_{y=0} = \mu \left. \frac{\partial \langle v_z \rangle}{\partial y} \right|_{y=0} \equiv \tau_0. \qquad (13.2\text{-}8)$$

Thus, the more rapid momentum transfer from the fluid to the wall in turbulent pipe flow must be associated with a much larger velocity gradient at the wall than in laminar flow. For a given mean velocity, it follows that the remainder of the velocity profile must be much flatter. This is shown in Fig. 13-5, in which velocity profiles for laminar and turbulent pipe flow are compared. Turbulent velocity profiles flatten progressively, approaching the ideal of plug flow, as Re is increased. The calculation of these profiles is discussed in Section 13.4.

In summary, cross-stream transport by the turbulent eddies can be orders of magnitude more rapid than the purely diffusive mechanisms present in fully developed laminar flow. Consequently, the entrance lengths and the time-smoothed velocity, temperature, and concentration profiles tend to be very different in turbulent than in laminar flow.

Scales in Turbulence

It has been mentioned that the fluctuating part of a velocity field may be described in terms of a population of eddies with a range of sizes and lifetimes. Useful insights into the nature of turbulence are provided by examining the velocity, length, and time scales associated with the largest and smallest of these eddies. Following Tennekes and Lumley (1972, pp. 19–24), these scales will be inferred by using dimensional analysis in conjunction with two key physical assumptions. Much of the underlying physics was revealed by the work of G. I. Taylor in the 1930s and A. N. Kolmogorov in the 1940s [e.g., see McComb (1990)].

   The physical assumptions are based on mechanical energy considerations. Consider the steady flow of an incompressible fluid in which gravitational effects can be ignored and there is negligible net inflow or outflow of kinetic energy. In such a flow the energy that is provided by an external source (e.g., a pump) to sustain the flow is dissipated ultimately as heat. As mentioned in connection with flow in pipes, turbulent flow requires a greater rate of energy input than a hypothetical laminar flow under the same conditions. Accordingly, turbulence involves additonal energy dissipation. The *additional* rate of energy dissipation in turbulence per unit mass is customarily denoted as ε, which has SI units of $m^2 s^{-3}$.

   A basic concept in turbulence theory is that mechanical energy is transferred from larger to successively smaller eddies, in a kind of cascade. Thus, the additional energy input associated with the turbulence first supplies the kinetic energy of the large eddies. Because the Reynolds number based on the size and velocity of these eddies is large, viscous effects are unimportant and energy dissipation at this level is negligible. Instead of the energy being converted immediately to heat, the large eddies give rise to smaller ones until the eddy Reynolds number is roughly unity. It is at that scale of motion that the mechanical energy is dissipated.

   The velocity, length, and time scales of the largest eddies, which are called here the *macroscales* of the turbulence, are denoted as u_1, L_1, and t_1, respectively. The key assumption concerning the largest eddies is that they are relatively short-lived. Specifically, it is assumed that their kinetic energy per unit mass, u_1^2, is dissipated in roughly one characteristic time, L_1/u_1. This allows ε to be estimated as

$$\varepsilon \sim \frac{u_1^2}{L_1/u_1} = \frac{u_1^3}{L_1}. \tag{13.2-9}$$

The key assumption for the smallest eddies is that their properties are determined only by the local flow conditions. Specifically, their scales (u_2, L_2, t_2) are assumed to depend on ε and $\nu = \mu/\rho$, but not on such factors as the system geometry. Using dimensional analysis now, we write that, for example,

$$L_2 = \nu^a \varepsilon^b [=] m. \tag{13.2-10}$$

where [=] indicates equality of units. Substituting in the units for ν and ε leads to two algebraic equations for the unknown constants a and b, from which it is found that $a = \frac{3}{4}$ and $b = -\frac{1}{4}$. Doing the same for u_2 and t_2, we find that

$$L_2 = \left(\frac{\nu^3}{\varepsilon}\right)^{1/4}, \qquad u_2 = (\nu\varepsilon)^{1/4}, \qquad t_2 = \left(\frac{\nu}{\varepsilon}\right)^{1/2}. \tag{13.2-11}$$

These are referred to here as the *microscales* of the turbulence; they are often called the *Kolmogorov scales*. Notice that Eq. (13.2-11) implies that the Reynolds number for the smallest eddies is unity, as suggested above; that is, $Re_2 = u_2 L_2 / \nu = 1$.

The microscales and macroscales are related now by using Eq. (13.2-9) to evaluate ε in Eq. (13.2-11). The result is

$$\frac{L_2}{L_1} \sim Re_1^{-3/4}, \qquad \frac{u_2}{u_1} \sim Re_1^{-1/4}, \qquad \frac{t_2}{t_1} \sim Re_1^{-1/2}, \qquad (13.2\text{-}12)$$

where Re_1 is the Reynolds number based on the turbulence macroscales, or

$$Re_1 = \frac{u_1 L_1}{\nu}. \qquad (13.2\text{-}13)$$

The most noteworthy conclusion from Eq. (13.2-12) is that the microscales and macroscales diverge more and more as Re_1 is increased. In other words, turbulence develops a progressively "finer-grained" structure as the Reynolds number is increased. There is much experimental evidence to support this conclusion.

As a numerical example, we consider pipe flow again at $Re = 1 \times 10^5$. The largest eddy size is limited by the cross-sectional dimension of the system, so that $L_1 = d$, the pipe diameter. Equating u_1, the largest velocity scale for the turbulence, with the rms fluctuation of the axial velocity, we infer from Eq. (13.2-2) that $u_1 \sim 0.1\ U$. It follows that $Re_1 \sim (0.1)\ Re = 1 \times 10^4$. Using Eq. (13.2-12) to estimate the length and time scales for the smallest eddies, we find that $L_2 \sim (0.001) d$ and $t_2 \sim (0.1)(d/U)$.

Suppose now that the system consists of air at 20°C in a 10-cm diameter pipe, for which $\nu = 1.5 \times 10^{-5}$ m²/s and $d = 0.1$ m. With $Re = 1 \times 10^5$, this implies that $U = 15$ m/s. (Although a large velocity, this is still only about 5% of the speed of sound.) It follows from the results in the preceding paragraph that $L_2 \sim 10^{-4}$ m $= 100$ μm. The mean free path in a gas under these conditions is $\ell \cong 0.1$ μm (Table 1-9), so that $L_2/\ell \sim 10^3$. The fact that the smallest eddies are orders of magnitude larger than the mean free path confirms that turbulence is a continuum phenomenon; this is true for all circumstances that are likely to be encountered. Finally, we calculate that $t_2 \sim 10^{-3}$ s $= 1$ ms. This is an estimate of the duration of the most rapid fluctuations and suggests the time response needed in equipment intended to record those events.

13.3 TIME-SMOOTHED EQUATIONS

It was emphasized in the preceding section that high-frequency, small-amplitude fluctuations in velocity are a hallmark of turbulent flow. Because the pressure, temperature, and species concentration fields all depend on the velocity, those variables also fluctuate. Nonetheless, a turbulent flow will be adequately characterized for most purposes if the time-smoothed velocity field can be determined. Calculations of the fluctuating part of the velocity are not routinely practical at present and, from an engineering viewpoint, would provide a level of detail that is usually unnecessary. The same is true for the other field variables. Accordingly, the usual approach is to rewrite the governing differential equations in terms of time-smoothed variables. This approach was pioneered by Reynolds (1895) and is usually called *Reynolds averaging*. In this section the time-smoothed conservation equations for mass, momentum, energy, and chemical species are derived for Newtonian fluids with constant physical properties. Although in time-smoothing one hopes to avoid dealing with fluctuating quantites altogether, it will be seen that this is not possible.

Basics

In introducing the concept of a time-smoothed function in Section 13.2, it was pointed out that one needs to average over an interval (t_a) that is large compared to the period of the fluctuations (t_f), but small compared to the macroscopic or process time scale (t_p). For a function $F(\mathbf{r}, t)$, the operation of time-smoothing is defined as

$$\langle F \rangle \equiv \frac{1}{t_a} \int_t^{t+t_a} F(\mathbf{r}, \tau) \, d\tau \qquad \left(\frac{t_f}{t_a} \to 0, \frac{t_a}{t_p} \to 0 \right), \qquad (13.3\text{-}1)$$

where τ serves as the dummy time variable in the integration. One must bear in mind that the time-smoothed function, $\langle F \rangle$, can still depend on t. The dual limits in Eq. (13.3-1) express the aforementioned restrictions on the time constants. With very strongly fluctuating functions (e.g., time derivatives of fluctuating quantities), the integration may need to be repeated before these limits will exist. The angle-bracket notation is used hereafter with the understanding that the integration has been repeated as necessary to suppress the rapid oscillations while preserving any slow variations over time.

The principal field variables are expressed as

$$\mathbf{v} = \langle \mathbf{v} \rangle + \mathbf{u}, \qquad \langle \mathbf{u} \rangle \equiv \mathbf{0}, \qquad (13.3\text{-}2)$$

$$\mathscr{P} = \langle \mathscr{P} \rangle + p, \qquad \langle p \rangle \equiv 0, \qquad (13.3\text{-}3)$$

$$T = \langle T \rangle + \theta, \qquad \langle \theta \rangle \equiv 0, \qquad (13.3\text{-}4)$$

$$C_i = \langle C_i \rangle + \chi_i, \qquad \langle \chi_i \rangle \equiv 0, \qquad (13.3\text{-}5)$$

where the fluctuations are represented by the functions $\mathbf{u}$, p, θ, and χ_i. As indicated, each of the time-smoothed fluctuations vanishes, by definition.

One important rule in manipulations with smoothed quantities is that *repeated smoothing of any time-smoothed variable has no effect*. This is shown formally by using the fact that smoothing, which involves an integration over time, is distributive. Taking the velocity as an example and smoothing both sides of Eq. (13.3-2), we obtain

$$\langle \mathbf{v} \rangle = \langle \langle \mathbf{v} \rangle + \mathbf{u} \rangle = \langle \langle \mathbf{v} \rangle \rangle + \langle \mathbf{u} \rangle = \langle \langle \mathbf{v} \rangle \rangle. \qquad (13.3\text{-}6)$$

Thus, the "twice-smoothed" velocity does not differ from the "once-smoothed" velocity, as asserted above. A second rule is that *the order of smoothing and differentiation can be interchanged*. This is clear for differential operators such as ∇ and ∇^2, because the differentiations involve spatial variables and the smoothing involves integration over a different variable, namely, time. Thus, for example,

$$\langle \nabla \cdot \mathbf{v} \rangle = \nabla \cdot \langle \mathbf{v} \rangle, \qquad \langle \nabla^2 \mathbf{v} \rangle = \nabla^2 \langle \mathbf{v} \rangle. \qquad (13.3\text{-}7)$$

The proof for time derivatives is given in McComb (1990, p. 34). Using the velocity again as an example, it is found that

$$\left\langle \frac{\partial \mathbf{v}}{\partial t} \right\rangle = \frac{\partial}{\partial t} \langle \mathbf{v} \rangle. \tag{13.3-8}$$

Conservation of Mass

In transforming the differential conservation equations to ones involving smoothed variables, we begin with the continuity equation,

$$\boldsymbol{\nabla} \cdot \mathbf{v} = 0. \tag{13.3-9}$$

Time-smoothing both sides of the equation results in

$$\langle \boldsymbol{\nabla} \cdot \mathbf{v} \rangle = \boldsymbol{\nabla} \cdot \langle \mathbf{v} \rangle = 0. \tag{13.3-10}$$

Thus, the smoothed velocity satisfies the same continuity equation as the instantaneous velocity. Important information about the fluctuating part of the velocity is obtained by using Eq. (13.3-2) to expand Eq. (13.3-9), which gives

$$\boldsymbol{\nabla} \cdot \mathbf{v} = \boldsymbol{\nabla} \cdot (\langle \mathbf{v} \rangle + \mathbf{u}) = \boldsymbol{\nabla} \cdot \langle \mathbf{v} \rangle + \boldsymbol{\nabla} \cdot \mathbf{u} = 0. \tag{13.3-11}$$

Comparing Eqs. (13.3-10) and (13.3-11), we conclude that

$$\boldsymbol{\nabla} \cdot \mathbf{u} = 0. \tag{13.3-12}$$

It is seen that the instantaneous velocity fluctuation also obeys the usual continuity equation.

Conservation of Momentum

The Navier–Stokes equation is

$$\rho \frac{D\mathbf{v}}{Dt} = \rho \left(\frac{\partial \mathbf{v}}{\partial t} + \mathbf{v} \cdot \boldsymbol{\nabla} \mathbf{v} \right) = -\boldsymbol{\nabla} \mathcal{P} + \mu \nabla^2 \mathbf{v}. \tag{13.3-13}$$

The key consideration in time-smoothing is the linearity or nonlinearity of the individual terms. Except for $\mathbf{v} \cdot \boldsymbol{\nabla} \mathbf{v}$, all of the terms are linear in $\mathbf{v}$ or $\mathcal{P}$. Applying the rules given by Eqs. (13.3-7) and (13.3-8), the effect of smoothing each of the linear terms is simply to replace the original (instantaneous) variable by the smoothed variable, as happened with the continuity equation. However, the nonlinear inertial term requires special attention. It is first expanded as

$$\mathbf{v} \cdot \boldsymbol{\nabla} \mathbf{v} = \langle \mathbf{v} \rangle \cdot \boldsymbol{\nabla} \langle \mathbf{v} \rangle + \mathbf{u} \cdot \boldsymbol{\nabla} \langle \mathbf{v} \rangle + \langle \mathbf{v} \rangle \cdot \boldsymbol{\nabla} \mathbf{u} + \mathbf{u} \cdot \boldsymbol{\nabla} \mathbf{u}. \tag{13.3-14}$$

Time-smoothing does not alter the first term on the right-hand side, whereas the second and third terms vanish because $\langle \mathbf{u} \rangle = \mathbf{0}$. The fourth term does not necessarily vanish. Thus, smoothing the inertial term gives

$$\langle \mathbf{v} \cdot \boldsymbol{\nabla} \mathbf{v} \rangle = \langle \mathbf{v} \rangle \cdot \boldsymbol{\nabla} \langle \mathbf{v} \rangle + \langle \mathbf{u} \cdot \boldsymbol{\nabla} \mathbf{u} \rangle. \tag{13.3-15}$$

What is important is that a new term has arisen, $\langle \mathbf{u} \cdot \boldsymbol{\nabla} \mathbf{u} \rangle$, which involves the fluctuating part of the velocity. Thus, as mentioned earlier, it is not possible to avoid fluctuating quantities altogether. The smoothed Navier–Stokes equation at this stage is

$$\rho \frac{D\langle \mathbf{v} \rangle}{Dt} = -\boldsymbol{\nabla} \langle \mathcal{P} \rangle + \mu \nabla^2 \langle \mathbf{v} \rangle - \rho \langle \mathbf{u} \cdot \boldsymbol{\nabla} \mathbf{u} \rangle, \tag{13.3-16}$$

where the new term has been placed on the right-hand side. Notice that pressure fluctuations do not appear in the time-smoothed momentum equation, so that p need not be evaluated. Only the velocity fluctuations are important.

It is helpful to rewrite the right-hand side of the smoothed Navier–Stokes equation in terms of stress tensors. Equation (5.6-8) gives

$$\mu \nabla^2 \mathbf{v} = \nabla \cdot \boldsymbol{\tau},$$ (13.3-17)

from which it follows that

$$\mu \nabla^2 \langle \mathbf{v} \rangle = \langle \mu \nabla^2 \mathbf{v} \rangle = \langle \nabla \cdot \boldsymbol{\tau} \rangle = \nabla \cdot \langle \boldsymbol{\tau} \rangle.$$ (13.3-18)

The turbulent contribution to the right-hand side of Eq. (13.3-16) can also be expressed as the divergence of a tensor. First, note from identity (12) of Table A-1 and Eq. (13.3-12) that

$$\nabla \cdot (\mathbf{uu}) = (\nabla \cdot \mathbf{u})\mathbf{u} + \mathbf{u} \cdot \nabla \mathbf{u} = \mathbf{u} \cdot \nabla \mathbf{u}.$$.(13.3-19)

Accordingly, the turbulent term in the momentum equation can be rewritten as

$$\rho \langle \mathbf{u} \cdot \nabla \mathbf{u} \rangle = \nabla \cdot \left(\rho \langle \mathbf{uu} \rangle \right).$$ (13.3-20)

The *turbulent stress* or *Reynolds stress* is defined as

$$\boldsymbol{\tau}^* \equiv -\rho \langle \mathbf{uu} \rangle.$$ (13.3-21)

It is apparent that $\boldsymbol{\tau}^*$, like $\boldsymbol{\tau}$, is a symmetric tensor. Using these results, the smoothed Navier–Stokes equation is written finally as

$$\rho \frac{D\langle \mathbf{v} \rangle}{Dt} = -\nabla \langle \mathscr{P} \rangle + \nabla \cdot \left(\langle \boldsymbol{\tau} \rangle + \boldsymbol{\tau}^* \right).$$ (13.3-22)

It is this form of the Navier–Stokes equation which is the usual starting point for solving problems involving turbulent flow.

It must be emphasized that although the same Greek letters are used, the stresses in the last term of Eq. (13.3-22) have very different physical origins. Whereas $\boldsymbol{\tau}$ and $\langle \boldsymbol{\tau} \rangle$ represent viscous transfer of momentum and are therefore molecular or diffusive in nature, $\boldsymbol{\tau}^*$ is due to the velocity fluctuations and is fundamentally convective. Given that distinction, it is all the more noteworthy that the effect of turbulence is formally equivalent to augmenting the viscous stress tensor. Of course, the manipulations leading to $\boldsymbol{\tau}^*$ have not addressed the basic issue of how the Reynolds stress is to be evaluated. Strategies for that are discussed in Section 13.4.

Conservation of Energy

For a single-component, Newtonian fluid with constant thermophysical properties, conservation of energy is expressed as

$$\rho \hat{C}_p \frac{DT}{Dt} = \rho \hat{C}_p \left(\frac{\partial T}{\partial t} + \mathbf{v} \cdot \nabla T \right) = k \nabla^2 T + \mu \Phi,$$ (13.3-23)

where Φ is the viscous dissipation function (see Section 9.6). The time-smoothing of the energy equation is very similar to that of the Navier–Stokes equation. Because T

depends on $\mathbf{v}$, the $\mathbf{v} \cdot \nabla T$ term is nonlinear and requires special attention. By analogy with Eqs. (13.3-15) and (13.3-16), time-smoothing of this term gives

$$\langle \mathbf{v} \cdot \nabla T \rangle = \langle \mathbf{v} \rangle \cdot \nabla \langle T \rangle + \langle \mathbf{u} \cdot \nabla \theta \rangle \tag{13.3-24}$$

and the smoothed energy equation is

$$\rho \hat{C}_p \frac{D \langle T \rangle}{Dt} = k \nabla^2 \langle T \rangle - \rho \hat{C}_p \langle \mathbf{u} \cdot \nabla \theta \rangle + \mu \langle \Phi \rangle. \tag{13.3-25}$$

Once again, a new term has arisen which contains only fluctuations. In this case both velocity and temperature fluctuations are involved. Although not written out, $\langle \Phi \rangle$ also contains time-smoothed products of fluctuations. This is because Φ is a nonlinear function of $\mathbf{v}$ (see Table 5-10).

The conduction and turbulence terms on the right-hand side of Eq. (13.3-25) are expressed now as the divergence of a sum of fluxes, similar to what was done with stresses in the momentum equation. In a pure fluid, the energy flux relative to the mass-average velocity is given by Fourier's law as $\mathbf{q} = -k \nabla T$. Accordingly, the smoothed conduction term is rewritten as

$$k \nabla^2 \langle T \rangle = \langle k \nabla^2 T \rangle = - \langle \nabla \cdot \mathbf{q} \rangle = - \nabla \cdot \langle \mathbf{q} \rangle. \tag{13.3-26}$$

The relation analogous to Eq. (13.3-19) is

$$\nabla \cdot (\mathbf{u} \theta) = (\nabla \cdot \mathbf{u}) \theta + \mathbf{u} \cdot \nabla \theta = \mathbf{u} \cdot \nabla \theta, \tag{13.3-27}$$

which motivates defining the *turbulent energy flux* as

$$\mathbf{q}^* \equiv \rho \hat{C}_p \langle \mathbf{u} \theta \rangle. \tag{13.3-28}$$

As shown below, $\mathbf{q}^*$ acts as an additional energy flux relative to the mass-average velocity. The turbulent energy flux defined here and the Reynolds stress defined in Eq. (13.3-21) are analogous. (The difference in algebraic sign is due to the sign convention we adopted for the stress.) The final form of the time-smoothed energy equation is

$$\rho \hat{C}_p \frac{D \langle T \rangle}{Dt} = - \nabla \cdot (\langle \mathbf{q} \rangle + \mathbf{q}^*) + \mu \langle \Phi \rangle. \tag{13.3-29}$$

Thus, the main effect of turbulence on energy transfer is formally equivalent to augmenting the energy flux relative to the mass-average velocity. As with the Reynolds stress, the evaluation of $\mathbf{q}^*$ remains to be discussed.

Conservation of Chemical Species

The time-smoothing of the species conservation equation is closely analogous to that for energy. For a dilute solution, the diffusive contribution to the flux of species i is given by Fick's law as $\mathbf{J}_i = -D_i \nabla C_i$. The *turbulent flux of species i* (relative to the mass-average velocity) is

$$\mathbf{J}_i^* \equiv \langle \mathbf{u} \, \chi_i \rangle \tag{13.3-30}$$

and the smoothed conservation equation is

$$\frac{D \langle C_i \rangle}{Dt} = - \nabla \cdot (\langle \mathbf{J}_i \rangle + \mathbf{J}_i^*) + \langle R_{Vi} \rangle. \tag{13.3-31}$$

Once again, the effect of turbulence is to augment the flux relative to the mass-average velocity.

The effects of turbulence on the calculation of reaction rates are illustrated by considering two simple rate laws. For a first-order reaction with $R_{Vi} = -k_1 C_i$, the time-smoothed rate is

$$\langle R_{Vi} \rangle = -k_1 \langle \langle C_i \rangle + \chi_i \rangle = -k_1 (\langle C_i \rangle + \langle \chi_i \rangle) = -k_1 \langle C_i \rangle. \tag{13.3-32}$$

Thus, the form of the rate expression is unaffected by smoothing. However, for a second-order reaction with $R_{Vi} = -k_2 C_i^2$, the smoothed rate becomes

$$\langle R_{Vi} \rangle = -k_2 \langle (\langle C_i \rangle + \chi_i)^2 \rangle = -k_2 (\langle C_i \rangle^2 + 2\langle C_i \rangle \langle \chi_i \rangle + \langle \chi_i^2 \rangle) = -k_2 (\langle C_i \rangle^2 + \langle \chi_i^2 \rangle). \tag{13.3-33}$$

The additional term involving $\langle \chi_i^2 \rangle$ indicates that, in this case, using the time-smoothed concentration in the original rate law would systematically underestimate the rate at which species i is consumed. In general, any nonlinear rate law will lead to additional terms when the concentrations are smoothed.

Turbulent Fluxes

The quantities τ^*, q^*, and J_i^* are referred to here as the *turbulent fluxes*. Considering transport in the y direction, examples of these fluxes are

$$-\tau_{yz}^* = \rho \langle u_y u_z \rangle, \tag{13.3-34}$$

$$q_y^* = \rho \hat{C}_p \langle u_y \theta \rangle, \tag{13.3-35}$$

$$J_{iy}^* = \langle u_y \chi_i \rangle. \tag{13.3-36}$$

The turbulent shear stress is viewed as a flux because τ_{yz}^* can be interpreted as a flux of z momentum along the y axis. It is seen that each flux is proportional to the time-smoothed product of two fluctuating variables.

The ability of fluctuations to yield net fluxes, despite the fact that the average of any single fluctuating quantity is zero, is illustrated by applying a simple model to heat transfer. In this model the velocity and temperature fluctuations are assumed to be sinusoidal in time; this is a more regular pattern than in actual turbulence. If the velocity and temperature fluctuations have the same period (t_f), but are out of phase by an angle $2\pi\beta$, then

$$u_y = u_0 \sin \left[2\pi \left(\frac{t}{t_f} \right) \right] \tag{13.3-37}$$

$$\theta = \theta_0 \sin \left[2\pi \left(\frac{t}{t_f} + \beta \right) \right], \tag{13.3-38}$$

where u_0 and θ_0 are constants. Because these idealized fluctuations are exactly periodic, their time-averaged product can be calculated by intergrating over any one period. Choosing $0 \leq t \leq t_f$, we obtain

$$\frac{q_y^*}{\rho \hat{C}_p} = \langle u_y \theta \rangle = \frac{1}{t_f} \int_0^{t_f} u_y \theta \, dt = \frac{u_0 \theta_0}{2} \cos 2\pi\beta. \tag{13.3-39}$$

It is seen that the energy flux has its greatest positive or negative value when the velocity and temperature fluctuations are in phase ($\beta=0$) or 180° out of phase ($\beta=\frac{1}{2}$), respectively. There is no average flux if the phase angle is 90° or 270° ($\beta=\frac{1}{4}$ or $\frac{3}{4}$). For $\beta=0$, note that u_y and θ are both positive for $0<t<t_f/2$ and both negative for $t_f/2<t<t_f$. This indicates that relatively warm fluid is transferred in the $+y$ direction during the first half of the period, and relatively cool fluid is transferred in the $-y$ direction during the second half. This amounts to positive heat transfer in the $+y$ direction at all times, which is why $q_y{}^*$ has its largest positive value for $\beta=0$. For any other phase angle, there is movement of cool fluid in the $+y$ direction and warm fluid in the $-y$ direction at least part of the time. A similar model was used by Kays and Crawford (1993) to illustrate the origin of the Reynolds stress.

An important aspect of the turbulent fluxes is their behavior at solid surfaces. At any such surface, moving or stationary, the no-slip and no-penetration conditions imply that the velocity fluctuations vanish. For example, at a stationary solid surface located at $y=0$, the no-penetration condition is that $v_y=0$ at every instant, which implies that $\langle v_y \rangle=0$ and $u_y=0$. In three dimensions, the requirement that $\mathbf{v}=\mathbf{0}$ at a surface indicates that both $\langle \mathbf{v} \rangle=\mathbf{0}$ and $\mathbf{u}=\mathbf{0}$. Given that $\mathbf{u}=\mathbf{0}$, we see from Eqs. (13.3-21), (13.3-28), and (13.3-30) that $\boldsymbol{\tau}^*=\mathbf{0}$, $\mathbf{q}^*=\mathbf{0}$, and $\mathbf{J}_i{}^*=\mathbf{0}$ at any solid surface. Thus, the shear stress at a surface is calculated simply by using the time-smoothed velocity in the usual viscous stress formula, as indicated in Eq. (13.2-8); there is no turbulent contribution at the surface. Similarly, the energy and species fluxes are calculated from gradients in the smoothed temperature and concentration, respectively. Because the turbulent fluxes may be orders of magnitude larger than their diffusive counterparts in much of the fluid (Section 13.2), they are evidently strong functions of position. This is true even in the simplest flows (e.g., fully developed flow in a tube). The fact that the turbulent fluxes are always zero at solid surfaces must be accounted for in any scheme used to evaluate $\boldsymbol{\tau}^*$, $\mathbf{q}^*$, and $\mathbf{J}_i{}^*$, such as the eddy diffusivity models discussed in Section 13.4.

Closure Problem

Although the fluxes at a fluid–solid interface can be calculated by knowing only the smoothed field variables, the smoothed fields cannot be determined without evaluating the fluctuations, at least in some approximate manner. When we review the relationships derived above, a problem becomes evident. Namely, the number of smoothed conservation equations only equals the number of smoothed field variables, whereas the velocity, temperature, and concentration fluctuations ($\mathbf{u}$, θ, χ_i) are also unknowns. This shortage of governing equations is the most basic form of what is called the *closure problem*. As discussed in Section 13.5, *any* attempt to use smoothed variables in turbulent flow leads to a formulation in which the number of rigorously derivable equations is less than the number of unknowns. Thus, the cost of averaging is the need to find additional relationships among the unknowns, the physical basis for which is not obvious.

The simplest way to overcome the closure problem is to postulate algebraic relationships between the turbulent fluxes ($\boldsymbol{\tau}^*$, $\mathbf{q}^*$, $\mathbf{J}_i{}^*$) and the corresponding smoothed field variables, based on empirical observations and/or physical reasoning. Thus, the fluctuations per se are ignored, and the problem is reduced to specifying what may be thought of as "turbulence constitutive equations." The best-known approach involves *eddy diffusivity models*, which are introduced in Section 13.4.

13.4 EDDY DIFFUSIVITY MODELS

The eddy diffusion approach described in this section has been extremely influential in the development of turbulence theory and has been widely used in engineering. We will focus initially on eddy diffusivity models in their most basic forms, leaving more elaborate approaches to be discussed in Section 13.5. Hereafter, to shorten equations, overbars are frequently used in place of angle brackets to denote smoothed values of *single* quantities. Thus, for example, $\bar{v}_z = \langle v_z \rangle$ and $\bar{T} = \langle T \rangle$. When *products* of quantities are to be smoothed, angle brackets are always used.

Eddy Diffusion Concept

It was shown in Section 13.3 that time-smoothing of the momentum, energy, and species conservation equations leads to additional flux terms for turbulent flow. An idealized model for the turbulent energy flux was presented, in which the flux was related to time-periodic fluctuations in the velocity and temperature. That model suggests that the turbulent fluxes might be independent of *gradients* in the time-smoothed field variables, in that no spatial variations were considered. This is true to a certain extent. Nonetheless, as emphasized in Section 13.2, it is well known that turbulence enhances cross-stream transport in a manner that speeds equilibration. That is, turbulent eddies tend to transfer momentum, energy, and chemical species from regions of higher to regions of lower concentration of the given quantity. Apparently, the turbulent fluxes act much like enhanced forms of diffusion.

The resemblance between the turbulent fluxes and diffusion is explained by a conceptual model in which an eddy moves an element of fluid from, say, a hotter to a colder region. After arrival at the colder region, the fluid element equilibrates with its new surroundings. Given the rotational nature of turbulent eddies, this is accompanied by the movement of some other element of fluid in the opposite direction. The net effect of such movements is heat transfer from the hotter to the colder region. This conceptual model is much the same as the lattice model for molecular diffusion in gases (Section 1.5), except that the motion is on an eddy rather than a molecular scale. For that reason, the mechanism just described is called *eddy diffusion.*

Throughout this section we will suppose that the main flow is in the z direction, and that y is distance from a tube wall or other solid surface. It is assumed that $\bar{v}_z$ depends mainly on y, as is true for fully developed flow or two-dimensional boundary layer flow. According to the eddy diffusion concept, the turbulent fluxes in the y direction are expressed as

$$\frac{\tau_{yz}^*}{\rho} \equiv \varepsilon_M \frac{\partial \bar{v}_z}{\partial y}, \qquad (13.4\text{-}1)$$

$$\frac{q_y^*}{\rho \hat{C}_p} \equiv -\varepsilon_H \frac{\partial \bar{T}}{\partial y}, \qquad (13.4\text{-}2)$$

$$J_{iy}^* \equiv -\varepsilon_i \frac{\partial \bar{C}_i}{\partial y}, \qquad (13.4\text{-}3)$$

where ε_M, ε_H, and ε_i are the *eddy diffusivities* for momentum, heat, and species i, respectively. The eddy diffusivities have the same units as molecular-scale diffusivities (m²/s).

On a fundamental level, Eqs. (13.4-1)–(13.4-3) do no more than define the respective ε's. However, to the extent that the underlying concept is correct and the eddy diffusivities are predictable, these algebraic relationships between the turbulent fluxes and the smoothed field variables provide a simple and effective solution to the closure problem.

Two general implications of this type of model are noteworthy. The first is that the eddy diffusivities for all quantities in a given flow should be very similar. This is because each of the turbulent fluxes is convective in origin and arises from the same eddy motion. By analogy with Eq. (1.5-4), we expect each of the eddy diffusivities to scale as $u\ell$, where u is a characteristic eddy velocity and ℓ is an eddy length scale. Taken to the extreme, this reasoning suggests that

$$\varepsilon_M = \varepsilon_H = \varepsilon_i, \tag{13.4-4}$$

which is called *Reynolds' analogy*. The usual approach in heat or mass transfer calculations is to evaluate ε_M from velocity measurements and to assume that Reynolds' analogy is correct (or nearly so). As will be discussed, experimental findings tend to support the assumption that the eddy diffusivities are approximately equal. To the extent that Eqs. (13.4-1)–(13.4-4) are correct, the problem of evaluating the turbulent fluxes is reduced to that of evaluating ε_M.

The second implication of Eqs. (13.4-1)–(13.4-3) is that the eddy diffusivities must vanish at solid surfaces. This follows from the fact that each turbulent flux vanishes (Section 13.3), whereas, in general, the gradients in the field variables do not. Thus, for a surface at $y=0$, any model must give $\varepsilon_M \rightarrow 0$ as $y \rightarrow 0$. We conclude that, unlike their molecular counterparts, the eddy diffusivities are strong functions of position; this closure scheme is not as simple as it first appears. The situation may be summarized by stating that the molecular diffusivities (ν, α, D_i) are properties of the *fluid*, whereas the eddy diffusivities (ε_M, ε_H, ε_i) are properties of the *flow*.

Wall Variables

As with laminar boundary layers, much of the important action in a turbulent shear flow occurs very near the bounding surfaces. Accordingly, much attention has been devoted to the behavior of the velocity and eddy diffusivities in the vicinity of a wall. As will be shown, $\bar{v}_z$ and ε_M both have a nearly universal character near a solid surface, largely independent of the overall flow geometry. The velocity scale which is found to be pertinent to the wall region is $\sqrt{(\tau_0/\rho)}$, where τ_0 is the shear stress at the wall; a corresponding length scale constructed from the fluid properties and τ_0 is $\nu/\sqrt{(\tau_0/\rho)}$. Accordingly, the dimensionless velocity and distance from the wall are expressed as

$$v^+_z \equiv \frac{\bar{v}_z}{\sqrt{\tau_0/\rho}}, \qquad y^+ \equiv \frac{\sqrt{\tau_0/\rho}}{\nu} y. \tag{13.4-5}$$

These *wall variables* are ubiquitous in the turbulence literature.

The velocity and length scales just introduced are peculiar, and it may be helpful to relate them to more familiar quantities. Steady, fully developed flow in a pipe of radius R and length L is used as an example. Whether the flow is laminar or turbulent, the overall force balance is $2\pi R L \tau_0 = \pi R^2 |\Delta \bar{\mathcal{P}}|$. Thus, the pressure drop and wall shear stress are related as

TABLE 13-1
**Friction Factor and Ratios of Scales for Turbulent
Pipe Flow**

Re	f	$\sqrt{f/2}$	$y_c^+ = \mathrm{Re}\sqrt{f/8}$
1×10^4	0.00773	0.0622	311
1×10^5	0.00450	0.0474	2,370
1×10^6	0.00291	0.0382	19,100
1×10^7	0.00203	0.0318	159,000

$$\frac{|\Delta\overline{\mathcal{P}}|}{L} = \frac{2\tau_0}{R}.$$

(13.4-6)

An alternate form of Eq. (13.2-3) is then

$$f = \frac{2\tau_0}{\rho U^2},$$

(13.4-7)

from which it follows that the ratio of the wall velocity scale to the mean velocity U is

$$\frac{\sqrt{\tau_0/\rho}}{U} = \sqrt{f/2}.$$

(13.4-8)

Again using Eq. (13.4-7), the ratio of the wall length scale to the radius is found to be

$$\frac{\nu/\sqrt{\tau_0/\rho}}{R} = \frac{1}{\mathrm{Re}\sqrt{f/8}}.$$

(13.4-9)

How small the wall scales are relative to U and R is illustrated by some numerical examples. The full range of the wall coordinate is $0 \leq y^+ \leq y_c^+$, where y_c^+ corresponds to the tube center (i.e., $y = R$). Some representative values of $y^+{}_c = \mathrm{Re}\sqrt{f/8}$ are given in Table 13-1. Also shown are values of f and $\sqrt{f/2}$, the latter being the wall velocity scale relative to the mean velocity. It so happens that the rms fluctuations in axial velocity in a pipe are about one to two times the wall velocity scale, except very near the wall, where the fluctuations are damped (Laufer, 1954; Durst et al., 1993). Accordingly, the turbulence intensity (based on U) is typically $\sim\sqrt{f}$. This is why the right-hand side of Eq. (13.2-2) is written more accurately as $\sqrt{f}$, as stated earlier.

Evaluation of Eddy Diffusivities

The most influential of the various approaches for evaluating ε_M is the *mixing length* model proposed by Prandtl in 1925 [see Goldstein (1938, pp. 207–208)]. In this model it is assumed that

$$\varepsilon_M = \ell^2 \left| \frac{\partial \overline{v}_z}{\partial y} \right|,$$

(13.4-10)

where ℓ is the "mixing length." The mixing length and $\ell \left|\partial\overline{v}_z/\partial y\right|$ are viewed as length and velocity scales, respectively, for eddy motion. Based on dimensional reasoning, Prandtl concluded that near any surface $\ell = \kappa y$, where κ is a constant. Using the dimensionless wall variables, Eq. (13.4-10) is rewritten as

$$\frac{\varepsilon_M}{\nu} = (\kappa y^+)^2 \left| \frac{\partial v_z^+}{\partial y^+} \right|.$$ (13.4-11)

Comparisons of velocities predicted by this model with those measured by numerous investigators yield a best-fit value of $\kappa = 0.40$ (see Example 13.4-1).

Although the mixing-length model with $\ell = \kappa y$ is remarkably accurate for most of the region near a wall, it breaks down for $y^+ < 30$. The problem is that the eddy diffusivity given by Eq. (13.4-11) does not decrease fast enough as $y^+ \to 0$. One solution to this problem has been to postulate a *laminar sublayer* adjacent to the wall, in which $\varepsilon_M = 0$; outside the sublayer, Eq. (13.4-11) is assumed to hold. An intermediate or "buffer" layer, in which a third expression is used for ε_M, is sometimes invoked to improve the fit to the velocity data. More plausible are any of several models in which ε_M is a continuous function of position. A useful expression of this type is that of Van Driest (1956),

$$\frac{\varepsilon_M}{\nu} = (\kappa y^+)^2 (1 - e^{-y^+/A})^2 \left| \frac{dv_z^+}{dy^+} \right|.$$ (13.4-12)

where κ is the same as before and A is another constant. Here the mixing length has been modified using a decaying exponential. Equation (13.4-12) is identical to Eq. (13.4-11) for $y^+ \gg A$, but the new term speeds the decrease of ε_M for $y^+ \to 0$. A good fit to velocity data for pipe flow is obtained with $A = 26$ (see Example 13.4-1). The optimal value of A decreases slightly if the pressure gradient is less favorable (Kays and Crawford, 1993, pp. 211–212); for the turbulent boundary layer on a flat plate parallel to the approaching flow (no pressure gradient), $A = 25$.

Other expressions for the eddy diffusivity near a wall are proposed, for example, in Deissler (1955), Notter and Sleicher (1971), and Petukhov (1970). The alternative models tend to give equivalent results for the velocity, within experimental error. What is most sensitive to the functional form of ε_M for $y^+ \to 0$ is the Nusselt number at large Pr or the Sherwood number at large Sc. The Van Driest expression is accurate enough for calculating temperature profiles at moderate Pr, but it appears to be inadequate at large Pr (see Problem 13-4).

Equations (13.4-11) and (13.4-12) do not apply very far from a wall. Although, for example, $y^+ = 100$ seems large, for pipe flow at $\text{Re} \geq 1 \times 10^5$ it is still $\leq 4\%$ of the distance from the wall to the center (Table 13-1). As recognized by Prandtl himself (Goldstein, 1938, p. 357), the assumption that $\ell = \kappa y$ begins to break down for $y/R > 0.1$; the slope of a plot of ℓ versus y decreases progressively toward the tube center. An analogous bending over of the ℓ versus y curve is seen in turbulent boundary layers (Andersen et al., 1975). In cylindrical tubes, the behavior of ε_M outside the sublayer may be approximated using

$$\frac{\varepsilon_M}{\nu} = \frac{\kappa y^+}{6} \left[1 + \left(\frac{r}{R} \right) \right] \left[1 + 2 \left(\frac{r}{R} \right)^2 \right].$$ (13.4-13)

This expression, due originally to Reichardt (1951), is written here in the form given by Kays and Crawford (1993, p. 246). It implies that $\varepsilon_M/\nu \to \kappa y^+/6$ as $r/R \to 0$ and $\varepsilon_M/\nu \to \kappa y^+$ as $r/R \to 1$. As shown in Example 13.4-2, the latter form matches the limiting behavior of the mixing-length expressions for large y^+. This is reminiscent of the asymptotic matching used in boundary layer theory; indeed, y^+ and r/R act much like boundary layer and outer coordinates, respectively. An ingenious aspect of Eq. (13.4-13) is that it yields an analytical expression for the velocity outside the sublayer (see Problem 13-2).

We consider now a series of examples which illustrate the use of eddy diffusivity models in turbulent velocity and temperature calculations. All four examples involve macroscopically steady, fully developed flow in a cylindrical tube. Certain well-known results are derived and compared with experimental data.

Example 13.4-1 Velocity Near a Wall   This example focuses eventually on the region near the wall, but we begin by characterizing the shear stress throughout a cylindrical tube. From Eq. (13.3-22), conservation of momentum is expressed as

$$0 = -\frac{d\bar{\mathcal{P}}}{dz} + \frac{1}{r}\frac{d}{dr}[r(\bar{\tau}_{rz} + \tau^*_{rz})]. \tag{13.4-14}$$

Integrating and requiring that the shear stress be finite at $r = 0$, we obtain

$$\bar{\tau}_{rz} + \tau^*_{rz} = \frac{d\bar{\mathcal{P}}}{dz}\frac{r}{2}, \tag{13.4-15}$$

which is analogous to Eq. (6.2-15) for laminar flow. From Eq. (13.4-6), the relationship between the pressure gradient and the wall shear stress is

$$\frac{d\bar{\mathcal{P}}}{dz} = -\frac{2\tau_0}{R}. \tag{13.4-16}$$

Eliminating the pressure gradient in favor of τ_0, and changing from the usual radial coordinate to the wall coordinate using $y = R - r$ and $\tau_{yz} = -\tau_{rz}$, Eq. (13.4-15) becomes

$$\bar{\tau}_{yz} + \tau^*_{yz} = \tau_0\left(1 - \frac{y}{R}\right). \tag{13.4-17}$$

Using the constitutive equation for a Newtonian fluid and Eq. (13.4-1) for the turbulent stress, the total stress is given also by

$$\bar{\tau}_{yz} + \tau^*_{yz} = \mu\left(1 + \frac{\varepsilon_M}{\nu}\right)\frac{d\bar{v}_z}{dy}. \tag{13.4-18}$$

Equating the last two expressions, we obtain

$$\frac{d\bar{v}_z}{dy} = \frac{\tau_0}{\mu}\frac{[1 - (y/R)]}{[1 + (\varepsilon_M/\nu)]}, \tag{13.4-19}$$

which is a general result for fully developed turbulent flow in a pipe.

In the region near the wall, where $y/R \ll 1$, the total shear stress is essentially constant at τ_0. For this region we will use the dimensionless wall variables defined by Eq. (13.4-5), and we will evaluate the eddy diffusivity using the Van Driest expression. For convenience, Eq. (13.4-12) is rewritten as

$$\frac{\varepsilon_M}{\nu} = g^2(y^+)\left|\frac{dv^+_z}{dy^+}\right|, \tag{13.4-20}$$

$$g(y^+) = \kappa y^+(1 - e^{-y^+/A}), \tag{13.4-21}$$

where the function $g(y^+)$ is a dimensionless mixing length. Evaluating the eddy diffusivity in this manner and noting that $dv^+_z/dy^+ \geq 0$, Eq. (13.4-19) becomes

$$\frac{dv^+_z}{dy^+} = \frac{1}{1+(\varepsilon_M/\nu)} = \frac{1}{1+(dv^+_z/dy^+)g^2}. \tag{13.4-22}$$

The best-known results for the velocity near a wall are obtained as asymptotic solutions of Eq. (13.4-22). For $y^+ \to 0$, Eqs. (13.4-20) and (13.4-21) indicate that $\varepsilon_M/\nu \ll 1$. This region, where the *turbulent* stress is negligible, is the "laminar sublayer" mentioned earlier. Using the no-slip condition at the wall, the simple result is

$$v^+_z = y^+ \qquad (y^+ \to 0). \tag{13.4-23}$$

At large values of y^+, we enter a region where $\varepsilon_M/\nu \gg 1$ and the *viscous* stress is negligible. In this case Eq. (13.4-22) is rearranged to

$$\frac{dv^+_z}{dy^+} = \frac{1}{g}. \tag{13.4-24}$$

If y^+ is large enough that $g = \kappa y^+$, then this can be integrated to give

$$v^+_z = \frac{1}{\kappa} \ln y^+ + c \qquad (y^+ \to \infty). \tag{13.4-25}$$

The constant c cannot be determined from the asymptotic solution alone, because there is no applicable boundary condition. Nonetheless, the prediction of a region where there is a logarithmic dependence of the velocity on y^+, confirmed by numerous experimental studies, is a notable achievement of the mixing length theory. The value of the integration constant which best fits the velocity measurements is usually given as $c = 5.5$; as stated earlier, the best-fit value of the other constant is $\kappa = 0.40$. Accordingly, Eq. (13.4-25) becomes

$$v^+_z = 2.5 \ln y^+ + 5.5, \tag{13.4-26}$$

which is called the *Nikuradse equation*.

The asymptotic results derived for the sublayer and logarithmic regions depend only on the limiting behavior of ε_M/ν for small and large y^+, which is the same for both the Prandtl and Van Driest versions of the mixing-length theory. Thus, Eqs. (13.4-23) and (13.4-25) are obtained also using the original Prandtl model, Eq. (13.4-11). The advantage of the Van Driest expression is its ability to interpolate between the two asymptotic solutions. Solving Eq. (13.4-22) as a quadratic for the velocity gradient and integrating, we obtain

$$v^+_z = \int_0^{y^+} \frac{\sqrt{1+4g^2}-1}{2g^2} \, ds, \tag{13.4-27}$$

where s is a dummy variable and $g = g(s)$ in the integrand. Figure 13-6 shows results obtained by evaluating the integral numerically, with $\kappa = 0.40$ and $A = 26$. It is seen that the Van Driest approach provides a transition between the asymptotic results which is in excellent agreement with representative data. The sublayer result is accurate for $y^+ < 5$ and the logarithmic profile is reasonably good for $y^+ > 30$. The two asymptotic solutions intersect at $y^+ = 11.6$. There is a slight offset between the Van Driest and Nikuradse results in the logarithmic region; with the parameter values used, the results computed from Eq. (13.4-27) imply that $c = 5.2$ rather than 5.5.

A remarkable aspect of the results in Fig. 13-6 is their universal character. The Reynolds number does not appear as a parameter in this plot, suggesting that the effects of Re are captured

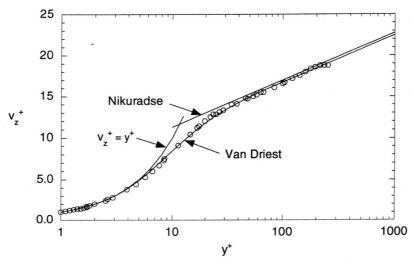

Figure 13-6. Velocity near a wall for turbulent flow. The curves labeled "Nikuradse" and "Van Driest" are based on Eqs. (13.4-26) and (13.4-27), respectively. The symbols represent measurements by Durst et al. (1993) using laser-Doppler velocimetry.

fully by the way in which the dimensionless wall variables are defined.[1] Moreover, essentially the same wall-region velocity profile is obtained in conduits with other shapes (e.g., parallel-plate channels) and in boundary layers on submerged objects [see, for example, Kim et al. (1987) and Kays and Crawford (1993, p. 211)]. The universality of this profile has led it to be termed *the law of the wall.* Whatever name is attached to it, this form of the time-smoothed velocity near a wall is so well-established that a requirement of any new experimental technique or theory for turbulent flow is consistency with this result.

Example 13.4-2 Complete Velocity Profile   This is an extension of Example 13.4-1 to include the entire tube cross section. Now the appropriate length and velocity scales are R and U, and the dimensionless variables are chosen as

$$\chi \equiv \frac{y}{R}, \qquad V(\chi) \equiv \frac{\bar{v}_z}{U}. \tag{13.4-28}$$

Equation (13.4-19) is rewritten as

$$\frac{dV}{d\chi} = \mathrm{Re}\,\frac{f}{4}\left[\frac{1-\chi}{1+(\varepsilon_M/\nu)}\right]. \tag{13.4-29}$$

Using the no-slip condition, this is integrated to obtain

$$V = \mathrm{Re}\,\frac{f}{4}\int_0^\chi \frac{1-s}{1+(\varepsilon_M/\nu)}\,ds. \tag{13.4-30}$$

The velocity is calculated by evaluating this integral using suitable choices for ε_M.

---

[1] The Reynolds number does enter indirectly, in that we assumed that $y/R \ll 1$ or (equivalently) $y^+/y_c^+ \ll 1$. Because Re determines, y_c^+, it sets the upper limit for the y^+ axis.

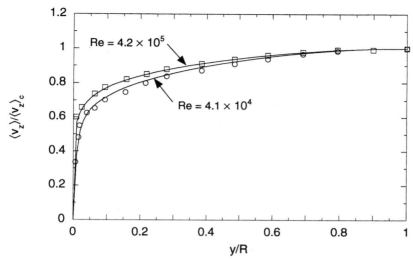

Figure 13-7. Velocity in a tube relative to the centerline value, at two Reynolds numbers. The curves were calculated numerically using a combination of the Van Driest and Reichardt eddy diffusivity models. The symbols represent the hot-wire anemometry data of Laufer (1954).

In Example 13.4-1 it was shown that the Van Driest expression for the eddy diffusivity gives accurate velocity results near the wall. In considering what to do for the core region, we note from Eq. (13.4-24) that the Van Driest model implies that $dv_z{}^+/dy^+ \to (\kappa y^+)^{-1}$ for large y^+. Thus, using Eq. (13.4-12), it is found that $\varepsilon_M/\nu \to \kappa y^+$ for large y^+; the Prandtl model has the same behavior. As mentioned in connection with Eq. (13.4-13), the eddy diffusivity expression of Reichardt yields $\varepsilon_M/\nu \to \kappa y^+$ for $r/R \to 1$ (or $\chi \to 0$). Thus, the Van Driest and Reichardt expressions are equivalent just outside the sublayer, and they may be used in a complementary manner. The integral in Eq. (13.4-30) was evaluated numerically in two parts, switching from the Van Driest to the Reichardt expression where the eddy diffusivities became equal; this occurred at values of y^+ on the order of 100. As shown in Fig. 13-7, the calculated results are in good agreement with data reported by Laufer (1954) at two Reynolds numbers. Normalizing the velocity with the centerline value, as done here, shows that the velocity profile becomes more blunt as Re is increased.

Figure 13-8 compares velocity profiles computed at $Re = 1 \times 10^5$ using four methods. The curve labeled "Van Driest and Reichardt," which should be the most accurate, was calculated numerically as just described. The other curves are based on: the Nikuradse equation [Eq. (13.4-26)]; an analytical expression derived from the Reichardt eddy diffusivity (Problem 13-2); and a purely empirical power-law fit given by

$$\frac{\langle v_z \rangle}{U} = \frac{60}{49}\left(\frac{y}{R}\right)^{1/7} = 1.22\left(\frac{y}{R}\right)^{1/7},\qquad(13.4\text{-}31)$$

which applies only for Re between about 10^4 and 10^5. All four methods yield very similar velocity profiles, as shown. Perhaps most remarkable is the accuracy of the Nikuradse equation. Although restricted in principle to the wall region, it provides an excellent approximation to the entire profile. The accuracy of Eq. (13.4-26) over most of the cross section is confirmed by comparisons with the extensive data of Nikuradse (Nikuradse, 1932; Goldstein, 1938, pp. 334–335). Recall that the logarithmic velocity profile was derived by assuming that the stress is approximately constant

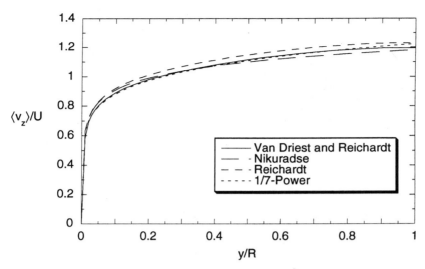

Figure 13-8. Turbulent velocity profiles in a tube at $Re = 1 \times 10^5$, calculated using four methods. The curve labeled "Van Driest and Reichardt" is a numerical result based on the eddy diffusivities given in Eqs. (13.4-12) and (13.4-13); "Nikuradse" is from Eq. (13.4-26); "Reichardt" is from an analytical expression based on Eq. (13.4-13), as given in Problem 13-2; and "$\frac{1}{7}$-Power" is from Eq. (13.4-31).

and that $\varepsilon_M/\nu = \kappa y^+$; both conditions are violated far from the wall, so that there is evidently a cancellation of errors. Because it is simple and, unlike the power-law expression, accounts for the effects of Re, the Nikuradse equation is often preferred for velocity calculations in pipes.

Example 13.4-3 Temperature Profiles   We now consider turbulent pipe flow in which the inlet temperature is T_0 and the heat flux at the wall is constant at q_w for $z>0$. The objective is to calculate the radial temperature variations for large z, in the thermally fully developed region. It is assumed that viscous dissipation and axial "conduction" (i.e., $\bar{q}_z + q_z^*$) are negligible. There are a number of parallels with the laminar case analyzed in Example 9.5-2, as will be seen.

The problem statement in dimensional form is

$$\bar{v}_z \frac{\partial \bar{T}}{\partial z} = \frac{1}{r} \frac{\partial}{\partial r}\left[r(\alpha + \varepsilon_H) \frac{\partial \bar{T}}{\partial r}\right], \qquad (13.4\text{-}32)$$

$$\bar{T}(r, 0) = T_0, \qquad \frac{\partial \bar{T}}{\partial r}(0, z) = 0, \qquad \frac{\partial \bar{T}}{\partial r}(R, z) = -\frac{q_w}{k}. \qquad (13.4\text{-}33)$$

As in Example 9.5-2, the dimensionless coordinates, temperature, and Péclet number are defined as

$$\eta \equiv \frac{r}{R}, \qquad \zeta \equiv \frac{z}{R\,\mathrm{Pe}}, \qquad \Theta \equiv \frac{\bar{T} - T_0}{q_w R/k}, \qquad \mathrm{Pe} \equiv \frac{2UR}{\alpha}. \qquad (13.4\text{-}34)$$

Defining the velocity V as in Eq. (13.4-28), the dimensionless problem statement is

$$\frac{V(\eta)}{2} \frac{\partial \Theta}{\partial \zeta} = \frac{1}{\eta} \frac{\partial}{\partial \eta}\left[\eta\left(1 + \frac{\varepsilon_H}{\alpha}\right) \frac{\partial \Theta}{\partial \eta}\right], \qquad (13.4\text{-}35)$$

$$\Theta(\eta, 0) = 0, \qquad \frac{\partial \Theta}{\partial \eta}(0, \zeta) = 0, \qquad \frac{\partial \Theta}{\partial \eta}(1, \zeta) = -1. \tag{13.4-36}$$

For laminar flow, where $V(\eta) = 2(1 - \eta^2)$ and $\varepsilon_H = 0$, Eq. (13.4-35) reduces to Eq. (9.4-6). Equations (13.4-35) and (13.4-36) constitute a Sturm–Liouville problem, which could be solved in principle using the finite Fourier transform method. However, the complicated forms of the functions $V(\eta)$ and $\varepsilon_H(\eta)$ require that the eigenfunctions be computed numerically; an extensive set of such results is given in Notter and Sleicher (1972).

In the present analysis the eigenvalue problem is avoided by writing the temperature as the sum of three functions, one of which vanishes for large ζ. As in Example 9.5-2, the asymptotic form of the temperature is expressed as

$$\lim_{\zeta \to \infty} \Theta(\eta, \zeta) = \psi(\eta) + \phi(\zeta). \tag{13.4-37}$$

The overall energy balance from the laminar flow example, Eq. (9.5-35), also still applies. From the energy balance and Eq. (13.4-37), we conclude that

$$\lim_{\zeta \to \infty} \frac{\partial \Theta}{\partial \zeta} = \frac{d\phi}{d\zeta} = \frac{d\Theta_b}{d\zeta} = -4. \tag{13.4-38}$$

Thus, the axial temperature gradient is a known constant, independent of radial position.

Using Eqs. (13.4-35)–(13.4-38), the governing equations for the radially varying part of the temperature are found to be

$$\frac{1}{\eta} \frac{d}{d\eta} \left[\eta \left(1 + \frac{\varepsilon_H}{\alpha} \right) \frac{d\psi}{d\eta} \right] = -2V(\eta), \tag{13.4-39}$$

$$\frac{d\psi}{d\eta}(0) = 0, \qquad \frac{d\psi}{d\eta}(1) = -1. \tag{13.4-40}$$

Changing to the wall coordinate used in Example 13.4-2, $\chi = 1 - \eta$, the problem for $\psi(\chi)$ is

$$\frac{1}{1 - \chi} \frac{d}{d\chi} \left[(1 - \chi) \left(1 + \frac{\varepsilon_H}{\alpha} \right) \frac{d\psi}{d\chi} \right] = -2V(\chi), \tag{13.4-41}$$

$$\frac{d\psi}{d\tau}(0) = 1, \qquad \frac{d\psi}{d\chi}(1) = 0. \tag{13.4-42}$$

Integrating Eq. (13.4-41) once gives

$$\frac{d\psi}{d\chi} = \frac{1}{(1 - \chi)[1 + (\varepsilon_H/\alpha)]} \left[1 - 2 \int_0^\chi V(s)(1 - s) \, ds \right], \tag{13.4-43}$$

which satisfies the boundary condition at the wall, $\chi = 0$. To simplify the analysis, the velocity integral is evaluated next by assuming plug flow. This has the effect of underestimating the temperature gradient, but the errors are small (see below). Setting $V = 1$, we find that

$$1 - 2 \int_0^\chi V(s)(1 - s) \, ds = (1 - \chi)^2. \tag{13.4-44}$$

The expression for the temperature gradient reduces now to

$$\frac{d\psi}{d\chi} = \frac{(1-\chi)}{1+(\varepsilon_H/\alpha)},$$
(13.4-45)

which is analogous to Eq. (13.4-29) for the velocity gradient.
    The eddy diffusivity term in Eq. (13.4-45) is rewritten as

$$\frac{\varepsilon_H}{\alpha} = \left(\frac{\varepsilon_H}{\varepsilon_M}\right)\left(\frac{\nu}{\alpha}\right)\left(\frac{\varepsilon_M}{\nu}\right) = \frac{\mathrm{Pr}}{\mathrm{Pr}_t}\frac{\varepsilon_M}{\nu}, \qquad \mathrm{Pr}_t \equiv \frac{\varepsilon_M}{\varepsilon_H},$$
(13.4-46)

where Pr_t is called the *turbulent Prandtl number*. Although the symbol and name for this ratio are based on the analogy with $\mathrm{Pr} = \nu/\alpha$, it is important to remember that, like the eddy diffusivities, Pr_t is a property of the flow. The value of the turbulent Prandtl number is not necessarily independent of position in a given flow, but a variety of experimental and theoretical findings reviewed by Kays and Crawford (1993, pp. 259–266) suggest that $\mathrm{Pr}_t \cong 0.85$. This supports the basic premise of the eddy diffusion concept that the diffusivities for all quantities should be very similar, and it indicates that Reynolds' analogy is approximately correct. Integrating Eq. (13.4-45), we obtain finally

$$\psi - \psi_w = \int_0^{\chi} \frac{1-s}{1+(\mathrm{Pr}/\mathrm{Pr}_t)(\varepsilon_M/\nu)}\, ds = \Theta - \Theta_w,$$
(13.4-47)

where $\psi_w = \psi(0)$ is the value at the tube wall. As indicated, the radial variations in ψ are the same as those in the overall temperature, Θ.
    The temperature differences given by Eq. (13.4-47) were evaluated numerically, using a combination of the Van Driest and Reichardt expressions for ε_M/ν and with $\mathrm{Pr}_t = 1$. Temperature profiles computed for $\mathrm{Re} = 1 \times 10^4$ and Pr ranging from 0.5 to 5 are given in Fig. 13-9. Also shown is the result for laminar flow, from Eq. (9.5-37). In this plot, which emphasizes the region near the wall, the errors in $\Theta - \Theta_w$ caused by the plug-flow approximation for turbulence are $\leq 3\%$; at the tube centerline, the errors for these moderate values of Pr are $\leq 6\%$. It is seen that,

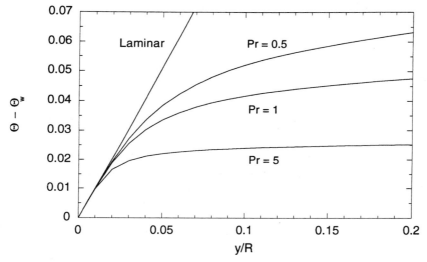

Figure 13-9. Temperature profiles near the wall in turbulent and laminar pipe flow, for a constant heat flux at the wall. The turbulent curves for $0.5 \leq \mathrm{Pr} \leq 5$ were calculated using $\mathrm{Re} = 1 \times 10^4$ and $\mathrm{Pr}_t = 1$.

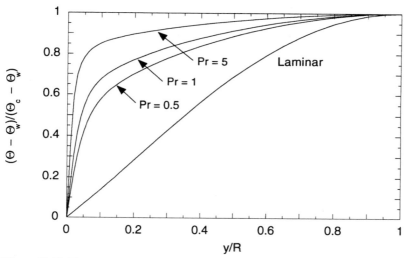

Figure 13-10. Temperature profiles in turbulent and laminar pipe flow, for a constant heat flux at the wall. The temperatures are normalized by their centerline values. The turbulent curves for $0.5 \leq \mathrm{Pr} \leq 5$ were calculated using $\mathrm{Re} = 1 \times 10^4$ and $\mathrm{Pr}_t = 1$.

unlike the laminar case, the temperature profiles in turbulent flow are sensitive to Pr; the temperature increase is smallest at large Pr. Also, as Pr is increased, the temperature changes occur over a progressively narrower region. This is shown more clearly in Fig. 13-10, in which the temperatures are normalized by their centerline values. In this plot the relative slopes of the curves at $y=0$ indicate the relative values of the Nusselt number. Whereas Nu in laminar tube flow is independent of Pr in the fully developed region, in turbulent flow it increases with Pr.

    The effects of Pr on the temperature profile and Nu in turbulent pipe flow are qualitatively similar to the effects of Pr in laminar boundary layers. As discussed in Section 10.3, Pr determines the relative thicknesses of the thermal and momentum boundary layers. In turbulent pipe flow, the region near the wall (say, $0<y^+<30$) is analogous to a momentum boundary layer. For $\mathrm{Pr} \to 0$, almost all of the radial temperature variation is outside the wall region, and $\mathrm{Nu} \sim 1$ as for laminar tube flow. For $\mathrm{Pr} \to \infty$, however, the entire temperature change occurs within the laminar sublayer (i.e., for $y^+<5$, where $v_z^+ \cong y^+$), and Nu is extremely large. Thus, as already noted, Nu increases with Pr. Because the thickness of the sublayer or wall region varies inversely with Re, Nu increases also with Re. The functional form of Nu in turbulent pipe flow is examined in more detail in the next example.

Example 13.4-4 Nu in Turbulent Pipe Flow for Pr $\sim$ 1 Certain results from the preceding examples are used now to predict Nu for turbulent pipe flow when Pr is close to unity (i.e., for gases). It is assumed that the velocity and temperature fields are fully developed and that the heat flux at the wall is constant. The approach is based on the fact that the turbulent velocity and temperature profiles have very similar shapes for Pr $\sim$ 1. Although several approximations are needed to obtain a simple analytical result, the final expression is remarkably accurate.

    We begin by recalling that the dimensionless velocity and temperature gradients in the direction normal to the wall are given by Eqs. (13.4-29) and (13.4-45), respectively; the latter was derived by assuming plug flow in the energy equation. Dividing the temperature gradient by the velocity gradient yields

$$\frac{d\psi/d\chi}{dV/d\chi} = \frac{d\psi}{dV} = \frac{4}{f \, \text{Re Pr}} \left(\frac{\nu + \varepsilon_M}{\alpha + \varepsilon_H} \right). \tag{13.4-48}$$

A key observation is that the ratio of the total diffusivities is approximately constant, such that

$$\left(\frac{\nu + \varepsilon_M}{\alpha + \varepsilon_H} \right) \cong 1. \tag{13.4-49}$$

For $\text{Pr} \sim 1$ most of the temperature change occurs outside the sublayer, where the eddy diffusivities are dominant. Accordingly, Eq. (13.4-49) depends mainly on the assertion that $\text{Pr}_t = \varepsilon_M/\varepsilon_H \cong 1$. With the diffusivity ratio a constant, Eq. (13.4-48) is readily integrated from the wall to the tube center. The result is that the temperature and velocity are related by

$$\psi_c - \psi_w \cong \frac{4V_c}{f \, \text{Re Pr}} \cong \Theta_c - \Theta_w, \tag{13.4-50}$$

where the subscripts c and w indicate quantities evaluated at the center and wall, respectively. The second equality is based on the fact that the differences in ψ over the tube cross section are the same as differences in the overall temperature, Θ.

The final assumption, which is based on the similarity in the shapes of the velocity and temperature profiles, is that

$$V_c \cong \frac{\Theta_c - \Theta_w}{\Theta_b - \Theta_w}, \tag{13.4-51}$$

where Θ_b is the bulk value of Θ (i.e., the velocity-weighted average). In words, it is assumed that the centerline velocity is to the mean velocity what the centerline temperature is to the bulk temperature. This assumption is somewhat less extreme than the plug-flow approximation. Combining Eqs. (13.4-50) and (13.4-51), the difference between the bulk and wall temperatures is found to be

$$\Theta_b - \Theta_w \cong \frac{4}{f \, \text{Re Pr}}. \tag{13.4-52}$$

Evaluating the Nusselt number as in the laminar flow analysis [see Eq. (9.5-38)], we conclude that

$$\text{Nu} = \frac{2}{\Theta_b - \Theta_w} \cong \frac{f}{2} \, \text{Re Pr}. \tag{13.4-53}$$

This expression clearly illustrates the strong dependence of Nu on Re. For turbulent flow with $\text{Re} \le 1 \times 10^5$, where the Blasius equation gives $f \propto \text{Re}^{-1/4}$, it implies that $\text{Nu} \propto \text{Re}^{3/4}$. This dependence on Re is essentially the same as that in the Colburn equation, Eq. (13.2-7). Indeed, Eq. (13.4-53) gives results which are within 6% of the Colburn equation for $\text{Pr} = 1$ and $1 \times 10^4 \le \text{Re} \le 1 \times 10^6$.

Many correlations have been published for Nu in turbulent pipe flow, as reviewed in Bhatti and Shah (1987). Most have some theoretical basis, but all contain numerical constants which have been fitted to experimental data. Several are considerably more accurate than the Colburn equation, especially for large Pr. In turbulent flow the thermal boundary condition usually has little effect on Nu; for $\text{Pr} \ge 0.7$ and $\text{Re} \ge 10^4$, the constant-flux and constant-temperature values differ by $\le 4\%$ (Petukhov, 1970), as compared with about 20% in laminar flow (Chapter 9). Given the scatter in the available data for Nu in turbulent flow, most correlations are viewed as being applicable to any boundary condition. Bhatti and Shah recommend an expression proposed by Gnielinski (1976),

$$Nu = \frac{(f/2)(Re - 1000)Pr}{1 + 12.7(f/2)^{1/2}(Pr^{2/3} - 1)}.$$ (13.4-54)

This correlation, which applies to constant-property fluids with $0.5 \leq Pr \leq 2000$ and $2300 \leq Re \leq 5 \times 10^6$, is accurate to within about $\pm 10\%$. Note that it is identical to Eq. (13.4-53) for $Re \gg 1000$ and $Pr = 1$. Correlations which account for temperature-dependent physical properties are discussed in Petukhov (1970), Sleicher and Rouse (1975), and Gnielinski (1976).

13.5 OTHER APPROACHES FOR TURBULENT FLOW CALCULATIONS

Need for Improved Models

Any mathematical representation of a real system, however exact or approximate, is properly termed a "model." In the turbulence literature, though, a "model" is usually something more specific, namely, a set of approximations invoked on physical grounds to overcome the closure problem or otherwise make calculations feasible. In this section we usually have in mind the more restricted usage of the term.

The eddy diffusion concept leads to the simplest type of model which provides useful results for turbulent flow. The closure problem is overcome by assuming that the Reynolds stress is related to the smoothed rate of strain in the same way that the viscous stress is related to the instantaneous rate of strain; only the coefficients differ. In the models described in Section 13.4, the function $\varepsilon_M(y)$ is specified. Such models are very successful in some respects (e.g., the law of the wall), but they have serious limitations. One problem is that the values of κ or other parameters in $\varepsilon_M(y)$ must be determined by fitting velocity measurements. Thus, prior experimental information is needed for even the simplest flow calculations. This is different than having to measure a material property such as the Newtonian viscosity. Once determined, the same value of μ applies to any flow, whereas the eddy diffusivities are flow-specific (except near a wall). Thus, however useful they may be as interpolation tools, the models of Section 13.4 are not truly predictive. Another problem is that the eddy diffusivity approach in its basic form fails to describe key features of a number of flows. Examples include boundary layers with adverse pressure gradients, "nonequilibrium" boundary layers which result from sudden changes in the pressure gradient along a surface (Kays and Crawford, 1993, pp. 215–216), and boundary layers with accelerations so severe that the flow again becomes partially laminar (Jones and Launder, 1972). A specific aspect of mixing-length models, in which all eddy diffusivities are proportional to the local velocity gradient, is their prediction that the turbulent energy and species fluxes will vanish in homogeneous turbulence (uniform mean velocity) or where there is a symmetry line or plane for the velocity (i.e., the center of a tube); this is clearly incorrect. It is evident that more powerful approaches are needed for engineering calculations, especially where complex geometries are involved.

Mixing-length and other models were fine for Prandtl's time, one might argue, but with modern supercomputers, why use models? In other words, why not just compute turbulent flows by numerical solution of the (unsmoothed) continuity and Navier–Stokes equations? In fact, this is a promising area of research; the foundations of purely computational approaches are reviewed in Rogallo and Moin (1984). Where practical, such simulations are like extremely well-instrumented experiments, in that they endeavor to provide the full details of the turbulent fluctuations. To cite one success, direct numerical simulations of fully developed flow in a parallel-plate channel, using a spectral method, have been able to accurately reproduce the law of the wall and certain other experimental findings (Kim et al., 1987). Some examples of more recent work are simulations of boundary layers (Spalart and Watmuff, 1993) and channel flow of polymer solutions (Sureshkumar and Beris, 1997). Direct simulations have been limited, however, to moderate Reynolds numbers ($\sim$3000 in channel flow) and simple geometries. The upper limit on Re is due to the number of grid points needed to fully resolve the Kolmogorov scales; as discussed in Section 13.2, the structure of turbulence becomes progressively finer-grained as Re increases. This problem will diminish as computer power increases, but will not disappear in the forseeable future. In weather forecasting and air pollution modeling, for instance, it is desired to compute turbulent flows in large parts of the atmosphere, for which there are enormous differences between the turbulence macroscales and microscales. Concerning flow geometry, even cylindrical tubes are difficult at present. These limitations have led many researchers to pursue *large eddy simulations,* in which most of the computational effort is focused on the large-scale flow that is peculiar to a given problem, and models are used to describe the smaller scales, where the dynamics are more universal. Thus, the majority of numerical simulations lie somewhere between the classical mixing-length approach and the ideal of "model-free" computations.

Types of Models

Accepting that neither mixing-length models nor direct numerical simulations are able to address many engineering needs, we briefly survey other approaches. All of these are *statistical* models, which are based on the view that turbulent fluctuations are essentially random and (at least practically) not predictable in detail. The governing equations are formulated using the Reynolds averaging discussed in Section 13.3 or other averaging procedures. Numerous statistical models for turbulence are discussed in the review by Speziale (1991) and in the texts by Hinze (1975), McComb (1990), and Tennekes and Lumley (1972). Only a few approaches are highlighted here.

A key quantity in the time-smoothed Navier–Stokes equation is the Reynolds stress, which is proportional to $\langle \mathbf{uu} \rangle$. Time-smoothed products such as $\langle \mathbf{uu} \rangle$ are called *correlations* or *moments,* and statistical models for turbulence are classified partly according to the types of correlations used. The correlations, in turn, are categorized according to the number of quantities involved and their extent of localization in space and time. The tensor $\langle \mathbf{uu} \rangle$ is a *second-order* correlation. As used here, both quantities (e.g., both $\mathbf{u}$'s) are evaluated at the same time and position. It has been argued that turbulence should be described using more general types of correlations, in which the individual quantities are evaluated at two or more points in space and/or instants in time. Although such models have been developed (e.g., two-point models), they tend to be difficult to use and have not penetrated into engineering practice. Thus, we are concerned here entirely with single-point or local models.

All statistical formulations for turbulence encounter some form of the closure problem, which was introduced in Section 13.3. That is, the number of unknowns in statistical formulations always exceeds the number of equations which are "fundamental" or "exact" (e.g., the continuity or Navier–Stokes equations). Consequently, the set of exact equations must be augmented with approximate relationships based on models. What distinguishes the various (single-point) formulations is how early or late such approximations are introduced. In what we will call *first-order* methods, the smoothed continuity and Navier–Stokes equations are solved for $\langle \mathbf{v} \rangle$ and $\langle \mathcal{P} \rangle$, and a model is invoked for the Reynolds stress, or $\langle \mathbf{uu} \rangle$. The most familiar examples are those in Section 13.4, in which the model for $\langle \mathbf{uu} \rangle$ consists mainly of the algebraic equation specified for ε_M. The eddy diffusivity idea is preserved in certain other first-order methods, but ε_M is evaluated by solving one or more additional differential equations. First-order methods are commonly labeled as *zero-equation*, *one-equation*, or *two-equation*, according to the number of differential equations which must be solved to compute ε_M. In this terminology the Prandtl mixing-length approach is a first-order, zero-equation model. What is perhaps the most widely used approach for engineering calculations in turbulent flow is a two-equation method called the *K–ε model*. In *second-order* methods, a fundamental differential equation is solved also for $\langle \mathbf{uu} \rangle$; the third-order correlation $\langle \mathbf{uuu} \rangle$, which appears in that equation, is modeled. Second-order methods appear to be advantageous for certain flows (Speziale, 1991), but they are not yet widely used.

The remainder of this chapter is devoted largely to discussing certain differential equations which are common to several of the methods mentioned above. We begin by deriving the differential equation for the Reynolds stress, expressed in component form as $\langle u_i u_j \rangle$. This equation, which is central to second-order methods, illustrates the persistence of the closure problem. It also leads to the differential equation for the turbulent kinetic energy, which is at the heart of most first-order, one- and two-equation methods. A one-equation approach based on the kinetic energy equation is described briefly, and finally, the *K–ε* model is discussed.

Reynolds Stress Equation

In Section 13.3 it was shown that the velocity fluctuation satisfies continuity in the form of Eq. (13.3-12), but no attempt was made to derive a differential equation for $\mathbf{u}$ from the Navier–Stokes equation. To obtain such a relationship we expand each term of the original (instantaneous) Navier–Stokes equation [see, for example, Eq. (13.3-14)]. Subtracting from that result the time-smoothed Navier–Stokes equation gives

$$\rho \left[\frac{\partial \mathbf{u}}{\partial t} + \bar{\mathbf{v}} \cdot \nabla \mathbf{u} + \mathbf{u} \cdot \nabla \bar{\mathbf{v}} + \nabla \cdot (\mathbf{uu} - \langle \mathbf{uu} \rangle) \right] = -\nabla p + \mu \nabla^2 \mathbf{u}, \tag{13.5-1}$$

where $\bar{\mathbf{v}} = \langle \mathbf{v} \rangle$. This and continuity are the governing equations for the instantaneous fluctuations, $\mathbf{u}$ and p. These equations are at least as difficult to solve as the original problem for $\mathbf{v}$ and $\mathcal{P}$, so that little has been gained at this stage. If Eqs. (13.3-12) and (13.5-1) are smoothed, then all terms vanish. Once again, little is learned. Important information is obtained, however, if Eq. (13.5-1) is multiplied by $\mathbf{u}$ and then smoothed. Performing the multiplication in different ways leads to differential equations for both the second-order correlation and the turbulent kinetic energy, as will be shown.

Before multiplying it by the velocity fluctuation, it is helpful to divide Eq. (13.5-1) by ρ and to write the result in component form as

$$\frac{\partial u_i}{\partial t} + \bar{\mathbf{v}} \cdot \nabla u_i + \mathbf{u} \cdot \nabla \bar{v}_i + \nabla \cdot (\mathbf{u} u_i - \langle \mathbf{u} u_i \rangle) = -\frac{1}{\rho} \frac{\partial p}{\partial x_i} + \nu \nabla^2 u_i. \tag{13.5-2}$$

The gradient and divergence terms are expanded as

$$\bar{\mathbf{v}} \cdot \nabla u_i = \sum_k \bar{v}_k \frac{\partial u_i}{\partial x_k}, \qquad \mathbf{u} \cdot \nabla \bar{v}_i = \sum_k u_k \frac{\partial \bar{v}_i}{\partial x_k}, \tag{13.5-3}$$

$$\nabla \cdot (\mathbf{u}\, u_i - \langle \mathbf{u}\, u_i \rangle) = \sum_k \frac{\partial}{\partial x_k} (u_k u_i - \langle u_k u_i \rangle). \tag{13.5-4}$$

Using these expressions in Eq. (13.5-2) and multiplying the result by u_j gives

$$u_j \frac{\partial u_i}{\partial t} + u_j \sum_k \bar{v}_k \frac{\partial u_i}{\partial x_k} + u_j \sum_k u_k \frac{\partial \bar{v}_i}{\partial x_k} + u_j \sum_k \frac{\partial}{\partial x_k} (u_k u_i - \langle u_k u_i \rangle)$$

$$= -\frac{u_j}{\rho} \frac{\partial p}{\partial x_i} + \nu u_j \nabla^2 u_i. \tag{13.5-5}$$

A complementary equation is obtained by interchanging the subscripts i and j. The two equations are then added term-by-term, the sums are rearranged by repeated use of the rule for differentiating a product, and the result is time-smoothed. These operations lead to

$$\frac{\partial}{\partial t} \langle u_i u_j \rangle + \sum_k \bar{v}_k \frac{\partial}{\partial x_k} \langle u_i u_j \rangle + \sum_k \left(\langle u_j u_k \rangle \frac{\partial \bar{v}_i}{\partial x_k} + \langle u_i u_k \rangle \frac{\partial \bar{v}_j}{\partial x_k} \right) + \sum_k \frac{\partial}{\partial x_k} \langle u_i u_j u_k \rangle$$

$$= -\frac{1}{\rho} \left(\left\langle u_j \frac{\partial p}{\partial x_i} \right\rangle + \left\langle u_i \frac{\partial p}{\partial x_j} \right\rangle \right) + \nu \left(\langle u_j \nabla^2 u_i \rangle + \langle u_i \nabla^2 u_j \rangle \right). \tag{13.5-6}$$

The viscous terms are now expanded and rearranged as

$$\langle u_j \nabla^2 u_i \rangle + \langle u_i \nabla^2 u_j \rangle = \sum_k \frac{\partial^2}{\partial x_k^2} \langle u_i u_j \rangle - 2 \sum_k \left\langle \frac{\partial u_i}{\partial x_k} \frac{\partial u_j}{\partial x_k} \right\rangle. \tag{13.5-7}$$

Using this in Eq. (13.5-6), the final differential equation for $\langle u_i u_j \rangle$ is

$$\frac{\partial}{\partial t} \langle u_i u_j \rangle + \sum_k \bar{v}_k \frac{\partial}{\partial x_k} \langle u_i u_j \rangle$$

$$= -\sum_k \left(\langle u_j u_k \rangle \frac{\partial \bar{v}_i}{\partial x_k} + \langle u_i u_k \rangle \frac{\partial \bar{v}_j}{\partial x_k} \right) + \sum_k \frac{\partial}{\partial x_k} \langle u_i u_j u_k \rangle$$

$$- \frac{1}{\rho} \left(\left\langle u_j \frac{\partial p}{\partial x_i} \right\rangle + \left\langle u_i \frac{\partial p}{\partial x_j} \right\rangle \right) + \nu \sum_k \frac{\partial^2}{\partial x_k^2} \langle u_i u_j \rangle - 2\nu \sum_k \left\langle \frac{\partial u_i}{\partial x_k} \frac{\partial u_j}{\partial x_k} \right\rangle, \tag{13.5-8}$$

which is equivalent to Eq. (1.20) of McComb (1990, p. 10). Among other complications, note that $\langle u_i u_j \rangle$ depends on $\langle u_i u_j u_k \rangle$. This indicates that Eq. (13.5-8) cannot be solved for the second moment (the Reynolds stress) without information on the third moment. Thus, the closure problem persists. In second-order methods, the additional relationships needed for closure are provided by models [see Speziale (1991) for information on those approaches].

Turbulent Kinetic Energy

A key part of most one- and two-equation models is a differential equation involving kinetic energy. The *turbulent kinetic energy* (per unit mass) is defined as

$$K \equiv \frac{1}{2}\langle u^2 \rangle = \frac{1}{2}\sum_i \langle u_i^2 \rangle. \tag{13.5-9}$$

Recall that in Chapter 9 an expression involving the overall kinetic energy (i.e., $v^2/2$) was obtained by forming the dot product of $\mathbf{v}$ with the Cauchy momentum equation [see Eq. (9.6-10), the "mechanical energy equation"]. The desired equation for K is found in similar fashion, by forming the dot product of $\mathbf{u}$ with its own "momentum equation," Eq. (13.5-1). This dot product is evaluated using Eq. (13.5-8), by setting $j = i$ and summing over i. After rearrangement, along with conversion back to Gibbs notation, the result is

$$\frac{DK}{Dt} = -\nabla \cdot \mathbf{j_K} - \langle \mathbf{uu} \rangle : \nabla \bar{\mathbf{v}} - \varepsilon, \tag{13.5-10}$$

$$\mathbf{j_K} \equiv -\nu \nabla K + \left\langle \frac{u^2}{2}\mathbf{u} \right\rangle + \left\langle \frac{p}{\rho}\mathbf{u} \right\rangle, \tag{13.5-11}$$

$$\varepsilon \equiv \nu \langle (\nabla \mathbf{u}) : (\nabla \mathbf{u})^t \rangle = \nu \sum_i \sum_j \langle (\partial u_i / \partial x_j)^2 \rangle. \tag{13.5-12}$$

Equation (13.5-10) is a conservation equation for turbulent kinetic energy. In this and subsequent relations, the material derivative is interpreted as $D/Dt = (\partial/\partial t) + \bar{\mathbf{v}} \cdot \nabla$. The left-hand side represents accumulation and convection, whereas $\mathbf{j_K}$ is the flux of turbulent kinetic energy relative to $\bar{\mathbf{v}}$. This flux, defined by Eq. (13.5-11), was identified by seeing which terms on the right-hand side of Eq. (13.5-10) could be written as a divergence. (The same approach was used in Section 11.8 to identify the entropy flux.) Of the remaining terms, one is a source of kinetic energy and the other a sink. The source in Eq. (13.5-10) is the term involving the second-order correlation. Often called the "production term," it is interpreted as the rate at which turbulent kinetic energy is derived from the main flow (Hinze, 1975; McComb, 1990). The sink is ε, which is the *turbulent dissipation rate*. Mentioned in a more qualitative context in Section 13.2, turbulent dissipation is defined now more precisely by Eq. (13.5-12).

The equation for the turbulent kinetic energy is applied often to boundary layers in which the flow is macroscopically steady and two-dimensional. To simplify Eq. (13.5-10) for boundary layers, let x and y be coordinates parallel to and normal to the surface, respectively. Assuming that the dominant contributions to $\nabla \cdot \mathbf{j_K}$ and the production term involve the y derivatives, we obtain

$$\bar{v}_x \frac{\partial K}{\partial x} + \bar{v}_y \frac{\partial K}{\partial y} = \frac{\partial}{\partial y}\left(\nu \frac{\partial K}{\partial y} - \left\langle \frac{u^2}{2}u_y \right\rangle - \left\langle \frac{p}{\rho}u_y \right\rangle \right) - \langle u_x u_y \rangle \frac{\partial \bar{v}_x}{\partial y} - \varepsilon. \tag{13.5-13}$$

This is as far as we can go without making closure approximations. The quantities which need to be modeled are ε, $\langle u_x u_y \rangle$, and the flux terms containing u^2 and p. The u^2 and p terms are evidently turbulent contributions to the "diffusive" flux of kinetic energy. By analogy with what was done with the turbulent fluxes in Section 13.4, it is customary to assume that

$$\left\langle \frac{u^2}{2}u_y\right\rangle + \left\langle \frac{p}{\rho}u_y\right\rangle = \left\langle \left(\frac{u^2}{2}+\frac{p}{\rho}\right)u_y\right\rangle = -\varepsilon_K\frac{\partial K}{\partial y},\tag{13.5-14}$$

where ε_K is the *eddy diffusivity for kinetic energy*. To the extent that the eddy diffusion concept is correct, ε_K should be very similar in magnitude to the other eddy diffusivities. As in Section 13.4, the Reynolds stress is evaluated using

$$\left\langle u_x u_y\right\rangle = \left\langle u_y u_x\right\rangle = -\varepsilon_M\frac{\partial \bar{v}_x}{\partial y}.\tag{13.5-15}$$

The modeling of ε is discussed later. With these approximations, Eq. (13.5-13) becomes

$$\bar{v}_x\frac{\partial K}{\partial x}+\bar{v}_y\frac{\partial K}{\partial y}=\frac{\partial}{\partial y}\left((\nu+\varepsilon_K)\frac{\partial K}{\partial y}\right)+\varepsilon_M\left(\frac{\partial \bar{v}_x}{\partial y}\right)^2-\varepsilon.\tag{13.5-16}$$

It is now much clearer that what we have been calling the production term does, indeed, act as a source for turbulent kinetic energy (i.e., the term containing ε_M is positive). Equation (13.5-16) and variations of it are widely used in calculations involving turbulent boundary layers (Jones and Launder, 1972; Kays and Crawford, 1993, p. 217; Launder and Spalding, 1974; Speziale, 1991).

One-Equation Model

Equation (13.5-16) can be used to create a one-equation model for turbulent boundary layers, as described by Kays and Crawford (1993, pp. 216–220). In this approach it is assumed that ε_M is a function of K and a turbulence length scale, ℓ_t. Although ℓ_t is like a mixing length, involving K instead of the time-smoothed velocity gradient is a definite departure from the Prandtl and Van Driest models discussed in Section 13.4. This approach can be rationalized by postulating that eddy-related transport depends on the turbulence intensity and by observing that K is more closely related to the turbulence intensity than is $\partial \bar{v}_x/\partial y$. On dimensional grounds, ε_M must then be of the form

$$\varepsilon_M = aK^{1/2}\ell_t,\tag{13.5-17}$$

where a is a constant. Applying similar reasoning to the dissipation rate, it is inferred that

$$\varepsilon = \frac{bK^{3/2}}{\ell_t},\tag{13.5-18}$$

where b is another constant. Finally, it is assumed that $\varepsilon_M/\varepsilon_K=1$; this ratio could also be treated as an empirical constant. Calculations using this model are performed by solving the smoothed continuity and momentum equations simultaneously with Eq. (13.5-16). The empirical information that is needed consists of a, b, and $\ell_t(y)$.

As explained by Kays and Crawford (1993), this one-equation model is not a significant improvement over the traditional mixing-length approach. Its key limitation is having to specify the function $\ell_t(y)$, just as it was necessary in Section 13.4 to specify the mixing length. This problem is overcome in the two-equation, K–ε model which follows.

Two-Equation Model

The K–ε model has much in common with the one-equation model just described. Its main additional feature is a differential equation for ε. For boundary layers this equation is (e.g., Kays and Crawford, 1993, p. 221)

$$\bar{v}_x\frac{\partial \varepsilon}{\partial x}+\bar{v}_y\frac{\partial \varepsilon}{\partial y}=\frac{\partial}{\partial y}\left((\nu+\varepsilon_\varepsilon)\frac{\partial \varepsilon}{\partial y}\right)+C_1\frac{\varepsilon}{K}\left[\varepsilon_M\left(\frac{\partial \bar{v}_x}{\partial y}\right)^2\right]-\frac{C_2\varepsilon^2}{K}, \tag{13.5-19}$$

where ε_ε is the eddy diffusivity for the dissipation rate and C_1 and C_2 are constants. The effect of having differential equations for both K and ε is that Eqs. (13.5-17) and (13.5-18) can be combined to eliminate the objectionable function $\ell_t(y)$. The resulting expression for ε_M is

$$\varepsilon_M=\frac{C_0 K^2}{\varepsilon}, \tag{13.5-20}$$

which involves only the constant C_0. The eddy diffusivity is able now to vary freely, according to the spatial variations in the turbulent kinetic energy and dissipation rate. Calculations using the K–ε model are performed by solving the smoothed continuity and momentum equations simultaneously with Eqs. (13.5-16) and (13.5-20). The empirical information that is needed consists of the constants C_0, C_1, C_2, $\varepsilon_M/\varepsilon_K$, and $\varepsilon_M/\varepsilon_\varepsilon$.

In a number of situations where the mixing-length approach does poorly (e.g., adverse pressure gradients or strong accelerations), the K–ε model has been found to perform quite well (Jones and Launder, 1972; Kays and Crawford, 1993; Launder and Spalding, 1974). The underlying physical assumptions are evidently fairly accurate, and versatility is gained by not having to specify a priori functional forms for $\ell(y)$ or $\ell_t(y)$. Based on a large body of experimental evidence, the best-fit values for the five constants are $C_0=0.09$, $C_1=1.44$, $C_2=1.92$, $\varepsilon_M/\varepsilon_K=1.0$, and $\varepsilon_M/\varepsilon_\varepsilon=1.3$ (Launder and Spalding, 1974). As one would hope, the three eddy diffusivities are in fact quite similar. See Launder and Spalding (1974) and Speziale (1991) for more information on the implementation, strengths, and limitations of the K–ε model.

References

Andersen, P. S., W. M. Kays, and R. `J. Moffat. Experimental results for the transpired turbulent boundary layer in an adverse pressure gradient. *J. Fluid Mech.* 69: 353–375, 1975.

Bhatti, M. S. and R. K. Shah. Turbulent and transition flow convective heat transfer in ducts. In *Handbook of Single-Phase Convective Heat Transfer.* Kakaç, S., R. K. Shah, and W. Aung, Eds. Wiley, New York, 1987, pp. 4.1–4.166.

Deissler, R. G. Analysis of turbulent heat transfer, mass transfer, and friction in smooth tubes at high Prandtl and Schmidt numbers. *NACA Tech. Rept.* No. 1210, 1955.

Durst, F., J. Jovanovic, and J. Sender. Detailed measurements of the near wall region of turbulent pipe flows. In *Ninth Symposium on Turbulent Shear Flows,* Kyoto, Japan, 1993.

Gnielinski, V. New equations for heat and mass transfer in turbulent pipe and channel flow. *Int. Chem. Eng.* 16: 359–368, 1976.

Goldstein, S. *Modern Developments in Fluid Dynamics.* Clarendon Press, Oxford, 1938 [reprinted by Dover, New York, 1965].

Hinze, J. O. *Turbulence,* second edition. McGraw-Hill, New York, 1975.

Jones, W. P. and B. E. Launder. The prediction of laminarization with a two-equation model of turbulence. *Int. J. Heat Mass Transfer* 15: 301–314, 1972.

Kays, W. M. and M. E. Crawford. *Convective Heat and Mass Transfer,* third edition. McGraw-Hill, New York, 1993.

Kim, J., P. Moin, and R. Moser. Turbulence statistics in fully developed channel flow at low Reynolds number. *J. Fluid Mech.* 177: 133–166, 1987.

Koury, E. Drag reduction by polymer solutions in riblet pipes. Ph. D. thesis, Department of Chemical Engineering, Massachusetts Institute of Technology, 1995.

Laufer, J. The structure of turbulence in fully developed pipe flow. *NACA Tech. Rept.* No.1174, 1954.

Launder, B. E. and D. B. Spalding. The numerical computation of turbulent flows. *Computer Meth. Appl. Mech. Eng.* 3: 269–289, 1974.

McComb, W. D. *The Physics of Fluid Turbulence.* Clarendon Press, Oxford, 1990.

Moody, L. F. Friction factors for pipe flow. *Trans. ASME* 66: 671–678, 1944.

Nikuradse, J. Gesetzmäßigkeiten der turbulenten Strömung in glatten Rohren. *VDI Forschungsheft* 356, 1932.

Nikuradse, J. Laws of flow in rough pipes. *NACA Tech. Memo.* No.1292, 1950. [This is an English translation of a German paper published in 1933.]

Notter, R. H. and C. A. Sleicher. The eddy diffusivity in the turbulent boundary layer near a wall. *Chem. Eng. Sci.* 26: 161–171, 1971.

Notter, R. H. and C. A. Sleicher. A solution to the turbulent Graetz problem—III. Fully developed and entry region heat transfer rates. *Chem. Eng. Sci.* 27: 2073–2093, 1972.

Petukhov, B. S. Heat transfer and friction in turbulent pipe flow with variable physical properties. *Adv. Heat Transfer* 6: 503–564, 1970.

Reichardt, H. Die Grundlagen des turbulenten Wärmeüberganges. *Arch. ges. Wärmetechn.* 6/7:129–142, 1951.

Reynolds, O. On the dynamical theory of incompressible viscous fluids and the determination of the criterion. *Philos. Trans. R. Soc. Lond. A* 186: 123–164, 1895.

Rogallo, R. S. and P. Moin. Numerical simulation of turbulent flows. *Annu. Rev. Fluid Mech.* 16: 99–137, 1984.

Rott, N. Note on the history of the Reynolds number. *Annu. Rev. Fluid Mech.* 22: 1–11, 1990.

Schlicting, H. *Boundary-Layer Theory,* sixth edition. McGraw-Hill, New York, 1968.

Sleicher, C. A. and M. W. Rouse. A convenient correlation for heat transfer to constant and variable property fluids in turbulent pipe flow. *Int. J. Heat Mass Transfer* 18: 677–683,1975.

Spalart, P. R. and J. H. Watmuff. Experimental and numerical study of a turbulent boundary layer with pressure gradients. *J. Fluid Mech.* 249: 337–371, 1993.

Speziale, C. G. Analytical methods for the development of Reynolds-stress closures in turbulence. *Annu. Rev. Fluid Mech.* 23: 107–157, 1991.

Sureshkumar, R., R. A. Handler, and A. N. Beris. Direct numerical simulation of the turbulent channel flow of a polymer solution. *Phys. Fluids* 9: 743–755, 1997.

Tennekes, H. and J. L. Lumley. *A First Course in Turbulence.* MIT Press, Cambridge, MA, 1972.

Van Driest, E. R. On turbulent flow near a wall. *J. Aero. Sci.* 23: 1007–1011, 1956.

Problems

13-1. Time-Smoothing

As a model for a fluctuating quantity in turbulent flow, consider the function

$$F(t) = \left(1 - \frac{t}{t_p}\right)\left(1 + k \sin\left(\frac{2\pi t}{t_f}\right)\right), \qquad 0 \le t \le t_p,$$

where t_p, t_f, and k are constants, with $t_p \gg t_f$ and $k < 1$. A realistic aspect of $F(t)$ is that rapid oscillations (with period t_f) are superimposed on a slowly varying baseline (with time constant t_p). However, the period and amplitude of the fluctuations are more regular than in actual turbulence.

(a) Sketch $F(t)$ for $k \cong 0.1$, and use Eq. (13.3-1) to determine $\langle F \rangle$. To simplify the calculations, let $t_a = nt_f$, where n is any positive integer. Show that

$$\langle F \rangle = \left(1 - \frac{t}{t_p}\right)$$

as expected from the sketch of $F(t)$.

(b) Calculate $d\langle F\rangle/dt$ and $\langle dF/dt\rangle$ with $t_a = nt_f$, and show that

$$\frac{d\langle F\rangle}{dt} = \left\langle \frac{dF}{dt}\right\rangle = -\frac{1}{t_p}.$$

In other words, confirm that the order of smoothing and time-differentiation can be interchanged. (*Hint:* To smooth dF/dt, an additional time integration is needed.)

(c) Given that turbulent fluctuations are not exactly periodic, a more realistic model problem is created by assuming that $t_a \ne nt_f$. Show that this more general choice of averaging time does not affect the results of parts (a) and (b).

13-2. Analytical Velocity from the Reichardt Model

An analytical expression which has been proposed for the velocity in turbulent pipe flow is

$$v_z^+ = 2.5 \ln\left[\frac{3}{2}y^+ \frac{1 + (r/R)}{1 + 2(r/R)^2}\right] + 5.5.$$

This result, based on the eddy diffusivity model of Reichardt (1951), is given in Kays and Crawford (1993, p. 247). It is intended to apply only in the core region, outside the laminar sublayer.

(a) Show that this expression follows from conservation of momentum and Eq. (13.4-13), if it is assumed that $\varepsilon_M/\nu \gg 1$. (*Hint:* Consider beginning with v_z^+ and working backwards.)

(b) Notice that this expression for v_z^+ is identical to the Nikuradse equation, Eq. (13.4-26), for $r/R \to 1$. Show that, unlike the Nikuradse equation, it satisfies the symmetry condition at $r = 0$.

13-3. Power-Law Velocity Profile and the Blasius Equation

The velocity in turbulent pipe or boundary layer flow is sometimes approximated using power-law expressions of the form

$$v_z^+ = C(y^+)^{1/n},$$

where C and n are constants; n is usually an integer such that $n \ge 7$. In pipe flow, at Reynolds numbers for which the Blasius equation for f is valid (about 10^4 to 10^5), $n = 7$ gives good results (see Fig. 13-8). Note, however, that this expression implies that $dv_z^+/dy^+ = \infty$ at $y^+ = 0$, so that it cannot be used to calculate the wall shear stress.

(a) Assuming that $n = 7$, show that the above expression leads to Eq. (13.4-31).
(b) Determine the value of C that is consistent with the Blasius equation, Eq. (13.2-6). (*Hint:* Start by evaluating v_z^+ at the tube centerline.)

13-4. Heat Transfer for Pr ≫ 1

For turbulent pipe flow with Re → ∞ and Pr → ∞, Eq. (13.4-54) reduces to

$$\text{Nu} = 0.079\sqrt{f/2}\ \text{Re}\ \text{Pr}^{1/3}.$$

Petukhov (1970) recommends a coefficient of 0.0855 instead of 0.079; another closely related expression is given in Notter and Sleicher (1971). Although various authors differ slightly, it is now generally agreed that $\text{Nu} \propto \text{Pr}^{1/3}$ for Pr → ∞. The objective of this problem is to explore what this implies for the functional form of ε_M near a wall.

(a) As mentioned in Example 13.4-3, almost the entire temperature change occurs within the laminar sublayer when Pr is large. In this region the eddy diffusivity has the general form

$$\frac{\varepsilon_M}{\nu} = a(y^+)^n \qquad (y^+ \to 0),$$

where the values of the constants a and n depend on the particular model. Use this functional form to derive a general expression for Nu that is valid for Pr → ∞. (*Hint:* The plug-flow assumption is unnecessary.)
(b) Using the result of part (a) and the empirical expression given above for Nu, determine the values of a and n.
(c) Show that the Van Driest model, Eq. (13.4-12), is incompatible with the asymptotic form for Nu, if Pr_t is a constant.

13-5. Calculation of Eddy Diffusivity from Velocity

(a) Suppose that you have measured a velocity profile for fully developed flow in a pipe and have found a function $v_z^+(y^+)$ which represents the data very accurately. Show how you would compute $\varepsilon_M(y^+)$ from these results.
(b) It has been suggested that the velocity in turbulent pipe flow be represented as

$$v_z^+ = \begin{cases} y^+, & 0 \le y^+ \le 5, \\ 5 \ln y^+ - 3.05, & 5 < y^+ < 30, \\ 2.5 \ln y^+ + 5.5, & y^+ \ge 30. \end{cases}$$

Sketch this velocity profile, calculate the eddy diffusivity it implies, and sketch $\varepsilon_M(y^+)$. Discuss the strengths and weaknesses of this form of velocity profile in comparison with other approaches.

13-6. Nikuradse Velocity Profile and the Kármán–Nikuradse Equation

Assume that the Nikuradse equation adequately represents the turbulent velocity profile throughout a tube. Show that it implies a relationship between f and Re which is very similar to the Kármán–Nikuradse equation. [*Hint:* Begin by using Eq. (13.4-26) to compute the cross-sectional average velocity. The values of the constants in your final result for $f(\text{Re})$ will differ somewhat from those in Eq. (13.2-5).]

13-7. Momentum Equations for Turbulent Boundary Layers

(a) By analogy with the arguments leading to Eq. (8.2-32) for laminar flow, show that the smoothed momentum equation for a turbulent boundary layer is

$$\bar{v}_x\frac{\partial \bar{v}_x}{\partial x}+\bar{v}_y\frac{\partial \bar{v}_x}{\partial y}=\bar{U}\frac{d\bar{U}}{dx}+\frac{\partial}{\partial y}\left((v+\varepsilon_M)\frac{\partial \bar{v}_x}{\partial y}\right),$$

where $\bar{U}$ is the "outer" velocity evaluated at the surface. (An uppercase letter is used here to avoid confusion with the velocity fluctuation, u.)

(b) Show that the integral form of the momentum equation for a turbulent boundary layer is

$$\frac{\tau_0}{\rho}=\frac{d}{dx}\int_0^\delta \bar{v}_x\left(\bar{U}-\bar{v}_x\right)\,dy+\frac{d\bar{U}}{dx}\int_0^\delta\left(\bar{U}-\bar{v}_x\right)\,dy,$$

which is analogous to the Kármán integral equation for laminar flow.

13-8. Turbulent Boundary Layer on a Flat Plate

In this problem the integral momentum equation from part (b) of Problem 13-7 is used to analyze turbulent flow parallel to a flat plate. For this geometry, a velocity profile which is found to be a good approximation for most of the boundary layer is (Kays and Crawford, 1993, p. 207)

$$v_x^+=8.75(y^+)^{1/7}.$$

This is analogous to the 1/7-power expression used for tubes [see Eq. (13.4-31)].

(a) Using the integral momentum equation and the 1/7-power approximation for the velocity, show that the wall shear stress is related to the boundary layer thickness by

$$\tau_0=\frac{7}{72}\rho\bar{U}^2\frac{d\delta}{dx}.$$

(b) What is needed now is an independent expression for τ_0. The 1/7-power velocity profile implies an infinite velocity gradient at the surface, so that the direct approach will not work. Show, however, that evaluating the velocity at the outer edge of the boundary layer gives

$$\tau_0=0.0225\,\rho\bar{U}^2\left(\frac{v}{\bar{U}\delta}\right)^{1/4}.$$

(c) Determine the boundary layer thickness, $\delta(x)$. Let $\delta(x_0)=\delta_0$, where x_0 is the position at which the laminar–turbulent transition occurs. For $x\gg x_0$, show that

$$\delta=0.370\left(\frac{vx^4}{\bar{U}}\right)^{1/5}.$$

Comparing this with Eq. (8.4-54), we see that the boundary layer on a flat plate grows much faster in turbulent than in laminar flow ($x^{4/5}$ versus $x^{1/2}$).

(d) If the plate dimension in the direction of flow (L) is large enough, the fraction of the surface for which there is laminar flow will be negligible. Show that the drag coefficient is given then by

$$C_D=\frac{0.0720}{\mathrm{Re}^{1/5}},$$

where Re is based on L. This may be compared with Eq. (8.4-18), the exact result for laminar flow.

Appendix

VECTORS AND TENSORS

A.1 INTRODUCTION

This appendix provides general information on vector and tensor manipulations, as used in derivations throughout the book. It is designed for the reader who would benefit from a review of basic vector concepts and who may have had little or no prior exposure to tensors. Some of the material is elementary and some (e.g., the discussion of surface geometry) is more advanced.

Many books have been devoted entirely to vector and tensor analysis, but space limitations require that the coverage here be quite limited. This appendix focuses more on operational aspects (i.e., how to perform manipulations) than on basic theory. In many instances the reader is referred to other sources for proofs. The content has been influenced most by the appendices in Bird et al. (1960, 1987), the summary of vector analysis in Hildebrand (1976), and the book by Brand (1947). Other useful discussions of vector and tensor analysis are given in Aris (1962), Hay (1953), Jeffreys (1963), Morse and Feshbach (1953), and Wilson (1901).

Section A.2 introduces the notation used to represent vectors and tensors and describes the more elementary operations involving those quantities (e.g., addition). The various types of vector and tensor multiplications are discussed in Section A.3. Vector differential operators are introduced in Section A.4, with an emphasis on rectangular coordinates. A number of useful integral theorems are given in Section A.5, and the special properties of position vectors are discussed in Section A.6. Section A.7 extends the treatment of differential operators to any system of orthogonal curvilinear coordinates and provides details for the important special cases of cylindrical and spherical coordinates. The differential geometry of surfaces, including normal and tangent vectors, surface gradients, and surface curvature, is reviewed in Section A.8.

A.2 REPRESENTATION OF VECTORS AND TENSORS

A *scalar* is a quantity that has a magnitude only (e.g., temperature), whereas a *vector* is characterized by both a magnitude and a direction (e.g., velocity). Vectors are often represented as arrows which have a certain length and spatial orientation. Two vectors are said to be equal if they are parallel, pointed in the same direction, and of equal length (magnitude). Vectors do not usually have a particular spatial location, and therefore they need not be collinear to be equal. (Position vectors, which extend from a reference point to some other point in space, are the exception.) A heavy reliance on the geometrical interpretation of vectors is confining from an analytical viewpoint, but the concept of an arrow in space reminds us that a vector is an entity which is independent of any coordinate system. That is, no matter what coordinate system is used to represent it, a vector must have the same length and direction. Likewise, physical laws involving directional quantities are described most naturally using vectors (and tensors), because those relationships too must be independent of any coodinate system that is chosen.

A vector in physical (three-dimensional) space may be viewed also as an ordered set of three numbers, each of which is associated with a particular direction; the directions usually correspond to a set of orthogonal coordinates. The fact that a vector is an independent entity implies that once the three numbers (or *components*) are known for one coordinate system, they are uniquely determined for all other coordinate systems. In other words, vector components must obey certain rules in coordinate transformations. The directionality of a vector and the transformation rules are what distinguish it from a row or column matrix. Two vectors are equal only if each of the corresponding components is equal.

A *second-order tensor* in three-dimensional space is an entity that can be represented as an ordered set of nine numbers, each of which is associated with *two* directions. The most prominent example in transport theory is the stress in a fluid. With second (or higher)-order tensors the "arrow-in-space" concept ceases to be helpful. An nth-order tensor has 3^n components, each corresponding to a set of n directions. As with vectors, the components of tensors must obey certain transformation rules [see, for example, Prager (1961)]. In general discussions of tensor analysis, scalars and vectors are regarded as zeroth-order and first-order tensors, respectively. In this text we have little need for tensors with $n > 2$, and the term "tensor" without a modifier is used as a shorthand for "second-order tensor."

Two systems of notation for vectors and tensors are in common use. This text employs *Gibbs notation,* developed by J. W. Gibbs (who was responsible also for much of the basic theory of chemical thermodynamics). The classical presentation of vector analysis in this notation is the book by Wilson (1901). In this system, vectors and tensors are each represented by single boldface letters, usually without subscripts. With occasional exceptions, we represent scalars as italic Roman letters (e.g., f), vectors as boldface Roman letters (e.g., $\mathbf{v}$), and tensors as boldface Greek letters (e.g., $\boldsymbol{\tau}$); each may be either uppercase or lowercase. The magnitudes of vectors and tensors are denoted usually by the corresponding italic letter; occasionally, absolute-value signs are used for clarity. Thus, the magnitude of $\mathbf{v}$ is written either as v or $|\mathbf{v}|$. Unit vectors corresponding to coordinate directions are denoted as $\mathbf{e}_i$, where the subscript is used to identify the coordinate. The advantage of Gibbs notation is that it allows most equations to be written in a simple and general form, without reference to a particular coordinate system.

The other major system, called *Cartesian tensor notation,* is based on vector and tensor components, which are identified using subscripts. It has the advantage of showing more explicitly the results of vector and tensor manipulations, and it is used here to evaluate various products and derivatives. One disadvantage is that the component representations of differential operators are valid only for rectangular coordinates. This is not a hindrance in proving general identities, because (as already noted) the essential features of vectors and tensors are independent of any coordinate system. Thus, an identity involving vectors and tensors which can be proved using one coordinate system will be valid for all others.

A vector **v** can be represented using rectangular (Cartesian) coordinates as

$$\mathbf{v} = v_x \mathbf{e}_x + v_y \mathbf{e}_y + v_z \mathbf{e}_z = \sum_{i=1}^{3} v_i \mathbf{e}_i, \tag{A.2-1}$$

where $(\mathbf{e}_x, \mathbf{e}_y, \mathbf{e}_z)$ are vectors of unit length which parallel the respective coordinate axes and (v_x, v_y, v_z) are the corresponding (scalar) components of **v**. Labeling the coordinates as $(1, 2, 3)$ instead of (x, y, z) leads to a more compact representation involving a sum, as shown. The economy of the summation notation makes it preferable to represent the base vectors as $(\mathbf{e}_1, \mathbf{e}_2, \mathbf{e}_3)$ instead of $(\mathbf{i}, \mathbf{j}, \mathbf{k})$, as is sometimes done. Because vectors in physical space are understood to have three components, the limits of the summations are omitted hereafter. In the research literature the notation is often pared down even further by omitting both the summation symbol and the unit vectors, so that **v** is represented simply as v_i. When this is done a *repeated* subscript or other index in an algebraic expression is used to *imply* certain summations. A standard treatment of vector and tensor analysis using implied summations is given in the small book by Jeffreys (1963). Whether the summation symbols are included or omitted, it is important to recognize that any quantity being summed over is a *dummy* index; changing each i in Eq. (A.2-1) to j does not alter the meaning of the expression.

A tensor $\boldsymbol{\tau}$ can be represented in component form as

$$\boldsymbol{\tau} = \sum_i \sum_j \tau_{ij} \mathbf{e}_i \mathbf{e}_j = \begin{pmatrix} \tau_{11} & \tau_{12} & \tau_{13} \\ \tau_{21} & \tau_{22} & \tau_{23} \\ \tau_{31} & \tau_{32} & \tau_{33} \end{pmatrix}. \tag{A.2-2}$$

In this case each scalar component, τ_{ij}, is associated with a *pair* of unit vectors, $\mathbf{e}_i\mathbf{e}_j$; the paired unit vectors are called *unit dyads.* When implied summations are used, the entire tensor is represented simply as τ_{ij}. The double sum yields nine terms, and the nine components of a tensor can be displayed as a 3×3 matrix, as shown. A matrix lacks the directional character of a tensor, so that the matrix in Eq. (A.2-2) does not literally equal $\boldsymbol{\tau}$. The matrix analogy is quite useful, however, in that several operations involving vectors and tensors follow the rules of matrix algebra. In such manipulations a vector can be represented as a row or column matrix. For example, $\mathbf{v} = (v_1, v_2, v_3)$ and $\mathbf{e}_2 = (0,1,0)$.

Analogous to the transpose of a square matrix, the *transpose* of the tensor $\boldsymbol{\tau}$, written as $\boldsymbol{\tau}^t$, is given by

$$\tau^t = \sum_i \sum_j \tau_{ji} e_i e_j = \begin{pmatrix} \tau_{11} & \tau_{21} & \tau_{31} \\ \tau_{12} & \tau_{22} & \tau_{32} \\ \tau_{13} & \tau_{23} & \tau_{33} \end{pmatrix}. \qquad (A.2-3)$$

Comparing Eqs. (A.2-2) and (A.2-3), it is seen that the transpose is obtained by replacing τ_{ij} by τ_{ji}, or (equivalently) interchanging the off-diagonal elements. If $\tau_{ij} = \tau_{ji}$, so that the tensor and its transpose are equal, τ is said to be *symmetric*; if $\tau_{ij} = -\tau_{ji}$, then τ is *antisymmetric*. A tensor can be antisymmetric only if the diagonal elements are all zero (i.e., $\tau_{ii} = 0$).

The addition and subtraction of vectors and tensors is the same as with matrices. That is, one adds or subtracts the corresponding elements. Likewise, multiplication (or division) of a vector or tensor by a scalar entails multiplication (or division) of each component. Two vectors **a** and **b** are equal if $\mathbf{a} - \mathbf{b} = \mathbf{0}$, where each component of the vector **0** is zero. Thus, as already mentioned, $\mathbf{a} = \mathbf{b}$ implies that each of the corresponding components of **a** and **b** are equal. Likewise, the equality of two tensors implies equality of their components. The zero vector and zero tensor are both represented by the symbol **0**; which one is meant will be clear from the context.

Several of the basic operations are illustrated by the decomposition of the tensor τ into symmetric and antisymmetric parts:

$$\tau = \frac{1}{2}(\tau + \tau^t) + \frac{1}{2}(\tau - \tau^t) = \frac{1}{2}\sum_i \sum_j (\tau_{ij} + \tau_{ji})e_i e_j + \frac{1}{2}\sum_i \sum_j (\tau_{ij} - \tau_{ji})e_i e_j, \quad (A.2-4)$$

$$\tau + \tau^t = \begin{pmatrix} 2\tau_{11} & \tau_{12} + \tau_{21} & \tau_{13} + \tau_{31} \\ \tau_{21} + \tau_{12} & 2\tau_{22} & \tau_{23} + \tau_{32} \\ \tau_{31} + \tau_{13} & \tau_{32} + \tau_{23} & 2\tau_{33} \end{pmatrix}, \qquad (A.2-5)$$

$$\tau - \tau^t = \begin{pmatrix} 0 & \tau_{12} - \tau_{21} & \tau_{13} - \tau_{31} \\ \tau_{21} - \tau_{12} & 0 & \tau_{23} - \tau_{32} \\ \tau_{31} - \tau_{13} & \tau_{32} - \tau_{23} & 0 \end{pmatrix}. \qquad (A.2-6)$$

Equation (A.2-5) shows that adding a tensor and its transpose creates a symmetric tensor, whereas Eq. (A.2-6) shows that subtracting the transpose leads to an antisymmetric tensor. *Any* tensor can be expressed as the sum of symmetric and antisymmetric tensors, in the manner of Eq. (A.2-4). This fact is exploited in Chapter 5 in describing the stress in a fluid.

A.3 VECTOR AND TENSOR PRODUCTS

It was pointed out in the preceding section that the addition or subtraction of vectors or tensors, and the multiplication or division of a vector or tensor by a scalar, are straightforward in that they follow the same rules as with matrices. The various products which can be formed using vectors and tensors require special attention, however, as discussed now.

Scalar Product of Vectors

The *scalar* (or *dot*) *product* of **a** and **b** is defined as

$$\mathbf{a} \cdot \mathbf{b} = ab \cos \varphi_{ab}, \tag{A.3-1}$$

where φ_{ab} is the angle between the vectors ($\leq 180°$) [see Fig. A-1(a)]. The operation $\mathbf{a} \cdot \mathbf{b}$ results in a scalar, hence the name. Noticing that $b \cos \varphi_{ab}$ equals the magnitude of **b** as projected on **a**, it is seen that $\mathbf{a} \cdot \mathbf{b}$ equals the magnitude of **a** times the projection of **b**, and vice versa.

If the vectors are perpendicular, then $\cos \varphi_{ab} = 0$ and the scalar product vanishes. If a vector is dotted with itself, then $\cos \varphi_{ab} = 1$ and the result is the square of the magnitude. Thus, the magnitude of **a** is given by

$$a = |\mathbf{a}| = (\mathbf{a} \cdot \mathbf{a})^{1/2}. \tag{A.3-2}$$

The scalar product of two vectors is *commutative*,

$$\mathbf{a} \cdot \mathbf{b} = \mathbf{b} \cdot \mathbf{a}, \tag{A.3-3}$$

and also *distributive*,

$$\mathbf{a} \cdot (\mathbf{b} + \mathbf{c}) = (\mathbf{a} \cdot \mathbf{b}) + (\mathbf{a} \cdot \mathbf{c}). \tag{A.3-4}$$

In computing dot products it is efficient to dispense with the angles between vectors and to work directly with the vector components. This approach exploits the special properties of the dot products of the base vectors. For an *orthogonal* (mutually perpendicular) set of unit vectors, it follows from Eq. (A.3-1) that

$$\begin{aligned}
\mathbf{e}_1 \cdot \mathbf{e}_1 &= 1, & \mathbf{e}_2 \cdot \mathbf{e}_2 &= 1, & \mathbf{e}_3 \cdot \mathbf{e}_3 &= 1, \\
\mathbf{e}_1 \cdot \mathbf{e}_2 &= 0, & \mathbf{e}_2 \cdot \mathbf{e}_3 &= 0, & \mathbf{e}_3 \cdot \mathbf{e}_1 &= 0, \\
\mathbf{e}_2 \cdot \mathbf{e}_1 &= 0, & \mathbf{e}_3 \cdot \mathbf{e}_2 &= 0, & \mathbf{e}_1 \cdot \mathbf{e}_3 &= 0.
\end{aligned} \tag{A.3-5}$$

These nine equations are summarized as

$$\mathbf{e}_i \cdot \mathbf{e}_j = \delta_{ij}, \tag{A.3-6}$$

where δ_{ij} is the *Kronecker delta,* given by

$$\delta_{ij} = \begin{cases} 1, & i = j, \\ 0, & i \neq j. \end{cases} \tag{A.3-7}$$

Thus, $\delta_{ij} = 1$ or 0, according to whether the subscripts are the same or different. The scalar product of **a** and **b** is calculated now as

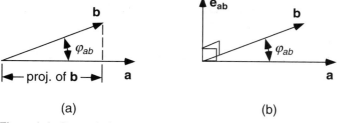

(a) (b)

Figure A-1. Geometrical representations of (a) the scalar or dot product, $\mathbf{a} \cdot \mathbf{b}$, and (b) the vector or cross product, $\mathbf{a} \times \mathbf{b}$.

$$\mathbf{a}\cdot\mathbf{b}=\left(\sum_i a_i\mathbf{e}_i\right)\cdot\left(\sum_j b_j\mathbf{e}_j\right)=\sum_i\sum_j a_ib_j(\mathbf{e}_i\cdot\mathbf{e}_j)=\sum_i\sum_j a_ib_j\delta_{ij}=\sum_i a_ib_i. \quad \text{(A.3-8)}$$

This illustrates the four main steps in evaluating a product involving two or more vectors. First, the vectors are each represented as summations. Second, all quantities are grouped inside the multiple summations. Notice that any quantity can be brought inside a summation involving a *different* index, as done in the second equality.[1] Third, the operations involving the base vectors ($\mathbf{e}_i$) are performed. Fourth and last, summations which have been made redundant by the Kronecker deltas are eliminated. As will be shown, the same steps are used to evaluate other products involving vectors and tensors, although different operations involving the base vectors are required.

Equation (A.3-8) shows that the dot product of two vectors is just the sum of the products of the respective components. The commutative property expressed by Eq. (A.3-3) is now more obvious, and the distributive property in Eq. (A.3-4) is easily confirmed by evaluating the left- and right-hand sides of that equation as done in Eq. (A.3-8). In the most pared-down notation, the dot product is written simply as a_ib_i, and the repeated index i is understood to imply summation over i. The less compact notation of Eq. (A.3-8) is employed here because it is more explicit.

A component of a vector is just its projection on the corresponding coordinate axis. Thus,

$$a_i=\mathbf{e}_i\cdot\mathbf{a}=\mathbf{a}\cdot\mathbf{e}_i \quad \text{(A.3-9)}$$

as may be confirmed by replacing $\mathbf{b}$ by $\mathbf{e}_i$ in Eq. (A.3-8). Likewise, the component (or projection) of a vector in an arbitrary direction is given by the dot product of the vector with a unit vector pointed in that direction. For example, if $\mathbf{n}$ is a unit vector that is normal to a surface, then $\mathbf{n}\cdot\mathbf{a}$ is the normal component of $\mathbf{a}$. Using Eqs. (A.3-2) and (A.3-8), the magnitude of $\mathbf{a}$ is expressed in terms of its components as

$$a=|\mathbf{a}|=\left(\sum_i a_i^2\right)^{1/2}. \quad \text{(A.3-10)}$$

A *normalized* vector (one with unit magnitude) is obtained by dividing each component by the magnitude of the vector, as calculated from Eq. (A.3-10).

Vector Product of Vectors

The *vector* (or *cross*) *product* of $\mathbf{a}$ and $\mathbf{b}$ is defined as

$$\mathbf{a}\times\mathbf{b}=ab\sin\varphi_{ab}\mathbf{e}_{ab}, \quad \text{(A.3-11)}$$

where $\mathbf{e}_{ab}$ is a unit vector that is perpendicular to the plane formed by $\mathbf{a}$ and $\mathbf{b}$ [see Fig. A-1(b)]. The direction of $\mathbf{e}_{ab}$ is such that $(\mathbf{a},\mathbf{b},\mathbf{e}_{ab})$ is a right-handed set of vectors; that is, if the fingers of the right hand are curled from $\mathbf{a}$ toward $\mathbf{b}$, then $\mathbf{e}_{ab}$ points in the direction of the extended thumb. The magnitude of $\mathbf{a}\times\mathbf{b}$ equals the area of a parallelogram which has $\mathbf{a}$ and $\mathbf{b}$ as adjacent sides.

---

[1] The second equality in Eq. (A.3-8) is analogous to

$$\left(\int a(x)\,dx\right)\left(\int b(y)\,dy\right)=\int\int a(x)b(y)\,dx\,dy.$$

Note that because $\mathbf{a} \times \mathbf{b}$ is perpendicular to the plane containing $\mathbf{a}$ and $\mathbf{b}$, it is perpendicular to both of those vectors. Thus, if $\mathbf{a}$ and $\mathbf{b}$ are also perpendicular to one another, then $(\mathbf{a}, \mathbf{b}, \mathbf{a} \times \mathbf{b})$ is an *orthogonal* set. If a vector is crossed with itself (or with any parallel vector), then $\sin \varphi_{ab} = 0$ and the product is zero. The right-hand rule implies that reversing the order of the vectors in the cross product reverses the direction of the resultant. That is,

$$\mathbf{a} \times \mathbf{b} = -\mathbf{b} \times \mathbf{a}, \tag{A.3-12}$$

indicating that the cross product is *not commutative*. It is, however, *distributive:*

$$\mathbf{a} \times (\mathbf{b} + \mathbf{c}) = (\mathbf{a} \times \mathbf{b}) + (\mathbf{a} \times \mathbf{c}). \tag{A.3-13}$$

Representations of cross products in terms of vector components make use of the special properties of the cross products of the base vectors. Those cross products are given by

$$
\begin{aligned}
&\mathbf{e}_1 \times \mathbf{e}_1 = \mathbf{0}, &&\mathbf{e}_2 \times \mathbf{e}_2 = \mathbf{0}, &&\mathbf{e}_3 \times \mathbf{e}_3 = \mathbf{0}, \\
&\mathbf{e}_1 \times \mathbf{e}_2 = \mathbf{e}_3, &&\mathbf{e}_2 \times \mathbf{e}_3 = \mathbf{e}_1, &&\mathbf{e}_3 \times \mathbf{e}_1 = \mathbf{e}_2, \\
&\mathbf{e}_2 \times \mathbf{e}_1 = -\mathbf{e}_3, &&\mathbf{e}_3 \times \mathbf{e}_2 = -\mathbf{e}_1, &&\mathbf{e}_1 \times \mathbf{e}_3 = -\mathbf{e}_2.
\end{aligned}
\tag{A.3-14}
$$

These nine equations are summarized as

$$\mathbf{e}_i \times \mathbf{e}_j = \sum_k \varepsilon_{ijk} \mathbf{e}_k, \tag{A.3-15}$$

where ε_{ijk} is the *permutation symbol,* given by

$$
\varepsilon_{ijk} = \begin{cases}
0, & i=j,\, j=k,\text{ or } i=k, \\
1, & ijk = 123,\, 231,\text{ or } 312, \\
-1, & ijk = 132,\, 213,\text{ or } 321.
\end{cases}
\tag{A.3-16}
$$

Thus, $\varepsilon_{ijk} = 0$, 1, or -1, according to whether two or more indices are alike, the indices increase in cyclic order, or the indices decrease in cyclic order, respectively. Expanding $\mathbf{a}$ and $\mathbf{b}$ as sums and using Eq. (A.3-15), it is found that

$$\mathbf{a} \times \mathbf{b} = \sum_i \sum_j \sum_k \varepsilon_{ijk} a_i b_j \mathbf{e}_k, \tag{A.3-17}$$

which may be used to confirm Eqs. (A.3-12) and (A.3-13). Writing out all of the terms in Eq. (A.3-17), we obtain

$$\mathbf{a} \times \mathbf{b} = (a_2 b_3 - a_3 b_2)\mathbf{e}_1 + (a_3 b_1 - a_1 b_3)\mathbf{e}_2 + (a_1 b_2 - a_2 b_1)\mathbf{e}_3, \tag{A.3-18}$$

which is equivalent to the determinant,

$$
\mathbf{a} \times \mathbf{b} = \begin{vmatrix}
\mathbf{e}_1 & \mathbf{e}_2 & \mathbf{e}_3 \\
a_1 & a_2 & a_3 \\
b_1 & b_2 & b_3
\end{vmatrix}.
\tag{A.3-19}
$$

Multiple Products

Several useful identities involving two or more dot or cross products are

$$\mathbf{a} \cdot (\mathbf{b} \times \mathbf{c}) = \mathbf{b} \cdot (\mathbf{c} \times \mathbf{a}) = \mathbf{c} \cdot (\mathbf{a} \times \mathbf{b}) = (\mathbf{b} \times \mathbf{c}) \cdot \mathbf{a} = (\mathbf{c} \times \mathbf{a}) \cdot \mathbf{b} = (\mathbf{a} \times \mathbf{b}) \cdot \mathbf{c}, \tag{A.3-20}$$

$$\mathbf{a} \times (\mathbf{b} \times \mathbf{c}) = (\mathbf{a} \cdot \mathbf{c})\mathbf{b} - (\mathbf{a} \cdot \mathbf{b})\mathbf{c}, \tag{A.3-21}$$

$$(\mathbf{a} \times \mathbf{b}) \times \mathbf{c} = (\mathbf{a} \cdot \mathbf{c})\mathbf{b} - (\mathbf{b} \cdot \mathbf{c})\mathbf{a}, \tag{A.3-22}$$

$$(\mathbf{a} \times \mathbf{b}) \cdot (\mathbf{c} \times \mathbf{d}) = (\mathbf{a} \cdot \mathbf{c})(\mathbf{b} \cdot \mathbf{d}) - (\mathbf{a} \cdot \mathbf{d})(\mathbf{b} \cdot \mathbf{c}). \tag{A.3-23}$$

The expressions in Eq. (A.3-20) are called *scalar triple products;* the scalar which results from any of these corresponds to the volume of a parallelepiped which has **a**, **b**, and **c** as adjacent sides. The first three expressions involve a cyclic rearrangement of the vectors, whereas the last three follow from the commutative property of the dot product. Comparing, for example, the first and last expressions in Eq. (A.3-20), it is seen that interchanging the dot and cross does not affect the result. Note also that no ambiguity would be created by omitting the parentheses in the scalar triple products, because any of the six expressions can be evaluated in only one way. Equations (A.3-21) and (A.3-22) are expansion formulas for what are called *vector triple products;* here the parentheses are essential. A relationship involving the permutation symbol that is very helpful in proving identities such as these is

$$\sum_k \varepsilon_{ijk} \varepsilon_{mnk} = \delta_{im}\delta_{jn} - \delta_{in}\delta_{jm}. \tag{A.3-24}$$

Dyadic Product

The formal multiplication of the vectors **a** and **b,** without a dot or cross, results in the *dyad,* **ab.** This dyad is represented as

$$\mathbf{ab} = \sum_i \sum_j a_i b_j \mathbf{e}_i \mathbf{e}_j = \begin{pmatrix} a_1 b_1 & a_1 b_2 & a_1 b_3 \\ a_2 b_1 & a_2 b_2 & a_2 b_3 \\ a_3 b_1 & a_3 b_2 & a_3 b_3 \end{pmatrix}. \tag{A.3-25}$$

It is seen that a dyad is simply a tensor constructed by pairing the components of two vectors. Unlike scalar multiplication, the order of **a** and **b** must be preserved. That is, unless the dyad happens to be symmetric, $\mathbf{ab} \neq \mathbf{ba}$. A dyad will be symmetric if **a** and **b** are parallel (i.e., if they differ at most by a multiplicative scalar constant). Dyadic products have a distributive property analogous to Eqs. (A.3-4) and (A.3-13). Manipulations involving dyads follow the same rules as with tensors, which are discussed next.

Products Involving Tensors

Using an approach like that in Eq. (A.3-8), dot products involving tensors are reduced to the following operations involving unit vectors and unit dyads:

$$\mathbf{e}_k \cdot \mathbf{e}_i \mathbf{e}_j = \delta_{ik} \mathbf{e}_j, \tag{A.3-26}$$

$$\mathbf{e}_i \mathbf{e}_j \cdot \mathbf{e}_k = \delta_{jk} \mathbf{e}_i, \tag{A.3-27}$$

$$\mathbf{e}_i \mathbf{e}_j \cdot \mathbf{e}_m \mathbf{e}_n = \delta_{jm} \mathbf{e}_i \mathbf{e}_n, \tag{A.3-28}$$

$$\mathbf{e}_i \mathbf{e}_j : \mathbf{e}_m \mathbf{e}_n = \delta_{jm} \delta_{in}. \tag{A.3-29}$$

It is seen that the dot product of a vector with a tensor results in a vector and that the dot product of two tensors yields another tensor. The *double-dot product* of two tensors gives a scalar, as shown in Eq. (A.3-29). These four relations can be remembered by noting that they are equivalent to forming dot products of the unit vectors on either side of the dot symbol, while keeping the other vectors in proper order. In the case of Eq. (A.3-29), the two dot products are performed sequentially.[2]

Using Eqs. (A.3-26) and (A.3-27) to evaluate $\mathbf{a} \cdot \boldsymbol{\tau}$ and $\boldsymbol{\tau} \cdot \mathbf{a}$, it is found that

$$\mathbf{a} \cdot \boldsymbol{\tau} = \sum_i \sum_j \sum_k a_k \tau_{ij} (\mathbf{e}_k \cdot \mathbf{e}_i \mathbf{e}_j) = \sum_i \sum_j \sum_k a_k \tau_{ij} \delta_{ik} \mathbf{e}_j = \sum_i \sum_j a_i \tau_{ij} \mathbf{e}_j, \qquad (A.3\text{-}30)$$

$$\boldsymbol{\tau} \cdot \mathbf{a} = \sum_i \sum_j \sum_k \tau_{ij} a_k (\mathbf{e}_i \mathbf{e}_j \cdot \mathbf{e}_k) = \sum_i \sum_j \sum_k \tau_{ij} a_k \delta_{jk} \mathbf{e}_i = \sum_i \sum_j \tau_{ji} a_i \mathbf{e}_j. \qquad (A.3\text{-}31)$$

In the first case the jth component of the product is $\sum_i a_i \tau_{ij}$, whereas in the second it is $\sum_i a_i \tau_{ji}$. These results are different unless $\boldsymbol{\tau}$ is symmetric. Thus, the dot product of a vector with a tensor is *not commutative*, in general. Likewise, the single-dot product of two tensors is *not commutative*, unless the tensors are symmetric. It is worth noting that these single-dot products are formally similar to matrix multiplication, which also is commutative only if a matrix is symmetric. Note also that, unlike multiplication of a vector by a scalar, the product $\boldsymbol{\tau} \cdot \mathbf{a}$ (or $\mathbf{a} \cdot \boldsymbol{\tau}$) has a different effect on each component of the vector. Consequently, dotting a vector with a tensor changes its direction; the vector is "twisted" or "deflected."

Equations (A.3-30) and (A.3-31) are easily extended to show that dot multiplication of a tensor by pre- and post-factors is *associative*,

$$(\mathbf{a} \cdot \boldsymbol{\tau}) \cdot \mathbf{b} = \mathbf{a} \cdot (\boldsymbol{\tau} \cdot \mathbf{b}) = \sum_i \sum_j a_i \tau_{ij} b_j. \qquad (A.3\text{-}32)$$

Accordingly, no ambiguity results if the parentheses are omitted and the expression is written simply as $\mathbf{a} \cdot \boldsymbol{\tau} \cdot \mathbf{b}$. Also, the double-dot product of two tensors is *commutative*:

$$\boldsymbol{\alpha} : \boldsymbol{\beta} = \boldsymbol{\beta} : \boldsymbol{\alpha} = \sum_i \sum_j \alpha_{ij} \beta_{ji}. \qquad (A.3\text{-}33)$$

This last operation leads to a useful defintion of the *magnitude* of the tensor $\boldsymbol{\alpha}$. Following Bird et al. (1987), we let

$$\alpha = \left(\frac{1}{2} \boldsymbol{\alpha} : \boldsymbol{\alpha}^t \right)^{1/2} = \left(\frac{1}{2} \sum_i \sum_j \alpha_{ij}^2 \right)^{1/2}, \qquad (A.3\text{-}34)$$

which is analogous to Eqs. (A.3-2) or (A.3-10) for a vector. If $\boldsymbol{\alpha}$ is symmetric and its only nonvanishing components are $\alpha_{12} = \alpha_{21}$, then its magnitude is $\alpha = |\alpha_{12}| = |\alpha_{21}|$. This is analogous to a vector $\mathbf{v}$ with a single nonzero component v_1, where the magnitude is $v = |v_1|$. The factor $\frac{1}{2}$ in Eq. (A.3-34) is needed to preserve this analogy.

Cross products involving tensors, which are encountered infrequently in transport theory, are evaluated in a similar manner. That is, the cross products involving the unit dyads are derived by evaluating the products of the unit vectors adjacent to the multiplication symbol, analogous to Eqs. (A.3-26)–(A.3-28) [see Brand (1947, pp. 187–189)].

---

[2] According to the original definition of the double-dot product (Wilson, 1901), the right-hand side of Eq. (A.3-29) would be written as $\delta_{jn} \delta_{im}$. The definition used here is a more natural extension of the single-dot product of Eq. (A.3-28) and is preferred now by many authors.

Identity Tensor

The *identity tensor* or *unit tensor*, which is analogous to the identity matrix, is given by

$$\boldsymbol{\delta}=\sum_i\sum_j\delta_{ij}\mathbf{e}_i\mathbf{e}_j=\sum_i\mathbf{e}_i\mathbf{e}_i=\begin{pmatrix}1 & 0 & 0\\0 & 1 & 0\\0 & 0 & 1\end{pmatrix}. \qquad (\text{A.3-35})$$

The symbol $\boldsymbol{\delta}$ has been chosen because, as shown, the components of the identity tensor are given by the Kronecker delta. The essential properties of $\boldsymbol{\delta}$ are that

$$\left(\begin{array}{l}\boldsymbol{\delta}\cdot\mathbf{v}=\mathbf{v}\cdot\boldsymbol{\delta}=\mathbf{v}, \qquad\qquad\qquad\qquad (\text{A.3-36})\\[2mm]\boldsymbol{\delta}\cdot\boldsymbol{\tau}=\boldsymbol{\tau}\cdot\boldsymbol{\delta}=\boldsymbol{\tau} \qquad\qquad\qquad\qquad (\text{A.3-37})\end{array}\right.$$

for any vector $\mathbf{v}$ or tensor $\boldsymbol{\tau}$. Thus, dot multiplication of $\boldsymbol{\delta}$ with a vector or tensor, in any order, returns that vector or tensor.

Alternating Unit Tensor

Just as a second-order tensor ($\boldsymbol{\delta}$) is constructed from the Kronecker delta, a third-order tensor ($\boldsymbol{\varepsilon}$) is derived from the permutation symbol. Called the *alternating unit tensor*, it is represented as

$$\boldsymbol{\varepsilon}=\sum_i\sum_j\sum_k\varepsilon_{ijk}\mathbf{e}_i\mathbf{e}_j\mathbf{e}_k. \qquad (\text{A.3-38})$$

The group of three unit vectors shown here is called a *unit triad*. Operations involving third (or higher)-order tensors are performed in a manner analogous to that shown above for second-order tensors. In particular, the dot products of unit *polyads* are governed by the mnemonic rules mentioned below Eq. (A.3-29).

A.4 VECTOR DIFFERENTIAL OPERATORS

In transport problems we are interested in evaluating various quantities, called *field variables*, which are functions of position. Two examples are temperature and fluid velocity. According to the directional nature of a particular field variable (or lack thereof), the value of the function is a scalar, a vector, or a tensor. The necessary information on the spatial rates of change of such quantities is obtained by using the differential operators discussed in this section. Field variables may, of course, be functions of time as well as position. However, the derivatives of vectors or tensors with respect to time are calculated simply by differentiating each component, and therefore they do not require special attention.

Unlike the results for vector and tensor products in Section A.3, each of the component (or summation) expressions presented in this section is valid only for *rectangular* coordinates. A unique aspect of rectangular coordinates is that the base vectors are independent of position. This allows the $\mathbf{e}_i$ to be treated as *constants* when differentiating (or integrating), and it greatly simplifies the derivation of results involving vector-differential operators. Although the component representations are valid only for rectangular coordinates, the vector–tensor identities presented in the examples and table (Table

A-1) at the end of this section are invariant to the choice of coordinate system. The component representations of the various differential operations in cylindrical and spherical coordinates are discussed in section A.7.

Gradient

Spatial derivatives of field variables are computed using the *gradient operator* or "del," which is represented in rectangular coordinates as

$$\nabla = \mathbf{e}_1 \frac{\partial}{\partial x_1} + \mathbf{e}_2 \frac{\partial}{\partial x_2} + \mathbf{e}_3 \frac{\partial}{\partial x_3} = \sum_i \mathbf{e}_i \frac{\partial}{\partial x_i}. \tag{A.4-1}$$

In derivations, ∇ may be treated much like an ordinary vector. However, because $x_j \partial/\partial x_i$ obviously differs from $\partial x_j/\partial x_i$, care must be taken to keep ∇ in the proper order relative to other quantities. In other words, none of the operations involving ∇ are commutative.

The direct operation of ∇ on any field variable (without a dot or cross) yields the *gradient* of that quantity. Thus, the gradient of the scalar-valued function (or "scalar function") f is a vector, represented as

$$\nabla f = \frac{\partial f}{\partial x_1} \mathbf{e}_1 + \frac{\partial f}{\partial x_2} \mathbf{e}_2 + \frac{\partial f}{\partial x_3} \mathbf{e}_3 = \sum_i \frac{\partial f}{\partial x_i} \mathbf{e}_i, \tag{A.4-2}$$

and the gradient of the vector function $\mathbf{v}$ is a tensor (or dyad), given by

$$\nabla \mathbf{v} = \sum_i \sum_j \frac{\partial v_j}{\partial x_i} \mathbf{e}_i \mathbf{e}_j = \begin{pmatrix} \partial v_1/\partial x_1 & \partial v_2/\partial x_1 & \partial v_3/\partial x_1 \\ \partial v_1/\partial x_2 & \partial v_2/\partial x_2 & \partial v_3/\partial x_2 \\ \partial v_1/\partial x_3 & \partial v_2/\partial x_3 & \partial v_3/\partial x_3 \end{pmatrix}. \tag{A.4-3}$$

Divergence

The dot product of ∇ with a field variable (vector or tensor) gives what is called the *divergence.* The divergence of the vector $\mathbf{v}$ is a scalar, represented as

$$\nabla \cdot \mathbf{v} = \frac{\partial v_1}{\partial x_1} + \frac{\partial v_2}{\partial x_2} + \frac{\partial v_3}{\partial x_3} = \sum_i \frac{\partial v_i}{\partial x_i}, \tag{A.4-4}$$

and the divergence of the tensor τ is a vector,

$$\nabla \cdot \tau = \sum_i \sum_j \frac{\partial \tau_{ij}}{\partial x_i} \mathbf{e}_j. \tag{A.4-5}$$

In some books $\nabla \cdot \mathbf{v}$ is written as "div $\mathbf{v}$."

Curl

The cross product of ∇ with a field variable gives what is termed the *curl.* For example, the curl of $\mathbf{v}$ is the vector represented by

$$\nabla \times \mathbf{v} = \sum_i \sum_j \sum_k \varepsilon_{ijk} \frac{\partial v_j}{\partial x_i} \mathbf{e}_k. \tag{A.4-6}$$

Writing out all terms, we have

$$\nabla \times \mathbf{v} = \left(\frac{\partial v_3}{\partial x_2} - \frac{\partial v_2}{\partial x_3}\right)\mathbf{e}_1 + \left(\frac{\partial v_1}{\partial x_3} - \frac{\partial v_3}{\partial x_1}\right)\mathbf{e}_2 + \left(\frac{\partial v_2}{\partial x_1} - \frac{\partial v_1}{\partial x_2}\right)\mathbf{e}_3. \tag{A.4-7}$$

Some authors write $\nabla \times \mathbf{v}$ as "curl $\mathbf{v}$" or "rot $\mathbf{v}$."

Laplacian

The dot product of the gradient operator with itself results in what is called the *Laplacian*,

$$\nabla \cdot \nabla = \nabla^2 = \frac{\partial^2}{\partial x_1^2} + \frac{\partial^2}{\partial x_2^2} + \frac{\partial^2}{\partial x_3^2} = \sum_i \frac{\partial^2}{\partial x_i^2}, \tag{A.4-8}$$

which can operate on a scalar, a vector, or a tensor. [As with the other component representations in this section, it is emphasized again that Eq. (A.4-8) is valid only for rectangular coordinates.] Given that the Laplacian is a scalar, it does not change the tensorial order of the field variable. For example, the Laplacian of the vector $\mathbf{v}$ is a vector written as

$$\nabla^2 \mathbf{v} = \mathbf{e}_1 \nabla^2 v_1 + \mathbf{e}_2 \nabla^2 v_2 + \mathbf{e}_3 \nabla^2 v_3 = \sum_i \sum_j \frac{\partial^2 v_j}{\partial x_i^2} \mathbf{e}_j. \tag{A.4-9}$$

In some books the symbol Δ is used instead of ∇^2.

Material Derivative

One other differential operator which is used very frequently in transport analysis is the *material derivative*. It is written as

$$\frac{D}{Dt} = \frac{\partial}{\partial t} + \mathbf{v} \cdot \nabla = \frac{\partial}{\partial t} + \sum_i v_i \frac{\partial}{\partial x_i}, \tag{A.4-10}$$

where t is time and $\mathbf{v}$ is now the local fluid velocity. As discussed in Chapter 2, the material derivative provides the rate of change of any field variable as perceived by an observer moving with the fluid. In particular, $\mathbf{v} \cdot \nabla$ gives the contribution to the apparent rate of change that is made by the motion of the observer.

Example A.4-1 Proof of a Vector–Tensor Identity As an example of how the representations of differential operators may be used to prove a vector–tensor identity, we will show that

$$\nabla \cdot [\nabla \mathbf{v} + (\nabla \mathbf{v})^t] = \nabla^2 \mathbf{v} + \nabla(\nabla \cdot \mathbf{v}). \tag{A.4-11}$$

Evaluating the left-hand side, we obtain

$$\nabla \cdot [\nabla \mathbf{v} + (\nabla \mathbf{v})^t] = \sum_i \sum_j \sum_k \frac{\partial}{\partial x_k}\left(\frac{\partial v_j}{\partial x_i} + \frac{\partial v_i}{\partial x_j}\right)(\mathbf{e}_k \cdot \mathbf{e}_i \mathbf{e}_j) = \sum_i \sum_j \left(\frac{\partial^2 v_j}{\partial x_i^2} + \frac{\partial^2 v_i}{\partial x_i \partial x_j}\right)\mathbf{e}_j. \tag{A.4-12}$$

Comparing with Eq. (A.4-9), we see that the first term on the far right equals $\nabla^2 \mathbf{v}$. The second term on the far right is rewritten as

$$\sum_i \sum_j \frac{\partial^2 v_i}{\partial x_i \partial x_j} \mathbf{e}_j = \sum_j \mathbf{e}_j \frac{\partial}{\partial x_j}\left(\sum_i \frac{\partial v_i}{\partial x_i}\right) = \nabla(\nabla \cdot \mathbf{v}). \tag{A.4-13}$$

This completes the proof of Eq. (A.4-11), which is used in deriving the Navier–Stokes equation in Chapter 5.

Example A.4-2 Proof of a Vector–Tensor Identity   As a second example, we will show that

$$\nabla \cdot (\boldsymbol{\tau} \cdot \mathbf{v}) = \boldsymbol{\tau} : \nabla \mathbf{v} + \mathbf{v} \cdot (\nabla \cdot \boldsymbol{\tau}),\tag{A.4-14}$$

provided that $\boldsymbol{\tau}$ is *symmetric*. We start again by evaluating the left-hand side to give

$$\nabla \cdot (\boldsymbol{\tau} \cdot \mathbf{v}) = \sum_i \sum_j \sum_m \sum_n \frac{\partial}{\partial x_i}(\tau_{jm} v_n) \mathbf{e}_i \cdot (\mathbf{e}_j \mathbf{e}_m \cdot \mathbf{e}_n)$$

$$= \sum_i \sum_j \frac{\partial}{\partial x_i}(\tau_{ij} v_j) = \sum_i \sum_j \left(\tau_{ji}\frac{\partial v_i}{\partial x_j} + v_i\frac{\partial \tau_{ji}}{\partial x_j}\right).\tag{A.4-15}$$

Obtaining the last equality required two steps, expanding the derivative of the product and interchanging the indices. The indices were switched to facilitate comparisons with the other expressions below. The first term on the right-hand side of Eq. (A.4-14) gives

$$\boldsymbol{\tau} : \nabla \mathbf{v} = \sum_i \sum_j \sum_m \sum_n \tau_{ij}\frac{\partial v_n}{\partial x_m}\mathbf{e}_i\mathbf{e}_j : \mathbf{e}_m\mathbf{e}_n = \sum_i \sum_j \tau_{ij}\frac{\partial v_i}{\partial x_j}.\tag{A.4-16}$$

For symmetric $\boldsymbol{\tau}$ (i.e., for $\tau_{ij} = \tau_{ji}$), this matches the first term on the far right of Eq. (A.4-15). Expanding the remaining term in Eq. (A.4-14), we obtain

$$\mathbf{v} \cdot (\nabla \cdot \boldsymbol{\tau}) = \sum_i \sum_j \sum_m \sum_n v_i \frac{\partial \tau_{mn}}{\partial x_j}\mathbf{e}_i \cdot (\mathbf{e}_j \cdot \mathbf{e}_m\mathbf{e}_n) = \sum_i \sum_j v_i \frac{\partial \tau_{ji}}{\partial x_j},\tag{A.4-17}$$

which matches the last term in Eq. (A.4-15). This completes the proof of Eq. (A.4-14), which is used in deriving the general form of the energy equation for a pure fluid in Chapter 9.

A number of other useful identities involving vector-differential operators are given in Table A-1. These identities are taken mainly from Brand (1947) and Hay (1953); proofs for most are given in those sources. All may be confirmed as in the preceding examples. Although the expressions in this section which involve vector and tensor components are generally valid only for rectangular coordinates, the identities in Table A-1 (and other expressions in Gibbs form) are independent of the coordinate system.

A.5 INTEGRAL TRANSFORMATIONS

General conservation statements for transported quantities are derived as integral equations involving the field variables. To put these equations in the most helpful forms, a number of integral transformations are needed. Some of the more useful ones are summarized in this section. These results are mainly from Brand (1947), which contains an exceptionally complete treatment of integral transformations; proofs of most are given in that source. In the general formulas in this section (and elsewhere in this book), all volume, surface, and contour (line) integrals are written using a single integral sign. The type of integration is identified both by the differential element inside the integral (dV, dS, or dC, respectively) and by the domain indicated under the integral sign (usually just V, S, or C). Although usually not shown explicitly, in all formulas in this section the limits of integration are allowed to depend on time.

TABLE A-1
Identities Involving Vector-Differential Operators[a]

(1)	$\nabla(f\mathbf{a})=(\nabla f)\mathbf{a}+f\nabla\mathbf{a}$
(2)	$\nabla\cdot(f\mathbf{a})=(\nabla f)\cdot\mathbf{a}+f\nabla\cdot\mathbf{a}$
(3)	$\nabla\times(f\mathbf{a})=(\nabla f)\times\mathbf{a}+f\nabla\times\mathbf{a}$
(4)	$\nabla(\mathbf{a}\times\mathbf{b})=(\nabla\mathbf{a})\times\mathbf{b}-(\nabla\mathbf{b})\times\mathbf{a}$
(5)	$\nabla\cdot(\mathbf{a}\times\mathbf{b})=\mathbf{b}\cdot(\nabla\times\mathbf{a})-\mathbf{a}\cdot(\nabla\times\mathbf{b})$
(6)	$\nabla\times(\mathbf{a}\times\mathbf{b})=\mathbf{b}\cdot\nabla\mathbf{a}-\mathbf{a}\cdot\nabla\mathbf{b}+\mathbf{a}(\nabla\cdot\mathbf{b})-\mathbf{b}(\nabla\cdot\mathbf{a})$
(7)	$\nabla\times\nabla f=\mathbf{0}$
(8)	$\nabla\cdot(\nabla f\times\nabla g)=0$
(9)	$\nabla\cdot(\nabla\times\mathbf{a})=0$
(10)	$\nabla\times(\nabla\times\mathbf{a})=\nabla(\nabla\cdot\mathbf{a})-\nabla^2\mathbf{a}$
(11)	$\nabla(\mathbf{a}\cdot\mathbf{b})=\mathbf{a}\cdot\nabla\mathbf{b}+\mathbf{b}\cdot\nabla\mathbf{a}+\mathbf{a}\times(\nabla\times\mathbf{b})+\mathbf{b}\times(\nabla\times\mathbf{a})=(\nabla\mathbf{a})\cdot\mathbf{b}+(\nabla\mathbf{b})\cdot\mathbf{a}$
(12)	$\nabla\cdot(\mathbf{ab})=\mathbf{a}\cdot\nabla\mathbf{b}+\mathbf{b}(\nabla\cdot\mathbf{a})$
(13)	$\nabla\cdot(f\boldsymbol{\delta})=\nabla f$
(14)	$\boldsymbol{\delta}{:}\nabla\mathbf{a}=\nabla\cdot\mathbf{a}$

[a] In these relationships f and g are any differentiable scalar functions and $\mathbf{a}$ and $\mathbf{b}$ are any differentiable vector functions.

Volume and Surface Integrals

Three of the more useful transformations involving volume and surface integrals are

$$\int_V \nabla f\, dV = \int_S \mathbf{n} f\, dS, \tag{A.5-1}$$

$$\int_V \nabla\cdot\mathbf{v}\, dV = \int_S \mathbf{n}\cdot\mathbf{v}\, dS, \tag{A.5-2}$$

$$\int_V \nabla\cdot\boldsymbol{\tau}\, dV = \int_S \mathbf{n}\cdot\boldsymbol{\tau}\, dS. \tag{A.5-3}$$

In each of these equations, S is a surface that completely encloses the volume V, and the functions (f, $\mathbf{v}$, $\boldsymbol{\tau}$) are assumed to be continuous and to have continuous partial derivatives in V and on S. The bounding surface may have two or more distinct parts. For example, V may be the volume between two concentric spheres, in which case S consists of the surface of the inner sphere plus the surface of the outer sphere. In any event, at any point on S there are definite "inward" and "outward" directions of any vector function of position, according to whether the vector is pointed toward or away from V, respectively. The vector $\mathbf{n}$ is a *unit outward normal,* meaning that it has unit magnitude, points outward, and is normal (perpendicular) to the surface. It is assumed that $\mathbf{n}$ is a *piecewise continuous* function of position. That is, S must be divisible into a finite number of sections, in each of which **n** is a continuous function. Such a surface is *piecewise smooth.*

Although written for a scalar, Eq. (A.5-1) applies to functions of any tensorial order. Equation (A.5-2), which is associated with Gauss, is usually called the *divergence theorem.* Its generalization for tensors, Eq. (A.5-3), is referred to here as the *divergence theorem for tensors.* Formulas analogous to Eqs. (A.5-2) and (A.5-3) exist also for cross products; in these, each dot symbol is replaced by a cross.

Green's Identities

Other useful relationships follow directly from the divergence theorem. For example, let $\mathbf{v} = \phi \, \nabla \psi$ in Eq. (A.5-2), where ϕ and ψ are scalar functions with continuous first and second partial derivatives in V. Then, with the help of identity (2) of Table A-1, we obtain

$$\int_V (\nabla\phi\cdot\nabla\psi + \phi\nabla^2\psi)\,dV = \int_S \phi\frac{\partial\psi}{\partial n}\,dS, \qquad (A.5\text{-}4)$$

where $\partial\psi/\partial n = \mathbf{n}\cdot\nabla\psi$, the outward-normal component of the gradient. This is called *Green's first identity.* An analogous result is obtained by interchanging ϕ and ψ. Subtracting that from Eq. (A.5-4), we obtain *Green's second identity,*

$$\int_V (\phi\nabla^2\psi - \psi\nabla^2\phi)\,dV = \int_S \left(\phi\frac{\partial\psi}{\partial n} - \psi\frac{\partial\phi}{\partial n}\right)dS, \qquad (A.5\text{-}5)$$

which has a symmetric form.

Surface and Contour Integrals

There are many relationships also involving integrals over a surface S and a contour C that bounds that surface. In such equations S is not a closed surface, in general. These relationships generally involve a unit vector **t** that is tangent to the curve C. As shown in Fig. A-2, when **n** is pointed upward and S is viewed from above, **t** is taken to be in the *counterclockwise* direction along C. A third unit vector is $\mathbf{m} = \mathbf{t}\times\mathbf{n}$, which is tangent to the surface and directed away from S (i.e., outwardly normal to C).

Two of the more useful relationships between surface and contour integrals are

$$\int_S (\mathbf{n}\times\nabla)\cdot\mathbf{v}\,dS = \int_C \mathbf{t}\cdot\mathbf{v}\,dC = \int_S \mathbf{n}\cdot(\nabla\times\mathbf{v})\,dS, \qquad (A.5\text{-}6)$$

Figure A-2. Unit normal and unit tangent vectors for a surface S bounded by a curve C.

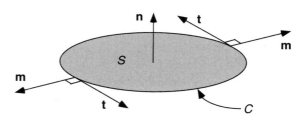

$$\int_S (\mathbf{n} \times \nabla) \times \mathbf{v} \, dS = \int_C \mathbf{t} \times \mathbf{v} \, dC, \tag{A.5-7}$$

which require that $\mathbf{v}$ have continuous derivatives over S. Equation (A.5-6), which is usually written in the second form, is called *Stokes' theorem;* it applies also to tensors (i.e., one can replace $\mathbf{v}$ by $\boldsymbol{\tau}$). The use of Eq. (A.5-7) is discussed further in Section A.8.

Leibniz Formulas for Differentiating Integrals

We often need to differentiate integrals which have variable limits of integration. Suppose, for example, that we have an integral of the function $f(x,t)$ over x, where x has the range $A(t) \leq x \leq B(t)$. The derivative with respect to t is given by

$$\frac{d}{dt} \int_{A(t)}^{B(t)} f(x, t) \, dx = \int_{A(t)}^{B(t)} \frac{\partial f}{\partial t} \, dx + \frac{dB}{dt} f(B(t), t) - \frac{dA}{dt} f(A(t), t), \tag{A.5-8}$$

which is often called the *Leibniz rule* (e.g., Hildebrand, 1976, pp. 364–365). Using $\mathbf{r}$ to denote position in space (Section A.6), two generalizations of this formula to volume integrals are

$$\frac{d}{dt} \int_{V(t)} f(\mathbf{r},t) \, dV = \int_{V(t)} \frac{\partial f}{\partial t} \, dV + \int_{S(t)} (\mathbf{n} \cdot \mathbf{v}_S) f \, dS, \tag{A.5-9}$$

$$\frac{d}{dt} \int_{V(t)} \mathbf{v}(\mathbf{r},t) \, dV = \int_{V(t)} \frac{\partial \mathbf{v}}{\partial t} \, dV + \int_{S(t)} (\mathbf{n} \cdot \mathbf{v}_S) \mathbf{v} \, dS, \tag{A.5-10}$$

where $\mathbf{v}_S$ is the velocity of the surface, a function of position and time. Thus, $\mathbf{n} \cdot \mathbf{v}_S$ is the outward-normal velocity of a given point on the surface at a given instant. All three of these equations imply that unless the boundary is stationary and/or the integrand vanishes at the boundary, it is not permissible to simply move the t derivative inside the integral.

If x and t in Eq. (A.5-8) are interpreted as position and time, respectively, then dA/dt and dB/dt represent the velocities of the endpoints of the interval. The "outward" velocities at the upper and lower limits of integration are in the positive and negative x directions, respectively; hence, the different algebraic signs of the terms containing dA/dt and dB/dt. In Eqs. (A.5-9) and (A.5-10), the direction of surface motion is accounted for automatically by $\mathbf{n} \cdot \mathbf{v}_S$, which may be positive or negative. Although only the most common special cases are shown, both the one-dimensional and three-dimensional forms of the Leibniz formula apply to functions of any tensorial order.

A.6 POSITION VECTORS

Definition

A *position vector,* denoted in this book as $\mathbf{r}$, is a vector that extends from an arbitrary reference point to some other point in space. Thus, $\mathbf{r}$ identifies the location of the latter point, and we often speak of "the point $\mathbf{r}$" as if the point and its position vector were one and the same. The position vector designates locations in three-dimensional space without reference to a coordinate system. For example, scalar, vector, and tensor functions of position are written simply as $f(\mathbf{r})$, $\mathbf{v}(\mathbf{r})$, and $\boldsymbol{\tau}(\mathbf{r})$, respectively. As with other vectors, $\mathbf{r}$ is invariant to the choice of coordinate system and may be represented using any coordinates which are convenient. If the reference point is the origin of a *rectangular* coordinate system, the position vector is given by

$$\mathbf{r} = x\mathbf{e}_x + y\mathbf{e}_y + z\mathbf{e}_z = \sum_i x_i \mathbf{e}_i. \qquad \text{(A.6-1)}$$

Thus, in this representation the components of $\mathbf{r}$ correspond to the rectangular coordinates of a point. The magnitude of $\mathbf{r}$ (denoted as r) is identical to the radial coordinate in a spherical coordinate system (see Section A.7). As shown in the summation in Eq. (A.6-1), it is convenient to represent the rectangular coordinates as x_i. (Thus, another commonly used symbol for the position vector is $\mathbf{x}$.) As in Section A.4, all of the relations in this section involving summations are restricted to rectangular coordinates, whereas those in Gibbs form are general.

Space Curves

Let s be arc length along some curve in three-dimensional space, such that the vector function $\mathbf{r}(s)$ represents all points on the curve. If $\Delta\mathbf{r}$ is a vector directed from one point to another on the curve, and Δs is the corresponding arc length, then the limit of $\Delta\mathbf{r}/\Delta s$ as $\Delta s \to 0$ equals $d\mathbf{r}/ds$. Moreover,

$$\frac{d\mathbf{r}}{ds} = \sum_i \frac{dx_i}{ds}\mathbf{e}_i = \mathbf{t}, \qquad \text{(A.6-2)}$$

where $\mathbf{t}$ is a unit vector that is tangent to the curve and directed toward increasing s.

The rate of change of a scalar function f along such a curve is given by

$$\frac{\partial f}{\partial s} = \mathbf{t} \cdot \nabla f. \qquad \text{(A.6-3)}$$

This result is obtained by writing

$$\frac{\partial f}{\partial s} = \sum_i \frac{\partial f}{\partial x_i}\frac{dx_i}{ds} = \left(\sum_i \frac{dx_i}{ds}\mathbf{e}_i\right) \cdot \left(\sum_j \frac{\partial f}{\partial x_j}\mathbf{e}_j\right). \qquad \text{(A.6-4)}$$

In the first equality we have used the "chain rule" for differentiation. The first and second quantities in parentheses represent $\mathbf{t}$ and ∇f, respectively. It follows that the change in f corresponding to a differential displacement $d\mathbf{r} = (dx_1, dx_2, dx_3)$ is given by

$$df = d\mathbf{r} \cdot \nabla f. \qquad \text{(A.6-5)}$$

Although written for the scalar function f, Eqs. (A.6-3) and (A.6-5) apply to functions of any tensorial order.

Taylor Series

Taylor-series expansions of functions can be written in terms of position vectors and vector-differential operators. For example, Eq. (A.6-5) indicates that the first correction to $f(\mathbf{r})$ at a position $\mathbf{r}+\mathbf{r}'$ is given by $\mathbf{r}'\cdot\nabla f$, where the gradient is evaluated at position $\mathbf{r}$. Including also the second-derivative terms, the value of a function at position $\mathbf{r}+\mathbf{r}'$, written as a Taylor expansion about position $\mathbf{r}$, is given by

$$f(\mathbf{r}+\mathbf{r}')=f(\mathbf{r})+\mathbf{r}'\cdot\nabla f+\frac{1}{2!}\mathbf{r}'\mathbf{r}':\nabla\nabla f+\cdots, \tag{A.6-6}$$

$$v(\mathbf{r}+\mathbf{r}')=v(\mathbf{r})+\mathbf{r}'\cdot\nabla v+\frac{1}{2!}\mathbf{r}'\mathbf{r}':\nabla\nabla v+\cdots, \tag{A.6-7}$$

$$\tau(\mathbf{r}+\mathbf{r}')=\tau(\mathbf{r})+\mathbf{r}'\cdot\nabla\tau+\frac{1}{2!}\mathbf{r}'\mathbf{r}':\nabla\nabla\tau+\cdots, \tag{A.6-8}$$

where all derivatives are evaluated at $\mathbf{r}$. Notice that the same formula applies to functions of any tensorial order. It is readily confirmed that Eq. (A.6-6) reduces to the usual one- or two-dimensional Taylor expansions of scalar functions presented in most calculus books. Terms beyond those shown may be calculated as described in Morse and Feshbach (1953, p. 1277).

Differential Identities

The relationship between the components of $\mathbf{r}$ and the coordinates leads to special properties for the gradient, divergence, and curl of $\mathbf{r}$. Recognizing that $\partial x_i/\partial x_j=\delta_{ij}$, we obtain

$$\nabla\mathbf{r}=\sum_i\sum_j\frac{\partial x_j}{\partial x_i}\mathbf{e}_i\mathbf{e}_j=\sum_i\mathbf{e}_i\mathbf{e}_i=\boldsymbol{\delta}, \tag{A.6-9}$$

$$\nabla\cdot\mathbf{r}=\sum_i\sum_j\frac{\partial x_j}{\partial x_i}\mathbf{e}_i\cdot\mathbf{e}_j=\sum_i\delta_{ii}=3, \tag{A.6-10}$$

$$\nabla\times\mathbf{r}=\sum_i\sum_j\sum_k\varepsilon_{ijk}\frac{\partial x_j}{\partial x_i}\mathbf{e}_k=\sum_i\sum_k\varepsilon_{iik}\mathbf{e}_k=\mathbf{0}. \tag{A.6-11}$$

Integral Identities

It is sometimes necessary to calculate the integral of a position vector or position dyad over a surface. The integral of any vector or tensor is calculated by integrating each component, including the unit vector or unit dyad; in rectangular coordinates, the constancy of the unit vectors and unit dyads allows them to be carried through the integration unchanged. Let S_R represent a spherical surface of radius R. Then, the integrals of $\mathbf{r}$ and $\mathbf{r}\mathbf{r}$ over that surface are

$$\int_{S_R}\mathbf{r}\,dS=\mathbf{0}, \tag{A.6-12}$$

$$\int_{S_R}\mathbf{r}\mathbf{r}\,dS=\frac{4\pi R^4}{3}\boldsymbol{\delta}. \tag{A.6-13}$$

The integral of $\mathbf{r}$ vanishes due to the spherical symmetry. That is, there are negative values of the integrand to cancel each positive contribution. The integral of $\mathbf{r}\mathbf{r}$ is evaluated by converting its components (but not the unit dyads) to spherical coordinates, and integrating each over the two spherical angles (see Section A.7). The off-diagonal terms in Eq. (A.6-13) vanish, again due to the symmetry.

A.7 ORTHOGONAL CURVILINEAR COORDINATES

Enormous simplificatons are achieved in solving a partial differential equation if all boundaries in the problem correspond to *coordinate surfaces,* which are surfaces generated by holding one coordinate constant and varying the other two. Accordingly, many special coordinate systems have been devised to solve problems in particular geometries. The most useful of these systems are *orthogonal;* that is, at any point in space the vectors aligned with the three coordinate directions are mutually perpendicular. In general, the variation of a single coordinate will generate a curve in space, rather than a straight line; hence the term *curvilinear.* In this section a general discussion of orthogonal curvilinear systems is given first, and then the relationships for cylindrical and spherical coordinates are derived as special cases. The presentation here closely follows that in Hildebrand (1976).

Base Vectors

Let (u_1, u_2, u_3) represent the three coordinates in a general, curvilinear system, and let $\mathbf{e}_i$ be the unit vector that points in the direction of increasing u_i. A curve produced by varying u_i, with u_j ($j \neq i$) held constant, will be referred to as a "u_i curve." Although the base vectors are each of constant (unit) magnitude, the fact that a u_i curve is not generally a straight line means that their direction is variable. In other words, $\mathbf{e}_i$ must be regarded as a function of position, in general. This discussion is restricted to coordinate systems in which $(\mathbf{e}_1, \mathbf{e}_2, \mathbf{e}_3)$ is an *orthonormal* and *right-handed* set. At any point in space, such a set has the properties of the base vectors used in Section A.3, namely,

$$\mathbf{e}_i \cdot \mathbf{e}_j = \delta_{ij}, \tag{A.7-1}$$

$$\mathbf{e}_i \times \mathbf{e}_j = \sum_k \varepsilon_{ijk} \mathbf{e}_k. \tag{A.7-2}$$

Recalling that the multiplication properties of vectors and tensors are derived from these relationships (and their extensions to unit dyads), we see that *all of the relations in Section A.3 apply to orthogonal curvilinear systems in general,* and not just to rectangular coordinates. It is with spatial derivatives that the variations in $\mathbf{e}_i$ come into play, and the main task in this section is to show how the various differential operators differ from those given in Section A.4 for rectangular coordinates. In the process, we will obtain general expressions for differential elements of arc length, volume, and surface area.

Arc Length

The key to deriving expressions for curvilinear coordinates is to consider the arc length along a curve. In particular, let s_i represent arc length along a u_i curve. From Eq. (A.6-2), a vector that is tangent to a u_i curve and directed toward increasing u_i is given by

$$\mathbf{a}_i = \frac{\partial \mathbf{r}}{\partial u_i} = h_i \mathbf{e}_i, \tag{A.7-3}$$

where $h_i \equiv ds_i/du_i$ is called the *scale factor*. In general, u_i will differ from s_i, so that $\mathbf{a}_i$ is not a *unit* tangent (i.e., $\mathbf{a}_i \neq \mathbf{e}_i$). The relationship between a coordinate and the corresponding arc length is embodied in the scale factor, which generally depends on position. For an arbitrary curve in space with arc length s, we find that

$$t = \frac{d\mathbf{r}}{ds} = \sum_i \frac{\partial \mathbf{r}}{\partial u_i} \frac{du_i}{ds} = \sum_i h_i \mathbf{e}_i \frac{du_i}{ds}. \tag{A.7-4}$$

The properties of the unit tangent imply that

$$\mathbf{t} \cdot \mathbf{t} = 1 = \sum_i h_i^2 \left(\frac{du_i}{ds}\right)^2 \tag{A.7-5}$$

or, after rearranging, that

$$ds = \left[\sum_i h_i^2 (du_i)^2\right]^{1/2}. \tag{A.7-6}$$

The fact that a space curve has an independent geometric significance indicates that the quantity in brackets must be invariant to the choice of coordinate system.

Volume and Surface Area

For a given coordinate system, the differential volume element dV corresponds to the volume of a parallelepiped with adjacent edges $\mathbf{e}_i ds_i$. From Eq. (A.7-3) and the definition of h_i, the edges can be represented also as $\mathbf{a}_i du_i$. As noted in connection with Eq. (A.3-20), the volume is given by the *scalar triple product,* or

$$dV = (\mathbf{a}_1 du_1) \times (\mathbf{a}_2 du_2) \cdot (\mathbf{a}_3 du_3) = h_1 h_2 h_3 \, du_1 \, du_2 \, du_3. \tag{A.7-7}$$

Expressions for differential surface elements are obtained in a similar manner, by using the geometric interpretation of the *cross product.* Thus, letting dS_i refer to a surface on which the coordinate u_i is held constant, we obtain

$$dS_1 = |\mathbf{a}_2 \times \mathbf{a}_3| \, du_2 \, du_3 = h_2 h_3 \, du_2 \, du_3, \tag{A.7-8a}$$

$$dS_2 = |\mathbf{a}_3 \times \mathbf{a}_1| \, du_1 \, du_3 = h_1 h_3 \, du_1 \, du_3, \tag{A.7-8b}$$

$$dS_3 = |\mathbf{a}_1 \times \mathbf{a}_2| \, du_1 \, du_2 = h_1 h_2 \, du_1 \, du_2. \tag{A.7-8c}$$

Gradient

An expression for the gradient is obtained by examining the differential change in a scalar function associated with a differential change in position. Letting $f = f(u_1, u_2, u_3)$, we have

$$df = \sum_i \frac{\partial f}{\partial u_i} \, du_i. \tag{A.7-9}$$

From Eqs. (A.6-5) and (A.7-4), df can also be written as

$$df = \mathbf{dr} \cdot \boldsymbol{\nabla} f = \left(\sum_i h_i \mathbf{e}_i \, du_i \right) \cdot \left(\sum_j \lambda_j \mathbf{e}_j \right) = \sum_i h_i \lambda_i \, du_i, \qquad \text{(A.7-10)}$$

where the quantities λ_i are to be determined. Comparing Eqs. (A.7-9) and (A.7-10), we see that $\lambda_i = (1/h_i) \, \partial f / \partial u_i$ and

$$\boldsymbol{\nabla} = \sum_i \frac{\mathbf{e}_i}{h_i} \frac{\partial}{\partial u_i}. \qquad \text{(A.7-11)}$$

This is the general expression for the gradient operator, valid for any orthogonal, curvilinear coordinate system.

Several identities involving u_i, h_i, and $\mathbf{e}_i$ are useful in deriving expressions for the other differential operators. From Eq. (A.7-11) we obtain

$$\boldsymbol{\nabla} u_i = \sum_j \frac{\mathbf{e}_j}{h_j} \frac{\partial u_i}{\partial u_j} = \sum_j \frac{\mathbf{e}_j}{h_j} \delta_{ij} = \frac{\mathbf{e}_i}{h_i}. \qquad \text{(A.7-12)}$$

From this and identity (7) of Table A-1, it follows that

$$\boldsymbol{\nabla} \times \boldsymbol{\nabla} u_i = \boldsymbol{\nabla} \times \frac{\mathbf{e}_i}{h_i} = \mathbf{0}. \qquad \text{(A.7-13)}$$

Using Eqs. (A.7-2) and (A.7-12) we find, for example, that

$$\frac{\mathbf{e}_1}{h_2 h_3} = \frac{\mathbf{e}_2}{h_2} \times \frac{\mathbf{e}_3}{h_3} = \boldsymbol{\nabla} u_2 \times \boldsymbol{\nabla} u_3. \qquad \text{(A.7-14)}$$

Two analogous relations are obtained by cyclic permutation of the subscripts. From these and identity (8) of Table A-1 it is found that

$$\boldsymbol{\nabla} \cdot \frac{\mathbf{e}_1}{h_2 h_3} = \boldsymbol{\nabla} \cdot \frac{\mathbf{e}_2}{h_3 h_1} = \boldsymbol{\nabla} \cdot \frac{\mathbf{e}_3}{h_1 h_2} = 0. \qquad \text{(A.7-15)}$$

Divergence

To evaluate the divergence of the vector $\mathbf{v}$, we first consider just one component. Bearing in mind that the unit vectors are not necessarily constants, and using identity (2) of Table A-1, the divergence of $v_1 \mathbf{e}_1$ is expanded as

$$\boldsymbol{\nabla} \cdot (v_1 \mathbf{e}_1) = \boldsymbol{\nabla} \cdot \left[(h_2 h_3 v_1) \left(\frac{\mathbf{e}_1}{h_2 h_3} \right) \right] = \boldsymbol{\nabla} (h_2 h_3 v_1) \cdot \frac{\mathbf{e}_1}{h_2 h_3} + h_2 h_3 v_1 \boldsymbol{\nabla} \cdot \left(\frac{\mathbf{e}_1}{h_2 h_3} \right).$$
$$\text{(A.7-16)}$$

It is seen from Eq. (A.7-15) that the term on the far right of Eq. (A.7-16) is identically zero. Using Eq. (A.7-11) to evaluate the gradient in the remaining term, we find that

$$\boldsymbol{\nabla} \cdot (v_1 \mathbf{e}_1) = \sum_i \frac{\mathbf{e}_i}{h_i} \frac{\partial}{\partial u_i} (h_2 h_3 v_1) \cdot \frac{\mathbf{e}_1}{h_2 h_3} = \frac{1}{h_1 h_2 h_3} \frac{\partial}{\partial u_1} (h_2 h_3 v_1). \qquad \text{(A.7-17)}$$

Treating the other components in a similar manner results in

$$\nabla \cdot \mathbf{v} = \frac{1}{h_1 h_2 h_3} \left[\frac{\partial}{\partial u_1} (h_2 h_3 v_1) + \frac{\partial}{\partial u_2} (h_3 h_1 v_2) + \frac{\partial}{\partial u_3} (h_1 h_2 v_3) \right], \qquad \text{(A.7-18)}$$

which is the general expression for the divergence.

Curl

The curl of $\mathbf{v}$ is evaluated in a similar manner. Expanding one component, we obtain

$$\nabla \times (v_1 \mathbf{e}_1) = \nabla \times \left[(h_1 v_1) \left(\frac{\mathbf{e}_1}{h_1} \right) \right] = \nabla (h_1 v_1) \times \frac{\mathbf{e}_1}{h_1} + h_1 v_1 \nabla \times \frac{\mathbf{e}_1}{h_1}. \qquad \text{(A.7-19)}$$

Again, the term on the far right vanishes [see Eq. (A.7-13)]. The remaining term is expanded further as

$$\nabla \times (v_1 \mathbf{e}_1) = \sum_i \frac{\mathbf{e}_i}{h_i} \frac{\partial}{\partial u_i} (h_1 v_1) \times \frac{\mathbf{e}_1}{h_1} = \frac{1}{h_1 h_2 h_3} \left(h_2 \mathbf{e}_2 \frac{\partial}{\partial u_3} - h_3 \mathbf{e}_3 \frac{\partial}{\partial u_2} \right) (h_1 v_1).$$

$$\text{(A.7-20)}$$

When all components are included, the curl is written most compactly as a determinant:

$$\nabla \times \mathbf{v} = \frac{1}{h_1 h_2 h_3} \begin{vmatrix} h_1 \mathbf{e}_1 & h_2 \mathbf{e}_2 & h_3 \mathbf{e}_3 \\ \partial/\partial u_1 & \partial/\partial u_2 & \partial/\partial u_3 \\ h_1 v_1 & h_2 v_2 & h_3 v_3 \end{vmatrix}. \qquad \text{(A.7-21)}$$

Laplacian

The Laplacian for curvilinear coordinates is derived from Eq. (A.7-18) by setting $\mathbf{v} = \nabla$, or $v_i = (1/h_i) \partial/\partial u_i$. The result is

$$\nabla \cdot \nabla = \nabla^2 = \frac{1}{h_1 h_2 h_3} \left[\frac{\partial}{\partial u_1} \left(\frac{h_2 h_3}{h_1} \frac{\partial}{\partial u_1} \right) + \frac{\partial}{\partial u_2} \left(\frac{h_3 h_1}{h_2} \frac{\partial}{\partial u_2} \right) + \frac{\partial}{\partial u_3} \left(\frac{h_1 h_2}{h_3} \frac{\partial}{\partial u_3} \right) \right].$$

$$\text{(A.7-22)}$$

Material Derivative

The material derivative is given by

$$\frac{D}{Dt} = \frac{\partial}{\partial t} + \mathbf{v} \cdot \nabla = \frac{\partial}{\partial t} + \sum_i \frac{v_i}{h_i} \frac{\partial}{\partial u_i}. \qquad \text{(A.7-23)}$$

This completes the general results for orthogonal curvilinear coordinates. The remainder of this section is devoted to the three most useful special cases.

Rectangular Coordinates

In rectangular coordinates, which we have written as either (x, y, z) or (x_1, x_2, x_3), the position vector is given by Eq. (A.6-1). In this case it is readily shown from Eq. (A.7-3) that $h_x = h_y = h_z = 1$, and it is found that the general expressions for the differential operators reduce to the forms given in Section A.4. For convenient reference, the principal results are summarized in Table A-2.

TABLE A-2
Differential Operations in Rectangular Coordinates[a]

(1)
$$\nabla f=\frac{\partial f}{\partial x}\mathbf{e}_x+\frac{\partial f}{\partial y}\mathbf{e}_y+\frac{\partial f}{\partial z}\mathbf{e}_z$$

(2)
$$\nabla\cdot\mathbf{v}=\frac{\partial v_x}{\partial x}+\frac{\partial v_y}{\partial y}+\frac{\partial v_z}{\partial z}$$

(3)
$$\nabla\times\mathbf{v}=\left[\frac{\partial v_z}{\partial y}-\frac{\partial v_y}{\partial z}\right]\mathbf{e}_x+\left[\frac{\partial v_x}{\partial z}-\frac{\partial v_z}{\partial x}\right]\mathbf{e}_y+\left[\frac{\partial v_y}{\partial x}-\frac{\partial v_x}{\partial y}\right]\mathbf{e}_z$$

(4)
$$\nabla^2 f=\frac{\partial^2 f}{\partial x^2}+\frac{\partial^2 f}{\partial y^2}+\frac{\partial^2 f}{\partial z^2}$$

(5)
$$(\nabla\mathbf{v})_{xx}=\frac{\partial v_x}{\partial x}$$

(6)
$$(\nabla\mathbf{v})_{xy}=\frac{\partial v_y}{\partial x}$$

(7)
$$(\nabla\mathbf{v})_{xz}=\frac{\partial v_z}{\partial x}$$

(8)
$$(\nabla\mathbf{v})_{yx}=\frac{\partial v_x}{\partial y}$$

(9)
$$(\nabla\mathbf{v})_{yy}=\frac{\partial v_y}{\partial y}$$

(10)
$$(\nabla\mathbf{v})_{yz}=\frac{\partial v_z}{\partial y}$$

(11)
$$(\nabla\mathbf{v})_{zx}=\frac{\partial v_x}{\partial z}$$

(12)
$$(\nabla\mathbf{v})_{zy}=\frac{\partial v_y}{\partial z}$$

(13)
$$(\nabla\mathbf{v})_{zz}=\frac{\partial v_z}{\partial z}$$

[a] In these relationships, f is any differentiable scalar function and $\mathbf{v}$ is any differentiable vector function.

Cylindrical Coordinates

Circular cylindrical coordinates, denoted as (r, θ, z), are shown in relation to rectangular coordinates in Fig. A-3(a). Using

$$x=r\cos\theta,\quad y=r\sin\theta,\quad z=z,\tag{A.7-24}$$

the position vector is expressed as

$$\mathbf{r}=r\cos\theta\mathbf{e}_x+r\sin\theta\mathbf{e}_y+z\mathbf{e}_z.\tag{A.7-25}$$

Alternatively, the position vector is given by

$$\mathbf{r}=r\mathbf{e}_r+z\mathbf{e}_z,\tag{A.7-26}$$

where $\mathbf{e}_r$, the unit vector in the radial direction, is given below. Equation (A.7-25) is more convenient for derivations involving differentiation or integration because it involves only constant base vectors. Whichever expression is used, note that in cylindrical coordinates there is an irregularity in our notation, such that $|\mathbf{r}|=(r^2+z^2)^{1/2}\neq r$.

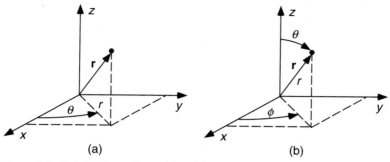

Figure A-3. Cylindrical coordinates (a) and spherical coordinates (b). The ranges of the angles are: cylindrical, $0 \le \theta \le 2\pi$; spherical, $0 \le \theta \le \pi$ and $0 \le \phi \le 2\pi$.

To illustrate the derivation of scale factors and base vectors, consider the θ quantities in cylindrical coordinates. From Eqs. (A.7-3) and (A.7-25) it is found that

$$\frac{\partial \mathbf{r}}{\partial \theta} = h_\theta \mathbf{e}_\theta = -r \sin \theta \mathbf{e}_x + r \cos \theta \mathbf{e}_y, \qquad (A.7\text{-}27)$$

$$h_\theta = \left| \frac{\partial \mathbf{r}}{\partial \theta} \right| = \left(r^2 \sin^2 \theta + r^2 \cos^2 \theta \right)^{1/2} = r, \qquad (A.7\text{-}28)$$

and $\mathbf{e}_\theta = (1/h_\theta)(\partial \mathbf{r}/\partial \theta)$. Repeating the calculations for the r and z quantities, the scale factors and base vectors for cylindrical coordinates are found to be

$$h_r = 1, \qquad h_\theta = r, \qquad h_z = 1, \qquad (A.7\text{-}29)$$

$$\mathbf{e}_r = \cos \theta \mathbf{e}_x + \sin \theta \mathbf{e}_y, \qquad (A.7\text{-}30a)$$

$$\mathbf{e}_\theta = -\sin \theta \mathbf{e}_x + \cos \theta \mathbf{e}_y, \qquad (A.7\text{-}30b)$$

$$\mathbf{e}_z = \mathbf{e}_z. \qquad (A.7\text{-}30c)$$

The dependence of $\mathbf{e}_r$ and $\mathbf{e}_\theta$ on θ is shown in Eq. (A.7-30); the expression for $\mathbf{e}_r$ confirms the equivalence of Eqs. (A.7-25) and (A.7-26). Inverting the relationships in Eq. (A.7-30) to find expressions for the rectangular base vectors, we obtain

$$\mathbf{e}_x = \cos \theta \mathbf{e}_r - \sin \theta \mathbf{e}_\theta, \qquad (A.7\text{-}31a)$$

$$\mathbf{e}_y = \sin \theta \mathbf{e}_r + \cos \theta \mathbf{e}_\theta, \qquad (A.7\text{-}31b)$$

$$\mathbf{e}_z = \mathbf{e}_z. \qquad (A.7\text{-}31c)$$

The differential volume and surface elements are evaluated using Eqs. (A.7-7) and (A.7-8) as

$$dV = r \, dr \, d\theta \, dz, \qquad (A.7\text{-}32)$$

$$dS_r = r \, d\theta \, dz, \qquad dS_\theta = dr \, dz, \qquad dS_z = r \, dr \, d\theta. \qquad (A.7\text{-}33)$$

A summary of differential operations in cylindrical coordinates is presented in Table A-3. Several quantities not shown, including $\nabla^2 \mathbf{v}$, $\nabla \cdot \boldsymbol{\tau}$, and $\mathbf{v} \cdot \nabla \mathbf{v}$, may be obtained from the tables in Chapter 5.

TABLE A-3
Differential Operations in Cylindrical Coordinates[a]

(1)
$$\nabla f = \frac{\partial f}{\partial r}\mathbf{e}_r + \frac{1}{r}\frac{\partial f}{\partial \theta}\mathbf{e}_\theta + \frac{\partial f}{\partial z}\mathbf{e}_z$$

(2)
$$\nabla \cdot \mathbf{v} = \frac{1}{r}\frac{\partial}{\partial r}(r v_r) + \frac{1}{r}\frac{\partial v_\theta}{\partial \theta} + \frac{\partial v_z}{\partial z}$$

(3)
$$\nabla \times \mathbf{v} = \left[\frac{1}{r}\frac{\partial v_z}{\partial \theta} - \frac{\partial v_\theta}{\partial z}\right]\mathbf{e}_r + \left[\frac{\partial v_r}{\partial z} - \frac{\partial v_z}{\partial r}\right]\mathbf{e}_\theta + \left[\frac{1}{r}\frac{\partial}{\partial r}(r v_\theta) - \frac{1}{r}\frac{\partial v_r}{\partial \theta}\right]\mathbf{e}_z$$

(4)
$$\nabla^2 f = \frac{1}{r}\frac{\partial}{\partial r}\left(r\frac{\partial f}{\partial r}\right) + \frac{1}{r^2}\frac{\partial^2 f}{\partial \theta^2} + \frac{\partial^2 f}{\partial z^2}$$

(5)
$$(\nabla \mathbf{v})_{rr} = \frac{\partial v_r}{\partial r}$$

(6)
$$(\nabla \mathbf{v})_{r\theta} = \frac{\partial v_\theta}{\partial r}$$

(7)
$$(\nabla \mathbf{v})_{rz} = \frac{\partial v_z}{\partial r}$$

(8)
$$(\nabla \mathbf{v})_{\theta r} = \frac{1}{r}\frac{\partial v_r}{\partial \theta} - \frac{v_\theta}{r}$$

(9)
$$(\nabla \mathbf{v})_{\theta\theta} = \frac{1}{r}\frac{\partial v_\theta}{\partial \theta} + \frac{v_r}{r}$$

(10)
$$(\nabla \mathbf{v})_{\theta z} = \frac{1}{r}\frac{\partial v_z}{\partial \theta}$$

(11)
$$(\nabla \mathbf{v})_{zr} = \frac{\partial v_r}{\partial z}$$

(12)
$$(\nabla \mathbf{v})_{z\theta} = \frac{\partial v_\theta}{\partial z}$$

(13)
$$(\nabla \mathbf{v})_{zz} = \frac{\partial v_z}{\partial z}$$

[a] In these relationships, f is any differentiable scalar function and $\mathbf{v}$ is any differentiable vector function.

Spherical Coordinates

Spherical coordinates, denoted as (r, θ, ϕ), are shown in relation to rectangular coordinates in Fig. A-3(b). Note that $0 \le \theta \le \pi$ and $0 \le \phi \le 2\pi$. Using

$$x = r \sin \theta \cos \phi, \qquad y = r \sin \theta \sin \phi, \qquad z = r \cos \theta \qquad \text{(A.7-34)}$$

the position vector is expressed as

$$\mathbf{r} = r \sin \theta \cos \phi \mathbf{e}_x + r \sin \theta \sin \phi \mathbf{e}_y + r \cos \theta \mathbf{e}_z. \qquad \text{(A.7-35)}$$

The position vector is also given by

$$\mathbf{r} = r\mathbf{e}_r. \qquad \text{(A.7-36)}$$

Either expression indicates that $|\mathbf{r}| = r$, consistent with our usual notation.

Employing Eq. (A.7-35) in the general relationships for curvilinear coordinates, the scale factors and base vectors for spherical coordinates are evaluated as

$$h_r = 1, \qquad h_\theta = r, \qquad h_\phi = r \sin \theta, \tag{A.7-37}$$

$$\mathbf{e}_r = \sin \theta \cos \phi \mathbf{e}_x + \sin \theta \sin \phi \mathbf{e}_y + \cos \theta \mathbf{e}_z, \tag{A.7-38a}$$

$$\mathbf{e}_\theta = \cos \theta \cos \phi \mathbf{e}_x + \cos \theta \sin \phi \mathbf{e}_y - \sin \theta \mathbf{e}_z, \tag{A.7-38b}$$

$$\mathbf{e}_\phi = -\sin \phi \mathbf{e}_x + \cos \phi \mathbf{e}_y. \tag{A.7-38c}$$

The dependence of the three base vectors on θ and ϕ is shown in Eq. (A.7-38); the expression for $\mathbf{e}_r$ confirms the equivalence of Eqs. (A.7-35) and (A.7-36). The complementary expressions for the rectangular base vectors are

$$\mathbf{e}_x = \sin \theta \cos \phi \mathbf{e}_r + \cos \theta \cos \phi \mathbf{e}_\theta - \sin \phi \mathbf{e}_\phi, \tag{A.7-39a}$$

$$\mathbf{e}_y = \sin \theta \sin \phi \mathbf{e}_r + \cos \theta \sin \phi \mathbf{e}_\theta + \cos \phi \mathbf{e}_\phi, \tag{A.7-39b}$$

$$\mathbf{e}_z = \cos \theta \mathbf{e}_r - \sin \theta \mathbf{e}_\theta. \tag{A.7-39c}$$

Finally, the differential volume and surface elements are evaluated as

$$dV = r^2 \sin \theta \, dr \, d\theta \, d\phi \tag{A.7-40}$$

$$dS_r = r^2 \sin \theta \, d\theta \, d\phi, \qquad dS_\theta = r \sin \theta \, dr \, d\phi, \qquad dS_\phi = r \, dr \, d\theta. \tag{A.7-41}$$

A summary of differential operations in spherical coordinates is presented in Table A-4. As already mentioned, several other quantities, including $\nabla^2 \mathbf{v}$, $\nabla \cdot \boldsymbol{\tau}$, and $\mathbf{v} \cdot \nabla \mathbf{v}$, may be obtained from the tables in Chapter 5.

Many other orthogonal coordinate systems have been developed. A compilation of scale factors, differential operators, and solutions of Laplace's equation in 40 such systems is provided in Moon and Spencer (1961).

A.8 SURFACE GEOMETRY

In Section A.5 a number of integral transformations were presented involving vectors which are normal or tangent to a surface. The objective of this section is to show how those vectors are computed and how they are used to define such quantities as surface gradients. This completes the information needed to understand the integral forms of the various conservation equations. Much of the material in this section is adapted from Brand (1947).

Normal and Tangent Vectors

We begin by assuming that positions on an arbitrary surface are described by two coordinates, u and v, which are not necessarily orthogonal. The position vector at the surface is denoted as $\mathbf{r}_s(u, v)$. From Eq. (A.7-3), two vectors that are tangent to the surface are

$$\mathbf{A} = \frac{\partial \mathbf{r}_s}{\partial u}, \qquad \mathbf{B} = \frac{\partial \mathbf{r}_s}{\partial v}. \tag{A.8-1}$$

Specifically, $\mathbf{A}$ is tangent to a "u curve" on the surface (i.e., a curve where v is held constant) and $\mathbf{B}$ is tangent to a "v curve." In general, these vectors are not orthogonal to one another and they are not of unit length. However, their cross product is orthogonal

TABLE A-4
Differential Operations in Spherical Coordinates[a]

(1)
$$\nabla f = \frac{\partial f}{\partial r}\mathbf{e}_r + \frac{1}{r}\frac{\partial f}{\partial \theta}\mathbf{e}_\theta + \frac{1}{r \sin \theta}\frac{\partial f}{\partial \phi}\mathbf{e}_\phi$$

(2)
$$\nabla \cdot \mathbf{v} = \frac{1}{r^2}\frac{\partial}{\partial r}(r^2 v_r) + \frac{1}{r \sin \theta}\frac{\partial}{\partial \theta}(v_\theta \sin \theta) + \frac{1}{r \sin \theta}\frac{\partial v_\phi}{\partial \phi}$$

(3)
$$\nabla \times \mathbf{v} = \frac{1}{r \sin \theta}\left[\frac{\partial}{\partial \theta}(v_\phi \sin \theta) - \frac{\partial v_\theta}{\partial \phi}\right]\mathbf{e}_r + \left[\frac{1}{r \sin \theta}\frac{\partial v_r}{\partial \phi} - \frac{1}{r}\frac{\partial}{\partial r}(rv_\phi)\right]\mathbf{e}_\theta + \left[\frac{1}{r}\frac{\partial}{\partial r}(rv_\theta) - \frac{1}{r}\frac{\partial v_r}{\partial \theta}\right]\mathbf{e}_\phi$$

(4)
$$\nabla^2 f = \frac{1}{r^2}\frac{\partial}{\partial r}\left(r^2\frac{\partial f}{\partial r}\right) + \frac{1}{r^2 \sin \theta}\frac{\partial}{\partial \theta}\left(\sin \theta\frac{\partial f}{\partial \theta}\right) + \frac{1}{r^2 \sin^2 \theta}\frac{\partial^2 f}{\partial \phi^2}$$

(5)
$$(\nabla \mathbf{v})_{rr} = \frac{\partial v_r}{\partial r}$$

(6)
$$(\nabla \mathbf{v})_{r\theta} = \frac{\partial v_\theta}{\partial r}$$

(7)
$$(\nabla \mathbf{v})_{r\phi} = \frac{\partial v_\phi}{\partial r}$$

(8)
$$(\nabla \mathbf{v})_{\theta r} = \frac{1}{r}\frac{\partial v_r}{\partial \theta} - \frac{v_\theta}{r}$$

(9)
$$(\nabla \mathbf{v})_{\theta\theta} = \frac{1}{r}\frac{\partial v_\theta}{\partial \theta} + \frac{v_r}{r}$$

(10)
$$(\nabla \mathbf{v})_{\theta\phi} = \frac{1}{r}\frac{\partial v_\phi}{\partial \theta}$$

(11)
$$(\nabla \mathbf{v})_{\phi r} = \frac{1}{r \sin \theta}\frac{\partial v_r}{\partial \phi} - \frac{v_\phi}{r}$$

(12)
$$(\nabla \mathbf{v})_{\phi\theta} = \frac{1}{r \sin \theta}\frac{\partial v_\theta}{\partial \phi} - \frac{v_\phi \cot \theta}{r}$$

(13)
$$(\nabla \mathbf{v})_{\phi\phi} = \frac{1}{r \sin \theta}\frac{\partial v_\phi}{\partial \phi} + \frac{v_r}{r} + \frac{v_\theta \cot \theta}{r}$$

[a] In these relationships f is any differentiable scalar function and $\mathbf{v}$ is any differentiable vector function.

to both, and hence is normal to the surface. It follows that a *unit normal* vector is given by

$$\mathbf{n} = \frac{\mathbf{A} \times \mathbf{B}}{|\mathbf{A} \times \mathbf{B}|}. \tag{A.8-2}$$

Depending on how u and v are selected, it may be necessary to reverse the order of $\mathbf{A}$ and $\mathbf{B}$ (thereby changing the sign of $\mathbf{n}$) to make the unit normal point *outward* from a closed surface.

Suppose now that $u = x$, $v = y$, and the surface is represented as the function $z = F(x, y)$. In this case the position vector at the surface is given by

$$\mathbf{r}_s = x\mathbf{e}_x + y\mathbf{e}_y + F(x, y)\mathbf{e}_z. \tag{A.8-3}$$

The two tangent vectors are found to be

$$\mathbf{A} = \frac{\partial \mathbf{r}_s}{\partial x} = (1)\mathbf{e}_x + (0)\mathbf{e}_y + \frac{\partial F}{\partial x}\mathbf{e}_z, \tag{A.8-4}$$

$$B = \frac{\partial \mathbf{r_s}}{\partial y} = (0)\mathbf{e}_x + (1)\mathbf{e}_y + \frac{\partial F}{\partial y}\mathbf{e}_z, \qquad (\text{A.8-5})$$

and the unit normal is computed as

$$\mathbf{n} = \frac{\mathbf{A} \times \mathbf{B}}{|\mathbf{A} \times \mathbf{B}|} = \frac{-(\partial F/\partial x)\mathbf{e}_x - (\partial F/\partial y)\mathbf{e}_y + \mathbf{e}_z}{\left[(\partial F/\partial x)^2 + (\partial F/\partial y)^2 + 1\right]^{1/2}}. \qquad (\text{A.8-6})$$

An alternate way to compute a unit normal is to represent the surface as $G(x, y, z)$ = 0 and to use (Hildebrand, 1976, p. 294)

$$\mathbf{n} = \frac{\boldsymbol{\nabla} G}{|\boldsymbol{\nabla} G|}. \qquad (\text{A.8-7})$$

It is readily confirmed that if $G(x, y, z) \equiv z - F(x, y)$, this formula yields the same result as in Eq. (A.8-6). As already mentioned, $\mathbf{n}$ may have to be replaced by $-\mathbf{n}$ to give the outward normal.

Reciprocal Bases

Until now we have employed base vectors, denoted as $\mathbf{e}_i$, which form orthonormal, right-handed sets. The special properties of such sets, as given by Eqs. (A.7-1) and (A.7-2), make them very convenient for the representation of other vectors. However, base vectors need not be orthonormal, or even orthogonal. Indeed, an arbitrary vector can be expressed in terms of any three vectors which are not coplanar or, equivalently, not linearly dependent (Brand, 1947). The volume of a parallelepiped which has the three vectors as adjacent edges will be nonzero only if the vectors are not coplanar; as discussed in Section A.3, that volume is equal to the scalar triple product. Thus, any three vectors with a nonvanishing scalar triple product constitute a possible *basis*. In describing the local curvature of a surface and other surface-related quantities, it proves convenient to employ two complementary sets of base vectors, neither of which is orthogonal. The special properties of these base vectors leads them to be termed *reciprocal bases*. Accordingly, a brief discussion of the properties of reciprocal bases is needed.

One basis of interest is the set of vectors defined above, $(\mathbf{A}, \mathbf{B}, \mathbf{n})$. Their scalar triple product is

$$H = \mathbf{A} \times \mathbf{B} \cdot \mathbf{n}. \qquad (\text{A.8-8})$$

Because $\mathbf{A}$ and $\mathbf{B}$ are tangent to the surface and $\mathbf{n}$ is normal to it, they are obviously not coplanar; thus, $H \neq 0$. Suppose now that $(\mathbf{A}, \mathbf{B}, \mathbf{n})$ and a second set $(\mathbf{a}, \mathbf{b}, \mathbf{c})$ are reciprocal bases. Then, by definiton, they satisfy

$$\begin{array}{lll} \mathbf{a} \cdot \mathbf{A} = 1, & \mathbf{a} \cdot \mathbf{B} = 0, & \mathbf{a} \cdot \mathbf{n} = 0, \\ \mathbf{b} \cdot \mathbf{A} = 0, & \mathbf{b} \cdot \mathbf{B} = 1, & \mathbf{b} \cdot \mathbf{n} = 0, \\ \mathbf{c} \cdot \mathbf{A} = 0, & \mathbf{c} \cdot \mathbf{B} = 0, & \mathbf{c} \cdot \mathbf{n} = 1. \end{array} \qquad (\text{A.8-9})$$

Notice that each base vector is orthogonal to two members of the *reciprocal* set. It is straightforward to verify that these relationships will hold if

$$\mathbf{a} = \frac{\mathbf{B} \times \mathbf{n}}{H}, \qquad \mathbf{b} = \frac{\mathbf{n} \times \mathbf{A}}{H}, \qquad \mathbf{c} = \frac{\mathbf{A} \times \mathbf{B}}{H} = \mathbf{n}. \qquad (\text{A.8-10})$$

An orthonormal basis is its own reciprocal; compare Eqs. (A.8-9) and (A.7-1). For an orthonormal basis we have $H = 1$.

For a surface represented as $z = F(x, y)$, it is found that

$$H = \left[\left(\frac{\partial F}{\partial x} \right)^2 + \left(\frac{\partial F}{\partial y} \right)^2 + 1 \right]^{1/2}, \tag{A.8-11}$$

$$\mathbf{a} = \frac{1}{H^2} \left\{ \left[1 + \left(\frac{\partial F}{\partial y} \right)^2 \right] \mathbf{e}_x - \frac{\partial F}{\partial x} \frac{\partial F}{\partial y} \mathbf{e}_y + \frac{\partial F}{\partial x} \mathbf{e}_z \right\}, \tag{A.8-12}$$

$$\mathbf{b} = \frac{1}{H^2} \left\{ -\frac{\partial F}{\partial x} \frac{\partial F}{\partial y} \mathbf{e}_x + \left[1 + \left(\frac{\partial F}{\partial x} \right)^2 \right] \mathbf{e}_y + \frac{\partial F}{\partial y} \mathbf{e}_z \right\}. \tag{A.8-13}$$

Surface Gradient

The key to describing variations of geometric quantities or field variables over a surface is the *surface gradient* operator, denoted as ∇_s; this operator is for surfaces what ∇ is for three-dimensional space. To derive an expression for ∇_s we consider a surface curve with arc length s. Evaluating the unit tangent using Eq. (A.6-2) and employing the coordinates u and v, we obtain

$$\mathbf{t} = \frac{d\mathbf{r}_s}{ds} = \frac{\partial \mathbf{r}_s}{\partial u} \frac{du}{ds} + \frac{\partial \mathbf{r}_s}{\partial v} \frac{dv}{ds} = \mathbf{A} \frac{du}{ds} + \mathbf{B} \frac{dv}{ds}. \tag{A.8-14}$$

From Eqs. (A.8-14) and (A.8-9) it is found that

$$\mathbf{a} \cdot \mathbf{t} = \frac{du}{ds}, \qquad \mathbf{b} \cdot \mathbf{t} = \frac{dv}{ds}. \tag{A.8-15}$$

Accordingly, the rate of change of a scalar function f along the curve is given by

$$\frac{\partial f}{\partial s} = \frac{\partial f}{\partial u} \frac{du}{ds} + \frac{\partial f}{\partial v} \frac{dv}{ds} = \mathbf{t} \cdot \left(\mathbf{a} \frac{\partial f}{\partial u} + \mathbf{b} \frac{\partial f}{\partial v} \right). \tag{A.8-16}$$

By analogy with Eq. (A.6-3), we define the quantity in parentheses as $\nabla_s f$. Accordingly, the surface gradient operator is

$$\nabla_s = \mathbf{a} \frac{\partial}{\partial u} + \mathbf{b} \frac{\partial}{\partial v}. \tag{A.8-17}$$

For a surface described by $z = F(x, y)$, the result is

$$\nabla_s = \frac{1}{H^2} \left\{ \left[1 + \left(\frac{\partial F}{\partial y} \right)^2 \right] \mathbf{e}_x - \frac{\partial F}{\partial x} \frac{\partial F}{\partial y} \mathbf{e}_y + \frac{\partial F}{\partial x} \mathbf{e}_z \right\} \frac{\partial}{\partial x}$$

$$+ \frac{1}{H^2} \left\{ -\frac{\partial F}{\partial x} \frac{\partial F}{\partial y} \mathbf{e}_x + \left[1 + \left(\frac{\partial F}{\partial x} \right)^2 \right] \mathbf{e}_y + \frac{\partial F}{\partial y} \mathbf{e}_z \right\} \frac{\partial}{\partial y} \tag{A.8-18}$$

where H is given by Eq. (A.8-11). As a simple example, consider a planar surface corresponding to a constant value of z. In this case $H = 1$ and the surface gradient operator reduces to

$$\nabla_s = \mathbf{e}_x \frac{\partial}{\partial x} + \mathbf{e}_y \frac{\partial}{\partial y} \qquad (z = F = \text{constant}), \tag{A.8-19}$$

which is simply a two-dimensional form of ∇ involving the surface coordinates x and y.

As shown in Brand (1947, p. 209), the gradient and surface gradient operators are related by

$$\nabla = \nabla_s + \mathbf{nn} \cdot \nabla, \tag{A.8-20}$$

so that an alternative expression for ∇_s is

$$\nabla_s = (\boldsymbol{\delta} - \mathbf{nn}) \cdot \nabla. \tag{A.8-21}$$

Subtracting the term $\mathbf{nn} \cdot \nabla$ from ∇ has the effect of removing from the operator any contributions which are not in the tangent plane.

Mean Curvature

The unit normal vector will be independent of position only for a planar surface. The rate of change of $\mathbf{n}$ along a surface is clearly related to the extent of surface curvature: The greater the curvature, the more rapid the variation in $\mathbf{n}$. A measure of the local curvature of a surface that is useful in fluid mechanics is the *mean curvature*, $\mathcal{H}$, which is proportional to the *surface divergence* of $\mathbf{n}$. Specifically,

$$\mathcal{H} \equiv -\frac{1}{2}\nabla_s \cdot \mathbf{n}. \tag{A.8-22}$$

For a surface expressed as $z = F(x, y)$, the mean curvature is given by

$$2\mathcal{H} = \frac{1}{H^3}\left\{\left[1 + \left(\frac{\partial F}{\partial y}\right)^2\right]\frac{\partial^2 F}{\partial x^2} - 2\frac{\partial F}{\partial x}\frac{\partial F}{\partial y}\frac{\partial^2 F}{\partial x \partial y} + \left[1 + \left(\frac{\partial F}{\partial x}\right)^2\right]\frac{\partial^2 F}{\partial y^2}\right\}. \tag{A.8-23}$$

For a surface with $z = F(x)$ only, this simplifies to

$$2\mathcal{H} = \left[\left(\frac{\partial F}{\partial x}\right)^2 + 1\right]^{-3/2}\frac{\partial^2 F}{\partial x^2}. \tag{A.8-24}$$

Two important special cases are cylinders and spheres. For a cylinder of radius R, with the surface defined as $z = F(x) = (R^2 - x^2)^{1/2}$, Eq. (A.8-24) gives

$$\mathcal{H} = -\frac{1}{2R} \quad \text{(cylinder).} \tag{A.8-25}$$

For a sphere with $z = F(x, y) = (R^2 - x^2 - y^2)^{1/2}$, it is found from Eq. (A.8-23) that

$$\mathcal{H} = -\frac{1}{R} \quad \text{(sphere).} \tag{A.8-26}$$

Unlike most surfaces, the curvature of cylinders and spheres is uniform (i.e., independent of position). For any piece of a surface described by $z = F(x, y)$, it is found that $\mathcal{H} < 0$ when $\mathbf{e}_z$ points away from the local center of curvature (as in these examples) and $\mathcal{H} > 0$ when $\mathbf{e}_z$ points toward the local center of curvature. For a general closed surface, $\mathcal{H} < 0$ when the outward normal $\mathbf{n}$ points away from the local center of curvature, and $\mathcal{H} > 0$ when $\mathbf{n}$ points toward the local center of curvature. In other words, $\mathcal{H}$ is negative or positive according to whether the surface is locally convex or concave, respectively.

Integral Transformation

One additional integral transformation is needed, which involves the surface gradient and surface curvature. Setting $\mathbf{v} = \mathbf{n}f$ in Eq. (A.5-7) gives

$$\int_S (\mathbf{n} \times \nabla) \times (\mathbf{n}f) \, dS = \int_C \mathbf{t} \times \mathbf{n}f \, dC. \tag{A.8-27}$$

As discussed in connection with Fig. A-2, the unit vector $\mathbf{m} = \mathbf{t} \times \mathbf{n}$ is tangent to the surface S and outwardly normal to the contour C. The left-hand side is rearranged by first noting that $\mathbf{n} \times \nabla = \mathbf{n} \times \nabla_s$; this is shown using Eq. (A.8-20). After further manipulation of the left-hand side, it is found that

$$\int_S (\nabla_s f - \nabla_s \cdot \mathbf{n}\mathbf{n}f) \, dS = \int_C \mathbf{m}f \, dC \tag{A.8-28}$$

or, using the symbol for mean curvature,

$$\int_S (\nabla_s f + 2\mathscr{H}\mathbf{n}f) \, dS = \int_C \mathbf{m}f \, dC. \tag{A.8-29}$$

This result is used in Chapter 5 in deriving the stress balance at a fluid–fluid interface, including the effects of surface tension.

References

Aris, R. *Vectors, Tensors, and the Basic Equations of Fluid Mechanics.* Prentice-Hall, Englewood Cliffs, NJ, 1962 (reprinted by Dover, New York, 1989).

Bird, R. B., R. C. Armstrong, and O. Hassager. *Dynamics of Polymeric Liquids,* Vol. 1, second edition. Wiley, New York, 1987.

Bird, R. B., W. E. Stewart, and E. N. Lightfoot. *Transport Phenomena.* Wiley, New York, 1960.

Brand, L. *Vector and Tensor Analysis.* Wiley, New York, 1947.

Hay, G. B. *Vector and Tensor Analysis.* Dover, New York, 1953.

Hildebrand, F. B. *Advanced Calculus for Applications,* second edition. Prentice-Hall, Englewood Cliffs, NJ, 1976.

Jeffreys, H. *Cartesian Tensors.* Cambridge University Press, Cambridge, 1963.

Moon, P. and D. E. Spencer. *Field Theory Handbook.* Springer-Verlag, Berlin, 1961.

Morse, P. M. and H. Feshbach. *Methods of Theoretical Physics.* McGraw-Hill, New York, 1953.

Prager, W. *Introduction to Mechanics of Continua.* Ginn, New York, 1961 (reprinted by Dover, New York, 1973).

Wilson, E. B. *Vector Analysis.* Yale University Press, New Haven, 1901.

AUTHOR INDEX

Page numbers are given only for first authors; bold numbers indicate where full citations appear.

SUBJECT INDEX